Night Trains

THE PULLMAN SYSTEM   IN THE GOLDEN YEARS OF AMERICAN RAIL TRAVEL

Peter T. Maiken

Chesapeake & Ohio Historical Society

Two young gentlemen bid mock farewell to friends in this 1935 publicity shot of the Chesapeake & Ohio's *George Washington* at White Sulphur Springs, W.Va.

The Johns Hopkins University Press

Baltimore and London

© 1989 by Peter T. Maiken
All rights reserved
Printed in the United States of America on acid-free paper
01 00 99 98 97 96 95 94   5 4 3 2

Originally published in 1989 in a hardcover edition by
Lakme Press, Beloit, Wisconsin 53512

Johns Hopkins Paperbacks edition, 1992

The Johns Hopkins University Press
2715 North Charles Street
Baltimore, Maryland 21218-4319
The Johns Hopkins Press Ltd., London

Designer: Franz Altschuler
Map maker: Carol Renaud
Body type: Baskerville, set by Chief City Graphics, Pontiac, Ill.

Library of Congress Cataloging-in-Publication Data

Maiken, Peter T., 1934-
    Night trains: the Pullman system in the golden years of American
rail travel / Peter T. Maiken.
        p.   cm.
    Originally published: Chicago : Lakme Press, 1989.
    Includes bibliographical references and index.
    ISBN 0-8018-4503-3 (pbk.)
    1. Railroads—United States—Pullman cars.   I. Title.
TF459.M35 1992
385'.0973—dc20   92-12586
                                                          CIP

A catalog record for this book is available from the British Library.

Table of Contents

List of Abbreviations

A&StAB Atlanta & St. Andrews Bay Railway

A&W Ahnapee & Western Railway

A&WP Atlanta & West Point Railroad

AA Ann Arbor Railroad

AB&C Atlanta, Birmingham & Coast Railroad

ACL Atlantic Coast Line Railroad

AT&SF Atchison, Topeka & Santa Fe Railway

B&A Boston & Albany Railroad

B&M Boston & Maine Railroad

B&O Baltimore & Ohio Railroad

BAR Bangor & Aroostook Railroad

Big Four Cleveland, Cincinnati, Chicago & St. Louis Railway

BR Blue Ridge Railway

BR&P Buffalo, Rochester & Pittsburgh Railway

C&EI Chicago & Eastern Illinois Railroad

C&G Columbus & Greenville Railway

C&IM Chicago & Illinois Midland Railway

C&NW Chicago & North Western Railway

C&O Chesapeake & Ohio Railway

C&S Colorado & Southern Railway

C&WC Charleston & Western Carolina Railway

C&WI Chicago & Western Indiana Railroad

CB&Q Chicago, Burlington & Quincy Railroad

CGW Chicago Great Western Railway

CIL Chicago, Indianapolis & Louisville Railway (Monon)

CM Colorado Midland Railway

CNJ Central Railroad of New Jersey

CNR Canadian National Railways

CofG Central of Georgia Railway

CPR Canadian Pacific Railway

CR Copper Range Railroad

CStPM&O Chicago, St. Paul, Minneapolis & Omaha Railway

CV Central Vermont Railway

CW California Western Railroad

D&H Delaware & Hudson Railroad

D&M Detroit & Mackinac Railway

D&RGW Denver & Rio Grande Western Railroad

D&SL Denver & Salt Lake Railway

DL&W Delaware, Lackawanna & Western Railroad

DSS&A Duluth, South Shore & Atlantic Railroad

FEC Florida East Coast Railway

FJ&G Fonda, Johnstown & Gloversville Railroad

FS&W Fort Smith & Western Railway

FW&DC Fort Worth & Denver City Railway

G&F Georgia & Florida Railroad

G&SI Gulf & Ship Island Railroad

GM&N Gulf, Mobile & Northern Railroad

GM&O Gulf Mobile & Ohio Railroad

GN Great Northern Railway

GR&I Grand Rapids & Indiana Railway

GS&F Georgia Southern & Florida Railway

GT Grand Trunk Railway System

GTW Grand Trunk Western

HPRA&S High Point, Randleman, Asheboro & Southern Railroad

IC Illinois Central Railroad

IPS Interstate Public Service Company

IT Illinois Terminal Railroad System

KCS Kansas City Southern Railway

KGB&W Kewaunee, Green Bay & Western Railroad

L&A Louisiana & Arkansas Railway

L&HR Lehigh & Hudson River Railway

L&N Louisville & Nashville Railroad

L&NE Lehigh & New England Railroad

LA&SL Los Angeles & Salt Lake Railroad

LIRR Long Island Rail Road

LV Lehigh Valley Railroad

LV&T Las Vegas & Tonopah Railroad

M&NA Missouri & North Arkansas Railway

M&O Mobile & Ohio Rail Road

M&StL Minneapolis & St. Louis Railway

M&WR Montpelier & Wells River Railroad

MC Michigan Central Railroad

MEC Maine Central Railroad

MILW Chicago, Milwaukee, St. Paul & Pacific Railroad (Milwaukee Road))

MKT Missouri-Kansas-Texas Railroad (Katy)

MO&G Missouri, Oklahoma & Gulf Railway

MP Missouri Pacific Lines

MT Midland Terminal Railway

N&W Norfolk & Western Railway

NC&StL Nashville, Chattanooga & St.Louis Railway

NdeM National Railways of Mexico

NH New York, New Haven & Hartford Railroad

NJI&I New Jersey, Indiana & Illinois Railroad

NJR Napierville Junction Railway

NKP New York, Chicago & St. Louis Railroad (Nickel Plate Road)

NN Nevada Northern Railway

NP Northern Pacific Railway

NS Norfolk Southern Railway

NWP Northwestern Pacific Railroad

NYC New York Central System

NYO&W New York, Ontario & Western Railway

OE Oregon Electric Railway

OR&N Oregon Railway & Navigation Co.

OSL Oregon Short Line Railroad

OT Oregon Trunk Railway

P&LE Pittsburgh & Lake Erie Railroad

PM Pere Marquette Railway

PRR Pennsylvania Railroad

QA&P Quanah, Acme & Pacific Railway

RDG Reading Railway System

RF&P Richmond, Fredericksburg & Potomac Railroad

RGS Rio Grande Southern Railroad (narrow)

RI Chicago, Rock Island & Pacific Railroad

RUT Rutland Railway

RW&O Rome, Watertown & Ogdensburg Railroad

SAL Seaboard Air Line Railroad

SI Spokane International Railroad

SLSF St. Louis-San Francisco Railway (Frisco)

Soo Minneapolis, St. Paul & Sault Ste. Marie Railroad (Soo Line)

SOU/SR Southern Railway System

SP Southern Pacific Lines

SP&S Spokane, Portland & Seattle Railway

SPdeM Southern Pacific Railroad Company of Mexico

StLSW St. Louis Southwestern Railway (Cotton Belt)

StJ&GI St. Joseph & Grand Island Railway

StJ&LC St. Johnsbury & Lamoille County Railroad

T&G Tonopah & Goldfield Railroad

T&P Texas & Pacific Railway

T&T Tonopah & Tidewater Railroad

TC Tennessee Central Railway

TH&B Toronto, Hamilton and Buffalo Railway

TP&W Toledo, Peoria & Western Railroad

UP Union Pacific Railroad

V&T Virginia & Truckee Railway

VGN Virginian Railway

W&NB Williamsport & North Branch Railway

WAB Wabash Railroad

WM Western Maryland Railway

WofA Western Railway of Alabama

WP Western Pacific Railroad

WPR West Point Route

WS West Shore Railroad

WSS Winston-Salem Southbound Railway

YV Yosemite Valley Railroad

Index of Maps

Preface

Night Trains is intended to be a general interest exposition of the "where" element of Pullman sleeping car travel: the routes these moveable hotels traveled, the trains that carried them, and the hundreds of towns and cities this remarkable system served. The book's main focus is on the late heavyweight and streamliner eras, from 1920 to 1955, but the time frame occasionally extends beyond these bounds.

The text of *Night Trains* is organized geographically—mostly by individual states, but regionally in one case and by major passenger train routes in several others. In each chapter, I've tried to give a capsule history of the main railroad lines within the states and the important trains operating over them. Almost without exception, these rail lines and trains and the major Pullman lines coincided. (A Pullman "line" is a specific selection of sleeping car accommodations operating on one or more specific trains, between two places). Inasmuch as the book is based on the interrelationship of all these lines, there is a certain amount of intended duplication because some material applies to both origin and destination.

In researching *Night Trains,* I relied almost exclusively on secondary sources. Most of the information on Pullman lines comes from *The Official Guide of the Railways,* monthly compilations of all passenger train schedules published from the late 1860's on. *Official Guides* were the bibles that travel agents and ticket vendors consulted to properly direct travelers over the nation's quarter-million-mile rail network. These voluminous periodicals, some of which amounted to more than 1,750 pages, were filled with timetables and other text that was often graphically numbing, haphazardly organized, confusingly cluttered, and maddeningly cryptic. The route maps of many of the railroads sacrificed the forest to the trees and were legible only to those with jewelers loupes. With their antediluvian covers, one issue looked just like its forebears of forty years earlier. On top of all their other faults, *Official Guides* were absolutely fascinating.

I made basic line compilations from four *Official Guides* and used fifteen others between 1890 and 1971 as supplements (see Bibliography for listing of volumes used). Because this was only a sampling, I make no claim to comprehensiveness. A full exploration of the field from 1890 to the beginning of Amtrak would require combing nearly a thousand issues of the *Official Guide,* which would have been an expenditure beyond the resources available for this project. Still, I'm sure that the major Pullman lines of the modern era are represented and those that eluded the sampling were fairly short-lived.

Furthermore, there are much additional data about Pullman lines that, for space reasons, *Night Trains* does not attempt to include. To keep track of its operations, the Pullman Company maintained "Line Books," in later years loose-leaf ledgers with histories of each line entered on one or more pages. Each page listed the dates of inauguration and termination (even some of the major lines were on-again, off-again affairs that underwent changes for one or more reasons, such as strike interference, seasonal traffic fluctuations, competitor realignment, etc.). They also listed origins and destinations (these changed as lines were shortened or lengthened), mileage (for accounting purposes), and descriptions of the equipment (which also varied with passenger demand, car availability, and other factors). Earlier in the century, most of this material was published by the company and bound into books called *Schedule of Lines.* These volumes offer much valuable data, but since they were not issued monthly, as were the *Official Guides,* they are probably not so comprehensive a catalog of the actual lines themselves. Doubtless someday all this information, together with car assignments, will be integrated, and it will fill a library shelf. For the moment, *Night Trains* opts for the more general-interest approach.

Because the Pullman system was a dynamic institution that constantly changed to meet demands, one must "stop" the action at some point to see how the various parts fit together. I have done this in the appendix titled *Midnight Sleepers,* where the entire system, as it existed at the height of the post-war streamliner era, is frozen in time. Through this example, the reader is able to see how every single car in the entire Pullman system was routed on a typical night.

Although most Pullman lines were bidirectional, a few were one-way only. In some cases, it was more advantageous to operate from A to B, then return to A from nearby C, in a triangular routing. In other cases, train schedules provided adequate service with through sleepers in one direction but not the other. If the only available train came through town at, say, 3:35 a.m., then a dedicated setout for later pickup would be required if passengers boarding there were to get a decent night's sleep. (A "dedicated" sleeping car serves its origin and destination specifically.) In still other cases, train schedules or passenger demand might require a car to operate as a sleeper one way and as a parlor car on the daylight run in the other direction. Whatever the case, when the Pullman system was healthy, virtually all communities served by the few one-way lines had adequate round-trip service by way of through cars.

In the main text, to keep relative antiquity in perspective, I have indicated lines that were primarily pre-1900 by putting in parentheses the year of the schedules in which they were listed. All other lines were of the Twentieth Century. Railroad names within parentheses indicate participating interline carriers. I use corporate names as they existed in mid-Twentieth Century, and they can be assumed to apply to any predecessor companies. (I've used ampersands wherever the word "and" appeared in the titles.) Although many of the lower-cost "tourist" Pullman lines duplicated the movements of the regular ones, others didn't, and those are indicated in parentheses as "tourist" cars.

I've also used the word "Pullman" as it came to be construed by most Americans, that is, in the generic sense to mean "sleeping car." Although Pullman operated numerous parlor cars—the daytime, first-class equivalent of the sleeper—*Night Trains* deals only with sleeping cars. As for these conveyances, certain railroads—the Milwaukee Road and Soo Line, for instance—operated their own for many years. Until it absorbed the Wagner company in 1899, Pullman had other competitors, though, except for Wagner, none of any significance in the 1890's. In its later years, too, the nature of the Pullman Company changed, with revised ownership and operating agreements brought on by court decree and changing economic conditions. Administrative matters, however, are not the focus of this book, and so I deal with sleeping car lines per se and differentiate between Pullman and others only where it would have been significant to the traveler.

In addition to data from the *Official Guides,* I used compilations made by various railroad historical societies and scholars, and information from public timetables, books, and articles (see Bibliography). In the first category, I'm particularly grateful for the contributions of Thomas W. Dixon Jr., Chesapeake & Ohio Historical Society, and Dr. John I. Dale, Southern Railway scholar.

Additionally, I greatly appreciate the contributions and suggestions made by the following:

Representing rail historical societies and museums: Henry F. Burger, Ann Arbor Railroad Technical & Historical Association; Frank C. Hoffman, Anthracite Railroads Historical Society; J. W. Barnard Jr., Affiliation for Baltimore & Ohio System Historical Research; Donald S. Robinson, Boston & Maine Railroad Historical Society; Rod Masterson, Burlington Route Historical Society; Art Million, Chesapeake & Ohio Historical Society; Ray Curl, Chicago & Eastern Illinois Railroad Historical Society; Jim Bennett, Cotton Belt Rail Historical Society; Harold H. Weber, Gulf, Mobile & Ohio Historical Society; Norman C. Keyes Jr., Great Northern Railway Historical Society; Dr. J.W. Laude, Illinois Central Historical Society; William E. Poynter, Missouri Pacific Historical Society; Edward J. Lewnard, Monon Railroad Historical-Technical Society; Jim Nichols, Norfolk & Western Historical Society; Alvin A. Lawrence and Charles A. Brown, New Haven Railroad Historical & Technical Association; Willard A. Harvey, Nickel Plate Road Historical & Technical Society; Jack Farley, Northwestern Pacific Railroad Historical & Technical Society; Charles M. Smith, New York Central System Historical Society; Joe Bux, Ontario & Western Railway Historical Society; James J. D. Lynch Jr., Pennsylvania Railroad Technical & Historical Society; Ken McElreath, Frisco Modelers Information Group; Joseph L. Oates, Southeastern Railroad Technical Society; Oscar W. Kimsey Jr., Southern Railway Historical Society; David E. Cline, Western Maryland Railway Historical Society; Norman W. Homes, Feather River Rail Society; Richard C. Datin, Nevada State Railroad Museum; Arthur M. Bixby Sr., Roanoke Chapter, National Railway Historical Society.

Authors and scholars: Tom Greco, Colorado Midland and B&O; Seth H. Bramson, Florida East Coast; William H. Patterson, Seaboard Air Line; Don L. Hofsommer, Southern Pacific; Robert Goldsborough, railroad art; William W. Kratville, Union Pacific; and Arthur D. Dubin, for resources and wise counsel.

State or local historical society members or affiliates: Gene Hull, Arkansas; Gregg M. Turner, Connecticut; Sam Boldrick, Florida; Jim Witherell, Idaho; Robert M. Sutton and George H. Douglas, Illinois; Howard B. Morris, Indiana; Harry Frye, New Hampshire; Sam Breck, Michigan; Vernon Glover, New Mexico; Walter E. Zullig Jr., New York; Preston George, Oklahoma; Bob Bell, Tennessee; Jim Fredrickson, Washington; James L. Ehrenberger, Wyoming.

Railroad corporation officers or their representatives: P. W. Stafford, Burlington Northern; J. Norman Lowe, Canadian National Rail; Craig MacQueen, Conrail; Bill Howes, CSX; Charles Castner, Louisville & Nashville; Walter A. Fussner, Missouri Pacific System; J. Donlan Piedmont, Norfolk & Western; Richard W. Harris, Norfolk Southern; William E. Griffin, Richmond, Fredericksburg & Potomac; Robert E. Gehrt, Santa Fe; George R. Cockle and William J. Fox, Union Pacific; Martin Rock, Pullman Company.

I appreciate the assistance given by the Chicago Public Library, Newberry Library, Chicago Historical Society, and most of all the Northwestern University Library, specifically Mary Roy, transportation librarian, and her staff.

Finally, my thanks to my late father, John A. Maiken, for his critiques, and to my wife, Margaret Maiken, for all her help, ranging from editing to the vital, detailed tasks that are part of preparing any book for publication. Naturally, I alone am responsible for the book's contents.

It's unfortunate that many night trains seemed to elude the lenses of our best rail photographers. The short-run trains departed late and arrived early, and night-time photographs of some of them are remarkably scarce. Many of the historic trains, of course, operated before the era of fast films, and that hurt, too. All of this points to the fact that while I have used the best night shots available to me, the vast majority of the train photographs in the book were taken during daytime. And why not? There is, after all, nothing wrong with showing a train in its best light. In any case, I'm grateful to all the people, especially the individual photographers, who answered the call.

Chicago, Illinois
July 1989

The Pullman enterprise

George Mortimer Pullman did not invent the sleeping car, nor did the Japanese invent the camera, the automobile, or the video tape recorder. Pullman simply took a good idea, built upon it, and shrewdly turned it into the most remarkable transportation system in the world. Simple testimony, it would seem, to the notion that if you can't be the first, be the best.

At the height of its operation in the 1920's, the Pullman system welcomed to its berths and rooms more than 50,000 people nightly, a traveling contingent that would have filled Chicago's Hilton Hotel and Towers—once the world's largest inn—twenty times over. Its 9,800 Pullman cars operated over trackage that, if put into a single line, would have girdled the globe five times and then some. But even though the company was a gargantuan hotel-keeping operation whose inventory of bed sheets and towels ran into the millions, Pullman's uniqueness lay in the fact that its customers went to bed in one place and awakened in another, often hundreds of miles of away.

For the nation's business travelers and tourists, the railroads provided virtually all intercity transportation until automobiles and buses began to seriously erode their monopoly in the 1920's. And if one wanted to travel as comfortably as possible over this network of rails, the Pullman sleeper was the way to go.

To the distant observer of a night train, a string of cheerfully lighted passenger cars moving across the dark landscape gave only a glimmer of the wondrous world that lay within: fashionably dressed urban travelers sipping cocktails and wines, or dining at tables set with starchy linen and gleaming crystal and silver, while other people reposed in private sleeping compartments paneled with exotic tropical woods, or perhaps engaged the services of a manicurist, barber, or train secretary.

This was first-class rail travel at its most elegant, the sort that one experienced by riding a *Super Chief* or *Twentieth Century Limited*. This was the Pullman system at the pinnacle.

George M. Pullman entered the sleeping car business in 1858, twenty years after the first sleeper was introduced, on the Cumberland Valley Railroad (which became part of the Pennsylvania). And it was not until the production of his legendary *Pioneer*, which he completed in 1865, that he decided to devote his full energies to this particular enterprise. By that time a dozen car builders had entered the field, and Pullman obviously had to offer something extraordinary to compete with two very well entrenched

competitors. Accordingly, he spared no effort in designing *Pioneer* to be his piece de resistance in selling the railroads on large-scale sleeper operation.

As John H. White Jr. points out in his monumental study, *The American Railroad Passenger Car*, the legend of *Pioneer* quickly outran the vehicle itself[1]. Traditional accounts tell of the rebuilding of railroad bridges and platforms around Chicago so that the oversized car could be included in Abraham Lincoln's funeral train. The association of *Pioneer* with the martyred president enhanced the legend, and soon George Pullman was getting unwarranted credit for playing obstetrician to the entire sleeping car industry.

In the formative years, competition among the three dozen declared car builders, coupled with the Victorian era's fascination with ornamented works, all but ensured that the various contestants would try to outdo one another in the craftsmanship of their products. The early cars were made of wood, and their interiors were finished by skilled cabinet makers with Circassian walnut, mahogany, amaranth, and ebony, among dozens of fine woods. Upholsterers chose velvets, velours, or plush for sleeper seats and often leathers for those in dining and lounge cars. Crystal lamps, leaded glass windows, marble wash basins, and silver faucets were not uncommon. It was what Lucius Beebe called "the most sumptuously upholstered landfaring in the history of human movement."

When Pullman took over its last competitor, the Wagner Palace Car Company in 1899, the sleeping car cartel was a fait accompli. A few railroads continued to run their own sleeper operations, but everything else was in the hands of the smoothly functioning machine called the Pullman Palace Car Company. This gigantic firm, which would soon be streamlined to just the Pullman Company, built sleepers at its factory next to its own model community on the south side of Chicago, and it operated the cars not just in the United States but into Canada and Mexico as well.

This scale of operation, however, brought opposition in the growing anti-monopolistic climate of the late 19th Century. Pullman, moreover, in reponse to the business depression of 1893, laid off workers and reduced wages, which led to the infamous and bitter strike of the following year. This darkened the image of the company—which to outsiders had seemed so paternalistic—for years to come. So while Pullman entered the 20th Century an all-powerful giant, it was ever alert to the need for defending its own interests.

In this new era, the fussy, ornate Victorian styles soon faded and yielded to simpler forms, from Spanish Mission to later Art Deco and finally to what

William F. Howes Jr. collection

A smartly uniformed Pullman crew for a Baltimore & Ohio train stands for a portrait in Chicago's Grand Central Station.

might be called Formica Moderne. A more significant change in the early years, however, was the conversion to all-steel cars, spurred by mounting concern for passenger safety and mandated for use on trains operating through the underground tunnels in and out of New York City's two new terminals. Initially, Pullman was reluctant to give up wooden car technology, in which its craftsmen were certainly preeminent, but once it saw that many railroads were rushing to reequip their trains with—and benefit from the promotion of—the safer new cars, the company moved aggressively into all-steel construction.

This ushered in the so-called "standard era," the twenty-five year period starting in 1910, when the pioneering all-steel *Carnegie* went into service. Before car building slowed to a trickle during the Depression, 8,000 of these dark olive-green, standard sleepers were built, half of them in the bread-and-butter, open-section, 12 and 1 design.

This great workhorse that dated well back into the wooden car era had a dozen pairs of facing seats that converted into lower berths, above each of which an upper berth folded down out of the wall and ceiling. The 12 sections yielded a dozen uppers and a dozen lowers, each outfitted by the porter with little green net hammocks for passengers to stow their clothing in. A drawing room at one end of the car provided for those who demanded and could afford greater privacy.

Only hanging curtains separated a passenger in an open section berth from the public at large, and the arduous exercise of undressing in or making dignified exit from an upper berth—with one's rump swaddled only in a flimsy nightshirt—proved a continuing challenge. The typical open-section Pullman car quickly

established itself in American comedic lore, becoming, in motion pictures, a well disciplined art form, with mistaken identities, inappropriate intrusions, ingenuous goosings, and predictable pratfalls all being part of the repertoire.

In the 1880's, The Mann Boudoir Car Company had attempted to spare the traveling public these indignities by offering all-private-room sleepers, with accommodations that presaged those in lightweight Pullmans by a half century. But the higher costs of Mann's cars, plus the public imagination that scandalous things took place in *boudoirs,* for heaven's sake, prematurely doomed the effort. This reinforced Pullman's resolve to put his money on the more economical open-section car, which America rode to the sunrise until private rooms became the accommodations of choice in the post-war era.

But just as the 12 and 1 served as Pullman's cash cows, they also became its millstones. These indestructible battlewagons with the Marx Brothers' dream interiors ultimately became a bore to the lively generation that set the style of the 1920's, people who had a hankering to breeze off to the mystical "somewhere west of Laramie" in an open roadster, people who were leading a wholesale defection to the highways, primitive as they were. The decline in rail traffic that began as a slow leak in the Roaring Twenties bordered on blowout proportions by the depths of the Depression. The railroads and Pullman now had to fight for their survival, and lightweight streamliners, with an emphasis on private rooms—mostly bedrooms and compartments for two persons, roomettes and duplex roomettes for singles—began rolling from the shops.

In 1940, the ultimate challenge came, when the gov-

Burlington Northern Railroad

This smiling model ornaments a display of the *North Coast Limited's* lower berth accommodations in 1936.

ernment charged Pullman with violating both the Sherman and Clayton antitrust acts by creating an illegal monopoly. The complaint said that the public interest was thwarted by the company's refusal to operate cars not of its own manufacture, giving passengers no choice but to ride in Pullman's superannuated fleet of sleepers rather than the modern ones that would presumably be available with competition among manufacturers. The suit aimed to separate the production of Pullman cars from their operation.

There was truth on both sides. Pullman had made great strides in air-conditioning its fleet and replacing heavyweight sleepers with new, streamlined versions, but it had also used its substantial muscle to push aside competing manufacturers, such as Budd and American Car & Foundry, through punitive contracts with the railroads.

After the judgment was rendered nearly four years later, Pullman chose to sell its operating division, and after a variety of plans were discussed, a consortium of fifty-seven railroads finally bought the assets of the sleeping car business in 1947. It was a decision Pullman could live with and, in short order, be grateful for, given the economics of passenger train operation.

Pullman had encountered various charges of monopoly for nearly three-quarters of a century, and its usual defense was that it was acting in the interest of the public, vis a vis the railroads: Its system provided a seamless flow of traffic across the nation, a standardized service that individual railroads would not and could not otherwise offer. Moreover, Pullman was able to maintain a pool of cars far beyond the resources of any single road, accommodations that could be dispatched wherever needed: to Florida in the winter, to the north and west in the summer, or for special events such as the Kentucky Derby, World Series, Rose Bowl, presidential inaugurations, and major conventions. Pullman marshaled nearly 1,200 sleepers for travel to the Eucharist Congress in Mundelein, Ill., near Chicago, in 1926—a pool record.

But these movements were just interesting footnotes to the daily routine of the Pullman system as it provided communities, large and small, with reliable and economical public transportation. It blanketed the country like no other entity has since. Towns of as little as five or ten thousand population had dedicated Pullman service (whereby the community was either the origin or destination of the car) to cities in their regions, and tens of thousands of towns had through service.

If eight or ten people, on a nightly average, traveled from one place to another, resulting in a gross sleeping car revenue of, say, $30, that would be sufficient to support a dedicated Pullman "line" between the two towns, and the railroad could request the company to put such a sleeper into service. (For their cars, Pullman provided all interior maintenance and furnishings, such as linens, soap, and paper cups, as well as the labor of porters and supervising conductors. The railroads hauled the sleepers, provided electricity and heating, and did routine exterior maintenance while the cars were in service.)

Encouraged by the exceptional though thoroughly misleading traffic surge during World War II, the railroads embarked on the last hurrah of private-enterprise, passenger train operation. Despite continuing drops in ridership, despite the advances and growing competition of commercial aviation, despite the growing disenchantment within the railroad community toward the passenger train, many of the key players nonetheless began reequipping their finest trains with state-of-the-art, streamlined equipment. The result was a memorable, though inevitably short, Indian Summer of rail passenger travel that lasted about ten years before the railroads' ailments became their terminal illness.

Apart from the hardware, the single most important ingredient in the Pullman Company's success was the service rendered by its employees, notably the porters, most of whom were black men. Economic opportunities never abounded for blacks during the lifetime of the Pullman Company, and employment as a porter, despite the originally modest wages, gave these men a degree of job security and respectability that most of their peers could only dream of.

The Pullman porter was subjected to a training regimen and rule book that left nothing to chance (a commissary department instruction devoted five pages of text to the proper serving of a bottle of beer, from the icing of the glass to the issuing of the receipt).

Whether a porter's particular line was on a one-, two-, or even three-night schedule, he was always on call. He reported to his assigned sleeper several hours before train departure to make up the beds, stock the supplies, and make sure the temperature was at the proper level. He carried and stowed the passengers' baggage, delivered meals and beverages to those who chose to remain in their accommodations, ran errands such as the forwarding of telegrams, and awakened guests in the early morning to ensure their getting off at the proper station. In between fitful late-night naps, he shined the shoes his passengers had deposited in little lockers in the wall.

For all this, the Pullman porter might or might not collect a tip to add to the basic wages the company paid him. But he did get a kind of mocking recognition in the form of a nickname that stuck with him until the civil rights movement. After black waiters and porters were hired en masse by Pullman in the 1880's, the mostly white passengers soon began calling the porters "George," as in "let George do it," no doubt in dubious honor of the company founder. (The practice was so widespread that a wealthy Iowan—named George—later formed a society of card-carriers sworn to do away with the obnoxious habit.)

This great, black fraternity, which, during World War II, reached a membership of 12,000 men, was honored on a 1989 U.S. postage stamp, commemorating A. Phillip Randolph and the Brotherhood of Sleeping Car Porters, the union he founded in 1925.

For those who remember the crowded trains of World War II, when passenger traffic soared to stratospheric levels, with servicemen and -women and civilians sitting on their suitcases in the aisles, the fate that awaited the Pullman Company just a quarter century later was unthinkable. On January 1, 1969 the one-time "world's largest hotelkeeper" quit running Pullman cars—there were now only 350 left—and assumed the role of merely supplying and maintaining those still in service. Now the death-rattles of this once great company and all of American passenger railroading sounded across the land, and it remained for Amtrak to pick up the broken pieces.

Night Trains, however, does not deal with the sad ending but rather chooses to look at the golden years of American passenger railroading, when Pullmans proliferated across America's darkened landscape. Some traveled in the best of company, on crack overnight trains like the *Broadway Limited* or the *Panama Limited* that were solid with sleeping cars. Others departed in the early morning hours as the last car in a mail or newspaper train and perhaps traveled

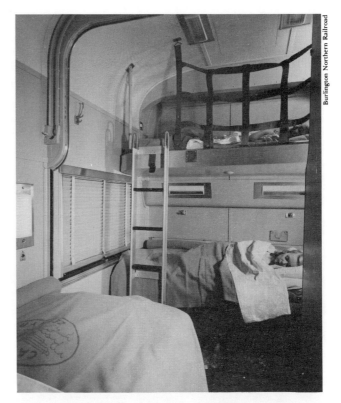

On a more modern note, these women demonstrate the comforts of a double bedroom, aboard the *California Zephyr.*

the final leg on a branch line accommodation run scheduled just for that purpose.

Whatever their pedigree, they delivered their passengers reliably, safely, and in hotel-like comfort to destinations in every corner of the land from Van Buren, Me., to Key West, Fla., to San Diego, to Bellingham, Wash. They were—*sic transeunt in gloria*—the night trains of America.

1. *John H. White Jr., The American Railroad Passenger Car, Part 1* (Baltimore: Johns Hopkins University Press, softcover edition, 1985) p. 248.

1
New England

A Boston lady once traveled by train to San Francisco.
"By what route did you come?" her hosts asked.
"Via Worcester," she replied.

Boston legend

Boston was not just the hub of the universe, it was also the center of the entire New England rail network that spread like a giant fan from New York City to the maritime provinces of Canada. This was befitting, inasmuch as Boston merited rail service on a scale that complemented its long-standing importance as a bustling port of entry. Just like other major eastern seaports, Boston was eager to tap the commerce flowing on the Great Lakes, and it enjoyed the advantage of having all-weather navigability.

Canada's major cities did not have this advantage, and their shippers needed an outlet that Boston was happy to provide. As a result of these circumstances, a number of rail lines radiated from Boston to the north and northwest, wending their way along river valleys and through the granite notches of New England, carrying freight and passengers to several Canadian interchanges.

At the beginning of the railroad era, one of the first orders of business was to connect Boston and New York City. Surveys of Connecticut were not encouraging; the topography was thought to be unfavorable for an all-rail route, and so the planning focused on joint boat-train service. Trains were scheduled to leave Boston in late afternoon for the wharves in such cities as Fall River, Mass.; Providence and Newport, R.I.; and Stonington, Norwich, and New London, Conn. At dockside, passengers transferred to steamships that made overnight passages through the protected waters of Long Island Sound and docked early in the morning in New York. The process was reversed for eastbound travel. It was a short-term solution, but once the rails were in place, the joint service proved so popular that remnants endured to the Air Age.

One early scheme sought to minimize the slower travel over water. The Long Island Rail Road was originally planned, not as the commuter road it eventually became, but as part of a New York-Boston route. Rather than leaving by steamboat from New York harbor, passengers rode the relatively swift, bright-yellow coaches of the Long Island to Greenport, where they boarded a ferry for a short ride to Stonington, near the Connecticut-Rhode Island line. They finished their trip on what later became the Shore Line of the New York, New Haven & Hartford. This arrangement worked well but briefly, for by mid-Nineteenth Century the all-rail route between Boston and New York was being laid, despite the early misgivings (though it would suffer for lack of two key bridges for several more decades). The Long Island,

A Canadian National 4-8-4 pauses with Central Vermont's *Washingtonian* at Essex Junction, Vt., in March 1954. Boston & Maine will take over this southbound version of the *Montrealer* at White River Junction, Vt.

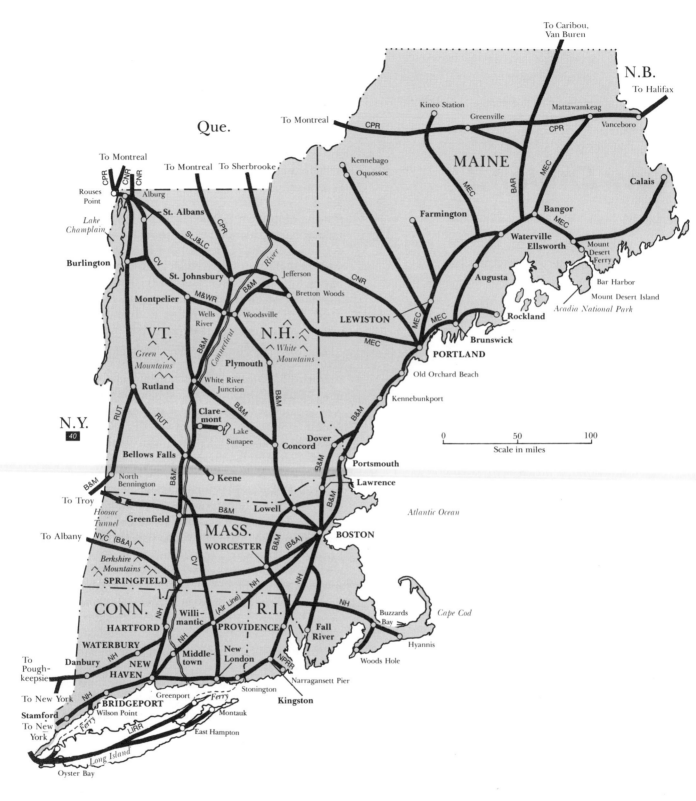

faced with the new competition and internal financial woes, went into receivership.

New Haven Railroad predecessors were involved with other shortlived boat-train operations between New York and Boston, as well as a backwoods all-rail route. Passengers left Brooklyn on Long Island tracks to Oyster Bay, from which the cars were ferried to Wilson Point, Conn. On the all-rail route, the *New York & Boston Pullman Limited* ran via what became the New York Central's Putnam Division to Brewster, N.Y., whereupon it was switched to the New York & New England line of the New Haven.[1]

The Fall River Line was the ranking survivor of the boat-train service. Starting in May 1847, the ancestor of the New Haven's boat-train departed Boston, bound for the wharf at Fall River. In 1937, just a decade short of a century, it was retired as the country's oldest passenger operation in continuous service.

Despite the loyal boat-train clientele, the all-rail route between New York and Boston made its mark with superior speed, and soon a fleet of trains, both day and night, was serving the two large cities. Eventually, control of the various railroads that laced the territory—there were four principal ones—passed to the modern New Haven. After an agreement with the Boston & Maine shortly before the turn of the century, essentially to split New England between the two companies, the New Haven enjoyed an almost complete monopoly of rail, interurban, and steamship transportation in New England south of Boston.[2]

After the New Haven's consolidation, completed in 1898, most trains between Boston and New York City traveled over one of two main routes: the Shore Line, via Providence and New London, or the joint Boston & Albany/New Haven route via Springfield and Hartford. These lines, together with the once-favored Air Line through Willimantic and Middletown, Conn., all came together at the city of New Haven.

The Shore Line was not completely bridged until 1889, which subjected passengers to the delays of ferrying, but after it became all-rail it never lost its dominance. Being the "shore line," of course, meant it was always vulnerable to severe weather off Long Island Sound. Much of its trackage was destroyed by the extraordinary hurricane that struck New England on September 21, 1938, and service was not restored for nearly three weeks.

It was over the Air Line that the fabled *New England Limited,* also known as the *White Train* or the *Ghost Train,* ran in the 1890's. This crack express was painted white and gold—on the inaugural runs even the coal that fueled its locomotive reportedly was whitewashed—which gave the train its ghostly appearance as it raced through the Connecticut uplands at twilight.

The *Merchants Limited,* inaugurated in December 1903, was the premier all-parlor-car, extra-fare train the New Haven operated. Others included the *Bay State,* the *Knickerbocker Limited,* on which afternoon tea was served, and the Depression-born *Yankee Clipper.* Bankers and brokers, however, favored the *Merchants Limited,* which carried two diners stocked with foodstuffs supplied by the famed Boston firm of S.S. Pierce. The train, which kept its extra-fare status until 1949, was the last daily all-parlor car limited in America.[3]

The *Owl* was a solid train of sleeping cars, and on this nocturnal operation the New Haven distinguished itself with a flair. In 1910, just before the railway quit operating its own sleepers and signed a contract with Pullman, the New Haven established what was reported to be the most luxurious sleeping car service ever seen, short of the sybaritic excesses found in the fraternity of private varnish. The Pullman company built special compartment cars, and each room contained a full-length, four-foot-wide brass bed, dresser, table, and chairs. Each unit was thermostatically controlled and each had its own toilet. In a day when the open-section sleeper was the norm, the New Haven's service was wildly extravagent.

In addition to operating trains from Boston into New York's Grand Central Terminal and through trains to Washington, Philadelphia, and Pittsburgh *(Federal, Quaker, William Penn),* the New Haven shared operation of the *Montrealer, State of Maine,* and seasonal *Bar Harbor Express.* Post-war, in the Indian summer of passenger travel, the New Haven went to a colorful orange, black and white livery. For many years, it was among the few railroads whose passenger trains earned more than the freight.

The New Haven carried sleepers from Boston to both New Haven and New York, and over the years it forwarded Pullmans to the Pennsylvania Railroad for Philadelphia, Pittsburgh, Chicago, St. Louis, Cincinnati, Washington, and New Orleans (Southern/Louisville & Nashville), in addition to cars for the Florida cities (all Richmond, Fredericksburg & Potomac) of Jacksonville (Atlantic Coast Line), Miami (both ACL/Florida East Coast and Seaboard Air Line), Palm Beach (ACL/FEC), St. Petersburg (ACL or SAL), and Tampa/Sarasota (ACL).

In the 1890's, the New Haven carried sleepers on its line to Beacon, N.Y. (then known as Fishkill Landing), where they were ferried across the Hudson to a connection with the Erie Railroad at Newburgh. They included cars for Chicago[4] on the *Isabella Express* (a Columbian Exposition promotion), and for Hornell, N.Y., and Washington (Baltimore & Ohio).[5]

New Haven trains operated out of Boston's huge South Station after it was dedicated in 1898, immediately becoming the largest and busiest terminal in the country, with more than 800 trains a day, two-thirds of them commuter. This landmark, with its weathered face of New England granite fronting on Dewey Square, had a 600-foot concourse, and its 28 stub tracks were covered by a huge, arched shed consisting of three spans.[6] Although the train shed was ripped down in the 1930's, and numerous tracks removed and the building truncated in the early 1970's, the rounded portion of the historic facade, with a sculpted eagle perched above the clock, was allowed to continue holding court over the square and the remaining portion of the station was given a tasteful restoration.

The New Haven's fellow tenant in South Station was a New York Central lessee that for most of its career maintained its own identity as the Boston & Albany until it was merged in the early 1960's. The B&A linked with the Central's main line at Albany—before New York City was similarly connected—and it became Boston's sole passenger window to the west in later years.

For many years the Central ran an exclusive, all-Pullman section of the *Twentieth Century Limited* from Boston to Albany, where it was coupled into one of the main sections. In 1938, Bostonians got their own train as a successor, the *New England States*, eliminating the switching delay in Albany that the *Century* had been subjected to. For a time all-Pullman, the *New England States* later took on reserved-seat coaches. Other Central trains or sections serving Boston included the *New England Wolverine, Berkshire,* and *North Shore Limited* to Chicago, the *Ohio State Limited* to Cincinnati, and *Southwestern Limited* to St. Louis.

The B&A scheduled numerous trains to Albany, and there it picked up many Boston-bound Pullmans from upstate New York cities: Buffalo, Niagara Falls, Albany, Clayton, Lake Placid, Rochester, Syracuse, and Utica. Boston however, had sleeper service over the full span of the New York Central system, with cars to Chicago (via both Detroit and Cleveland, and including dedicated Pullmans to those cities, too), Cincinnati, French Lick, Ind. (Monon), Louisville (L&N at Cincinnati), Pittsburgh (Pittsburgh & Lake Erie), St. Louis, Toronto (Toronto, Hamilton & Buffalo/Canadian Pacific), and from Toledo. The B&A also participated in an interline sleeper to New York City, turning the cars over to the New Haven in Springfield.

Boston's other major railroad terminal, North Station, was also deserving of superlatives in that it was the country's busiest station serving one railroad, in this case the Boston & Maine. The original North Station was built in 1894, about a mile north of South Station, on the Charles River, just across from Cambridge. The new North Station, built in 1928, was distinguished by being the home of the Boston Garden, New England's largest sports palace. This arena for the Bruins and Celtics was built atop the station, which also included a hotel and numerous shops within its spacious precincts.

The B&M's Fitchburg line ran to Troy, N.Y., across the Hudson River from Albany, and provided Boston with substantial sleeping car service to the west in the years before World War I. Virtually all of this was gone by the Depression, however, except for a Chicago Pullman. But in 1907, for example, B&M trains carrying Chicago accommodations outnumbered their B&A counterparts,[7] and the service was by no means limited to Chicago. The most important advantage the Fitchburg line had over the B&A was its Hoosac Tunnel.

All east-west traffic through Massachusetts must

New Haven's *Fall River Boat Express* races a local near Boston Oct. 11, 1930. This train-boat operation lasted 90 years.

cross the formidable barrier rising between the Hudson and Connecticut River valleys, the Berkshire Mountains. Contrary to its lessor's water-level image, the B&A went up and over, eventually honoring the heavy class of steam locomotives that did so much of the work by naming them for the mountains. The B&M, on the other hand, went through, rather than over, by building the Hoosac Tunnel, dug by hand through 4.75 miles of shale and granite.

When it was opened in 1875, after more than two decades of construction, the tunnel was hailed as an engineering wonder, which, for the times, it truly was. It was the longest such bore and would be surpassed in the United States by only two other tunnels, both built with the benefit of Twentieth Century technology. Unfortunately, the Hoosac was not achieved without heavy human cost: 195 men lost their lives during its construction.[8]

Despite a ventilation shaft to the center of the tunnel, smoke from steam locomotives proved to be a great hazard, and the railroad decided to electrify the Hoosac and several miles of approaches in 1910. After its two tracks were put under wire, the engineers could bank their fires, and electric locomotives towed the steam engines through. (Some of the work was later undone. When diesels came in after World War II, the B&M abandoned the electrification and later removed one of the tunnel's tracks.) Pullmans traveled over the Fitchburg line through the Hoosac Tunnel to Troy (a dedicated sleeper to Troy/Albany operated early in this century). The cars were taken across the Hudson to Albany by the New York Central for forwarding by its trains, or, for a time around the turn of the century, given over to the Delaware & Hudson. Before the New York Central leased the B&A, it carried the B&M's through cars, but afterward it pushed them to its secondary West Shore line between Albany and Buffalo. Much of this joint service was terminated by the government when it operated the nation's railroads during World War I. Thereafter, the few surviving B&M Pullmans to the west traveled on the Central's main line trains.

Pullman service to Chicago on the B&M was indicative of the several arrangements that were in effect. Around the turn of the century, sleepers moved via West Shore to Buffalo, west of which some were forwarded by Nickel Plate, others by the Wabash. The D&H, meanwhile, moved B&M sleepers to Binghamton, N.Y., whence the Erie took them west. With the *Minute Man* service, introduced in May 1926, Chicago sleepers moved on the Central's *Lake Shore Limited.* This was the sole surviving Pullman on the line to Troy.

During the West Shore era, Boston sleepers also traveled to Buffalo, Detroit (Wabash), and St. Louis (Wabash), and service supposedly included a twice-a-week tourist car to the West Coast.[9] In addition to the Chicago sleeper, the D&H/Erie handled cars to Cincinnati and Cleveland.

Before the New Haven/Pennsylvania Hell Gate route was opened through New York City, Boston-Washington trains moved over a variety of routes involving the New Haven's railroad bridge across the Hudson at Poughkeepsie, N.Y. The B&M participated in at least two joint movements of these sleepers to the nation's capital.

By way of the former Boston & Lowell, the B&M reached into New Hampshire, dividing its line in two at the capital city, Concord. One portion went to the interchange with the Central Vermont at the busy rail hub of White River Junction, Vt., and the other went to Wells River, Vt., where it connected with Canadian Pacific tracks.

Some of the country's most unusual sleeping car lines used the B&M's White Mountains Division route to Wells River. Early in the century a tourist sleeper operated from Boston all the way to Vancouver, B.C. (Canadian Pacific), the longest transcontinental line in the country. Only slightly less remarkable was the sleeping car line, inaugurated in 1889, to Minneapolis and St. Paul; the CPR delivered the sleeper to the Soo Line at Sault Ste. Marie. The White Mountains line was one of three sleeper routes between Boston and Montreal (CPR), and it also saw dedicated service via Quebec Central to Levis, Sherbrooke, and Quebec. The overnight *Red Wing* and the daylight *Alouette* were the major trains.

On a smaller scale, the B&M carried a Pullman to the junction of Woodsville and one to Bretton Woods via the railroad's own route from the north. It also once turned over a sleeper for Montpelier to the Montpelier & Wells River Railroad at Wells River.

The daylight *Ambassador* and overnight *New Englander* traveled another major route to Montreal, this one passing through White River Junction. There the B&M turned over to the CV a Chicago-bound sleeper that continued west on the lines of the Canadian National, which controlled the CV. On its own tracks, the CV carried Pullmans to Burlington, Montpelier, and

St. Albans, Vt., and with the CNR to Montreal and Depot Harbor, Ont.

The third Boston-Montreal line, which predeceased the other two, followed the B&M's Fitchburg route, then branched off to Bellows Falls, where the Rutland picked up the sleepers. This was the route of the daytime *Green Mountain Flyer* and the overnight Boston section of the *Mount Royal*, which carried Pullmans for Ogdensburg, N.Y., and Montreal, the latter of which the Rutland turned over to the Canadian National at Rouses Point.

The original Boston & Maine line went north through Lawrence and Dover. It was on these tracks that railroad history was made in 1848, when the B&M's little thirty-five horsepower locomotive *Antelope* made the first mile-a-minute train run, from Boston to Lawrence.[10]

After the B&M acquired the competing Eastern line through Portsmouth, the railroad had two routes to Portland. Initially it split the through passenger traffic between the two lines and even operated the international *Gull* as well as Boston-Portland sleepers northbound on one line and southbound on the other. In time, however, the Dover line got most of the traffic.

The B&M gave Boston year-round Pullman service, with Maine Central, to the Maine destinations of Bangor, Calais, Waterville, as well as the Canadian cities of Saint John (CPR), Moncton (CPR/CNR), and Halifax (CPR/CNR). In the summertime, it carried sleepers to the Maine resorts and vacation gateways of Farmington, Kineo Station, Mount Desert Ferry, and Rockland.

In the northern potato country of Maine, the little Bangor & Aroostook ruled, picking up Boston Pullmans for Van Buren, Greenville, and Caribou from the Maine Central. The *Potatoland Special* was the night train to Van Buren, the *Aroostook Flyer* the daytime accommodation.

Charles B. Chaney/Smithsonian Institution

Boston & Maine's *Minute Man* raises the dust in East Deerfield, Mass., July 1939. B&M carried Boston Pullmans for the west.

The trains serving Boston dominated New England passenger travel, but they were by no means the whole story. Both Springfield, Mass., and Portland, Me., were rail centers in their own right, and each one had a fair amount of dedicated Pullman service. Furthermore, a substantial volume of passenger traffic moved between New York and northern New England and Canada, bypassing Boston.

In Springfield, the New Haven delivered sleepers to the Boston & Maine, whose Connecticut River line served the White Mountains region of New Hampshire and connected with the major Canadian lines. This, plus the Boston & Albany main line through town, gave Springfield good train service, from the New Haven/B&M *Montrealer* and *Washingtonian,* to the Central's *New England States, Wolverine,* and *Southwestern Limited.*

The B&M operated dedicated sleepers from Springfield to the Canadian cities of Sherbrooke (CPR/QC) and Montreal (CV/CNR). A Pullman operated on the New York Central to Syracuse. The remainder of Springfield's service, via New Haven, included Pullmans to New York and (with the Pennsylvania) Philadelphia, Pittsburgh, Washington, Miami (RF&P/ACL/FEC), and St. Petersburg (RF&P/SAL).

Worcester, a major rail junction with a well-designed, elevated Union Station, saw mostly through Pullmans, either on the B&A's Boston trains, or on the New Haven/B&M expresses to Maine. Worcester, however, did have a dedicated sleeper to New York, which the B&A turned over to the New Haven in Springfield.

In addition to being an important rail terminus, Portland was also an all-weather seaport that challenged Boston. In the 1850's, when the future Grand Trunk Railway Company of Canada's line between Montreal and Portland was being constructed, the investors were not sure whether the railhead should be located at the Maine seaport or Boston. They conducted a test, sending mail pouches from Liverpool to Montreal by way of both Portland and Boston; the pouch that reached the Canadian city first would determine the choice of seaport. Thanks to a well planned relay of horses, the sleigh driver from Portland entered Montreal, an American flag fluttering beside him, hours before his Boston competitor arrived.[11] Portland got the railroad and became a major transshipper of Canadian freight.

Between 1890 and World War I, three sleeping car lines operated between Portland and Chicago, including one by Canadian National, and two others that departed from Portland on the Maine Central line into New Hampshire.[12] The CNR carried sleepers to Levis and Montreal, and MEC and CPR offered joint service to the latter.

Portland had a nine-track Union Station, where the B&M and MEC connected. Many of the important trains operated straight through Portland, and the cars of the two roads wore the same tuscan red in later years, when the B&M managed the MEC (after this arrangement ended in the 1950s, MEC went to a bright green). Northbound on the *Gull,* the Maine Central carried a Pullman to Vanceboro, Me., on the Canadian border. South on the B&M, Portland had Pullman service to Boston, New York (NH), and Philadelphia and Washington (NH/PRR).

Trains, or sometimes just individual cars, came to Maine from as far south as Washington, via the Pennsy, New Haven, and Boston & Maine. The *State of Maine* was a year-round train that carried sleepers from Washington and New York to both Portland and Concord and to some of the northern resorts in its swollen summertime consists. Avoiding Boston, where there never was a direct link between South and North stations, the train traveled via Worcester and Nashua, N.H. (later, Lowell, Mass.). A train with a similar route but far more exclusive pedigree was the *Bar Harbor Express.*

In the 1880's the Maine Central began ferry service to Mount Desert Island off the Maine Coast and contructed a branch from the main line down to the ferry landing. New England's first limited train, the *Boston & Mount Desert Limited,* carrying elegant umber-colored sleepers with gilt trim and lettering, was inaugurated in June 1887.[13] This train lasted only two years, but meanwhile the island's posh resort of Bar Harbor was attracting an expanding circle of wealthy families, and not just Bostonians.

The *Bar Harbor Express,* the first through train to the resort area from New York, was introduced in the summer of 1902, and, except for wartime suspensions, it carried America's well-to-do to their summer homes in Maine for nearly six decades. Numerous "camp extras," solid trains of as many as twenty Pullmans each, were scheduled north in June on both the *Bar Harbor* and Connecticut River valley routes, carrying the young scions to meet their waiting counselors. The flow was reversed in August.[14] After the traffic crested in both directions, the baggage room at Portland Union Station was filled to the ceiling with the children's trunks, waiting to be forwarded.

The New Haven originated the *Bar Harbor Express,* and for the first few years, the Boston & Albany carried it between Springfield and Worcester, before handing it over to the B&M. The B&A ended its participation, however, and a succession of three different New Haven routings followed. In Portland the cars were turned over to the Maine Central. After the opening of the Hell Gate route through New York, separate sections of the *Bar Harbor Express* originated in both Washington and Philadelphia on the Pennsylvania Railroad.

Otto C. Perry/Denver Public Library

The Boston & Albany's *Lake Shore Limited* steams into Springfield, Mass., on Aug. 22, 1933 with a six-car consist.

A former Pullman conductor recalled working the train when it carrried the cream of Philadelphia society: "The first trains of each season were filled with housekeepers, cooks, maids, butlers, gardeners, and other domestics being dispatched north to prepare the summer mansions for the arrival of their employers around the first of July. All went ahead except, of course, the chauffeurs, who did not depart until they had safely installed their employers aboard the train at the appointed time; then there was an interstate parade as they raced those custom-bodied Rolls-Royces north on Route 1 as fast as machinery and law allowed. Their masters were not ones to be kept waiting."[15]

The *Bar Harbor* reached full flowering in the 1920's, when it operated daily with cars to Mount Desert Ferry, Ellsworth, and several other resorts, and a separate section going to Rockland. On Labor Day 1923, the Maine Central turned over 102 cars to the Boston & Maine in Portland,[16] but the public accommodations were only part of the story. In one day, the *Bar Harbor*, operating in three sections, carried twenty-one private cars.[17]

The patrician era ended with the onset of the Depression, and in 1931 the island's exclusivity was diminished with the opening of a causeway and the resulting termination of the ferry service. The creation of Acadia National Park on Mount Desert Island in 1916 had no doubt helped open the gates to ordinary mortals, and in the late 1930's the place was further popularized by one of the special trains of the era designed for the budget-conscious traveler. In June 1940, the deluxe, canary-yellow, all-coach *East Wind* went into service between Washington and Maine, operating on a daytime, seasonal schedule until it was terminated during World War II.

With the return of peace, the *Bar Harbor* resumed service several times a week. Highways and airlines, however, continued to take away its patronage, and the celebrated *Bar Harbor Express* made its final run on Labor Day 1960. Within a year, Maine saw no more rail passenger service of its own north of Portland.

The three principal destinations of the *Bar Harbor*—Rockland, Ellsworth, and Mount Desert Ferry—all had dedicated sleepers from Washington, Philadelphia, and New York. Over the years the train also carried New York Pullmans to Farmington, Kennebago, Kineo Station, Van Buren (BAR), and Waterville, and a Washington car to Oquossoc.

The *Bar Harbor Express*, of course, skirted Boston, but the regular trains of the MEC/B&M carried many Pullmans there from Maine. Sleepers regularly operated to Boston from Bangor, Calais, Waterville, and Van Buren (BAR). In the summertime, Farmington, Kineo Station, Mount Desert Ferry, Rockland, and (with BAR) Caribou and Greenville all had dedicated Pullmans to Boston.

Several Maine resorts were especially popular with Montreal people, and early in the century the Grand Trunk Railway offered service, with B&M, to Kennebunkport and Old Orchard Beach, as did the CPR to the latter (MEC/B&M).

In one case, the Canadian Pacific ran its tuscan red sleeping cars through Maine entirely on its own tracks, except for the rights it exercised on the MEC line from Mattawamkeag to Vanceboro (which it later purchased). Despite their stateside sojourn, both origin and destination of the cars were Canadian. They were the sleepers operated between Montreal and Saint John, N.B., on the CPR line through Greenville. In the early 1890's, the CPR and MEC carried a Bangor-Saint John sleeping car.

With the exception of Concord's service on the B&M/New Haven to New York and Washington (PRR), virtually all of New Hampshire's Pullmans were summer-seasonal. Bretton Woods, the resort in the midst of the Presidential Range, near the famous Mount Washington cog railway, received sleepers on the *Night White Mountains Express* (NH/B&M/MEC) from New York, Philadelphia (PRR), and Washington (PRR). Sleepers came from New York (NH/B&M) to Claremont, Jefferson, Keene, Lake Sunapee, Plymouth, and Woodsville (MEC), and a Washington car (PRR) ran to Plymouth. Boston Pullmans operated to Bretton Woods and Woodsville on the B&M. In the 1880's, a sleeper from Syracuse came to nearby Fabyan's station and resort via NYC/Rutland/CV/Montpelier & Wells River/B&M.

Except for Boston sleepers from Burlington, Montpelier, and St. Albans (all CV/B&M) and another from Montpelier (M&WR/B&M), Vermont's other dedicated Pullman service was directed to New York City, and the two railroads that were virtually the state's own—and onetime bitter rivals—carried most of it.

Vermont's premier railway was the Central Vermont, a line that cut diagonally across the state after leaving the Connecticut River Valley at White River Junction, then traversing the main pass of the Green Mountains, through the capital of Montpelier on the way to its headquarters city of St. Albans. The CV's Pullman trade was mostly with Boston, but it did carry two New York sleepers from St. Albans, one via B&M/NH, and an earlier one in company with the Rutland/B&M/NYC. (St. Albans, scene of the most northerly raid in the Civil War—a group of Confederates in mufti robbed all the banks in town—was also known for its picturesque trainshed. CV trains once entered the cavernous interior of this fortress against New England winters through an ornate facade that had brick-arch portals above each of the four tracks.)

The B&M carried a sleeper from White River Junction to New York.

The Rutland, which drew its name from the Marble City, was Vermont's other name railway, running the length of the state from Lake Champlain south along the flank of the Green Mountains. It served as a bridge route for Montreal traffic and also had a line to Ogdensburg, N.Y., where it interchanged Great Lakes freight. The Rutland terminated passenger operations in the early 1950's, and after bitter labor disputes abandoned its whole system a decade later. In better times, however, it carried Pullmans to New York (B&M/NYC) from Alburg, Burlington, and Rutland.

The Maine Central's Mount Desert Island section of the *Bar Harbor Express* unloads at Ellsworth, Me., Aug. 10, 1951.

Some sizable communities of southern New England had relatively little dedicated Pullman service, but the New Haven's monopoly was probably not a factor. The travel distances to Boston and New York were short and the train service to them frequent, and so there was not a great need for dedicated sleepers from these cities.

Pullmans left Providence on the New Haven for New York, Philadelphia (PRR), and Washington (PRR). Hartford, whose historic station was destroyed by fire in 1914 and rebuilt within the existing walls, had sleeping cars to New York (NH) and Washington via the Poughkeepsie Bridge route. New Haven had Pullman service to Boston. Other southern New England sleepers were seasonal.

The *Night Cape Codder* (NH) brought Pullmans from New York to Woods Hole, from which ferries to Martha's Vineyard and Nantucket operated, and to Hyannis, the cape's summer business center. Sleeping cars for Hyannis also originated in Washington (PRR).

For a few seasons before World War I, the Central

Jim Shaughnessy

Canadian Pacific's Montreal-Saint John, N.B., trains crossed northern Maine. Here overnighter No. 39 pauses at Brownville Jct.

Vermont operated the *Seashore Express,* an all-Pullman train running from Montreal (Grand Trunk) to New London.

Before the Depression, a Sunday evening sleeper left the summer colony of Narragansett Pier, R.I., on the railroad of the same name, bound for New York. The New Haven picked up the car at Kingston, R.I.

Apparently the infrequency of Pier service resulted in occasional lapses. One night, according to a historian's account, the NPRR train roared off to Kingston, only to have to return because the chagrined crew had forgotten to attach the sleeper.[18]

1. Warren Jacobs, "The Story of the New England," *RLHS Bulletin,* No. 1 (1921), pp. 17-18.

2. George H. Drury, *The Historical Guide to North American Railroads* (Milwaukee: Kalmbach, 1985), p. 225.

3. Arthur D. Dubin, *Some Classic Trains* (Milwaukee: Kalmbach, 1964), p. 54.

4. Charles E. Fisher, "Through Car Service From New England," *RLHS Bulletin,* No. 87 (October, 1952), p. 51.

5. Willard H. Hart, "...Of Trains That Once Ran," *NRHS Bulletin,* XXII, No. 1 (1st Quarter, 1957), pp. 24-5.

6. John A. Droege, *Passenger Terminals and Trains* (Milwaukee: Kalmbach, 1969. Reprint of original, New York: McGraw-Hill, 1916), p. 114.

7. Fisher, p. 56.

8. F. M. Cramer, "Hoosac Tunnel," *Trains,* Vol. 2, No. 4 (February 1942), p. 30.

9. Details are sketchy. A Nickel Plate schedule for its Train No. 1 in the *Official Guide* of August 1909 (p. 299) lists such a car as leaving Boston on the B&M, probably traveling West Shore to Buffalo, then NKP to Chicago (No. 2 carried the car eastbound). I could find no corroborating details, however, from the schedules of other railroads. A similar listing of even earlier service appears in "Westbound Nickel Plate Express Passenger Trains—1902," a tabular accommodation summary published in John A. Rehor's *The Nickel Plate Story* (Milwaukee: Kalmbach, 1965), p. 47. That tourist sleeper also was carried on No. 1, with the same routing and via "connecting railroads at Chicago."

10. Edward P. Sanderson, "Railroads of Massachusetts," *Trains,* Vol. 4, No. 5 (March 1944), p. 14.

11. Federal Writers' Project of the Works Progress Administration, *Maine: A Guide Down East* (Boston: Houghton Mifflin, 1937), p. 180.

12. Fisher, p. 57. One line, operated in the summer of 1890, included stops at both Fabyan, N.H., and Niagara Falls (MEC/St. Johnsbury & Lake Champlain/CV-Rutland/NYC-Michigan Central). The other line was listed in a June 1916 schedule (MEC/CPR/MCRR).

13. Dubin, p. 41.

14. Letter of March 19, 1988 received from Donald S. Robinson, Boston & Maine Railroad Historical Society.

15. William Moedinger, "Diary of a Pullman Conductor—2," Trains, Vol. 30, No. 5 (March 1970), p. 43. Copyright Kalmbach Publishing Co.; reprinted with permission.

16. Dubin, p. 42.

17. Lucius Beebe and Charles Clegg, *The Trains We Rode* (Berkeley: Howell-North, 1965), p. 959.

18. James N.J. Henwood, *A Short Haul to the Bay: A History of the Narragansett Pier Railroad* (Brattleboro, Vt.: The Stephen Greene Press, 1969), p. 22.

2
Going to Chicago

Under private enterprise, American passenger trains offered travelers almost 120 years of service between the country's two largest cities, and, but for the last few, the accommodations were plentiful and frequently luxurious. Naturally the competition pitted the two giant passenger carriers, the New York Central and the Pennsylvania Railroad, against each other, and both of them committed entire fleets of crack express trains to the 900-mile-plus journey. Until the final few years, the railroads' two premier passenger trains, the Central's *Twentieth Century Limited* and the Pennsy's *Broadway Limited,* ran the race in the manner of noble sportsmen, never allowing the other as much as a minute's advantage in time, never conceding a single perquisite of luxury rail travel.

Those two railroads accounted for most of the accommodations between New York and Chicago, yet various other lines were involved in the competition at one time or another: the Baltimore & Ohio and the Erie Railroad ran their own through service, and the Delaware, Lackawanna & Western, the Lehigh Valley, the New York, Ontario & Western, and the West Shore Railroad (a New York Central leasee) operated joint service with western affiliates that included the Nickel Plate, Wabash, Michigan Central, and Canadian National/Grand Trunk Western. Shortly after the Civil War the Central of New Jersey advertised through service also involving the Reading and Pennsylvania. Over the years, all these combinations created more than a dozen and a half sleeping car lines between the two metropolises.

The New York Central instigated the competition, when, in 1852, both the Michigan Central and the Lake Shore & Michigan Southern—two future components of the NYC system—entered Chicago. In Upstate New York, the Central went through the Mohawk River valley, thus availing itself of the only total cleavage in the Appalachian Mountains south of Vermont. From this came its slogan, the Water Level Route.

By 1855, passengers were traveling between New York and Chicago in 36 hours, including ferry trips at Albany, Niagara, and Detroit,[1] but not until after the Civil War could they enjoy the comfort of Pullman sleepers over portions of the route. A quickening of pace as well a break from the bright crimson livery of the era came with the introduction of the *Fast Mail* in 1875. This briefly operated train, painted a gleaming white and buff with gold ornamentation, cut running time to slightly more than 27 hours and introduced the practice of postal crews' picking up and throwing off mail bags at speed.[2]

In 1893, the Central set a 20-hour schedule for the *Exposition Flyer,* which catered to passengers traveling to the world's fair in Chicago. The train operated just for the season, but in 1897 the railroad introduced its luxurious *Lake Shore Limited,* an all-sleeper consist pro-

The *Twentieth Century Limited* departs Chicago in 1935 under power of the shrouded *Commodore Vanderbuilt* locomotive. The famous train's more familiar livery (inset) was the Henry Dreyfuss-designed streamlining introduced in 1938 on its 36th anniversary.

Main photo copyright Chicago Tribune; inset New York Central photo by Arnold Hass, collection of Howard W. Ameling.

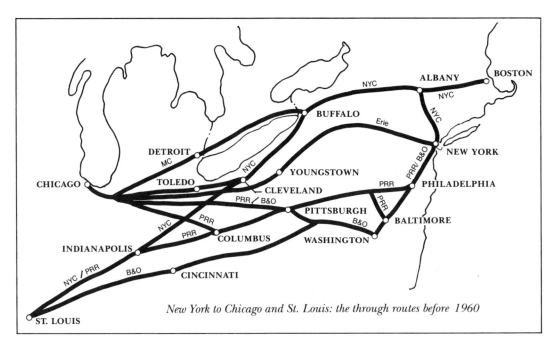

New York to Chicago and St. Louis: the through routes before 1960

duced by the Wagner Palace Car Company (NYC's exclusive supplier of sleeping cars until Wagner was absorbed by Pullman in 1899). When the *Twentieth Century Limited* was introduced on June 15, 1902 it also ran on a 20-hour schedule that over the years was pared to 16, and sometimes less during the streamline era.[3]

The *Twentieth Century* was the most famous of all American trains. It was spotlighted on stage (Ben Hecht and Charles MacArthur's "Twentieth Century") and in films ("North By Northwest" and "The Hucksters") and in countless journalistic renderings whenever city editors yearned for evidence of overindulgent travel spiced with the words of the rich and famous. The all-Pullman *Century* spared no efforts or cost in providing the perks of Brahmin travel, starting with the ritual unfurling of the red carpet down the platform in Grand Central Terminal.

Each train required a crew of nearly 70 persons, which meant on the average fewer than two passengers for each crew member. In addition to the liberal staffing of sleeper and dining car attendants, the train carried maids, barbers, manicurists, and valets, as well as male secretaries who, for eastbound runs, were sent ahead to Elkhart, Ind., where they transcribed closing stock prices, then boarded the *Century* to convey the market news they had gathered.[4] Men and women prominent in theater, sports, banking, formal society, and so forth, could avail themselves of functionaries who arranged meetings "to accommodate travelers who desired the pleasure or business convenience of one another's company en route."[5] The train's floral bill for complimentary corsages and boutonnieres and table arrangements ran to mob funeral proportions.

For such an exclusive train, the *Twentieth Century Limited* was immensely popular. Said a New York Central advertisement in early 1930: "The *Twentieth Century Limited* long since ceased to be a 'train.' It is a daily fleet of trains. In the past year this famous standard bearer of New York Central passenger service has been operated as 2,153 trains and has carried more than 240,000 passengers." The record load came in January 1929, when the Central filled seven separate sections of the train, carrying 822 passengers to New York for the national automobile show. To maintain the service, the Central kept 24 locomotives and a fleet of 122 identical luxury sleepers on call,[6] and the railroad operating instructions specified that the cars always be emplaced so that all private room space would face the Hudson River on the scenic main line north of New York City.[7] After it changed from electric motive power at Harmon, N.Y., the stops that the *Century* made before reaching Englewood Station in Chicago were both few and brief.[8]

The most spectacular train sets for the *Century* were designed by Henry Dreyfuss and introduced June 15, 1938, the train's 36th birthday. The streamlined equipment, painted two shades of gray, with blue and silver accents, contained all-room sleeping cars—no more open-section Pullmans for this great limited that was now carded at 16 hours flat. Ten Hudson locomotives, in bullet-nosed, streamlined shrouds, their drive wheels painted a silvery aluminum, were reserved to power the superlative new trains.

A newspaper editor once recalled his experience as a schoolboy, when he and friends crept down to the main line near Peekskill, N.Y., and awaited the passage of the westbound *Century*, listening to its whistle grow louder as it echoed off the mountains along the

Hudson:

"Finally out of the night she came, the *Century* . . . only the stoutest of young hearts could survive the absolute agony of excitement on that quaking road-bed. First, the appalling white headlight, jerking and leaping in a kind of mad disembodiment in the night, and then the awful noises of a devil's thunder symphony rushing down, all crescendo. There simply never was anything like the shattering explosion of sound when that great 4-6-4 Hudson yanked the *Century* past . . . It was all over in one flashing spasm of delight . . ."[9]

With the *Twentieth Century Limited's* ten-dollar extra fare (a princely sum, considering that until the early 1930's, ten dollars was the equivalent of half an ounce of gold) the train was a highly profitable operation in its heyday. (Traditionally, the Central refunded a dollar to each passenger for every hour the *Century* was late.) In its lifetime, the train grossed more than $200 million, a billion and then some by 1980's standards.

Unfortunately the legend proved mortal. In April 1958, the financially ailing New York Central stripped the train of many of its luxury features and added coaches. But even spartan measures proved insufficient to restore the train's profitability. On December 4, 1967, the future spy novelist Bill Granger wrote an obituary for the Chicago *Tribune:* "The *Twentieth Century Limited,* one of the world's most famous passenger trains, died yesterday in La Salle Street Station after a lingering illness. It was 65." Derailment of a freight train had delayed its arrival from New York nine hours.

The New York Central's Great Steel Fleet accounted for about a sixth of Pullman's entire business (as did the Pennsylvania), and the New York-Chicago

traffic was a great part of the total.[10] Even as late as April 1950, by which time the airlines were carrying more passengers than were traveling by Pullman, there were 37 trains containing 412 sleeping cars, providing more than 6,000 rooms, west of Albany, on the Chicago to New York City main line on a typical midnight.[11] That would have been equivalent to double the capacity of Chicago's giant Stevens/Conrad Hilton Hotel, then the world's largest.

The Central's purchase of more than 700 new passenger cars after World War II no doubt contributed heavily to its later financial hardship, but for at least a decade they provided superb accommodations on its trains of the night between New York and Chicago.

In addition to the *Twentieth Century Limited,* the New York Central operated a fleet of trains between the nation's two largest cities, the following of which were all-Pullman during their golden years:

• *Commodore Vanderbilt,* named for founding father Cornelius when the train was introduced in 1929 as a virtually identical, later section of the *Twentieth Century Limited* that developed its own loyal clientele. Both the *Century* and the *Commodore* also had Advance versions, exact replicas in equipment and service of the originals.

• *Lake Shore Limited,* the illustrious predecessor of the *Century* that continued to garner acclaim as an extra-fare train offering early evening departures from both termini, as well as Chicago-Boston service. The post-war *Lake Shore* carried San Francisco sleepers for both the Overland and the Zephyr routes.

• Such other onetime, all-sleeper trains as the *Wolverine* and *North Shore Limited* via Detroit, the *Niagara,* whose sleepers were spotted for daytime viewing of

Three miles into steam territory, the westbound all-Pullman *Commodore Vanderbilt* hurtles through Oscawanna, N.Y., in 1946.

The Pennsylvania Railroad's prime flyers: At left, the *Broadway Limited* briefly pauses in Harrisburg on an August night in 1965. In this 1948 rendering (above), the *General* steams into Chicago on the Pennsy's broad, elevated main line.

the falls; and the eastbound *Fast Mail, Fifth Avenue Special,* and *Mohawk,* and the westbound *Iroquois* and *Chicagoan.*

The New York Central offered Pullman service from New York's Grand Central Terminal in addition to operating sleepers jointly with several other lines (see below).

The Pennsylvania Railroad had the advantage of a shorter route to Chicago, but its trains faced the arduous climb to its crest of the Alleghenies at Gallitzin, Pa. Despite the formidable engineering task of building such a mountainous right-of-way, the railroad completed its line from Philadelphia to Pittsburgh in 1854. By way of the railroad it later leased, it reached Chicago in 1858.

In the autumn of 1881, the Pennsy inaugurated the country's first extra-fare, all-Pullman train, which was to become known as the *Pennsylvania Limited.*[12] The flyer, which made the New York-Chicago run in just under 26 hours, would also become the first all-vestibuled train in America as well as the Pennsylvania's oldest name train. The smart-looking consist got a further adornment in 1898, when it was painted in a green, cream, and red livery, colors presumably inspired by a Mexican presidential train that railroad officials had chanced to see.

On the day the New York Central first dispatched its *Twentieth Century Limited,* June 15, 1902, the Pennsy countered with its new *Pennsylvania Special.* Three years later a speed war ensued, when the *Special* reduced its running time to 18 hours and the *Century* took up the gauntlet. On its first outing on the new schedule, June 11, 1905, the *Special* lost time to a

hotbox, but with a substitute freight locomotive, Engineer Jerry W. McCarthy drove the train to a record speed of 127.1 miles an hour near Elida, Ohio. Because passengers tended to confuse the *Limited* with the *Special,* the railroad in 1912 renamed the latter the *Broad Way Limited* in honor of its six main-line tracks between New York and Philadelphia. (In time, the one-word usage became standard, paying tribute, as far as the public was concerned, to the famous Manhattan thoroughfare.)

Although the *Broadway* never attained the fame or popularity that the *Twentieth Century Limited* did, it nonetheless kept pace by offering everything expected of a luxury train. Designer Raymond Loewy oversaw the streamlining of the *Broadway* in 1938. The new Tuscan red cars were pulled by Brunswick green GG1's under catenary and the faithful K4 Pacific locomotives west of Harrisburg. Like the new *Twentieth Century,* which bowed on the same day, June 15, the *Broadway* was an all-room train.

One of the great traditions of the two palladins was the unofficial race they staged after leaving Englewood Station on Chicago's South Side. The Chicago *Tribune's* late columnist Will Leonard recalled watching the contests from his father's office south of Englewood.

"By the time they passed 75th Street," Leonard wrote, "they were at full speed, on tracks only a few feet apart, with New York a thousand miles away. The railroads always said the trains were not engaged in a race, but you couldn't tell that to the men in my father's office. Work stopped every afternoon about 3:15, and the entire office force gathered to watch the giants charge by, the *Broadway Limited* on the right-hand track closest to the windows, the *Twentieth Century* on the other. . . . Usually, as we recall, it was a dead heat."[13]

In the end, the *Broadway* outlasted its worthy rival. Nine days after the *Century's* ignominious death, the *Broadway* made the last all-Pullman run in the country. The next day, December 13, 1967, it acquired the coaches of the Pennsy's second-ranked *General,* itself a onetime all-Pullman limited. And if it was any consolation at all, when Amtrak took over on May 1, 1971, it was the *Broadway* name and route that survived.

The *Broadway Limited,* of course, was the flagship, but the numerous fine trains of the Pennsylvania made its four-track main line near the crest of the Alleghenies no less exciting to watch than the New York Central's. In the Pennsy's namesake state, the famous Horseshoe Curve, which helped westbound trains climb the summit, was the place to be. On the eve of World War II, a marvelous midnight parade of twenty-seven limiteds, starting with the *General* and ending with the *Broadway,* negotiated the curve within a five-hour period ending at 3:41 a.m., "sparks flying from stacks of westbound locomotives, sparks from brakes of eastbound trains."[14]

The procession included other Pennsylvania trains in the New York-Chicago fleet: the *Golden Arrow,* the road's second-ranked train from its introduction in 1929 until the birth of the *General* eight years later; the all-Pullman *Manhattan Limited; Admiral; Gotham Limited;* and longest-running train, the *Pennsylvania Limited.* West of Pittsburgh, the *Metropolitan* and *Commercial Express* also operated Pullman-only.

In later years, all the Pennsylvania's New York-Chicago sleepers moved by way of the Fort Wayne line, but at one time a trip via the Panhandle route through Columbus, Ohio, was possible. Trains departed from Pennsylvania Station in midtown Manhattan.

In the post-Civil-War years, when numerous car manufacturers were tooling up for the growing sleeping car trade, the Pennsylvania participated in a New York-Chicago line on which Silver Palace cars were operated over what the CNJ advertised as its "Allentown Route." According to schedules announced for the spring of 1869, the anthracite-hauling Central of New Jersey took the sleepers to Allentown, where the Reading picked them up and forwarded them to the Pennsy at Harrisburg. Although sleeping cars were then operating on the modern route through Philadelphia, the Pennsylvania did not yet have a controlling connection to New York.

The Baltimore & Ohio offered New York-Chicago Pullman service by way of Washington, Pittsburgh, and Akron, until its Royal Blue train service ended in April 1958, and Baltimore became the road's eastern terminus. The through trains often didn't take their final shape until they reached Washington, where additional cars were put aboard such trains as the *Capitol Limited* and *Shenandoah.* Except for a brief period during and after World War I, when they used Penn Station in New York, B&O trains departed from the Jersey City terminal of the Central of New Jersey. Trains of the B&O operated on CNJ to Bound Brook, N.J., then over the Reading to their own tracks at Park Junction, Philadelphia. (See District of Columbia chapter for more on B&O service.)

The Erie Railroad was the fourth solo Pullman-carrier between New York and Chicago. In 1851 this 446-mile line had the distinction of being the first single railroad to connect the Atlantic Ocean and the Great Lakes; it was also the country's longest and most important single line at the time.[15] The Erie got access into Chicago in 1880, the same year it converted to standard gauge from the monstrous six-foot track width it had used during its first four decades of operation. Erie iron was laid between the main lines of the New York Central and Pennsylvania and avoided any large cities; its route was longer than the competitors' and its green and gray trains were more leisurely. The *Erie Limited,* introduced in June 1929 on a twenty-five-hour schedule, somehow managed to justify its $4.80 extra fare with the quality of service it offered. The *Lake Cities* and a third train also carried sleepers from Jersey City to Chicago. In addition to

The fury of an early snow storm shows on the B&O's Capitol Limited *in Chicago's Grand Central Station, December 1962.*

Richard J. Cook

Donald Sims

Two smaller railroads offered New York-Chicago travel alternatives. The Erie's *Lake Cities Limited* (bottom) makes time through Indiana in the 1940's; and the eastbound *Nickel Plate Limited* (top) crosses bridge at Painesville, Ohio, in 1948.

Before 1960, however, DL&W had several partners in the Hoboken-Chicago trade, the longest standing of which was the Nickel Plate, whose tracks went west from Buffalo to Chicago, Peoria, and St. Louis. The premier *Lackawanna Limited's* midwestern partner was the *Nickel Plate Limited*, the cars of which, in the streamline era, wore blue and stainless dress. In the standard era, the DL&W ran four trains with sleepers to Chicago, including the *Lackawanna Limited*, *Chicago Limited*, *Western Special*, and *Whitelight Limited*, with departures spaced throughout the day. Not all the Pullmans, however, were switched over to NKP. Some finished the trip on the rails of the Michigan Central and others went via Wabash.

In November 1949 the Lackawanna resurrected an old and honored name for its new daytime, maroon, yellow, and gray streamliner to Buffalo: *Phoebe Snow* (a woman who, advertising legend had it, kept her white gown spotless by riding the road of clean-burning anthracite). The westbound *Phoebe Snow* was one of the Lackawanna's two trains carrying Chicago sleepers (the *Westerner* was the other), both with NKP connections. The streamliner was extended to Chicago after the EL merger, suspended, then revived for several more years before expiring in 1966. The Erie's old *Lake Cities* was the last sleeper-carrying, long-distance EL passenger train.

The name of its premier train, the *Black Diamond*, spoke volumes about the prime cargo of the Lehigh Valley, another anthracite hauler that terminated at Buffalo. The train was inaugurated in May 1896 as a daytime express from Jersey City to Buffalo, with a connection from Philadelphia. As the Lehigh Valley's passenger-carrying ambitions expanded, sleepers were added to the train. From 1918 on, LV trains operated into and out of Penn Station in New York. For many years the Canadian National picked up sleeping cars from LV trains at Niagara Falls. A Chicago sleeper, switched in this manner, rode the *Black Diamond*, and the *New Yorker* and *Chicagoan* offered through Pullman service between the two namesake cities. The Lehigh Valley also teamed with Michigan Central, turning over another Chicago sleeper in Buffalo. Unfortunately, the railroad's freight revenues ebbed as coal traffic declined, hindering its efforts to support an increasingly marginal passenger operation. A stream-styled *Black Diamond* in Cornell red and black, designed by Otto Kuhler, was part of the road's 1940 refurbishing, but after a wartime boom, it was all downhill in the scenic LV terrain. The *Black Diamond* died in May 1959, and all Lehigh Valley passenger service followed two years later, making it the first major eastern road to become freight-only.

One more marginal carrier struggled for passenger business in this region, but the New York, Ontario &

running on its own, the Erie in 1890 handed off Chicago sleepers to the Canadian National at Niagara Falls and of course offered consolidated service after merging with the Lackawanna in 1960.

Before the Erie Lackawanna merger, Buffalo was the western end of the Delaware, Lackawanna & Western Railroad. The EL consolidation made good sense inasmuch as the Lackawanna, another six-foot gauge railroad at birth, was an anthracite road whose scenic right-of-way through the Delaware Water Gap and the Poconos was virtually parallel to the Erie's.

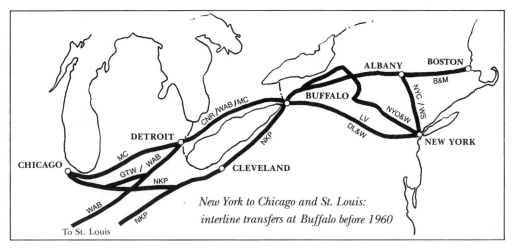

New York to Chicago and St. Louis: interline transfers at Buffalo before 1960

Western Railway—dubbed, in its geriatric years, the "Old Woman" by Wall Street—had the further distinction of going across the grain of the mountains in its trek from nowhere to nowhere. Wall Street, of course, does inflate; in fact, the OW ran from Weehawken, across from New York City, to Utica and Oswego, N.Y. This ocean-to-lake freight line had a brief taste of big-time passenger activity early in the century, when it operated New York-Chicago Pullmans over two routes: It transferred cars at Oswego to the New York Central's Rome, Watertown & Ogdensburg line and, later, at Earlville, N.Y., to the Central's West Shore; Wabash provided the Buffalo-Chicago connection in both cases. The OW once had a lucrative passenger business to the Catskills (a Kuhler streamstyled maroon, orange, and black *Mountaineer* served this trade starting in 1938), but even that traffic evaporated, and its freight business never fulfilled expectations. In March 1957, the railroad ceased all operation, in the largest abandonment of a major system up to that time.

A final minor-league conveyor of sleeping passengers between New York and Chicago was the nuisance railroad that became part of the New York Central, itself, the West Shore Railroad, which had been built up the left bank of the Hudson on its way to Buffalo by rivals of the Vanderbilts. Before and after the turn of the century, when it was leased by NYC, the West Shore carried Chicago sleepers that it turned over at different times to three railroads: Canadian National, Nickel Plate, and finally Wabash.

1. Alvin F. Harlow, *The Road of the Century: The Story of the New York Central* (New York: Creative Age Press, 1947), p. 228.
2. "Lake Shore & Michigan Southern," *Trains*, Vol. 2, No. 5 (March 1942), p. 19.
3. The most-quoted passenger on the maiden run was John W. ("Bet-a-Million") Gates, whom Lucius Beebe called "a vulgarian of the first chop." The barbed-wire magnate, it seems, told reporters in New York that the *Twentieth Century Limited* would make Chicago a suburb of Gotham; at the other end of the line, naturally, he reversed the story.
4. Beebe, *Twentieth Century* (Berkeley: Howell-North Books, 1962) p. 47.
5. Beebe, p. 86.
6. Arthur Dubin, "Twentieth Century Limited," *Trains*, Vol. 22, No. 10 (August 1962), p. 26.
7. Beebe, p. 47.
8. "You could not board the *Century* at Utica and be dignified," *Trains* reader Ralph W. Shaw Jr. wrote in the July 1965 issue (p. 47) of the magazine. Shaw recounted how a traveler booking space to Chicago was escorted to whichever of the ten tracks his section of the *Century* was to arrive on, just minutes before the headlight of the locomotive appeared. The train, he wrote, didn't stop—it just slowed to maybe 10 mph. Red caps shot the man's baggage into the one open door, and then a couple of burly trainmen picked up the passenger and heaved him in next. "Two hoots from the locomotive," said Shaw, "and you were highballing to Syracuse."
9. David R. Dear, "So Long, Century," *Passenger Train Journal*, Vol. 9, No. 12 (February 1978), p. 17. This was a reprint of a piece originally published December 5, 1967, in the Elizabeth City (N.C.) *Daily Advance* by Mr. Dear, its editor.
10. When the New York Central announced in January 1958 that it would terminate its contract with Pullman and operate its own sleepers, it started a trend that would ultimately cripple the once-great sleeping car company.
11. "The Return of the Lake Shore," *Passenger Train Journal*, Vol. 8, No. 2 (Fall 1975), pp. 8-9.
12. Bert Pennypacker, "'The Grandest Railway Terminal in America,'" *Trains*, Vol. 44, No. 2, (December 1983), p. 45.
13. *Chicago Tribune*, December 9, 1967.
14. Harry T. Sohlberg, "Horseshoe Curve," *Trains*, Vol. 1, No. 5 (March 1941), p. 29.
15. Wallace W. Abbey, "Route of the Flying Saucers," *Trains*, Vol. 11, No. 7 (May 1951), p. 24.

Metro-North Commuter Railroad

William J. Brennan

Mid-day sunbeams slant through the south windows of New York's Grand Central Terminal (top), illuminating bygone passengers and their ghosted images in this classic time-exposure. Representing all the fine trains that arrived at and departed from the Jersey side of the Hudson River (bottom), the eastbound *Phoebe Snow*, now in its last year of service for the Erie Lackawanna, moves through Newark at dusk in late 1966.

3
New York City

Water has always been the special concern of railroads serving New York City; travel to and from Manhattan invariably has involved ferries, bridges, and tunnels. The predecessors of the New York Central had the least trouble laying tracks out of Manhattan because the Harlem River was easily spanned. Still, Central trains could go far into Upstate New York and never be out of sight of or much higher than the water whose course they followed. By 1845, predecessors of the New York, New Haven & Hartford had built an all-rail route to Boston, but for more than half a century, the Central and the New Haven—because of water—were the only railroads directly serving Manhattan.

The Hudson River was the most formidable barrier, and all railroads terminating on the New Jersey side for many years carried their passengers to and from New York with fleets of ferries. The terminals, all of which developed substantial commuter business, were in Jersey City, Hoboken, and Weehawken. The Pennsylvania, Central of New Jersey, and Erie all maintained terminals in Jersey City, which, of the three communities, saw the most long-distance traffic. (Reading and Baltimore & Ohio trains also used the CNJ station.)

The giant of the west shore railroads was the Pennsylvania. Trains began running between Jersey City and Philadelphia in 1840 in twice-a-day trips that took six hours in each direction, and within a decade, commutation tickets were being sold on the line.[1] The Philadelphia-based Pennyslvania eventually brought the railroad into its camp and by 1873 was offering its own through service to Washington. The gateway at the nation's capital permitted the Pennsy to make important connections to cities all over the South, in addition to its regular main line service to Chicago and St. Louis and intermediate points. The railroad's growth was swift, and by 1908 the Jersey City terminal had 101 Pullman lines originating and terminating within its precincts.[2]

There was talk of bringing this traffic into Manhattan by tunnel as early as 1874, but steam locomotives were obviously unsuited for the job and electromotive technology was not sufficiently advanced.[3] And so for the moment, the huge volume of New York passengers either began or completed their trip on one of the numerous ferries that served Manhattan and Brooklyn. (Entire trains, with their passengers, also made the trip by water. Through Boston-Washington expresses, which first saw service in 1876, were carried via steamer from the New Haven's Harlem station, around the Battery, to Jersey City.)

The Central of New Jersey had a two-level ferry terminal, many of whose twenty tracks served the railroad's substantial commuter operation, but several platforms were restricted to the intercity trains of the three inhabiting railroads. Immediately after the Civil

War, the CNJ participated in joint sleeping car service to Chicago and Pittsburgh (RDG and PRR), but for most of its life confined its interstate service to trains serving eastern Pennsylvania. The Reading operated some crack Philadelphia trains, notably the stainless steel, streamlined *Crusader,* but the sovereign tenants of the station were the twenty-one trains that operated between Washington and New York.

The Royal Blue Line, as it was called, was established in 1890, with the B&O, Reading, and CNJ the contracted operators using custom-built Pullman equipment painted royal blue. In time, the B&O used the color scheme for all its passenger equipment and began using the generic name for specific trains, finally inaugurating in 1935 the famous train of the modern era called the *Royal Blue.*[4] (Trains of this pedigreed service traveled the four-track CNJ main line to Bound Brook, then the Reading to Philadelphia, where they picked up the B&O main line. The routing led to a proprietary relationship, under which the B&O acquired control of the Reading, which in turn acquired control of the CNJ.)

B&O trains couldn't match the schedules of the competing Pennsylvania flyers, however, nor in later years could they offer the convenience of midtown arrival and departures. A long-held dream of gaining access to Manhattan by way of a B&O-controlled transit line on Staten Island and subsequent crossing of New York Harbor through municipal subway tubes eventually foundered.[5] Special trainside motor coaches from various points in New York City had to suffice instead.

Once the consists of the through trains were fleshed out in Washington, the *National Limited, Diplomat,* and *Metropolitan* continued on to St. Louis, and the *Capitol Limited, Shenandoah,* and all-coach *Columbian* to Chicago. But by 1958, the overhead costs of operating into the New York market had become prohibitive, and the B&O terminated all service east of Baltimore. Now, with long-distance service gone, the Jersey City facility had less than a decade of life ahead for its remaining intercity and commutation business. Operation of the terminal and ferry service to Liberty Street in New York ceased in April 1967, when trains were shifted over to Newark's Pennsylvania Station.[6]

Passengers took the Chambers Street ferry to the Jersey City terminal of the Erie, which served a substantial commuter operation and a handful of intercity trains, such as the *Erie Limited* and *Lake Cities*—slower and less expensive conveyances to Chicago, but ones not lacking in congeniality. The Erie was designed and for several decades operated as a six-foot-gauge railroad. Before it was converted to standard gauge in 1880, passengers could travel from the Hudson to the Mississippi River across from St. Louis on the Erie and its affiliates' extra-wide track (which, in turn, accommodated more spacious sleeping cars).

The Hoboken Terminal of the Delaware, Lackawanna & Western opened in February 1907, after a fire destroyed the previous facilities two years earlier. The new terminal, with Tiffany glass skylighting, had sixteen tracks and six ferry slips in which the boats from Christopher and Barclay streets unloaded their cargoes of passengers and vehicles on two levels. In addition to heavy commutation business, Lackawanna trains, led by the *Lackawanna Limited* and later the streamlined *Phoebe Snow,* carried Pullmans to Buffalo, where cooperating railroads took over cars going further west. In the 1930's, the trackage west of Hoboken, one of the oldest parts of the Lackawanna system, was electrified for suburban multiple-unit coaches. DL&W trains in the postwar era included the *Westerner* and *New Yorker,* with Chicago sleepers, the *Twilight* and *Pocono Express,* which carried interline Pullmans with the Michigan Central for Detroit, and the premier *Phoebe Snow.* In the late 1950s the Erie moved its trains to the DL&W terminal, and the two roads merged in 1960.

For years, the ferry houses at West Twenty-third Street in New York had resounded with the scuffling of passengers and baggage and the docking of the boats in the buildings' six slips. In its heyday, the facility was virtually a union ferry terminal, serving Baltimore & Ohio, Reading, Jersey Central, Lackawanna, and Erie passengers. Then one by one, the terminals, including Twenty-third Street, closed. Finally, on November 22, 1967, Erie Lackawanna boats made their last calls at Barclay Street and Hoboken, officially ending all ferry service across the Hudson.

Train passengers could still ride "the tubes," a subway built by the Hudson & Manhattan Railroad in 1909, connecting New York with the terminals at Hoboken, Jersey City, and Newark. But the early morning river trips, with the redolence of salt-air and fresh ink from ten-cent copies of the *Times* and *Herald-Tribune,* the sounds of city and seagulls, and the sight of the looming Manhattan skyline backlit by the rising sun, were over.

Two railroads originated at Weehawken, which had been accessible from New York by the West Forty-second and Cortlandt Street ferries. The West Shore Railroad, which followed the left bank of the Hudson to Albany before heading west to Buffalo, was taken over by the New York Central shortly after its opening in 1884. Before long-distance service was eliminated during World War I, West Shore trains carried sleepers to the Midwest and Canada. The railroad ended all passenger service in 1959.

The New York, Ontario & Western, a redundant Hudson-to-Lake Ontario railroad that expired two years earlier, once operated Pullman service as far west as Chicago from Weehawken.

By the turn of the century, technology was able to

Jim Shaughnessy

A southbound New York Central train slips into Harmon for a change to third-rail power that will take it into Grand Central.

furnish the Pennsylvania Railroad just what it needed to bring its trains onto Manhattan Island. A construction railroad was chartered to build a line from northeast of Newark across the Jersey Meadows, tunnel under the Hudson River and beneath Manhattan, where a huge new station would be built, then continue under the East River to Long Island, where the new Sunnyside Yard were to be located. The project, launched in 1902, became "the most massive and expensive privately financed engineering feat ever accomplished."[7]

In a neat balancing act made possible by its acquisition of the Long Island Rail Road in 1900, the project opened Manhattan Island to both the west and east. Owing to the nature of island travel, LIRR passengers had to cross water, in their case the East River, by ferry, to get from the railhead at Long Island City to downtown New York. With the four East River tunnels, Long Island trains could go by way of Sunnyside Yard straight into the new Pennsylvania Station in Manhattan. With future construction of the Hell Gate Bridge, through Boston-Washington trains could be similarly handled.

Pennsylvania Station occupied two square blocks and was modeled after the Baths of Caracalla and the Basilica of Constantine. The concourse leading to the station's 21 tracks was enclosed by arched girders, all covered by skylights. The colonnaded waiting room had a 150-foot ceiling.[8] After it opened in November 1910, the station was soon accommodating 750 trains a day.

Trains coming in on the main line from Philadelphia stopped at the new Manhattan Transfer near Newark, where crews switched from steam to electric locomotives for tunnel operation. This procedure continued until the main line was electrified to Philadelphia in 1933, whereupon the electric locomotives ran straight through. Manhattan Transfer was closed and the Hudson & Manhattan tube service to it ended several years later.

During World War I, when American railroads were under federal administration, the government ordered the Pennsylvania to accommodate the passenger trains of both the Lehigh Valley and Baltimore & Ohio in the new station. The Royal Blue fleet returned to Jersey City in 1926, but Lehigh Valley trains, such as the *Black Diamond* and *Maple Leaf,* used the station until the end of the railroad's passenger service.

Probably the most far-reaching consequence of the Penn Station project was the decision of Pennsylvania President Alexander J. Cassatt to operate only steel coaches through the tunnels. Since the beginning of car building, railroad coaches had been made of wood. Although iron and steel cars had been manufactured as early as the 1840's, and newspaper editori-

This steel- and glass-covered train concourse at New York's Pennsylvania Station was demolished in the 1964 remodeling.

als had continually inveighed against the splinter and fire hazards of wooden coaches, most of the cars continued to be made of wood.

In 1907 the Pennsylvania completed plans for a new, 88-passenger, steel coach, designated the P-70, which would soon become ubiquitous. Later that year it ordered 200 steel cars, against an estimate that more than 1,500 would be required to reequip all the trains serving Penn Station, and in so doing tipped the balance in favor of the heavier but safer type of construction.[9] Other railroads joined the movement (as did the Pullman Company, though somewhat reluctantly) and the standard era of heavyweight passenger cars, which would last until the streamliners came on the scene, began.

Trains went through Pennsylvania Station in an almost seamless flow. The *Broadway Limited* carried the standard, but it was not alone among all-Pullman trains heading west: *Manhattan Limited, Metropolitan, Commercial Express, "Spirit of St. Louis," American, Red Knight, Red Arrow, Golden Arrow.* The fleet to the South was equally impressive: *Orange Blossom Special,*

Miamian, Gulf Coast Limited, Havana Special, Seaboard Florida Limited, Crescent Limited, Everglades. Pennsylvania Station was where you caught "the Chattanooga Choo-Choo" and, seemingly, trains to everywhere else.

While the Pennsylvania was building in and across Manhattan, its rival New York Central was by no means idle. Since 1871, when Commodore Vanderbilt opened his Grand Central Depot on East Forty-second Street, the railroad had been very well positioned to exercise its monopoly of intercity train service in and out of Manhattan Island. But by the 1890s, the depot was inadequate to handle the traffic, which also included trains of the New Haven, as well as heavy commuter trade from both railroads, and a substantial remodeling was begun. But even after this was completed, in 1898, traffic soon outgrew the station, and the railroad decided to build, for the ages, the present Grand Central Terminal.

Unlike Penn Station, this construction work went on while an unabated flow of train and passenger traffic continued. Three million cubic yards of earth and rock and thirteen years later, in February 1913, the new Grand Central Terminal, with two levels of electrified, subterranean tracks, was opened. Over the depressed tracks north of GCT, a new Park Avenue was created where once an appalling smoke nuisance had existed, and electrification was extended all the way to Harmon, thirty-two miles up the Hudson. Suburban trains used the lower platforms, intercity trains the upper ones; a total of 48 tracks led to the terminal gates. To the south, underground loops around the station allowed for the quick turning of trains after they had deposited their passengers.

Beaux Arts facades enclosed the spectacular main concourse, 300 feet long with a 125-foot-high vaulted ceiling containing a gold and blue zodiac mural, whose constellations appeared backwards. One authority surmises that the artist took his inspiration from a manuscript of Medieval times, "when the convention was to depict the heavens as they would be seen from outside the celestial sphere."[10] In any case, it was a heavenly sight, and with the sun filtering through the terminal's south windows onto the Tennessee marble floor, the concourse was a great and spectacular interior space.

Grand Central was a city unto itself, with such amenities as a theater, art gallery, oyster bar, dressing rooms, and numerous shops, as well as direct entrances to the nearby Commodore, Biltmore, and Roosevelt hotels.[11] For those departing on Mr. Pullman's movable hotels, the premier accommodations of which were found just off the red carpet on the platform serving the *Twentieth Century Limited,* there was a choice over the years of more than 100 sleeper lines serving nearly that number of cities.

Although the Pennsylvania Railroad offered Pullmans to more cities from New York than the Central did, the inclusion of New Haven sleepers made Grand Central's total selection of destinations almost equal to Penn Station's. And despite its shorter list of destinations, the Central carried more Pullman passengers than the Pennsy.[12] Both railroads intensively worked the area between New York, Chicago, and St. Louis; both cooperated in post-war expansion of Pullman service into Texas and to the West Coast. But over the years, the Pennsylvania's longer menu resulted from traffic coming to it through the gateway of Washington.[13]

No train in the country, however, had any advantage over the *Twentieth Century Limited,* whose departure for Chicago every afternoon, often in several sections, was frequently a news event because of the train's clientele. Still, there were other one-time all-Pullman notables in the Central's Great Steel Fleet that ran up the Hudson before heading west: the *Commodore Vanderbilt,* Advance editions of both the *Century* and *Commodore, Wolverine, Lake Shore Limited, North Shore Limited, Iroquois,* and *Chicagoan.* To St. Louis the Central dispatched the premier *Southwestern Limited,* and to their namesake destinations, the *Cleveland Limited, Detroiter, Montreal Limited, Ohio State Limited,* and *Toronto Limited.*

These and other New York Central trains carried Pullmans between New York City and the remarkable number of thirty-two different Upstate communities, many of them summer or winter resorts. No other state could boast of such numerous, dedicated accommodations within its borders. The Central hauled sleepers to an almost equal number of out-of-state cities, from Canada to Mexico.

North of Troy, the Delaware & Hudson took over two New York-Montreal trains, the *Montrealer* and daytime *Laurentian.* The D&H continued the water-level trip up its Champlain & Saratoga Division to Rouses Point, where the road's Canadian subsidiary, Napierville Junction Railway, took over. Over the years, D&H passenger livery went from bright yellow, to Pullman green, to bottle-green trimmed in cream and red, to blue and yellow. The railroad was an active partner with the Central in the delivery of many sleepers to upstate resorts in the Adirondack and Lake Champlain regions.

What the new Manhattan terminals did not provide for was through service between New England and the south, but this was remedied with the opening in 1918 of the Hell Gate Bridge, a joint project of the New Haven and Pennsylvania railroads. This huge, steel-arch span, at 977 feet reported to be the largest in the world when built, crossed a narrow portion of the East River named by early Dutch settlers for its dangerous reefs. Anchored by two massive masonry piers, the four-track bridge connected the New Haven main line with the Pennsylvania's Sunnyside Yard, which then made possible the routing of Boston trains through Penn Station. (New Haven trains terminating in New York continued to use Grand Central.) The opening of Hell Gate put an end to barging cars from Harlem to Jersey City and to Pullman lines that crossed the Hudson on the Poughkeepsie Bridge, upriver.

The New Haven had a hand in virtually all passenger operations from New York into New England because it monopolized the territory. Its trains from New York traveled a four-track main line to New Haven, which was electrified in 1914, the first trunk line of any substantial distance to be wired. In addition to carrying thousands of commuters in both New York and Boston, the New Haven ran virtually hourly service between the two cities.

In the age of Amtrak, New York's Pennsylvania Station garnered the greater share of intercity passenger traffic, but the facility itself suffered an ignominious destruction just before the railroad went into the ill-fated Penn Central company. Despite vociferous, organized protest, the station was razed in 1964 and a new Madison Square Garden constructed above it. The girders and skylights, the majestic waiting room, the statuary—all of it was dismantled, and thereafter, the trains arrived and departed in a nondescript, underground context of noveau subway.

Grand Central was affected, though not irretrievably scarred, by the construction of the Pan Am Building over tracks north of the terminal. But despite the esthetically damaging, revenue-raising remedies the private railroads employed in their fight for survival, there is little to defend in the incredibly destructive public policy that hastened their journey to hard times.

As late as 1956, according to New York Central President Alfred E. Perlman, his railroad was paying taxes of seven million dollars a year on Grand Central Terminal alone,[14] an ignorant levy on a vital public accommodation. Across the Hudson, the state of New Jersey had virtually taxed the CNJ into oblivion.

At the end of the road, the railways that for more than a century had given New York City such incom-

parable passenger service were in worse shape than most other lines in the country. The three largest carriers (PRR, NYC, and NH) were by then part of Penn Central, which, in 1970, became the biggest bankruptcy in the nation. After the deluge, it was still possible to catch a Metroliner from the deflated Pennsylvania Station, but the grandeur was gone forever.

1. W R. Osborne, "Greatest Show on Earth," *Trains*, Vol. 2, No. 3 (January 1942), p. 23.
2. Carl W. Condit, *The Port of New York*, Vol. II (Chicago: University of Chicago Press, 1980) p. 274.
3. George H. Burgess and Miles C. Kennedy, *Centennial History of the Pennsylvania Railroad Company* (Philadelphia: Pennsylvania Railroad Company, 1949) p. 469.
4. Letter of April 21, 1988 received from J.W. Barnard Jr., Affiliation for Baltimore & Ohio System Historical Research.
5. Barnard, "Passenger Train Cars of the Staten Island Lines," *Handbook* (AB&OSHR, 1988), pp. 1-7.
6. George H. Drury, *The Historical Guide to North American Railroads* (Milwaukee: Kalmbach, 1985), p. 58.
7. Ron Ziel and George H. Foster, *Steel Rails to the Sunrise* (New York: Duell, Sloan and Pearce, 1965), p. 184.
8. Zeil, p. 186.
9. John H. White Jr., *The American Railroad Passenger Car*, Part 1 (Baltimore: Johns Hopkins University Press, softcover edition, 1985), pp. 137-8.
10. William D. Middleton, "The Grandest Terminal of Them All," *Trains*, Vol. 35, No. 7 (May 1975), pp. 22-35.

Sleeping car service

Following is a compilation of New York City's sleeping car service.

Central of New Jersey Pullman service included Jersey City to: Philadelphia, Pottsville, and Williamsport, Pa., and from Atlantic City, N.J.

The **Baltimore & Ohio's** dedicated Pullman service from Jersey City: Baltimore, Washington, Richmond (Richmond, Fredericksburg & Potomac/Chesapeake & Ohio), Cumberland, Md., Fairmont, Parkersburg, and Wheeling, W.Va., Cincinnati, Louisville, Memphis (Illinois Central), New Orleans (IC), St. Louis, Oklahoma City (Frisco), Pittsburgh, Cleveland, Columbus, Toledo (Hocking Valley-C&O), Chicago (via either Akron, Newark, or Wheeling), and New Orleans (Norfolk & Western/Southern, 1898).

The **Erie** carried Pullmans from New York to: Binghamton, Hornell, Salamanca, and Buffalo in Upstate New York, and Bradford, Meadville, and Oil City, Pa., Cincinnati, Cleveland, and Marion, Ohio, and Chicago. Erie service to New York City ca. 1890 included sleepers from Rochester, and (with Canadian National predecessors) Chicago and Toronto, and (with a B&O predecessor) St. Louis and Louisville.

Lackawanna Pullman service from New York City included: Scranton, Pa., and the New York communities of Binghamton, Elmira, Ithaca, Oswego, Richfield Springs, Syracuse, Utica, and Buffalo. West of Buffalo, partners moved cars to: Chicago (three routes—Nickel Plate, Michigan Central, Wabash), Cleveland (NKP), Detroit (MC), and St. Louis (NKP or WAB). After the Erie Lackawanna merger, the new railroad carried a sleeper to Youngstown and took the Chicago Pullman the entire distance.

The **West Shore** turned over its Chicago sleepers to the Wabash and, before the turn of the century, CNR and NKP. A St. Louis Pullman went via MC and WAB until the latter got trackage rights to Suspension Bridge. CNR also picked up a Toronto sleeper (1890).

The **New York, Ontario & Western** carried a Pullman to Liberty, N.Y.; additional cars went to Suspension Bridge (1890) via the NYC's Rome, Watertown & Ogdensburg or West Shore line, and the Chicago cars, which traveled both routes, continued west by way of the Wabash.

The **Lehigh Valley's** *Black Diamond*, a Buffalo train, carried a Pullman for Chicago (NYC); another LV Chicago sleeping car and one for Toronto were joint operations with CNR. Lehigh Valley's remaining sleeper service was to cities in its region: Buffalo, Ithaca,

Niagara Falls, and Rochester, N.Y., and Wilkes-Barre, Pa. The railroad also carried a Pullman from upstate Geneva to New York.

The **Pennsylvania Railroad's** Pullman service was extensive.

In the wake of the *Broadway Limited*, Pennsylvania trains carried New York sleepers on the main line west to Philadelphia, Harrisburg, Pittsburgh, and Chicago, others branching off for Niles, Akron, Cleveland, Youngstown, and Toledo, Ohio, and Detroit. At Chicago, in the post-war years, the Pennsy turned over Pullmans for Los Angeles (Chicago & North Western/Union Pacific, or Rock Island/Southern Pacific, or Santa Fe) and San Francisco (Burlington/Denver & Rio Grande Western/Western Pacific, or C&NW/UP/SP).

At Harrisburg, **PRR** trains headed north up the Susquehanna River with sleepers from New York for Williamsport, Emporium, Bradford, Oil City, and Erie, Pa. South from Harrisburg, they carried a car for York, Pa., and, among others on the Cumberland Valley line, a sleeper for Hagerstown, Md., where the N&W picked up the Pullmans for Roanoke and Bristol, Va., and Bluefield and Williamson, W.Va., Winston-Salem, N.C., Jacksonville (Winston-Salem Southbound Railway/Atlantic Coast Line), and (with Southern) Knoxville (1890), Chattanooga (1882), Memphis (1890), and New Orleans (ca. 1891).

Over the old **Panhandle** line from Pittsburgh to St. Louis, **Pennsy** trains carried New York sleepers for Columbus and Dayton, Ohio, Richmond (1890) and Indianapolis, Ind., and St. Louis, and to Wheeling, Cincinnati, and Chicago. From Indianapolis, a Pullman went to Louisville and another to the Indiana resort of French Lick on the Monon. At Cincinnati the Louisville & Nashville picked up sleepers for Louisville, Nashville, Memphis, and Birmingham. At St. Louis Union Station, the *Penn-Texas* delivered Pullmans that went aboard the *Texas Eagle* (Missouri Pacific/Texas & Pacific) for El Paso, Houston, San Antonio, and Mexico City (Nacional de Mexico). Another San Antonio car rode the *Texas Special* (Frisco/Missouri-Kansas-Texas) and a sleeper for Tulsa went aboard Frisco's *Meteor*.

The **Pennsylvania** carried Pullmans for Cape Charles, Va., Baltimore, and Washington, where it turned over sleepers bound for Florida and most of its other destinations in the Deep South. From the nation's capital, the **RF&P** carried a New York sleeper to Richmond, Va., as well as all the cars bound for **Atlantic Coast Line:** Norfolk (N&W) and Virginia Beach, Va. (N&W/Norfolk South-

Pennsylvania Railroad/courtesy Conrail

Soon after it opened in 1910, Pennsylvania Station was handling 750 trains a day coming in under both Hudson and East rivers.

11. A.C. Kalmbach, "Grand Central," *Trains*, Vol. 3, No. 5 (March 1943), p. 10.
12. The *Night Trains* compilation shows this breakdown of New York Pullman service over the years: New York Central, 72 lines, 62 destinations; Pennsylvania, 144 lines, 102 destinations; New Haven, 34 lines, 33 destinations. (The number of lines exceeds the number of destinations because some cities were served by way of two or more different routes.)

13. Contrary to what might be expected, the prime competitors for New York's Pullman business, within their own territories, were not the Pennsylvania and New York Central, but rather the Pennsy and the B&O, which served fifteen common destinations compared to the NYC and PRR's nine. The B&O and NYC competed in six cities, all of which the Pennsy served, as well.
14. Chicago *Tribune*, May 25, 1956.

ern), Wilmington, N.C., Charleston, Summerville (SOU), and Myrtle Beach, S.C., Thomasville, Augusta, Savannah, and Brunswick, Ga., and the Florida communities of Jacksonville, Clewiston, Fort Myers, St. Petersburg, Tampa/Sarasota, Sebring, and (with Florida East Coast) St. Augustine (1890), Palm Beach, Miami, Knights Key, and Key West.

RF&P also carried **PRR's** New York Pullmans bound for **Seaboard Air Line** points: Raleigh, Pinehurst (NS), and Hamlet, N.C., Columbia, S.C., Atlanta, Brunswick (Atlanta, Birmingham & Coast), and Columbus (West Point Route/Central of Georgia), Ga., Birmingham, Memphis (Frisco), and the Florida resorts of Naples, Boca Grande, St. Petersburg, Venice/Sarasota, Palm Beach, and Miami. The RF&P also forwarded a sleeper for Old Point Comfort, Va., to the C&O.

The **Southern Railway** participated in the most interline Pullman routes with the **Pennsy,** including New York sleepers for: Danville, Va. (1890), Greensboro, Charlotte, Durham/Raleigh, Salisbury, Winston-Salem, Asheville, and Hot Springs (1893), N.C., Aiken, Columbia, and Greenville, S.C., Augusta, Atlanta, Brunswick (1898), Macon, and Columbus, Ga., Anniston and Birmingham, Ala., Knoxville, New Orleans, and Jacksonville.

In cooperation with other railroads, **Southern** forwarded the **Pennsylvania's** New York Pullmans to: Montgomery (WPR), Mobile and New Orleans (WPR/L&N), St. Augustine (FEC, 1898), Tampa (either ACL or SAL, 1898), and (via N&W) Bristol, Va., Bluefield and Williamson, W.Va., and (via N&W/SOU) Birmingham, Chattanooga, Knoxville, Memphis, New Orleans, Nashville (NC&StL), Shreveport, La. (IC), and Vicksburg, Miss. (IC), and (via Frisco at Birmingham) Memphis (1898), and (via Asheville and NC&StL) Nashville.

Finally, the **Pennsy** delivered to the **C&O** in Washington Pullmans from New York bound for Hot Springs, Clifton Forge, and Roanoke (N&W), Va., White Sulphur Springs and Huntington, W.Va., Ashland, Lexington, and Louisville, Ky., and Cincinnati.

The **New York Central** carried Pullmans within its home state to: Albany, Buffalo (three routes: via main line, Lockport, and West Shore), Cape Vincent, Clayton, Fort Edward (Delaware & Hudson), Fort Ticonderoga (D&H), Lake George (D&H), Lake Placid, Lockport, Loon Lake, Malone, Massena, Newton Falls, Niagara Falls, North Creek (D&H), Ogdensburg, Oswego, Paul Smith's Hotel, Plattsburg (D&H), Racquette Lake, Rochester, Rouses Point

(D&H), Sacandaga Park (Fonda, Johnstown & Gloversville), Saranac Lake, Saratoga Springs (D&H), Syracuse, Thendara (Fulton Chain), Tupper Lake Junction, Utica, Watertown, and Westport (D&H).

Pullmans rode in **Central** trains to the Vermont communities (with Boston & Maine/Rutland) of Alburgh, Burlington, Rutland, and St. Albans (Central Vermont), and Canadian cities of Montreal (B&M/Rutland/CNR or D&H), Ottawa (B&M/Rutland/CNR or D&H/Canadian Pacific), Quebec (D&H/CPR), and Toronto (Toronto, Hamilton & Buffalo/CPR).

New York Central trains carried sleepers west for Bay City, Mich., Chicago (via Cleveland or Detroit or Niagara Falls), Cleveland, Toledo, Detroit, Grand Rapids, Los Angeles (C&NW/UP, or RI/SP, or Santa Fe), and San Francisco (Burlington/D&RGW/WP or C&NW/UP/SP).

For destinations in the lower Midwest, St. Louis, and Southwest, **NYC** originated Pullmans to Cincinnati, Columbus, Indianapolis, Dallas (Frisco/MKT), Dallas/Fort Worth (MP/T&P), French Lick, Ind. (Monon), Houston (MP/T&P/MP), Mexico City (MP/T&P/MP/Nacional de Mexico), Oklahoma City (Frisco), Peoria, San Antonio (Frisco/MKT), St. Louis, Tulsa (Frisco), and Waco (Frisco/MKT).

From New York, the **New Haven** carried sleepers to Hartford, Providence, Springfield, and Boston, in addition to summertime Pullmans to Hyannis and Woods Hole, Mass. It carried cars to Worcester and Boston jointly with the Boston & Albany. Aside from a seasonal sleeper to New York that it picked up from the Narragansett Pier (R.I.) Railroad, its other Pullman service to the north was conducted jointly with the Boston and Maine:

Interline **NH/B&M** sleepers departed year-round for Portland, Me., Concord, N.H., and White River Junction and St. Albans, Vt. (Central Vermont); and to Canadian destinations of Sherbrooke (CPR/Quebec Central), Levis, Montreal, and Ottawa (CV/CNR); and Halifax (Maine Central/CPR/CNR)

Summer season movements of the **New Haven/B&M** served the Maine communities of (all via **Maine Central**) Rockland, Kineo Station, Mount Desert Ferry, Waterville, Kennebago, Farmington, Ellsworth, and Van Buren (Bangor & Aroostook), and the New Hampshire stations of Jefferson, Plymouth, Lake Sunapee, Keene, Claremont, and (with Maine Central) Bretton Woods and Woodsville.

William D. Middleton

Jim Shaughnessy

4
New York State

A New York Central Niagara 4-8-4 (top) pounds its way up Albany hill with the *Mohawk* in May 1948. It's 4 o'clock on an April 1955 morning (bottom) in Troy, and the Delaware & Hudson has just turned over its Montreal sleeper to the Central for the final leg into New York City. The GP-7 is now under way.

The New York Central was the indisputable main line of the Empire State. The famed water-level route went up the Hudson River from New York City to Albany, from which it headed west through the Mohawk River valley to Buffalo. Of all the railroads connecting New York and Buffalo, the Central traveled the greatest distance, yet none of the others could ever match its swiftness or volume of traffic. After the *Empire State Express* established a world speed record of 112.5 miles an hour in 1891, the six tracks across much of the state remained a raceway for the passenger trains of the Great Steel Fleet, led by the *Twentieth Century Limited,* and the freights that occupied the remaining iron. The Erie, the Delaware, Lackawanna & Western, and the Lehigh Valley all made the journey to Buffalo—more directly, but over mountainous terrain. Always, the Central was dominant.

The Erie Canal, the historic waterway that first connected the Hudson River with the Great Lakes, at Buffalo, naturally was built along a route most suitable for canal operation, one with a minimum grade. This was the case except for the portion between Albany and Schenectady, where numerous locks were required to raise the bed to the Mohawk River level. What suited canal construction was equally favorable to railroads, and eventually ten short lines were constructed in a chain along the same route between Albany and Buffalo. The Erie Canal had been open but just six years when, in 1831, the locomotive *DeWitt Clinton* began rail service on the Mohawk & Hudson Railroad between Albany and Schenectady. Being spared the greater distance the canal had to travel to make this one major ascent, the railroad proved far more expeditious. Soon the railroads were cooperating in through service to Buffalo that was much faster than canal boat travel.

Despite competition of the steamboats on the Hudson (except when the river iced over), a rail line was built up the east bank of the river, and it reached Rensselaer in 1851. Passengers had to be ferried over to Albany, but with this exception, the rail route from New York to Buffalo was in place. Two years later, after unsuccessful efforts to protect its investment in the Erie Canal, the state approved consolidation of the ten Albany-Buffalo railroads, and the New York Central was born. In the next decade, the bridge across the Hudson was completed and more important consolidations were made. Commodore Cornelius Vanderbilt, who had gained control of the Hudson River line and its connection into New York City, merged this and the upstate line into one system in 1869.

Meanwhile, expansion and consolidations were going on west of Buffalo, too. Oddly, the spectacular gorge of the Niagara River was bridged a full decade before the far simpler crossing of the Hudson (which steamboat interests, including the Commodore, had successfully stalled) was made.

Since Michigan was already a state and the West was opening up to more and more immigration, the need was recognized for a through line, fed by traffic from both New York City and Boston, from Albany to Buffalo that continued across Ontario to Detroit. In 1847, a company was formed to construct a suspension bridge across the roiling water just north of Niagara Falls. To get himself started, the builder offered local boys five dollars to lay a kite string across the gorge. Finally, one youth succeeded, but the contractor never went any further than constructing a foot bridge. After this failure, the company turned to a

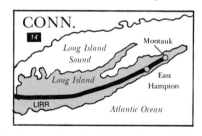

previous bidder, John A. Roebling, then the reigning expert on suspension bridge construction, an engineer who would later distinguish himself by designing the great Brooklyn Bridge.

Inasmuch as several other suspension bridges elsewhere had been broken up by wind and other misfortunes, it was a brave undertaking to build across waters as turbulent as the Niagara River's an 821-foot span, a structure whose thousands of tiny wires must collectively support a heavy locomotive and an entire train of cars. Roebling nonetheless persisted and on March 8, 1855 opened his graceful "Great International Railway Suspension Bridge." The designated locomotive engineer suddenly decided against going down in history (with the bridge, he was convinced), but officials found a substitute, who made a successful crossing with a 23-ton locomotive. The remarkable structure was used by numerous railroads for through travel until it was replaced by a steel-arch span in 1897. It remained to the end the only successful suspension bridge mainly built for a railroad.[1]

With the opening of both the north and south shores of Lake Erie to busy rail lines, Buffalo became an important terminus and would be just the sort of vital center for the New York Central that Pittsburgh

was for the Pennsylvania Railroad. Buffalo was served by more than 300 passenger trains a day, and of course the Central's traffic was dominant.

Between Buffalo and New York, four railroads offered sleeping car service over six different lines, three of which were NYC: the main line route, a slight variation via Lockport early in the century, and the West Shore route. The West Shore, which accounted for the fifth and sixth of the Central's main line tracks across New York state, was a nuisance road built in the 1880's up the left bank of the Hudson. The parallel Central leased it in 1885.

The Erie Railroad, another competitor, completed its trackage from tidewater to Lake Erie in 1851, but without the benefits of the water-level routing of the Central. The Erie's wide-gauge route was more direct, but also more mountainous, because by its charter it was intended to serve the Southern Tier communities that had not benefited from the Erie Canal.

The Lehigh Valley and the Lackawanna, two railroads whose routes went through Pennsylvania's anthracite region, also provided Pullman service between Buffalo and New York. During the standard era, the Lackawanna had three eastbound trains carrying sleepers to New York: the *Whitelight Limited, Garden State Limited,* and *Lackawanna Special.* The westbound versions of the trains that superseded them in the 1930s were known as the *Buffalonian* and *Owl.* Post-war, the *Owl* and *New Yorker* carried the sleepers between New York City and upstate destinations, including Buffalo.

In Buffalo, the Lackawanna also handed over the interline sleepers operating between New York and cities in the Midwest to its western partners, principally the Nickel Plate.

Equally competitive with the DL&W, the Lehigh Valley also offered a choice of three eastbound trains carrying Pullmans to New York City, including the *Lehigh Limited.* This and the *Star* provided the westbound accommodations. The trains of this picturesque route also offered Buffalo the additional dividend of direct service to and from Philadelphia. As late as the early 1950's, the Lehigh Valley was still scheduling two trains with New York City-Buffalo Pullmans—one was the Toronto-bound *Maple Leaf;* the other was still carrying a Philadelphia sleeper.

The Erie was the first of this threesome to quit the Buffalo-New York City competition, but in its day it offered overnight accommodations on both the *Lake Cities Express* and the *Atlantic/Pacific Expresses.*

All three lines gave Buffalo additional sleeping car service: the Lackawanna to Scranton; the Lehigh Valley, of course, to Philadelphia (with Reading, in a joint service the Erie had once participated in) and Washington (Reading/Baltimore & Ohio); and, early in the century, the Erie to Cincinnati.

The Buffalo, Rochester & Pittsburgh (later B&O) carried a sleeper to Pittsburgh, and Buffalo had Pullman service on the Nickel Plate to Chicago and, briefly, Hot Springs, Va. (C&O).

The Canadian National offered sleeping car service from the west that came to Buffalo from Suspension Bridge by way of either the New York Central or the Lehigh Valley. In addition to cars that it and subsidiary Grand Trunk Western hauled to Chicago and

A conductor scans the ample consist of the New York Central's westbound *North Shore* at Fonda, in February 1957.

Detroit, CNR carried seasonal sleepers to the Ontario stations of Kingston Wharf, Muskoka Wharf, and Temagami (Temiskaming & Northern Ontario). When the Wabash had trackage rights over CNR rails to Suspension Bridge, it carried Pullmans for Buffalo from Chicago and St. Louis.

The Pennsylvania Railroad gave Buffalo sleeping car connections with Pittsburgh, Philadelphia, and Washington, in addition to summertime service to Atlantic City and, according to 1929-30 schedules, winter service to Florida via Washington and the Richmond, Fredericksburg & Potomac. Pullmans departed for Miami (Atlantic Coast Line/Florida East Coast or Seaboard Air Line), St. Petersburg (ACL or SAL), and Tampa (ACL).

To the east, the New York Central carried sleepers to Clayton, Lake Placid, Massena, Montreal, Albany, New York City, and Boston (in the 1890's the NYC's West Shore also provided Boston service jointly with the Boston & Maine). To the west, the Michigan Central ran Pullmans to Detroit, Grand Rapids, Chicago, and, before the turn of the century, St. Louis (Wabash). On the Lake Shore line through Ohio, sleepers traveled from Buffalo to Pittsburgh (Pittsburgh & Lake Erie), Cleveland, Toledo, Cincinnati, St. Louis, and Chicago. In Cincinnati, the Central turned over Pullmans to the Southern Railway for Jacksonville, Miami (FEC), and St. Petersburg (SAL).

In a memorable dedication ceremony on June 22, 1929, waiters served luncheon to 2,200 guests of the

The bridge across the Hudson River at Poughkeepsie served several through Boston-Washington Pullman lines that bypassed New York City. Here, a freight train crosses in 1936.

New York Central in the passenger concourse of its elegant new Central Terminal, two and a half miles from downtown Buffalo. The 225-foot hall, the most spectacular room of the new station, had a vaulted ceiling, Art Deco detailing, and a terrazzo floor. A fifteen-story office tower crowned the project. The concourse leading to the trains was half again as long as a football field, spanning seven platforms and fourteen through tracks used by 200 trains a day, including those of the Pennsylvania and Toronto, Hamilton & Buffalo. Entering service as it did on the eve of the stock market crash and great Depression and being somewhat remote from downtown Buffalo, Central Terminal did not have an entirely happy life, but after abandonment by Amtrak, the heroic structure at least gained entry on both state and federal Registers of Historic Places.[2]

The capital, Albany, ranked third among New York cities in the Pullman service it obtained. Day and night, mostly night, the Central's four-track main line saw an almost constant rush of limiteds moving between New York and Chicago or St. Louis and intermediate cities. These offered a great variety of accommodations to all the major cities of Upstate New York, but of course one gauge of a community's importance was the number of dedicated sleepers the Pullman company provided it. Being a nationally visible state capital helped.

Until 1968, Albany passengers were served by a two-level, French Renaissance Union Station built in 1900. Not as busy as the Buffalo terminal, it nonetheless saw movements of about 150 trains a day. The eight tracks on the upper level carried Central traffic; the lower ones accommodated Delaware & Hudson and West Shore trains.

Albany had dedicated sleepers to more than a dozen cities from Boston to Chicago. Both the New York Central and D&H (with Erie) provided the Chicago service; Boston cars moved via the Central's Boston & Albany and, earlier, also on the B&M to Troy, across the Hudson River.

The Delaware & Hudson, the nation's first north-south railroad, had its general offices in an ornate, turreted building in Albany. Over the years the D&H carried sleepers from the capital to Binghamton, Hornell (Erie), Plattsburg, and Montreal (Napierville Junction and Canadian Pacific). The remainder of Albany's Pullman service was provided by the Central to New York, Buffalo, Cleveland, Pittsburgh (P&LE), and Niagara Falls, and from Rochester and Detroit.

Syracuse had the dubious honor of having the New York Central's main line tracks running down Washington Street, one of its busy thoroughfares, for nearly a century until finally they were rerouted just before World War II. Although the continuing passage of the Great Steel Fleet occasioned some amusing interplay between passenger and townsfolk, mostly it was a smoke and traffic nuisance the city was glad to be rid of.

From Syracuse a branch of the Central ran north to Oswego and Watertown, and the secondary (although once main line) "Auburn Road" went west to Rochester by way of Auburn and Geneva. The Lackawanna's branch to Oswego passed through Syracuse, and over this the road scheduled Pullmans to New York, Philadelphia (Central of New Jersey/Reading), and Washington (CNJ/Reading/B&O). The NYC provided the remainder of Syracuse's sleeping car service, to New York, Boston, Springfield, Cleveland, Detroit, and Chicago.

Rochester, whose downtown was graced by picturesque falls on the Genesee River, also saw all the Central's main line traffic, in addition to serving as the western terminus of the Auburn Road. Rochester had Pullman service to New York via both routes, in addition to Boston, Cleveland, and Detroit. Sleepers to Philadelphia and Washington left town on the Auburn Road and were picked up by the Pennsylvania at Canandaigua (which service the Pennsy maintained after the Rochester link was dropped). The Lehigh

In September 1941, the *Ohio State Limited* hustles up the Central's four-track main along the Hudson near Oscawanna.

Howard W. Ameling collection

Valley also carried Pullmans to New York, Philadelphia (Reading), and Washington (Reading/B&O). The BR&P offered sleepers to Pittsburgh, and in the 1890's, Rochester had a fourth sleeping car route to New York, via the Erie.

Utica's Pullman service was limited and, in later years, all New York Central: to New York, Boston, and Chicago. The Lackawanna also carried a New York sleeper, according to a 1917 schedule. Utica's ornate station, the last survivor of the Central's big upstate depots, was refurbished for the winter games of the 1980 Olympics.[3]

New York had more communities with dedicated Pullman service than any other state in the union.

Ithaca, a gateway to the beautiful Finger Lakes country and home of Cornell University, originally had Pullman service to New York via both the Lehigh Valley and the Lackawanna. The DL&W quit during the Depression, but it is remembered for its unusual entry into Ithaca. Coming into town several hundred feet above Cayuga's waters, Lackawanna trains had to get down to the station level by negotiating a switchback—a construction shortcut entailing two wyes that permits quicker, though more cumbersome, changing of elevation in a restricted area.[4]

The Long Island Rail Road, originally conceived as part of a joint boat-train route to Boston and later a mostly electric, commuter giant, carried sleepers from Pittsburgh to both East Hampton and Montauk.

Lake Placid, scene of the winter games of both the 1932 and 1980 Olympics, had Pullman service, via the NYC (over a branch it acquired from D&H), to New York, Boston, Chicago, and Philadelphia (PRR). It was one of the few northern resorts to have both winter and summer sleeping car service. The NYC's *Whiteface Mountain Special* brought the heavy consists of Pullmans up the Adirondack Division from Utica.

Niagara Falls, of course, was the spectacular vista without which no new marriage stood a chance of succeeding. Over the years the railroads, principally the Central, paused their trains at the Falls, routed Pullmans there on side trips from Buffalo, or scheduled sleepers with Niagara Falls as the actual destination. These included the New York Central's service from New York, Boston, and Chicago, and Lehigh Valley Pullmans from New York and Philadelphia (Reading).

Being at the vital center of the Erie system, Hornell had the benefit of through sleeping car service to New York, Boston (New Haven), Cincinnati, and Detroit (CNR). One of the Erie's Depression-era innovations was the *Southern Tier Express*, which offered air-conditioned Pullman service from Hornell to New York, giving such communities as Corning, Elmira, and Binghamton convenient evening departures. It was perhaps the Erie's most profitable Pullman operation during those lean years.

Several other New York communites had dedicated sleepers to large cities out of state: Auburn (NYC), Jamestown and Salamanca (Erie), and Kingston (West Shore/Wabash) to Chicago. Clayton (NYC) and Ogdensburg (Rutland/B&M) to Boston. Auburn (LV/Reading) and Binghamton (DL&W/CNJ/Reading) to Philadelphia. And Massena (NYC/P&LE), Mayville (PRR), and Olean (PRR) to Pittsburgh.

After the Civil War, when the arbiters of Nineteenth Century society decided that Saratoga Springs, in the Adirondack foothills, was the proper place to spend summers, an exodus began from New York's Grand Central Depot, as the predecessor facility was then known. The migration resumed every summer, and the list of destinations expanded. Not all upstate resorts offered the posh accommodations of Saratoga Springs, with its horse racing, gambling, and obligatory social calendar, but vacation spots were soon developed for all tastes and budgets, for summer as well as winter pleasure.

In 1906, Paul Smith opened a seven-mile electric railroad from Lake Clear Junction to his hotel on Lower St. Regis Lake, replacing the stage coach service to several area rail stations. An electric car with seats for twenty-four coach passengers shuttled from the New York Central station at the junction to the hotel, but through Pullmans and private cars were pushed to the resort by the doughty little interurban. Service ended in the late 1920's.[5]

The Adirondack Mountains, with their healthful air and numerous lakes for fishing, were popular, but so too were Lake Ontario destinations, the Thousand Islands, and Lake Champlain resorts. As one would expect, the New York Central was the main provider of this intrastate Pullman service. (See listings at the end of the New York City chapter for a compilation of accommodations.)

1. Richard Sanders Allen, "Over Niagara on a Wire," *Trains,* Vol. 18, No. 2 (December 1957), pp. 44-47.
2. Garnet R. Cousins, "A Station Too Late, Too Far," *Trains,* Vol. 45, No. 11 (September 1985), pp. 20-33.
3. Tom Nelligan, "Empire Corridor," *Passenger Train Journal,* Vol. 14, No. 2 (August 1982), p. 36.
4. Edwin A. Wilde, "The Ithaca Branch of the Lackawanna," *NRHS Bulletin,* 23, No. 2 (1958), p. 14.
5. Michael Kudish, "The Paul Smith's Electric Railway," *Franklin Historical Review,* Vol. 13, 1976, pp. 20-29.

Streamstyled 4-6-2 No. 1136 (top) with a nine-car *Lackawanna Limited* at Binghamton, Sept. 5, 1937. Nearby, on the same day (lower), this K4 brought the *Erie Limited* to town with a ten-car consist. Neither the Erie nor Lackawanna had the New York Central's water-level profile.

5
New Jersey

New Jersey figured prominently in early railroad development. The Garden State claimed the first locomotive built and operated in this country, an experimental "steam waggon" Col. John Stevens demonstrated on a loop of track at his estate in Hoboken in 1825. The more famous locomotive *John Bull* saw its first main line use in New Jersey in 1831.

Great rivalry attended the building of the modern railroad expressway across the central portion of the state. By leasing an entrenched line that had a state monopoly on New York-Philadelphia traffic, the Pennsylvania Railroad for a time had it all, but after years of rampant political maneuvering and bribery, the state finally cleared the way for the Central of New Jersey and Reading predecessor to build a competitive route, whose elements were connected at Bound Brook. Not that the Baltimore and Ohio, which eventually used the CNJ/RDG route, and the Pennsylvania ever let the competitive spirit lapse while their trains charged through the corridor, and of course it didn't until the B&O quit in 1958.

The Pennsylvania's four-track main line across New Jersey was reputedly the busiest stretch of trunk-line trackage in the world, averaging a train movement

Pennsylvania Railroad/courtesy Conrail

Workmen put the finishing touches on the Pennsylvania's Newark station, a joint city-railroad project of the 1930's

every five minutes, around the clock. And this magnificent right of way saw virtually everything the lordly Pennsylvania had to offer: suburban locals, Philadelphia clockers, Washington expresses, the fleet of western flyers led by the *Broadway Limited* and *"Spirit of St. Louis,"* and the New York accommodations for most of the great trains of the South. Whether under the authoritative tow of a Pacific K4 out of Manhattan Transfer or gliding behind a powerful GG1 electric locomotive under wire, the Pennsy's trains on this New York Division established a remarkable standard for passenger travel.

Nearly two dozen all-Pullman trains rolled south over these tracks on a typical night during the winter of 1929-30, and this was counting just the scheduled movements; extra sections frequently increased the number. Eight of the trains were Florida-bound.

Against this formidable competition, the B&O was always a worthy and valiant contender. Even though outgunned by the powerful Pennsylvania, it operated as many as 21 passenger trains between New York and Washington on its Royal Blue Line service, offering through Pullmans on the *Capitol Limited* and the *National Limited* as well as other Chicago and St. Louis trains. The B&O, moreover, had its own clientele, people who preferred Royal Blue service to the rival's tuscan red. They rode B&O trains, and that was the way it was.

After New York and Philadelphia were connected by rail, promoters turned their attention to linking the anthracite fields and carriers in eastern Pennsylvania with the New York market and its tidewater shipping facilities. (For passengers, these lines gave the best account of New Jersey's natural beauty as they went through the mountainous northwestern part of the state.) All these roads coming in from the west funneled into a heavy concentration of Hudson River terminals, which ultimately gave New Jersey more rail mileage per land area than any other state.

The metropolis of Newark had next to no dedicated Pullman service but was more than adequately served by the endless parade of sleepers passing through town on the Pennsylvania, Lehigh Valley, and

Lackawanna. In the mid 1930s, the Pennsy and the city of Newark built a new station with a terminal for public transit lines, including the Hudson & Manhattan "tubes" to New York. The Lehigh Valley also used the facility, but the other roads had their own. Although it did not directly serve Newark, the B&O did provide a setout sleeper to Washington that left the CNJ station in time to be put aboard the overnight train from Jersey City to the capital.

The resort and amusement center of Atlantic City had New Jersey's only sizable selection of dedicated sleepers. The Pennsylvania ran a train called the *Sea Gull* that brought Pullmans from as far as Chicago to this town known for its 60-foot-wide boardwalk. Most of the sleepers arrived once a week in winter and daily during warm weather. Over the years, Atlantic City also had Pullman service, via the Pennsy, from Akron, Buffalo, Cincinnati, Cleveland, Harrisburg, Memphis (L&N), Pittsburgh, and Wheeling. The CNJ carried a sleeper to New York briefly before and after World War I.

Overnight service, however, was not so important to Atlantic City, which had frequently scheduled day trains from New York and Philadelphia.

Between 1929 and 1941, the CNJ operated a special daytime train called the *Blue Comet* down its 136-mile line from New York to Atlantic City. The train, painted royal blue and cream, pioneered the concept of the all-coach, economy streamliners that became popular a decade later.

Both the Pennsylvania and the Reading operated fast trains from Camden, and the trip, including the ferry ride from Philadelphia, took just an hour. With the competition becoming ruinous, the two railroads merged operations in 1933 into the Pennsylvania-Reading Seashore Lines, which opened an eight-track union station in Atlantic City the following year. Trains from New York left the Pennsy main line at Trenton; Philadelphia trains operated from Camden and from Broad Street via North Philadelphia. PRSL service once exceeded 175 daily trains. Although the last remnants came to rest in June 1982, Amtrak undertook a revival in 1989.

Autumn 1963, and Pennsy trains to the south already are showing longer consists, such as this under GGI tow at Kilmer Tower, N.J.

6
Pennsylvania

The *Trail Blazer* (top), shown here in Philadelphia's Broad Street Station, was a night train, but a different sort from most in this book. Inaugurated in the summer of 1939 to serve coach travelers going to New York World's Fair, the *Trail Blazer* was typical of the new economy trains of its era: they were fast, generally streamlined, and had commodious coach seats. If they carried sleepers, they were usually the lower-priced, tourist variety. Horseshoe Curve (bottom) was the Pennsy's great engineering landmark; here, the ten-car *Metropolitan* climbs the hill in May 1934.

On a map of the Class I railroads in Pennsylvania, the state looked like a rectangular snapshot mounted at its four corners by, on the upper left, the New York Central, Nickel Plate, and Erie; on the upper right by the Lehigh Valley, Lackawanna, and Erie; and on the two lower corners by the B&O and B&O/Reading. The picture itself, only minor hyperbole intended, was of the Pennsylvania Railroad going everywhere else in the state.

The mighty Pennsylvania did not wear its symbolic keystone or respected name for either light or transient reasons: It *was* the railroad of Pennsylvania. It was an archetypal Philadelphia corporation, dignified and circumspect. As the largest—and self-proclaimed "standard"—railroad of the world, it was also a 1,200-pound gorilla. Nonetheless, it was a model of financial prudence and had the longest continuous record of dividend payment in United States corporate history. Despite the shambles that resulted from the Pennsylvania-New York Central merger in 1968, the Pennsy left behind enduring monuments to its dominating presence, and we are richer today for them. It all began with Manifest Destiny.

In the early 19th Century, settlers were moving west in increasing numbers, generating their own demand for improved transportation facilities. In 1818, the National Road reached Wheeling, W.Va. (at the time, part of Virginia), and in 1825 the Erie Canal opened an all-water route between the Hudson River and the Great Lakes. At this point, America's three most important mid-coastal seaports were New York, Philadelphia, and Baltimore. New York, of course, was the main beneficiary of the Erie Canal, just as Baltimore had the most to gain from the National Road, by which it could tap the surging trade of the Ohio River valley. Pennsylvania was losing out to the competition.

There were, however, suitable options. Pittsburgh, like Wheeling, also provided access to the Ohio River trade, just as the port of Erie was able to tap the lake commerce. The question was: What was the best way to breast the formidable Appalachian barrier and connect Philadelphia with the state's two western outposts?

After building a primitive road from Philadelphia to Pittsburgh, Pennsylvania embarked on a joint rail-canal project, which it completed in 1834. The journey required four days and entailed a canal trip through 108 locks up the Susquehanna and Juniata rivers, then a trip over the crest of the Alleghenies by way of a series of ten inclines on the Portage Railroad (vehicles were pulled up the tracks by ropes spooled by stationary engines). After crossing the summit, travelers resumed canal travel from Johnstown to Pittsburgh. For a time, canal boats were portaged in sections over the inclines. Despite the cumbersome arrangement and even though canal traffic was al-

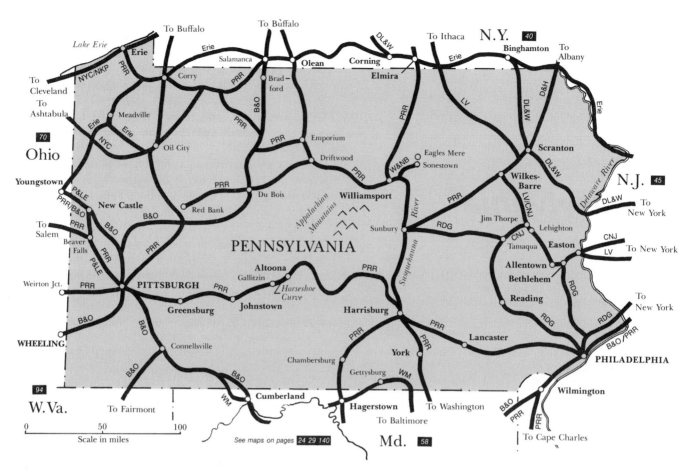

ready drying up, the state wanted to protect its investment in waterways, and it refused to invest further in an all-rail line. The Baltimore & Ohio Railroad, meanwhile pushing west through Maryland and Virginia, was seeking entry into Pittsburgh, and this spurred the Pennsylvania legislators to action. They granted a charter in 1846 to a new corporation to build a line to Pittsburgh from Harrisburg, which already had a connection with Philadelphia via Lancaster, and it was to include a branch to Erie. The new line to Pittsburgh, the main artery of the modern Pennsylvania Railroad, opened in 1854.

The Pennsy fleshed out its western system to Chicago and Cincinnati before acquiring what became its most important and enduring avenue of passenger travel, the main line to New York. The railroad made a connection with Chicago in 1858 via the Pittsburgh, Fort Wayne & Chicago, which it gained control of eleven years later. In 1868, it began putting together and leasing the roads through Ohio that would take it to St. Louis. It was not until 1871, however, that it gained access to New York (at Jersey City) by leasing several New Jersey railroads.

In 1873, the Pennsylvania established its own through service from Jersey City to Washington, using the existing Philadelphia, Wilmington & Baltimore Railroad and by building a new road south of Baltimore, with a branch to Washington, thus breaking the B&O's monopoly between those two cities. The Pennsy gained majority control of the PW&B eight years later, and by this time, other portions of the giant system were rapidly falling into place.

Although New York quickly became the new eastern terminus of the system, the Pennsylvania was still a Philadelphia corporation, where, some said, it's heart always remained. To give its citizens a suitably commodius terminal, the Pennsylvania built Broad Street Station, which it opened several weeks before Christmas in 1881. By bridging the Schuylkill River and elevating the trackage on what came to be known as the Chinese Wall, the railroad could operate all its trains into the heart of the city, rather than just to its west side terminal.

Twelve years later, a new ten-story head house and office building, an ornate granite and brick Gothic structure, were added and an arched trainshed built to cover the sixteen tracks. The Victorian showplace offered marble floors and staircases, walls covered with walnut wainscoting, and heavy brass hardware.[1] The lordly office tower accommodated the railroad's corporate offices.

Despite its grandeur, however, Broad Street was very much a creation of its times, and the downtown stub terminal never was able to fully meet the needs of the 20th Century, such as accommodating the faster, through trains.

Charles B. Chaney/Smithsonian Institution

Lehigh Valley's *Black Diamond* in an almost trademark setting east of Ox Bow Curve in the Lehigh River Valley. The train carried sleepers for Chicago.

An increasing number of New York-Washington trains stopped only at West Philadelphia Station, and through trains to and from the west, such as the *Broadway Limited* and *Pennsylvania Limited*, used the North Philadelphia Station, built in 1904. Broad Street, however, continued to handle some through trains, short-haul traffic such as the hourly "Clockers" to New York, and suburban service.

After two major fires at the old station, railroad and city officials finally drew up a master terminal plan to address all the various problems. It provided for a bi-level, union station at 30th Street, which would replace the west side terminal, and a twenty-two-story commuter station—reached by underground trackage—just north and west of Old Broad Street. In 1933, the 30th Street Station, flanked with stately Corinthian columns, opened, but hard times scuttled the proposed move of the B&O and Reading to the new terminal as well as the timetable for razing the Chinese Wall and Old Broad Street.

The venerable Victorian station, which had once handled a train every two minutes, continued to function for almost twenty more years until the end came in 1952. On April 27, Eugene Ormandy conducted a brief concert of his Philadelphia Orchestra in the station concourse before the musicians boarded the last scheduled train from the terminal. As a crowd of several thousand onlookers sang "Auld Lang Syne," the train pulled out into the rainy night, and Old Broad Street was history.[2]

On the line to New York, which was completely electrified in 1933 and became a raceway for express trains under the tow of the Raymond Loewy-designed GG1's, Philadelphians had Pullman service to New York, Providence, Boston, and Springfield, Mass. In the summertime, when the *Bar Harbor Express* was the Main-Line-to-Maine connection, the Pennsy originated sleepers for Ellsworth, Mount Desert Ferry, Rockland, and Portland, as well as Bretton Woods, N.H. (They moved via New York, New Haven & Hartford-/Boston & Maine/Maine Central.) In the summer of 1909, the railroad turned a sleeper over to the New York Central in New York for Lake Placid, N.Y.

Philadelphia, America's No. 3 city for many years, did not have as much dedicated Pullman service as some smaller communities had. But because it was less than two hours from New York, most of the through sleeper schedules proved convenient.

Over the Pennsylvania's main line south, which carried the Atlantic Coast Line and Seaboard Air Line's Florida trains and Southern Railway and Chesapeake & Ohio's fleet of expresses, Philadelphia had dedicated sleeper service to Baltimore, Washington, Richmond, Va. (Richmond, Fredericksburg & Potomac), Cincinnati (C&O), and from Memphis (SOU/Norfolk & Western/SOU), and Key West (Florida East Coast/ACL/RF&P). On the Delmarva line, which branched off at Wilmington, Dela., the Pennsy ran a Pullman to Cape Charles, Va., connected by ferry to Norfolk. Westward, the railroad ran sleepers from Philadelphia to Altoona, Pittsburgh, Cleveland, and Chicago.

Some of Philadelphia's service operated over scenic secondary routes. The Cumberland Valley line ran from Harrisburg to Hagerstown, Md., where it connected with the Norfolk & Western. The Pennsylvania scheduled sleepers from Philadelphia to Hagerstown and, with N&W, to both Roanoke and Gary, W.Va. Before the turn of the century, service included cars to both Knoxville (SOU) and New Orleans (SOU-/Louisville & Nashville).

Pullmans from Philadelphia traveled north up the Susquehanna River to Williamsport. From this one-time lumbering center, other cars for Canandaigua, N.Y., and Rochester (NYC) continued on through Elmira, N.Y. The train carrying sleepers for Erie set out a Red Bank car at Driftwood. From Emporium, which had its own service from New York, Philadelphia Pullmans went on to Buffalo and Toronto (Toronto, Hamilton & Buffalo/Canadian Pacific), and to Erie.

The Baltimore & Ohio, which maintained its own station at Chestnut Street on the east bank of the Schuylkill River, gave Philadelphia more varied Pullman service to the west, although it tended to be one-directional. The B&O carried sleepers to Washington, Cumberland, Md., Pittsburgh, Cleveland (alone and also with Pittsburgh & Lake Erie/Erie), Detroit, Louisville, and Chicago, and to Philadelphia from Parkersburg, W.Va., Cincinnati, and St. Louis. The B&O's partners on Royal Blue service to the north, Reading and Central of New Jersey, carried Pullmans to New York. All B&O service through Philadelphia terminated in April 1958.

The Reading Company was another Philadelphia-born railroad, but its domain was the eastern part of the state, where it moved anthracite coal from the mines to tidewater markets. Its landmark station in Philadelphia opened in 1893 at 12th and Market streets, a few blocks east of the Pennsylvania's Broad Street station. Beneath its 13 tracks, Reading Terminal had the largest retail farmer's market in the country, and commuters customarily availed themselves of the fresh meats and produce.

By way of its charter main line through Reading, the railroad carried sleepers to Williamsport and, during summers until the mid 1920s, to Sonestown (Williamsport & North Branch Railroad) for the benefit of tourists catching the narrow gauge train to Eagles Mere. The remainder of the Reading's Pullman movements from Philadelphia traveled the line to Bethlehem.

At this milltown on the Lehigh River, the CNJ picked up sleepers for Scranton, and (both Delaware, Lackawanna & Western) Binghamton and Syracuse, N.Y. The Reading, however, turned over the majority of its Pullmans to the Lehigh Valley, which carried them to such destinations as Wilkes-Barre, and Auburn, Rochester, Buffalo, and Niagara Falls, N.Y. The Erie briefly cooperated in handling a Buffalo car it picked up at Elmira. The New York Central took Pullmans on to Detroit and Chicago, and the Canadian National/Grand Trunk Western picked up sleepers for Toronto and, before the turn of the century, Chicago. The Reading's trains on this line included a Philadelphia section of the LV's *Black Diamond* as well

as the Pullman workhorse *Interstate Express.*

Reading's most photogenic train did not carry sleeping cars. It was the the Budd-built *Crusader,* a stainless steel, steam-powered streamliner that began making two trips daily between Philadelphia and New York (CNJ) in December 1937.

Pennsylvania's second city, the steel-making metropolis of Pittsburgh, had a greater selection of Pullman service than Philadelphia. The dominant presence in both cities was, of course, the Pennsylvania Railroad, and in both places the main line split into two new trunks. From Pittsburgh's Pennsylvania Station, the Fort Wayne line went to Chicago, and the Panhandle line to St. Louis (the name derived from a predecessor's crossing of the West Virginia panhandle between the Pennsylvania state line and the Ohio River).

The main line from the east brought sleepers from New York—on the all-Pullman *Pittsburgher* and *Red Knight,* among others—as well as through cars from Boston (New Haven) and Springfield, Mass. (NH); the Long Island communities of Montauk (Long Island Rail Road) and East Hampton (LIRR); Atlantic City, N.J., and Harrisburg. Another Pittsburgh train, the *Statesman,* carried sleepers from Washington and Baltimore, and by way of this route the city had brief seasonal service to Miami, St. Petersburg, and Tampa (RF&P/ACL or SAL; FEC to Miami). Pullmans came by way of Harrisburg and Sunbury from Wilkes-Barre and Scranton (DL&W). Shortly after the Civil War, the Pennsy brought a New York sleeper that had come via Central of New Jersey and Reading at Harrisburg.

Led by the all-Pullman *American* and *"Spirit of St. Louis,"* as many as eight trains carried sleepers through and from Pittsburgh to St. Louis. In addition, the Pennsy offered dedicated service to Columbus, Cincinnati, Louisville, Indianapolis, Chicago, and New Orleans (L&N). At one time, the PRR carried sleepers to Pittsburgh from Dayton and Wheeling.

Over the line to Chicago, which saw the likes of the *Broadway Limited, General, Pennsylvania Limited, Golden Arrow, Manhattan Limited,* and the Pittsburgh-Chicago *Golden Triangle,* the Pennsy provided its main Pullman service to the Windy City as well as sleepers to Cleveland, Akron, Toledo, Detroit,[3] and Grand Rapids. Pittsburgh had sleeping car service from Fort Wayne, Ind., and Niles, Ohio.

Although most of the traffic was main line, the Steel City also had Pennsylvania connections to Upstate New York, with sleepers to Olean and Buffalo, and summertime service to Mayville (for Lake Chautauqua) and Muskoka Wharf, Ont. (CNR).

After the Baltimore & Ohio broke the Pennsylvania's monopoly and entered Pittsburgh in 1871, the B&O was able to broaden service and eventually com-

The skyline of Pittsburgh is backdrop for the B&O's *Capitol Limited* before the train departs P&LE station for Chicago.

pete with the PRR for business to many major cities; additionally it offered considerable service into West Virginia.

The B&O main line over Sand Patch from Cumberland and the East brought dedicated sleepers from New York, Baltimore, Washington and Richmond (RF&P), in addition to cars from Charleston, Clarksburg, Elkins (Western Maryland), and Fairmont, W.Va. Over the former Buffalo, Rochester & Pittsburgh line came sleepers from the two Upstate New York cities. The line to Cincinnati provided Pullman service to the West Virginia communities of Huntington, Kenova, Charleston (C&O), and Hinton (C&O), and from Parkersburg, as well as sleepers to Columbus, Cincinnati, Louisville, and St. Louis.

Over the B&O main line through northern Ohio, the route of the all-Pullman *Fort Pitt Limited* and *Capitol Limited,* and the *Shenandoah* and *Ambassador,* the B&O carried sleepers to Akron, Detroit, and Chicago. In 1934, the B&O speeded up trains on this route by almost an hour after acquiring trackage rights over the flatter and shorter Pittsburgh & Lake Erie line through the Steel City. B&O trains, now more competitive with the Pennsy's, used Pittsburgh's P&LE station from that time on.

Although the P&LE eventually came to be its own road, it was controlled by the New York Central for most of its life, giving Pittsburgh Pullman connections with many cities that the Central served. The Erie was a popular bridge route for sleepers originating in Pittsburgh; P&LE trains took cars to Youngstown, where the Erie picked them up for the 66-mile trip into Cleveland Union Terminal. In addition to the Cleveland car, Pullmans for Chicago, Detroit, Toledo, and St. Louis made the same trip, which they completed on one or another line of the New York Central system. Cars going east went via P&LE to Youngstown, then by Central trains to Ashtabula, Ohio; they included sleepers to Buffalo, Albany, and Massena, N.Y.; Boston, and Toronto (TH&B/CPR).

The P&LE's other association with joint Pullman service came in the years before World War I, when the Western Maryland was originating sleepers to Chicago. The P&LE was one of the partners in this routing—as was the Erie—and the service included a Pittsburgh-Baltimore car the WM carried east from Connellsville.

Pittsburgh briefly had a Pullman connection with Chicago via the Wabash, when George Jay Gould was attempting to put together a transcontinental system, which the WM would have been part of. Unfortunately, building his Wabash Pittsburgh Terminal Railway and downtown station was one of the financial strains that caused the collapse of the Gould empire.[4]

With three divisions and one branch of the Pennsylvania converging at Harrisburg—not to forget substantial Reading activity—the state capital hummed with passenger train activity, to the extent of more than 200 movements a day during World War II.[5] Under wire from Philadelphia came the main line trains that paused briefly while switching motive pow-

William J. Brennan

The westbound *Penn Texas*, with Indianapolis and St. Louis sleepers, bids farewell to catenary country in Harrisburg.

er before moving on and crossing the Susquehanna River at Rockville, on what reputedly was the largest stone-arch bridge in the world, and continuing their journey west. Down the river from Williamsport came the trains from Erie and Upstate New York. The *Liberty Limited* and other trains carrying the through Washington Pullmans (and a New York sleeper from York) came into town from the south. Finally, from the Cumberland Valley line to Hagerstown, came trains carrying sleepers from the Norfolk & Western, some of which originated as far south as Florida. It was on this line to Hagerstown that the sleeping car was born.

In the 1830's, stage coach passengers arriving in Chambersburg after a grueling trip over the Alleghenies from Pittsburgh were accustomed to riding the Cumberland Valley Railroad to Harrisburg, where they could board one of the state's trains to Philadelphia. The Cumberland Valley's connection departed late at night, and the weary travelers complained about the spartan coaches with low-back seats. The railroad, later part of the Pennsylvania, responded by inaugurating what is credited with being the first sleeping car, between Chambersburg and Harrisburg in early 1838.[6] This primitive but certainly recognizable conveyance was not the creation of George Pullman, but of a Philadelphia car manufacturer named Richard Imlay. Thanks to this innovator, Philadelphia became a center of early sleeping car activity, and by the autumn of 1838 another Imlay sleeper was running, between Philadelphia and a terminal on the Susquehanna River, across from Havre de Grace, Md. Both cars, running on future PRR lines, were honored pioneers whose progeny would someday reach astounding numbers on America's trains of the night.

Harrisburg saw and sorted so many through Pullman sleepers that it hardly needed its own, but it did have dedicated service to New York and Pittsburgh,

as well as sleeping cars to Atlantic City and from Washington.

It was at Altoona, with the Pennsylvania's huge yards and multitude of red-brick shops, that train crews prepared for the climb to the Allegheny summit at Gallitzin. The enabling contruction on the eastern slope was the famous Horseshoe Curve of the four-track main line just west of Altoona. Completed in 1854, this engineering marvel made the Allegheny Portage Railroad and its incline planes obsolete with the new 4,300-foot, sweeping curve that brought about a rise of 72 feet in grade. For decades, it was the most spectacular train-watching spot on the entire Pennsy system as one after another Pullman-laden limited made its nightly passage. (Between Harrisburg and the state line on the west, 26 trains carrying sleepers were plying the Pennsy main line on a typical midnight in March 1952.) On the night of October 21, 1954, the railroad and the Sylvania company momentarily illuminated Horseshoe Curve with 6,000 flash bulbs in honor of its 100th anniversary.[7]

Altoona itself saw through Pullmans and helper locomotives galore but little dedicated sleeper service except for a pre-1900 car to Chicago.

The cities of Pennsylvania's anthracite region, aside from their sleeper connections with Pittsburgh, had modest Pullman service elsewhere, mostly to New York. Scranton, in hard coal's heyday the state's third largest city, got most of its sleeping car accommodations from the Delaware, Lackawanna & Western, which came into Pennsylvania from New Jersey through the Delaware Water Gap and the Poconos, then transited the northeastern corner of the state en route to Binghamton. This service was appropriate, inasmuch as two of the railroad's predecessors were joined at Scranton in 1853 to form the Lackawanna itself.[8]

The callboard at Scranton's classic French Renaissance station[9] listed the arrivals of the *Lackawanna Limited, Whitelight Limited, Westerner, Phoebe Snow, Lake Cities* and others down through the years. The Lackawanna provided the city dedicated Pullman service to New York and Buffalo, to Pittsburgh with the Pennsylvania, and, via Nickel Plate, to Cleveland and Chicago. On the railroad's scenic route north of Scranton, DL&W trains crossed the famous Tunkhannock Viaduct, which its builders claimed to be the largest concrete-arch structure of its kind when completed in 1915.

Two other coal communities saw Pullman service to New York: Wilkes-Barre by way of the Lehigh Valley, a pioneering anthracite hauler that got its start in Jim Thorpe (ex-Mauch Chunk), and Pottsville (the model for "Gibbsville" in John O'Hara's "Appointment in Samarra") via Reading and CNJ.

Both the PRR and Reading scheduled Pullmans from Williamsport to Philadelphia and New York (Reading jointly with CNJ), and the Pennsy carried a sleeper to Washington.

Despite the fact that many trains of the New York Central's Great Steel Fleet thundered through Erie, the busy lakeport got its dedicated Pullman service from the Pennsylvania, to Philadelphia, Washington, and New York. Erie's service was limited but long-lived; all three lines were still running in the 1950s. In the early days, Erie tried a power play by enacting an ordinance in 1851 requiring railroads entering the town from the east to be of different gauges from those lines coming in from the west. Presumably this would mandate a change of trains and offer employment opportunities for local freight handlers. The governors of Erie came to their senses and repealed the law two years later.[10] In modern times, the Central and PRR shared a Union Station; the Nickel Plate had its own facility.

Oil City, once a major petroleum center whose oil exchange controlled world prices long before the Arabs took charge, had dedicated Pullman service by way of three railroads, but this waned as the wells dried up. The New York Central offered berths to Chicago, the Erie and the PRR to New York, and the Pennsy to Washington.

Bradford, famed for its backyard oil wells, had scheduled sleepers from New York via both Erie and Pennsylvania.

Meadville, home of the Talon slide fastener, saw Pullmans from Chicago and New York brought to town via the Erie.

The industrial and rail center of New Castle had dedicated sleepers that traveled on the Pennsy by way of Youngstown and Alliance to Chicago. Pullmans for Columbus and Cincinnati went via the direct route from Orrville, Ohio.

Of all the what-might-have-been stories in railroading, the tale of the South Pennsylvania Railroad is unique. When it appeared that Philadelphia interests were soon to dominate railroad building in western Pennsylvania, the patriarch of the New York Central, Cornelius Vanderbilt, had a route surveyed across southern Pennsylvania that was much shorter than the PRR's main line. The survey lay dormant until Vanderbilt's son, William Henry—suspecting that the giant Pennsylvania Railroad was backing construction of the NYC's rival, the West Shore, up the Hudson River to Albany and Buffalo—dusted it off in 1883. Sufficiently provoked by this supposed intrusion, Vanderbilt sent construction crews to Pennsylvania, where they began tunneling and grading the right-of-way for the South Pennsylvania. The struggle between the titans went on for two years and involved the Rockefeller and Carnegie names before a settlement was worked out by J. P. Morgan, when he convened a summit meeting one foggy morning on his yacht. The palladins came to terms and agreed to get out of each other's territories.

In the late 1930's, the old South Pennsylvania tunnels were drained, the graded embankments were reclaimed from the woodland growth that covered them, and miles of four-lane concrete pavement were laid. In 1940, the South Penn route was finally opened, as one of the most modern roads of its kind, the Pennsylvania Turnpike.

1. Bert Pennypacker, "'The Grandest Railway Terminal in America,'" *Trains,* Vol. 44, No. 2 (December 1983), pp. 40-57.
2. Pennypacker, p.57.
3. Before the Pennsyvlania got entry into Detroit, New York Central handled the sleeper north of Toledo.
4. George H. Drury, *The Historical Guide to North American Railroads* (Milwaukee: Kalmbach, 1985), p. 343.
5. Richard H. Steinmetz, "Mainline Hot Spot," *Trains,* Vol. 5, No. 4 (February 1945), pp. 16-22.
6. John H. White Jr., *The American Railroad Passenger Car, Part 1* (Baltimore: Johns Hopkins University Press, softcover edition, 1985), pp. 204-5. White offers convincing, though circumstantial, evidence disputing other accounts that have dated the service as early as 1836.
7. Chicago *Tribune,* October 22, 1954.
8. Drury p. 108.
9. The station was restored, and its two-and-a-half-story waiting room, which featured a barrel-vaulted ceiling of leaded glass, walls of Siena and Alpine marble, and terrazo floors laid in mosaic patterns, became the lobby of a Hilton hotel. (New York *Times,* August 4, 1985.)
10. Alvin F. Harlow, *The Road of the Century: The Story of the New York Central* (New York: Creative Age Press, 1947), p. 268.

Charles B. Chaney/Smithsonian Institution

Pennsylvania Railroad/courtesy Conrail

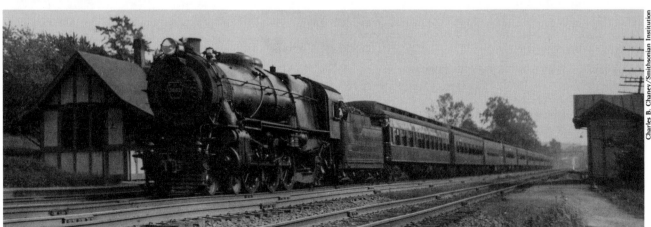

Charles B. Chaney/Smithsonian Institution

7
Maryland Delaware

It was truly an old-fashioned Fourth of July, when, on Independence Day 1828 in the seaport of Baltimore, a group of perspiring dignitaries broke ground for the new Baltimore & Ohio Railroad. In a nice salute to tradition, the 90-year-old gentleman chosen to lay the first stone was none other than Charles Carroll, who 52 summers earlier had set his name to the Declaration of Independence. Now standing at the dawn of the railroad era, here was the last surviving signatory of the nation's birth certificate, linking our past and future.

But it was soon down to business for the fledgling railroad, for there was competition to meet. The immediate concern of the B&O and other railroad investors of the period was not to connect existing centers of population, as was later the case, but rather to give their home cities exclusive avenues to the west, to either the Ohio River valley or the Great Lakes. As a result, there was a rivalry among the great seaports, and Baltimore interests were quick to sense the need to compete with the Chesapeake & Ohio canal, whose builders were making their way up the Potomac River from Georgetown, near Washington.

Accordingly, the Baltimore & Ohio was chartered to build a railroad from the Maryland metropolis to a suitable railhead on the Ohio River, where it could tap that waterway's increasing flow of commerce. And so construction of the country's first common-carrier railroad began, the first leg destined to reach the Potomac River at Point of Rocks. By 1834, the crews had lain tracks to Harpers Ferry, W.Va. (then part of Virginia); by 1842 they were in Cumberland, and finally in late 1852 they reached Wheeling, W.Va., on the Ohio River.

In 1835, the B&O completed the Thomas Viaduct, its famed 700-foot stone bridge across the Patapsco River at Relay, and opened its line to Washington. After the Metropolitan Branch from Washington connected with the old main line at Point of Rocks in 1873, the B&O's modern passenger train routing to the west—via the nation's capital—was established. Meanwhile, it was apparent that lucrative traffic was moving between the cities of the eastern seaboard.

Starting in the late 1860s, the B&O was one of the carriers to participate in joint service advertised as the New York & Washington Air Line Railway. By terms of its charter, the railroad enjoyed a monopoly on Baltimore-Washington travel, so it operated the trains on that portion of the route. The Philadelphia, Wilmington & Baltimore carried through cars on to Philadelphia, and from there, the future affiliates of the Pennsylvania Railroad took over for the remainder of the trip to Jersey City.

Friction developed, however, during the Pennsylvania's consolidation phase, and cooperation ceased. The Pennsy acquired a dormant charter, built a line south from Baltimore, and made an end-run around

B&O's eastbound *Diplomat* (top) crosses the Thomas Viaduct at Relay, Md., May 30, 1932. Locomotive is the President Arthur. Baltimore's Pennsylvania station (middle). One the the Pennsy's workhorse K4's (bottom) speeds Washington's own *Liberty Limited* through Halethorpe, Md., in May 1931.

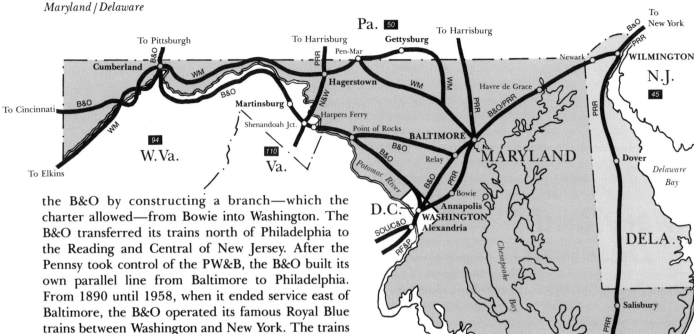

See maps on pages 24 29 140

0   50

Scale in miles

the B&O by constructing a branch—which the charter allowed—from Bowie into Washington. The B&O transferred its trains north of Philadelphia to the Reading and Central of New Jersey. After the Pennsy took control of the PW&B, the B&O built its own parallel line from Baltimore to Philadelphia. From 1890 until 1958, when it ended service east of Baltimore, the B&O operated its famous Royal Blue trains between Washington and New York. The trains were fewer and somewhat less convenient than the Pennsy's, yet the B&O operation had a certain panache that had its own loyal following.

As one might expect, the B&O provided the largest share of Baltimore's dedicated Pullman service, which for a city of its size and importance, was not voluminous. Like Philadelphia, Baltimore was on the busy Northeast Corridor and was well served by numerous through trains. New York and Washington had substantial service, and much of that was available to the two big intermediate cities.

The B&O carried Baltimore-New York sleepers, as did the Pennsylvania, but all the rest of its service was to the west, through Washington: Grafton, Fairmont, and Parkersburg, W.Va., Cincinnati, Louisville, and St. Louis (with C&O); and Pittsburgh, Akron, Cleveland, Columbus, Toledo (Hocking Valley-C&O), Detroit, and Chicago (via either Newark, Akron, or Indianapolis with the Monon). The railroad also carried Pullmans to Baltimore from St. Louis (B&O only) and Wheeling.

Although the Pennsylvania scheduled minimal dedicated Pullman service from Baltimore (New York, Philadelphia, and Pittsburgh), it did offer through accommodations on the full spectrum of trains operating south of Washington, such as Seaboard Air Line's *Silver Star, Comet,* and *Meteor;* Atlantic Coast Line's *East Coast* and *West Coast Champions, Miamian,* and *Havana Special;* Southern Railway's *Crescent* and *Peach Queen;* the Southern/Norfolk & Western's *Tennessean, Birmingham Special,* and *Pelican;* and Chesapeake & Ohio's *Fast Flying Virginian* and *George Washington.* The same applied to the northbound flyers, such as the *Montrealer,* the summertime *Bar Harbor Express,* and all the Boston trains.

The Western Maryland, another Baltimore railroad, and one the city itself had been principal owner of, expanded its passenger service briefly before

World War I. The WM, whose tracks had carried the special train on which President Lincoln composed his Gettysburg Address, generally paralleled the B&O on its short run into West Virginia and Pennsylvania. Although known for being a heavy coal hauler, the Western Maryland was no stranger to passenger business in its early days.

In 1883, the railroad built a hotel, the Blue Mountain House, near Pen-Mar, a resort in the Blue Ridge Mountains on the Pennsylvania/Maryland border. The WM did a brisk excursion business to this scenic area from Baltimore.[1]

Long-distance passenger travel on the WM, sleepers from Baltimore to both Pittsburgh and Chicago, was inaugurated in June 1913, after the railroad made connection with the Pittsburgh & Lake Erie the year before. The Pullmans moved via P&LE to Youngstown, Ohio, Erie to Cleveland, and New York Central to Chicago.[2] Although the schedules were competitive with the B&O's, the service didn't last and was terminated four years later. The Western Maryland, which had the difficult task of operating the highest railroad east of the Rockies, was ultimately absorbed by the Chesapeake & Ohio/B&O.

With the opening of its Howard Street tunnel under Baltimore in 1895, the B&O demonstrated the first practical electrification of a steam trunk line in America. Eastbound trains were pulled through the one-and-a-half-mile tunnel by electric locomotives, which were then cut off "on the fly," whereupon the steam locomotives resumed their smoky operation. Westbound steam trains, rolling through the tunnel on the downgrade, emitted little smoke and didn't require electric helpers.[3]

Mount Royal Station, the B&O's uptown facility, stood at the north end of the tunnel. This beautifully landscaped facility, with its graceful, Romanesque clock tower, was opened in September 1896. A mile and a half to the south was the railroad's historic Camden Station, which, when opened in 1853, was reported to be the largest rail terminal in America. Camden, which was rebuilt the year Mount Royal opened, was the B&O's principal station and continued to be used as a commuter facility even after private rail service ended.[4]

Baltimore's pink granite Pennsylvania Station, opened in 1911, also was used by Western Maryland trains. North of the station, the Pennsy's line to Harrisburg, Pa., branched off; this was the route used by the *Liberty Limited* and Washington sections of the major east-west passenger trains. Baltimore's sleeper for Pittsburgh traveled these tracks.

Cumberland, at one time Maryland's second city, gained great importance in 1806, when Congress passed legislation designating it as the starting point of the new national road to Wheeling. Chesapeake & Ohio canal builders headed for Cumberland, but the B&O beat them into town by eight years, and it was a railroad stronghold forever after.

The B&O maintained shops and a huge yard in Cumberland, where its main line to the West forked. The older trunk to Cincinnati and St. Louis, the route of the *National Limited, Diplomat,* and *Metropolitan Special,* continued to follow the Potomac River, south and west. The Pittsburgh and Chicago main,

which hosted the *Capitol Limited, Columbian* and *Shenandoah,* headed north through the Narrows, a break in the mountains shared by a creek, a roadway, and tracks once used by four railroads, including the Western Maryland.[5]

Despite all its rail activity, Cumberland's dedicated sleeping car service was all gone by the Depression: cars from New York, Philadelphia, Washington, and Chicago via the B&O, and WM's Elkins, W.Va., sleeper for Baltimore, which laid over for several hours in Cumberland.

Four trunk railroads visited Hagerstown, which was the southern terminus of the Pennsy's Cumberland Valley branch and the northern terminus of the Norfolk & Western's Shenandoah division. As a result of this connection and the through service the two roads offered, some unusual Pullmans passed through town, including a New York-Jacksonville car. Hagerstown had its own service to Philadelphia and New York on the Pennsylvania.

1. The Western Maryland operated the *Blue Mountain Express,* carrying cafe/observation/parlor cars, and the *Pen-Mar Express,* which often ran in four or five sections of the road's olive-green coaches, around the turn of the century. Roger Cook and Karl Zimmermann, *The Western Maryland Railway* (San Diego: Howell-North Books, 1981), pp. 239-66.
2. The Western Maryland split into two separate lines at Emory Grove. The through trains generally followed the most direct route, to Hagerstown.
3. E. L. Thompson, "Railroads of Maryland," *Trains,* Vol. 4, No. 7 (May 1944), pp. 31-2.
4. Camden also served the interurban Baltimore & Annapolis Railroad before it terminated passenger operations in 1950. The electric B&A, which ran half-hourly trains between Baltimore and Washington before that service ended in 1935, had the distinction of being the only railroad serving Annapolis, which in turn had the distinction of being the only state capital that wasn't on a steam rail line. There were some exceptions: The B&O ran football specials to Annapolis over the interurban's tracks for transporting midshipmen to Philadelphia for Army-Navy football games.
5. Thompson p. 34.

Delaware

Wilmington, the largest city in Delaware, in fact the only community in the state to exceed a population of 5,000 until the mid-20th Century, was well served by Pennsylvania and Baltimore & Ohio trains of the Northeast Corridor, with their numerous through Pullmans.

The Pennsylvania offered frequent service from New York to Washington, and the great trains to the South—such as Southern Railway's *Crescent,* Seaboard Air Line's *Silver Star,* Atlantic Coast Line's *Champions,* Chesapeake & Ohio's *Fast Flying Virginian*—made at least conditional stops for passengers going beyond the Pennsy's territorial limits of Washington.

The B&O also provided New York-Washington ser-

vice as well as through sleepers on such trains as the *Capitol Limited* and *National Limited* to Chicago and St. Louis, respectively.

What was once the Delaware Railroad, but in modern times part of the Pennsylvania system, branched off the main line at Wilmington and headed south through Dover, the state capital, continuing down the Delmarva peninsula to Cape Charles, Va. Several trains on this route bore New York or Philadelphia sleepers for Cape Charles, where passengers transferred to ferries for Old Point Comfort and Norfolk.[1]

1. No Delaware community had any dedicated Pullman service, as far as this study was able to determine.

8
District of Columbia

President Kennedy liked to tell the old joke that Washington was a city where northern charm met southern efficiency. He was speaking facetiously, of course, but there was a shred of truth in the remark, at least as it applied to the railroad systems that met at the Potomac River. The Royal Blue Line of the Baltimore & Ohio Railroad had a well deserved element of charm, and the efficiency of the roads approaching the capital from the South was hardly ever in doubt.

Whatever else Washington was, it was a unique rail metropolis, the easternmost of four major cities (including Louisville, Cincinnati, and St. Louis) in which the railroads of the North and the South came together. The nation's capital continues to be the southern terminus of the Northeast Corridor, the busy stretch of trackage running to New York and Boston that once tested the rivalry between the Pennsylvania Railroad and the B&O and later became the racetrack of the Metroliner.

Washington was also the northern terminus of both the giant Southern Railway system and the small but important bridge route of the Richmond, Fredericksburg & Potomac. RF&P handled the through traffic of the Seaboard Air Line and Atlantic Coast Line railroads north of Richmond, and Southern rails also accommodated Chesapeake & Ohio trains, by a trackage arrangement, north of Orange, Va. Through cars from the Norfolk & Western came to Washington on both the Southern and ACL/RF&P.

The several systems came together in ominously narrow straits, the Pennsylvania's bridge over the Potomac River into Virginia, on which all eastern trains to the South traveled. This caused defense authorities no little concern during World War II, and emergency bridge plans were supposedly available. But strategic drawbacks aside, there was hardly a more scenic spot—with the various edifices and memorials of Washington as a backdrop—to watch a greater concentration of some of America's premier passenger trains.

These included the great Florida fleet of the ACL: *Miamian, Florida Special, Havana Special,* the *Champions;* SAL's *Orange Blossom Special,* the *Silver Comet, Star* and *Meteor;* Southern's *Crescent, Piedmont Limited, Southerner, Peach Queen, Tennessean;* and the C&O's *Fast Flying Virginian (F.F.V.)* and *George Washington.*

Across the river in Virginia, trains entered the giant Potomac Yard, where the interchange of freight between North and South took place. Through passenger trains, of course, had to undergo a similar sorting, but that took place in the remarkable structure of Washington Union Station.

At the dawn of the Twentieth Century, when it became clear that the capital of the newly emerging world power needed a major rail facility befitting its importance, attention turned to a plot of land north of the Capitol not unlike much of the acreage in the

Now in the third decade of their service, three GG1's (top) repose in Ivy City engine terminal in Washington on a June night in 1960. America was not yet out of the horse and buggy era when Washington Union Station (bottom) opened in 1907. Pennsylvania and B&O tracks from New York stubbed in the rear of the station; trains to the south continued on through a tunnel under Capitol Hill, which lay just a half mile beyond the photographer's right shoulder.

Baltimore & Ohio's Detroit train, the *Ambassador*, pulls out of Washington with eleven cars on August 3, 1939.

District of Columbia: It was swampland. Nonetheless, plans went ahead, and in 1901, the Pennsylvania and B&O laid aside at least one element of their keen rivalry and jointly formed the Washington Terminal Company for the purpose of building the new Union Station. The planners chose a renowned architect from Chicago, Daniel Burnham, who designed a gleaming structure of white marble and granite with interior space of immense proportions.

To accommodate the huge crowds expected for presidential inaugurations, the station featured a concourse 760 feet long and 130 feet wide, covered by an arched roof with no supporting pillars. To accommodate the president or visiting heads of state, a suite of rooms was appropriately furnished for their exclusive use.[1] (Apparently the comings and goings of members of the legislature and judiciary were regarded as nothing more than routine.)[2]

And so a half mile from the U.S. Capitol, where Massachusetts and Delaware avenues and a few other streets converged at a public fountain and monument known as Columbus Circle, the new Washington Union Station opened on November 17, 1907.

Passengers from the north and northwest arrived on trains of the Pennsylvania and B&O, which used the 20 stub tracks at the station's ground level. Through cars to the South were spotted on one of nine tracks on the lower level. These rails converged into a subway, then a tunnel running for a length of three-quarters of a mile beneath Capitol Hill before exiting on the south slope and continuing on to the Potomac River bridge. Unfortunately, clearances in the tunnel later proved inadequate for the postwar dome coaches.

Its few inadaquacies notwithstanding, Washington Union Station has served its purpose well over the years, hosting citizens and sovereigns alike. After an inept conversion to a Bicentennial tourist center, followed by a period of alarming neglect, Union Station reopened in 1989, handsomely refurbished as both terminal and commercial center.

At one time or another passengers could travel, without changing sleepers, from Washington to more than ninety-five destinations in all parts of the country except the Pacific Northwest. Washington was the eastern terminus of the longest, long-running sleeper route in the country, the tourist Pullmans that traveled the Sunset Route to Los Angeles and San Francisco.

Despite the fact that only five major railroads operated passenger trains in and out of the nation's capital, the service was competitive. At various times, Washington had sleeper connections over three or more routes with eleven cities. Fifteen destinations were served by both the Pennsylvania and B&O, and seven by the B&O and C&O.[3]

While the rail systems of the East were taking shape, the Pennsylvania and B&O actually cooperated in the movement of passengers between Washington and New York. The Pennsylvania had trackage north of Philadelphia, but no Baltimore-Washington connection. It was just the opposite with the B&O, and the vital center of this route—provided by the Philadelphia, Wilmington & Baltimore—accommodated both railroads. After the Pennsy built its own line into Washington, the B&O formed an alliance with the Reading and Central of New Jersey to forward its trains to New York. The Pennsylvania then acquired the PW&B, whereupon the B&O built its own line to Philadelphia, which it completed in 1886. Joint operation was now a memory, and 70-odd years of tough competition lay ahead.

In 1890, the B&O/Reading/CNJ inaugurated the famous Royal Blue Line service between Washington and New York, which, during its heyday, consisted of ten trains a day in each direction. The equipment, all painted royal blue until the industry succumbed to Pullman-green uniformity early in the Twentieth Century, was contributed by the three participating carriers. On each of the coaches was painted the coat of arms of the state in which its owner-railroad was chartered: Maryland for the B&O, Pennsylvania for the Reading, and New Jersey for the CNJ.

Before the tunneling under the Hudson River was opened in 1910, Pennsylvania trains from Washington to New York terminated in Jersey City. Starting in 1918, B&O trains were allowed to operate to and from New York's Pennsylvania Station by order of the United States Railroad Administration, but they returned to Jersey City in 1926 rather than continuing to pay the Pennsy's exorbitant rents. The B&O compensated for the loss of convenience by offering passengers direct-to-trainside bus service from several locations in Manhattan and Brooklyn. Over the Royal Blue main line, the B&O operated sleepers to Washington from New York, Newark, and Philadelphia.

At Philadelphia, where the northbound B&O trains went onto Reading tracks, several Washington sleepers for upstate New York were cut out. Before World War I, Pullmans for both Buffalo and Rochester traveled north via the Reading and Lehigh Valley. Briefly during the 1920s, a Washington-Syracuse car moved on the *Interstate Limited* (RDG/CNJ/Lackawanna).

After several decades of Pullman green, the B&O's true colors returned to the Royal Blue Line. During the 1930s all railroads were facing a precipitous decline in passenger travel, and they spared no efforts in getting their business back on track. In 1931, the B&O scored a coup by putting into service between New York and Washington two air-conditioned train sets of the *Columbian,* making it the country's first completely air-conditioned train.

A Royal Blue train made of lightweight aluminum, the East Coast's first non-articulated streamliner,

On its 25th anniversary, The C&O's *George Washington* prepares to depart from Union Station, April 24, 1957.

made its bow in June 1935, under the power of a streamstyled Hudson named Lord Baltimore. Later that summer the Electro-Motive Corporation's first single-unit diesel made trial runs with the train, starting the tradition of locomotives in blue and gray that would haul B&O streamliners until the logos of the Chessie era came along.

The competitive spirit in the Northeast Corridor endured until the B&O capitulated to the Pennsylvania's faster, more frequent service into downtown New York. For the Royal Blue Line, which had always prided itself in doing business on a personal scale, the end of more than two-thirds of a century's service came in April 1958, by which time the six daily trains in each direction were running with unprofitably light loads. Thereafter, Baltimore, the city of the B&O's historic birth, was the road's eastern terminus.

West of Washington, the competition with the Pennsy, both to Chicago and St. Louis, was no less intense. Despite the disparity in scale and manner, the two roads geographically were remarkably similar, and it was small wonder that they both offered Pullman service to the same fifteen cities. The B&O differed, however, in that its main line from Washington went right to the point, by heading up the Potomac River to a much more difficult crossing of the Alleghenies than the Pennsylvania had to make.

The *Capitol Limited* was the flagship of the B&O's Chicago trains. Inaugurated in May 1923, the *Capitol* was the first all-Pullman train operating between Washington and Chicago, and in 1937 it became the first train from the East Coast to Chicago hauled by

C&OHS

William F. Howes Jr. collection

Pullman cars also served as destination hotels. Here West Point cadets assemble for inauguration of President Truman in 1949.

diesel locomotive. Postwar, it offered a Strata Dome sleeper (none of the B&O's competitors had dome cars) and a rare Pullman for San Diego, which it turned over to the Santa Fe in Chicago. The B&O carried two other West Coast sleepers: a San Francisco car for the Chicago & North Western/Overland route and a Los Angeles Pullman for the Santa Fe Chief. The *Capitol Limited* remained an exclusive Pullman train until the 1960's, when the mostly coach *Columbian* was combined with it. Other key trains on this route were the *Shenandoah* and *Washington-Pittsburgh-Chicago Express,* the *Ambassador* to Detroit, and the *Cleveland Night Express.*

Before the turn of the century, the B&O had a short involvement—in both time and distance—in sleeper service to the South. At Shenandoah Junction, W. Va., B&O turned over cars from Washington to the Norfolk & Western, which took one to Roanoke, and the others to Bristol, on the Virginia/Tennessee line, where the Southern reclaimed them for the remainder of their journeys to Nashville (Nashville, Chattanooga & St. Louis), Memphis, and New Orleans. The Southern also took over Memphis sleepers at Cincin-

nati. In addition to Pullmans for Chicago on this main line, the B&O also carried sleepers from Washington to Cumberland, Pittsburgh, Akron, Columbus, Cleveland, and Detroit.

Slightly younger and not exclusively all-Pullman for as long a time as the *Capitol Limited,* the *National Limited* nonetheless proudly carried the B&O's train numbers 1 and 2, a reminder of the seniority of the road's main line to St. Louis. In addition to sleepers for Cincinnati and St. Louis, the *National* once carried a Pullman to the Illinois capital, Springfield, and after World War II it participated in sleeping car lines to Oklahoma City (Frisco), Fort Worth (Missouri Pacific-/Texas & Pacific), Houston (MP/T&P/MP), and San Antonio (Frisco/Missouri-Kansas-Texas). The *Diplomat* and the two-night *Metropolitan Express* also operated on this route to St. Louis, which before the turn of the century also formed part of a joint Pullman route to Chicago. The B&O and a predecessor turned over these sleepers to the Monon in Indianapolis. Washington also had service from Louisville and Newark, Ohio, and regionally from the West Virginia communities of Wheeling, Parkersburg, Grafton, and

Fairmont.[4]

After the Chesapeake & Ohio got access into the nation's capital, it originated sleepers in Washington and New York City, in addition to Newport News, Va. This expanded service began with the *F.F.V.* in 1889. Over the years the railroad carried sleepers from Washington to destinations ranging from White Sulphur Springs, W.Va., to St. Louis (with New York Central, later B&O) and San Antonio (NYC/MP/T&P/MP). On its southerly route through West Virginia to the Ohio River valley, the C&O offered Pullman service to Chicago (NYC), Cincinnati, Columbus, Detroit, Indianapolis (NYC), Lexington, Ky., and Louisville. Additionally, it carried dedicated sleepers to the West Virginia cities of Hinton, Charleston, and Huntington, and also to Ashland, Ky.

In 1963, the C&O gained control of the B&O, and a decade later the two became subsidiaries of the new Chessie System. The two roads jointly operated a Washington-St. Louis car for a time, and on the eve of Amtrak an emaciated *Capitol Limited* and *George Washington* were the sole survivors providing "Chessie Sleeper" service.

The Pennsylvania Railroad had a dominating grip on service to New York that it never lost, at least while it was its own railroad. It offered virtually hourly service, and, with the completion of electrification into Washington, further speed records. The deluxe *Congressional*, which in the Gay Nineties had sported an elegant red, green, and cream-colored livery, celebrated its 50th anniversary in 1935, and on Feb. 10 that year inaugurated completely electrified service between the two cities. From this time on, the Northeast Corridor accommodated an almost endless parade of swift limiteds powered by the distinctive GG1 locomotives that Raymond Loewy had designed in a dark-Brunswick-green, gold-pin-striped, streamlined shroud. These mighty workhorses served well into the Amtrak era, hauling virtually all the major trains between Washington and New York.

New York and Boston expresses were plentiful and mostly carried governmental names, such as *Legislator, Executive, Judiciary, Mount Vernon, Federal,*[5] and *Patriot.* Before the all-but-seamless flow of trains to Boston was possible, however, the crossing of waterways around New York City was an impediment. Until the tunnels were drilled under both the Hudson and the East rivers, Hell Gate Bridge built, and Pennsylvania Station opened, two alternate solutions were used. The Pennsylvania Railroad barged its through cars across New York Harbor to the Bronx on the steamer *Maryland* for an interline transfer with the New York, New Haven & Hartford. The other alternative involved what was known as the Poughkeepsie Bridge Route.

In 1890, a high, cantilevered steel bridge was completed across the Hudson River at Poughkeepsie, N.Y., making possible a freight route between New England and the coal fields of Pennsylvania. Passenger service was soon instituted between Washington and Boston over the bridge, which was owned by several railroads throughout its history. The "bridge route" had a few variations beyond the actual crossing of the Hudson itself. At the southern end, the B&O/Reading family were at first involved, as were the Boston & Maine and New Haven (and their predecessors) in New England. The middle ground in New Jersey was variously occupied by the Lehigh & Hudson River, Lehigh & New England, and Central of New Jersey.

Between 1912, when the Pennsy began having mechanical problems with the *Maryland,* and 1917, when the Hell Gate Route opened, the railroad sent the *Federal Express,* the premier Washington-Boston train, by way of the bridge route, turning it over to the L&HR at Phillipsburg, N.J. Although through passenger service on the Poughkeepsie route ended with the opening of Hell Gate, the bridge itself bore the weight of many freight trains until it burned in 1974.

Apart from the Boston trains of the Pennsylvania/New Haven, a more international air was achieved on the *Montrealer,* which was inaugurated in June 1924. Although officially named the *Washingtonian,* the southbound version of the train became jocularly referred to as the *Bootlegger.* Prohibition was the law of the land then in the United States—but not in Canada—and American passengers riding the *Washingtonian's* Pullmans from Montreal, Quebec, and Murray Bay often returned to the States with their illegal alcoholic spirits ingeniously concealed. The train was a joint operation of Canadian National/Central Vermont/Boston & Maine/New Haven/PRR.

Another old reliable was the year-round *State of Maine,* which carried Washington sleepers for Portland, Me., and Concord, N.H., skirting Boston in the process. Most passenger travel to coastal New England came in the summertime, however, and the venerable, all-Pullman *Bar Harbor Express* was the conveyance of choice for well-heeled seasonal travelers. This posh train, a longtime New York tradition, carried sleepers to resorts in New Hampshire and Maine by way of the New Haven, Boston & Maine, and Maine Central railroads. In 1917, after the opening of the Hell Gate Bridge route in New York, Washingtonians could book through accommodations to such destinations as Ellsworth, Mount Desert Ferry, Oquossoc, and Rockland, Me., and Bretton Woods and Plymouth, N.H. So heavy was the traffic at certain times, especially in the 1920s, when the train generally ran daily, that separate sections of the *Bar Harbor Express* were operated just from Washington. For summer resorters traveling to Hyannis, Mass., the New Haven's *Night Cape Codder* carried the Washington sleepers.

In addition to this seasonal service, the Pennsy routinely carried Washington Pullmans up the corridor to Philadelphia and New York, and with the New Haven to Providence, Springfield, and Boston.

Pullmans on the Pennsylvania bound for the West also moved northeast as far as Baltimore, where they left the main line and were hauled north to Harrisburg, Pa. At this busy rail center, crews switched them to the through trains from New York. Many of the Pennsy's name trains had their own sections serving Washington: the *Gotham Limited*, "*Spirit of St. Louis*," *Clevelander*, *Red Arrow*, *Admiral*, *Pennsylvania Limited*, and *Penn Texas* among them.

The Pennsylvania, however, gave Washington its own Chicago train, the *Liberty Limited*, in the mid 1920s after operating a D.C. section of the *Broadway Limited* for a couple of years. The *Liberty* operated all-Pullman until the late 1930s.

After World War II, the Pennsy shared in transcontinental Pullman service to the West Coast, carrying Washington sleepers to Chicago for both San Francisco and Los Angeles via the Overland Route. Sleepers bound for Harrisburg, Pittsburgh, Cleveland, Detroit, and Chicago also moved on this line.

At Pittsburgh, the "Panhandle" line of the Pennsylvania branched off for St. Louis. The railroad carried Pullmans for the Missourigateway, where it also turned over sleepers for Dallas/Fort Worth, Galveston, and Houston to the Missouri Pacific. Cincinnati, Columbus, and Indianapolis were intermediate destinations with dedicated service from the capital.

The Pennsylvania offered Washington one final sector of travel, north of Harrisburg, to cities of central Pennsylvania and western New York. Through Williamsport, Pa., which had its own sleeping car, passed such trains as the *Dominion Express* with Washington sleepers for Erie and Oil City, Pa., and Buffalo, Canandaigua, and (with NYC) Rochester, N.Y.

South of Washington, the Richmond, Fredericks-

Three diesel units in Seaboard Air Lines' citrus colors, here at Ivy City in 1939, powered the railroad's posh *Orange Blossom*.

burg & Potomac moved the great parade of trains that served Florida winter and summer. Both the Atlantic Coast Line and Seaboard Air Line carried Washington Pullmans to Jacksonville, St. Petersburg, and Miami; ACL had cars to Orlando, Tampa/Sarasota, and Key West (FEC), and SAL scheduled dedicated sleepers to Palm Beach (FEC). Both railroads, moreover, carried Washington Pullmans to many intermediate cities: the ACL to New Bern (Norfolk Southern) and Wilmington, N.C., Charleston, S.C., Augusta (1890, Southern) and Savannah, Ga.; and SAL to Hamlet, N.C., Columbia, S.C., Atlanta and Columbus, Ga. (Central of Georgia), and Birmingham, Ala. The RF&P carried Richmond sleepers.

Washington Union Station's lower-level tracks also accommodated trains of the Southern Railway and the C&O. Southern trains traveled the main line to the historic canal town of Lynchburg, Va. (which had a dedicated sleeper in the 1890's). From there the *Crescent* route went south through Atlanta and on to New Orleans via the West Point Route and Louisville & Nashville. The Washington-Sunset tourist cars to the West Coast once operated over these tracks.

At Lynchburg, the Norfolk & Western Railway took

On a prewar Sunday in February 1940, RF&P 4-8-4 No. 605 approaches the Capitol Hill tunnel with the *Florida Special*.

over other Southern Pullmans, for Roanoke and Bristol, Va., and Bluefield and Williamson, W.Va., as well as those that returned to Southern tracks west of Bristol. This was the route of the *Birmingham Special, Pelican,* and *Tennessean,* trains that carried sleepers to Knoxville, Chattanooga, Nashville (NC&StL), Memphis, Tenn., and Shreveport, La. (Illinois Central). Passengers from Washington could travel by Pullman on either the all-Southern or the joint Southern/N&W routes to Birmingham, and by both those lines plus the joint Southern/WPR/L&N route to New Orleans.

Over the *Crescent* route, the Southern carried sleepers to Asheville, Charlotte, Southern Pines, Hot Springs, (1890), Winston-Salem, and Raleigh, N.C., Columbia and Charleston, S.C. (1882), Atlanta, Augusta, Macon, and Columbus, Ga., Montgomery (WPR) and Selma, Ala., and Memphis (1893, Frisco).

Northern charm, southern efficiency: It all came together in Washington Union Station.

1. For at least a century, trains were the primary means of long-distance transportation of American presidents, and in the final years of such travel the Pullman company provided for this purpose an impressive piece of varnish. Officially known as U.S. Car No. 1, this armor-plated observation/sleeper was more familiarly known as the *Ferdinand Magellan.* Stored under guard at the Washington Naval Shipyard, the car was thoroughly inspected before each presidential trip. The extra-heavyweight *Magellan,* which had an open platform that was ideal for campaign whistlestops, was used chiefly by presidents Franklin D. Roosevelt and Truman, and to a limited degree by Eisenhower.
2. The author's first impression of Washington Union Station was anything but routine for him, when, in his tenth year, he arrived in the company of an uncle after a trip from Jersey City in a B&O parlor car (actually a section sleeper). It was at dusk on a spring evening in 1944, and the memory of the throng of people, many of whom were Army and Navy officers, queuing up for taxicabs and streetcars, was unforgettable. After following the required Washington protocol of "rank has its privilege," his father, a wartime naval officer, but a junior one, finally hailed a cab for the party. Never before having seen such a galaxy of stars, stripes, bars, and oak leaves, the author was still taking inventory as he felt his arm being pulled into the waiting taxi.
3. At one time or another, sleepers move by at least three different railroads or routings from Washington to Birmingham, Boston, Chicago, Cincinnati, Columbus (Ohio), Detroit, Los Angeles, Memphis, New Orleans, San Francisco, St. Louis.

   Both the Pennsylvania and B&O provided Pullman service from Washington to Boston, Buffalo, Chicago, Cincinnati, Cleveland, Columbus (Ohio), Detroit, Houston, Los Angeles, New York, Philadelphia, Pittsburgh, Rochester, St. Louis, and San Francisco.

   The B&O and C&O both offered sleeping car accommodations from Washington to Chicago, Cincinnati, Columbus (Ohio), Detroit, Louisville, San Antonio, St. Louis.
4. For a few years after World War II, this line saw the daytime passage of a smart little all-coach streamliner called the *Cincinnatian.* On the expectation that the C&O was going to inaugurate a deluxe, Washington-to-Cincinnati train, the B&O remodeled several heavyweight coaches and streamlined four of its Pacific locomotives in a two-tone blue, black, and silver dress. The five-car *Cincinnatian* made its bow in January 1947, but owing to a lack of traffic—which caused the C&O to cancel its plans—the *Cincinnatian* ran the race alone before becoming a Cincinnati-Detroit train in 1950.

   The *Cincinnatian's* C&O adversary was to have been called the *Chessie,* but instead it became known as the train that never was. Robert R. Young, chairman of the C&O, announced plans for the daylight, Washington-Cincinnati streamliner, saying that it would run on a 12-hour schedule and include such amenities as two dome cars, a soda fountain, library, telephones, cartoon theater, and playroom. The orange and silver Chessie was to be powered by a revolutionary steam turbine/electric, coal-burning locomotive built by Baldwin. By the time the coaches began arriving from Budd, the manufacturer, in mid-1948, it was apparent that daytime traffic was not sufficient for two competing trains, and the cars for the stillborn streamliner were either sold or reassigned.
5. The *Federal Express* made all the newspapers in early 1953. The southbound train with 16 cars was routinely entering Washington at about 70 miles an hour on January 15, when the engineer of the GG1 discovered with alarm that his brakes had failed. He reversed the engines, but with only partial success. The *Federal* hit the bumper of Track 16 in Washington Union Station at about 35 miles per hour, then lurched across the concourse before plunging through the floor into a baggage room. Fifty-one persons were injured.

Southern No. 25, the *Memphis Special,* leaves Charlottesville, Va., in May 1938. The *Tennessean* replaced this train in 1941.

W.H. Thrall/Frank Ardrey collection

9
Ohio

During the first half of the Nineteenth Century, Cincinnati was the ruling American metropolis west of the Appalachians, and it was from the Ohio River that the Queen City derived its bustling trade. Day and night, freight-laden steamboats and luxurious passenger packets whistled their sonorous calls and docked along the city's six-mile waterfront. But while the steamboat was king, the railways were coming.

By 1846, the future Pennsylvania Railroad line had connected Cincinnati with Columbus. By mid-century, early New York Central lines had made through trav-el possible to Cleveland as well as Kankakee, Ill., south of Chicago. In the late 1850's, the Atlantic Sea-board and Mississippi River were finally connected by rails, and these became part of the Baltimore & Ohio's line to St. Louis, passing through Cincinnati.

By the Civil War, Ohio had the most railroad mile-age of any state in the Union and was crisscrossed by all the modern eastern trunk lines. For Cincinnati, of course, the shift to the railroads meant the demise of much of the river traffic that had made it prosper. The rail promoters and merchants, however, quickly built upon the trade that was long established with cities as far downriver as New Orleans. Cincinnati, as a result, became to the Great Lakes region what Washington, D.C., was to the East: a great rail gate-way between North and South.

The rival port of Louisville had a head start in tapping the commerce of the southern states, by way of the Louisville & Nashville Railroad, but its rail connections to the North were not well established until after the Civil War. Moreover, as a north-south line whose physical plant became the target of numer-ous raids, the L&N suffered from the war more than most railroads. Accordingly, there was still time for Cincinnati to assert itself, and the city responded by building its own railroad.

Local boosters had talked about a rail line running south through Kentucky and Tennessee well before the Civil War, but for a variety of reasons—mostly lack of capital—it never got built. By the late 1860's, with the city of Cincinnati's financial backing, the pro-ject finally got off the ground.

The Cincinnati Southern, as the railroad was named, was owned by the city of Cincinnati and even-tually found its way, by lease, into the Southern Rail-way system.[1] The Cincinnati, New Orleans & Texas Pacific leased the Cincinnati Southern. The CNO&TP became part of the Queen & Crescent Route of the Southern Railway System. The L&N also acquired a line into Cincinnati, giving the city a sec-ond connection to the south.

From the north and the Atlantic Seaboard came the east-west trunk lines—New York Central, Pennsylva-nia, Baltimore & Ohio, and Erie—as well as the big coal-carriers, Chesapeake & Ohio and Norfolk & Western. With all these connections, Cincinnati be-

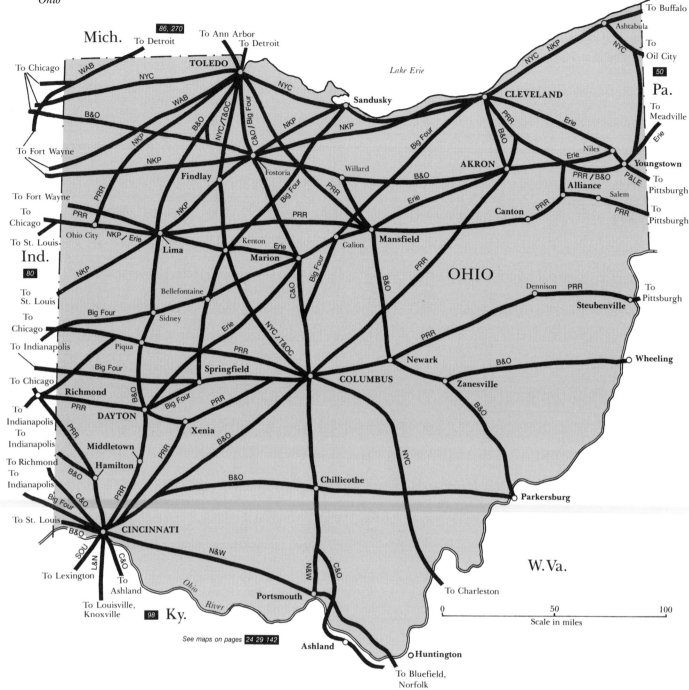

See maps on pages 24 29 142

came a busy gateway, where numerous sleepers were interchanged.

The Erie provided Cincinnati with limited sleeping car service, but all of it was gone by the 1920's. Sleepers operated to Youngstown, Hornell, N.Y., Buffalo, New York City, and Boston (Delaware & Hudson/Boston & Maine).

The first eastern trunk line to serve Cincinnati carried Pullmans in five directions. On the B&O's main line west, route of the all-Pullman *National Limited, Diplomat,* and *Metropolitan Special,* the railroad scheduled sleepers to St. Louis and Louisville, where it made a connection with the Illinois Central. The IC forwarded the cars to Memphis, Gulfport, Miss., New Orleans (over both its routes), and Los Angeles and

San Francisco (tourist sleepers that traveled Southern Pacific from New Orleans).

Over its line to Hamilton and Indianapolis, the B&O scheduled Pullmans to Springfield, Ill., and Chicago (Monon). Before the turn of the century, service included sleepers to Jacksonville, Ill., (Wabash) and St. Louis (PRR).

The B&O carried Pullmans north to Toledo and Detroit and, via C&O in summers, to the Michigan resorts of Bay View and Harbor Springs (PRR).

Cincinnati had sleepers to Columbus, Wheeling, and Pittsburgh, and, over the B&O main line through Parkersburg, W.Va., to Washington, Baltimore, Philadelphia, and New York. This was the route of the *Cincinnatian,* a steam-powered, all-coach streamliner the B&O inaugurated after World War II, anticipating C&O competition that never developed. In Chillicothe, the B&O turned over interline Pullmans to the Norfolk & Western for that road's namesake city.

Following the river of its name, the Chesapeake & Ohio laid tracks into Cincinnati in 1888 and the next year premiered the crack *Fast Flying Virginian (F.F.V.)*. This venerable express carried sleepers from New York (PRR) and Newport News, Va., and led a stable of thoroughbreds that would eventually include the *George Washington* and *Sportsman*.

The C&O originally went no further west than Cincinnati, but early in this century it acquired a line to Chicago, the C&O of Indiana, over which it ran a through train from Old Point Comfort, Va., from 1911 to 1917.[2] Called the *Old Dominion Limited*, this was among the few trains to operate through Cincinnati. At other times, sleepers from the C&O continued on to Chicago and St. Louis by way of the Big Four. (Before Pullman green won the day, the cars of both the C&O and Big Four wore colors of maroon and orange.) The C&O also briefly scheduled Cincinnati-Chicago sleepers on its own line through Indiana.

From the East, the C&O brought Pullmans to Cincinnati from Ashland, Ky.; Huntington, Charleston, and Hinton, W.Va.; Clifton Forge, Charlottesville, Richmond, and Newport News, Va.; Washington, and, with the Pennsylvania, Philadelphia and New York.

The Norfolk & Western, which gained entry into Cincinnati in 1901, was the road of the *Pocahontas* and *Cavalier,* and the post-war, all-coach *Powhatan Arrow*. The N&W carried Pullmans from Cincinnati for Norfolk, and Winston-Salem, N.C., and through cars from Chicago (PRR) to the East. In later years, the railroad's attractive tuscan red and black streamlined J class locomotives and red coaches graced the rail scene in Cincinnati.

Seven express trains, led by the New York-bound, all-Pullman *Ohio State Limited*, worked the Big Four main line to Cleveland in the 1920's. Cincinnati had sleepers to Columbus, Cleveland, Buffalo, New York, and Boston via this route. Trains branched off at Springfield with cars for Toledo, Sandusky, Detroit, and, in summers, Mackinaw City, Mich.

The *Carolina Special* and daytime *James Whitcomb Riley* operated on the main line through Indianapolis, once a speedway for a half dozen trains between Cincinnati and Chicago. Sleepers for Chicago, St. Louis, and Peoria traveled via this route.

The Pennsylvania scheduled fewer trains to Chicago, but its service also included sleepers. From Richmond, Ind., a line ran north through Fort Wayne into Michigan. This was the route of the summertime *Northern Arrow,* a train that gathered Pullmans from several midwestern cities and delivered them to resorts in Michigan. Cincinnati had seasonal sleepers to

V-J Day was a month away when this photograph was taken in the soaring rotunda of Cincinnati Union Terminal.

Harbor Springs, Mackinaw City, Muskegon, Traverse City, and from Petoskey, in addition to year-round service to Grand Rapids.

From Chicago, the Pennsy's *Southland* traveled to Cincinnati through Fort Wayne, and the Central's *Royal Palm* came by way of Indianapolis. The Florida-bound Pullmans these trains brought from northern cities were switched to either Southern or L&N rails in the rail yards of Cincinnati.

The *Cincinnati Limited* was the Pennsylvania's principal train to the East. Over the line through Xenia, the railroad scheduled Pullmans to Cleveland, New Castle, Pa., Pittsburgh, Washington, Atlantic City, N.J., New York, and Boston (New Haven), and from Columbus. The Pennsy and L&N also exchanged many through cars to and from southern cities in Cincinnati.

The L&N had two main lines going south from Cincinnati. The western line went through its namesake cities then south into Alabama to the Gulf Coast and New Orleans. The eastern line went to Knoxville and Atlanta.

Over the western route, the L&N carried Pullmans from Cincinnati to Louisville, Nashville, Montgomery, Mobile, Pensacola, and New Orleans, as well as Memphis. This was the way of the *Pan-American,*

Humming Bird, and *Azalean.*

After its inauguration in 1921, the L&N's *Pan-American* immediately became the flagship of the line to New Orleans. In 1925, the train was reequipped, this time as an all-Pullman limited, and passengers were offered maid and valet service, shower baths, and the new entertainment sensation of radio.

A few years later, the luxurious *Pan-American* itself became a radio star. From 1933 to the end of World War II, the clear-channel voice of Nashville radio station WSM broadcast live the sounds of the train as it passed a tower on the edge of town, sending to all the Southland the news of the wonderful clangor of steam locomotive and heavy Pullmans trucking over steel rails on their way to Canal Street in the Crescent City.

The train took on coaches during the Depression, and after World War II the L&N inaugurated the all-coach, streamlined *Humming Bird,* but the *Pan-American* retained its traditional position of honor. When the end came in 1971, just a few months short of its 50th birthday, the train was the sole survivor on this historic route.

Around midnight for many years, the southbound *Pan-American* passed through Georgiana, Ala., birthplace of country music star Hank Williams Sr. He wrote a song in its honor, and no doubt it was this nightly passage, among all the rest, that he had in mind when he sang: ". . . a midnight train is whining low."

The sprightly blue and silver *Humming Bird* ran on a faster schedule than the *Pan-American.* Although originally all coach, it soon was offering Pullman sleepers between Cincinnati and New Orleans.

The *Southland, Flamingo,* and pre-World War II *Georgian* were the prime runners on the L&N line to Atlanta, carrying through cars from Great Lakes cities to Florida as well as sleepers from Cincinnati to Knoxville; Atlanta; Bristol (Southern) and Norton, Va.; and jointly with Central of Georgia to Macon and from Savannah, Ga.; and to Jacksonville and Tampa (Atlantic Coast Line) and Miami (ACL/Florida East Coast).

The heaviest concentration of Pullmans departing from Cincinnati over the years traveled on the Southern Railway, which handled the heavy Florida traffic from the Midwest on such trains as the *Suwanee River Special, Ponce de Leon,* and the *Royal Palms.*

From Cincinnati, the Southern carried sleepers to Knoxville; Atlanta, Macon, Augusta, and Rome, Ga.; Charleston, S.C. (Georgia Railroad/Southern); and Jacksonville, St. Augustine (FEC), Miami (FEC or Seaboard Air Line), Sarasota, and (all SAL) St. Petersburg, Tampa (1898), and Venice, Fla. But there were more—trains to the Carolinas and New Orleans also departed from Cincinnati on the same tracks.

Service to the southeast included sleepers to Asheville, Greensboro, Winston-Salem, Goldsboro, and Hot Springs, N.C.; Spartanburg, Columbia, and Charleston, S.C., as well as a more round-about routing through the Carolinas to Jacksonville (Southern or Southern/SAL), St. Petersburg (SAL), and Tampa (ACL). Many of these cars moved on either the *Carolina Special* or the *Skyland Special.*

The Southern's line to New Orleans was another major route from Cincinnati to the Crescent City. Its premier train was named the *Queen & Crescent Limited.* (Oddly, the Southern briefly operated its own *Pan-American Special* before the famous L&N rival was inaugurated.) Over this route the Southern carried sleepers from Cincinnati to Chattanooga; Birmingham, Mobile (Mobile & Ohio in 1898, or later Southern), and Livingston, Ala.; Meridian, Miss.; and Shreveport and New Orleans, La. Schedules of 1895 also show a Cincinnati-San Francisco tourist sleeper operating over this route, presumably going to the Southern Pacific at New Orleans.

Though not the largest city in Ohio, Cincinnati handled by far the greatest Pullman volume. Over the years, the railroads provided sleeping car service to more than seventy-five destinations. Until World War I, five different railroads carried sleepers to New York. Four provided accommodations to Chicago, Jacksonville, and New Orleans, and three to Columbus, Boston, St. Louis, and Washington. The through cars further compounded the flow. To handle all this traffic, the seven railroads still serving Cincinnati with passenger trains in the 1920's pooled resources to build a commodious new station to replace the five then in use.

The entryway of the new Union Terminal was a soaring, Art Deco half-dome fronting on a park with terraced fountains. Inside, the ticket windows and other service facilities were arranged in a semi-circular rotunda, above which giant mosaic murals depicted the development of America, and specifically the Cincinnati region, with a focus on transportation. Extending off the rotunda to the west, a concourse of more than 400 feet spanned the sixteen tracks below it. This huge space, with arched ceiling, red marble walls interspersed with other murals, and a boldly striped terrazzo floor, was also the main waiting room. Leather seats were grouped near the ramps and stairways leading to the eight platforms.

Cincinnati Union Terminal opened in 1933, one of the last three great railroad stations built in America (the other two: Philadelphia's Thirtieth Street and Los Angeles's Union Passenger Terminal). But despite all its utilitarian beauty, the station was built too late and too commodiously. The 200-odd daily train movements of the war years had dropped to half that number by the early 1950's, and the decline never stopped. Its days as a passenger terminal ended in 1972, but in the late 1980's it was undergoing restoration for use as a museum.

Richard J. Cook

New York Central No. 9 pulls out of Cleveland Union Terminal in March 1953. The mail train cut in a Toledo-Chicago Pullman.

The successor to the title of Ohio's largest city, meanwhile had just put an even bigger mark on its skyline with a gargantuan project completed in 1930. Cleveland's Union Terminal, which was built at more than four times the cost of Cincinnati's, involved construction of not just a terminal, but of a commercial complex, a rapid-transit line, and two suburban stations, as well as electrification of seventeen miles of track. When it was finished, Terminal Tower was the dominant Cleveland landmark, at fifty-two stories the tallest building west of New York City.[3]

Union Terminal was a joint project of the New York Central (including its Big Four subsidiary) and the Nickel Plate. Ever since the Central had sold it in 1916, the Nickel Plate had been in the hands of the Van Sweringen brothers—given names Oris Paxton and Mantis James—Cleveland men who had made their fortune in real estate. They were trying to build a rapid transit line from downtown to their posh Shaker Heights development, in the process promoting the concept of a union station for all the steam railroads and interurbans serving Cleveland. (Ohio was laced with railroads of both kinds: The state's interurban mileage exceeded 2,800 at its high point.) After long negotiations, an agreement was reached,

whereby the Nickel Plate would have a seven per cent share, and the Vanderbilt roads the remainder. Construction of the terminal, which fronted on Cleveland's Public Square, began in 1922.

The Central was able to agree to the project because now it considered the Nickel Plate to be in friendly hands, which was originally not the case. A little more than a decade after the Central acquired its Lake Shore line across northern Ohio in 1869, the Nickel Plate extended its tracks from Cleveland east to Buffalo, paralleling the Central all the way. To keep this nuisance road away from owners it might deem even more undesirable, the Vanderbilts simply bought it. In 1916, with the possibility of anti-trust action looming, the Central decided that the Van Sweringens' ownership would be tolerable and sold the NKP to them.

Before the diesel era, steam trains had to switch to electric locomotion to go from the outskirts of Cleveland to the twenty-eight stub tracks of Union Terminal. Not all trains of the Great Steel Fleet, however, called at the station. The *Twentieth Century Limited, New England States,* and *Commodore Vanderbilt* all passed through Cleveland like . . . trains in the night. But the city was well served by those that did stop, including the *Lake Shore Limited, Southwestern Limited,*

On a rainy March night in 1949, Nickel Plate No. 9, the road's Cleveland-St. Louis sleeper train, pauses in Lorain.

Chicagoan, Iroquois, and *Fifth Avenue Special.*

The all-Pullman *Cleveland Limited* ran to the East, where the Central scheduled dedicated sleepers to Buffalo, Rochester, Syracuse, Albany, Boston, New York, and Toronto (Toronto, Hamilton & Buffalo-/Canadian Pacific). To the west, Cleveland had Pullman service to Detroit and Chicago, on the *Forest City* to the latter.

The Big Four's main line from Cleveland to the southwest carried a great variety of Pullman traffic, much of it joint service with other roads. Over the line connecting Ohio's "Big C" cities, the railroad scheduled Pullmans to Columbus, Dayton and Cincinnati. At Columbus, the Norfolk & Western took over cars for Roanoke and Norfolk, and the Chesapeake & Ohio picked up Pullmans for Huntington and White Sulphur Springs, W.Va., and Clifton Forge and Charlottesville, Va. At Galion, the all-Pullman *Southwestern Limited, Knickerbocker,* and *Missourian* continued on the main line to Indianapolis and St. Louis, both of which cities had sleeping car service from Cleveland.

The Big Four turned over Florida sleepers to the Southern at Cincinnati, cars for Jacksonville, St. Augustine (FEC), Miami (FEC or SAL), Tampa (SAL), and St. Petersburg (SAL). The L&N picked up a Jacksonville (CofG/Southern/ACL) Pullman, and the Southern also carried a Cleveland sleeper to New Or-

leans.

The Nickel Plate, which terminated at Buffalo, had lines going west to Chicago, Peoria, and St. Louis. Its Pullman service was not extensive, but NKP trains had a down-home friendliness, and a post-war order of fifteen stainless-steel and blue lightweight sleepers thoroughly modernized the fleet.

The NKP generally paired with the Lackawanna at Buffalo on through trains from Chicago to the East Coast and scheduling sleepers from Cleveland to Scranton, Pa., and New York. The *Nickel Plate Limited* carried Pullmans to Fort Wayne and Chicago, and the railroad also scheduled a Cleveland-St. Louis sleeper. At Fostoria, during the Depression years, NKP turned over to the C&O Pullmans for Clifton Forge, White Sulphur Springs, and Hot Springs; later the Big Four delivered them to the C&O at Columbus.

The Pennsylvania, which was the one passenger railroad in Cleveland that never used the new Union Terminal, carried sleepers on its line through Akron to Chicago, Columbus, St. Louis, and Cincinnati, where it also delivered St. Petersburg Pullmans for the *Southland.*

To the east, the premier trains to and from New York were the *Buckeye Limited* and *Clevelander.* The Pennsy scheduled service via both Salem and Youngstown, with sleepers for Pittsburgh, Philadelphia,

Washington, and Atlantic City. In 1929-30, the railroad scheduled sleeping cars to Miami, St. Petersburg, and Tampa by way of Washington (Richmond, Fredericksburg & Potomac/etc.).

On such trains as the *Cleveland Night Express*, the Baltimore & Ohio brought Pullmans into Union Terminal from Washington, Baltimore, Philadelphia, and New York. The railroad moved seasonal sleepers to Washington for St. Petersburg and Tampa (RF&P, etc.) and scheduled a sleeper to Chicago.

The Erie's *Lake Cities* carried Pullmans to New York, but the railroad conducted the rest of its service from Cleveland jointly. Around the turn of the century, a Boston sleeper moved via Binghamton, N.Y. (Delaware & Hudson/Boston & Maine). At Youngstown, the Pittsburgh & Lake Erie picked up Pullmans for Pittsburgh and Philadelphia (B&O).

For a time, Cleveland saw no Amtrak trains, but that was remedied with a modicum of east-west service operating from a tiny lakefront station. In the end, Union Terminal was shorn of its twenty-eight tracks and reverted to the sort of facility the Van Sweringens had sought before they became railroad men: a transit terminal, no more, no less, but in a downtown that was now fighting for its life. In this limited role, it continued to see many intercity passengers—except that they came just a short distance by rail, over the relatively few miles on the transit line from Cleveland's airport.

Four decades earlier, another air strip, known as Port Columbus, had figured into the railroad picture, when the Pennsylvania's *Airway Limited* stopped to put passengers aboard Transcontinental Air Transport's Ford trimotors in the nation's most elaborate coast-to-coast air/rail service (See chapter on The Transcontinentals). This experiment was short-lived, however, and Columbus residents settled down for at least another quarter century of first-rate rail service from their wonderfully convenient Union Station, less than a mile north of the state Capitol, before the airlines took it all away.

Although Columbus's rapid growth rate came later, it was well served during the golden years of rail travel by five major roads that offered four different sleeper lines to Chicago, three to New York and Cincinnati, and two to Cleveland, Detroit, Indianapolis, Mackinaw City, and St. Louis.

The callboards at Union Station noted the arrivals and departures of the Pennsy's all-Pullman *"Spirit of St. Louis"* and *American*, the *Penn Texas*, *St. Louisan*, and all-coach *Jeffersonian;* the New York Central's solid-sleeper *Ohio State Limited*, Norfolk & Western's *Pocahontas*, and the C&O's *Sportsman*.

Pennsylvania trains brought Pullmans from New York, Washington, Pittsburgh, New Castle, Cleveland,

and Akron. West of Columbus, the railroad carried sleepers to Cincinnati, Indianapolis, St. Louis, Chicago, and, in summers, Mackinaw City.

The Central gave Columbus slightly less and, except for its New York, Cincinnati, and Cleveland sleepers, far more roundabout—and sometimes exotic—Pullman service. St. Louis cars traveled by way of Cincinnati and Indianapolis; a Detroit sleeper went to Cincinnati, then backtracked to Springfield and traveled north via Toledo. In Cincinnati, the Big Four turned over to the Southern its Pullmans bound for St. Petersburg and Miami (both SAL).

Over the former Peoria & Eastern line from Springfield to Indianapolis, Big Four trains once carried sleepers to both Indianapolis and Chicago. Doubtless the most unusual service was the sleeping car line that operated from Columbus to St. Paul, Minn., by way of Peoria and the Minneapolis & St. Louis Railway. This improbable routing was devised to spare passengers the trouble of changing trains in Chicago.[4]

The tracks of the Toledo & Ohio Central, which the New York Central acquired early in the century, extended north from Columbus through Findlay to the road's namesake city. The Central scheduled sleepers to Chicago and, in summers, Mackinaw City on this line.

The Chesapeake & Ohio brought Pullmans to Columbus from Charleston, Hinton, and Clifton Forge. Schedules from 1883 show a sleeper from Washington and a car to White Sulphur Springs. On the former Hocking Valley line north, C&O trains carried sleepers to Toledo, Detroit, Frankfort, Mich. (with Ann Arbor Railroad, summer 1907), and Chicago (Erie).

The Norfolk & Western, which served the same eastern region as the C&O, carried sleepers from Columbus to Roanoke, Richmond (ACL), and Norfolk, Va., and Durham, N.C.

Columbus was on the B&O's line from Pittsburgh to Cincinnati, which generally had both a daytime and night train. The city had Pullman service to Cincinnati, Louisville, Chicago, Pittsburgh, Washington, and Baltimore, and from New York.

Toledo, which once was caught in a tug of war between Michigan and Ohio, went on to become the nation's third-ranking railroad center.[5] This important coal port and glass making center had thirteen trunk lines and one terminal railroad serving it in 1930, when it was still a thriving interurban hub in the sunset of that era. Because of its location, Toledo was the main gateway to Detroit from the south. All of Toledo's passenger railroads except the Pennsylvania used Union Station.

With all the important midwestern components of the Vanderbilt lines connecting at Toledo—the New York Central, Michigan Central, and Big Four, as well

The once-premier *Pennsylvania Limited* was an octogenarian when it made this July 1964 run under tow of an E8 through Irwin, Pa.

as the T&OC—it was not surprising that the Central provided the major share of the city's Pullman service. This included sleepers to Buffalo, Pittsburgh (Erie/P&LE), Boston, and New York, to Mackinaw City in summers, to Indianapolis, and to Chicago by way of two routes, the NYC main line and the T&OC/B&O. The Big Four turned over Pullmans to the Southern in Cincinnati for forwarding to Jacksonville and Miami (SAL).

In June 1947, the Pere Marquette Railway, a mostly Michigan system, became part of the C&O, with which it had long been affiliated and had interchanged cars at Toledo. Pre-merger, the PM carried a sleeper from Toledo to Bay View, Mich. The remainder of the C&O service was south over the former Hocking Valley line, to Huntington, Charleston, and Hinton, and from Clifton Forge. The Hocking Valley as an independent participated with the B&O on interline Pullmans to Baltimore and New York.

Toledo had Pullman service to St. Louis by way of both the Nickel Plate and the Wabash, and before the turn of the century the Wabash ran a sleeper to Decatur, Ill. The B&O offered a Louisville Pullman, and the Pennsy carred a sleeping car to Pittsburgh (and briefly one from New York). Summer schedules of 1909 show the Ann Arbor carrying a sleeper from Toledo to Frankfort.

By the time the joint air/rail service was operating via Columbus, Dayton had already become known as the cradle of aviation, a technology partially spawned in the local bicycle shop owned by Wilbur and Orville Wright. At this stage, however, Dayton still relied on Pullman travel, and three railroads provided the accommodations in Union Depot.

Dayton was on the Pittsburgh-St. Louis "Panhandle" line of the Pennsylvania, though not all trains on this route came through town—some went by way of Piqua. Before World War I, the Pennsy carried sleepers to Pittsburgh and to Mackinaw City, and from New York. In later years, the Dayton-Chicago Pullman was the railroad's sole contribution.

The Cincinnati, Hamilton & Dayton, a steam railway that became part of the B&O, also offered a sleeper to Chicago, this jointly with the Monon at Indianapolis. Both the B&O and New York Central carried Pullmans from Dayton to Detroit.

Just as Dayton was a hotbed for the development of aviation, Akron became a key supplier to the automobile industry in its role as the world's largest rubber manufacturing center. With the exception of the Erie's Pullman to Chicago, Akron's service was about equally divided between the Pennsylvania and B&O, which shared a union station.

The B&O brought its crack Chicago trains through town: *Capitol Limited, Fort Pitt Limited, Shenandoah, Washington-Pittsburgh-Chicago Express,* and the coach train *Columbian.* On these and other trains, the railroad scheduled dedicated sleepers from Akron to Pittsburgh, Washington, Baltimore, Detroit, and Chicago.

The Pennsylvania, which served Akron with the *Akronite* as well as several Cleveland trains, offered Pullmans to Chicago, Pittsburgh, and New York, and, on its resort train the *Sea Gull,* a car to Atlantic City.

Youngstown saw most of the same through passenger trains that stopped in Akron, including the Erie's *Lake Cities Limited, Erie Limited,* and *Atlantic/Pacific*

Expresses, as well as the *Phoebe Snow* after the Lackawanna merger. Of all the major Ohio cities, Youngstown had the least service of the greatest variety: five lines, each operated by different single or paired railroads: Pullmans to Chicago traveled by either B&O or Erie, those to New York via either Erie Lackawanna or Pennsylvania. The Erie and New York Central jointly carried a Youngstown sleeper to Detroit.

Several other communities in Ohio had dedicated sleeping car service, probably the most unusual being the Lake Erie & Western's line from Sandusky to Peoria, which operated in the early 1890's. The LE&W, which went by the graceless slogan of "The Natural Gas Route," became part of the Nickel Plate.

The Pennsy's great tuscan red fleet of trains to Chicago—the *Broadway Limited, Golden Triangle, Manhattan Limited, Pennsylvania Limited,* and their compatri-ots—passed through northern Ohio mostly at night. Three of the cities they visited had dedicated Pullman service, Alliance and Mansfield to Chicago, and Canton from Chicago. Springfield also had its own sleepers on the Pennsy to Chicago, and the railroad scheduled cars from Niles to Pittsburgh and New York.

The B&O scheduled Pullmans from Newark to Washington and from Zanesville to Chicago. Marion had a New York sleeper via the Erie.

---

1. The Cincinnati, New Orleans & Texas Pacific leased the Cincinnati Southern. The CNO&TP became part of the Queen & Crescent Route of the Southern Railway System.
2. Carl W. Condit, *The Railroad and the City* (Columbus: Ohio State University Press, 1977), p. 123.
3. John A. Rehor, *The Nickel Plate Story* (Milwaukee: Kalmbach, 1965), pp. 183-9.
4. Frank P. Donovan Jr., *Mileposts on the Prairie* (New York: Simmons-Boardman, 1950), p. 114.
5. Willard V. Anderson, "The Princely New York Central," *Trains,* Vol. 9, No. 1 (November 1948), p. 22.

The B&O's westbound *Shenandoah* departs Deshler, Ohio, in February 1965.

10
Indiana

The Monon was Indiana's own railroad, and one of its most picturesque activities was bringing sleepers into the hotel tracks at the resort of French Lick. In the last year of regular service, 1949, the process started with No. 23 (top, left) backing up with the Chicago Pullman at Orleans to the wye leading to French Lick. The cars were parked in a woodland setting (top right and middle) on one of two tracks leading to the grounds of the seven-story hotel (below). Monon Pullman Sir Henry W. Thornton (middle) probably arrived as a special movement.

Indiana had two levels of passenger railroading. The focal point of the first was Indianapolis, and this was best illustrated during the era of the electric interurbans, when the Hoosier capital was also the nation's capital of intercity trolley travel. From the imposing downtown Traction Terminal, interurban cars fanned out in every direction, connecting Indianapolis with all the principal cities and towns in the state. This service also had a counterpart in the steam railroads, inasmuch as Indianapolis was the hub of numerous trunk lines whose trains departed from its Romanesque, red brick and granite Union Station to destinations in Indiana and nearby states.

The second level of passenger travel was represented by the sort of traffic patterns that were determined by outside forces. Indiana was crisscrossed by numerous railroads, and for these lines, the state was little more than another bridge across which the trains pounded day and night en route to destinations elsewhere. Cities far more sizable than Indianapolis were the controlling forces. Chicago's magnetism for railroads was irresistable, but St. Louis, Cincinnati, Detroit, and the great population centers of the East all exerted the sort of influence that determined much of the passenger traffic flow across Indiana.

From early times, Indianapolis was handicapped in not being situated on any body of navigable water. As a result, the industries—notably those consuming coal and iron ore—that depended on the cheap transportation that boats offered were located elsewhere. And the cities that had the boat traffic in the first place, such as St. Louis and Cincinnati, had an unsurpassable head start as commercial centers.

As a result, many interstate passenger trains that traversed Indiana, such as the numerous flyers connecting New York and Chicago, served Hoosier communities only incidentally. But the cities and towns that were on the principal trunk routes—and Indianapolis was on a fair number—certainly profited from the service.

Indianapolis Union Station, which has been put on the National Register of Historic Places and artfully restored, was opened in 1888. The terminal replaced the nation's first union station, which dated back to 1853. Architecturally, the newer one was very much a structure of its times, with a tall, turreted clock tower, a barrel-vaulted ceiling in the waiting room, and two walls containing large, stained-glass, rose windows. The facilities underwent renovation during World War I, when the main line tracks were elevated and a new train shed erected over the twelve passenger tracks.[1]

Six railroads provided passenger service: the New York Central's Big Four, Pennsylvania, Baltimore & Ohio, Monon, Illinois Central, and Nickel Plate. Somewhat over a hundred trains arrived and departed daily just before World War II.

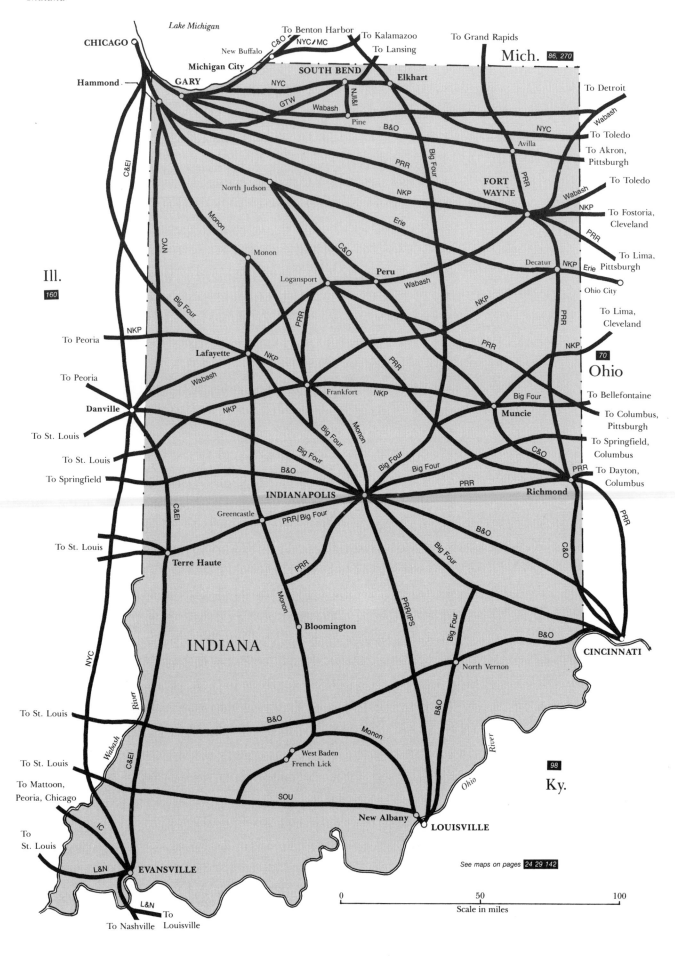

Lake Michigan

CHICAGO

To Benton Harbor
To Kalamazoo
To Lansing
To Grand Rapids

Mich.   86, 270

New Buffalo
C&O
NYC/MC

Michigan City
Hammond
GARY
SOUTH BEND
Elkhart

To Detroit

NYC
GTW
NJI&I
Wabash
Pine
B&O
Avilla
NYC
Wabash
To Toledo

To Akron,
Pittsburgh

To Toledo

FORT
WAYNE
PRR
NKP
Wabash
To Fostoria,
Cleveland

North Judson

C&EI

Monon

PRR

Big Four

Erie

PRR

Decatur
NKP
Erie
To Lima,
Pittsburgh

Ill.   160

Monon

Monon

Logansport

Peru
Wabash

Ohio City

NYC

Big Four

NKP

NKP

PRR

To Lima,
Cleveland

70

To Peoria
NKP

Lafayette
NKP

PRR

PRR

NKP

Ohio

To Peoria

Wabash

Frankfort
NKP

Big Four

Muncie
Big Four
To Bellefontaine

Danville
NKP

Big Four

Monon

Big Four

To Columbus,
Pittsburgh

To St. Louis

To St. Louis

B&O

Big Four

Big Four

C&O

To Springfield,
Columbus

To Springfield

C&EI

INDIANAPOLIS

PRR

Richmond

PRR
To Dayton,
Columbus

Greencastle
PRR/Big Four

B&O

To St. Louis

PRR

Big Four

C&O

PRR

Terre Haute

Monon

PRR/IPS

Big Four

INDIANA

Bloomington

North Vernon

B&O

CINCINNATI

NYC

River

Wabash

C&EI

B&O

Monon

West Baden
French Lick

Ohio   River

98

SOU

Ky.

New Albany
LOUISVILLE

To St. Louis

IC

See maps on pages   24 29 142

To
St. Louis

L&N   EVANSVILLE

L&N
To Louisville

To Nashville

0   50   100
Scale in miles

The Monon carried sleepers to Chicago, including a setout from its north suburban Boulevard Station, but all other Pullman service in and out of Indianapolis Union Station was in the hands of two railroads, and large ones at that: the Big Four, with six rail lines radiating out of the city, and the Pennsylvania, with five. The same carriers provided most of the name trains serving the Indiana capital.

Both of the railroads' main lines from the East to St. Louis passed through Indianapolis, bringing the Central's *Southwestern Limited, Knickerbocker,* and *Missourian,* and the Pennsy's *"Spirit of St. Louis," American, Penn-Texas, St. Louisan,* and *Jeffersonian.* Over the Chicago-Cincinnati line of the Big Four came the *Carolina Special, James Whitcomb Riley, Royal Palm,* and *Sycamore.* The *South Wind, Kentuckian,* and *Union* operated on the Pennsy's Chicago-Louisville line.

Passengers could travel by sleeper to eighteen destinations from Indianapolis, nine of which were served by both the Big Four and the Pennsylvania: Chicago, Columbus, Detroit (PRR with Wabash), Evansville (PRR with Chicago & Eastern Illinois), Grand Rapids (NYC with Chesapeake & Ohio, 1890), Jacksonville (PRR/Louisville & Nashville/Central of Georgia/Atlantic Coast Line; and NYC/Southern), New York, St. Louis, and Washington (NYC with C&O).

The New York Central scheduled Pullmans to Cleveland, Toledo, and Peoria. On the Pennsy, sleepers traveled from Indianapolis during summertime to Mackinaw City, Northport, and Traverse City, Mich., and year-round to Pittsburgh and Louisville.

The Pennsylvania, however, had competition for several years between Indianapolis and Louisville, but not from the rival Central. Until early in the Depression, the Interstate Public Service Company operated one of the nation's few interurban sleepers, a red and orange, open-section car, between the two cities. Promptly at 11:30 every night, the train swung out of the Indianapolis Traction Terminal onto the street and departed under trolley wire for the Kentucky city.

Indiana had a railroad that bespoke the character of the state like none other, and, more than any of its competitors, it fit the description of both levels of passenger operation. The Chicago, Indianapolis & Louisville was informally and finally officially known as the Monon—the name came from the Indiana town in which its two main lines crossed. The railroad said that Monon was an Indian word for "swift running," and if that didn't always describe the service the trains offered, it at least expressed the goals of the road's able, post-war president, John W. Barriger III, who had an unfulfilled dream of transforming the Monon into the country's first super railroad, with trains operating at 100 miles an hour.

The Monon's ancestor was chartered in 1847 to

The Interstate Public Service company had its own equipment trust subsidiary, for which this special stock (shown in a detail) was issued. The certificates lists two buffet parlor cars and three sleepers as among its shareholders' assets.

build a railway the length of Indiana, from the booming Ohio River town of New Albany—then the largest settlement in the state—to Michigan City. It was a potentially lucrative route: Michigan City, at that time, had a good chance of becoming a major port on the southern shore of Lake Michigan. History, of course, proved otherwise, and the Monon, despite various reorganizations, spent most of its life as a secondary carrier.

Over its other line running from Chicago through Monon to Indianapolis, the railroad scheduled sleepers to the Indiana capital. It eventually extending the Chicago service, via the B&O, to Cincinnati and, briefly around the turn of the century, to Washington and Baltimore. During this period the Monon carried sleepers from Chicago to Cincinnati for Jacksonville (B&O/Southern/ACL) and St. Augustine (same, with FEC) and via Louisville to Burgin, Ky., where passengers transferred to Southern trains from Cincinnati.

The Monon was probably best remembered by generations of college students who rode its trains between Chicago and Louisville to Purdue University, Indiana University, and other schools. After World War II, the railroad paid the Big Ten universities the honor of painting its coaches in their colors: Red and gray for Indiana, and, later, black and gold for Purdue.

Perhaps the Monon's most colorful service in Indiana was the Pullmans it switched in and out of the resorts of French Lick and West Baden. Several artesian mineral springs, the most famous yielding Pluto Water, were the attractions that prompted the building of the first hotel at French Lick, in the middle of the Nineteenth Century. The Monon came to town in 1887, and the resort eventually grew to accommodate a seven-story hotel of multiple wings, situated on sev-

The Pennsylvania's *Northern Arrow* assembled in Richmond with sleepers from St. Louis, Cincinnati, and other cities bound for Michigan summer resorts. The Chicago section, shown upon its arrival home on Aug. 10, 1939, left the main stem at Fort Wayne.

eral thousand acres. Pullman cars were spotted in a sylvan setting on the house tracks just a short walk from the hotel entrance.

West Baden, which was just a mile away, boasted a circular hotel with an interior atrium enclosed by what was reported to be the largest unsupported dome in America until the Houston Astrodome was built.[2] The two big resort hotels were connected with door-to-door trolley service. Thanks to local officials who winked at the state laws prohibiting gambling, numerous casinos sprung up in the two communities, and Chicago gangsters were frequent callers.

Despite the waggish bumper stickers of later years proclaiming that "French Lick isn't as exciting as it sounds," both of the resorts accommodated a substantial flow of tourists, especially in spring and fall. The weekend of the Kentucky Derby was the big event of the year. On Friday evening the house tracks began filling up with Pullmans and private cars, and their occupants gambled and drank the night away before boarding the trains for the short run to Louisville on Saturday. After the race, the trains returned and the party resumed until departure time on Sunday afternoon.

The Monon's *Red Devil*, an all-Pullman train operating from Chicago before World War I, served the resorts, and over the years the railroad scheduled sleepers that it cut out of its Louisville trains. The Monon also picked up Pullmans from New York City, from both the Pennsylvania and New York Central, as well as a car from Boston (NYC). French Lick had sleepers from St. Louis, via both the Southern Rail-

way and the B&O.

In early autumn 1967, the last Monon passenger train, the *Thoroughbred,* quit serving the college towns and the country of covered bridges and limestone quarries. Four years later, the railroad completely lost its Hoosier identity and became part of the L&N.

The once-bustling Ohio River port of Evansville, in addition to its C&EI/PRR and Big Four Pullman connections with Indianapolis, had service to five other midwestern cities. At Evansville, the Chicago to Florida "Dixie" fleet of trains were exchanged by the C&EI and L&N. Additionally, C&EI trains brought to Evansville sleepers from Chicago and Detroit (Wabash). Earlier in this century, Evansville also had Pullman service to Chicago on both the New York Central and the Illinois Central. In the 1890's, the Peoria, Decatur & Evansville, which became part of the IC, carried sleepers between the terminal cities of its name. The L&N scheduled Pullmans from Evansville to both St. Louis and Louisville.

Richmond was a community with a large Quaker population and the Gennett recording studio, which gave the world some of its earliest jazz records, notably those of Bix Beiderbecke. The town, moreover, was an important junction of several lines of the Pennsylvania Railroad, including Chicago-Cincinnati, Pittsburgh-St. Louis, and a line going north through Fort Wayne into Michigan.

The Grand Rapids & Indiana (later Pennsylvania) scheduled a Pullman over these tracks through Fort Wayne from Richmond to Mackinaw City. In later years, the summertime *Northern Arrow* traveled this route after assembling sleepers cut out of various trains passing through Richmond. In the 1890's, Richmond had its own sleeping car to New York. Both the Pennsylvania and, briefly, the C&O carried Pullmans to Chicago.

The B&O main line between Cincinnati and St. Louis carried substantial Pullman traffic on such trains as the *National Limited, Diplomat,* and *Metropolitan Special,* and at North Vernon the accommodation trains from Louisville delivered cars to be put aboard the through trains to the east. The B&O always had a reputation for excellence in service, and one example from its employee instructions reflected the lengths it was prepared to go to in serving its Louisville customers.[3]

The eastbound *National Limited* was ordered to be held for up to fifteen minutes at North Vernon if the Louisville train bringing a sleeper for New York was delayed. If the connection was not made, the accommodation train was to continue on to Cincinnati, trying to overtake the *National*—which was to be held another fifteen minutes in the Queen City, if necessary. Failing that connection, the train from Louisville was to attach a diner and coach and be prepared to go as far as Washington trying to overtake the *National Limited.* Such was the level of service on the Royal Blue trains.

Fort Wayne, the reputed birthplace of night baseball, was a busy junction of two Pennsylvania lines, the Wabash, and Nickel Plate. Here the *Broadway Limited, General,* and all the other crack Pennsy flyers to Chicago crossed the paths of the *Nickel Plate Limited* and the daytime Detroit-St. Louis *Wabash Cannon Ball.* The Pennsylvania carried Pullmans from Fort Wayne to Chicago and Pittsburgh, and the Nickel Plate scheduled a Cleveland sleeper.

For Gary, through sleepers from Chicago had to suffice, but the big three eastern carriers at least conditionally stopped many of their trains in this steel city. The New York Central and B&O shared a facility; the Michigan Central and Pennsylvania had their own stations.

Gary was the principal stop in Indiana for the B&O's *Capitol Limited, Shenandoah, Fort Pitt,* and *West Virginia Night Express.* The New York Central, which also served South Bend and Elkhart, put Gary on the cards of its *Fifth Avenue Special* and *New England States,* and the Michigan Central's *New York Special* and *Niagara* stopped at their station about a mile away.

Such Erie trains as the *Lake Cities, Erie Limited,* and the *Atlantic/Pacific Limiteds* stopped at Hammond. Michigan City was the major Indiana stop for the Pere

The B&O's *National Limited,* with its Colonial diner, was one of the more cosmopolitan carriers on the Indiana rail scene.

Marquette/Chesapeake & Ohio Chicago-to-Grand Rapids fleet.

Both Muncie and Peru had sleeping car service to Chicago, via the C&O, and the C&EI carried a Pullman between Chicago and Terre Haute.

The New Jersey, Indiana & Illinois, a tiny railroad with a carload of a name,[4] scheduled a sleeping car run over the entire length of its system—twelve miles—from South Bend to a station called Pine. Passenger service on the NJI&I was exclusively Owl. Train No. 1 left South Bend with its Pullman at 1:15 a.m., passed the Studebaker facility two miles down the line, and arrived at Pine thirty minutes later. Before long, a Wabash train came by, attached the sleeper, and departed for Detroit. The NJI&I crew waited. In a couple of hours, another Wabash train came through Pine from the opposite direction and cut out the westbound sleeper. The crew of what was now NJI&I train No. 2 had the car safely back in South Bend before the Fighting Irish were off to early mass.

1. John Uckley, "Hard Times for a Dowager," *Passenger Train Journal,* Vol. 12, No. 3 (July 1980), pp. 24-31.
2. Stan Conyer, "The French Lick Springs Valley and the Railroads That Made It Prosper," *The Hoosier Line,* Vol. 4, No. 3 (December 1984), pp. 4-9.
3. "Consolidated Schedule and Instructions for Make-Up of Through Passenger Trains," Baltimore & Ohio Railroad Transportation Department, December 1, 1928, p. 22. Collection of Howard B. Morris.
4. Mr. Morris reports that the naming of the New Jersey, Indiana & Illinois was a mouse-that-roared situation involving the Singer sewing machine company, which had a plant in South Bend and built the line after experiencing difficulties with local railroads. Because Singer also had plants in New Jersey and Illinois, company officers chose the name as a tongue-in-cheek reminder to railroads in those states that Singer would also lay rails in their precincts if they ever stifled competition. (Letter of February 14, 1988, from Mr. Morris, of the Indiana Historical Society).

11
Michigan

Michigan is called the "Great Lakes State" and for good reason: Four of five of these freshwater inland seas wash its shores. During the first two centuries of the white man's presence in this region—the explorer, fur trapper, missionary, soldier—water was the principal medium of travel. This did not quickly change as lumbermen, farmers, fishermen, and others began settling what was called the Northwest Territory.

When the railroads came in, they had a formidable task of wresting traffic away from this long-entrenched lake competition. Rail trips across southern Michigan, of course, were far shorter than lake voyages around the entire Lower Peninsula. In the wintertime, the railroads prospered, but when the ice left the lakes in spring, the ship owners regained much of the bulk freight business because they could offer lower rates. Gradually, however, as the population on inland farms and communities grew and the tempo of the times increased, the railroads came into their own.

Railroad building began in Michigan just as the territory was preparing for statehood, which was granted in 1837. By 1849, the original charter line of the Michigan Central extended across the southern part of the state, from Detroit to New Buffalo.[1] When it became clear that Chicago, not Michigan City, Ind., was the preferred destination, the railroad engaged in some horse-trading with a Monon ancestor and thereby gained the right to use the Indiana railway's charter to lay rails across the northwest corner of the state.

The Michigan Central completed the last lap into Chicago by arranging trackage rights on the Illinois Central, winning its race with the rival Lake Shore & Michigan Southern line, built west from Toledo through Michigan's southern tier of counties. (Both railroads eventually became part of the New York Central system. The MC, however, maintained a contractual relationship with the IC for more than a century, using its Chicago terminal, but for few exceptions, rather than the NYC's La Salle Street Station.)

By the mid-1850's, after construction of the Suspension Bridge across the Niagara River below the Falls and completion of a railway west to Windsor, Ont., one could travel from New York to Detroit entirely by rail, with two exceptions: the ferry trips across the Hudson at Albany and the Detroit River at Windsor. The latter proved to be the more troublesome barrier, and after more than four decades of planning and intermittent activity, a tunnel was finally opened in 1910.

In conjunction with the tunnel project the Michigan Central constructed a large new station, slightly more than two miles from downtown Detroit, just west of where the tracks from Windsor emerged into daylight. The station was opened in late 1913, serving Michigan Central/New York Central and Canadian

B&O's *Capitol Detroit* nee *Ambassador* (top photo, right) will soon be cut into the regular *Capitol* (to its left) here at Willard, Ohio, for the trip east. Meanwhile, busy platform activity eludes this nighttime exposure. Engine crew of Pere Marquette's venerable summertime *Resort Special* (bottom) checks out the scenery at Bay View in 1934.

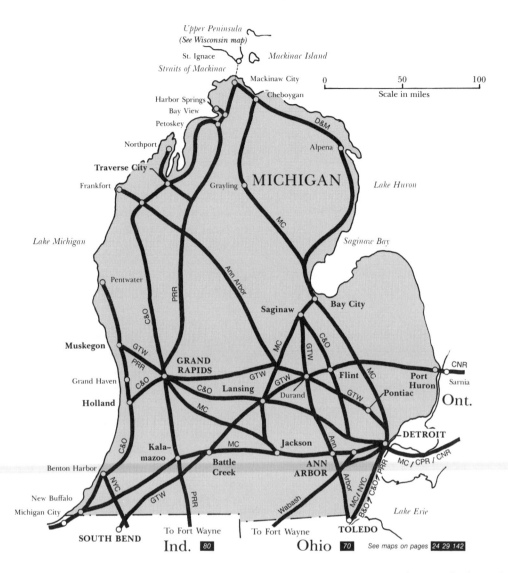

Pacific trains, and for a time after World War II, those of the Baltimore and Ohio.

The MC's through trains included the all-Pullman *Wolverine, Niagara,* and *North Shore Limited,* as well as Detroit's own daytime *Chicago Mercury,* all-parlor *Twilight Limited,* and *Motor City Special* among the eight or nine daily expresses to Chicago. The *Detroiter,* an all-Pullman, extra-fare train for much of its life, which began in 1911, and the reserved-seat, all-coach *Empire State Express* went to New York. The *Flamingo, Ponce de Leon,* and *Royal Palm* offered through service to Florida.

Over the MC line east through Ontario, Detroit had Pullman service to Buffalo, Syracuse, Albany, Boston, New York (from Buffalo via either the NYC main line or the Lackawanna), Philadelphia (Lehigh Valley/Reading); and from Rochester. To the west the MC carried sleepers to Grand Rapids and Chicago.

The largest selection of the Central's sleepers, however, went south through Toledo to Cincinnati, Cleveland, Columbus, Dayton, Indianapolis, and St. Louis.

In conjunction with other roads the MC/NYC carried Pullmans to Norfolk (Norfolk & Western at Columbus), Youngstown (Erie), and Pittsburgh (Erie/Pittsburgh & Lake Erie), and, early in the century, Pennsylvania Railroad at Toledo).

Florida service, via Cincinnati, included sleepers for Jacksonville, Fort Myers, and Orlando (Louisville & Nashville/Central of Georgia/Atlantic Coast Line) and, via the Southern Railway, to Jacksonville, St. Augustine (Florida East Coast), Miami (FEC or Seaboard Air Line), Tampa (SAL), and St. Petersburg (either SAL or ACL). The Southern also picked up Pullmans for Atlanta and New Orleans.

The MC handled the Canadian Pacific's through Pullmans from Montreal to Chicago. From Detroit, the CPR scheduled sleepers to Toronto and Montreal.

Fort Street Union Depot, a stub terminal on the edge of downtown, accommodated trains of the Baltimore & Ohio, Pennsylvania, Chesapeake & Ohio, and Wabash. All but the Wabash were late-comers to the Detroit passenger market. B&O and Pennsy service

Paul Moffitt/NRHS, Danville Junction

Michigan Central's premier *Wolverine* **was an extra-fare, all-Pullman train from Chicago to New York and Boston. This is a 1939 edition.**

started in 1920 and the C&O's began several years later with joint operations involving the Pere Marquette, which eventually merged into the C&O.

In 1946, the B&O moved its passenger trains from Fort Street to Michigan Central Station, where the longer platforms expedited the loading of its trains. The *Ambassador* was the B&O's flagship between Detroit and eastern cities; the streamlined *Cincinnatian*, a daytime coach train with a stewardess, carried the standard going south.

Detroit had Pullman service on the B&O from Akron, Pittsburgh, Wheeling, Washington, Baltimore, and Philadelphia. The railroad carried sleepers south to Dayton, Louisville, and Cincinnati, where it turned over Florida Pullmans: Jacksonville (Southern), St. Augustine (Southern/FEC), and St. Petersburg (L&N/CofG/ACL or Southern/SAL).

The C&O became an official force on the scene after the Pere Marquette merger in 1947. The PM itself, which had been controlled by the C&O for two decades, resulted from a consolidation of three Michigan railroads at the turn of the century. In the 1960's, the C&O gained control of the B&O, which then returned to Fort Street.

Over its historic main line from Virginia, the C&O brought to Detroit Pullmans from Washington; Newport News, Richmond, Charlottesville, Gordonsville, and Clifton Forge, Va.; Hinton and Huntington, W.Va.; and Columbus, Ohio. It also carried sleepers to Charleston, W.Va. The *Sportsman*, which was designed to connect resort areas of the Great Lakes with those in the eastern mountains and the Virginia tidewater region, was the C&O's premier train on this route.

The Pennsy had a limited presence in Detroit, but it ran a substantial overnight service to Washington and New York on the all-Pullman *Red Arrow.* It also carried a sleeper to Pittsburgh and participated with the Wabash, which it gained control of in 1928, on other Pullman lines serving Detroit.

The Wabash was a unique railroad that straddled east and west, with a main line running from Buffalo to Kansas City and Omaha. Its entry into Detroit led to extension eastward, through trackage rights on the Grand Trunk Western, to Buffalo. Until World War I, the Wabash had great passenger-carrying ambitions and participated in service, through Detroit, from the East Coast to Chicago and St. Louis. Thereafter, however, it retrenched and confined itself to the Midwest. The daytime Detroit-St. Louis train, wearing the railroad's customary blue livery, took the name *Wabash Cannon Ball.*

Over its own lines the Wabash carried sleepers from Detroit to Decatur, Kansas City, and St. Louis. In company with the Pennsy it offered Pullmans to Chicago, Indianapolis, and the Florida resorts of Sarasota, St. Petersburg, and Tampa by way of the *Southland* at Fort Wayne (PRR/L&N/CofG/ACL). Service to Indiana included Evansville (Chicago & Eastern Illinois) and South Bend (New Jersey, Indiana & Illinois). Before World War I, the Wabash carried a sleeper from Boston (West Shore-NYC/Boston & Maine). The Wabash was leased to the Norfolk & Western in 1964 and later merged.

Detroit's third railway terminal, Brush Street Station, gave way to Renaissance Center, but during its lifetime it served passengers of the Canadian National's Grand Trunk Western. The GTW (the CNR company operating in the midwestern United States) scheduled sleepers to both Muskegon and Chicago.

Detroit was not on the main line of the GTW; its sleepers and coaches from the Motor City went to Durand, where they were put onto the through trains coming west from Port Huron and Canada. This was the first all-rail line to cross the water barrier between Michigan and Canada, and it did so a full twenty years before the MC's tunnel between Detroit and Windsor was completed. Railroad officials organized a construction company in 1884 to build a tunnel under the St. Clair River, between Sarnia, Ont., and Port Huron. Despite several setbacks, work continued, and on August 23, 1890, the crew on the American side drilled through to the bore coming over from Canada. At the height of the celebration, one of the crews shoved through the opening the first freight that the first international underwater tunnel accommodated: a plug of tobacco.[2]

The premier train on the line, the Montreal-to-Chicago *International Limited,* was inaugurated in 1900. The Canadian National made a big commitment to radio in 1924, when it set up eleven stations along its train routes. To acquaint passengers with this remarkable invention, the railroad staffed the *International* with radio operators, who supervised the use of headsets in the observation cars.

In addition to through service from Canadian cities in the East and the feeder traffic from Detroit, the GTW provided several Michigan communities with sleepers to Chicago: Bay City, Flint, Lansing, Pontiac, and Port Huron.

The northern part of Michigan's Lower Peninsula was covered with virgin timber when settlers arrived, and lumbering interests were quick to exploit it, often very rapaciously. Much of the wood initially moved by water, but as the shoreline areas were depleted and the lumberjacks harvested further inland, away from the rivers, railways inherited the job. Five lines extended their fingers into the portion of the state northwest of Saginaw Bay.

Thanks to land grants, both the Michigan Central and Grand Rapids & Indiana lines to Mackinaw City were in operation by 1883, the MC's running north from Bay City, and the GR&I going up the western portion of the state.[3]

One of the C&O's Pere Marquette routes paralleled the GR&I north from Grand Rapids to Petoskey and Bay View. The Ann Arbor angled diagonally across the state from Toledo to Frankfort. The easternmost line, the Detroit & Mackinac, started as a narrow-gauge logger and followed the Lake Huron shoreline from Bay City to Cheboygan, which it finally reached in 1904.

By then the logging boom was over, leaving much of the once-forested land barren, but tourism was in vogue, and well-to-do citizens of St. Louis and Cincinnati and other midwestern cities began seeking summertime relief in what was left of the northern woods.

William J. Miller

The eastbound *Red Arrow*, the Pennsylvania Railroad's overnighter to New York City, pulls out of Detroit in 1941.

Ready to begin its run to the Virginia tidewater, C&O's *Sportsman* waits in Detroit's Fort Street Station in the mid-1950's.

Mackinac Island, once a military stronghold from which the French commanded the straits separating Lake Huron from Lake Michigan, became an attractive tourist destination in the late Nineteenth Century. The Michigan Central and GR&I railroads joined with a steamship line to build the palatial Grand Hotel, which opened in 1887. Until passenger train service ended in 1961, both the Central and Pennsy brought trains full of summer tourists, who alighted from coaches and sleepers in Mackinaw City and clambered aboard the ferries for the breezy ride across the straits to the picturesque island with the horse-drawn carriages. Among all Michigan destinations, Mackinaw City was second only to Detroit in the number of Pullman lines terminating there—over the years at least a dozen and a half—and in summertime the yards were swollen with the two-toned gray and tuscan red sleepers of the two big passenger carriers.

The Grand Rapids & Indiana billed itself as "The Fishing Line" in its whimsical logo that pictured a jumping, but well hooked fish. The GR&I, whose line to Mackinaw City was leased by the Pennsylvania for many years and finally absorbed, operated two summertime name trains made up of cars delivered by the Pennsy. The *Northland Limited* carried Pullmans to Mackinaw City from Grand Rapids and cities in and near the Ohio River valley: Cincinnati, Columbus, Dayton, Indianapolis, Louisville, and Richmond, Ind., where most of the final version of the train was assembled. In Kalamazoo, the Michigan Central dropped off sleepers from Chicago and St. Louis (Illinois Central—St. Louis Pullmans also traveled an all-PRR route, via Richmond, Ind.). These cars went into the GR&I train called the *North Michigan Resorter*. The *Northern Arrow* was the Pennsy's successor train, which handled the Mackinaw City run from the mid-1930's until its last trip in 1961.

Detroit had Pullman service to Mackinaw City via the Michigan Central, which over the years forwarded sleepers from Toledo, Pittsburgh (P&LE), Columbus, Cincinnati, Jackson, and, briefly before World War I, Bay City. The MC also carried a Detroit Pullman to Grayling.

In addition to terminating heavy summertime passenger traffic, Mackinaw City was strategically connected, by railroad ferry, with the Upper Peninsula. The Duluth, South Shore & Atlantic Railroad was an ore and lumber carrier whose tracks extended across the U.P. from St. Ignace, opposite Mackinaw City, and Sault Ste. Marie west to Duluth. Train ferries crossed the Straits of Mackinac, exchanging cars between the upper and lower portions of Michigan. The DSS&A brought sleepers to Mackinaw City from Calumet and Duluth and returned with Michigan Central Pullmans from Detroit bound for Marquette, Calumet, and Sault Ste. Marie (Soo).

Although Mackinac Island was Michigan's major tourist destination, it was not the whole story. After the logging business faded, the Detroit & Mackinac made some gains during summertime by scheduling tourist outings. The D&M carried sleepers from Detroit to Alpena (C&O, later MC, to Bay City), and after winning a brawl with MC crews over the right to cross their tracks and enter the town of Cheboygan, the railroad entered the Mackinac Island trade. The D&M and C&O ran a Pullman line from Detroit to Cheboygan, where passengers boarded a steamer to the island.

The C&O/Pere Marquette lines did a good resort business over much of their history, even with the Pennsylvania's competition. For nearly sixty summers, the *Resort Special* made its northerly trek with Pullmans delivered to Grand Rapids. C&O rails went to Bay View, which was founded on Little Traverse Bay as an assembly of the Methodist Church in 1876. Bay View had through sleepers from Grand Rapids, Chi-

Canadian National set up its own radio stations when the medium was young; *Maple Leaf* passengers tuned in for this PR shot.

cago, St. Louis (Alton or C&EI), Detroit, Toledo, and Cincinnati (B&O).

From nearby Petoskey, Pullmans departed for Chicago (C&O), and (PRR) Grand Rapids and Cincinnati. Further around Little Traverse Bay, Harbor Springs was served early in the century by various combinations—the C&O with a sleeper from Cincinnati (B&O), a joint Michigan Central/Pennsylvania Pullman from Detroit—and later by the Pennsy with sleepers from Cincinnati, Chicago (MC), and St. Louis (either exclusively PRR, or IC/MC).

Traverse City had Pullmans from Detroit, Grand Rapids, and Chicago via C&O, and from Indianapolis and Cincinnati by way of the Pennsylvania. Northport, at the end of the line of the Manistee & Northeastern on the Leelanau Peninsula, had service from Chicago on the C&O and from Indianapolis on the Pennsylvania. The GR&I originally built the Northport branch in hopes of getting freight traffic by ferry from Manistique, in the Upper Peninsula.[4] The line never generated much business, however, and for passenger service was used only incidentally. Further south on the Lake Michigan shore, C&O Pullmans from Chicago terminated in Pentwater.

For a few summers after the turn of the century, the Ann Arbor carried sleepers to its railhead at Frankfort, from Toledo and Columbus, Ohio (C&O), as well as a car that came on the C&O from Chicago and finished the last few miles of its trip on the Ann Arbor.

Travel to the north country mostly ended on Labor Day, but this was not the case in the southern part of the state, especially in Grand Rapids, Michigan's second largest city and for years the nation's principal producer of furniture. In the lumbering era, logs were floated down the Grand River for processing in the mills at Grand Rapids. It was from this background that the city's furniture industry developed, which in turn meant substantial Pullman traffic bringing buyers to town. The three railroads carrying sleepers to Grand Rapids all shared Union Station; the Grand Trunk Western had its own facility.

In addition to summertime resort service, Grand Rapids had Pullman accommodations by way of the C&O to Chicago, Detroit, Richmond, Va., and, in the 1890's, via the Big Four at Benton Harbor, to Indianapolis. The line to Detroit was the route of the *Pere Marquettes,* the first all-new streamlined trains after World War II. These blue and maize daylight flyers made three trips a day between Grand Rapids and the Motor City.

The Michigan Central scheduled sleepers from Grand Rapids to Buffalo, New York, Jacksonville (Southern), and Miami (Southern/FEC). The Pennsylvania carried Pullmans from Grand Rapids to Cincinnati, Indianapolis, Pittsburgh, Jacksonville (L&N/CofG/Southern/ACL), and St. Petersburg (L&N/CofG/ACL). The PRR/MC formed a joint line to Chicago, via Kalamazoo.

Bay City, another onetime mill town, had Pullman service on the Michigan Central to New York, on both MC and C&O to Detroit, and by a variety of connections to Chicago: MC, Grand Trunk Western, and an early C&O/GTW line via Flint. Nearby Saginaw had service to Chicago on the MC and from Detroit on the C&O.

The C&O and GTW both scheduled sleepers from Muskegon to Detroit. The C&O provided Muskegon with a Pullman to Chicago, and the Pennsylvania carried one to Cincinnati.

The Michigan Central carried sleepers from Kalamazoo and Lansing to Chicago.

Michigan's Upper Peninsula was a booming region when copper- and iron-mining and lumbering were the economic mainstays. Starting in the 1880's, however, tourism became an increasingly important factor in the U.P. Despite the fact that this area is politically part of Michigan, in rail travel, at least, it had far greater affinity to other states west of Lake Michigan, principally Wisconsin. (See Wisconsin chapter for map of Michigan's Upper Peninsula.)

With the exception of Pullman service from Detroit across the Straits of Mackinac to Calumet, Marquette, and Sault Ste. Marie, and a Calumet-to-Sault Ste. Marie sleeper, all of the UP's other overnight service terminated in either Chicago, Milwaukee, Duluth, or the Twin Cities, in that order of importance.

The Upper Peninsula had two principal east-west railroads, the Duluth, South Shore & Atlantic, and the Soo Line, which the South Shore became part of in 1961 (as separate roads, both had been controlled by Canadian Pacific). The Soo's line, which went from the Twin Cities to Sault Ste. Marie, where it connected with the Canadian Pacific, had originally been built as an alternate route for shipping flour east, to avoid the higher rates charged by railroads serving Chicago. Both the South Shore and Soo formed part of the CPR's alternate routes for freight, south of Lake Superior.

The South Shore carried sleepers from Duluth to Calumet, Marquette, and Sault Ste. Marie. The Soo scheduled service from the Twin Cities to Sault Ste.

Marie. Both railroads operated their own sleeping cars, generally in the wine red livery of the CPR. Both also carried Pullmans jointly with the Chicago & North Western and the Milwaukee Road. From Chicago, the North Western's *Iron and Copper Country Express* and daytime *Peninsula 400* provided name-train service. The Milwaukee's venerable *Copper Country Limited* and *Chippewa Hiawatha* ruled its iron.

Five UP communities had Pullman service from Chicago at one time or another via both C&NW and Milwaukee Road: Calumet (both with DSS&A), Iron Mountain, Iron River, Marquette (DSS&A), and Sault Ste. Marie (both with Soo). Calumet also had a sleeper operating by way of the Copper Range Railroad and the Milwaukee.

The North Western carried Pullmans from both Chicago and Milwaukee to Iron Mountain, Ironwood, Watersmeet, and Escanaba. The Milwaukee Road brought sleepers to Champion and Iron River from both cities.

From Cisco Lake, Ishpeming, Marenisco, and Menominee, the North Western scheduled Pullmans for Chicago. The Milwaukee Road carried a Pembine sleeper to Chicago and cars for Milwaukee from Channing and Ontonagon.

1. Alvin F. Harlow, *The Road of the Century: The Story of the New York Central* (New York: Creative Age Press, 1947), p. 220.
2. Jerry A. Pinkepank, "A Tale of Two Tunnels," *Trains,* Vol. 24, No. 11 (September 1964), p. 40.
3. Willis Frederick Dunbar, *All Aboard! A History of Railroads in Michigan* (Grand Rapids: William B. Eerdmans, 1969), p. 163.
4. Dunbar, p. 266.

An eight-car *Resort Special* of a generation removed from the opening photo, zips along Grand Traverse Bay in 1911.

12
West Virginia

Norfolk & Western's eastbound *Cavalier* (top), behind a Roanoke-shrouded 4-8-4, stops at Bluefield, W.Va., in May 1958. A brakeman's lantern produces intermittent light streaks in this time exposure. The *Fast Flying Virginian* (bottom), the C&O's senior flyer, nears Charleston, W.Va., June 2, 1951.

Development of West Virginia's railroads started in the north and moved south. The three main east-west trunk lines—Baltimore & Ohio, Chesapeake & Ohio, and Norfolk & Western—were laid down at about twenty-year intervals. The original goal of early builders was to reach the Ohio River, which formed West Virginia's northwestern boundary and provided the region's greatest avenue of commerce before the era of railroad construction began.

The B&O was the first railroad in the state to reach the Ohio, which it did at Wheeling in 1852 and Parkersburg four years later. (West Virginia, of course, was part of Virginia until 1863.) After the Civil War, it was obvious that the Ohio was no longer the frontier it once had been, and the B&O management saw the necessity of pushing on to acquire an all-rail connection with St. Louis. While several eastern roads, the B&O among them, were aiming for this booming community on the Mississippi, however, Chicago emerged as the priority destination in the Midwest.

The Appalachian coal fields, meantime, were providing a source of freight all their own and would even prompt the building of one major road, the Virginian, just for hauling the fuel to an ocean terminal. When it became clear that there was a demand for coal on both sides of the mountains, two of the coal-carriers continued their expansion westward.

After the Civil War, the newly formed C&O reached the Ohio River, and its terminus was located a few miles east of where the Big Sandy River joined the mainstream. The C&O, an extension of a much older line called the Virginia Central, was under the control of Collis P. Huntington, one of the builders of the first transcontinental line. Huntington's choice of terminus outraged the citizens of nearby Guyandotte, a thriving sawmill-cum-saloon community, whose newspaper editor argued the case for locating the railhead there. A reader, however, responded that while Guyandotte might have had certain advantages, it would "take a search warrant to find them in the annual crop of dog fennel, popcorn, and empty pint bottles."[1] The rail baron wasn't swayed either, and the first C&O train steamed into the new community, named for Huntington, in January 1873.

West Virginia's third major trunk line railroad, the Norfolk & Western, which followed the contour of the state's southwestern boundary from Bluefield to Kenova, reached the Ohio River in 1892. A fourth railroad, the Virginian, was completed to Deepwater in 1909. This well-engineered road, which became part of the N&W fifty years later, was built primarily to be a coal carrier, and it scheduled only minimal passenger service in West Virginia.

Wheeling, a bustling outpost on the Ohio River, acquired new importance when the National Road reached it in 1818, permitting the transfer of freight between wagons from the east and boats from the

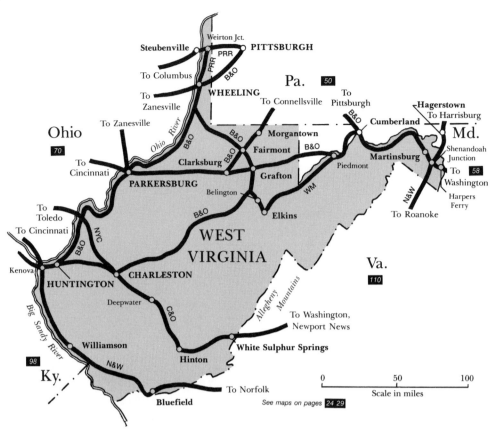

See maps on pages 24 29

west. At the time, Wheeling was engaged in an intense rivalry with Pittsburgh to be the head of navigation on the Ohio River. The flow of goods through Wheeling became so heavy that Congress made it a port of entry in 1831, complete with customs facilities. This pattern of trade, of course, changed with the coming of the railroads, and Wheeling acquired its later importance as a steel manufacturing and commercial center.

The B&O was Wheeling's major provider of Pullman service, on such trains as the *Chicago/West Virginia Night Expresses.* The city had been the western terminus of the railroad's original line built through Cumberland, Md., and Grafton, as designated in its charter. Over these tracks the railroad ran dedicated sleepers to Wheeling from Washington, Baltimore, and New York. On its line going down the Ohio River, it carried a Pullman to Huntington. The B&O's Pittsburgh-Cincinnati line also went through town, and this brought to Wheeling sleepers from Cincinnati, Detroit, and Chicago.

Wheeling also was served by a branch off the Pennsylvania's Panhandle route (the name derived from its crossing of West Virginia's Panhandle) to St. Louis. The Pennsy offered dedicated sleepers to Pittsburgh, New York, and, on the *Sea Gull,* a popular resort train of the heavyweight era, to Atlantic City, N.J.

The capital city, Charleston, in the western foothills of the Appalachians, was almost exclusively a C&O preserve, except for a Pullman line to Chicago, listed in New York Central schedules for 1909. In 1917,

however, the B&O acquired its line into Charleston, which gave the city dedicated sleepers to Clarksburg and Pittsburgh (via Grafton). Charleston also had joint service to the two cities by way of a C&O/B&O interchange at Huntington. The C&O had two joint Pullman lines with the N&W via Kenova: Charleston to Bluefield and Roanoke. On its main line trains, which included the *F.F.V., Sportsman,* and *George Washington,* C&O carried dedicated sleepers from Cincinnati, Columbus, Toledo, Washington, and Detroit.

Huntington, like Charleston, was another C&O stronghold except for the B&O's dedicated sleepers to Parkersburg and Pittsburgh (a sleeping car for Pittsburgh also originated in the rail junction of Kenova, where the C&O and N&W mains crossed, eight miles west of Huntington.) The C&O carried Pullmans to Huntington from Cincinnati, Cleveland (NYC), Detroit, and Toledo in the west, and from White Sulphur Springs (1880's); Clifton Forge (1880's), Old Point Comfort, and Richmond, Va.; Washington, and New York in the east. A sleeper to Louisville operated briefly after World War I. The C&O had its main shops at Huntington.

Parkersburg had two B&O stations, one on Sixth Street serving main line trains to Cincinnati and St. Louis, such as the *National Limited, Diplomat,* and *Metropolitan Special,* and the Ann Street depot, a half mile away, for Ohio River Division trains. A Pullman made the trip to Pittsburgh over the latter, and a Chicago sleeper left town on the branch line to Zanesville, Ohio. Pullmans were scheduled on the main line east

Ray Grant/William F. Howes Jr. collection

Sporting C&O power and by now, in 1970, serving primarily as a mail train, the B&O's eastbound *Metropolitan* stops at Martinsburg, W.Va. The train once carried St. Louis-New York Pullmans, among others.

to Washington, Baltimore, Philadelphia, and New York.

Grafton, where the newer trackage to Parkersburg and St. Louis departed from the old main line to Wheeling, had its own sleeping cars to both Baltimore and Washington.

Fairmont had sporadic Pullman service via Grafton to Washington, Baltimore, and New York, and to Chicago by way of Wheeling. Night trains provided dedicated sleepers to Pittsburgh from both Clarksburg and Fairmont over the line to Connellsville, Pa.

White Sulphur Springs was an elegant Allegheny spa, especially popular during the heat of summer long before the railroad arrived at its doorstep. Southern plantation society originally traveled to the Springs in horse-drawn coaches. Early on, the resort acquired the reputation of being a marriage mart for the well-to-do. Cottages were the original dwellings on the 5,000-acre compound; a hotel was built in 1854. The addition of a golf course in the 1880s made it a popular year-round resort.

The C&O began purchasing an interest in the company in 1910, replaced the hotel with the palatial Greenbrier two years later, and acquired full ownership in 1918. Over the years the railroad carried great numbers of tourists and conventioneers to White Sulphur Springs, and it was the summer White House of choice for three presidents. Just as the original hotel had been used during the Civil War, the Greenbrier served as a government hospital in World War II. Such C&O trains as the *F.F.V.*, *George Washington*, and *Sportsman* brought Pullmans from Cleveland (Nickel Plate or NYC), Columbus (1883), New York (PRR), Old Point Comfort (1890), Richmond, and Washington to the white colonial station at the entrance to the immaculately groomed grounds.

Because of its proximity to the mountain springs resorts and the fact that it was a division point on the C&O, Hinton saw a substantial amount of Pullman activity. Dedicated sleepers departed for Cincinnati, Columbus, Pittsburgh (B&O), Toledo, Detroit, Richmond, Va., Old Point Comfort (1893), Washington, and New York, and to Hinton from Chicago. Northwest of Hinton, C&O trains traversed one of the most scenic areas on the whole system, the New River Gorge. Ten miles east of town lay Big Bend Tunnel, scene of John Henry's legendary but mortal efforts against the steam drill.

The Norfolk & Western, whose crack trains in this region included the *Pocahontas*, *Cavalier*, and the postwar, all-coach *Powhatan Arrow*, had very little dedicated sleeper service to West Virginia cities. N&W, with the Southern, ran Pullmans to both Bluefield and Williamson from New York (PRR) and Washington. In addition, a Williamson car once operated to New York via the Cumberland Valley route to Harrisburg, Pa. (PRR), as did a Gary-Philadelphia sleeper.

The Western Maryland Railway, during its brief fling with bigtime passenger operation before World War I, ran sleepers from Elkins to Baltimore over its own lines, and to Pittsburgh (B&O at Belington) for a time in the late '20s.

Piedmont had its own dedicated sleeper to Chicago on the B&O in 1894 to accommodate the traffic fed to it by the Western Maryland.

Ultimately, coal proved to be the more enduring source of revenue for West Virginia railroading, but in the golden years of passenger travel, the scenic mountains and valleys of the state made some memorable vistas outside the windows of passing Pullmans cars.

---

1. Federal Writers' Project of the Works Progress Administration, *West Virginia: A Guide to the Mountain State* (New York: Oxford University Press, 1941), p. 241.

Gene Huddleston/H.H. Harwood collection

William D. Middleton

13
Kentucky

On a July morning in 1948 (top), the Louisville section of the C&O *George Washington* blasts up the grade from the Ohio River Valley at Ashland. The two light-colored cars are a lounge and children's car from the stillborn *Chessie*. In western Kentucky, the IC's *City of Miami* (bottom) stops in Fulton on a balmy April 20, 1956.

In the Great Lakes states to the north, the flow of railroad traffic was mostly east and west, but in Kentucky, the major lines ran north and south, much like tributaries emptying into the Ohio River, the state's naturally defined northern boundary and its original avenue of commerce. Two railroads dominated Kentucky's freight and passenger travel, and they were the main lines of Dixie: the Louisville & Nashville and the Southern Railway.

West of Washington, D.C., there were only two major gateways between the North and the Deep South, Louisville and Cincinnati, and the lines of the two key southern systems radiating out from them created the dominant design of Kentucky railroading. Both cities were served by these railways.

There were other crossings and interchanges at the Ohio River, but none that achieved the cosmopolitan mixing that went on in Louisville and Cincinnati. In western Kentucky, where the Mississippi River washes the toes of the Bluegrass State, the main lines of the Illinois Central and the Gulf, Mobile & Ohio briefly cut through on their way south. At Henderson, the L&N line from St. Louis came across the Ohio after picking up the *Georgian* and the *Dixie* fleet of Florida trains that the Chicago & Eastern Illinois delivered to Evansville, Ind.

Kentucky's other river gateway for passenger traffic was near Ashland, where the Chesapeake & Ohio rails entered from the East via Huntington, W. Va., crossing over the Big Sandy River then continuing down the Ohio River to Cincinnati. This, plus the L&N lines paralleling the river, and the IC's line from Louisville southwest through Paducah, completed the state's list of major rail thoroughfares.

Railroading in Kentucky began in the summer of 1832 with the opening of a short line between Lexington and the capital, Frankfort, on tracks that eventually became part of the L&N system. This portion, however historic, was overshadowed by the chartering of the L&N itself in 1850, with the mandate to build a line between the terminal cities of its name, with a branch to Memphis. The line to Nashville was opened in 1859, and the one from Bowling Green to Memphis two years later, just in time to be of inestimable value to Union forces—as well as a frequent target of Confederate raiders—during the Civil War.[1]

To meet the competition of Louisville as a rival for interchanging rail traffic between north and south, Cincinnati interests embarked on building a railroad of their own. Investors had failed in their attempt to push a railroad south through Kentucky and Tennessee, but in the post-war years, with the city itself behind the financing, the project went ahead . . . almost.

Kentucky refused to grant the railroad a charter, once again sidetracking the construction, but several years later the legislators had a change of heart, and by 1881 the railroad, the Cincinnati Southern, had been built to the outskirts of Chattanooga. Throughout its history, Cincinnati maintained ownership of the CS, which became part of the Queen & Crescent Route, and a principal main line of the Southern Railway System.

The topography of the CS route made for rugged railroading. A portion of the line south of Danville that initially coursed through twenty-three tunnels was given the infamous name "Rat Hole" Division. Working this stretch, steam engine crews constantly faced the danger of being parboiled or asphyxiated in the underground bores by their own locomotives' exhaust. Eighteen miles north of Danville, the railroad built a 306-foot-high bridge in 1911 to replace the one built thirty-four years earlier. The structure is the highest double-track railway bridge in the United States.[2]

Like some of the early lines, the L&N originally was to have been built in the extra-wide, six-foot gauge, but the developers finally settled on the five-foot width common in the South. This worked well enough until it became clear, after the Civil War, that to facilitate the interchange of cars, one standard gauge was necessary. Both the Southern and L&N made a prodigious one-day conversion in May 1886.

In time, the Southern acquired a line from Danville through Louisville to St. Louis, and the L&N put together its main line south from Cincinnati to Knoxville and Atlanta, rounding out the railways' modern route structure through Kentucky.

Louisville had Kentucky's largest Pullman selection, with service to most of the large cities east of the Mississippi on one or more of its eight steam railroads. The L&N, whose corporate headquarters were in Louisville, built Union Station, a Romanesque showplace with a four-story-high, barrel-vaulted ceiling

with skylight, large stained-glass rose windows, and Corinthian columns rising from a mosaic tile floor.[3] Opened in 1891, the stub-end terminal, called "Tenth Street," served such trains as the L&N's all-Pullman *Pan-American, Humming Bird,* and *Azalean,* the Pennsylvania's *South Wind, Kentuckian,* and *Union,* and the Monon's *Thoroughbred* and *Bluegrass.*

The IC used Central Station, as did the Baltimore & Ohio, Chesapeake & Ohio, New York Central, and Southern Railway. Sections of various name trains, such as the C&O's *George Washington* and B&O's *National Limited,* operated to and from Louisville, but Central never had the number of pedigreed flyers that Union Station counted. Central Station closed in 1963, whereupon trains of the C&O moved to Union.

On its "Short Line" to Cincinnati, the L&N carried Pullmans from Louisville both for the Queen City and for interchange with railroads serving the northeast. The *Pan-American* customarily hauled New York sleepers that the Pennsy forwarded. Before the turn of the century, the L&N picked up a sleeper from Boston delivered by the New York Central.

The original L&N charter specified construction of a branch to Lebanon, which later connected with the Cincinnati-Knoxville main line. Over this route the railroad scheduled Pullmans to Knoxville, Atlanta, and Jacksonville (Central of Georgia/Southern/Atlantic Coast Line). Local service included sleepers to Middlesboro (1890) and Lynch, and to Bristol (Southern) and Norton, Va.

L&N trains traveling the main line to Bowling Green carried Pullmans from Louisville to Memphis, Nashville, Montgomery, and Pensacola. In the 1890's, the railroad jointly operated sleeping car lines to Atlanta with the Nashville, Chattanooga & St. Louis, connecting at Nashville, and to Knoxville with the

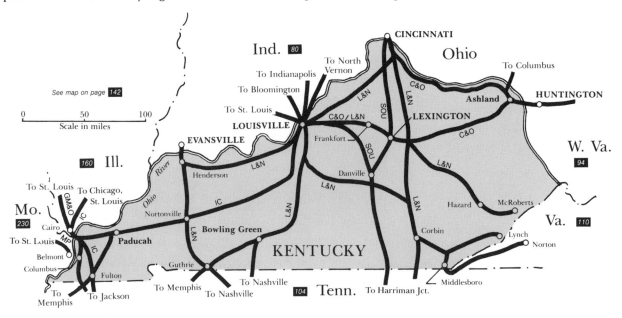

Louisville & Nashville's *Flamingo* loads a moderate Depression summer consist of seven cars in Louisville's Union Station.

Southern, via Jellico, Tenn.

On the line through Henderson, the L&N carried Pullmans to Evansville and St. Louis.

The Pennsylvania's passenger trains from Louisville started their journeys on the line to Indianapolis, where sleepers to the East were cut out. Pennsy service from Louisville included Pullmans to Indianapolis, Chicago, Pittsburgh, New York, and, in summertime, Mackinaw City, Mich.

For several years, until early in the Depression, travelers had their choice of another sleeper to Indianapolis, via the Interstate Public Service Company's electric interurbans.

The other tenant in Louisville's historic Union Station, the Monon, carried Pullmans to Chicago on its night train, the *Bluegrass.*

If it lacked some of Union Station's cachet, Central Station over the years at least gave the city a larger selection of sleeping car service. Although Louisville was essentially a branch line operation for the B&O—the connection with the main line to St. Louis was at North Vernon, Ind.—the level of Pullman service the railroad provided did not show it.

The B&O carried sleepers from Louisville to St. Louis, and north and east to Cincinnati, Toledo, Detroit, Pittsburgh, Washington, Baltimore, Philadelphia, and New York. (In the 1890's, the B&O and Erie jointly offered a sleeper to New York.) The B&O also brought Pullmans to Louisville from Columbus, Ohio. The Big Four, which had trackage rights on the B&O line to North Vernon, briefly carried a Chicago

sleeper.

The C&O entered Louisville by way of trackage rights over the L&N line from Lexington. All the C&O's Pullman traffic into Louisville came from the east: Ashland; Hinton, W.Va.; Clifton Forge, Charlottesville, Old Point Comfort, and Richmond, Va.; Washington; and New York (PRR).

Louisville lay on the Southern Railway's line to St. Louis, a secondary route that saw nothing of the SR's name trains except for connecting Pullmans for the *Carolina Special.* This line joined the Cincinnati-Chattanooga main at Danville. The Southern carried sleepers from Louisville to Knoxville, Chattanooga, Asheville, Columbia, Atlanta, Macon, and Jacksonville, Tampa (Seaboard Air Line), and west to St. Louis.

The Illinois Central and the B&O exchanged Pullmans in Louisville, generally cars from Cincinnati destined for cities in the south. From Louisville itself, the IC scheduled sleepers for Paducah, Memphis, Vicksburg, Miss. (1890), and New Orleans.

Altogether, during the prime years of rail travel, Louisville had Pullman service to more than three dozen destinations, including a choice of four different routes to New York, and three each to Chicago, Atlanta, and St. Louis. The usual workaday lineup, however, was inconsequential compared to the surge of traffic that visited Louisville every year on the first Saturday in May for the traditional running of the Kentucky Derby.

Special trains laden with Pullmans came to town

The IC's *Irvin S. Cobb* awaits departure from Louisville Central Station with Memphis and Paducah sleepers.

An Atlantic Coast Line power unit of the *South Wind* gets its windows washed in Louisville, June 1957.

and parked on station tracks and in coach yards around the city, the sleepers serving as hotel rooms during the visitors' stay in Louisville. During the 1920's, the number of these trains sometimes reached fifty, the number of extra cars 500. And because the sleepers were occupied, the logistics involved far more than a parking job: Auxiliary steam, electricity, and icing were also required.

After the throngs of Derby fans saluted an Eddie Arcaro or Willie Shoemaker, a Black Gold, Whirlaway, or Citation, they returned to the train yards, hoping the special numbering system would correctly guide them to their Pullman cars. Soon, the great parade of departing varnish began, the sounds of panting steam locomotives or growling diesels reverberating throughout the city and river valley before fading into the night.

Less than a hundred miles east of Louisville is another community in which the horse is at the top of the agenda. Lexington, in the heart of Kentucky's bluegrass country, is the capital of thoroughbred horse-breeding and buying, and as such it was important enough to have dedicated Pullman service to several eastern cities.

The C&O brought sleepers from Richmond, Washington, and New York (PRR) into Lexington's Union Station, which the road shared with the L&N. Pullman service on the L&N consisted of sleepers to both Hazard and McRoberts.

The Southern had its own facility in Lexington, where the trains of the old Queen & Crescent Route stopped. This important north-south main line hosted a diversity of passenger traffic, including such Cincinnati-New Orleans trains as the *Queen & Crescent Limited;* the Florida-bound *Ponce de Leon, Suwanee River Special,* and *Royal Palms,* carrying sleepers to and from several cities in the Great Lakes region; and the *Carolina Special,* connecting Chicago with North and South Carolina.

Ashland, the iron, coal, and oil center of eastern Kentucky, was an exclusively C&O city, where the line to Lexington branched off the main tracks from the east to Cincinnati. All the C&O thoroughbreds—*F.F.V., George Washington,* and *Sportsman*—came through Ashland, which had Pullman service to Louisville, Cincinnati and Chicago (Hocking Valley/Erie, 1890) and Richmond and Newport News, Va., Washington, and New York (PRR).

To provide connections with main line trains coming south from Cincinnati, the Southern carried a Pullman from St. Louis to Danville. In 1889, the Monon was scheduling a sleeper from Chicago to Louisville, whence the Southern took it to Burgin, another main line junction, just north of Danville.

In western Kentucky, Paducah was an important water and rail crossroads and freight gateway to the

R.D. Sharpless/Frank Ardrey collection

A Southern FP7 leads a heavy vacation consist of the northbound *New Royal Palm* at Austell, Ga., on New Year's Day, 1953.

north. A railroad bridge shared by the NC&StL, IC, and Burlington crossed the Ohio River at Paducah, just west of where the Tennessee River flowed into the waterway. Aside from its sleeper to Louisville, Paducah had Pullman service to both Chicago and St. Louis via the Illinois Central.

Fulton, where the IC line from Louisville and Paducah joined the main line to Memphis, was probably the most important junction of the railroad's entire system. All the important north-south trains—the *Panama Limited, City of New Orleans, Seminole, Floridan, City of Miami*—came down the main line of mid-America to Fulton. From there, the Florida trains traveled through Jackson, Tenn., to Birmingham, and the New Orleans trains continued south through Memphis. Fulton had dedicated sleepers to both St. Louis and Birmingham, typical of the once-common service to major rail junctions.

---

1. George H. Drury, *The Historical Guide to North American Railroads* (Milwaukee: Kalmbach, 1985), p. 178.
2. J. David Ingles, "Second Section," *Trains*, Vol. 46, No. 12 (October 1986), p. 64.
3. Edison H. Thomas, "A Tale of Two Union Stations," *Trains*, (May 1972), pp. 26-37.

The southbound *Royal Palm* clatters over the large drawbridge across the Tennessee River at Chattanooga, in August 1963.

J. Parker Lamb

14
Tennessee

All four of the settlements that grew into Tennessee's largest cities were no strangers to the steamboat whistle; they all had good water transportation from the earliest days. Travel inside the long and narrow state, however, was complicated by the fact that the four major cities lay on three different river systems and a formidable barrier of mountains cut through the central portion of the state.

Knoxville and Chattanooga were the major settlements in the Tennessee River valley, which was bracketed by the Appalachian Mountains on the east and the Cumberlands on the west. Beyond the Cumberland ridge lies the state capital, Nashville, which was the major port on the Cumberland River. The waters of the Tennessee and Cumberland Rivers eventually poured into the Ohio River, but this did nothing to facilitate navigation between Tennessee's larger cities.

The Tennessee metropolis of Memphis, of course, was a thriving port on the great waterway of mid-America, the Mississippi River, but it was not connected by water with the rest of the state. And so it remained for the railroads to provide what the rivers couldn't.

The decade of the 1850's saw remarkable progress in the construction of several of the modern main lines. The Southern Railway's route from Bristol, on the Virginia line, through Knoxville to Chattanooga was completed, as was the line from Chattanooga west to Memphis. The rails basically followed the Tennessee River valley from east to west.

By 1854, the Nashville, Chattanooga & St. Louis's ancestor had crossed the Cumberlands to connect the Tennessee cities of its name and provide through service, via another predecessor, the Western & Atlantic, to Atlanta.[1] The intersection of all these lines established the importance of Chattanooga as a railroad center, making it a prime target of Union forces during the Civil War. After long and costly fighting at Chickamauga Creek and on Lookout Mountain, Union troops captured Chattanooga, which became the staging area for General Sherman's campaign on Atlanta.

When the Southern Railway finally emerged as a large regional system in 1894, its vital center was the city on Moccasin Bend of the Tennessee River, Chattanooga. And even if the NC&StL served Chattanooga well with the *Dixie* fleet of Chicago-Florida passenger trains, it was very much a Southern Railway city with five major lines fanning out in nearly starlike angularity.

The Cincinnati Southern's line north brought traffic from midwestern cities, trains and Pullmans that had funneled through the gateway on the Ohio River: the Florida-bound *Royal Palms, Ponce de Leon, Suwanee River Special;* the *Queen & Crescent Limited* for New Orleans. Over this route, Chattanooga had dedicated sleepers to Cincinnati and Louisville.

The Southern line to Knoxville had an outlet to Asheville and the Carolinas as well as a connection at Bristol with the joint Norfolk & Western/Southern line to Washington, D.C. Through trains carried Pullmans from Chattanooga to Asheville, Salisbury, and Raleigh, N.C. (1898), Washington, and New York (Pennsylvania Railroad). An early (1882) sleeping car traveled to New York via the Cumberland Valley line (N&W/PRR).

Southern rails also went to Atlanta and continued on to Florida; the *Royal Palm* carried a Miami sleeper (Florida East Coast). Most of Chattanooga's service on this line, except for the Florida car, ended before the turn of the century: sleepers to Atlanta, Macon, and Brunswick, Ga., and from the junction of Rome, Ga.

The Alabama Great Southern line went southwest to New Orleans, the route of the *Birmingham Special* and the *Pelican*. In addition to sleepers for Birmingham (1891) and New Orleans, the Southern exchanged cars at Meridian, Miss., with other lines: for Mobile, via Mobile & Ohio, and for Vicksburg (1882) and Shreveport by way of the Illinois Central.

Along the Tennessee River to the west, the Southern's trains rode the NC&StL tracks into northern Alabama before continuing on to Memphis. The streamlined *Tennessean* and, earlier, the *Memphis Special* traveled this route. Chattanooga had its own Pullmans to Memphis.

South of Chattanooga, NC&StL trains traveled the historic stretch of tracks in Georgia made famous by "the great locomotive chase" during the Civil War. A Union agent, James J. Andrews, had orders to disrupt the line to Atlanta to hinder Confederate attempts to defend Chattanooga. Andrews failed, and the locomotive he had seized to help accomplish his mission, the *General*, was later enshrined in Chattanooga's NC&StL station.

In addition to the numerous through Pullmans on the Florida trains and the *Georgian*, the railroad scheduled sleepers from Chattanooga to Nashville and Atlanta. Around the turn of the century, it delivered a sleeper to the IC at Martin. The car went to St. Louis, the western terminus the NC&StL targeted but never reached.

Further up the Tennessee River valley, west of the Great Smoky Mountains, lies Knoxville. Once known for having the nation's only streetcars with mailboxes on them, Knoxville was on the joint Southern/Norfolk & Western line across the Appalachians to Washington. In addition to the New Orleans and Memphis trains, the *Carolina Special* also stopped in Knoxville before heading up tracks along the French Broad River to Asheville.

The Louisville & Nashville's line from Cincinnati to Atlanta also went through Knoxville, carrying the Florida-bound *Southland* and *Flamingo* and the pre-World War II *Georgian*. The L&N scheduled Pullmans from Knoxville to both Cincinnati and Louisville, as did the Southern, and in the late 1890's both railroads participated in joint service.

Southern Railway offered a greater selection of sleeper service from Knoxville, but, like Chattanooga's, much of it was early and short-lived. On the Norfolk & Western joint line, the Southern carried sleeping cars to Bristol and Lynchburg, Va., and (via Asheville) Goldsboro, N.C., and New York (PRR), all before the turn of the century. More recent service on the line included Pullmans to Salisbury, N.C., Washington, and New York. In 1890, travelers from Knoxville could book berths on Philadelphia and New York sleepers going by way of the Cumberland Valley route (N&W/PRR), and on cars heading in opposite directions, to Atlanta and Mobile. Several years later, the Southern scheduled a sleeper to St. Louis.

If Chattanooga and Knoxville were Southern Railway strongholds, the second city of Tennessee and

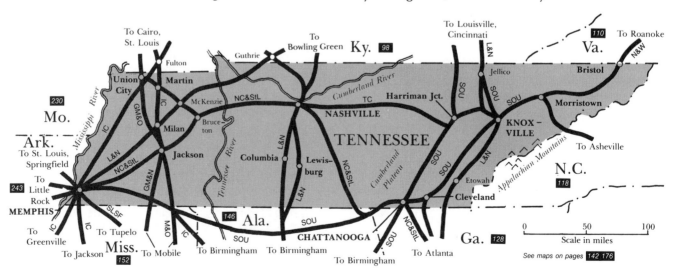

J. Parker Lamb

Laden with head-end cars, the northbound *Dixie Flyer* drifts to a stop after coming down the hill at Cowan, Tenn., June 1964.

first city of country music was, with one exception, firmly in the camp of the Nashville roads, the NC&StL and its parent, L&N.

Both railroads shared Nashville Union Station, a Romanesque structure replete with ornamental ironwork and glass, which opened in 1900.

At this Nashville landmark, the L&N handed over to the NC&StL the *Dixie Flagler* and all the other Florida trains with the *Dixie* names, as well as the streamlined *Georgian* for their trip to Chattanooga and Atlanta. L&N trains from Cincinnati and Louisville, such as the all-Pullman *Pan-American, Azalean,* and streamlined *Humming Bird,* continued on to Birmingham and New Orleans. The Chicago-Florida *South Wind* followed this route as far as Montgomery.

On this line, the L&N carried Pullmans from Nashville to Birmingham, Montgomery, and New Orleans, and north to Louisville, Cincinnati, and New York (PRR). By way of Evansville, Ind., the L&N offered sleepers to St. Louis and Chicago (Chicago & Eastern Illinois).

In the last century, the NC&StL had visions of expansion and began acquiring other railroads. L&N regarded this activity as a threat and acquired majority control of NC&StL stock. But even though the NC&StL was basically a Memphis-Atlanta line, it put together some remarkably varied sleeper routings, even after L&N took charge.

From Nashville, NC&StL scheduled on-line Pullmans south to Chattanooga and Atlanta and west to Memphis. The latter was the route of the daytime *City of Memphis,* a shortlived, steam-powered streamliner that in 1947 replaced the Depression-era *Volunteer* on the run from Nashville.

The other NC&StL sleeping car lines, of which there were nearly a dozen, required the cooperation of other railroads. The Southern was a good neighbor and from Chattanooga moved Nashville sleepers to Knoxville, Bristol, Washington (N&W/Southern), and New York (N&W/Southern/PRR). In addition, it carried New York Pullmans by way of Asheville, which had its own sleepers from Nashville. An 1893 schedule showed alternate service to Washington, in which the Norfolk & Western turned over sleepers to the Baltimore & Ohio at Shenandoah Junction, W.Va.

The NC&StL scheduled sleeping cars for St. Louis, a terminus the L&N didn't want it to have, over two routes. The first through Pullman route between the two cities was established in the early 1880's, when the NC&StL carried a car to Union City, whence the Mobile & Ohio forwarded it to Columbus, Ky. From there the sleeper was ferried to Belmont, Mo., where a predecessor of the Missouri Pacific took it for its final lap. By the turn of the century, the Illinois Central was handling the car north of Martin. The NC&StL also made an end run around the L&N with a sleeper to Chicago, also via Martin and the IC.

About this time, the NC&StL teamed with the L&N on a Memphis sleeping car, connecting at McKenzie. Cotton Belt schedules of December 1897 mention a through sleeper from Nashville to Waco, Tex., but make no mention of routing east of Memphis.

In modern times, a Nashville Pullman was cut into the *Dixie Flagler* for Miami (Atlantic Coast Line/Florida East Coast). An 1895 line to Jacksonville involved Central of Georgia and ACL. The NC&StL, it seems, had no shortage of willing partners.

Nashville's only other Pullman service came from

Otto C. Perry/Denver Public Library

Powered by a streamlined, N&W J class 4-8-4, the Memphis-bound *Tennessean* hits 70 miles an hour near Salem, Va., in July 1953.

the Tennessee Central, a coal-hauling road built across the rugged Cumberland Plateau. The Southern had two partners in the Nashville-Knoxville sleeper trade; the NC&StL was one, the TC the other. The latter collaboration also included a Pullman from Nashville to Asheville.

Memphis, the state's most populous city, was a major railhead in its own right in country far removed from the surroundings of its eastern cousins. Memphis was a thriving Mississippi River port during the steamboat era, when the city looked north and south for its commerce. When the railroad arrived from the east in 1857, however, the perspective changed, and Memphis became not just a rail terminus at river's edge, but rather a major gateway to the west as well.

Two stations accommodated the trains of the eight railways providing Memphis passenger service. Most of the railroads terminating in the city used Union Station: Southern, L&N, NC&StL, and two from across the river, Missouri Pacific, and St. Louis Southwestern (Cotton Belt). Central Station served the Rock Island and the two lines that went through Memphis, the IC and the St. Louis-San Francisco (Frisco).

Union Station was a mixture of Southern's greens and the blues of the Nashville roads and the MoPac. The Southern brought to town the *Memphis Special* and streamlined *Tennessean* from New York, and the

L&N scheduled sections of the *Pan-American, Azalean,* and *Humming Bird* from Cincinnati. The MP's *Sunshine Special, Southerner, Texan,* and *Texas Eagle* and the Cotton Belt's *Lone Star Limited* bridged the Mississippi with Pullmans for the Southwest.

In Memphis's Central Station, the reds of the Rock Island and Frisco mingled with the brown, orange, and yellow of the IC. The all-Pullman *Panama Limited,* the coach train *City of New Orleans,* and *Louisiane* from Chicago and the *Chickasaw* from St. Louis were on the IC's card. The Frisco brought the Florida-bound *Kansas City-Florida Special* to town from Kansas City, the *Memphian* from St. Louis, and the *Sunnyland,* which had sections from both Missouri cities (in the final years the *Sunnyland* operated only south of Memphis, with cars to both Atlanta and Pensacola). The Rock Island offered the *Choctaw* and the *Cherokee,* the through cars of the latter connecting with the California-bound *Imperial.* Rail passenger service in Memphis was national in scope.

The Southern carried Pullmans from Memphis to Chattanooga and Knoxville; Asheville and Salisbury, N.C.; and Bristol and Richmond, Va. In company with the Norfolk & Western at Bristol, the railroad scheduled sleepers for Roanoke (1890), Norfolk, Washington (Southern), Philadelphia (Southern/PRR), and New York (same). Before the turn of the century, the Norfolk & Western carried a Washington sleeper on its Shenandoah Valley line for the

B&O. By a similar route, a New York car went via Hagerstown and Harrisburg to the Pennsy. The Southern and B&O also made an interline exchange of a Washington sleeper at Cincinnati.

The L&N scheduled Pullmans to Louisville and Cincinnati, where the Pennsylvania picked up through cars for New York and Atlantic City. The L&N cooperated with the NC&StL on sleeper lines to Atlanta and, around the turn of the century, to Nashville, by way of McKenzie. According to 1890 schedules, the L&N delivered a sleeper for St. Louis to the IC at Milan. From Memphis the NC&StL dispatched a Pullman to Nashville.

After consolidation of the southern portions of the Illinois Central, two lines came north from New Orleans to Memphis, the main passenger route via Jackson, Miss., and the other by way of Vicksburg. The IC carried Pullmans from Memphis to the Mississippi communities of Jackson and Gulfport on the former line and to Greenville, Yazoo City, Jackson (via Clarksdale), and Natchez on the latter; it served Greenwood via another branch. Sleepers traveled to New Orleans over both lines.

Northbound, the IC scheduled Pullman service from Memphis to Louisville, Cincinnati (B&O), and New York (B&O), and to St. Louis and Chicago.

Memphis's other through railroad, the Frisco, was a southwestern system with aborted West Coast ambitions. Around the turn of the century, however, it put together a line extending southeast to Birmingham and Pensacola, making it, like the Wabash, a major system serving both east and west.

Into its primary territory, the Frisco carried sleepers from Memphis to Kansas City, Oklahoma City, Springfield, Mo. (1890), and St. Louis. Eastbound, the railroad scheduled cars from Memphis to Mobile (GM&O), Pensacola, Birmingham, Atlanta (Southern or Seaboard Air Line), and Jacksonville.

Several sleepers from Memphis had a somewhat circuitous routing through Birmingham. In the 1890's, the Frisco exchanged Washington and New York (PRR) cars there with the Southern. Later, the Seaboard was the Frisco's partner carrying Pullmans for Portsmouth, Va., and New York (Richmond, Fredericksburg & Potomac/PRR).

The Missouri Pacific carried the largest schedule of sleepers from Memphis west of the Mississippi. MoPac offered Pullmans to St. Louis, Kansas City, and Pueblo, Colo., but most of the flow from Memphis was southwest, to Little Rock, Hot Springs, Texarkana (Cotton Belt, 1890), the Louisiana communities of Alexandria, Ferriday, Lake Charles, and Monroe, and (with Texas & Pacific) Dallas/Fort Worth, Houston (MP), and Los Angeles (Southern Pacific).

The Rock Island served the same territory, but not so extensively, scheduling sleepers from Memphis to Little Rock, Hot Springs, Oklahoma City, Amarillo, and Los Angeles (SP).

Yet another line, the Cotton Belt, worked the same region, gaining access to Memphis, first from the Missouri Pacific, then after 1912 from the Rock Island. Officially known as the St. Louis Southwestern, the railway was originally a narrow-gauge line built to haul cotton from Texas to St. Louis.[2] The Cotton Belt carried Pullmans from Memphis to Texarkana, Dallas, Fort Worth, Waco (1895), San Antonio (SP), and Shreveport.

Three other communities in Tennessee had dedicated Pullman service.

At Bristol, part of which lies in Virginia, part in Tennessee, the rails of the Southern and the Norfolk & Western met. Bristol had its own Pullmans to Washington and New York (N&W/Southern), as well as alternate New York service over the Shenandoah Valley route (N&W/PRR). Eastbound service from Bristol also included sleepers to Richmond (N&W/ACL) and Norfolk (N&W). The Southern carried cars to Knoxville (1898), Memphis, Cincinnati, and Louisville.

Jackson, a major rail junction that is also the final resting place of John Luther "Casey" Jones, had dedicated sleeper service courtesy of the Gulf, Mobile & Northern route to Jackson, Miss., Mobile, and New Orleans (the latter continuing under the successor Gulf, Mobile & Ohio).

The GM&N made news in July 1935, when the *Rebels*, the first diesel streamliners in the South, were inaugurated. Otto Kuhler designed the red and silver trains, which were also the first non-articulated streamliners, and the first trains to carry hostesses. The observation car offered six sleeper sections for travel between Jackson and New Orleans.

Jackson was also a regular stop for the Illinois Central's trains to from Chicago and St. Louis to Florida, including the *Floridan, Seminole,* and *City of Miami.* Through sleepers to and from Memphis also served Jackson on the NC&StL.

Morristown, where the Southern's tracks to Asheville branched off the Bristol line, had brief sleeping car service in the early '90s, via Asheville, to both Raleigh and Greensboro, N.C. About the same time, the junction town of Cleveland was the terminus for sleepers from Mobile.

---

1. Stuart Covington, "NC&StL: Dixie Success Story," *Trains* (January 1948), p. 44.
2. George H. Drury, *The Historical Guide to North American Railroads* (Milwaukee: Kalmbach, 1985), p. 289-90.

William D. Middleton

William E. Griffin Jr. collection

J. Parker Lamb

15
Virginia

Between the salt water of Chesapeake Bay and the foothills of the Blue Ridge Mountains in Virginia, four great railroad systems crossed one another. Two of them were parts of the main north and south trunks of the Eastern Seaboard; the other two were the hard-working coal railroads that ran from the seaports of Hampton Roads to the Ohio River valley. The passenger traffic from the north came through Washington, some of it all the way from Canada. From Norfolk and Newport News came trains traveling over the first major east-west routes north of Jacksonville. Like stars in a constellation, the four crossings—at Richmond, Petersburg, Lynchburg, and Charlottesville—lay in the Piedmont region of Virginia.

Richmond, the capital of Virginia, was the southern terminus of the 116-mile Richmond, Fredericksburg & Potomac Railroad's double-track main line from Washington and was the point where the Atlantic Coast Line and Seaboard Air Line railroads picked up the through cars to Florida. The Chesapeake & Ohio main line west from Newport News also passed through Richmond, and the resulting intersection was spectacular: At 17th and Dock Street, C&O and SAL lines passed over a switching track of the Southern Railway in the only tri-level crossing of three major American railroads.

In Charlottesville, home of presidents Jefferson and Madison and the University of Virginia, the C&O crossed the main line of the Southern Railway, the other major north-south trunk from Washington.

Lynchburg, which gained early importance as a canal town, became a key rail junction with the crossing of the Southern by the main line of the Norfolk & Western, as well as the James River line of the C&O. From Lynchburg, some of the Southern traffic went on to Atlanta over the railroad's own main line; the rest of it switched over to the N&W, which took the cars across the Blue Ridge and Appalachian mountains and returned them to the Southern at Bristol, a bistate town on the Tennessee line, for the remainder of the way to Chattanooga and Mississippi valley cities from Memphis to New Orleans.

Petersburg, the fourth star in the constellation, saw the crossing of the Norfolk & Western by the ACL and SAL, which there quit running in parallel and headed south their separate ways. The portion of the ACL to Weldon, N.C., opened in 1833, was the oldest segment of the railroad's modern network.

Over the years an unequaled parade of varnish passed through this rolling Virginia countryside: to and from Washington, the Southern's *Crescent, Piedmont Limited, Southerner,* and *Tennessean;* ACL's *Florida Special, Havana Special, Champions,* and *Everglades;* SAL's *Orange Blossom Special, Silver Comet, Silver Star,* and *Silver Meteor.* From Norfolk and Newport News came the N&W *Pocahontas, Cavalier,* and all-coach *Pow-*

Richmond's Broad Street Station (top) is aglow as Atlantic Coast Line takes over its twenty-car *East Coast Champion* on August 25, 1957. Inside, the station's architectural elegance is evident in the cool waiting room (bottom). Both ACL and Richmond, Fredericksburg & Potomac had their own train boards (this RF&P posting is of February 1927 vintage).

hatan Arrow; on the C&O the historic *Fast Flying Virginian, George Washington,* and *Sportsman.*

What a parade and what a rainbow!—the elegant green and gold of the Southern *Crescent,* N&W's tuscan red and gold, C&O's early orange and maroon, later olive green, and finally blue and yellow; the silver streamliners under diesel tow of RF&P's blue and gray, ACL's purple and SAL's "Orange Blossom Scheme" in citrus colors; in the distant past, the orange cars of the Virginian Railway.

Richmond, onetime capital of the Confederacy, and the ports of Hampton Roads were about equal in the level of dedicated Pullman service they enjoyed and together they accounted for the major portion of traffic in the state. Residents of northern Virginia, of course, could choose from the greater selection of sleepers operating in and out of Washington.

In 1836, Richmond saw its first passenger train, on what is now the RF&P. This city at the head of navigation on the James River quickly developed into an important rail center, a fact that figured prominently into Civil War strategy on both sides. The town of Fredericksburg, midway between Richmond and Washington on the RF&P, changed hands seven times in some of the war's bloodiest fighting, and near Manassas on the Southern Railway the two forces fought

the battles of Bull Run. In the last days of the Confederacy, Richmond was evacuated and burned by its citizens rather than having its stores and rail facilities fall into northern hands.

In 1888, a light-rail development took place in Richmond that quickly carried the urban transit systems of the world out of the horse- and cable-car era. Inventor Frank J. Sprague finally perfected a successful design of an electric traction motor for streetcars, and his trolleys went into operation in the Virginia capital that year. It was the breakthrough the industry had been waiting for, and within a few years electric streetcar and interurban systems, using the Sprague technology, proliferated all over America.

All of Richmond's north-south railroad passenger traffic went through Washington by way of the RF&P, a key bridge route whose ownership was shared by the major railroads serving the two cities: ACL, C&O, SAL, Southern, Pennsylvania, and Baltimore & Ohio. The RF&P tracks hummed with passenger activity, which accounted for more than 40 percent of the railroad's sales dollar. Sometimes as many as five of its passenger trains rolled into Richmond within an hour.

The RF&P participated in three different interline Pullman agreements, all involving the Pennsylvania Railroad, which carried the sleepers between Wash-

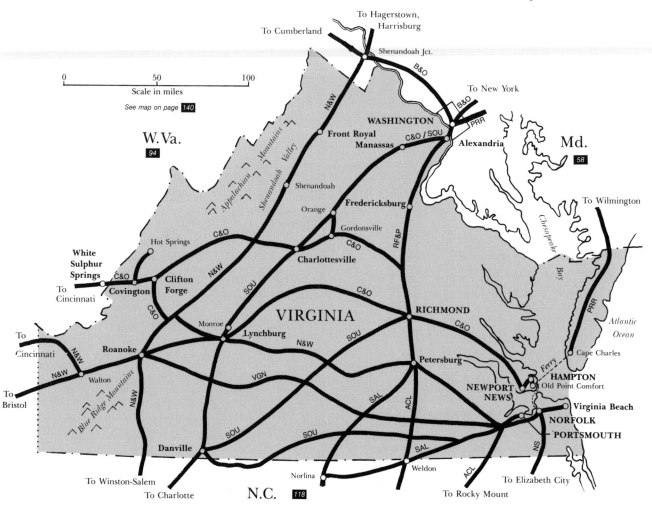

ington and New York. South of Richmond, the other participants were ACL/Florida East Coast, N&W (ACL handled these cars to Petersburg), and SAL. On the basis of route mileage, the railroads purchased their shares of the new equipment required, including coaches and revenue cars. For the ACL/FEC line, for instance, the four railroads ordered seventy-one new sleepers in 1946. The RF&P owned seven of these and three each for the SAL and N&W services.[1]

Both of Richmond's main passenger stations were architecturally notable. The city's Broad Street facility was designed by John Russell Pope, whose work also included the Jefferson Memorial in the nation's capital. The station, which opened in 1919, was distinguished by its shallow dome and pillared front. From the concourse, stairways descended to the platforms serving the through tracks. RF&P and ACL were the principal users of Broad Street until SAL moved its trains there for the last decade of private enterprise passenger service.

SAL and C&O formerly shared Main Street Station, which the two roads had built just after the turn of the century. The building's French Revival style, which gave it the look of a 17th Century chateau, and the large, elevated trainshed made it one of the South's most distinctive depots. Southern used the station for several years, but finally moved to its small Hull Street facility. Broad Street survived as a rail facility until 1975; it was later converted to a science museum.[2] Main Street was renovated and turned into a shopping mall.[3]

At Broad Street, Pullman service was mostly of the through variety. Main Street, thanks to the C&O, originated much more traffic, which was not surprising considering that Richmond was the fountainhead of the C&O system.

C&O's oldest predecessor, the Louisa Railroad, had extended its line fifty miles northwest of Richmond by 1837. The westward move continued, and the state of Virginia had a hand in building portions of the line through the Blue Ridge Mountains. The Civil War resulted in massive destruction, but building resumed afterwards and by 1873 the C&O had reached the Ohio River. By then the road was under the control of Collis P. Huntington, one of the builders of the Central Pacific, and he envisioned the C&O as part of a great transcontinental system, connecting with the Southern Pacific. Huntington manged to assemble the linkage, but his empire collapsed in the late 1880s.

Meanwhile, the C&O had pushed west to Cincinnati and east to Newport News and had negotiated trackage rights over the Southern Railway line into Washington, D.C. Although its principal trade was hauling coal from the booming mines of Appalachia to the docks at Newport News, the C&O was now equipped to introduce a luxurious long-distance train that would run for 79 years and a day and serve as the line's flagship for four decades.[4]

In May 1889, the *Fast Flying Virginian* made its bow with new vestibuled, electrically lighted cars, whose interiors were paneled in rosewood, cherry, and mahogany and whose exteriors carried the handsome color scheme of orange and maroon with gold lettering. The train was known as the *F.F.V.*, and the railroad had no objection at all to its association in the popular mind with the phrase "First Families of Virginia," for it was a prestigious consist in all respects. After the Pennsylvania Railroad agreed to move *F.F.V.* sleepers to and from New York, C&O was able to establish Pullman service that tied Richmond, Newport News, and New York with Cincinnati, Chicago, and St. Louis.

Over the years, the C&O provided dedicated Pullman service from Richmond to at least a dozen destinations. It offered three routes to Chicago: Before World War I, cars moved over the railroad's own line from Cincinnati through Indiana, but the New York Central's Big Four route eventually proved to be the more expeditious. In later years the C&O also turned Chicago sleepers over to the NYC at Toledo. The railroad had two routes to Clifton Forge, both offering Pullman service: the main line by way of Charlottesville, and the James River line via Lynchburg, which followed the route of the defunct James River and Kanawha Canal. A Lynchburg sleeper was operated on the mostly freight, river line in 1910-11. About that time, the C&O participated briefly in a Pullman line that presaged its future association with the B&O: a Richmond-New York car, with a little bridgework by RF&P to Washington.

All the other C&O sleeper service from Richmond moved west: White Sulphur Springs, Hinton, and Huntington, W.Va.; Ashland, Lexington, and Louisville, Ky.; Cincinnati, Detroit, Grand Rapids, and St. Louis (via NYC).

Although one train after another with Florida Pullmans came through Broad Street, relatively few dedicated sleepers departed from the station moving north over RF&P or south via ACL. The RF&P carried a Pullman to Washington, where it also turned over sleepers for Philadelphia and New York to the Pennsylvania, and one for Pittsburgh to the B&O. ACL service was limited to hauling cars to Petersburg, whence the N&W took them on to Lynchburg, Roanoke, Bristol, and Columbus, Ohio.

From Hull Street Station, Southern carried Pullmans to Danville; Greensboro (1890), Salisbury, Winston-Salem, Asheville, and Charlotte, N.C.; Atlanta and Augusta, Ga. (1898), Birmingham, and Memphis.

In addition to the dominating naval presence in their midst, the Hampton Roads cities, for a time after World War II, led all other American maritime cities in total export tonnage. Coal, the principal com-

J. Parker Lamb

From Richmond's Main Street Station, the C&O's *George Washington* (left) will leave for Charlottesville, where it joins the section from Washington. On this October afternoon in 1961, the SAL *Silver Star*, now using Broad Street, slides on by.

modity, came down from Appalachian mines to Newport News via the C&O, and to Norfolk by way of the Norfolk & Western and the Virginian Railway, a relative late-comer to the coal trade, but one known for having the best engineered line down the mountains.

The C&O and N&W accounted for the major share of Pullman traffic in and out of the area, and the level of their services was about equal. After the C&O built its Hampton branch from Newport News, it dispatched sleepers variously from three stations in the vicinity: Old Point Comfort, a posh beach resort near the military post of Fort Monroe; Phoebus, a nearby town (now part of Hampton) named for an early hotelkeeper at Old Point Comfort; and Newport News itself. (The branch, the outer few miles of which are now gone, was of lasting importance: In 1889, the C&O began renumbering all its mileposts to make Old Point Comfort the eastern end of the system, and that pattern is still followed.[5])

Although the national economy was in a steep decline in the early 1930s, the C&O went ahead with plans to introduce two new trains. The *Sportsman*, connecting Newport News and Detroit, was intended to link the vacation areas of Virginia and Michigan and the mountain springs resorts in between. The train was inaugurated in March 1930.

By 1932, the level of business activity was lower

than ever, but the C&O nonetheless made preparations for a train that would assume the *F.F.V.'s* premier title. When the new flyer was inaugurated in April, it was advertised as the first long-distance, air-conditioned, sleeping-car train. Dining car customers were seated in Duncan Phyfe chairs, served with specially patterned china on Irish linens, and given mountain water from White Sulphur Springs. Because the C&O's James River line had supplanted a canal that our country's first president had been involved in, and since this Depression year also happened to be the bicentennial of his birth, it was only fitting that the company that was now calling itself "George Washington's Railroad" named the train in honor of the Father of Our Country, a portrait of whom adorned the drumhead on the observation car.

The *George Washington*, which connected Newport News and Washington with Cincinnati and Louisville, was given the fastest schedule of all C&O trains, and, despite the economic adversity of the times, the train proved popular and enduring. (The next year, 1933, the railroad introduced in its advertising copy a figure that proved even more popular and enduring, the little animal that became its logo of Chessie the cat.[6])

In the streamlined, postwar era, the C&O equipped the *George Washington* with a "Chessie Theatre on Rails" in the diner, where, after the evening meal, first-run motion pictures were shown. After 1968, for the final three years of its life, the *George Washington* was the C&O's last surviving passenger train.[7]

At several times, the C&O offered sleepers from Old Point Comfort to New York (RF&P/PRR), but all the rest of its service from the Hampton area was to the west: White Sulphur Springs (1890), Hinton (1893), and Huntington, W.Va.; Ashland and Louisville, Ky.; Cincinnati, Detroit, St. Louis (NYC), and Chicago (via the three different routes).

While C&O held down the north shore of Hampton Roads, the area's other passenger service operated out of Norfolk and Portsmouth. Norfolk's Terminal Station had the most sleeper traffic and involved cars from the Virginian, Norfolk Southern, and Norfolk & Western. (This Norfolk Southern was a modest-size carrier in Virginia and North Carolina, and it eventually became part of, and gave its name to, the modern megarail system.) ACL had its own facility. Norfolk residents had to ride ferries to Portsmouth for SAL trains, to Pinners Point for those of the Southern, and to Newport News for the C&O's fleet (buses later provided some of the connections). Steamers crossed Chesapeake Bay to Cape Charles for the Pennsylvania Railroad's Delmarva service.

The Norfolk & Western, which began hauling its modern lifeblood, coal, to Norfolk in 1883, remained a no-frills freight mover without so much as a titled passenger train until the 1920's. N&W reached the Ohio River much later than the C&O did, and it didn't arrive in Cincinnati until 1901. The railroad, however, did make arrangements with the Pennsylvania (which had controlled it for many years) to run sleeping cars to both New York and Chicago and to cooperate on service through the Shenandoah Valley. The N&W also made a lasting agreement with the Southern to provide the Lynchburg-Bristol link in the two roads' joint line to Chattanooga and New Orleans.

Early in this century, N&W originated some sleepers in Norfolk for this route, cars for Memphis and New Orleans that it turned over to the Southern in Bristol. This service, however, was short-lived, and mostly N&W was content to be a bridge line for traffic to the south. The Southern was running a sleeping car from Portsmouth to New Orleans about this time, via Atlanta and the Crescent route, but neither did this last, and, oddly, there never was any regularly sustained Pullman service between these great seaports of Hampton Roads and the Gulf of Mexico.

Over its main line to Cincinnati, N&W ran several fine name trains, introducing the *Pocahontas* in 1926 and the *Cavalier* two years later. It added to the *Pocahontas's* consist a number of streamlined coaches in 1941 and refurbished them in 1946 for the newly created all-coach *Powhatan Arrow*, the N&W's entry in a Tidewater-to-Cincinnati train race that never fully developed.

Being a coal road, the N&W stayed with steam locomotives on its passenger trains until 1958, much longer than other major railroads, and among the most dazzling sights of the postwar era were its crack passenger trains running under power of the streamlined, bullet-nosed Class J Northern locomotives, shrouded in glistening black with a gold-accented horizontal stripe of tuscan red. The last of these locomotives, all of which were built in the railroad's own Roanoke shops, were also America's last steam locomotives produced for main-line passenger use.

N&W brought Pullmans to Norfolk from Roanoke, Bristol, Cincinnati, Columbus, Cleveland (NYC), Chicago (PRR), St. Louis (B&O), and New York (PRR/RF&P/ACL).

The other tenants in Norfolk's Terminal Station had more limited Pullman service. Norfolk Southern operated sleepers south to New Bern and Raleigh, N.C., but perhaps its most unusual contribution was in shuttling a Virginia Beach-New York Pullman over its electric division between the resort and Norfolk. The car continued its journey via N&W/ACL/RF&P/PRR.

The Virginian Railway, the third major coal hauler, was one of the last major rail systems to be built in America. Started in 1904 and finished five years later, it was quickly recognized as a major engineering achievement for the energy-conserving, gradual downgrades its coal trains rode from the Appalachian mines to the Norfolk docks. This felicitous route, however, avoided any population center of consequence, except for Roanoke, and so its passenger service was always minimal. Until early in the Depression, the Virginian competed with the N&W by carrying a Pullman to Roanoke. Although it was not sleeper country, VGN trains west of Roanoke rode under catenary after the line was electrified in 1926. In one of the first of the modern rail mergers, the Virginian became part of Norfolk & Western in 1959.

ACL provided Norfolk's remaining Pullman service, with sleepers south to Wilmington, N.C., and Jacksonville.

Portsmouth, across the Elizabeth River from Norfolk, had been Seaboard Air Line's historic railhead since its charter line to Weldon, N.C., was opened in 1837. SAL owned an interest in the Old Bay Line, which provided connecting steamship service to Norfolk, Old Point Comfort, and, overnight, to Baltimore. The steamers offered staterooms with private toilets and baths ranging from $1.25 to $5.00 a night. The existence of overnight boats to both Baltimore and Washington well into the post-war era helps explain the lack of any long-term dedicated Pullman service to those cities from the Norfolk area.

Although SAL was headquartered in Norfolk, its biggest car shops were in Portsmouth. In the early years, the railroad ran sleepers to Raleigh and Hamlet, N.C., and Birmingham, Ala.; in modern times, to Charlotte, N.C., Atlanta, Tampa, and Memphis (Frisco at Birmingham).

In addition to its early service to New Orleans, the Southern Railway ran sleepers from Portsmouth to Danville, and to Salisbury, Charlotte, Greensboro (1893), and Asheville, N.C., over the Atlantic & Danville line, which it leased for many years.

The Pennsylvania Railroad, by way of the Delmarva Peninsula, gave the Hampton Roads area an extended maritime Pullman connection. Two of the four daily trains that operated from Wilmington, Dela., to Cape Charles, on the Eastern Shore, carried sleepers to and from Philadelphia and/or New York. Morning and evening ferries calling at both Old Point Comfort and Norfolk met the trains, and passengers could have a leisurely breakfast or dinner in the ships' dining rooms while crossing a twenty-five-mile span of Chesapeake Bay. The train/steamer service originated in 1885.

Roanoke, an Indian word for "shell money," was known by the more prosaic name of Big Lick until 1882, when it was selected to be the junction of a railway running down the Shenandoah Valley and the Norfolk & Western main line. Roanoke quickly grew into an archetypal railroad town, and the N&W, which located both its shops and headquarters in this community framed by Allegheny and Blue Ridge Mountains, was unquestionably the railroad. The Shenandoah Valley line eventually became part of the N&W.

Forty miles southwest of Roanoke, at Walton, the N&W main line divided, the left fork going to Bristol, the right one becoming the main line to Cincinnati. From Roanoke a branch ran south to Winston-Salem, and the Shenandoah line ran north to Hagerstown, Md. These gave the city some unusual connections. In the early 1890s, the *Shenandoah Limited* was plying this route with through sleepers, and eventually Roanoke had cars of its own going both south and north. A Pullman to Jacksonville operated via N&W, Winston-Salem Southbound, and ACL. Sleepers ran north to Philadelphia and New York, via the Pennsylvania at Hagerstown, Md., and, before the turn of the century, to Washington, by way of the B&O at Shenandoah Junction, W.Va. Later versions of the Roanoke service took the N&W's New York Pullman via C&O and PRR through Washington, and the Washington car by way of Southern.

The N&W also scheduled Roanoke sleepers to Charleston, W. Va., (C&O), Columbus, Ohio, and Memphis (1890, Southern), and for a time the New York Central assisted in moving from Cleveland a car to Roanoke. It was fitting that N&W, which provided the city with nearly all its Pullman service, financed an architecturally elegant remodeling of the city's railway station, a postwar redesigning by Raymond Loewy.

In the mountain country north of Roanoke lay the resorts whose curative waters were eagerly sought even in the Colonial era: Hot Springs and, across the West Virginia line, White Sulphur Springs. Before moving westward with its sleepers from New York for

The *Silver Star* poses on the Rappahannock River bridge in Fredericksburg with two RF&P E8's for this 1950 publicity shot.

W.H. Thrall/Frank Ardrey collection

A Southern 4-6-2 steams past Alexandria's CR Tower in February 1937 with the northbound *Birmingham Special*.

White Sulphur Springs, the C&O's *F.F.V.* dropped Pullmans in Clifton Forge, from which they were shuttled to a branch line running from Covington to Hot Springs. This beautifully landscaped property with the Homestead Hotel was acquired by C&O interests in the early 1890s, and over the years the railroad—often in all-Pullman trains—brought in thousands of vacationers, who came for golfing, hiking, riding, and bathing. In addition to New York sleepers, the C&O carried Pullmans from Chicago (NYC), and from both Buffalo and Cleveland, via the Nickel Plate at Fostoria, Ohio.

The division point of Clifton Forge had substantial dedicated Pullman service because it was a convenient terminus for passengers visiting the springs resorts. Pullmans from New York and eight western cities terminated in this town of railroad shops and steep residential hillsides: Chicago (NYC), Cincinnati, Cleveland (both NKP and NYC), Columbus, Detroit, Huntington (1885), Louisville, and Toledo.

The C&O also operated sleepers from Charlottesville, their destinations including Chicago (NYC), Cincinnati, Cleveland (NYC), Detroit, and Louisville. A Pullman for Detroit originated in the junction of Gordonsville, where vendors renowned for their fried chicken once had fed hungry passengers on trains pausing for a meal stop.

For a brief time in the early 1890s, the Southern originated sleepers in Lynchburg for Washington, Knoxville (N&W/Southern) and Mobile (N&W/Southern). It was on the Southern's main line through Virginia that the wreck of "Old 97," the *Fast Mail* from Washington to Atlanta, occurred on Sept. 27, 1903 near North Danville.

Danville, which for a few days in April 1865 was the last capital of the Confederacy after Richmond was evacuated, saw numerous through Pullmans on the Southern's main line, as well as on its branches to Richmond (which handled much of the Civil War exodus from the capital) and Portsmouth (over the Atlantic & Danville). Apart from a sleeper to Portsmouth, its dedicated service from Augusta, Ga., and New York didn't survive the turn of the century.

In the coal town of Norton, where, it was said, residents dug their winter's fuel supply from their own cellars, a branch of the Louisville & Nashville Railroad met one line of the N&W. From this small mountain terminus, the L&N ran Pullmans to both Cincinnati and Louisville.

1. Letter of June 1, 1988, received from William E. Griffin Jr., director of personnel and labor relations for the RF&P, containing roster information on the railroad's passenger cars.

2. Griffin, *One Hundred Fifty Years of History Along the Richmond, Fredericksburg and Potomac Railroad* (Richmond, Va.: Whittet & Shepperson, 1984) pp. 47-55.

3. Letter of August 11, 1987, from Thomas W. Dixon Jr., president of the Chesapeake and Ohio Historical Society, Inc.

4. Dixon, "Along Came the F.F.V.," *Chesapeake and Ohio Historical Newsletter*, XI, No. 12 (December 1979), pp. 4-22.

5. Letter of July 31, 1987, received from Mr. Dixon.

6. See Dixon, *Chessie: The Railroad Kitten* (Sterling, Va.: TLC Publishing Co., 1988).

7. Dixon, "The George Washington: The Most Wonderful Train in the World," *Chesapeake and Ohio Historical Newsletter*, IX, No. 5 (May 1977), pp. 14-31.

16
North Carolina

Looking down on North Carolina from on high, the cosmic dispatcher might have concluded that most of the state's passenger train activity was concentrated on three main lines—the Southern Railway, the Atlantic Coast Line, and the Seaboard Air Line—that cut across the Piedmont from northeast to southwest like hash marks on a boatswain's jumper. This was true, but for a pleasant anomaly that was recorded over the years in the beautiful country known as the Land of the Sky.

Although Asheville is not even among the six largest cities of North Carolina, it surpassed all others in dedicated Pullman traffic. Remarkably, the movement of all this rolling hostelry was in the hands of one railroad, the Southern. Some travelers came to do business with the city's textile and furniture manufacturers, but it was Asheville's reputation as a health resort that was the basis for most of its sleeping car service. Visitors came to town on such trains as the *Skyland Special, Asheville Special,* and *Carolina Special,* in dark green Pullmans that had originated in Miami, New York, Chicago, and more than two dozen cities in between.

West of Asheville lies the eastern portal of Great Smoky Mountains National Park, which showcases one of the ranges of the Blue Ridge Mountains that surround the community. Recognizing the beauty of this countryside, the New York capitalist George Vanderbilt began developing the great Biltmore estate and village after his arrival in 1889. By the 1920's, the influx of tourists was reaching boom proportions, but, just as in Florida, the bubble burst, capping the growth. All the while, however, the trains came to town, the trains that Thomas Wolfe heard in his youth, their stacks shattering the mountain stillness as they struggled up the steep grades, their sanded drive wheels biting every inch of the glistening steel rails. Day and night the trains came to Asheville.

Up the water-level route alongside the French Broad River, came the trains from Knoxville and the west, from cities on the Great Lakes to the Gulf of Mexico. The *Carolina Special* began working this line in 1911 and didn't cease until 1968. Over the years it carried Pullmans principally from Cincinnati and, by extension of the New York Central, from Chicago, both for Asheville and beyond. In Asheville the train was split into two sections, the southern one bound for Spartanburg and Charleston, S.C., and the northern one carrying sleepers for Greensboro and Goldsboro.

Pullmans from St. Louis and Louisville also came to Asheville over this Cincinnati line. Trains from the South moved through Chattanooga and Knoxville with sleepers for Asheville from Birmingham, Chattanooga, Memphis, Mobile, Nashville (with either Nashville, Chattanooga & St. Louis or Tennessee Central), and New Orleans. These tracks along the

The eastbound *Carolina Special* eases downgrade through a cloud of brakeshoe smoke near Old Fort, N.C., in August 1961. This is the North Carolina section of the train, which divided in Asheville.   J. Parker Lamb

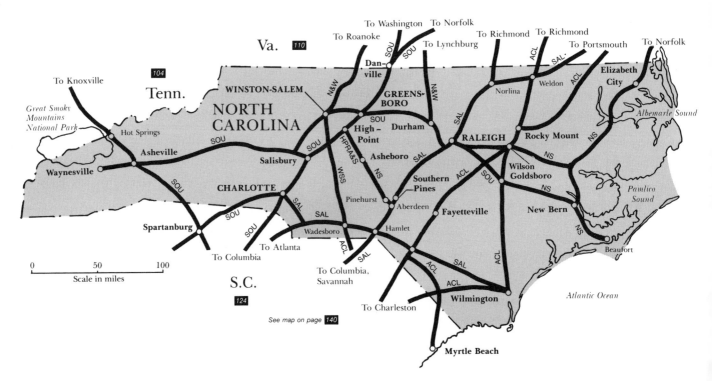

French Broad River were the easy way into town. The other two main routes presented the full challenges of mountain railroading.

. The steepest grades found on any main-line, standard-gauge American railroad are at Saluda Mountain, thirty-four miles southeast of Asheville, on the Southern's line to Spartanburg. Trains were routinely boosted by helper engines up the 4.7 per cent grade leading to this Blue Ridge summit, and extraordinary safety measures were instituted after several early runaway disasters. Two sets of safety tracks were installed so that any train that went out of control while traveling down the mountain would automatically be switched to an uphill, deadend spur and halted by gravity. The railroad stationed a switch tender where each of the safety spurs branched off the downhill main line. Only after hearing a particular whistle signal from an engineer, indicating that his train was under control, could the tender throw the switch, thus clearing the train for continued downhill movement.

Passengers on the Charleston section of the *Carolina Special* were routinely treated to these services, but over the years, many additional Pullmans traveling between Asheville and cities in the Southeast were hauled over Saluda Mountain. The line crossed the Washington-Atlanta main at Spartanburg, where sleepers for Atlanta and Macon, Ga., and New Orleans were switched to southbound trains. A Pullman for Augusta, Ga., also was cut out at Spartanburg and turned over to the Charleston & Western Carolina Railway; in more recent times the Southern handled it all the way, via Columbia, S.C.

At Columbia, which had its own sleeper from Ashe-

ville, Pullmans for Savannah, Ga., Jacksonville, and Miami (Florida East Coast) were switched to the Florida main line, and the Wilmington car to the Atlantic Coast Line. Sleepers for Charleston proceeded by way of Orangeburg and Branchville, S.C. The two sections of the *Skyland Special* formed up in Columbia for the run to Jacksonville with cars from Asheville and Cincinnati and others from Charlotte and Greensboro.

For a time in the 1920's, the Southern turned over Asheville's Savannah and Florida Pullmans to the Seaboard Air Line in Columbia, including sleepers to Jacksonville, Miami (FEC), Sarasota, and St. Petersburg.

The Southern had two more routes in and out of Asheville: The one to the southwest carried sleepers from Charleston to the health resort of Waynesville. The line to the northeast skirted Mount Mitchell—at 6,684 feet the tallest peak in the eastern United States—and connected the Land of the Sky with the cities of the Northeast. The trains reached the main line either at Salisbury or Greensboro. In later years, the Asheville sleepers for Washington and New York (Pennsylvania) operated on the *Asheville Special* by way of Winston-Salem rather than Salisbury. The Southern also scheduled Pullmans to Salisbury, Raleigh, and Norfolk, and, on the northern section of the *Carolina Special,* to Goldsboro and Richmond.

Southern Railway not only monopolized Asheville's Pullman traffic, but it dominated service in all of North Carolina. And nowhere was the parade of varnish more handsome than that flowing through the main-line cities of Charlotte and Greensboro. Seven

SAL No. 18 has arrived in Raleigh from Portsmouth, Va., on this night in 1961. The train carried sleepers for the *Silver Comet*.

daily express trains, their Pacific engines painted forest green, trimmed in gold and white with cab roofs a deep red, worked this Washington-Atlanta main line at the end of the Roaring '20's. The *Crescent Limited, Piedmont Limited, Aiken-Augusta Special*—these and four others called around the clock, carrying their own load of heavyweight sleepers as well as cutting in those that came down the mountains from Asheville, either at Spartanburg, Salisbury, or Greensboro. The parade continued as some of the classic limiteds were refitted—the *Crescent* in shining new streamlined dress—and joined by the likes of the *Peach Queen* and *New Yorker* and the stainless-clad, primarily coach *Southerner.*

Charlotte, the industrial giant of the Piedmont, was once known for having a United States mint in operation—it was the center of gold-mining in America until the clamor of the forty-niners diverted the action elsewhere. The Southern Railway's main line from the north split at Charlotte, the right fork continued on to Atlanta, the left to Columbia and Jacksonville. Seaboard Air Line extended one of its western legs to the city, and the old Norfolk Southern Railway made Charlotte its western terminus.

Southern Railway trains from Atlanta and New Orleans stopped at Charlotte's landmark station with its

graceful tower. Charlotte had dedicated Pullmans from Birmingham, Atlanta, and Columbus, Ga., and, in the other direction, from Goldsboro and Raleigh, Portsmouth and Richmond, Va., Washington, and New York (PRR). On the Florida line, Charlotte sleepers operated to Augusta (1895) and Savannah, Ga., and Jacksonville. Seaboard Air Line carried Pullmans to Hamlet and Portsmouth.

Pullman service to and from Greensboro—like Asheville's—was entirely in the hands of the Southern Railway. Greensboro was at the intersection of the Southern's main line and an east-west route that linked four of the state's five largest cities with Asheville.

Greensboro had Pullman service, via Asheville, to Morristown, Tenn. (1890), and, later on the *Carolina Special,* to Cincinnati and Chicago (NYC). To the east, the Southern carried sleepers to Raleigh, Goldsboro, Wilmington (ACL), and Beaufort (Norfolk Southern). Main-line trains hauled Pullmans from Greensboro north to Portsmouth (1893), Richmond (1890) and New York (PRR), and south to Atlanta, Augusta (1893), and Jacksonville.

Salisbury, a colonial era town founded in 1753, was another Southern sleeping car monopoly. Over its line to Asheville, the railroad carried Pullmans to Knox-

A pusher boosts the southern section of the *Carolina Special* up Saluda grade, the steepest on any main line in the United States.

ville, Chattanooga, and Memphis. Salisbury also had dedicated service from Portsmouth, Richmond, and New York (PRR).

The capital city, Raleigh, boasted North Carolina's first railroad, an experimental installation used to move stone for the contruction of the Capitol in the early 1830's. Although Southern provided most of Raleigh's dedicated Pullman service over the years, the city was a principal stop for Seaboard Air Line trains to Florida and Alabama. SAL's standard-era *Southern States Special* and *New York-Florida Limited* paused to take on passengers, as did the *Silver* streamliners—*Star, Comet,* and *Meteor*—the *Orange Blossom Special, Sunland,* and *Palmland.* Seaboard provided Raleigh with a car to Portsmouth in the 1880's and a New York sleeper in more recent times. The Norfolk Southern, which eventually gave its name to the modern mega-rail system it became part of, carried sleepers from Raleigh to Norfolk.

The Southern Railway, before the turn of the century, scheduled sleepers from Raleigh to Morristown and Chattanooga. In the modern era, it offered dedicated Pullman service to Atlanta, Washington, and New York (PRR) and to the home-state cities of Asheville, Charlotte, Greensboro, Winston-Salem, and Wilmington (ACL).

Wilmington was once North Carolina's largest city, but it lost its population leadership as the manufacturing communities of the Piedmont gained importance over the coastal settlements. Wilmington had been at the nucleus of the original Atlantic Coast Line system, but in 1892, with the completion of its new main line from Wilson south through Fayetteville, the ACL relegated Wilmington to secondary passenger train status, even though its headquarters were located there.

Nonetheless, Wilmington had a respectable level of sleeping car service on the ACL: south to Columbia, Augusta (1890, Southern), and Atlanta (Georgia Railroad), and north to Raleigh and Greensboro (Southern), Norfolk, Washington (Richmond, Fredericksburg & Potomac), and New York (RF&P/PRR). The Seaboard scheduled Pullmans to Charlotte and Atlanta, and these traveled to Hamlet over a 78.8-mile segment of tangent track, the longest in the country.

The new ACL main line went through none of North Carolina's major cities, and some of the crack Florida trains, especially the seasonal, all-Pullman numbers that mostly made nighttime transits, did not make scheduled stops in the state at all. Those that did included the *Everglades, Miamian, Palmetto,* both *Champions,* and the *Havana Special,* generally at Rocky Mount, Wilson, and Fayetteville.

Before the merger of ACL and SAL in 1967, there was a history of cooperation between the near-parallel rivals. Although SAL's line from Portsmouth to Weldon was completed in 1837, not until 1900 did it build a line to Richmond. Until then, SAL carried its main traffic to and from the Tidewater terminal and turned over the Washington and New York cars to ACL at Weldon.

The tobacco and textile manufacturing center of Winston-Salem was one of two termini of the Norfolk & Western Railway in the state, giving it and Durham passenger connections to the west otherwise provided only by the Southern's *Carolina Special*. The N&W, known for its smart tuscan red coaches trimmed in gold, served Winston-Salem with a Cincinnati Pullman on the *Pocahontas* and, earlier, with a New York sleeper when it and the Pennsylvania shared operation of the Shenandoah Valley route through Hagerstown, Md., and Harrisburg, Pa. The *Pocahontas* also carried sleepers from Durham to Columbus, Ohio.

An unusual extension of the Shenandoah Valley service was provided by the Winston-Salem Southbound Railway, a bridge line built early in this century to an interchange with ACL at Wadesboro. Both N&W and ACL controlled this coal-hauling short line, which once was part of a New York to Florida Pullman route (PRR/N&W/WSS/ACL). Winston-Salem briefly had its own sleeper to Jacksonville over this line and by way of the Southern Railway as well.

All of Winston-Salem's other Pullman service was courtesy of the Southern, which provided sleepers to Cincinnati, Richmond, Washington, and New York (PRR), and eastward to Raleigh, Goldsboro, and Beaufort (NS).

Goldsboro, terminus of the northern section of the *Carolina Special*, was also served by the ACL line to Wilmington and the Norfolk Southern line to New Bern; as a result, it hosted a bit of interline switching of sleepers. Goldsboro's own Pullman service, however, was entirely in the hands of the Southern: to Charlotte, Asheville, Knoxville (1891), Cincinnati, St. Louis, and Chicago (NYC).

The busy Seaboard junction of Hamlet, which counted twenty-two passenger trains a day in the late 1940's, had dedicated Pullman service to New York, Washington, and Portsmouth. The main trunk from the north fanned out at Hamlet into four parts: two main lines—one going to Atlanta and Birmingham, the other to Florida—the low-grade freight line via Charleston, and the arrow-straight line to Wilmington.

Just north of Hamlet is the Sandhills area of North Carolina, where numerous recreational facilities, including dozens of golf courses, were built in the rolling country noted for its long-needle pines. The community of Southern Pines is in the heart of this winter resort area.

Early in the century, the Southern carried sleepers from Washington to Southern Pines, in cooperation with the High Point, Randleman, Asheboro & Southern, and the original Norfolk Southern.

Several decades later, Seaboard put together an all-Pullman train called the *Carolina Golfer* between New York and Southern Pines/Pinehurst. The train, which left New York after work and arrived in the resort area early the next morning, made its first trip six days before Christmas 1929 and ran for that winter season.

Sleepers for the private development of Pinehurst made the last lap of the journey over the tracks of the Norfolk Southern. Crews of this railroad had to take care to spot the Pullmans during daylight, for the corporate rules governing the posh resort prohibited night operation of locomotives.

In the 1890's, another North Carolina resort had dedicated sleeper service, from New York (PRR), Washington, and Cincinnati: the little spa of Hot Springs on the Southern's line from Asheville to Knoxville. These particular springs remained a popular gathering point until the resort hotel burned in the 1920's.

The Norfolk Southern provided some of the state's more colorful Pullman service before abandoning its passenger trains in 1948. New Bern was a major port, shipbuilding town, and distribution point that saw its importance diminish after the railroads came in. Nonetheless, the Norfolk Southern served it, bringing through town Pullmans to and from Beaufort, and originating sleepers for both Norfolk and Washington. The Norfolk car went NS all the way, which involved a trek across the 5.8-mile trestle over Albemarle Sound before skirting the Dismal Swamp on the way into Virginia. The Washington sleeper was passed to the ACL at Goldsboro.

17
South Carolina

Behind a 4-6-2, the Southern's *Piedmont Limited* (top) turns heads at Spartanburg, S.C., in 1937. An RF&P E8 (bottom) witnesses the loading of ACL's *West Coast Champion* in Richmond's Broad Street Station in 1963.

Andrew Jackson was president of a young and exuberant republic, when a trial run near Charleston helped usher in American rail transportation. The year was 1830, and the occasion was the inauguration of the South Carolina Rail Road, an enterprise planned to link Charleston with Hamburg, a small settlement across the Savannah River from Augusta, Ga. The intention of the investors was to divert to the South Carolina seaport some of the lucrative freight traffic—cotton, lumber and other commodities-that was moving down-river by barge to Savannah. During the winter of 1830-1, the railroad began hauling primitive passenger coaches with the historic engine named *Best Friend of Charleston.* At this auspicious inaugural, however, the friendship was largely unrequited, for Charleston was one of many communities unwilling to let the huffing steam monsters operate within the city limits. It seems that the railroads had yet to endear themselves to the national consciousness.

South Carolina passenger railroading eventually came to be dominated by three carriers: Southern Railway, Atlantic Coast Line, and Seaboard Air Line. All three had main lines to Florida running through South Carolina, and Southern and SAL rails to Atlanta crossed the northwestern part of the state.

Charleston, a picturesque peninsular community noted for its finely crafted 18th Century architecture, was well served by the ACL's fleet of Florida trains. By the post-war years, most of the key players were night trains: the *Champions* to both coasts of Florida, *Palmetto,* and *Everglades.* The *Miamian* and *Vacationer* were non-stops, as was the southbound *Florida Special,* according to the schedules of early 1952. Only the *Havana Special* broke the time pattern with early afternoon arrivals in both directions.

Charleston had dedicated Pullman service over the years to more than a dozen destinations, a larger selection than any other city in South Carolina. Although SAL had a low-grade freight line through Charleston, it accommodated only incidental passenger traffic. Pullman service to and from the city moved via either ACL or Southern.

ACL carried sleepers from Charleston north to Washington (Richmond, Fredericksburg & Potomac) and New York (RF&P/Pennsylvania). A circuitously routed car for Atlanta started out northbound but was shunted over to Sumter and finally Augusta, where Central of Georgia took it for the remainder of the journey. A more direct interline car traveled ACL and Charleston & Western Carolina to Augusta, thence Georgia Railroad. ACL schedules of 1917 show a sleeper running to Montgomery, Ala., but the return to Charleston involved a two-night journey, with a day's layover in Savannah.

The far-flung Southern Railway System gave Charleston some unusual Pullman service. In 1911, the Southern inaugurated the *Carolina Special,* a Cin-

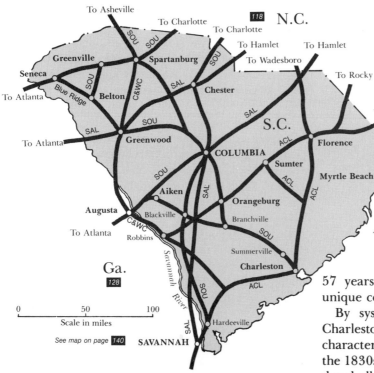

cinnati to Charleston train that later included through cars to and from Chicago via the New York Central. Despite its singular name, the train served both Carolinas, with a northern section going to Greensboro and Goldsboro, N.C., and the southern section serving Spartanburg, Columbia, and Charleston. The train was divided at Asheville. Over the years the consists varied (sleepers from Cincinnati to Columbia and Charleston were listed in 1917, Chicago-Columbia in 1925, Cincinnati-Spartanburg in 1930, and so forth), but typically they included a Chicago-Charleston car. Before the train was inaugurated, the Southern had run a Charleston-Cincinnati sleeper by way of Augusta and Atlanta (Georgia Railroad), but for the

57 years of its life, the *Carolina Special* offered a unique connection to the Cincinnati gateway.

By system standards, the Southern's line from Charleston to Branchville was itself almost of branch character, yet this historic route, which began life in the 1830s as the South Carolina Rail Road, accommodated all Southern sleepers serving Charleston. Pullmans for Augusta and Atlanta (with the Georgia Railroad) branched off to the south via Aiken, and sleepers for Columbia, Spartanburg, Greenville (via Belton), Asheville, Waynesville, N.C., and Atlanta (by an all-Southern route) took the north fork through Orangeburg. Before the turn of the century, the Southern carried a St. Louis sleeper via Augusta (Georgia Railroad/Nashville, Chattanooga & St. Louis/Louisville & Nashville) and a Charleston-Washington car the railroad handled all the way (1882). At Summerville, just northwest of Charleston, the Southern picked up a New York sleeper it turned over to the ACL.

The Southern, which had five lines radiating from Columbia, gave the capital city the majority of its dedicated Pullman service. By way of Spartanburg, sleepers rolled into Columbia from Asheville, Louisville, Cincinnati, and Chicago (NYC). Pullmans from Washington and New York (PRR) came through Charlotte. Columbia also had sleepers to Savannah and Jacksonville and several Pullman routes to Atlanta: one all-Southern by way of Greenwood and Greenville; a similar routing that used the Blue Ridge Railway from Belton to Seneca; and another via Augusta and the Georgia Railroad.

The *Aiken-Augusta Special*, which carried sleepers from New York to both of those popular winter retreats, paused in Columbia, and at one time, considerable Florida traffic passed through town on the Southern's line to Jacksonville.

Columbia, however, was served by the full flow of the Seaboard Air Line trains to Florida—*Orange Blos-*

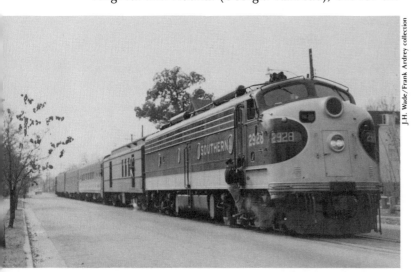

Aiken-Augusta Special departs Augusta in December 1968 with the usual half a dozen blocks of street running.

SAL's *Cotton States Special* at Birmingham in September 1946. The *Silver Comet* eventually displaced this train.

som Special, Seaboard Florida Limited, New York-Florida Limited, Silver Meteor, Silver Star, Palmland, and *Sunland.* Seaboard gave Columbia its own Pullman service, via RF&P, to Washington and New York (PRR). ACL's contribution to Columbia was a sleeper to Wilmington, N.C.

For six years, until 1899, the Southern and predecessor roads shared in the operation of through New York-Florida trains with the SAL's ancestor south of Columbia, where the interline transfer was made. This arrangement ended when the Seaboard took over the line to Savannah, but it was revived briefly in the 1920's, when the Southern and SAL exchanged Asheville-Florida Pullmans.

The Southern Railway's Washington-Atlanta trains—the *Crescent, Piedmont Limited, Southerner,* and *Peach Queen* among them—served Greenville and Spartanburg. Both cities had dedicated Pullman service to Charleston and Atlanta; in addition Greenville had a sleeper to New York, and Spartanburg one to Cincinnati.

A third Seaboard line through South Carolina, with through Pullmans between New York, Atlanta, and Birmingham, served Greenwood and Chester. This was the route of the Birmingham section of the *Southern States Special* and the streamlined *Silver Comet.*

The ACL, which maintained shops at Florence, served this major junction with a Pullman to Atlanta (Georgia Railroad).

When Myrtle Beach was still a quiet seaside community, visitors could book accommodations on a sleeper from New York delivered by the ACL.

In the ACL's busy yards at Florence, the *Everglades* gets a face wash before continuing south in August 1957.

18
Georgia

Atlanta's original designation on a railroad survey, Terminus, underscored its future significance as a transportation center. After construction of its first rail line in the 1840s, the newly founded settlement attracted numerous others, and soon tracks were radiating out of town like bicycle spokes. Two decades later, Atlanta's stature as the rail and commercial metropolis of the Southeast doomed it to the terrible siege and burning it suffered during the Civil War. The city, however, quickly rebuilt and reestablished its primacy, and as the nation's railways evolved from local to regional systems, Atlanta came to be the major crossroads of rail traffic between New York and New Orleans, and Chicago and Florida. Down through the years, passenger trains carrying first-class overnight accommodations departed from Atlanta in twelve different directions.

Atlanta had two main train stations. Terminal, with its tenants Central of Georgia, West Point Route, Seaboard Air Line, and Southern Railway, accommodated much of the New York-New Orleans traffic. The through trains at nearby Union Station, serving Atlantic Coast Line, Louisville & Nashville, the Nashville, Chattanooga & St. Louis (NC&StL), and the Georgia Railroad, had a predominately midwestern origin. Some of the Florida trains from Cincinnati and Chicago, such as the *Southland, Flamingo,* and *Dixie Flyer,* called at both stations to effect a transfer between CofG and the Nashville roads.

Terminal Station, an ornate, twin-towered structure, was opened in 1905 and served two decades before its Venetian design underwent substantial alterations. The great arched train shed was dismantled in 1925 and the towers truncated during World War II. Neither of Atlanta's two main stations survives.

Within Georgia itself, one could travel in an overnight sleeper from Atlanta to Albany, Augusta, Macon, and Waycross, and have a choice of alternate routes to Savannah, Brunswick, Thomasville, and Valdosta. From Savannah, Pullmans traveled to Augusta, Columbus, and Macon, and from Macon to Augusta and Columbus. The southern part of Georgia, especially, was interlaced with sleeping car lines as a result of the great variety of routes leading to the Jacksonville gateway.

The flagship of the Southern Railway's line to Washington was the *Crescent,* the New York to New Orleans all-Pullman limited, whose history under private operation spanned nine decades. Its precursor began service in the 1890's as an Atlanta train, but the route was soon extended to New Orleans. The *Southern Crescent,* as it was formerly named, was the South's first extra-fare train.

Southern Railway also operated the *Piedmont Limited* and the streamlined *Southerner* by way of Atlanta to the Crescent City, and it gave the Georgia capital its own trains, the *Peach Queen* and northbound *New*

May 1954, and an L&N E6 crosses a bridge near Chatsworth, Ga., with the *Southland* (top). The southbound *Ponce de Leon* (bottom) waits to leave Atlanta's Terminal Station while a Southern switcher goes about its chores. The *Flamingo* can be seen in the background.

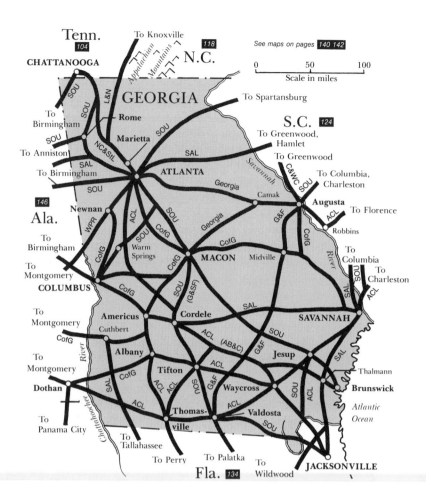

Yorker. Sleepers to Washington and New York (via Pennsylvania Railroad) were plentiful on this line, which also carried accommodations to Greenville, Spartanburg, Columbia, and Charleston, S.C., Asheville, Charlotte, Greensboro, and Raleigh, N.C., and Richmond, Va. In addition, through Pullmans operated to Boston (PRR/New Haven).

In Atlanta's Terminal Station, the Southern Railway turned over the *Crescent, Piedmont Limited,* and other New Orleans trains to the West Point Route, which moved them on to Montgomery, where the Louisville & Nashville took over. Atlanta had dedicated sleepers to Montgomery, New Orleans, and San Antonio (Southern Pacific) among these movements.

Not all Southern trains to New Orleans, however, traveled this route. The *Southerner,* a stainless steel streamliner, was a notable exception and followed Southern tracks all the way. (When the Southern continued operating its own trains rather than joining Amtrak, the *Crescent* also ran entirely on Southern iron.) At Meridian, Miss., the Southern turned over Atlanta sleepers to the Illinois Central for Jackson, Miss., Monroe and Shreveport, La., and Fort Worth (Texas & Pacific).

The Southern had a busy main line to Macon, which accommodated dedicated Pullman service

from Atlanta, in addition to substantial through traffic. South of Macon, the line divided into two routes to Jacksonville: One went via Jesup, from which the Southern had trackage rights on the ACL, and the other was the former Georgia Southern & Florida Railway route through Valdosta. By way of Jesup, Atlanta had sleeping car service to Savannah (ACL, 1893), Brunswick, and Jacksonville, as well as through service into Florida on the *Kansas City-Florida Special* and *Sunnyland* (before the latter was cut back to Atlanta). The railroad's Midwest-Florida trains-the *Royal Palms* and *Ponce de Leon*—originally moved by way of Jesup. For most of their years, however, they used the line through Valdosta, which community once had its own dedicated sleeping car from Atlanta. The Southern scheduled Atlanta-Miami Pullmans jointly with both the Florida East Coast and SAL.

Over its line to Birmingham, where it turned over the *Kansas City-Florida Special* and *Sunnyland* to the Frisco, the Southern offered Pullman service from Atlanta to Birmingham, Memphis, Kansas City, and Denver (with both Rock Island and Missouri Pacific/Denver & Rio Grande Western). Atlanta also had overnight accommodations to Greenville, Miss., via the Southern Railway and its former line across Mississippi that became the Columbus & Greenville.

In later years, the Southern's main line north to Tennessee saw little dedicated Pullman service from Atlanta, although at one time the railroad had operated cars to Chattanooga (1898), Knoxville (1890), Louisville, Cincinnati, Chicago (New York Central, 1893), and Detroit (NYC). Through traffic between Florida and the Great Lakes cities, however, remained heavy until long after World War II.

The Seaboard Air Line also carried sleepers between Atlanta and Washington (Richmond, Fredericksburg & Potomac) and New York (RF&P/PRR). A section of the standard *Southern States Special* worked this line, through to Birmingham, and with the streamline era came the *Silver Comet*. SAL also scheduled Pullmans from Atlanta to Portsmouth, Va., and Wilmington, N.C., and to Birmingham and Memphis (Frisco).

Less than two decades after its line from Atlanta to Augusta was completed in 1845, the Georgia Railroad suffered the destruction inflicted by Union forces, who removed, heated, and twisted its rails around trees, all within sight of the future Confederate monument at Stone Mountain. The railroad, whose properties were owned by a banking company, fortunately went on to more than a century of peaceful operation as a bridge route across the center of the state. The Georgia Railroad scheduled Pullmans from Atlanta to Augusta; in company with ACL to Charleston and Florence, S.C., and Wilmington, N.C.; and jointly with the Southern to Charleston and Columbia, S.C.

During the siege of Atlanta by Union troops in 1864, the Central of Georgia's line south to Macon was the last supply route into the city to remain open. Finally it, too, was captured, and the city was evacuated and burned.

South of Macon, the CofG main line headed to Americus, where such trains as the *Dixie Flyer, Dixie Limited, Flamingo,* and *Southland* briefly paused. There, the line from Birmingham joined the flow, bringing the *Floridan, Seminole,* and *City of Miami.* The CofG switched cars in Albany to the ACL for Florida points: Those going to the east coast went via the main line through Waycross (where the *South Wind* came into the main trunk over the ACL from Montgomery), and those for Gulf Coast resorts operated via Thomasville and the Perry (Fla.) Cutoff.

Atlanta had dedicated sleeping cars over CofG /ACL for Albany, Thomasville, Jacksonville, and Tampa/Sarasota. In addition, several branches off the CofG main stem took sleepers from Atlanta to Montgomery (via Americus), Panama City (Atlanta & St. Andrews Bay), Tallahassee (SAL at Cuthbert), and Savannah. The CofG and GS&F jointly carried sleeping cars to Valdosta and on into Florida, to Jacksonville (ACL), Palatka, and Tampa (with SAL in 1919

Tandem E units power a heavy holiday consist of Seaboard's *Palmland* near Savannah, July 4, 1948.

and, before the turn of the century, the ACL-predecessor Plant System).

Sleepers from the Georgia capital once ran on the former Atlanta, Birmingham & Coast Railroad by way of Cordele to Thomasville, Waycross, and Brunswick. This line became part of the ACL in 1945. Of all the Florida streamliners, the *Dixie Flagler* made exclusive use of this route, during early morning hours, in both directions.

An ancestor of the NC&StL, the Western & Atlantic, gained fame as host of the Great Locomotive Chase, a Civil War adventure that took place in the northwestern corner of Georgia. The drama started a few miles from Marietta, in April 1862, when a band of Union raiders, disguised as Confederates, stole an eight-wheel locomotive called the *General* and headed north. Their leader, a scout named James J. Andrews, was ordered to destroy the tracks behind him, thus severing the supply line from Atlanta to make the Union capture of Chattanooga easier. Soon, however, the raiders were being pursued, and after a harrowing, high-speed chase, they were eventually captured.

Over the years, the NC&StL and L&N jointly provided dedicated Pullman service from Atlanta to Chicago (Chicago & Eastern Illinois), Louisville (1895), Memphis, and St. Louis. And both carried substantial through traffic between the Midwest and Florida: The NC&StL was the vital center of the "Dixie" route, and the L&N moved the *Flamingo* and *Southland* over its Atlanta-Cincinnati main line. The NC&StL also served Atlanta with Pullmans to Chat-

R.D. Sharples/Frank Ardrey collection

The *Dixie Flyer* winds through a cut on the Nashville, Chattanooga & St. Louis near Emerson, Ga., in April 1953.

tanooga and Nashville, and the L&N scheduled sleepers to Louisville and Cincinnati. Atlanta was the southern terminus of the postwar, streamlined *Georgian,* a St. Louis train that was shifted to Chicago, through a C&EI connection. The *Georgian* carried sleepers from both cities to Atlanta.

Savannah, the oldest and for many years the largest city in Georgia, is a picturebook seaport whose brick houses and grilled gateways mirror the charm of Georgian England. The city, built on a bluff above the Savannah River, was well served by railroads and bore witness to most of the ACL and SAL trains to Florida, which included the streamlined *East Coast* and *West Coast Champions* of the ACL and the *Silver Star* and *Silver Meteor* of SAL. In the heyday of train travel, Savannah residents could book space on most of the through Pullmans operating between the Northeast and Florida resorts, from Quebec and Murray Bay, to Key West.

ACL, SAL, and Southern shared the city's Union Station. Not all the main trains stopped in Savannah, however, and SAL trains to Florida, which had to back into the terminal, in some cases used a West Savannah stop about two miles from downtown. Central of Georgia had its own station a quarter of a mile from Union.

The ACL's *Palmetto,* which once carried New York Pullmans to Brunswick and Thomasville and several Florida points, came to be mostly a New York-Georgia train, despite its apparent namesake, the Palmetto state of South Carolina. It offered Savannah sleepers to New York and Washington.

ACL carried dedicated Pullmans from Savannah to Montgomery, Charleston, and Jacksonville. In the early 1890's, the Plant System delivered sleepers to the Southern at Jesup for both Atlanta and Chicago (NYC).

Shortly after the Civil War, ACL and SAL predecessors participated in a Jacksonville line through Live Oak, Fla., until a more direct route was established. Before the turn of the century, SAL had a sleeper running from Savannah to Birmingham via Americus, where the car was switched over to CofG. SAL also delivered an Asheville sleeper to Southern at Columbia.

Over its nearly parallel route to the South Carolina capital, the Southern carried Savannah Pullmans to Columbia, Asheville, and Charlotte.

The Central of Georgia, which operated its own cars until contracting with the Pullman company in 1923, offered sleepers from Savannah to Augusta, Columbus, Macon, Birmingham, Chicago (via IC's *Seminole* at Columbus), and Cincinnati (L&N).

Augusta's long history of rail passenger service began in the early 1830's, after the South Carolina Rail Road's line (later the Southern Railway) from Charleston was completed to Hamburg, S.C., just across the Savannah River from Augusta. The Southern provided a major portion of Augusta's passenger accommodations, by way of Columbia. During the 1890's service included sleepers from Charlotte, Greensboro, and Wilmington, N.C. (ACL), Danville and Richmond, Va., and Washington (ACL). Over the SCRR charter line, the Southern carried Pullmans from Augusta to Charleston and cut out a Jacksonville car at Blackville, S.C. In more recent times, Augusta had dedicated sleepers to Asheville and Cincinnati, and, aboard the *Aiken-Augusta Special,* to Washington and New York.

The Charleston & Western Carolina's line went through Augusta, and the railroad participated in joint Pullman service to Asheville (Southern) and New York via the ACL's *Palmetto.*

The Georgia & Florida Railroad, which eventually became part of the CofG, operated the *Bon-Air Special* from Augusta to Valdosta, whence the Southern forwarded it to Jacksonville. The *Bon-Air* carried sleepers from Augusta to Jacksonville and for Miami (FEC). The Georgia Railroad carried Pullmans to Macon (1890) and through cars via Atlanta for both Chicago (NC&StL/L&N/C&EI) and St. Louis (NC&StL/IC).

Brunswick had Pullman service to New York (PRR) over the years via the three main north-south lines: Southern (1898), SAL, and ACL. In recent times Southern offered overnight accommodations to Atlanta on a line that had seen sleeping car service early this century to Birmingham, Chattanooga, and—

probably Georgia's most unusual connection—to Colorado Springs by way of the Frisco and Rock Island.

After the IC's fleet of Florida trains—the *Floridan, Seminole,* and *City of Miami*—crossed the Chattahoochee River on the CofG, they were in Columbus. These trains gave what is now the state's second largest city through Pullman service to Chicago (for a time, Columbus had its own car on the *Seminole*) and St. Louis, as well as the Florida resorts.

Columbus's other interstate sleepers operated by way of Atlanta. One line carried Pullmans to Charlotte, Washington, and New York via the Southern through Warm Springs (site of the winter White House of President Franklin D. Roosevelt, whose visits invariably brought a marvelous spectacle of all-Pullman trains). The other line, also to New York with a car for Washington, operated over a branch of the CofG to the West Point Route at Newnan, then by way of SAL. The CofG's two daily *Man o' War* stream-liners to Atlanta also followed this route.

In later years, Macon lost all its dedicated Pullman service, but as a hub of the Central of Georgia system, it saw much early morning activity in Terminal Station, where sleepers from Atlanta, Augusta, Columbus, Savannah, and Panama City were switched and put on connecting trains to their destinations. At one time the CofG scheduled sleepers from Macon to Columbus and Cincinnati (L&N).

The Southern Railway was once the principal provider of Pullman service in Macon, and this included cars to Atlanta, Chattanooga (1890), Louisville, Cincinnati, Chicago (NYC, 1893), Asheville, Washington, and New York (PRR). Sleepers from Macon to both Jacksonville and Palatka (1893) once traveled on the Southern's line through Valdosta.

Also in the distant past, the Southern Railway junction of Rome had its own sleeper service, according to 1890 schedules, to both Cincinnati and Mobile.

B.F. Roberts/Frank Ardrey Jr. collection

Carrying a flagman for Augusta street-running, ACL 50 pulls out of Union Station with a sleeper for New York, ca. 1936.

19
Florida

The Florida land boom is getting underway, and these baggage wagons and crews (top) at Miami's FEC depot bear witness to the throngs already mustering in late 1921. In a highly romantic, retouched rendering of the same period (lower left), an FEC train crosses Knights Key Bridge on the Oversea Extension to Key West. SAL's *Silver Meteor* (lower right) pauses with a load of returning Florida vacationers one night in 1961 in Raleigh, N.C.

Florida's railroad passenger service is remarkable in that it matured as late as it did. Although the northern part of the state had substantial railway activity at the time of the Civil War, the modern passenger trails weren't blazed for at least fifteen years after the entire country was spanned by rails in 1869.

In the mid-1880's Florida started attracting fugitives from northern winters, and over the years sleeping cars originating in at least forty-three different cities deposited their pallid inhabitants in the wondrously bright and plentiful sunshine of our southernmost seaboard state.

During the last hurrah of private-enterprise rail service, a host of silver streamliners and standard limiteds, some of the latter in continual operation since the previous century, raced down the Atlantic coast or made their way from the cities of midAmerica. For many of these cars, the first call in Florida was Jacksonville.

Built where the wide St. Johns River turns east toward the Atlantic, Jacksonville had been a transportation center long before the coming of the railroads. Ocean vessels served it, as did river steamers that ventured as far as 200 miles inland. When the Florida tourist flow started, Jacksonville was a ready gateway.

As Florida's northern metropolis, Jacksonville never had much in common with its resort siblings to the south. Its principal contribution to them was in sorting their clientele, switching cars among the fleets of trains that carried tourists to the winter playgrounds of both coasts.

Jacksonville Terminal, once billed as the busiest station south of Washington, D.C., was accustomed to handling more than 100 trains a day. In this 1919-vintage limestone structure, the emphasis was on switching—and not just the routine changing of locomotives from, say, Atlantic Coast Line purple to the red and yellow of the Florida East Coast Railway.

One train might roll in from the north with Pullmans from five different cities, and the cars were bound for several destinations on both coasts of Florida. Many of the sleepers had to be switched from one train to another—individually, gently, and with full services, for usually they were occupied. The process was reversed on the trip north. During World War II, the switching crews at Jacksonville handled, on the average, more than 1,200 cars a day. Holiday seasons, with their express shipments of fresh flowers and citrus fruits, put tremendous additional burdens on the station's staff.

In the winter of 1951-2, after the departure for Miami of the *Florida Special* at 7:40 a.m., fourteen more trains left Jacksonville Terminal in about a two-hour period. Most of the traffic was southbound, and, after a quiet spell in the middle of the day, the flow reversed itself. In the five hours starting about 9 p.m., seventeen trains departed. Within an hour, seven of

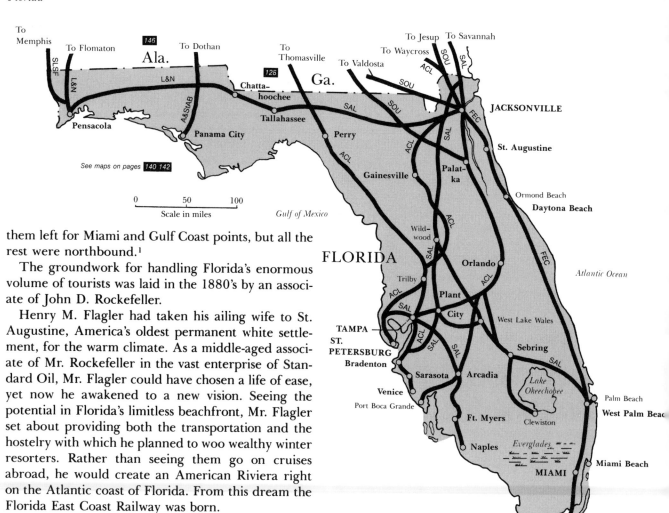

them left for Miami and Gulf Coast points, but all the rest were northbound.[1]

The groundwork for handling Florida's enormous volume of tourists was laid in the 1880's by an associate of John D. Rockefeller.

Henry M. Flagler had taken his ailing wife to St. Augustine, America's oldest permanent white settlement, for the warm climate. As a middle-aged associate of Mr. Rockefeller in the vast enterprise of Standard Oil, Mr. Flagler could have chosen a life of ease, yet now he awakened to a new vision. Seeing the potential in Florida's limitless beachfront, Mr. Flagler set about providing both the transportation and the hostelry with which he planned to woo wealthy winter resorters. Rather than seeing them go on cruises abroad, he would create an American Riviera right on the Atlantic coast of Florida. From this dream the Florida East Coast Railway was born.

In 1885 Mr. Flagler bought the little narrow-gauge line that ran from the south bank of the St. Johns River at Jacksonville to St. Augustine. There he built the Ponce de Leon and Alcazar hotels, purchased what became the Cordova, and began pushing his new railroad south. He widened the old line and in 1890 bridged the St. Johns, thus making possible through travel by rail from the north. At Ormond, which would become the winter home of Mr. Rockefeller, he bought another resort hotel.

Mr. Flagler continued to push south, reaching West Palm Beach in 1894. In the seaward community of Palm Beach, he built two more hotels: the Royal Poinciana and an ocean-front inn that became known as The Breakers. So that travelers of means would be only minimally inconvenienced, he constructed a wooden trestle for his trains across Lake Worth to Palm Beach. Sleeping cars from the north were spotted on the house tracks of each of the hotels, the entrances to which were just a brief stroll across a luxuriant carpet of grass. For several decades the shuttling of private cars and Pullmans from Boston, New York, Washington, Chicago, and St. Louis signaled the arrivals and departures of the cream of Palm Beach society. (With the Depression, the FEC fell into receivership and the once-lucrative traffic dwindled. The

branch to Palm Beach was abandoned in 1935, and the Royal Poinciana was razed the following year.)

By 1896, Mr. Flagler had extended his FEC to Miami, where, on Biscayne Bay, he built yet another hotel, the Royal Palm. Miami, which had been the site of an Indian mission under Spanish rule and a U.S. military fort during the Seminole Wars, was a community of just a handful of settlers until right before the turn of the century, when still it claimed a populace of no more than 2,000 souls. Yet now with Flagler's support system in place—a 366-mile railroad, whose main purpose was to bring the people to fill the luxury hotels along his new American Riviera—Miami witnessed rapid growth, and by 1926 its population had soared to 131,000.[2]

What Henry Flagler was to the development of Florida's Atlantic Seaboard, Henry B. Plant was to the growth of the Gulf Coast resort trade. By various means, including land grants and acquisitions, this dynamic owner of the Southern Express Company expanded his rail network across the northern part of the state. After building into Tampa in 1884, he began development of its port facilities, and in 1891, with an eye to his East Coast rival, he opened the opulent Tampa Bay Hotel. In what was doubtless the biggest event in the history of America's new cigar-making capital and Florida's third largest city, the grand opening attracted a couple thousand celebrities and members of royalty from abroad. Several years later, during the Spanish-American War, Col. Theodore Roosevelt and his Rough Riders put up at the hotel and used its grounds for training.

Around the turn of the century, numerous rail consolidations were taking place. In 1899, the Seaboard Air Line Railway acquired its line into Tampa, and three years later the Atlantic Coast Line Railroad took over the entire Plant System. With Henry Flagler's yellow and brown trains chuffing up and down Florida's East Coast, and SAL and ACL trains feeding the FEC and serving the Gulf Coast, the state's modern rail system was virtually in place, ready to handle the great migrations, seasonal and permanent, that already were forming.

Florida's rail system was *almost* in place, but for its most remarkable and regrettably short-lived component: the Oversea Extension of the FEC to Key West. Pirates were the first modern settlers to congregate on this coral island, which was well positioned to facilitate the efforts of those whose trade was commercial coitus interruptus. After the free-booters were chased out by the Navy in the 1820's, salvors moved in to avail themselves of the abundant shipwrecks on the surrounding coral reefs. This harvest reputedly made Key West, in the next decade, the wealthiest city per capita in all of America, and for much of the 1800's it was the metropolis of the Sunshine State as well. Next came Cuban cigar makers fleeing cruel Spanish overseers, fishermen from the Bahamas, the U.S. Navy, and, down through the years, a raffish assortment of inhabitants, from writers to rumrunners, that have made Key West our most free-spirited outpost.

Both the prospect of increased trade with the newly liberated Cuba and the impending construction of the Panama Canal stimulated Henry Flagler's interest in Key West because of its proximity to them. He was by then disappointed with the harbor facilities in Miami, and so he began to think very boldly about a railroad running across islands more than 100 miles out to sea, a line that would make Key West the primary gateway for Latin American freight and tourists bound for Havana.

Mr. Flagler had surveys run through the mosquito- and alligator-infested Everglades, and construction over the swampland began in 1904. Once at sea, crews built concrete bridges over 37 miles of open water, but not without great human cost: Hurricanes took the lives of more than 100 laborers. The railroad began freight service to Knights Key after opening a marine terminal there, and in 1908 passenger service was inaugurated. By then nearly 80 years old, Mr.

No. 192, the ACL's St. Petersburg section of the *West Coast Champion*, arrives in Jacksonville, July 1967.

Flagler wanted to see his project to completion and ordered his construction crews to redouble their efforts. Finally on Jan. 12, 1912, the old man rode a special train into Key West to participate in a triumphal opening. Regular service began the next day.

For 23 years, Florida East Coast trains, notably the *Havana Special* from New York, steamed down the 128 miles of the Oversea Extension. Passengers on the *Special* gathered on the observation car platform to watch the sun rise over the tiny necklace of coral islands they had just traversed and to sniff the balmy Gulf breezes that rippled the blue salt waters their Pullmans were now remarkably crossing. Upon early morning arrival in Key West, the trains pulled out onto a dock, where passengers could transfer to a waiting Peninsular & Occidental steamer bound for Havana.

With the advent of commercial air travel, the Seaboard Air Line Railway was finally able to compete for the Havana trade with its combined rail-air service via train over its new extension to Miami and the "boats with wings—Flying Yachts" of the New York Rio and Buenos Aires Line. For a time, the FEC and Pan American Airways offered almost-similar joint service. Aviation progress and the effects of the Depression, however, rapidly modified the nature of rail-air travel, and, as it happened, the Oversea Extension was running out of time anyway.

On the Labor Day weekend of 1935, the railroad that went to sea died in agony. A devastating hurricane struck the Florida Keys, killing several hundred people, many of them highway construction workers. Newspapers published a gruesome aerial photograph of an FEC train that had tried to go to their rescue, but had been blown off the tracks and its seven cars, locomotive, and tender were lying on their sides amidst matchstick debris. Forty miles of track and roadbed were in ruins. Rather than rebuilding the

Crowds cheer Henry Flagler (left center, holding straw hat) after he arrives on the first train into Key West.

exorbitantly expensive trackage, the railway abandoned it and sold the standing remains to the federal government for less than Mr. Flagler had spent constructing a mile of bridge. In 1938, the right-of-way was reopened as U.S. Highway 1, the world's longest oversea roadway. For all the saleable remnants of the forty-nine-million-dollar Oversea Extension, the FEC finally collected $804,000.[3]

In short order, the railways in Florida had set two records: The extension to Key West proved to be America's costliest short line, and the Seaboard Air Line's central Florida route to Miami had the distinction of being the last long line to be built in the country. The fact that Seaboard built on the scale it did in the 1920's—even though the country's railroad mileage was by then steadily declining—is explained by

the great Florida land boom. For years the state had been the winter retreat of the wealthy, but now growing numbers of people were seeking their place in the sun. The developers and speculators were waiting.

After World War I, traffic from the north increased enormously, nearly doubling from one season to the next, reaching a peak in the winter of 1924-5. Travelers wrote of snail-paced journeys south of Jacksonville, where FEC trains were operating with 20 or more Pullmans, frequently in four or five sections of similar consists, often stalled for hours waiting to "saw" past their counterparts coming from the opposite direction. The railroad's single-track main line was swamped, and because numerous sidings were jammed with unloaded freight cars, the company had to put an embargo on freight movements, which di-

Four sections of the *Florida Special*, the venerable ACL/FEC flyer, lined up at Miami in the winter of 1927-8.

minished the supply of construction materials the boom was devouring. Local legend has it that a desperate builder once had a carload of bricks shipped to Florida in a reefer, carded as lettuce, and iced regularly on its way down.[4]

FEC responded with a crash program to add a double track to its main line, and SAL, meanwhile, built a new line from just south of Wildwood, across the Ridge Section of central Florida, along the north shore of Lake Okeechobee, to West Palm Beach and Miami.

At the peak of Pullman traffic to Florida in the 1920s, three-quarters of a million visitors kept about 1,000 sleeping cars in constant use, many of which at the beginning and end of the season deadheaded against the predominant flow of traffic. In February 1926, the Florida East Coast was running 11 trains a day between Jacksonville and Miami.[5] Over the years, Florida drew Pullmans from places as far west as Los Angeles and West Yellowstone, Mont.,[6] and as far north as Quebec and Murray Bay.

Unfortunately, by the time the 600 dignitaries riding in five sections of the *Orange Blossom Special* could be assembled for the grand opening of the SAL's new line to Miami, in January 1927, the bubble had burst, and entire communities were disappearing in weeds. In three years' time, Pullman traffic diminished by almost 40 per cent, and the railroads themselves operated under increasing financial strain.

Despite the economic reversal, however, ACL went ahead and built its Perry Cutoff, which meant that trains from the Midwest could avoid Jacksonville, shaving hours off their travel time. The Seaboard also completed a bypass around Jacksonville for some of its premier trains from New York to use.

By the late 1930's, however, passenger traffic had returned to the point where new records were being set, and the eagerly awaited streamlined service whetted an appetite for Florida travel that proved voracious in the post-war years. The parade was long and enduring. The silvery streamliners and even the luxurious all-Pullman heavyweight trains continued to speed their way to the Sunshine State for many years after passenger service was written off elsewhere as a lost cause. The American Riviera of Mr. Flagler and the Gulf Coast resorts of Mr. Plant fulfilled their pioneering dreams many times over.

---

1. In addition to all the through sleepers it accommodated, Jacksonville had dedicated Pullman service, within Florida itself, to Arcadia/Boca Grande (SAL), Fort Myers (ACL and SAL), Key West (FEC), Miami (FEC and SAL) Orlando (ACL), Palm Beach and West Palm Beach (FEC), Pensacola (SAL/L&N), Sebring (ACL), St. Petersburg (ACL), Tallahassee (SAL), Tampa (SAL), Tampa/Sarasota (ACL).
2. Letter received April 6, 1988, from Sam Boldrick, Miami-Dade Public Library System.
3. Seth Bramson, *Speedway to Sunshine: The Story of the Florida East Coast Railway* (Erin, Ont.: Boston Mills Press, 1984), p. 121.
4. Boldrick letter.
5. Bramson, p. 104.
6. *The Official Guide of the Railways*, September 1925. This Jacksonville-to-West Yellowstone Pullman operated very briefly, via ACL/CofG/NC&StL/L&N to St. Louis, thence Wabash/Union Pacific.

William D. Middleton

William E. Griffin Jr. collection

Florida State Archives

20
Going to Florida

By the 1880's, America's burgeoning industrial order was producing sizable fortunes for many of its leaders, giving them, or at least their families, the means to enjoy substantial leisure. At the same time, the network of sleeping cars lines was growing rapidly, permitting overland travel in a degree of comfort not previously known. Both these influences combined to produce a vacation-travel boom, and in the wintertime, the exotic new destination of Florida exerted a powerful lure, especially to those seeking relief from pulmonary disorders.

Before the turn of the century and the full fleshing out of the nation's regional rail systems, the Richmond & Danville and the Florida, Central & Peninsular Railway operated through Pullman trains between New York and Tampa, with cars for St. Augustine. This arrangement, with the interline transfer in Columbia, S.C., continued for six years, until 1899, when the FC&P became part of the Seaboard Air Line system.[1]

The Southern Railway, which meanwhile had acquired the R&D and other roads, continued to offer sleeping car service between Washington and Florida on its own, but it soon bowed out and concentrated on other markets. This narrowed the field to the Atlantic Coast Line and Seaboard Air Line, both of which competed in this lucrative market until finally merging in 1967.

One minor variation on the Florida route structure went by way of the Pennsylvania Railroad through Harrisburg, Pa., down the Cumberland Valley via Norfolk & Western, then finally connecting with ACL by way of the Winston-Salem Southbound Railway. A New York-Jacksonville sleeper moved by way of this briefly used line around the time of World War I.

Over the years, ACL, which once advertised itself as "the Standard Railroad of the South," dominated the Florida trade in terms of numbers. But although ACL routinely operated its fleet of silvery streamliners at 100 mph on the double-tracked main line north of Jacksonville, SAL sometimes managed to card better times for its silver-sided flyers. The trains of both roads, in any case, were superlative examples of modern rail transportation. South of Washington the trains traveled over the Richmond, Fredericksburg & Potomac Railroad to Richmond, where the ACL and SAL took them to their own lines for the rest of the trip to Florida.

Although northern Florida was well traversed by railroads in the late Nineteenth Century, the resorts in the southern part of the state were just beginning to open for business. As a result, St. Augustine remained a major destination until early in this century. South of Jacksonville, the Florida East Coast Railway generally handled Pullmans destined for resorts along the Atlantic, at least until SAL built its new line to Miami in the 1920's and took over some of the business. Both SAL and ACL carried sleepers over their

ACL Engineer J. F. Brunson (top left) has the *Havana Special* Florida-bound south of Florence, S.C., on August 28, 1957. For years this ornate track gate sign (top right) at New York's Pennsylvania Station announced the departure of Seaboard's posh *Orange Blossom Special*. When the *Havana Special* ran to Key West, this could have been a typical scene (bottom) in its lounge car *Camaguey*.

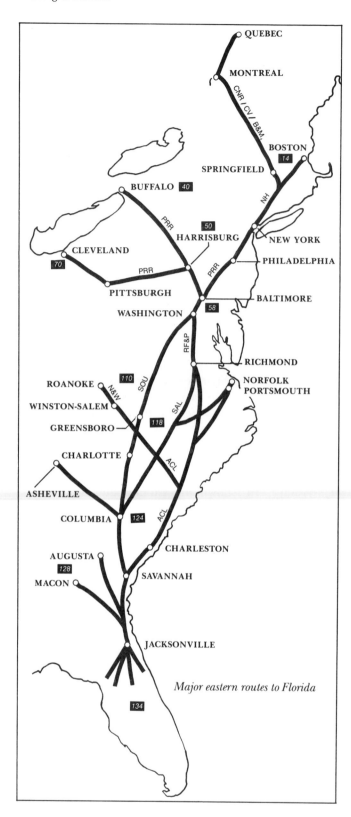

Major eastern routes to Florida

various branches to the Gulf Coast resorts.

North of Washington, the show was all the Pennsylvania Railroad's on its busy four-track main line to New York. The New York, New Haven & Hartford carried the cars to Boston, and through the additional efforts of the Boston & Maine, Central Vermont and

Canadian National Railways, Pullman service to and from Florida extended to such snowy outposts as Montreal, Quebec, and Murray Bay.[2] Cities in the northeastern United States with dedicated Pullman service to Florida included Boston and Springfield, Mass., New York, Philadelphia, Pittsburgh, Buffalo, and Washington. Southern cities with their own Pullman connections were: Norfolk, Portsmouth, and Roanoke, Va.; Asheville, Charlotte, Greensboro, and Winston-Salem, N.C.; Atlanta, Augusta, Macon, and Savannah, Ga.; Chattanooga, Memphis, and Nashville, Tenn.; Birmingham and Montgomery, Ala.; and New Orleans.

The monarch of the Florida trains was the ACL's deluxe all-Pullman express that began running in 1887-8 with the opening of Henry Flagler's Ponce de Leon Hotel in St. Augustine. This train, which came to be known as the *Florida Special,* was the first in the country equipped with both electric lighting and vestibules, and except for a brief suspension in World War II, it ran all-Pullman seasonally for most of its first 75 years. It was both fast (24 hours New York to Miami) and fun: Its consists, which numbered anywhere up to the record of seven separate sections set in February 1936, included Pullman recreation cars: hollowed out sleepers that featured bridge tables, an orchestra, and a dance floor. The cars' swaying potted palms more than sufficed until the real things came along.

The *Seaboard Florida Limited* began offering deluxe wintertime competition shortly after the turn of the century, and the SAL's similarly all-Pullman *Orange Blossom Special* made its debut in 1925, during the Florida land boom. This train, carrying many all-room cars, first rolled into Miami in early 1927 to great fanfare, celebrating the railroad's extension across central Florida to West Palm Beach and Biscayne Bay. The *Special* served resorts on both coasts with separate sections but in its later years concentrated on the Miami market. After World War II, during which time service was suspended, the *Special* was equipped with heavyweight Pullmans painted maroon and gray, sleepers that in the summertime made up the *Bar Harbor Express* to Maine.[3] The ACL may have had longevity on its side, but the SAL's luxury trains gave no quarter to any of its competitor's. "Less than 44 hours from New York to Havana," advised the timetables of late 1929 as they touted the combined service of the ACL/FEC *Havana Special* and the steamships of the Peninsular & Occidental lines. The *Havana Special* rode the rails of Henry Flagler's Oversea Extension of the FEC to Key West, where it offered dockside transfer to a steamship bound for Cuba. Pullman travel over the 100-odd miles of sea and keys was charming (Mr. Flagler once envisioned ferrying cars all the way to Havana, but decided

Having departed Richmond Broad Street, the northbound *Silver Star* heads for Washington on the RF&P. A SAL freight is at left.

against it), though on occasion, when the winds gave a preview of the horrible fate that awaited the Extension, the Caribbean roiled in gray, and train movements were canceled. After the Labor Day hurricane of 1935 destroyed the oversea trackage, the *Havana Special* was cut back to Miami. One of the train sets, stranded in Key West, which had escaped the main fury of the storm, was finally barged back to Miami to the impertinent newspaper headlines of "FEC's *Havana Special* Arrives From Key West Months Late."[4]

In the heyday of the oversea service, there was plenty of Havana-bound passenger traffic for everybody.

Seaboard's *Silver Meteor* carried this glass-roofed car as an answer to the inadequate dome clearances in the East.

Both the *Florida Special* and *Seaboard Florida Limited* carried Pullmans for Key West, and by the end of the Roaring '20s, both SAL and ACL were offering air/rail service to Cuba from Miami. The *Limited* arrived at 8:30 a.m., and passengers were shuttled over to Biscayne Bay, where they boarded a gently bobbing "boat with wings . . . capable of landing upon and riding any Sea, insuring perfect safety." This wondrous new seaplane took off for Havana at 9. The same season, ACL was advertising its all-Pullman *Everglades* as a fine, fast new train, connecting at Miami with Pan-American Airways for Havana, Nassau, and the West Indies.

And the parade of heavyweight luxury trains continued: the ACL/FEC *Florida East Coast Limited*, replaced in the late '20s by the *Miamian*, which commonly ran double sections through the winter; the ACL's *Gulf Coast Limited* and *Vacationer*, and numerous other trains of lesser credentials.

In February 1939, SAL inaugurated the first Florida streamliner, the *Silver Meteor*, a Budd-built all-coach consist of seven stainless steel cars hauled by EMD E4's dressed in a citrus-colored design of green, yellow, and orange. The competition was heating up, and by the end of the year ACL and FEC were offering the stainless steel *Champion*. Both trains stepped up their frequency, Pullmans were added, and service expanded to Gulf Coast resorts. ACL inaugurated a separate *West Coast Champion* in 1946, and SAL added the *Silver Star* to its stainless fleet.

The exclusive heavyweights faded in the 1960's, but the silver-sided Florida streamliners rolled into the age of Amtrak and kept right on going.

From Chicago

Over the years, Chicago-to-Florida trains followed four principal routes, with minor variations. Because no one railroad extended the entire distance, all were joint operations.

Of all the midwestern routes, the oldest and the busiest was the "Dixie" way: Chicago & Eastern Illinois to Evansville; Louisville & Nashville to Nashville; Nashville, Chattanooga & St. Louis to Atlanta; then ACL to Jacksonville (often with the help of Central of Georgia or CofG/Georgia, Southern & Florida in combination. The GS&F later became part of the Southern.). The flagship of this route was the *Dixie Flyer,* a Florida train whose components dated to 1885. In the 1920's, the *Dixie Flyer* ran as an all-Pullman special serving both coasts of Florida. The *Dixie Limited,* inaugurated in 1913, was another of the all-Pullman trains, and eventually this famous fleet grew to include the *Dixie Express,* the *Dixieland, Dixie Mail, Dixiana,* and the *Dixie Flagler.*

The Illinois Central, which for several years early in the century operated the Chicago and St. Louis legs of the *Dixie Flyer,* assembled a Florida route of its own by acquiring trackage rights into Birmingham in 1908 and control of the Central of Georgia the following year. That autumn, the *Seminole* became the first year-round train to offer sleeping car service between Chicago, St. Louis and Florida. ACL took the cars to Jacksonville and Gulf Coast points and handed off others to FEC. During the 1920's land boom, the luxurious all-Pullman *Floridan* went into wintertime service on this route, offering accommodations from Chicago and St. Louis to Miami, Palm Beach, and several Gulf Coast resorts.

Despite their essentially east-west character, both the New York Central and Pennsylvania were long involved in the Chicago to Florida passenger traffic. The Central turned over its Pullmans to the Southern at Cincinnati for the remainder of the trip to Jacksonville. Dating from before World War I, the Central's year-round *Royal Palm* at first provided service from Chicago, then, as traffic expanded, the list of cities grew to include Detroit, Cleveland, Buffalo, Grand Rapids, Indianapolis, and of course Cincinnati. In the 1920's, the *Royal Palm De Luxe* was added as a seasonal, all-Pullman second section, and the streamlined *New Royal Palm* bolstered post-World-War-II service. In 1921, the *Suwanee River Special* began operating from several midwestern cities to Florida's West Coast. The *Ponce De Leon* served Miami year-round.

The Pennsylvania began operating the *Southland* in

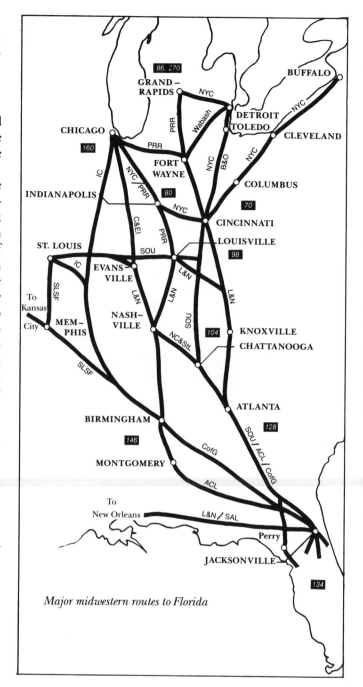

Major midwestern routes to Florida

December 1915; its primary routing took it from Chicago to Cincinnati (for a time through Fort Wayne rather than Logansport, Ind.), then via L&N to Atlanta, and CofG and ACL to Jacksonville. An Indianapolis car operated through Louisville. After the ACL's Perry cutoff was opened in 1928, the train focused on Gulf Coast service with sleepers from Chicago, Detroit (in cooperation with the Wabash), Grand Rapids, Cleveland, and Cincinnati. The Pennsy's section of the *Flamingo*—a joint operation with the New York Central—ran by way of Indianapolis and Louisville, whence the L&N forwarded it to Atlanta. The PRR handled Chicago Pullmans for Jacksonville and Mi-

Flanked by tracks of a local trolley line, the northbound *City of Miami* leaves Birmingham on July 4, 1950.

ami, and the Central forwarded sleepers from Detroit and Cleveland.

In 1940, nine railroads joined together to offer streamlined service over three different routes from Chicago to Miami. The trains, each of which was initially operated every third day, were created in response to criticism that midwestern service was inferior to that offered eastern cities. Although the streamliners originally were all-coach, they later incorporated Pullmans for first-class passengers. Generally they operated every third day off season, and sometimes two of three days or alternate days in wintertime.

The Illinois Central entered the Florida streamliner race in colorful beachwear with the new *City of Miami* done up in green and orange with red striping in an effect that made its EMD diesel appear to be breaking waves. After that splash, however, the IC opted for the more sedate, systemwide dress of brown and orange with yellow striping. Later in its career, the *City of Miami* carried dome cars borrowed from the Northern Pacific during wintertime. At the beginning and end of the season, the IC dutifully repainted the cars in the appropriate road's colors. The *Seminole* came to concentrate on the Gulf Coast resorts.

Another member of the 1940 trio was the *South Wind,* painted in the PRR's Tuscan red, and operated via Louisville, then over the L&N to Montgomery, where the ACL took over. The third streamliner was dressed in stainless steel but hauled by a C&EI steam locomotive in a custom-made shroud painted black with aluminum and red trim. This was the *Dixie Flagler,* which traveled the L&N/NC&StL route via Atlanta, then Atlanta, Birmingham & Coast/ACL-/FEC.

In addition to the C&EI, IC, PRR, and NYC, the Monon briefly scheduled Chicago-to-St. Augustine

sleeping car service in the early part of the century (B&O/Southern/FEC).

From St. Louis

Over the years, sleeping car passengers could leave St. Louis for Florida over four principal routes: L&N, Frisco, Southern, and the sole-surviving IC.

The L&N teamed with ACL for the Jacksonville service: around the turn of the century via Montgomery and later via NC&StL through Atlanta, with the Central of Georgia and the Southern (GS&F) sometimes assisting on both lines. St. Augustine, Palm Beach, and Miami sleepers also traveled via Atlanta and eventually FEC. Around the time of World War I, a Pullman operated from St. Louis to Pensacola all the way on the L&N.

Frisco carried St. Louis sleepers to Pensacola and exchanged cars for Jacksonville and St. Petersburg (SAL) with the Southern at Birmingham. The Southern also offered service from St. Louis to Jacksonville over its own system.

The IC was the last railroad hauling Florida-bound Pullmans out of St. Louis' mammoth Union Station. At one time, the IC had tapped into the Dixie route by turning sleepers over to the NC&StL at Martin, Tenn., to be forwarded to the *Dixie Flyer* at Nashville. But after getting its route through Birmingham, it put St. Louis cars aboard Chicago trains at Carbondale, Ill. St. Louis-Miami Pullmans rode the *City of Miami;* those to Tampa/Sarasota and St. Petersburg traveled on the *Seminole.*

From other cities

The Frisco's *Kansas City-Florida Special* and the *Sunnyland* turned over Kansas City Pullmans bound for the Sunshine State to the Southern at Birmingham. In 1926, the Frisco originated thrice-weekly sleepers to

Two other streamlined trains alternated with the *City of Miami* in Midwest-Florida service: the *South Wind* (above), here with L&N No. 275, and the *Dixie Flagler* (right), on the C&EI rails.

Jacksonville from, alternately, Oklahoma City and Wichita. Florida Pullmans came to Kansas City from as far west as Denver and Colorado Springs over the Rock Island.

Cincinnati was a great gateway where the midwestern lines turned over their Florida-bound Pullmans to either the L&N or the Southern, which carried the greater share of the traffic. In addition to its main routing by way of Chattanooga and Atlanta, the Southern also carried Florida sleepers through the Carolinas via Asheville.

Both Cleveland and Buffalo had eastern and midwestern sleeping car service to Florida. The New York Central served Cleveland by way of Cincinnati, turning over most of the traffic to the Southern, but at least in one instance to the L&N (CofG/GS&F/ACL). The Pennsy also offered connections to the Southern at Cincinnati, and both it and the B&O carried Cleveland-Florida sleepers by way of Washington. Buffalo residents mostly went south through Cincinnati via the Central, although the PRR once offered service by way of Washington.

From Detroit, Florida Pullman cars traveled on either the Central, B&O, or a combination of Wabash-PRR to Cincinnati, where the Southern took over.

Louisville had its own sleepers to Florida as well as a borrowed NP dome car in the final days of the *South Wind*.[5]

No doubt the most unusual service to the Sunshine State came on the heels of the great land boom of the 1920s.

A Jacksonville-Los Angeles sleeper, which was carried via SAL and L&N to New Orleans, then on the Southern Pacific's *Sunset Limited,* operated for several years in the late 1920's. (Jacksonville-New Orleans service, of course, was long-standing, and the streamlined *Gulf Wind* continued the tradition after World War II.)

Perhaps even more extraordinary was the Jacksonville-West Yellowstone Pullman provided in the summer of 1925 via the Dixie route to St. Louis, then Wabash and Union Pacific to Montana. Even in the prosperous 1920s, there must have been limited demand for its berths. One would think travelers would pause, at least briefly to see their hometown broker, before exchanging one season's playground for another.

---

1. William H. Patterson, "Through the Heart of the South: A History of the Seaboard Air Line Railroad Company, 1832-1950," Diss. University of South Carolina 1951.

2. Seth Bramson, *Speedway to Sunshine: The Story of the Florida East Coast Railway* (Erin, Ont.: Boston Mills Press, 1984), p. 211.

3. Arthur D. Dubin, *More Classic Trains* (Milwaukee: Kalmbach, 1974), p. 223.

4. Pat Parks, *The Railroad That Died at Sea: The Florida East Coast's Key West Extension* (Brattleboro, Vt.: Stephen Greene Press, 1968), p. 4.

5. Letter received April 8, 1988, from Sam Boldrick, Miami-Dade Public Library System.

21
Alabama

Alabama's main railroad passenger lines mostly followed the pattern of a giant "X" as the routes from the Northeast to New Orleans intersected those from the Midwest to the Gulf of Mexico and Florida. One exception was the Louisville & Nashville's main line from Cincinnati, which came straight out of the north like a Yankee frost, connecting Birmingham with Montgomery before curving west to pass through Mobile and continue along the Gulf Coast to New Orleans.

Birmingham, the state's largest city, was itself the product of a railroad intersection. Whereas the comparatively ancient seaport of Mobile was founded in 1699, and Montgomery in 1819, Birmingham was not even a dot on the maps until after the Civil War. In 1871, developers of the town started peddling lots in anticipation of the growth they expected to come with the crossing of the new L&N line by the Chattanooga-to-New-Orleans tracks of the future Southern Railway.

Steel was the industrial wonder of the era, and the rail entrepreneurs looked with great interest upon the abundance of its three principal ingredients, found in the footings of Birmingham: the iron ore of Red Mountain, and coal and limestone from the ground below. By 1872, the railroad X had marked the spot, and the rapid and boisterous growth of Birmingham, as the hub of Alabama railroading and "the Pittsburgh of the South," was under way.

From this city of Vulcan, six of its eight major rail roads carried sleepers to eleven different points on the compass, the L&N from its own station, and the Southern, Seaboard Air Line, Central of Georgia, St. Louis-San Francisco, and Illinois Central from Terminal Station on Fifth Avenue. Pullmans originated in Birmingham for destinations as far northwest as Denver and as far northeast as New York.

The flagship of the main line of the L&N from Nashville was the luxurious, onetime all-Pullman *Pan-American*. Other trains included the heavyweight *New Orleans Limited*, the post-war streamliner *Humming Bird*, the *Azalean*, and the Florida-bound *South Wind*, which turned over its cars to the ACL in Montgomery. The L&N brought Pullmans to Birmingham from New York (Pennsylvania Railroad), Nashville, Chicago (Chicago & Eastern Illinois), Mobile, Pensacola, and New Orleans.

As late as the early 1950's, travelers from Birmingham had a choice of four different Pullman lines on five trains to New York City, the L&N's *Azalean* being one.

The Southern Railway offered several alternatives to New York, including the fastest train of all, the streamlined *Southerner*, whose Birmingham sleeper was an exception to the original all-coach consist operating by way of Atlanta. A night train carried an Atlanta sleeper that also continued on to New York (PRR).

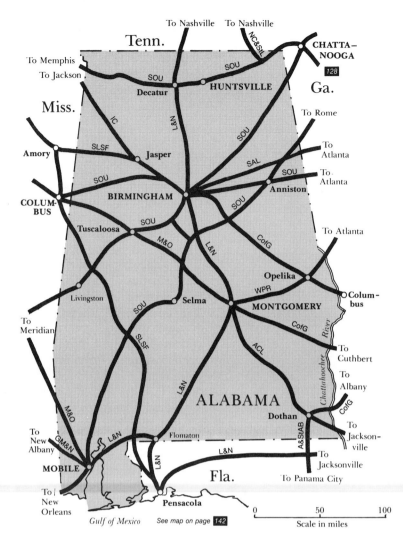

Gulf of Mexico   See map on page **142**

Scale in miles
0   50   100

Early in the century, service on the Southern's line through Atlanta included sleepers from Birmingham to Brunswick, Ga., Charlotte, N.C., and Richmond, Va. Pullmans for Washington, D.C., and Jacksonville operated in more recent times.

The Seaboard's *Silver Comet*, a train of stainless steel cars hauled by orange, yellow, and green diesels, also carried sleepers from Birmingham to Washington (Richmond, Fredericksburg & Potomac) and New York (RF&P/PRR). Previously on this route, SAL had operated a section of the *Southern States Special* and *Cotton States Special*. The railroad scheduled Pullmans from Birmingham to Atlanta and, around the turn of the century, Portsmouth, Va.

The Southern's other major trains to New York operated through Chattanooga and on the railroad's joint route with the Norfolk & Western. The *Birmingham Special* carried Pullmans from Birmingham for Washington and New York (PRR) and proved inspirational in the process. After one trip on the train, Harry Warren wrote a song with Mack Gordon that Glenn Miller made into the enormously popular hit of 1941: *Chattanooga Choo Choo.*

The *Pelican*, operating from New Orleans to New York, carried through sleepers. Birmingham also had Pullman service to Asheville, N.C., via this line.

North of Chattanooga, the Southern's Queen & Crescent Route continued to Cincinnati (the Queen City), and the premier limited working these tracks during the heavyweight era was given the Q&C name. The companion *New Orleans/Cincinnati Specials* carried sleepers from Birmingham to Cincinnati, and before the turn of the century Q&C trains incorporated sleeping cars to Chattanooga and Chicago (New York Central).

Both the *City of Miami* and *South Wind*, two of the three Chicago-to-Florida streamliners jointly created by nine railroads in 1940, operated by way of Birmingham, offering through Pullmans to Jacksonville and Miami. The *City* and the IC's *Seminole* left town on the Central of Georgia tracks, a route that early in the century offered Birmingham its own sleepers to Savannah and Jacksonville (Atlantic Coast Line) and later accommodated the luxurious, all-Pullman *Floridan* from Chicago and St. Louis. On its regular run to the Gulf Coast resorts of Florida, the *Seminole* also brought the IC's Birmingham-Chicago Pullman. For a time after the IC began Florida service in 1909,

J. Parker Lamb

The northbound *Crescent* rolls through Loachapoka, Ala., on the WPR; RPO drops the mail and collects the outgoing bag (right).

schedules included a sleeper to its junction at Fulton, Ky. (Many early sleeping car lines terminated at rather rustic crossings, where passengers transferred to other trains passing through.)

The *South Wind* operated on the L&N to Montgomery, Alabama's capital, where the ACL took over.

Frisco's entry into Birmingham in 1901 resulted in a unique linkage between cities in the Central Plains states and Florida. The *Kansas City-Florida Special* and the *Sunnyland*, jointly operated with the Southern Railway, which handled the trains east of Birmingham, gave the city dedicated Pullman service to Memphis, St. Louis, Springfield, Mo., Oklahoma City, Kansas City, and Denver (Rock Island), as well as through service to and from Miami and the Gulf Coast of Florida.

The Southern's line to New Orleans by way of Tuscaloosa saw Queen & Crescent Route trains from Cincinnati as well as through traffic from New York. The *Pelican* and predecessor *Washington / New Orleans Limited* worked this line, as did the famous *Crescent*, after the Southern chose to remain out of Amtrak and operated the train entirely on its own. Pullmans from Birmingham traveled this route to the Crescent City as well as several other destinations by way of Meridian, Miss., where the IC picked up cars for Jackson, Miss., and Shreveport, La., and, shortly after the turn of the century, the Mobile & Ohio handled those for Mobile. About the same time, the Southern also was carrying Pullmans to Mobile on its own line via Selma.

Before the Southern disposed of its Mississippi line that became the Columbus and Greenville Railway in the early 1920s, it ran a sleeping car from Birming-

ham through Columbus to Greenville and another to Winona, Miss. (1893).

The first capital of the Confederacy, Montgomery was an important junction on the main passenger route from New York to New Orleans. In 1836, Montgomery interests started to build a railroad to Georgia, which was finally completed in 1851. This was the Western Railway of Alabama portion of the West Point Route, which bridged the lines of the Southern and L&N, eventually carrying such trains as the *Crescent* and *Piedmont Limited*. In addition to moving numerous through sleepers, the WPR scheduled Pullmans from Montgomery to Atlanta and, with the

J. Parker Lamb

The Seaboard E6 will soon pull out with the *Silver Comet* from Birmingham's station, noted for its Moorish design. July 1959.

In Central of Georgia territory, the *Seminole* speeds through a cut north of Opelika, Ala., in February 1955.

J. Parker Lamb

Southern, to Washington and New York (PRR).

For a time, the Central of Georgia also carrried an Atlanta sleeper, by way of Americus and Macon, Ga. Over the ACL from Montgomery, the *South Wind* carried through Florida Pullmans, and the railroad also scheduled a sleeping car from Montgomery to Savannah.

Before the turn of the century, Montgomery had a short-lived sleeper service to Tampa by a round-about routing through Pensacola, involving the L&N and SAL. Accommodations on the L&N to Pensacola itself and Mobile were more enduring. On its line to the north, the L&N carried Pullmans from Montgomery to Nashville, Louisville, Cincinnati, St. Louis, and Chicago (C&EI). The Gulf, Mobile & Ohio also briefly offered sleepers between Montgomery and St. Louis.

Being on the L&N's Crescent route, Mobile had Pullman service to New York (L&N/WPR/Southern/PRR) and New Orleans as well as Montgomery, Birmingham, and Cincinnati. This major seaport of Alabama, moreover, once saw considerable activity on a secondary line of the Southern: Sleeping cars departed from Mobile for Selma, Birmingham, Rome, Ga., Knoxville, Cincinnati, Chicago (NYC), Asheville, N.C., and Lynchburg, Va. (N&W), although most of this service had ended by the turn of the century.

Mobile was the home office of the Gulf, Mobile & Ohio, a modern Gulf-to-St. Louis rail system born in 1940, with the consolidation of the Mobile & Ohio and Gulf, Mobile & Northern. The GM&O extended its reach to Kansas City and Chicago after the merger with the Alton Railroad in 1947. Thereafter, all the new road's passenger trains, whether dashing across the soybean fields of Illinois or gliding through the

Frank Ardrey

A doubleheaded *Kansas City-Florida Special* arrives in Birmingham on Southern iron against a backdrop of steel blast furnaces.

With the city skyline in the background, L&N's *Pan-American* departs Birmingham for the north on May 23, 1948.

pine woods of the Deep South, wore the classic Alton livery of red and maroon, with gold trim.

Both of the GM&O's predecessor southern lines carried overnight traffic to and from Mobile. The flagship of the Mobile & Ohio was the *Gulf Coast Special*, a Mobile-St. Louis train that also carried sleepers to Chicago (IC from Cairo). Under the GM&O colors, the *Special* became the *Gulf Coast Rebel*, and the Chicago cars moved via the railroad's former Alton line. The M&O also worked interline transfers with the Southern at Meridian, scheduling Pullmans from Mobile to Birmingham, Chattanooga, and Cincinnati (1898), and with the Frisco at Tupelo, Miss., for sleepers to Memphis. The GM&N, which had pioneered the *Rebel* streamliner in 1935, scheduled sleeping cars to both Jacksons—Mississippi and Tennessee.

Although Birmingham, Montgomery, and Mobile dispatched virtually all of Alabama's Pullman traffic, the *Birmingham Special* carried one exception when operating by way of Atlanta, early in the Depression: an observation-sleeper between Selma and Washington, D.C.

The Southern also provided Anniston its own Pullman to New York in the mid-1930's.

Several lines made brief sojourns into Alabama with mostly out-of-state traffic. The busiest was the main line of the Nashville, Chattanooga & St. Louis Railway between its first two namesake cities. Over those tracks, which creased the northeast corner of Alabama, rolled the Florida-bound Dixie trains.

By early 1952, Pensacola's sleeper service had dwindled down to just a Frisco car that crossed southwestern Alabama on its way to and from Memphis. The St. Louis sleeper over the SLSF was by then long

Mobile's ornate Mobile & Ohio—later GM&O—station, built in 1907, was remodeled into an office building.

gone. The last Birmingham-Pensacola run that February ended dedicated service on the L&N that had included Pullmans, operating via Flomaton, all the way to Chicago (C&EI), Cincinnati, Louisville, New Orleans, St. Louis, Montgomery, and Jacksonville.

The Southern's Memphis line went through Huntsville and Decatur carrying the *Memphis Special* and streamlined *Tennessean*, and in the state's southeastern corner, the tiny Atlanta & Saint Andrews Bay Railway and the Central of Georgia exchanged Panama City-Atlanta Pullmans in Dothan.

22
Mississippi

GM&O's southbound *Gulf Coast Rebel* (top left) heads out of Meridian behind a pair of Baldwin locomotives in May 1951. On a December day in 1958, the L&N's northbound *Azalean* (top right) eases into the Ocean Springs station. No. 102 (bottom), one of two northbound IC trains carrying Pullmans to the main line at Jackson, departs Gulfport in April 1932 with a Memphis sleeper.

For passenger travel, the "Main Line of Mid-America" was also the main line of Mississippi. On a map of the state, the Illinois Central tracks almost could have been mistaken for the line marking Longitude 90 west. The principal route of the IC came straight north from the Louisiana bayous to the capital city of Jackson, then continued on through central Mississippi to Memphis and beyond.

This was the speedway for such crack trains as the *Panama Limited* and *City of New Orleans.* It was also the main road north for blacks who migrated from Louisiana and Mississippi to Chicago and other midwestern cities during and after World War II. The IC was not the only passenger carrier in Mississippi, but it dominated train travel there like no other line.

Originally, as the nation's first land-grant railroad, the Illinois Central was confined to its home state. But with a unique north-south line among all the other, mostly east-west railroads, the IC quickly saw the advantages of expansion to the south.

From its terminus at Cairo, Ill., in 1858, the railroad inaugurated a fleet of packets that carried freight and passengers down the Mississippi to New Orleans. The steamboat line was set up by the IC's chief engineer, George B. McClellan, of eventual Civil War fame. One of its cub pilots would later earn his living writing under the name of Mark Twain.[1] (The IC must have been an inspiring place for name-droppers to have worked; Abraham Lincoln was an attorney for the road.)

Early rail construction in Mississippi, which dated from the 1830's, was like that in many other eastern states: Railways were constructed primarily as adjuncts to the riverboat routes, and of course the Mississippi River was the principal artery in this part of the Deep South. After the Civil War hiatus, steamboats again flourished, especially in states like Mississippi, where many of the railroads had been destroyed. But their revival was short-lived, and the railway revolution irresistable.

In the 1870's, the Illinois Central consolidated several southern railway lines, principally in Mississippi. After bridging the Ohio River at Cairo the following decade, the IC had an all-rail route running from the Great Lakes to the Gulf. One of the lines it acquired also traversed Mississippi from north to south, in the western part of the state near the river. With the large network of rails that came under IC management, most of the sleeping car business in Mississippi came with it.

Jackson, the state's largest city, had the most service, although it was not extensive by many standards. The IC brought dedicated sleepers from Chicago, Memphis (via the main line through Grenada, and by way of Clarksdale), and New Orleans. In addition, the IC delivered Birmingham and Atlanta Pullmans to the Southern at Meridian. Jackson's two other sleeper

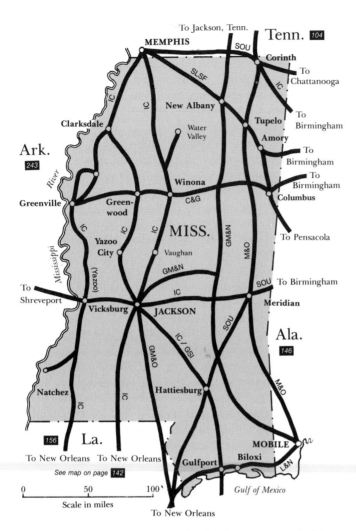

lines, to Jackson, Tenn., and Mobile, were carried by the Gulf, Mobile & Northern, the route of the *Rebel*.

The junction of Meridian was on the Southern's line to New Orleans, bringing through the city such trains as the *Queen & Crescent Limited*, *Pelican*, and the Amtrak-era version of the railroad's famed *Crescent*. For many years, the Southern and IC exchanged interline Pullmans in Meridian, notably New York cars to Vicksburg and Shreveport (Pennsylvania/South-

ern/Norfolk & Western/Southern/IC). Dedicated service to Meridian—sleepers on the Southern from Cincinnati and on the IC from Vicksburg and New Orleans—was gone by the early part of this century.

The Mobile & Ohio's main line in eastern Mississippi also came through Meridian. The M&O's *Gulf Coast Special* (and later, Gulf, Mobile & Ohio's *Gulf Coast Rebel*) paused on their way to Mobile.

The river port of Greenville, in the prime cotton-growing country, had Pullman service to both Memphis and Chicago on the IC, and to Birmingham and Atlanta on a line running across north-central Mississippi that once was part of the Southern Railway System. After the Southern disposed of it in the 1920's, the line began operation as the Columbus & Greenville, with passenger coaches of tuscan red. The Southern handled the C&G sleepers east of Columbus.

Before the Southern gave up the C&G line, it had scheduled a sleeper from Winona, on the IC main line, to Birmingham (1893).

Gulfport originated as a planned community, with wide streets lined with palms, when the harbor facilities were opened there in the early part of this century. The IC subsidiary Gulf & Ship Island Railroad served the new port with tracks from the main line at Jackson. During the 1920's, the railroad began Pullman service to Gulfport from Chicago, Memphis, Cincinnati (Baltimore & Ohio), and St. Louis, much of it on the *Panama Limited*. Motor bus service provided connections to many of the Gulf Coast resorts.

Vicksburg, the city on bluffs that was known as the Gibraltar of the Confederacy, had sleeping car service on the IC to New Orleans and Louisville (1890), in addition to the New York and Meridian sleepers. In 1882, the railroad also carried a sleeping car for Chattanooga that the Southern took over at Meridian.

The ubiquitous IC also scheduled Pullmans to Greenwood, Natchez, and Yazoo City from Memphis, and from Natchez and Water Valley to New Orleans.

Other trunk lines went through Mississippi, but with only through sleeping car service. The Southern's *Tennesseean* and the Florida trains of the Frisco and IC creased the northeast corner of the state, the *Kansas City-Florida Limited* and *Sunnyland* stopping in Tupelo, and the *Seminole* and *City of Miami* in Corinth. The Gulf Coast of Mississippi was served by the crack trains of the L&N going to and from New Orleans: the *Crescent*, *Pan-American*, *Humming Bird*, and *Piedmont Limited*.

The L&N was semi-nautical in these parts, making a mile-and-a-quarter crossing of Biloxi Bay, and one of two miles over St. Louis Bay on trestles consistently buffeted by wind and waves. Hurricane Betsy, in 1965, destroyed thirty-three miles of track, and Hurricane Camille four years later was even more destructive, carrying away entire bridges.[2]

Heading for the joint Southern/Norfolk & Western route to Washington, the *Pelican* departs Birmingham in April 1948.

The Gulf Coast has a surprise snowfall in which the combined *Gulf Wind/Pan-American/Piedmont Limited* makes swirls near Biloxi in December 1963.

The most famous night-train run in Mississippi, maybe in all the land, came on the last day of April 1900 on the main line of the Illinois Central. The express popularly known as the *Cannonball* was running more than an hour and half late out of Memphis, and its engineer, a last-minute fill-in doing double-duty, name of John Luther "Casey" Jones, had been ordered to get back on schedule before he reached Jackson. At the town of Vaughan, a freight train was to have moved onto a siding so the *Cannonball* could pass. Part of the freight, however, was still hanging out on the main track when Casey's train loomed on the scene. Too late to stop, the engineer ordered his fireman to jump before the *Cannonball* crashed into the rear of the freight. Casey died in the wreck, and the legend began.

1. Carlton J. Corliss, *Main Line of Mid-America* (New York: Creative Age Press, 1950), pp. 100, 154.
2. J.G. Lachaussee and J. Parker Lamb, "The Railroad That Walks on Water," *Trains*, Vol. 47, No. 3 (January 1987), p. 40.

Southern's *Washington-Atlanta & New Orleans Express* carried both sleepers and diner. Here it's leaving Alexandria, Va., in 1938.

J. Parker Lamb

J. Parker Lamb

23
Louisiana

The L&N (top) went to great lengths crossing the waters of the Gulf, not to mention repairing their damage when hurricane-force winds swept them up in a fury. The southbound *Humming Bird* here rumbles across the Pascagoula River drawbridge in December 1961. Two summers later in Shreveport (bottom), a Kansas City Southern switcher puts finishing touches on the consist of the northbound *Flying Crow*. IC's *Northeastern Limited* has just departed the station for Meridian.

What Chicago was on a larger scale, New Orleans was to the rail lines across the southern United States. Railroads fanned out from the Crescent City to both coasts and numerous northern cities, giving this great Gulf port direct passenger service to all parts of the country but the Pacific Northwest. In all cases, New Orleans was a terminus of the lines serving it. From the northeast came the Louisville & Nashville and Southern Railway; from the north the Gulf, Mobile & Ohio and Illinois Central; from the northwest the Kansas City Southern, Missouri Pacific, and Texas & Pacific; and over the Sunset Route from the west, the Southern Pacific.

Given this reach, New Orleans had Pullman service to more than fifty cities in a wide arc ranging from Jacksonville, Boston, Chicago, Denver, and San Francisco, to San Diego. Over the years, service involved at least six different routes to New York.

In 1891, a predecessor of the Southern inaugurated the *Washington and Southwestern Vestibuled Limited*, a luxurious extra-fare train carrying the new Pullman Palace cars with the enclosed ends. Originally terminating in Atlanta, the train was extended to New Orleans over the West Point Route and the L&N by way of Montgomery.

The *"Vestibule"* evolved into the *New York & New Orleans Limited* and finally, in 1925, the new *Crescent Limited*. About this time, Southern president Fairfax Harrison was taking a fancy to the colorful green locomotives of the London & North Eastern line in England. On his return home, he had some of the Southern's locomotives painted forest green and trimmed in gold, silver and red.[1] In 1929, Pullman delivered new equipment for the *Crescent Limited*, all of it painted in the same elegant green and gold. Interior amenities of the extra-fare, all-Pullman train included showers, maid and valet service, and an observation car with women's lounging rooms.

The *Crescent* (the name was shortened in 1938) got green and white diesel engines in 1941 and after World War II a string of stainless steel cars. For many years it remained all-Pullman between New York and Atlanta. The *Crescent* had the bittersweet distinction of being the last of America's historic passenger trains, when Southern chose to continue operating it (and did so until 1979) in lieu of joining Amtrak.

With the *Crescent* setting the tone, there was little doubt that the premier route between the eastern seaboard and New Orleans was this combination Pennsylvania/Southern/WPR/L&N right-of-way via Atlanta and Montgomery. Over its rails, starting in 1898, rode the country's longest-lived, coast-to-coast sleeper service, the tourist cars from Washington, D.C., that were turned over to the Southern Pacific in New Orleans for the rest of the trip to Los Angeles and San Francisco.

On the last lap of the Crescent line, along the Gulf

Under Alco PA power, the Southern Pacific's Los Angeles-bound *Argonaut* rolls through New Iberia in 1960.

Coast between Mobile and New Orleans, the environment was particularly maritime; wind and sea were constant concerns of the L&N. The more leisurely *Piedmont Limited* also plied these tracks with Pullmans from New York.

Some Southern trains from Washington continued on company rails through Atlanta and Birmingham all the way to New Orleans. Pullmans from New York to New Orleans traveled this line, as did the *Crescent* itself after the Southern chose to remain out of Am-

trak. The *Southerner*, a New York–New Orleans streamliner was inaugurated as an all-coach train on this route in 1941.

The Southern had another collaborator on service to Washington, the Norfolk & Western, which picked up cars at Bristol, Va., and returned them to the Southern at Lynchburg, Va. The *Pelican*, which carried Pullmans from New York to New Orleans and Shreveport (IC), traveled on this line. The joint Southern/N&W route figured into two earlier sleeping car lines to New York, both pre-turn of the century. The Norfolk & Western delivered cars (including one for Washington) to the Baltimore & Ohio at Shenandoah Junction, W.Va., and to the Pennsylvania's Cumberland Valley line at Hagerstown, Md.

Finally, the Illinois Central carried New Orleans sleepers to Louisville, where they were put aboard the B&O for New York.

The Southern also provided New Orleans with Pullman service to Birmingham, Chattanooga, Asheville, and Cincinnati, where it turned sleepers over to the New York Central for Cleveland and Detroit. The *Queen & Crescent Limited* traveled this route to Cincinnati. At Bristol, the Norfolk & Western picked up Pullmans for Norfolk and Washington (Southern). At Meridian, the Southern cut out a St. Louis sleeper to

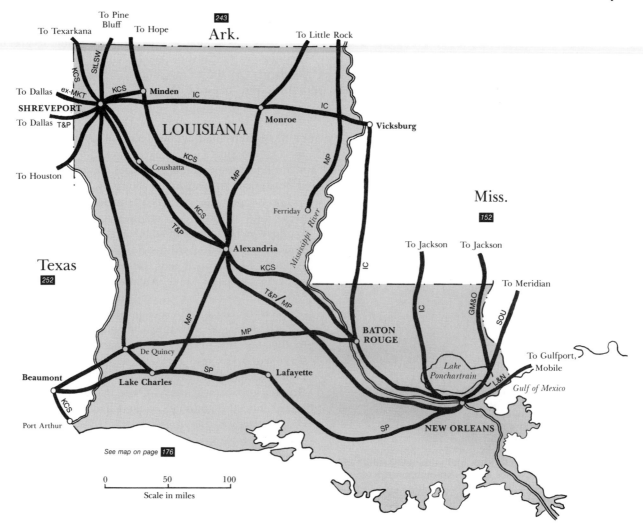

See map on page 176

be forwarded by the Mobile & Ohio.

Over its main line to Cincinnati, served by such trains as the *Pan-American, Azalean,* and *Humming Bird,* the L&N carried Pullmans to Mobile, Birmingham, Nashville, and Cincinnati. At Flomaton, Ala., it dropped off sleepers for Pensacola and Jacksonville (Seaboard Air Line), and in Montgomery turned over Pullmans to the WPR for Atlanta and (all via Southern) Asheville, Portsmouth, Washington, New York (PRR), and Boston (PRR/New Haven). Chicago cars traveled via Evansville, Ind. (Chicago & Eastern Illinois), and those for Pittsburgh via Cincinnati (PRR). An early (1890) line to Philadelphia tapped into the Southern at Calera, Ala., and continued via the N&W/PRR Cumberland Valley route.

The pride of the Illinois Central's main line to Chicago was the *Panama Limited,* a distinguished train named in honor of the canal, which was under construction when the train was introduced in 1911. The *Panama,* which also served St. Louis, was an all-Pullman, extra fare train for most of its life. The Depression brought about discontinuance for two and a half years, but it returned in 1934 as a faster, all-new, air-conditioned train. With brown and orange, streamlined equipment delivered in 1942, the *Panama Limited* was to be the last new pre-war train. The country's second-to-last all-Pullman train, the *Panama* succumbed to coach trade just months before the *Broadway Limited* did in December 1967.

Another IC streamliner that was entirely coach, the *City of New Orleans,* rode the fastest track to Chicago, with schedules of less than sixteen hours. The *City,* which was inaugurated in 1947, and other main line trains to the north skirted the western shore of Lake Pontchartrain on what was reputedly the longest continuous curve, almost nine and a half miles, on an American railroad.

South of Kentucky, the Illinois Central had alternate routes, and some Pullman service operated over both main line and the others, via Vicksburg or Jackson, Tenn. This included sleepers to Memphis, Chicago, and Cincinnati (B&O). Additionally, the IC carried Pullmans to Denver (Frisco/Union Pacific and Frisco/Rock Island), Kansas City (Frisco), Louisville, and St. Louis, plus the Mississippi communities of Jackson, Meridian, Natchez, Vicksburg, and Water Valley, and Monroe, La.

The Gulf, Mobile & Ohio carried modest sleeper traffic to New Orleans from the same neighborhood as the IC, with which it eventually merged. The GM&O, however, provided railroad-owned service in its articulated *Rebel* streamliner, from St. Louis, and later from Jackson, Tenn.

Another passenger carrier operating on the same scale as the GM&O was the Louisiana & Arkansas, which merged with the Kansas City Southern in 1939. This union resulted in creation of the New Orleans-

Texas & Pacific's nine-car *Louisiana Limited* treks through the outskirts of Shreveport in August 1938.

Kansas City *Southern Belle,* a streamliner done in the striking livery of blackish green with yellow and red horizontal striping. The *Belle* carried sleepers between its terminal cities. On the L&A portion of the line, the railroad scheduled New Orleans-Shreveport Pullmans.

For many years the Missouri Pacific controlled and had overlapping operations with the Texas & Pacific before the two were finally merged. As a result, three lines—one from the north through Monroe, the T&P main through Shreveport, and the Gulf Coast line from Beaumont in the west—brought Pullman traffic to New Orleans.

MoPac trains on the line north included the *Louisiana Sunshine Special* and *Southerner,* and to Texas the *Orleanean* and *Houstonian.* The railroad scheduled sleepers to Brownsville, Galveston, Hot Springs, Houston, Kansas City, Little Rock, Monroe, St. Louis, and (with Santa Fe) to Los Angeles and San Francisco.

The *Louisiana Limited* and later *Louisiana Eagle*

A Southern 4-6-2 steps out of New Orleans Terminal Station in January 1939 with the *Queen & Crescent.*

were name trains of the T&P, which carried Pullmans from New Orleans to Alexandria, Shreveport, Dallas-/Fort Worth, Abilene, El Paso, and Denver (Santa Fe or Burlington).

In 1881, the Texas & Pacific connected with the Southern Pacific's "pass of the Rio Grande" line from California, forming the nation's second transcontinental route, and two years later, the SP's Sunset Route was completed through to New Orleans. The venerable *Sunset Limited* was the premier train of this southernmost line to the West Coast. Over the years the *Sunset* carried Pullmans from New Orleans to Los Angeles, Santa Barbara, Pacific Grove, and San Francisco and to Globe, Ariz. The SP scheduled sleepers for San Diego (San Diego & Arizona) on its secondary *Argonaut.*

Southern Pacific also served New Orleans with Pullmans to Lafayette, Lake Charles, Galveston, Houston, San Antonio, Dallas, Fort Worth, and Denver (Burlington).

In later years, the Missouri Pacific was the main conduit of through service to Mexico, but the Southern Pacific originated the first regular luxury train in 1889, and it departed from New Orleans rather than St. Louis. This was the deluxe Pullman *Montezuma Special,* which traveled three times a month over SP to Eagle Pass/Piedras Negras, then via Mexican International and Mexican Central to the capital of Mexico.[2]

Before the era of transcontinental sleepers, passengers could travel coast to coast entirely under the flag of the Southern Pacific. Its "Morgan Line" steamship subsidiary offered weekly freighter sailings from New York to New Orleans, where passengers transferred to either the *Sunset Limited* or *Argonaut* for the remainder of their journey to the West Coast by rail. The steamships *Dixie, Creole,* and *Momus* offered two or three classes of accommodations for the six-day trip. According to advertisements, third-class fare in early 1930 was thirty-three dollars.

In the sunset years of train travel, New Orleans finally was able to abandon the five separate railroad stations when its new Union Passenger Terminal opened in 1954. No longer did L&N trains tie up traffic while crossing Canal Street or passengers wonder whether their MoPac train was leaving from old Union or the stately pillared structure the road shared with the T&P at Thalia and Annunciation streets. Now Rampart Street and the other stations were closed, and all lines were rerouted into the twelve-track, completely air-conditioned terminal built on a redeveloped strip west of Canal. Although

more than forty trains initially used the station, the attrition of rail service was by then well under way, and in just a few years it would more than suffice for the three arrivals and departures of the Amtrak system.

A more spectacular rail construction took place in the area during the mid-1930's, when the high-level Huey Long Bridge opened, carrying two tracks as well as a public roadway across the Mississippi. Trains of the MP/T&P and Southern Pacific used the bridge, which was built to a sufficient elevation to allow ocean-going shipping to pass beneath.

Six railroads provided Pullman service to Shreveport, principally the Illinois Central. One of the IC lines was unique in that it carried the only regularly scheduled sleeping cars operating between New York and a city west of the Mississippi River (excluding St. Louis) until the post-war transcontinental routes were established. The Pullmans traveled over the joint Southern/Norfolk & Western line from Meridian to Washington, which also had service from Shreveport. The IC also originated sleepers for Chicago and turned over Pullmans to the Southern for Birmingham, Atlanta, and Chattanooga.

The components of the modern Kansas City Southern came together at Shreveport, and the KCS scheduled sleepers to Lake Charles, Little Rock (MP), Fort Smith, St. Louis (MP), and Denver (UP). The *Flying Crow,* inaugurated in the late 1920's, worked the premerger KCS from Kansas City to Port Arthur. Before selling its line to one of the KCS predecessors in the 1920's, the Missouri-Kansas-Texas carried sleepers between Dallas and Shreveport.

Also from Shreveport, the Cotton Belt offered sleeping car service to Memphis and St. Louis, the Southern Pacific to Houston, and the T&P to Dallas-/Fort Worth.

With the exception of a Monroe-Atlanta line (IC-/Southern), the Missouri Pacific dominated Pullman service in Louisiana outside of New Orleans and Shreveport. The MP carried sleepers to Monroe from Memphis and St. Louis.

Travelers in Lake Charles could book berths to Little Rock, Memphis, St. Louis, and Chicago (Alton). The MoPac also carried Pullmans to Memphis from both Alexandria and Ferriday.

1. Burke Davis, *The Southern Railway: Road of the Innovators* (Chapel Hill: University of North Carolina Press, 1985), p. 63.
2. Arthur D. Dubin, "Mexicano de Lujo!" *Trains,* Vol. 27, No. 4 (February 1967), p. 34.

24
Illinois

Most railroads in Illinois led to one of two destinations, either Chicago in the north, or St. Louis in the south. Inasmuch as both cities were rail hubs for the entire nation, a great travel corridor developed between them.

Five railroads sponsored Chicago-St. Louis service, offering, over the years, at least seven different routes with overnight sleepers. Some of the trains had all-Pullman consists, touting buffets, showers, and valet service, and they made no intermediate passenger stops. The term "rainbow race" was used to characterize the competition among the last three surviving carriers: the Alton's reds, the Wabash blue, and the green of the Illinois Central.

In the early 1890's, Pullmans for St. Louis departed from Chicago on the Santa Fe, were switched over to the Toledo, Peoria & Western and completed the journey on several lines, part of whose tracks eventually became part of the Chicago & Illinois Midland. The service lasted only three years.[1] With its new line to Kansas City, the Santa Fe apparently had more important fish to fry as a transcontinental.

The Chicago & Eastern Illinois was also involved in the St. Louis trade, although it had to use New York Central iron for the last eighty-odd miles. The C&EI scheduled the imaginatively named overnight *Curfew* and *Silent Knight* and the non-stop *Dearborn*. The *La Salle* was a crack day train. The C&EI's withdrawal in April 1949 left the race to the rainbow trains.

The Illinois Central didn't acquire its own tracks into St. Louis until 1900; before then it transferred sleepers to the Pennsylvania Railroad at Effingham. Nonetheless, it was a strong contender for the St. Louis trade with a swift fleet of day and nighttime trains. The IC traded on its corporate logo for names, designating its overnighters the *Diamond Special* and, later, *Night Diamond.* In May 1936 the IC introduced its first diesel streamliner, the articulated *Green Diamond.*

The Wabash's *Banner Blue* carried the road's flag during daylight hours and was augmented by the post-war, streamlined *Blue Bird,* the only Chicago-St. Louis train offering a dome car. The *Midnight* carried sleepers, including setouts that awaited patrons at St. Louis' west-end Delmar Station.

In the end, the railroad that had started the competition by pioneering George Pullman's first efforts in 1859, was the last to withdraw. The Alton's elegant, old *Midnight Special* was the night runner of its fleet, which included the daytime *Abraham Lincoln, Ann Rutledge,* and *Alton Limited.*[2] After the last *Midnight* ride in 1968 (by then under the Gulf, Mobile & Ohio), the Chicago-St. Louis corridor was without sleeping car service for the first time in more than a century.

One Illinois city, at the time the state's second larg-

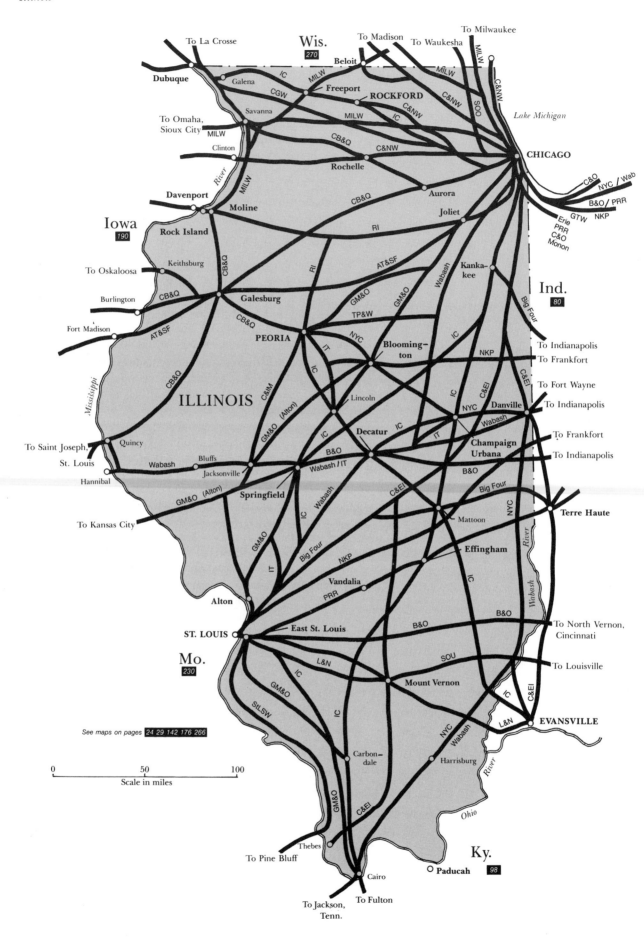

est,[3] became a miniature railroad hub of its own, independent of Chicago and St. Louis. Peoria never escaped the same sort of lampooning its fellow burghers in Dubuque, Iowa, suffered. Many a marketing campaign was, and still is, plotted on the basis of the old show-business concern of how something will "play in Peoria." But while its urbane cousins were laughing, Peoria was playing very well everywhere that whiskey was consumed. Ideally surrounded by a vast part of the nation's corn crop, Peoria had been a major distillery center since the mid-1880's. By this century it was producing a greater volume of corn squeezings than any other American city.[4] Louisville may have had its horsey sophistication, but Peoria was the life of the party.

Water was a big factor in the early development of Peoria, and not just the stuff the distillers poured into their vats. Situated on the Illinois River, the city had steamboat and barge connections with both Chicago and the Mississippi waterway. The early railroads recognized the importance of this, and ultimately fourteen lines came to town, an extraordinary number for a community whose population slightly exceeded 100,000.

At first Peoria was a rail/water transfer point, but as Chicago grew, Peoria acquired its own importance as a rail gateway between east and west. Shippers could avoid the congestion of Chicago and gain time by routing their goods via Peoria. Its importance as a freight handler also gave the city significant passenger service, although none of it truly main line.

Over the years Peoria had Pullman service to more than a dozen cities, from New York to Denver. Sleepers traveled to Chicago by way of four different routes, to St. Louis and the Twin Cities by way of two. Ten of its fourteen railroads scheduled sleeping cars, seven of them using Union Station.

Among these, the New York Central scheduled Pullmans to Indianapolis, Cincinnati, and New York by way of its Big Four lessee Peoria & Eastern. An ancestral line of the Nickel Plate made an unusual run with sleeping cars to Sandusky (1890) at the same time an IC predecessor was offering service to Evansville.

From the west, the Burlington delivered Pullmans to Peoria from Denver, Lincoln, Omaha, and Minneapolis/St. Paul, and the Alton scheduled service from Kansas City. A predecessor of the Chicago & Illinois Midland offered sleepers to St. Louis (1895).

Operating from its own station, the Rock Island also scheduled sleepers to the Twin Cities and one to Sioux Falls (1890). The Rock's tenant Minneapolis & St. Louis put its freight slogan, "The Peoria Gateway," to work with one of the country's most improbable sleeper lines, a Columbus, Ohio, car the M&StL picked up from the New York Central in Peoria for delivery to Minneapolis and St. Paul. The idea was to spare travelers the ordeal of changing trains in Chicago.

The Rock Island was the long-standing Pullman carrier between Peoria and Chicago, but several railroads previously entered into the mix. The Alton carried sleepers on its own and also in cooperation with the Toledo, Peoria & Western. Earlier (1890), the TP&W had worked jointly with the Wabash.

Until the summer of 1940, Peorians could book a bedroom berth to St. Louis on the last surviving electric interurban to offer sleeping car service, the Illi-

At the Elgin, Joliet & Eastern station at Rondout, north of Chicago, signals guarding the crossing of the Milwaukee Road twinkle in the darkness on this winter night in January 1958. This is a *Hiawatha* and *Pioneer Limited* place.

William D. Middleton

J. Parker Lamb

nois Terminal. The IT was part of a central-Illinois traction empire that once had sent sleepers under wire to St. Louis from Bloomington, Champaign, Decatur, and Springfield. It was one of only three electric interurbans in the country offering such service and was by far the largest.

The heavy, tangerine-colored cars offered berths claimed to be six inches longer than Pullman's, and the upper ones had their own windows. For $2.10 beyond the regular fare, travelers could sleep in the air-conditioned comfort of their nine-bedroom car and wake up in the IT's own terminal in downtown St. Louis.[5]

After World War II, when the few surviving interurban carriers were pulling down their trolley poles for the last time, the Illinois Terminal invested in three streamliners. The blue and silver trains offered a reserved-seat observation car and a la carte lounge and dining service on their runs from St. Louis to Decatur and East Peoria. Mechanical and other difficulties, however, hampered the trains' operation, and the biggest problem, lack of patronage, brought the IT's entire intercity passenger service to an end in 1956.

Of all the communities in Illinois that glorified their association with Abraham Lincoln, it was the capital city of Springfield that claimed him as its own and where his body was finally laid to rest. No doubt the tourism that grew from the Lincoln legend helped sustain a fair amount of the passenger service the city enjoyed. Aside from the IT sleepers to St. Louis, Pullmans operated from Springfield on the Baltimore & Ohio to both Cincinnati and Washington and to Kansas City on the Alton. Both the Alton and IC scheduled sleepers to Chicago, but faced with diminishing patronage in later years, the two roads worked out alternate-month service.

After Peoria, Rock Island and its sister cities on the Mississippi had the widest choice of Pullman service outside Chicago. Rock Island, Moline, and Davenport all had their own sleepers to Chicago on the Rock Island Line, which also carried Pullmans from its namesake city to Des Moines, Kansas City, Omaha, and Minneapolis/St. Paul. The Burlington scheduled sleepers from Rock Island to St. Louis (1890) and the Twin Cities. Rock Island had service to Milwaukee (1886) on the Milwaukee Road, and in more recent times Davenport had its own Pullman to Kansas City on the *Southwest Limited*.

Decatur was a main crossroads and repair facility of the Wabash, which gave the city sleepers, before the

Now with the woodland leaves and branches covered by an overnight snowfall, the combined *Georgian/Humming Bird* glides north over C&EI tracks near Danville in December 1959.

Rochelle was a train-watcher's paradise with both Overland and Burlington flyers; here's the *City of Denver* in 1950.

turn of the century, to Kansas City, Detroit, and Toledo. Both the Wabash and IC carried Pullmans to Chicago.

Like Galena, once Illinois's largest city, Cairo suffered declining fortunes when the railroads relegated the steamboat to the role of minor tourism. As the state's southernmost—and also remarkably "southern"—city, Cairo was the original railhead of the Illinois Central, whence steamboats departed with freight and passengers down the Mississippi.

One after another of the IC's main line trains paused in North Cairo before crossing the Ohio River into Kentucky and continuing south, and the railroad gave the town dedicated sleepers to St. Louis and from Chicago. In 1885, when the St. Louis Southwestern was still a narrow-gauge line, it scheduled a sleeper from Cairo to Waco, Texas. The New York Central offered Pullman service to Chicago on the *Egyptian.*

From Carbondale, where the IC's St. Louis trains entered the main line, the railroad operated dedicated sleepers to Chicago and New Orleans.

Quincy, a town on the Missisippi across from a major Burlington junction, had scheduled Pullmans on the CB&Q to both Chicago and St. Louis. Before the turn of the century, the Burlington crossed the river here with through Texas sleepers it turned over to the Missouri-Kansas-Texas. The Wabash also carried a Quincy-Chicago sleeping car.

A far busier Burlington junction, Galesburg had Pullman service on the CB&Q to both St. Louis (1890) and the Twin Cities.

Several other Illinois communities qualified for sleeping car lines, mostly to Chicago: Freeport (Chicago & North Western, 1876), Mattoon (IC), and (via Chicago & Eastern Illinois) Marion, Thebes, and Harrisburg (NYC). Jacksonville once had a sleeper to Cincinnati (Wabash/B&O, 1895), and the Wabash also carried a car (1890) from Bluffs, where the railroad's Keokuk branch joined the main line. The Burlington scheduled a Pullman from Savanna to St. Louis.

Chicago, of course, did not determine the whole pattern of Illinois railroading, and the passenger trains of ten major systems crossed the southern part of the state connecting St. Louis with destinations outside of Illinois. The most notable among these were the big eastern carriers, whose trains included the Pennsylvania's *"Spirit of St. Louis"* and *Penn Texas,* the New York Central's *Southwestern Limited* and *Knickerbocker,* and the Baltimore & Ohio's *National Limited* and *Diplomat.*

1. Richard R. Wallin, Paul H. Stringham, and John Szwajkart, *Chicago & Illinois Midland* (San Marino, Calif.: Golden West Books, 1979), p. 60.
2. The Alton also carried St. Louis sleepers on its line through Jacksonville.
3. Peoria lost the title of second-largest city to Rockford, which, according to schedules at hand, never had any dedicated Pullman service.
4. Federal Writers' Project of the Works Progress Administration, *The WPA Guide to Illinois* (New York: Pantheon, 1983), p. 357.
5. Gordon E. Lloyd, "Sleeping Under Wire on the Illinois Terminal," *Passenger Train Journal,* Vol. 12, No. 1 (May 1980) p. 23.

25 Chicago

Chicago was the nation's railroad capital. Any further description would be superfluous.

On the eve of World War I, nearly 600 through trains were serving the city.[1] Shortly after World War II the total was just under 400,[2] but twenty major railroad systems were still loading and unloading their trains day and night, still outdrawing their formidable new competitors at Midway Airport, ten miles to the southwest. Over the years, more than 500 Pullman lines radiated from Chicago's six major rail terminals to nearly half that number of destinations. No greater selection was available anywhere, an appropriate tribute to the home of the giant Pullman system itself.

For years, the politicians, the newspapers, and the traveling public fulminated over the inconvenience of Chicago's scattered terminal facilities. All twenty railroads terminated in Chicago, and more often than not passengers changing trains and railroads generally were forced to change stations as well. But despite all efforts, it was not until Amtrak that intercity operations were brought under one roof (if that's a proper description of Union Station's nouveau subway remodeling).

Until 1954, the major work of ferrying passengers was performed by a Chicago institution founded during the Civil War by Franklin Parmelee. In the early days, Parmelee stabled more than a thousand horses to haul his wagons and omnibuses. In later years, they gave way to a motorized fleet. But in exchange for the coupon that was part of the accordion-like rail tickets, Parmelee's burly draymen and drivers hauled baggage and passengers alike, for nearly a century, among the six great terminals of the Windy City.

Like Gaul of Caesar's time, commutation on the Chicago & North Western was, and still is, divided into three parts: west, northwest, and north. So was the railroad's intercity passenger service, which pattern stemmed from the lines that formed the modern system. The Galena & Chicago Union, which ran west from Chicago to Freeport, had the honor of introducing the first locomotive, the *Pioneer*, to the nation's future rail metropolis in 1848. This route, however, was eventually superseded by the railroad's newer west line to Omaha.

The railroad that gave its name to the modern system ran northwest into Wisconsin, through Janesville and Fond du Lac. The north line to Milwaukee was brought into the fold just after the Civil War.

Most of the glamor belonged to the line west, which accommodated the transcontinental flyers, from the *Overland Limited* to the great fleet of *City* streamliners. The north line handled the majority of the midwestern passenger business, which fanned out from Milwaukee into northern Wisconsin, Michigan's Upper Peninsula, and Minnesota. The northwest line, originally the primary route to the Twin Cities, principally accommodated trains through Madison to the Dakotas and Duluth. Although *400* streamliners served all three of the North Western's lines, most of these crack yellow and green trains traveled north.

The North Western occupied a series of stations in Chicago, unlike other railroads, all on its own. In 1911, it moved from north of the Chicago River to a new Italian Renaissance depot at Madison and Canal streets, vacating land that eventually became the site of the mammoth Merchandise Mart.

The railroads were then in the prime of their passenger-carrying business, and the station was equipped with such custom facilities as dressing rooms, baths, nurses and matrons rooms, and a doctor's office. Whether its patrons were transcontinental passengers waiting between trains or Lake Forest commuters dressing for the opera, North Western Station delivered the goods and services.

Green tile covered the barrel-vaulted ceiling in the cavernous waiting room. The adjoining concourse led to the sixteen stub tracks. The city's busiest and last unreconstructed terminal, North Western stood until it was demolished in the mid-1980's to make way for a new glass office tower.

In addition to running Chicago's largest commuter operation over its three lines, the North Western also operated numerous name trains. Until 1955, the fleet of *City* streamliners to Denver and the Pacific Coast traveled the west line to Omaha. Forerunners on this route included the *Columbine* to Colorado and the *Corn King Limited* to Omaha, and such California trains as *Los Angeles Limited*, *Gold Coast*, and *Overland*. Trains operating through Milwaukee included the *Iron and Copper Country Express*, *Ashland Limited*, *North Western Limited*, and *400's* bearing the names *Twin Cities*, *Commuter*, *Flambeau*, *Shoreland*, *Valley*, and *Peninsula*. Serving the line through Madison were the *Minnesota and Black Hills Express*, *Duluth-Superior Limited*, and *Dakota 400*.

Union Station, the last major Chicago terminal built, opened in July 1925 along the river, several blocks south of the North Western Station. It also became the city's first truly union station after Amtrak took over the nation's drastically pared schedule of passenger trains. Amtrak also acquired ownership of the facility, which had been built jointly by the Pennsylvania Railroad, the Chicago, Milwaukee, St. Paul & Pacific, and the Chicago, Burlington & Quincy (the Alton, later Gulf, Mobile & Ohio, was a tenant).

Union Station was designed as two back-to-back stub terminals with several through tracks. On the south side, the tuscan reds of the Pennsy mixed with the scarlet and wine-colored Alton-GM&O liveries and the silver-sided Burlington trains. With the orange and several greens of CB&Q's parent Great Northern and Northern Pacific trains, the spectrum was nearly complete. The Milwaukee's orange and maroon occupied the north tracks.

Burlington trains began rolling in 1850, not from Chicago, but from suburban Aurora, forty miles west, and the railroad began building and consolidating lines across Illinois. By the end of the century, it had long since backtracked to Lake Michigan and extended west to Denver and Billings, Mont. Its two major passenger routes from Chicago went to Denver and Minneapolis/St. Paul. Both routes included through service to the West Coast.

Burlington trains enjoyed a reputation for excellence, and in 1930 the road inaugurated its posh new (eightieth) "Anniversary Fleet," the *Aristocrat* to Denver, *Black Hawk* to the Twin Cities, and *Ak-Sar-Ben* (named for the Omaha booster organization, which spelled the home state backwards) to Omaha and Lincoln. Though luxurious, the CB&Q heavyweights were not particularly fleet of foot, but within four years, the railroad fired the shot that signaled a new age.

On May 26, 1934, a three-car, articulated, diesel-powered train called the *Zephyr*, streaked nonstop from Denver to Chicago in a daring, dawn-to-dusk run at an average speed of more than seventy-seven miles an hour (it had topped 112). The stainless-steel, Budd-built train rolled triumphantly to a lakefront stage at Chicago's Century of Progress exposition, ushering in with it the era of the streamliner.[3]

Soon, an entire fleet of these flyers, including the 1936 *Denver Zephyr*, were in operation. The post-war era saw an impressive array of freshly minted name trains plying the main line to Aurora: the *California Zephyr*, *Empire Builder* (GN), *North Coast Limited* (NP), and of course the *Twins* that had inaugurated streamlined service on the line in 1935, now called the *Morn-*

The westbound *National Parks Special* rolls through the suburbs of Chicago in 1951 on the C&NW main line to Omaha.

GM&O's *Abraham Lincoln*, which once carried Mobile and Oklahoma City sleepers, meets PRR's *General* near Union Station.

ing and *Afternoon Zephyrs.*

Every other day, one of the Q's neighbors delivered a Pullman car for the *California Zephyr's* consist, the Pennsylvania Railroad's contribution to this post-war New York-San Francisco sleeping car line. Over the Pennsy's tracks to the East ran the *Broadway Limited, General, Admiral, Pennsylvania Limited, Liberty Limited, Gotham Limited, Manhattan Limited, Golden Triangle,* and *Golden Arrow.* The *South Wind* and *Southland* departed for Florida, the *Kentuckian* for Louisville.

The Alton Railroad, which ran southwest from Chicago through Springfield to St. Louis and Kansas City, had a checkered history of financial legerdemain and red-ink operation until the Baltimore & Ohio took it over in 1929. Nonetheless, the railroad operated superior passenger trains and claimed credit for operating the first sleepers—remodeled day coaches—produced by George Pullman and his then-partner, Benjamin C. Field, in 1859.[4] Nine years later the Alton put the first bona fide dining car into service.[5]

The railroad's flagship in the busy Chicago-St. Louis corridor was the *Alton Limited,* inaugurated in 1899. The posh *Limited* was a favorite of politicians traveling between Chicago and the capital at Springfield. Among the Alton's other crack St. Louis trains were the daylight *Abraham Lincoln* and *Ann Rutledge,* and the elegant *Midnight Special,* a nonstop, all-Pullman train offering "Stag" sleepers as well as some of the earlier single-room cars. The overnight *Hummer,* inaugurated in 1880 between Chicago and Kansas City, was one of the country's oldest name trains.

The Mobile-based Gulf, Mobile & Northern and parallel Mobile & Ohio Rail Road merged into the Gulf, Mobile & Ohio in 1940, creating a system ex-

tending from Mobile to St. Louis. The GM&O extended its reach to Kansas City and Chicago by merging the Alton into the system in 1947. Through Pullmans from Chicago for the *Gulf Coast Rebel* now moved under the GM&O flag all the way.

The Milwaukee Road, which occupied the nine tracks on the north side of Chicago Union Station's vaulted concourse, began as a Lake Michigan-to-Mississippi River line chartered in Wisconsin. Expansion, however, brought it into Chicago, took it west to Omaha and Kansas City, northwest to the Twin Cities and grange country, and north into Michigan's Upper Peninsula. The culmination came in 1909 with the completion of its ambitious Pacific Coast Extension.

This costly and ultimately uneconomical project established the main line through Milwaukee to St. Paul and Minneapolis as the premier route of the system, with the transcontinental *Olympian* and *Columbian* joining the already prestigious overnight *Pioneer Limited* to the Twin Cities.[6] In the streamline age, the addition of the *Hiawathas* and initiation of the *Olympian Hi* into the family made for thrilling sights and speed records on the Milwaukee's fast tracks through Wisconsin.

The crack trains on the main line to Omaha included the *Arrow* and *Pacific Limited,* which carried through sleeping cars for the West Coast. With one bold move in 1955, the Milwaukee ceased playing second fiddle to the North Western's Overland operation and took over the entire fleet of *City* streamliners to the west. By then the Milwaukee's traffic on its Seattle/Tacoma line was withering, so the acquisition of the Overland trains put new life into the railroad's passenger service.

Hiawathas briefly served Michigan's Upper Peninsula, northern Wisconsin, and Omaha. The Milwaukee operated several other important name trains over the years: The *Copper Country Limited* to Michigan, the *Sioux* to South Dakota, and the *Southwest Limited* to Kansas City.

In 1969, the skylights and steel girders of the Union Station concourse fell to developers, who built a new office tower using air rights over the eastern portion of the terminal. Although the massive interior space of the waiting room on the west was only minimally changed, the concourse—which had esthetic similarity to Pennsylvania Station in New York—went the way of its eastern counterpart. Redevelopment turned it into an underground causeway with low ceiling, where, for several hours a day, intercity passengers—by then, clearly the pariahs of the new age—were thrust into the human sea of Burlington and Milwaukee commuters. The impediments of the supporting columns and profusion of retail shops closeted amidst cinder-block walls gave the concourse the no-frills ambience of a bus station.

Despite its geographically diverse traffic and all-embracing name, Union Station was only the third-busiest Chicago terminal in the prime years of rail travel. La Salle Street Station was not only second in total volume, but first in the number of intercity passengers it handled. Opened in 1903, La Salle Street was a joint project of the New York Central and Rock Island and the city's only station located on the Loop, the area defined by the downtown elevated railway. It was also among the trains of the Central's Great Steel Fleet from the east and the Rock Island's family of

westbound *Rockets* that the Nickel Plate made its Chicago home.

With the Central came the most prestigious of all trains from New York, the *Twentieth Century Limited*. Often arriving and departing in several sections, the *Century* was a regular beat for newspaper and radio reporters, who interviewed whatever prominent passengers they were able to beard. For a time, attendants even replicated the storied New York ceremony and rolled a red carpet down the platform for the elite clientele boarding this classic train.

Not all New York Central trains, however, departed from La Salle Street. With a few exceptions, the trains of the Michigan Central and Big Four used Central Station under an arrangement dating from 1852, when the MC got trackage rights into the city over the Illinois Central. But the lineup at La Salle was impressive enough, including the *Century, Commodore Vanderbilt, Wolverine, New England States, Lake Shore Limited, Fifth Avenue Special, Mohawk,* and *Forest City.*

Like many railroads, the Rock Island began as an adjunct to thriving waterways when it was chartered to connect the Illinois and Michigan Canal with the Mississippi River. In time, however, it grew into a major midwestern system that reached Colorado Springs by the late 1880's. An early episode, involving entrenched riverine interests, threatened the future of not just the Rock Island, but all railroads as well. To make a through route from Chicago to Council Bluffs, builders had to bridge the Mississippi, which, by the mid-1850's, still had not been done. The railroad built a crossing at Rock Island, Ill., to the great hostility of steamboat interests, who deemed the in-

The Santa Fe *Super Chief* arrives at Dearborn Street July 9, 1966, with a passenger load swollen by an airline strike.

William D. Middleton

The afternoon sun plays on the *Panama Limited*, departing for New Orleans in August 1948. At left, IC's suburban MU cars.

trusion a navigational hazard. The ramming of the bridge by a steamboat in May 1856 led to celebrated litigation between the rival camps (Abraham Lincoln argued for the railroads) that was not resolved until the Supreme Court, ten years later, ruled in favor of Lincoln's former clients.[7]

By the time the Rock Island emerged as a mature system, it was offering Pullman service from Chicago to destinations ranging in a wide arc from the Twin Cities, to Colorado, to the West Coast, to Texas. The *Golden State Limited* led the parade to California, which included the *Apache, Californian,* and *Imperial.* The *Rocky Mountain Limited,* inaugurated in 1898, served both Colorado Springs and Denver.

The Rock Island's response to the Depression's traffic loss was the introduction in 1937 of streamlined, Budd-built trains on five of its routes. Thus was born the famous family of *Rockets,* named in honor of the railroad's little engine that first steamed out of Chicago in 1852. In November 1939, with new stainless cars trimmed in red, the *Rocky Mountain Rocket* supplanted the namesake *Limited.* The tradition ended in 1978, when the bankrupt road terminated the last *Rockets,* to Peoria and Rock Island.

The *Nickel Plate Limited* was the premier train of that railroad's modest fleet, which generally consisted of two or three expresses between Chicago's La Salle Street Station and New York. The NKP also carried its blue and silver Pullmans to Cleveland.

On Chicago's South Side, the inhabitants of La Salle Street made convenience stops at an Englewood Station, which also served the Pennsylvania. For many years, New York Central and Pennsy trains squared off in a great race, the most recounted runnings of which involved one or another edition of the eastbound *Twentieth Century Limited* and the *Broadway Limited.* Whenever they happened to depart Englewood within seconds of each other on parallel tracks, a thrilling contest of thoroughbreds ensued.

Other trains wearing New York Central grays into Chicago made their calls at Central Station, a baronial structure on the near South Side. While the steam limiteds eased under the train shed, the adjacent electrified tracks hummed with the suburban traffic of both the Illinois Central and the interurban South Shore lines. Michigan Central trains using Central included the *North Shore Limited, Cayuga, Niagara, Motor City Special,* and at one time the *Wolverine.* The station also hosted the Big Four's *Carolina Special, James Whitcomb Riley,* and *Cincinnati Night Express,* as well as the Florida-bound *Royal Palms, Ponce de Leon,* and *Suwanee River Special.*

Tenants though they were (until the late 1950's), these roads did not set the distinctively southern tone of Central Station-that was the province of the Illinois

Howard W. Ameling collection

At Englewood Station on Chicago's South Side, the Pennsy's *New Yorker* gets the jump on Central's *Interstate Express*, 1940.

Central, the self-proclaimed "Main Line of Mid-America."

The IC—the nation's first land-grant railroad and for a time, upon completion, the largest—built its charter line from East Dubuque, through Illinois's onetime lead-mining metropolis of Galena, south to Cairo, where the Ohio flows into the Mississippi River. A branch was built to Chicago, where officialdom gave the railroad a 300-foot right of way along—and sometimes in—the lake, provided that it build and maintain a breakwater. And so began the railroad's unique and lucrative association with Chicago's lakefront.

Central Station was built well to the south of its predecessor, which was located about where the present Randolph Street suburban station occupies its subterranean quarters. The new terminal opened in 1893, in time to serve the tourists arriving for the Columbian Exposition from the railroad's distant outposts, which ranged from New Orleans to Omaha. Regarded an eyesore by many, Central Station put a dominant mark on the neighborhood, with its tall clock tower and huge electric green diamond touting IC passenger service. Central, with its eight tracks, was the only true through-type station in Chicago (although the tracks had nowhere to go but the yards to the north and the suburban electric terminal at Randolph Street).

Nonetheless, Central was the end of the line for many southern blacks, who migrated over the years to Chicago in search of better opportunity. Most of them fairly poor, they were riding the railroad from Louisiana and Mississippi long after more affluent Americans deserted the trains for airports and highways.

The Illinois Central dispatched crack trains to New Orleans: the all-Pullman *Panama Limited,* the coach streamliner *City of New Orleans,* and the more leisurely *Louisiane.* Vacationers traveled to Florida on the *City of Miami, Seminole,* and earlier *Floridan.* A fleet of trains, including the *Night Diamond,* worked the corridor to St. Louis, and the *Hawkeye* and daytime *Land o' Corn* the line west to Omaha.

Mindful of the criticism of its lakefront presence, the IC proposed a new terminal in the 1920's that it said would be in keeping with the classic lines of the nearby Field Museum of Natural History, Soldier Field, and other civic structures. The trackage was to be at three levels: suburban (which the railroad was then in the process of electrifying) the lowest, baggage and express next, and finally an upper, raised approach on which intercity trains would enter a twenty-five-track, stub terminal. The white-marble colossus never got beyond a striking rendering that made it appear to be yet another monumental Grant Park edifice.[8] A half-century later, the original, well-worn Victorian-era station was turned into the rubble its architectural critics had long favored.

Chicago had another Central station, this prefaced by a "Grand," and many thought the Norman structure with a soaring Victorian clock tower to be the most archtecturally distinguished terminal in the city. Opened in 1890, Grand Central flaunted the ornamental elegance of the period with synthetic marble columns topped with floral capitals, extensive iron grillwork, stained glass, and a graceful steel and glass train shed. Nonetheless, this station that spent its old age fronting on an expressway circle, also ultimately was reduced to brownstone and brick rubble.

It was unusual that such architectural finery was underwritten by two fairly marginal passenger carriers, including the Wisconsin Central. After financial defaults involving several railroads—including the Northern Pacific, which had leased the Wisconsin Central to gain an entrance to Chicago—the terminal and access trackage were sold to the Baltimore &

Ohio. The B&O had moved to Grand Central in 1891, rather than shifting operations to the IC's new Central station.

The Wisconsin Central's line went north through Fond du Lac, Wis., avoiding most sizeable communities on the way. In 1909, the Soo leased the railroad, giving it access to Chicago. Over the years, the Soo was an off-and-on tenant of Grand Central.

The Chicago Great Western also avoided populous communities on its line west into Iowa, and was generally regarded as an unnecessary project from the start. Nonetheless, it operated some memorable passenger service, and a favorite location for photographers was its half-mile-long tunnel in the rolling land of northwestern Illinois, the longest rail tunnel in the Prairie State.

The Pere Marquette, which became part of the Chesapeake & Ohio, was the other tenant in Grand Central, gaining access over B&O facilities.

Generally, the railroads using Grand Central had tough competitors in their territories, and so they stressed a more muted approach, such as the solid comfort of heavyweight Soo sleepers, the cosmopolitan charm of the Royal Blue trains. The B&O was the dominant carrier, with it *Capitol Limited, Fort Pitt Limited,* and *Shenandoah* traversing the historic line to Pittsburgh and the East.

The Soo offered sleeping car service to the Twin Cities and Lake Superior region as well as through service, via Canadian Pacific, to Vancouver on its one-time all-Pullman *Mountaineer,* a summertime train, and the year-round *Soo-Dominion.*

The Great Western was yet another of the seven railroads that together offered eleven different sleeper routes to the Twin Cities. The CGW's entries in this field were its luxurious *Great Western Limited* and *Legionnaire* and the Depression-era *Minnesotan.* From Chicago, the railroad carried Pullmans to Oelwein, Iowa, where it shuffled and dealt them to other midwestern destinations. The Great Western was merged into the Chicago & North Western system in 1968.

For many years, the marble floors of Grand Central were strewn with duffle bags of campers and fishermen riding the Soo into northern Wisconsin, or with the baggage of vacationers heading for the Michigan resorts on the PM/C&O's seasonal *Resort Special.* If the Pere Marquette's summer train conjured up visions of families in Brooks Brothers sportswear, the regular trains to Grand Rapids seemed to operate at a more meat-and-potatoes level: the *Shoppers', Business Man's,* and *Furniture City Specials.*

Over the years, Pullmans brought into town from C&O origins managed to dock in all six of Chicago's major terminals. Pere Marquette trains from Michigan used Grand Central. Through sleepers from the east arrived in either Central or La Salle, depending on whether they came Big Four from Cincinnati or

NYC via Toledo. Those delivered by the C&O of Indiana came to Dearborn Station. After Grand Central closed in 1969, C&O and B&O trains moved to North Western, and of course the Aquarian Age of Amtrak finally brought everything to Union Station.

Chicago's oldest terminal was its most ecumenical in terms of passenger service and architecturally its most eclectic. Dearborn Station, opened in 1885, hosted six railroads, more than any other facility in the city, and its callboards displayed such names as the Grand Trunk Western's *International Limited* from Montreal, the *Dixie* fleet of Florida expresses, and the star-studded *Super Chief* of the Santa Fe.

The Italianate structure of red brick originally had an imposingly top-heavy clock tower with dormers, designed in Flemish Revival style. The tower and main roof were damaged by a fire in the 1920's and were both flattened out in the refurbishing. Further interior remodeling brought a melange of Art Deco modernization, and butterfly platform shelters were extended beyond the reach of the historic train shed.

The landlord of it all, the Chicago & Western Indiana, was a terminal railroad owned by five of the participating lines, the Chicago & Eastern Illinois, Erie, GTW, Monon, and Wabash, all of which gained access to Dearborn over C&WI tracks. (C&WI also had an Englewood Station, used by C&EI, Erie, Monon, and Wabash.) The Santa Fe had a lease arrangement.

Though just a lessee, the Santa Fe, with its silvery streamliners brought to the post by throbbing diesels in yellow and red warbonnets, provided most of the glamor. Just as their colleagues in La Salle Street station worked the *Twentieth Century Limited,* reporters and photographers waited at Dearborn for the *Chief* or *Super Chief* to pull in with their usual consignment of Hollywood celebrities. Truly insignificant were the

The Alton's *Midnight Special* steams into Chicago in August 1939. Trolleys on bridge are passing, not running in tandem.

movie stars of mid-Twentieth Century whose arrivals in this ancient terminal weren't duly recorded on film and notepad by the Chicago press.

For ordinary mortals, the Santa Fe provided such additional trains to the west as the all-coach *El Capitan*, the *San Francisco Chief, Grand Canyon, California Limited, Scout, Navajo, Missionary,* and *Hopi*.

The C&EI, which traversed the Illinois-Indiana state line and coal mining areas, also ran a creditable passenger operation on the Chicago-Evansville leg of the Dixie route to Florida. Its fleet included the historic *Dixie Flyer,* the streamlined *Dixie Flagler,* and several similarly prefaced trains in between. The C&EI was also once a competitor for the St. Louis Pullman traffic, and it briefly operated the blue and orange, postwar *Wippoorwill* and *Meadowlark* streamliners in coach service downstate.

The neighboring Wabash was also a keen competitor for St. Louis business, which was its main focus in later years. The Wabash of the free-wheeling Gould era, however, sent its trains from Chicago in every direction but north, with sleeping car destinations ranging from Boston to Mexico City.[9]

Although the Wabash also once scheduled sleepers to New York, the Erie provided more durable service on such trains as the *Lake Cities* and *Erie Limited* and, after the Erie-Lackawanna merger, the *Phoebe Snow.*

After ambitious, though brief Pullman service to Florida and the East Coast, the Hoosier Line known as the Monon settled down to serving destinations in the home state, Ohio, and Kentucky, which it looked to when naming such trains as the daytime *Thoroughbred* and overnight *Bluegrass.*

The venerable, turn-of-the-century *International Limited* led the schedule of Grand Trunk Western trains from Montreal to Chicago, which also included the 1920's vintage *Inter-City Limited* and *Maple Leaf* and the *La Salle,* which bowed in 1937. The *New Yorker* and *Chicagoan,* with help of the Lehigh Valley, carried sleepers between their namesake cities. The GTW also scheduled Pullman-carrying trains from Chicago to Detroit and Toronto.

Although developers beat the preservationists and dismantled Dearborn Station's train shed, the clock tower and headhouse were restored in the late 1980's as part of a South Loop commercial redevelopment. Nearby, the engines of bulldozers continued to resound amidst the weeds and rubble of the once-great, though long-vacant rail yards south of downtown Chicago. Where steam and diesel locomotives had clat-

Sleeping car lines

Here is a listing, by railroads, of the more than 500 Pullman and other sleeping car lines that served Chicago over the years. Those that were principally 19th Century lines are indicated with the year, in parentheses, of the schedules in which they were listed.

Baltimore & Ohio: Akron, Columbus, Cleveland, Toledo (NYC), Youngstown, and Zanesville, Ohio; Fairmont, Parkersburg, Piedmont, and Wheeling, W.Va.; Pittsburgh and Philadelphia; Cumberland and Baltimore, Md.; Washington; and New York.

Chesapeake & Ohio: Grand Rapids, Bay City (also with CNR), Frankfort, Muskegon, Northport, Pentwater, Petoskey, and Traverse City, Mich.; Muncie, Peru, and Richmond, Ind.; Cincinnati; Hinton, W.Va.; Richmond and Old Point Comfort, Va.

Burlington: Quincy, Ill.; Burlington, Keokuk, and Marshalltown (M&StL, 1886), Iowa; Lincoln, Omaha, and Alliance, Neb.; Kansas City and St. Joseph, Mo.; Atchison, Kans. (1890); Edgemont, S.D.; Cody (NP) and Sheridan, Wyo.; Minneapolis/St. Paul; with NP to Billings, Bozeman, Gardiner, and Red Lodge, Mont.; with GN to Glacier Park, Great Falls, and Helena, Mont.; Portland (GN/SP&S or NP/SP&S); Seattle (GN or NP) and Tacoma (NP), Wash.; Colorado Springs (D&RGW) and Denver; with MKT to Houston (1895) and Galveston (SP, 1890); Los Angeles (tourist, various Colorado routes); San Francisco (D&RGW or Overland routes).

Chicago & Eastern Illinois: Marion, Thebes, and Harrisburg, Ill. (NYC); Evansville and Terre Haute, Ind., and St. Louis. **With L&N:** Nashville, Birmingham, Montgomery, New Orleans, and Pensacola. **With L&N and NC&StL:** Atlanta and Augusta, Ga.; Jacksonville (ACL or CofG/ACL), Miami (ACL/FEC), Palm Beach (CofG/ACL/FEC), Sarasota and St. Petersburg (CofG/ACL), and St. Augustine (CofG/SOU/ACL/FEC), Fla. **With Frisco:** Tulsa and Dallas.

Chicago Great Western: Des Moines, Fort Dodge, and Waterloo, Iowa; Minneapolis/St. Paul and Rochester, Minn.; Omaha; San Francisco (UP/SP).

Chicago & North Western: Freeport (1876) and East Dubuque (IC, 1869), Ill.; Ashland, Drummond, Eagle River, Fond du Lac (1869), Green Bay (3 routes, via Janesville 1869), La Crosse (1876), Madison, Mercer, Milwaukee, Oshkosh, Phelps, Rhinelander, Sturgeon Bay (KGB&W/A&W), and Wausau, Wis.; Calumet, Cisco Lake, Escanaba, Gogebic, Iron Mountain, Iron River, Ironwood, Ishpeming, Marenisco, Marquette (DSS&A), Menominee, Sault Ste. Marie (Soo), and Watersmeet, Mich.; Duluth (2 routes), Mankato, Minneapolis (2 routes), St. Paul (2 routes), Rochester, Tracy (1890), and Winona, Minn. (1876); Cedar Rapids, Des Moines, Eagle Grove, Hawarden, Mason City, and Sioux City, Iowa; Deadwood and Rapid City, S.D.; Omaha; Lander, Wyo., Portland, Ore. (NP/SP&S); Seattle (NP); Winnipeg (CNR) and Vancouver (CNR or Soo/CPR) **With UP:** Denver; Laramie, Wyo.; Long Beach, Los Angeles (UP only, or tourist cars via D&RGW/UP or SP), Pasadena, and San Francisco, Calif. (SP); Portland, Ore. (UP only or UP/SP); Seattle; West Yellowstone, Mont.

Erie Railroad: Akron, Youngstown, and Columbus, Ohio; Ashland, Ky. (C&O/N&W, 1890); Meadville, Pa.; Jamestown, Salamanca, Albany (D&H), and New York, N.Y.; and Boston (D&H/B&M, or NH 1893).

Grand Trunk Western/Canadian National: Detroit, Flint, Pontiac, Lansing, and Port Huron, Mich.; London and Hamilton, Ont., Toronto, Montreal; Portland, Me.; Boston (CV/B&M); Buffalo and New York (LV or Erie 1890, or NYC West Shore 1890).

Gulf, Mobile & Ohio: Mobile. All the following lines originally were scheduled by the Alton Railroad portion of the GM&O: Peoria (Alton alone or with TP&W) and Springfield, Ill.; St. Louis (2 routes) and Kansas City; Hot Springs, Ark. (MP); Oklahoma City (Frisco); Dallas/Fort Worth (MP/T&P), Houston (MP/T&P/MP), San Antonio (MKT), El Paso (MP/T&P, 1895); Lake Charles, La. (MP); and Los Angeles (RI/SP). The Chicago & Alton participated in a variety of pre-1900 lines to Los Angeles and San Francisco, via various connections at St. Louis and Kansas City.

Illinois Central: Mattoon, Decatur, Carbondale, Springfield, and Cairo, Ill.; Paducah, Ky.; Evansville, Ind.; Memphis and Nashville (NC&StL); Greenville, Jackson, and Gulfport, Miss.; Mobile (M&O) and Birmingham, Ala.; New Orleans (2 routes) and Shreveport, La.; Savannah, Ga. (CofG); all Florida with CofG/ACL—

tered over switches onto tracks leading to their appointed gates, a new neighborhood of townhouses and apartments was now taking root.

1. John A. Droege, *Passenger Terminals and Trains* (Milwaukee: Kalmbach, 1969. Reprint of original, New York: McGraw-Hill, 1916), p. 195.
2. Richard Caress, "The Terminals of Chicago," *National Railway Bulletin*, 45, No. 1, p. 41.
3. The Burlington retired the little train in 1960 after it completed 3.2 million miles of service. It was put on display at Chicago's Museum of Science and Industry.
4. John H. White Jr., *The American Railroad Passenger Car, Part 1* (Baltimore: Johns Hopkins University Press, softcover edition, 1985), p. 247.
5. Arthur D. Dubin, *More Classic Trains* (Milwaukee: Kalmbach, 1974), p. 192.
6. In railroad parlance, the term "transcontinental" is commonly used in referring to those lines that extended from Mid-America to the Pacific Ocean. This expression probably stems from the meeting in 1869 of the Union Pacific and Central Pacific at Promontory, Utah, which resulted in completion of America's first transcontinental rail line—even though neither UP nor CP, alone or together, were transcontinental railroads. The Chicago-to-Seattle/Tacoma Milwaukee Road, among several others, qualifies for the term as it is more loosely defined.
7. George H. Drury, *The Historical Guide to North American Railroads* (Milwaukee: Kalmbach, 1985), p. 88.
8. *Proposed Lake Front Passenger Terminal, Twelfth Street Boulevard and Indiana Avenue, Chicago, Illinois*, Illinois Central Railroad Company, 1925. A folio of architectural renderings in the collection of the Chicago Historical Society.
9. W.M. Adams, "The Sunshine Special," *National Railway Bulletin*, 41, No. 2 (1976), p. 7. The car for Mexico City went aboard a private tour train in St. Louis.

Union Station's concourse, inspired by that of New York's Pennsylvania Station, was lost to wreckers in the 1960's.

Jacksonville, Fort Myers, Miami (FEC), Palm Beach (FEC), St. Petersburg, and Tampa/Sarasota, Fla.; St. Louis (2 routes); Hot Springs, Ark. (RI); Oklahoma City (Frisco); via New Orleans to Brownsville (MP), Houston (SP), and San Antonio (SP), Tex., and San Francisco (SP); Waco, Tex. (StLSW, 1893); Dubuque, Cedar Falls, Fort Dodge, Waterloo, and Sioux City, Iowa; Minneapolis/St. Paul (M&StL); Sioux Falls, S.D.; and Omaha.

Milwaukee Road: Green Bay, Madison, Prairie du Chien (1890), Wausau, Minocqua, Star Lake, and Tomahawk, Wis.; Calumet (DSS&A or Copper Range), Champion, Iron Mountain, Iron River, Marquette (DSS&A), Pembine, and Sault Ste. Marie (Soo), Mich.; St. Paul/Minneapolis (3 routes), Austin, and Jackson, Minn.; Dubuque, Des Moines, Marion (1890), Mason City, Ottumwa, Sanborn (1895), and Sioux City, Iowa; Sioux Falls, Rapid City, Mitchell, and Canton, S.D.; Omaha; Kansas City; Denver (UP); Gallatin Gateway, Mont.; Portland, Ore. (UP or NP 1890); Spokane, Tacoma, and Seattle; Los Angeles and San Francisco (Overland and various Colorado routes).

Monon: Indianapolis and French Lick, Ind.; Louisville. **With B&O:** Cincinnati and Dayton, Ohio; Baltimore (1898), and Washington (1898). **With B&O/SOU:** Burgin, Ky. (1889); Jacksonville (ACL, 1890) and St. Augustine (FEC), Fla.

Nickel Plate: Cleveland, Buffalo, Scranton, Pa. (DL&W), Boston (NYC-WS, B&M), New York (DL&W or NYC-WS 1895)

New York Central: With PRR—Harbor Springs and Mackinaw City, Mich.; Toledo, Cleveland, and Columbus, Ohio; Charleston, W.Va.; Pittsburgh (Erie/P&LE) and Oil City, Pa.; Auburn, Buffalo, New York, Lake Placid, Syracuse, and Utica, N.Y.; Boston (also with B&M); with C&O at Toledo—Charlottesville, Clifton Forge, Richmond, and Phoebus, Va.; Baltimore (Erie/P&LE/WM); Cairo, Ill.; Evansville, Ind.

New York Central (Michigan Central): Bay City, Detroit, Grand Rapids, Kalamazoo, Lansing, and Saginaw, Mich.; with CPR, Toronto and Montreal; Albany, Buffalo, New York (via Buffalo or Niagara Falls, or also with DL&W or LV from Buffalo), and Niagara Falls, N.Y.; Philadelphia (LV/RDG); Boston; and Portland, Me. (CPR/MeC or NYC/Rutland/CV/StJLC/MeC 1890).

New York Central (Big Four): Indianapolis; Louisville; Cincinnati; with C&O—Washington; Charlottesville, Clifton Forge, Hot Springs, Newport News-Old Point Comfort-Phoebus, and Richmond, Va.; and all following with Southern Railway—Asheville, Goldsboro, and Greensboro, N.C.; Charleston and Columbia, S.C.; Atlanta (1893), Macon (1893), and Savannah (ACL, 1893), Ga.; Jacksonville, Miami (SAL or FEC), Palm Beach (FEC), St. Augustine (FEC), St. Petersburg (SAL), and Tampa (ACL), Fla.; and (both 1893) Birmingham and Mobile, Ala.

Pennsylvania (Fort Wayne route): Fort Wayne, Ind.; Detroit (Wabash); Akron, Alliance, Canton, Cleveland, and Mansfield, Ohio; Altoona (1890), New Castle, Philadelphia, and Pittsburgh, Pa.; Atlantic City, N.J.; Boston (NH); New York (also with RDG/CNJ, 1869); Washington. Via Logansport, Ind.: Richmond and Indianapolis, Ind.; Louisville; Cincinnati, Springfield, Columbus, and Dayton, Ohio; Norfolk, Va. (N&W); New York; Pittsburgh; via L&N routes—Jacksonville, Miami, St. Petersburg, Tampa/Sarasota, Fla.

Rock Island: Peoria, Moline, and Rock Island, Ill.; Cedar Rapids (1890), Des Moines, Iowa City, Davenport, and Washington, Iowa; Omaha and Lincoln, Neb.; Minneapolis/St. Paul (also with M&StL 1890); Kansas City and St. Joseph, Mo.; Atchison and Leavenworth, Kans.; Sioux Falls, S.D.; Brownsville (SP), Dallas/Fort Worth, and El Paso (SP), Tex.; Colorado Springs, Denver, and Pueblo, Colo.; and all following with SP: Los Angeles (also via Colorado routes), San Diego (SD&AE), San Francisco (also D&RGW/WP or D&RGW/SP), Santa Barbara; Phoenix and Tucson, Ariz.

Santa Fe: Fort Madison, Iowa (1890); Kansas City and St. Louis (TP&W/C&IM 1893); Wichita; Oklahoma City and Tulsa; Colorado Springs and Denver; Albuquerque; Phoenix; Los Angeles, San Diego, and San Francisco; Dallas/Fort Worth and Galveston.

Soo Line: Neenah, Ashland, Stevens Point, and Park Falls, Wis.; Minneapolis/St Paul and Duluth; Portland, Ore. (NP/1890); Winnipeg, Banff, and Vancouver, Canada.

Wabash: Detroit, Pittsburgh, Buffalo; Boston (WS/B&M), Kingston, N.Y., (WS), New York City (WS or MiC/NYO&W or DL&W); Toronto (CPR) and Montreal (CPR); Bluffs (1890), Decatur, Peoria (TP&W, 1890), and Quincy, Ill.; Hot Springs (MP), St. Louis, Kansas City (1890), and Moberley, Mo.; Laredo (MP, 1895) and San Antonio, Tex. (SLSF/MKT); Mexico City (MP/NdeM)

What was it the engines said,
Pilots touching—head to head,
Facing on the single track,
Half a world behind each back? . . .

—Bret Harte

26
Going to California

On May 10, 1869, the telegraph from Promontory, Utah, signaled the message: America's transcontinental railroad was completed, the rails of the Union Pacific had joined those of the Central Pacific, and the country was united by parallel ribbons of iron stretching from coast to coast. Within five days of the golden spike ceremony, sleeping car service was inaugurated, bringing San Francisco Bay within 130 hours travel time from Chicago.

The route of this new railroad had been known since early in the century by numerous travelers, from trappers and missionaries, to the Mormons and forty-niners.[1] But while there was great popular enthusiasm for building the nation's first transcontinental railway, the choice of route was complicated by sectional rivalries. Southerners and northerners alike wanted the tracks to run west from their own latitudes.

In 1853 Congress authorized the Pacific Railroad Surveys, under which Army engineers were sent out to find the best routes across the Rocky Mountains to the Pacific Ocean. They were under the direction of the secretary of war, Jefferson Davis. The surveys were completed in good time and the results published in 1855, but not until the southern states had seceded could a majority vote be obtained on any one route. On July 1, 1862, President Lincoln signed the Pacific Railroad Act, setting in motion the elements that would convene at Promontory seven years later.

From Omaha, Union Pacific crews built west, through Nebraska, to Cheyenne, Wyo., from which they laid their tracks over the high ground of Sherman Hill and the crossing of the Continental Divide at Creston, Wyo. Central Pacific crews, mostly Chinese coolies, built their far more mountainous line up from Sacramento over the Sierra Nevada. It all came together on a desolate plateau north of the Great Salt Lake, the Central Pacific's engine *Jupiter* facing the UP's No. 119, while several hundreds witnessed the meeting.

In the first two decades of passenger service on this Overland Route, as the Union Pacific called it, travel was fairly spartan, and traffic was not great. Travelers had to change cars, generally at Council Bluffs, Iowa, across the Missouri River from Omaha, and at either Ogden or Promontory. By the late 1880's, however, California was enjoying its greatest migration since the Gold Rush, and the railroads both spurred and profited from the boom.

In 1888, the UP inaugurated a train that brought a

The *eminence grise* of travel to the Golden State, the *Overland Limited* (top) makes a call around the turn of the century at Rocklin, Calif. The silvery eminence of modern day travel, the *Super Chief* (bottom) loads passengers in Chicago's dingy Dearborn station.

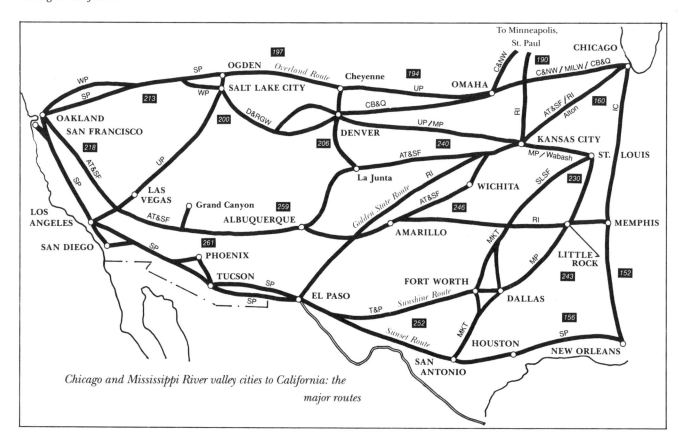

Chicago and Mississippi River valley cities to California: the major routes

standard of elegance as unique as it was brief. The *Golden Gate Special,* which operated only once a week from December to the following May, was an all-Pullman train that carried some of the manufacturer's earliest electrically lighted, vestibuled coaches. The cars were painted umber, with red and gold trim, and the lounge carried a brass-railed observation platform. Among interior amenities, the train carried the first bathtub available in a public sleeper. Fare from Council Bluffs to Oakland was one hundred dollars, berths and meals included.

After the short reign of the *Golden Gate Special,* the Union Pacific decided to upgrade a train it had inaugurated two years before, the *Overland Flyer.* Despite living temporarily in the shadow of the deluxe *Special,* the *Flyer* became a fine train in its own right, helping to establish the standards of service that would govern transcontinental travel well into the Twentieth Century. In the mid-1890's, the UP changed the name of the train, which would be its flagship until the streamline era, to the *Overland Limited.*

East of Omaha, part of the *Overland Limited* ran by way of the Chicago & North Western, part by way of the Milwaukee Road. In a strange bit of corporate rivalry, the Southern Pacific, which had assumed control of the Central Pacific in 1870, refused to acknowledge the name Overland—either for the route or the train—preferring to keep the original generic train names of *Atlantic* and *Pacific Expresses* until 1899.

For much of its career, the *Overland Limited* was all-

Pullman. As early as 1902, it offered pre-departure telephone service from its observation car, where a uniformed attendant assisted callers, and it was one of the first western trains on which an extra fare (ten dollars) was charged.[2]

In later years, it was called the *San Francisco Overland,* and although it remained a respectable all-Pullman train, it lost its premier status in the 1930's. After World War II, the train carried some of the transcontinental sleepers from New York, but in the mid-1950's it was shorn of all its through cars and later merged into its successor, the *City of San Francisco.*

When the bright yellow,[3] streamlined *City* started operation in June 1936, it proved immensely popular, but with just one eleven-car train set, it could offer only sixth-day departures. A new seventeen-car train was delivered in early 1938, permitting more frequent service, but not until September 1947 did it operate daily. In January 1952, the train made the news after a westbound section became snowbound in the Sierra Nevada. Two hundred passengers were rescued three days later.

Traffic to the West, like the topography, had its own peaks and valleys, and the railroads responded by introducing new trains as conditions warranted. In 1913, for the first time, the Overland route had three daily trains with the inauguration of the *San Francisco Limited* and *Pacific Limited.*[4] In the mid-1920's, the *Gold Coast* was introduced, and it provided through sleepers to both San Francisco and Los Angeles. By the end of the decade, four daily trains were plying the route.

The Depression brought cutbacks, but with some economic recovery, with the interest generated by the new streamliners, and with the attraction of San Francisco's Golden Gate Exposition, traffic again surged in the latter part of the decade. Two all-Pullman, extra-fare trains, neither daily, supplemented the regulars: the *Treasure Island Special* and the gray and gold-trimmed *Forty-Niner*, which had gone into service in 1937. Inspired by the success of the economy *Challenger* on the Los Angeles run, a San Francisco version was introduced and operated until 1947.

In the early days, before the railroads scheduled through sleepers from Chicago, passengers had the choice of numerous lines and trains serving Omaha. From the late 1880's on, however, through cars became more and more common, and although such railroads as the Burlington and Chicago Great Western offered service, two lines, the North Western and the Milwaukee Road, controlled most of it.[5]

The North Western, over the years, had the larger share of the Pacific Coast traffic, at least until the sudden announcement in 1955 that the Milwaukee Road would be the UP's new Omaha-Chicago connection and that the C&NW's historic involvement was terminating. In honor of its marriage, the Milwaukee even changed livery, from orange and maroon to the more exuberant UP yellow, red and gray.

The Overland Route was the original way to San Francisco, and for many years, at least as long as the city was the metropolis of the West, it seemed to be San Francisco's own route. But with the growth of the Union Pacific system, newer outlets permitted it to tap the business of both Southern California—which was burgeoning—and the Pacific Northwest.

Historically, the UP/SP had carried through Pullmans for Los Angeles by way of Sacramento or Oakland, service that was far slower than the Santa Fe's. But after completion of the San Pedro, Los Angeles & Salt Lake, which became a UP subsidiary just before World War I, the Union Pacific could offer more direct service.

The all-Pullman *Los Angeles Limited,* inaugurated early in this century, was the premier train on the new line, which also was served by the *Gold Coast* and *Continental Limited.* (These two trains and others carried sleepers between various western and midwestern destinations, in a manner typical of UP's mammoth Pullman-juggling operation.)

In 1935, the original *Challenger* began operation as a second section of the *Los Angeles Limited.* America's only all-coach, tourist-sleeper train, the *Challenger* was designed to attract travelers who, though still budget-minded, were beginning to venture out in increasing numbers.

The oeuvre of this segment of the Overland Route, however, was the *City of Los Angeles,* another train in the famous family of streamliners. The *City* began service in May 1936, getting a year's head start on its toughest competitor, the streamlined version of Santa Fe's *Super Chief.* A new, longer train set was added in late 1937, permitting more frequent scheduling, and,

The *Forty-niner*, an extra-fare, all-Pullman train, began its short-lived, pre-war career in 1937 with a gray livery trimmed in gold.

Union Pacific Railroad

like its San Francisco sister, the *City of Los Angeles* finally went daily in 1947. It made the trip in the mandatory new bench-mark time of thirty-nine hours, forty-five minutes.

Pullman service from Chicago to Los Angeles over the C&NW/UP route at one time also included separate sleepers for Pasadena and Long Beach. The North Western also carried sleepers for Los Angeles from Minneapolis/St. Paul.

The Oregon Short Line gave the Union Pacific access to Portland from its main line at Granger, Wyo. These tracks accommodated the *Portland Limited,* the Depression-born *Portland Rose,* and finally, the first arrival in the UP's streamliner fleet, the *City of Portland.* This experimental flyer, recognizable by the ominous scowl on its front grill, reduced travel time from Chicago by eighteen hours and pioneered routine schedules of less than forty hours. This earliest *City* was a seven-car, articulated train and included the first articulated Pullman cars. Its maiden run was in June 1935.

Even though Portland, Tacoma, and Seattle were well served by the northern transcontinentals from Chicago and the Twin Cities, the UP successfully competed with them. The North Western delivered Pullmans for Portland and Seattle to the UP at Omaha and in the early 1930's the two railroads briefly participated in a new line that went north on the Southern Pacific from Fernley, Nev. Pool service trains operated by UP/Great Northern/Northern Pacific carried sleepers from Portland to Seattle. The Milwaukee Road also once cooperated with the UP on sleeper service to Portland.

Both Portland and Seattle had Pullman service from St. Louis. The Alton provided an 1890 connection from St. Louis to the UP at Kansas City, but in modern times the Wabash handled the business, which also included sleepers for San Francisco and Los Angeles. The *City of St. Louis,* inaugurated in 1946, accommodated all the various Pullmans. Twelve years later, this *City* had the bittersweet honor of adding to its consist the last dome cars built for American railroads.

The *City* streamliners, their Electro-Motive diesels flaunting the logos of the participating railroads, were an unforgettable chapter in American railroading. Whether they were lined up on adjacent tracks in Chicago's North Western Station, ready to race out into the night just minutes apart, or streaking across Nebraska averaging more than 78 miles an hour, or breaking the darkened silence of a western desert, they were magnificent trains.

In a one of the last great flourishes of the stream-

line era, the *City* streamliners were enhanced with the addition of thirty-five dome cars—some coaches, some observation-lounges, some diners. The delivery came just before the switch to the Milwaukee Road. The dining cars on the new "Domeliners," as the trains were proudly billed, offered three different chambers: a main-floor area, a private room accommodating parties of up to ten, and the tour de force, the dome area itself.[6]

By 1936, Americans were well aware of the ferment in the passenger train business. Railroads, stung by the revenue loss of the Depression and the continuing defection to the highways, were going to great lengths to attract people back to the rails. As a result, some odd-looking conveyances were taking to the tracks, as railroads experimented with technology and studied consumer preferences. Passengers eagerly accepted the early streamliners, but entire fleets of trains couldn't be modernized overnight.

On May 12, 1936, a train of heavyweight Pullmans under power of a blunt-nosed diesel locomotive that resembled the rail cars known as doodlebugs pulled out of Chicago's Dearborn Station and began its run to Los Angeles. As it raced through the twilight down the Santa Fe main line along the Illinois waterway, the train wouldn't have appeared at all out of the ordinary were it not for the locomotive's rather peacock-like color scheme of cobalt and saratoga blue, olive, and scarlet.

By the time it finished its 2,239-mile journey, however, there was little doubt that this train was more than met the eye. It had just cut more than fifteen hours off the Santa Fe's previous best schedule, having made the trip in thirty-nine hours, forty-four minutes. This was the soon-to-be legendary *Super Chief,* though not the sleek, stainless steel consist under tow of a brace of diesels wearing the Santa Fe's blazing warbonnet of yellow and red. That train was a year in the

Seven sections of Santa Fe's *California Limited* line up for this classic publicity shot in Los Angeles of the 1920's.

future, and this was a temporary version. But the starting gun had been fired, and the race to Southern California, between the *Super Chief* and the *City of Los Angeles,* was on.

The old pattern of two days and three nights between Chicago and Los Angeles had now gone up in a wisp of diesel exhaust, as the new streamliners maintained under-forty-hour scheduling. Nor would the railroads rest their case on speed alone. The new trains were not ornate in the manner of the Gilded Age flyers, but they were not the least bit wanting in the realm of luxury, and the builders strove to deliver all the comforts that modern technology made possible.

The Atchison, Topeka & Santa Fe started construction in 1868, the year before Promontory, with the relatively modest ambition of hauling cattle in its home state of Kansas. But with a land grant and the prospects of Colorado minerals over the horizon, it soon raised its sights and, despite Indian thievery and stampeding buffaloes, continued pushing west.

In Colorado, the Santa Fe encountered the Denver & Rio Grande, whose goal was to reach El Paso. Both railroads wanted Raton Pass—on the border of Colorado and New Mexico—through which the historic trail to Santa Fe passed, but the AT&SF crews beat their rivals to it one day in 1878 and forged on, south and west. By 1883, the Santa Fe had reached California, and two years later, with the opening of its line to San Diego, it became a transcontinental with its eastern terminus in Kansas City.

Within the next few years, the railroad built a line to Chicago, straight and fast, and in 1888 began operating vestibuled trains from Lake Michigan through to the Pacific. The time was right, for Southern California was in the midst of a land boom. The year before, a rate war had broken out between the Santa Fe and Southern Pacific, and both sold tickets at bargain levels. Town sites were hurriedly laid out on the railroads' land-grant holdings, and passengers flocked West in every sort of conveyance, from luxury sleepers to Zulu cars, which were designed to carry the heads of families and their livestock. Los Angeles added 120,000 new landowners, San Diego 50,000.[7]

One of the great attractions of the Santa Fe line was a longtime railroad tradition, the Harvey Houses. During the early years, western cooks were generally innocent of what later became known as quality control. The well-to-do, eating in posh dining cars of the period, did quite well, stuffing themselves with the abundant prairie game and the already rich selection of food and beverage brought from the East. In lower circles, however, the food could be gritty and mean.

Before dining cars were common, railroads stopped their trains at appointed restaurants, usually their own, and allowed their passengers all of twenty or thirty minutes to wolf down a meal before the whistle sounded. Fred Harvey, a man of sensibilities, was appalled by what he saw the Santa Fe's dining halls dishing out, and he persuaded the railroad to give him control of all its restaurants, in return for which he would operate them under the same high standards set at his Harvey House in Topeka.

Harvey advertised in the eastern and midwestern press for well-mannered young women to come work as waitresses, thereby giving the restaurants the aura of gentility they had lacked. The quality of the food was high, and the prices reasonable.

Harvey Girls came west in ample numbers, and luckily so, for they were in great demand among the local men of means who wanted to marry well. The girls thrived and so did the houses, even though dining cars eventually reduced their numbers. But even with change, the Santa Fe continued to label its dining car cuisine with the proud hallmark of the Fred Harvey name.

Santa Fe Southern Pacific Corp.

The Santa Fe had trains that offered more luxury and panache, but none ever performed the sheer volume of yeoman service between Chicago and Los Angeles that the *California Limited* did. This all-Pullman train operated from 1892, with minor interruptions, until well into the post-World War II era. Although it boasted such amenities as valet, maid, barber, and bath service, the *California Limited* also wrote the book on doing volume business.

During summers of the 1920's, as many as seven sections of the train—each carrying eleven Pullmans—operated within a half hour of one another. The Santa Fe achieved a record with twenty-two westbound and twenty-three eastbound sections of the train under way at one time. Each section required at least fifteen locomotives and train crews over the length of the route.[8]

But while it relied on the *California Limited* to do the heavy work, the Santa Fe took pride in operating some of its trains for the select few. In December 1911, the railroad inaugurated its first train in this class, the *De Luxe*, which operated once weekly during winters on a sixty-three-hour schedule, five hours faster than usual. The service, which carried a twenty-five-dollar surcharge, was far beyond the usual. The *De Luxe* offered all manner of personal attendants, and at Cajon Pass in California, a flower boy boarded the westbound section and gave complimentary bouquets to the women and boutonnieres to the men. The luxurious Pullman accommodations for the train's maximum of sixty passengers included some of the earliest all-room sleepers. The *De Luxe* was discontinued during World War I.

In November 1926, the Santa Fe introduced the *Chief* as its new standard bearer. An all-Pullman train with a ten-dollar extra fare, the *Chief* immediately became the favorite of Hollywood celebrities, and *Variety* picked up the scent by publishing the train's daily "sailing" list. But despite the full retinue of servants and the comforts the heavyweight *Chief* provided, its fifty-eight-hour travel time would not be good enough in the new era that was dawning. Something even finer and more fleet-of-foot was necessary, and the *Super Chief* was the answer.

From the yellow and red herald on its diesel units to the purple and red drumhead on the observation car, the *Super Chief* was one shimmering, energized shaft of stainless steel. Its external dress was brash and exciting, as if the train were the sun's own centurion. Inside, it bore a muted elegance derived from tones of the earth, from veneers of ebony, rosewood, and teak, among other exotic woods used in the Pullmans, to the turquoise ceiling and copper walls of the lounge.[9]

While greeting passengers in the diner, where yellow roses graced the tables, the maître d'hotel always wore the appropriate morning coat or dinner jacket. Those who knew their Fred Harvey favored the freshly caught Colorado mountain trout. The extra fare from Chicago to Los Angeles was the considerable Depression-era sum of fifteen dollars. But the sort of person who rode the train was not the sort who would think twice about such an expenditure, for this *was* the *Super Chief*. If the New York Central's *Twentieth Century Limited* was the train of choice in the East, then the West belonged to the Santa Fe's new superstar.

The *Super Chief* became the country's first all-Pullman, diesel-powered streamliner when it started its maiden run west on May 18, 1937. The train was able to keep its thirty-nine-hour, forty-five-minute schedule, thanks to extensive improvements in the Santa Fe right of way. A second train was delivered the following February, permitting twice-weekly service. With new, post-war equipment, daily service began in February 1948. The Santa Fe put six completely new train sets into operation on the *Super Chief* in 1950-1 and refurbished them in 1958.

Far more than most railroads, the Santa Fe made great efforts to keep its fleet of trains in top condition. It ordered lightweight equipment for the *Chief* and gave the train distinctive, streamlined Hudson locomotives painted in two shades of blue. The *El Capitan*, the only all-coach train in the country with an extra fare, developed in parallel with the *Super Chief* and finally was merged with it as rail traffic waned.

The *El Capitan*, however, was most notable for the high-level coaches the Santa Fe equipped it with in the late 1950's, when investment in new passenger rail cars had dwindled to a trickle. The bold designs were the esthetic and structural forebears of modern Amtrak Superliner equipment, which followed two decades later. Over the years, the Santa Fe accounted for the largest share of the Chicago-Los Angeles passenger business, and its superlative trains at all levels of accommodation, were a prime reason.

The *Grand Canyon* held the record for the longest streamliner run in the country, with several Chicago-Oakland sleepers, until the Santa Fe inaugurated the *San Francisco Chief* from Chicago in June 1954. A section of this *Chief* offered connections between the Bay area and Texas cities.

The *Grand Canyon*, though originally an economy train born in the late 1920's, became all-Pullman and operated in northern and southern sections, one using the regular main line over Raton Pass, the other operating via the more southerly, mostly freight line through Clovis, N.M. The *Grand Canyons* carried Chicago-Los Angeles Pullmans that were cut out at Williams, Ariz., and hauled over the Santa Fe's branch line to the south rim of the canyon itself. There, passengers enjoyed a day's outing before returning in

Donald Sims

SP's venerable *Sunset Limited* heads west through Beaumont, Calif., with one of its five streamlined train sets delivered in 1950.

their sleepers at night to the main line and continuation of the journey.

Santa Fe was the only western railroad with its own tracks all the way from Chicago to California, so in modern times it routinely handled Pullman cars to Los Angeles, San Francisco, and San Diego over the entire route. For a few years after building its Chicago line in 1888, however, it continued working with the Alton and also the Frisco, which it briefly owned in the early 1890's.

The Alton delivered Chicago sleepers for Los Angeles and San Francisco to the Santa Fe at Kansas City and to the Frisco at St. Louis. The Frisco forwarded the cars to its interchange with the Santa Fe at Burrton, Kans.

From St. Louis, the Frisco also carried sleepers for Los Angeles, San Francisco, and San Diego via Burrton, and the Alton delivered Los Angeles and San Francisco cars to the Santa Fe at Kansas City. The more enduring St. Louis-Los Angeles service involved sleepers carried to Kansas City on a rotating basis by the Alton/Burlington, Missouri Pacific, and Wabash.

The Santa Fe provided early, brief Pullman service through to Minneapolis/St. Paul. Before the turn of the century, the Burlington forwarded a San Diego sleeper from Kansas City to Omaha, which continued to the Twin Cities via the Illinois Central and Minneapolis & St. Louis. Several decades later the Chicago Great Western carried a Twin Cities-Los Angeles Pullman to Kansas City.

At Houston, the Santa Fe picked up New Orleans sleepers from the MoPac for both Los Angeles and San Francisco. Dallas/Fort Worth Pullmans went Santa Fe all the way to Los Angeles.

America's second transcontinental route was its southernmost and for that reason was a Civil War casualty. It was eventually built, however, and opened in 1881. This route, the western part of which was all Southern Pacific, had three stems east of El Paso: the Sunset Route to San Antonio and New Orleans; what is sometimes called the Sunshine Route, with several variations, to St. Louis; and the Golden State Route to Kansas City and Chicago.[10] Each stem had a namesake, principal train.

The Gadsden Purchase of 1850, the wedge of Mexican land that enlarged southern Arizona and New Mexico, was made precisely with this rail route in mind because it afforded a favorable gateway for tracks that would be laid three decades later. Two charters were involved in the southern route: The Texas & Pacific was building west, the Southern Pacific east. Twelve years after Promontory, the two railroads met just east of El Paso, at Sierra Blanca, Tex., opening a route that passed through Dallas and Fort Worth. The same year, 1881, the Santa Fe joined with the SP at Deming, N.M., creating a brief alternate transcontinental route before its own tracks reached the Pacific. And in early 1883, the SP crew made the connection with the New Orleans, or Sunset leg, at the Pecos River in Texas. With subsequent consolidations, this became an all-SP route.

Through service on the Sunset Route began almost immediately, but its most famous train did not come along for more than a decade. The *Sunset Limited* had an interesting, though far from seamless history, and its destinations, both eastern and western, varied. The SP inaugurated the luxury Pullman train in 1894 on a once-weekly, wintertime schedule between New Orleans and San Francisco. In 1897-8 it operated to Chicago via St. Louis (T&P/MP/Alton), with a connecting train from El Paso to New Orleans. The next season it returned to its original route.[11] The *Sunset* ceased operation early in this century but was reinstated in 1911 and became a daily, year-round train from New Orleans two years later.

Over the next few decades, on its trek through Louisiana bayous and across plains, desert, and mountains to the California coast, the all-Pullman *Sunset Limited* carried sleepers from New Orleans to Los Angeles, Santa Barbara, Pacific Grove, and San Francisco. In World War II, its West Coast run was cut back to Los Angeles. In 1950, five new streamlined sets of fifteen cars each, in stainless steel with red letterboards, were put on a reduced schedule.

The good times were brief, however, and the end came for a sadly truncated *Sunset* in 1968.

The *Argonaut*, which began service on the Sunset Route in the mid-1920's, also carried Pullmans, including one from New Orleans to San Diego (San Diego & Arizona Eastern) and the Washington-Sunset tourist sleepers that operated from the nation's capital to Los Angeles and San Francisco.

Around the turn of the century, sleeping cars from several midwestern cities traveled to New Orleans, where they were cut into SP trains bound for California. The IC's young *Panama Limited* carried Pullmans from Chicago for San Francisco, and the railroad also scheduled similarly destined tourist cars from Chicago and Cincinnati. The Mobile & Ohio offered San Francisco sleepers from Chicago (Alton/M&O-/Southern) and St. Louis (M&O/Southern).

The Golden State Route to Chicago proved to be a more expeditious connection than the long, though colorful detour through New Orleans. The *Golden State Limited*, a handsome Pullman train in olive green with gold striping, began service for the winter season

Union Pacific PR, St. Louis

MoPac's *Sunshine Special* rolls north on a fine morning in 1948 through Carondelet Park in St. Louis.

Otto C. Perry/Denver Public Library

San Diego & Arizona Eastern No. 3, approaching its namesake city, carried sleepers brought from the east by Southern Pacific, the parent railroad. This July 1935 edition operates under both SP and SD&AE power.

in November 1902, inaugurating the new Chicago-to-Los Angeles route. The Rock Island and the Southern Pacific were the principals, and the Alton handled the Chicago-Kansas City leg. (Initially the partnership involved the tracks of the El Paso & Northeastern, which became part of the El Paso & Southwestern and, finally in 1924, the SP. Interline transfers between the SP and Rock Island were made at Tucumcari, N.M.)

The *Golden State Limited,* never as fast as its more direct competitors to the north, nonetheless became popular in its own right, in part because it served areas of the Southwest that had a salubrious winter climate. In 1924, the train set a record of nine eastbound sections.[12]

The *Californian,* which the Alton carried to Kansas City, and the *Apache* also operated over the Golden State Route and some of its variations west of El Paso. The *Imperial* was introduced in the late 1930's as a secondary train operating via El Centro in California's Imperial Valley.

Service from Chicago included Pullmans to San Diego (SD&AE), Los Angeles, Santa Barbara, and San Francisco. For a number of years, the Alton carried Los Angeles sleepers to the Rock Island at Kansas City.

With its own lines serving both Minneapolis/St. Paul and Memphis, the Rock Island also offered Pullman service from those cities to Los Angeles. The Rock Island and Alton both carried St. Louis-Los Angeles sleepers to Kansas City, but in later years the Missouri Pacific handled the connection.

In a brief period of optimism after World War II, the SP and Rock Island laid plans for a shiny new *Golden Rocket,* which would operate three times a week on schedules matching the under-forty-hour running time of the Santa Fe and Union Pacific's finest, far better timing than even the *Golden State Limit-*

ed had offered. Unfortunately, the SP waffled and finally canceled its order for new equipment, and the *Golden Rocket* never came into being. The equipment the Rock Island had ordered, plus some other rebuilt coaches went into the consists of a somewhat face-lifted *Golden State,* which first appeared in early 1948.[13] The train carried some of the coast-to-coast Pullmans from New York, but withdrew from that in 1951. After two-thirds of a century of service, the *Golden State* died in 1968.

The Sunshine Route to the West literally applied to the Missouri Pacific/Texas & Pacific, but it was sometimes used in describing the service of other railroads. Its principal origin was St. Louis, although before the turn of the century a few sleepers from Chicago were scheduled, and it included a connection to Memphis. Pullmans from these and intermediate cities traveled the MP/T&P main line to El Paso, where they were put aboard SP's Sunset Route flyers. The principal destination was Los Angeles, although original service included San Francisco.

As early as 1884, the Missouri Pacific predecessor Iron Mountain was scheduling through Palace sleepers from St. Louis to Los Angeles and California, as well as offering connections for travel to Mexico City. The MP and T&P, meanwhile, expanded on their numerous sleeping car lines into Arkansas, Louisiana, and Texas. The *Sunshine Special,* the longtime flagship that cruised this territory, often in many sections, was inaugurated in December 1915.

This handsomely equipped new train was designed to provide better service between Texas and the East so as to entice some of the traffic moving via New Orleans. In the process, it gave St. Louis and other cities in the mid-South better service to California. By the late 1920's, MoPac officials were trying to fill three Pullmans with passengers for Los Angeles over the "mild, low altitude, all-weather route" through the Southwest.[14]

To meet this competition, the Katy introduced its *Texas Special* the same December, but it couldn't match the schedule the *'Shine* offered. In 1917, under direction of the wartime United States Railroad Administration, the Katy made an agreement with the competing Frisco to jointly operate the *Texas Special*, taking advantage of the Frisco's shorter main line from St. Louis into Oklahoma. MKT trains tapped into the Sunset Line at San Antonio.

And so the Pullmans came south, in the *Sunshine Special*, or *Texas Special*, or other, earlier trains, to join the Sunset flow to the Pacific coast. The MoPac combination moved sleepers to Los Angeles from St. Louis, Memphis, and Chicago (Alton) and to San Francisco from Chicago (Alton) and St. Louis (both 1890). The Katy's service was from St. Louis, via the *Texas Special/Sunset Limited*, to Los Angeles, and a tourist sleeper (MKT only/SP) to San Francisco.

Nowhere was the crossing of the Rockies more difficult nor spectacular than in Colorado, which at various times accounted for one narrow- and three standard-gauge mountain passes and more Pullman lines than any of the other transcontinental routes.

Perhaps oblivious to this variety, the average passenger of modern times might remember Colorado railroading best from a trip he or she took on a silver train with the most publicized scenery of all, the *California Zephyr*. Not even the lordly *Super Chief* could offer the scenic equivalents of the Colorado Rockies and California's Feather River Canyons, through which the *CZ* was scheduled for the best possible daylight viewing. But long before this trip was possible, crossing the Divide in Colorado was a story of other routes and other trains.

The Denver & Rio Grande Western's original line across the Rockies, opened to Ogden in 1883, was narrow gauge. From Pueblo, the tracks went through the Royal Gorge of the Arkansas River and west from Salida via Gunnison over Marshall Pass to Grand Junction. Faced with the competition of the shorter, standard-gauge line of the Colorado Midland from Colorado Springs, the Rio Grande pushed construction of a new main line over Tennessee Pass to the valley of the Colorado River. This newer "Royal Gorge Route" was the road's standard line until it was displaced by the Moffat Tunnel/Dotsero Cutoff route, opened in 1934.

Though faced with stratospheric elevation, hairpin curves, and ruinous grades on its line through Leadville, the Colorado Midland ran several trains with through sleepers from Colorado Springs to Ogden, the westbound *Ute* and eastbound *Coyote* being the principals. The Rio Grande's *Scenic Special* was its turn-of-the-century workhorse.

The trains of both roads carried Pullmans, both standard and tourist, bound for Los Angeles and San Francisco from Chicago. Sleepers from Minneapolis/St. Paul moved via D&RGW en route to Los Angeles, and via the Midland route to San Francisco. The Rio Grande was the preferred connection for those originating in St. Louis.

From the east, the Midland took on sleepers delivered by the Burlington and Rock Island, some of the latter of which had originated on the Milwaukee Road. West of Grand Junction, the Rio Grande-/Southern Pacific handled the through cars. Railroading across the Continental Divide in Colorado became the sole province of the D&RGW in 1918, when the Midland came to its unhappy end, abandonment.

Over the years, the Rio Grande picked up West

Richard Kindig

Rio Grande's two prime name trains on the eve of World War II: (left) the westbound *Exposition Flyer*, which traveled the Moffat Tunnel route, is near Ralston, Colo., late in 1940; and (right) the *Scenic Limited*, which, despite the 1,000-foot climb to Tennessee Pass on the Royal Gorge line, is making 30 miles an hour with 13 cars on this July 3rd, 1941. The 2-8-8-2 makes the difference.

Coast Pullmans from a variety of midwestern railroads and combinations: Burlington, C&NW/UP (Tourist), Missouri Pacific, Milwaukee/MP, Rock Island, Alton/Santa Fe, and M&StL/IC/RI. At Salt Lake City or Ogden, the San Francisco cars went to Western Pacific or SP, Los Angeles Pullmans to UP or SP (via Sacramento).

In April 1915, the *Scenic Limited* replaced the *Scenic Special* on the Rio Grande, the new St. Louis-to-San Francisco train intended to accommodate travelers bound for the Panama-Pacific Exposition. The joint MP/D&RGW/WP flyer was aptly named, offering the scenery of both the Royal Gorge and the Feather River Canyon. In 1942, the new *Colorado Eagle* assumed duties of carrying through cars on the MoPac portion of the line; the Rio Grande portion of the *Scenic* eventually was renamed the *Royal Gorge*.

With the opening of the Moffat route, the D&RGW inaugurated passenger service with a new train called the *Panoramic*, which operated for five years. The *Exposition Flyer*, named in honor of the world's fairs of 1939, went into service the same year. This train, a joint effort of the Burlington, Rio Grande, and Western Pacific, operated for a decade, often in as many as eight sections during wartime, preparing the way for the most memorable train of all, the *California Zephyr*.

Six eleven-car sets of stainless steel coaches were manufactured for the *Zephyr*, which began its storied twenty-one-year career on March 20, 1949. The *Cali-fornia Zephyr* was justifiably acclaimed for being the first train to the West Coast to employ dome cars, and it used them to maximum advantage, with five in each consist. Under tow of the yellow-orange diesels of the western roads, the train moved at a respectable, though not urgent pace through the spectacular landscapes between Denver and San Francisco Bay.

It was fitting that the *California Zephyr* so valued the grand view, because Colorado's Glenwood Canyon is considered to be the birthplace of the postwar generation of dome cars. The manager of General Motors' Electro-Motive Division, Cyrus R. Osborn, got the idea while riding through the canyon in the fireman's seat of a Rio Grande diesel. Impressed by what he was able to see, Osborn wondered how to offer the rail passenger the same sensation. His interest and early sketches convinced both GM and the Burlington of the dome car's merit, and by 1945, the first one emerged from the railroad's shops.

The day after the *California Zephyr* died in 1970, the Rio Grande put its own version of the train into service between its eastern and western termini. Like a pleasant afterglow, it continued the foreshortened, thrice-weekly run for the first dozen years of the Amtrak era, but finally it, too, was gone.

Happily, Amtrak trains, at this writing, are still running on several of the great transcontinental routes, including the Rio Grande. Mostly the trains have shed the tattered liveries their predecessors handed down

Donald Sims

to them, but deep in the canyons and gorges and beneath the rocky escarpments, the ancestral echoes live on.

1. Ralph Budd, "The Conquest of the Rockies," *Trains*, Vol. 7, No. 12 (October 1947), p. 49.
2. Arthur D. Dubin, *Some Classic Trains* (Milwaukee: Kalmbach, 1964), p. 174.
3. The UP originally experimented with a yellow and brown combination, later choosing yellow and gray, with red accents.
4. Donald M. Steffee, "Times to Remember," *Trains*, Vol. 29, No. 7 (May 1969), p. 54C.
5. According to Lucius Beebe, the first through sleeper to San Francisco left Chicago, not on the North Western or Milwaukee Road, but on the Burlington, which never had much further involvement with the Overland Route. The car, inaugurated May 12, 1885, operated once a week. Through service on the North Western, said Beebe, did not begin until 1887. See Lucius Beebe, *The Overland Limited*, (Berkeley: Howell-North, 1963), p. 14.

   The Alton turned over through sleepers to the Union Pacific in Kansas City in 1890. The CGW carried a westbound car to Omaha, according to Phil Borleske, "Last Days of Great Western Passenger Trains," *Passenger Train Journal*, (October 1983), p. 27.
6. Karl Zimmermann, "From Salina to Everywhere," *Passenger Train Journal*, (March 1977), pp. 16-22.
7. Neill C. Wilson and Frank J. Taylor, *Southern Pacific: The Roaring Story of a Fighting Railroad* (New York: McGraw-Hill, 1952), p. 240.
8. James Marshall, *Santa Fe: The Railroad That Built an Empire* (New York: Random House, 1945), p. 290.
9. Stan Repp, "The Story of the Super Chief," *Trains*, Vol. 22, No. 7 (May 1962), pp. 30-41.
10. West of El Paso, there were three alternatives to the original route through New Mexico, Arizona, and California. In the first westbound alternative, the Golden State Route went through Douglas, the Sunset via Bowie. See chapters dealing with those states.
11. David F. Myrick, *Railroads of Arizona Volume I* (San Diego: Howell-North, 1981, second printing), p. 84.
12. Dubin, *More Classic Trains* (Milwaukee: Kalmbach, 1974), p. 370.
13. Don L. Hofsommer, *The Southern Pacific, 1901-1985* (College Station, Texas: Texas A&M University Press, 1986) p. 214.
14. W. M. Adams, "The Sunshine Special," *National Railway Bulletin*, 41, No. 2 (1976), p. 11.

There were trains to the west with more luxury and service, but none with the panache of the *California Zephyr*, here eastbound on the Western Pacific near Niles, Calif.

Stan Mailer collection

William D. Middleton

The great fleet of West Coast streamliners was building when this photograph of the *City of Los Angeles* (top) was taken at Clinton, Iowa, around
1940. Fewer name trains plied the north-south routes through Iowa, but
among them was the Rock Island's *Twin Star Rocket* (bottom), shown here
southbound on the approach to the Mississippi River bridge at Newport,
Minn., April 13, 1958.

27
Iowa

Iowa was interwoven with railroad tracks, the warp consisting of seven main lines that crossed the state from east to west, the woof being the tracks that fanned out south from Minneapolis and St. Paul. And like all granger states, Iowa was further threaded by branch lines crossing its cornfield fabric, calling at numerous centers of commerce consisting of not so much as a gas station or coffee shop, just the towering hulk of a grain elevator. No place in Iowa, it was said, was more than twelve miles from a railroad.

But while the rustic crop haulers pottered along and some of the country's last surviving interurbans

danced under the trolley wires between small towns seemingly locked in time, the heavy work was being done on the east-west main lines. Iowa was part of the great land bridge between the yards of Chicago and the gateway to the Overland Route at Council Bluffs. Long freight trains and crack transcontinental limiteds worked these lines almost with the frequency of crosstown streetcars. The east-west routes through the Hawkeye State were the fast tracks to the Pacific Coast.

For Iowa's early railroad builders, the prize target was Omaha/Council Bluffs, eastern terminus of the Union Pacific, which connected with the Central Pacific in 1869 to make the nation's first transcontinental rail line. The prospect of feeding traffic from Chicago and the East to the UP spurred many a construction plan, and land grants across the fertile prairie soil of Iowa facilitated the way.

The Chicago & North Western gained a major advantage by arriving in Council Bluffs, across the Missouri River from Omaha in 1867, two years before the ceremony at Promontory. The North Western established early and lasting ties, remaining the UP's favored connection to Chicago for passenger trains until 1955.

Other railroads, however, were soon in the picture, and two of them, the Rock Island and Chicago, Burlington & Quincy both reached the Missouri River by 1869. In time, the Illinois Central, Milwaukee Road, and Chicago Great Western followed, and the Wabash angled up from Missouri to make the seventh rail connection between Chicago and Omaha.

The resulting flow of traffic coming together at Council Bluffs was voluminous.[1] Four of the railroads—Milwaukee Road, C&NW, CB&Q, and Rock Island—were still carrying Pullmans from Chicago to Omaha well into the 1950's. Only the more indirectly routed IC and Great Western had, by then, dropped out.

Across the southern half of Iowa, the parade of passenger trains included many of the prime flyers to the West: the *Overland*, *Gold Coast*, and *Los Angeles Limited* and the later *Cities* streamliners traveled the North Western route through Cedar Rapids until 1955, when the Milwaukee Road took over. The Rock Island's line through Des Moines carried the *Rocky Mountain Limited*, *Colorado Express*, and the streamlined *Des Moines*, *Corn Belt*, and *Rocky Mountain Rockets*. The southernmost main line, through Ottumwa, was the route of the Burlington's *Fast Mail*, *Aristocrat*, *Ak-Sar-Ben*, *Exposition Flyer*, and the silvery *Zephyrs*: *California*, *Denver*, and *Nebraska*.

Creasing the southeastern corner of the state, en route to Kansas City, were the Milwaukee's *Southwest Limited* and the Rock Island's California fleet: *Golden State Limited*, *Imperial*, *Apache*, and *Californian*. After entering Iowa at Fort Madison, the Santa Fe's superla-

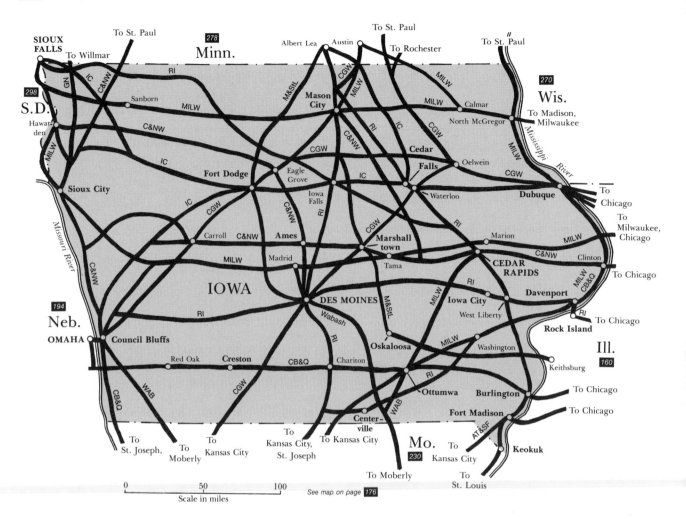

tive name trains—from the *De Luxe* and *California Limited* to the modern *Super Chief, Chief, El Capitan,* and many others—made their brief dash across the very tip of the state.

A similar pattern was once in evidence in the northeastern part of Iowa among the trains competing for the Chicago-Twin Cities trade. The Great Western, IC/Minneapolis & St. Louis, Rock Island, and even the Milwaukee Road, by way of its line up the Mississippi River from Dubuque, vied for this business until their rivals with more direct routes through Wisconsin won out. The IC, however, continued working its tracks to Sioux City with such trains as the *Land o' Corn* and overnight *Hawkeye.*

The northernmost line across Iowa belonged to the Milwaukee Road and connected with its original line from Milwaukee to Prairie du Chien. This was the route of the *Sioux,* which carried sleepers to Chicago from South Dakota and Minnesota.

For Minneapolis and St. Paul, several windows to the south opened up through Iowa by way of five railroads—CGW, C&NW, M&StL, Milwaukee Road, and Rock Island—although the name trains were not so prevalent as on the east-west routes. The Great Western's *Mill Cities* and *Tri-State Limiteds* paused in

Des Moines on their way to Kansas City, as did the Rock Island's *Mid-Continent Special* and later *Twin Star Rocket* en route to Texas. The *Zephyr Rocket,* jointly operated by the Rock Island and Burlington to St. Louis, passed through Cedar Rapids and Waterloo.

Although the capital city of Des Moines had the most Pullman lines serving it, the choice involved only seven destinations. Des Moines had four routes to Chicago (C&NW, CGW, RI, Milwaukee) and three to the Twin Cities (CGW, M&StL, RI) and Kansas City (CGW, RI, Wabash). The Rock Island carried Pullmans to Omaha and Rock Island, the Wabash to St. Louis, and the Milwaukee Road to Sioux City.

Sioux City, a major livestock market and site of a Corn Palace of seemingly Russian-Orthodox character, also had Pullman service to seven destinations, including three lines to Chicago: IC, C&NW, and Milwaukee Road (1890). The North Western scheduled sleepers to Minneapolis/St. Paul, Omaha, and Kansas City (Missouri Pacific). The Milwaukee Road, which operated a section of the *Arrow* to Sioux City, offered sleepers to Aberdeen and Murdo, S.D., in addition to Des Moines. Before the turn of the century, Sioux City also had service to the Twin Cities by way of the Great Northern.

The Quad Cities, of which Davenport was one, had a level of Pullman service similar to that of Des Moines and Sioux City. Rock Island, Ill., across the Mississippi River, had the greater share, although Davenport had dedicated sleepers to both Milwaukee and Kansas City on the Milwaukee Road and to Chicago on the Rock Island. The city of Rock Island also had Pullman service on the Burlington to the Twin Cities and St. Louis (1895) and on its namesake railroad to Des Moines, Minneapolis/St. Paul, and Omaha.

In the 1950's, the Rock Island Railroad made it all the way to Broadway, thanks to native Iowan Meredith Willson. The traveling salesman banter ("Ya gotta know the territory") in the opening scene of "The Music Man" was set in an imaginary Rock Island day coach traversing Iowa in the early part of this century.

Dubuque's best passenger connections were found across the Mississippi River in East Dubuque, Ill., where Burlington trains stopped briefly with their dome cars and Pullmans bound for the Pacific Northwest. For local sleeping car service, however, Dubuque depended on the IC and Milwaukee Road, both of which carried sleepers to Chicago (before its own line to Chicago was built, in the 1880's, the IC teamed with the North Western at Freeport, Ill.). The Milwaukee Road's service from Dubuque to the Twin Cities and Sioux Falls was pre-turn of the century, as was the IC's car to Sioux Falls.

Cedar Rapids was well served by the Overland Route trains, whether they came through town on the North Western, or stopped in nearby Marion on the Milwaukee Road line. The North Western carried a Pullman from Cedar Rapids to Chicago, and the Milwaukee Road scheduled a car for Kansas City that went aboard the *Southwest Limited* at Ottumwa. The Rock Island offered Pullman service to St. Louis (CB&Q) and Chicago (1890).

Pullmans came to Fort Dodge from Chicago on both the IC's *Hawkeye* and on the Great Western, whose main lines ran southwest to Omaha. In the 1920's, the *Nebraska Limited* and *Omaha Express* from Minneapolis and St. Paul plied the CGW line, which, in 1965, would be the last to see the wine-colored Great Western varnish. The M&StL carried a sleeper from Fort Dodge to the Twin Cities.

Waterloo had nightly visits of the *Zephyr Rockets,* but all the sleepers on the Rock Island were of the through variety. Waterloo, like Fort Dodge, had Pullman service to Chicago on both the IC and CGW.

To the northeast of Waterloo lay Oelwein, the heart of the Great Western system, where its rails from Chicago radiated into lines to the Twin-Cities, Omaha, and Kansas City. The Great Western always competed against stronger rivals, and until Depression-era

M&StL *North Star Limited,* which carried sleepers to the Twin Cities from St. Louis and, earlier, Chicago, at St. Paul in 1931.

cutbacks it conducted a creditable and innovative passenger service, dispersing Pullmans from its cornbelt hub at Oelwein. At the zenith of service in the 1920's, the *Legionnaire* carried sleepers from Chicago to Rochester and St. Paul/Minneapolis, and the *Mill Cities Limited* and *Tri-State Limited* worked the line between Kansas City and the Twin Cities.

Ottumwa lay on the main line of the Burlington to Omaha and was visited by all its crack trains to Denver and the West Coast. Before the turn of the century, the Wabash scheduled a sleeper to St. Louis, and the Milwaukee Road, together with the M&StL, offered a car to St. Paul. In more recent times, however, the Milwaukee provided Ottumwa's sole service, sleepers to both Chicago and Kansas City on the *Southwest Limited.*

Mason City was the major station in Iowa on the Milwaukee Road's line to South Dakota, and the railroad served the community with sleepers from Chicago, Milwaukee, and Mitchell, S.D. The North Western also scheduled a car to Chicago.

The Milwaukee Road, over the years, had more sleeping car lines terminating in Iowa than any other railroad.

Sanborn, a meal stop on the Milwaukee Road's line across Wisconsin and Iowa to Rapid City, S.D., had dedicated sleepers before the turn of the century to both Madison, Wis., and Chicago.

Burlington had Pullman service provided by its namesake railroad to both Chicago and St. Louis.

Several other cities and towns in Iowa had Pullman service to Chicago: Iowa City and Washington (RI), Cedar Falls (IC), Fort Madison (Santa Fe, 1890), Keokuk (CB&Q), Marion (Milwaukee, 1890), and Marshalltown (M&StL/CB&Q at Keithsburg, Ill., 1886). The North Western carried a Chicago sleeper to Hawarden. Two railroads scheduled sleeping cars to the Twin Cities: the M&StL from Oskaloosa and the Milwaukee Road from North McGregor (1890), where the charter line from Milwaukee crossed the Mississippi River.

---

1. The Burlington initially crossed down-river at Plattsmouth, but later routed its premier passenger trains through Council Bluffs, the UP's major transfer point.

28
Nebraska

The farmlands of eastern Nebraska were webbed with the lines of various railroads, but in the central and western part of the state, where the Great Plains rise into the hill country, the main lines of the Union Pacific and Burlington ruled the terrain.

From the great gateway of Omaha on the Missouri River, the tracks of the Union Pacific followed the gentle curves of the Platte River across Nebraska before dividing, one line for Denver, the other to continue west and cross the Continental Divide. This was the historic Overland Route that carried the yellow *City* streamliners in the twilight of UP passenger service.

At nighttime, during this last great era, Union Station in Omaha was the setting of an incomparable spectacle of arrivals and departures. In July 1951, within a period of less than seven hours, starting at 9:25 p.m., eight crack UP trains departed for the west against the flow of seven arrivals. After a 488-mile trip on the Chicago & North Western from Chicago, in as little as seven hours, forty-five minutes, the *City* streamliners paused for ten minutes of servicing, the other trains a bit longer, before continuing their journey.

The lineup of westbound departures consisted of the *Los Angeles Limited* (9:25 p.m.), seasonal *National Parks Special* (9:35), *Idahoan* (9:45), *City of Denver* (11:50), *San Francisco Overland* (12:55 a.m.), *City of Portland* (1:40), *City of San Francisco* (1:55), and the *City of Los Angeles* (2:10). Arriving from the west: *City of Denver* (12:35 a.m.), *City of Los Angeles* (1:40), *City of San Francisco* (1:50), *San Francisco Overland* (2:50), *City of Portland* (3:00), *Los Angeles Limited* (3:50), and *National Parks Special* (4:10).

All this took place in addition to the departures of the Rock Island's *Rocky Mountain Rocket* to Denver/ Colorado Springs (11:06 p.m.) and Chicago (11:59), plus those of the silvery flyers of the Burlington from its own station just across the tracks: the Chicago-bound *Ak-Sar-Ben* (10:30) and *Denver Zephyr* (1:15 a.m.), and the westbound *Nebraska Zephyr* (9:25), *Coloradoan* (10:20), *California Zephyr* (11:59), and *Denver Zephyr* (12:40). By the time the eastbound *California Zephyr* departed town at 5 a.m., Omaha was welcoming the dawn.

Passenger facilities in Omaha were attractive and convenient. Union Station, dedicated in 1931, served seven of the nine railroads providing the city with Pullman service: Union Pacific, C&NW, Milwaukee Road, Rock Island, Missouri Pacific, Wabash, and Illinois Central. The concourse that extended over the platforms, served by escalators, continued across the yards and connected with the Burlington's station just opposite. CB&Q and Chicago Great Western trains shared that facility. (UP donated Union Station to the city in 1973, and it became the Omaha History Museum.) A third station, Webster Street, served trains of

Omaha Union Station (top left), the street entrance of which is shown here, was alive all night long with the arrivals and departures of West Coast trains and many others. Burlington No. 42 (top right) takes on a Pullman for Omaha while being serviced in Alliance in 1965. The original *City of Portland* (center) puts on the speed near Grand Island in 1936. The *Rocky Mountain Rocket* (bottom), leaving Denver in December 1939 with five cars, soon will merge with the Colorado Springs section in Limon, Colo.

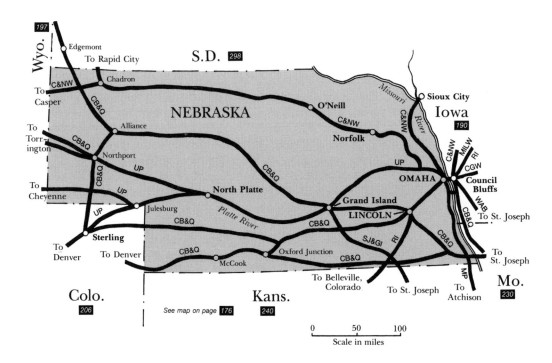

the Chicago, Minneapolis, St. Paul & Omaha operating on the west side of the Missouri River. The "Omaha Road" was owned by and operated as an integral part of the North Western system.

Of the two dominant passenger carriers in Nebraska, the Burlington offered a much longer Pullman menu, especially from Omaha. The UP emphasized through service from Chicago and scheduled few dedicated sleepers from Omaha. Exceptions were the cars to Denver, Cheyenne, South Torrington, Wyo., Salt Lake City (1890), and Los Angeles.

The Burlington, on the other hand, offered Pullman service from Omaha to nearly four times that number of destinations, most of them to the Northwest. The CB&Q completed this route through Alliance to Billings, Mont., in 1894, tapping the coal traffic from Wyoming and connecting with one of the road's future parents, the Northern Pacific, in Montana.[1]

Over the years, the Burlington scheduled sleeping cars to cities and towns as far northwest as Tacoma (NP), much of the service for the benefit of the vacation trade: Alliance; Edgemont and Deadwood, S.D.; Cheyenne (1890), Casper, Cody, and Sheridan, Wyo.; Billings, and, with the Great Northern, Shelby, Glacier Park, and Great Falls, Mont. Over its Missouri River line south, the Q carried Pullmans to St. Joseph, Kansas City, and St. Louis. On the Chicago-Denver main line, it scheduled sleeping cars to both those cities and Peoria.

The Burlington's tenant Great Western also dispatched Pullmans to Chicago, but it was the first to quit this stiff competition that once involved six railroads, including the C&NW, Milwaukee, IC, and Rock Island. The CGW was a latecomer in Midwestern passenger circles—it didn't enter the Omaha market until after the turn of the century—but its service was generally praiseworthy.

The Great Western also scheduled trains on the slightly less competitive (four-railroad) run to Minneapolis/St. Paul, with its *Twin City* and *Omaha Expresses* and *Twin City* and *Nebraska Limiteds*. The final remnants of these Corn Belt trains pulled into Omaha and Minneapolis on September 30, 1965, ending CGW passenger operations for all time. The railroad was merged into the Chicago & North Western three years later.

The North Western, longtime collaborator on the UP's Overland Route west, provided Omaha with a fair amount of dedicated Pullman service ranging from Wyoming to Minnesota. The C&NW had much of northern Nebraska to itself, and on this line ran sleepers to Chadron; Casper and Lander, Wyo.; and Deadwood. Trains on the Omaha Road carried Pullmans for Sioux City; Sioux Falls, Mitchell, Huron, and Aberdeen, S.D.; and Winona, Rochester, and Minneapolis/St. Paul, Minn. Schedules of the West Coast trains that came through town in the middle of the night, however, were not the most convenient for travel from Omaha to Chicago. The *Corn King*, inaugurated in the late 1920's, filled this void as the North Western's heavyweight overnighter to the Windy City.

The *Arrow*, dating to the early 1920's, was the Milwaukee Road's night train to Chicago that also cut in cars from Sioux City and, later, Sioux Falls. For a time the *Arrow* carried Omaha-Milwaukee sleepers that went aboard the *Southwest Limited* from Kansas City. The *Pacific Limited* maintained the Milwaukee's connection to the Overland Route, with through Pullmans from Chicago for San Francisco and Los Angeles, via Oakland.

In December 1940, the Milwaukee Road's mostly double-track main line got a premier daytime speedster with the inauguration of the *Midwest Hiawatha*, complete with the classic maroon and orange beavertail observation cars. The *Hiawatha*, which also had a Sioux Falls section, operated until the Milwaukee took over the Chicago leg of the Overland trains in 1955 and combined the *Hiawatha* with the *Challenger*.

Although the Rock Island's line through Omaha was renowned for service to Colorado on the venerable *Rocky Mountain Limited*, the railroad also carried through Pullmans for San Francisco (Denver & Rio Grande Western/Southern Pacific) on the *Colorado Express*. In November 1939, the *Limited* became the *Rocky Mountain Rocket*, the patriarch of the famous family that also gave Omaha the *Corn Belt Rocket*, a night train westbound, a day train returning to Chicago. The Rock Island scheduled Pullman service from Omaha to Des Moines, Minneapolis/St. Paul, Rock Island and Chicago, Ill., and west to Denver, and Wichita and Caldwell, Kans.

The Illinois Central's primary presence was elsewhere, but, like the Milwaukee Road, it funneled traffic from Omaha, Sioux City, and Sioux Falls onto a main line east to Chicago. Its overnight *Hawkeye Limited* carried Omaha-Chicago Pullmans but later focused solely on Sioux City. Early in the century, the IC traded interline Omaha-Twin Cities sleepers with the Minneapolis & St. Louis at Fort Dodge, Iowa.

Both the Wabash and Missouri Pacific carried Pullmans from Omaha to St. Louis. The MoPac, which inaugurated the daytime *Missouri River Eagle* on this route just before World War II, also scheduled sleepers from Omaha to Kansas City and Little Rock.

Nebraska's capital city, Lincoln, was a major center of Burlington activity, from which the lines to Denver, Billings, St. Joseph, and Chicago (via Omaha) radiated. And apart from the modest contribution of two other roads, Pullman movements in Lincoln were all Burlington.

The CB&Q's Nebraska service from Chicago—the 1930-vintage *Ak-Sar-Ben* (named for the Omaha booster organization) and post-war *Nebraska Zephyr*—terminated in Lincoln, except for the trains going through to Denver and the West Coast: the steam-era *Aristocrat* and *Exposition Flyer*, and the streamlined *Denver* and *California Zephyrs*. The famous first *Zephyr*, later prefaced with *Pioneer*, and its successor, the *Silver Streak Zephyr*, operated to Lincoln from Kansas City, by way of Omaha.

At Lincoln, the Burlington cut out sleepers from Omaha and Chicago that were bound for destinations on its line through Alliance to Billings. These and Lincoln setouts went aboard trains from Kansas City. Lincoln Pullmans included cars for Seattle (either GN or NP), Billings, Cody, Newcastle, Wyo., (1890), Deadwood, Edgemont, and Alliance.

Lincoln had sleeping car service to Chicago on both the Burlington and Rock Island, whose Colorado trains also operated through the Nebraska capital. In addition, the CB&Q scheduled sleepers to Kansas City and Peoria, the MoPac to Kansas City and St. Louis.

Another Burlington main line ran through eastern Nebraska, skirting the Kansas state line, before converging into the single trunk to Denver, at Oxford Junction. This was the route of the *Colorado Limited* and *St. Louis Limited*, which, by avoiding Omaha and Lincoln, provided more direct service between St. Louis and Denver, as well as to the intermediate stops of Kansas City and St. Joseph.

Pullman service from Alliance, in northwestern Nebraska, was all Burlington and included sleepers to Billings, Chicago, Kansas City, St. Louis, and Denver.

The *City* streamliners customarily made one-minute stops in the middle of the night at Grand Island, which had virtually all the UP's through Pullman service from Omaha to choose from. Here, too, the Burlington's trains from Lincoln to Alliance and Billings called, generally at night. The community's own sleeper, however, was a much smaller operation, provided by the tiny St. Joseph & Grand Island Railway, which carried the Pullman between its namesake cities. Early in the century, a sleeper operated through to Kansas City. For some residents of Kansas and southeastern Nebraska, the SJ&GI service to Grand Island provided convenient connections with the Overland and Burlington trains.

---

1. George H. Drury, *The Historical Guide to North American Railroads* (Milwaukee: Kalmbach, 1985), p. 72.

Richard Kindig

Otto C. Perry/Denver Public Library

The westbound *Pacific Limited* (top) departs Denver with 14 cars on March 16, 1941. Coming in out of the wildflowers with its bell chiming, North Western's No. 603 (bottom) enters Casper on Independence Day, 1932. The train delivered an Omaha sleeper to the Wyoming city.

29
Wyoming

The main line of Wyoming was the high road of America's first transcontinental railway. The Union Pacific entered Wyoming from Nebraska, joining the former Denver Pacific line at Cheyenne, before crossing the Overland summit at Sherman Hill on the way to Laramie. At Granger, the UP's Oregon Short Line departed for Portland from the main stem, which continued west to Ogden.

Although the railroad's crossing over Sherman Hill peaked at more than a mile and a half elevation, the climb was reasonably gradual and did not involve the usual mountainous conditions. And since the original UP tracks were laid, two relocations have reduced the climb over the top by several hundred feet. Throughout all their history, the tracks have been busy, and UP locomotive power, at least in this century, sufficient to the task.

All the West Coast *City* streamliners plied this Overland Route across the Rockies; Portland's branched off at Granger, and Los Angeles and San Francisco's rolled on to Ogden. Trains from Denver and Kansas City came north to Cheyenne, in later years mostly routed via the Borie Cutoff, which bypassed the capital city itself.

With such trains as the *Overland Limited, Portland Rose, Los Angeles Limited,* and numerous others hefting their consists over Sherman Hill, the main-line communities of Wyoming never suffered for lack of through Pullman connections. Dedicated service, however, was rare, except for a Cheyenne-Omaha sleeper and occasional through Pullmans for Portland that were listed as originating in Cheyenne or Laramie.

Early afternoons in South Torrington, a sleeper that had departed from Omaha on the *Columbine* arrived on the UP train that worked the south bank of the North Platte River. This branch left the Union Pacific main line just west of North Platte, Neb. After

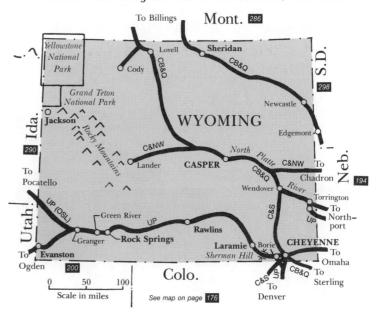

See map on page 176

Preston George

The westbound *City of San Francisco*, with its original E2 units, kicks up a cloud of dust east of Cheyenne, Nov. 30, 1941.

a notably quick turnaround, the Pullman was scheduled to depart fifteen minutes later on the eastbound train for the return to trip to Omaha.

The Burlington originated Wyoming's largest Pullman selection over the two lines leading to Montana. Much of this traffic consisted of summertime tourists traveling to Yellowstone National Park and other recreational areas.

The CB&Q was the last of the three major railroads serving Yellowstone, but once it established Cody as its gateway, it scheduled substantial service over the new line, giving the town the largest selection of Pullmans initiated in Wyoming.

Yellowstone Park, occupying the northwest corner of the state, was created in 1872, but it was not until the Northern Pacific's line was laid through southern Montana in 1883 that tourist travel to the new preserve was feasible. The NP was quick to exploit its advantage, and the railroad sponsored construction of hotels and lodges, including the famed Old Faithful Inn, to accommodate visitors. Popularity of the area increased, and, before automobiles took their toll, three other railroads offered substantial service to the park: CB&Q via Cody, the UP through Idaho, and the

Milwaukee Road at Gallatin Gateway, Mont.

The Burlington scheduled Pullmans to Cody from Omaha, Lincoln, Denver (Colorado & Southern), Billings, and Chicago (NP/CB&Q). Some of Cody's service in the 1920's permitted overnight travel to other scenic attractions in the region: Sleepers to Edgemont, S.D., served the Black Hills trade, and cars ran via the Great Northern to Glacier National Park in Montana, a GN-sponsored preserve.

In addition to Union Pacific trains, Cheyenne was also served by the Burlington, and later through the Colorado & Southern, which it controlled. The C&S/CB&Q scheduled Pullmans from Cheyenne to Billings, Casper, Omaha (1890), and Chicago (1888), and from Casper to both Denver and Omaha. The Burlington offered sleepers from Sheridan to Omaha and Chicago and from Torrington to Denver. A pre-1900 sleeping car from Lincoln terminated in Newcastle.

The most distant western outpost of the Chicago & North Western was Lander, the railroad's gateway to the national parks of Wyoming. Lander had summertime Pullman service to Chicago on the C&NW, and a Casper-Omaha sleeper also traveled this route.

Otto C. Perry/Denver Public Library

Colorado & Southern's No. 29 approaches Casper on May 30, 1935. The overnight train carried sleepers from Denver and Cheyenne.

30
Utah

The enterprise culminating at Promontory in 1869 had a freewheeling, all-American charm about it. The legislation that enabled the construction of the nation's first transcontinental railway was signed by a soon-to-be beloved president, who, given the choice, picked the wrong gauge. Abraham Lincoln decided that the tracks would be of five-foot width; Congress overruled him and specified standard (4 feet 8 1/2 inches). Once that was settled, construction began, again with an all-American cast: Irish crews pushing the tracks of the Union Pacific west from Omaha, Chinese gangs laying rails of the Central Pacific over the Sierra Nevada from Sacramento, the work of both lubricated by cash subsidies and western land grants.

That all of this effort came together at a certain point on a barren plateau north of Utah's Great Salt Lake was also the doing of Congress. Both companies, motivated by bonuses, passed each other like mule trains and were greedily laying rails in the other's territory until the legislative compromise was reached. Promontory it would be, and there on May 10, 1869, the golden spike ceremony took place as America awaited the telegraphic news signaling completion of the railway that unified the nation.

The ceremony itself had all the ingenuous qualities of amateur theater, a momentous pageant ordained to play with the limited props at hand in a rustic venue. Floods had delayed trains from the east, forcing several postponements. In true frontier spirit, final arrangements remained makeshift. Nonetheless, the show went on, played to a remarkably democratic and often inattentive assemblage of about six hundred, including dignitaries, construction workers, soldiers, tourists, strumpets, drifters . . . all the usual suspects.

The driving of the "last" spike involved a specially wired sledge hammer. Central Pacific President Leland Stanford was supposed to bring the maul down on the similarly wired spike, thus completing an electrical circuit and sending a telegraphic impulse to inform the anxious nation that the job was done. Stanford swung and missed. Next, the Union Pacific executive stepped up to bat, and he missed. Rather than allowing strike three to be called, the alert telegrapher completed the circuit with his key and fired off the message that America was spanned by rails. Whereupon the ample supplies of libations were uncorked. By nightfall, probably only the Mormons in all of northern Utah were drawing sober breaths.

For several years, until through sleeping car service was the rule, liquor flowed freely in this makeshift Utah rail junction of Promontory. Prostitution and gambling flourished, and while it wasn't exactly like the Parmelee transfer service in Chicago, it was all part of the ritual of changing trains.

In the end, however, Promontory proved to be one more disposable element in the American rail economy. The lines of both the CP and UP had been laid in

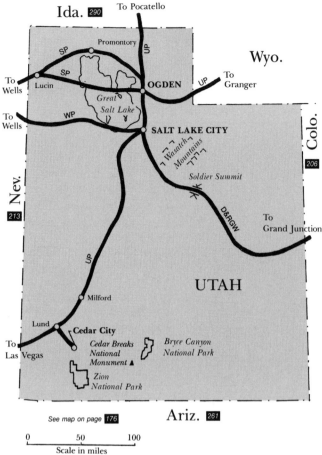

See map on page 176

```
0        50        100
Scale in miles
```

the route through Promontory and cut forty-four miles off the line west of Ogden. Thereafter, only occasional trains rolled through the historic community of the golden spike, the end for which came in 1942, when the rails on the old line were removed for the wartime scrap effort. A squat obelisk was erected on the spot where the pilots of the *Jupiter* and No. 119 had once touched head to head and thrilled a nation.

After the epic 1869 meeting of locomotives at Promotory, most of the workaday transfers took place at Ogden, Utah's second city and the nation's principal intermountain railroad center. There the line of the Union Pacific from Omaha met that of the Southern Pacific (which controlled the CP by lease) to Oakland. This key junction was additionally fed by the UP line to Pocatello, Idaho, and Butte, Mont., and the San Pedro, Los Angeles & Salt Lake Railroad, which came into the UP system shortly before World War I.

Historically, it was a day's trip west of Ogden; the streamliner schedules cut that to an overnight run.

The Southern Pacific took over trains bound for San Francisco: the *Overland, San Francisco Limited, Pacific Limited,* and the *City of San Francisco.* For the *Los Angeles Limited, Continental Limited,* and *City of Los Angeles,* it was UP all the way.

The flow of Pullmans through Ogden was enormous, but few sleepers terminated there. The SP, however, scheduled Pullmans from Ogden to San Francisco, and the Denver & Rio Grande Western carried sleepers to Denver via the Royal Gorge route in Colorado. Ogden's other service was tied into the operation of the Colorado Midland, which was abandoned in 1918. The Midland's service included through Pullmans from Chicago (1892) and Denver, and the Rio Grande carried them between Grand Junction, Colo., and Ogden. The D&RGW also sched-

haste, and by the turn of the century both were under the control of Edward H. Harriman, a rail baron determined to upgrade them into first-class properties. One of his achievements was the construction, across Great Salt Lake, of the Lucin Cutoff.

These 103 miles of track, part of which rode a twelve-mile trestle built on wooden piles driven into the lake bottom, eliminated the bothersome grades of

Union Pacific touring cars and passengers line up in front of the lodge at Bryce Canyon National Park at the start of an outing. UP and other western railroads actively promoted the development and use of national parks.

Bryce Canyon National Park

In an early publicity shot, the *Overland Limited* crosses the Lucin Cutoff when it was a wooden trestle across Great Salt Lake.

uled a sleeper from Glenwood Springs.

Although Salt Lake City lacked some of the through service that Ogden had, the UP's Los Angeles trains came through this scenic capital of Utah. Salt Lake City, also renowned as an interurban railway center, with four lines running as late as World War II, had far more of its own Pullman service than Ogden, most of it from the Union Pacific.

Much of the UP's service was to the north by way of such trains as the *Northwest Express, Northwest Special, Butte Express,* and *Butte Special.* Summertime traffic was especially heavy, with sleepers bound for Yellowstone Park and other recreational areas. Salt Lake City had Pullman service on this route to Ashton, Victor, Buhl, Idaho Falls (tourist), and Boise, Ida.; West Yellowstone and Butte, Mont.; Huntington and Portland, Ore., and Spokane.

On its line to Los Angeles, the UP carried sleepers to Cedar City; Caliente and Las Vegas, Nev., and Los Angeles. The Union Pacific opened its branch line to Cedar City in 1923, which provided a convenient gateway to Zion, Bryce Canyon, and Grand Canyon National Parks, as well as the later Cedar Breaks National Monument. The railroad operated tour buses from Cedar City and built lodges at the parks, the most notable of which was the one on the north rim of the Grand Canyon. In addition to the Salt Lake City cars, the UP carried through sleepers to Cedar City from Chicago and Los Angeles. The *National Parks Special,* a summertime train from Omaha, served this region.

The Union Pacific scheduled few Pullmans from Salt Lake City east: in modern times, Denver and Kansas City; before the turn of the century, Omaha and St. Louis (Wabash). UP carried a sleeper for Ely, Nev., to Ogden that the SP delivered to the tiny Nevada Northern.

The Denver & Rio Grande Western came into Salt Lake City from the southeast, after crossing the Wasatch Mountains at Soldier Summit. The Rio Grande offered Pullman service to Denver, via the Moffat Tunnel Route, most recently in the *Prospector.* In 1892, the Rio Grande had shared an interline Denver-Salt Lake City sleeper with the Colorado Midland/Santa Fe.

The D&RGW's principal connection at Salt Lake City was with the Western Pacific, a railroad it had helped finance shortly after the turn of the century in an arrangement that brought both roads into bank-

Tomorrow morning, Feather River Canyon—but for now the *California Zephyr* changes power and loads in Salt Lake City.

ruptcy. The WP was built to provide an alternate route to San Francisco, which it did by way of the scenic Feather River Canyon in the Sierra Nevada, to break the SP's monopoly.

Several memorable passengers trains made good promotional use of the natural wonders along their route. The *Panama-Pacific Limited,* inaugurated to serve visitors to the world's fair, was the WP's first name train and operated from 1912-15.[1] Thereafter, the durable *Scenic Limited* held sway alone until the opening of the Moffat Route in Colorado and inauguration of the *Panoramic* over those new tracks in 1934. The world's fairs of 1939 prompted the unveiling of the *Exposition Flyer,* a heavyweight that became the D&RGW's principal train and presaged the most memorable flyer of all, the *California Zephyr.* The famed *CZ* carried the West's first dome cars when it began its twenty-one-year career in 1949. During stops of these trains in Salt Lake City, Rio Grande and Western Pacific motive power was exchanged, and then they were on their way, westbound for Feather River Canyon, eastbound for the soaring Colorado Rockies.

Both Western Pacific and onetime archrival Southern Pacific carried sleepers for San Francisco.

---

1. Guy L. Dunscomb and Fred A. Stindt, *Western Pacific Steam Locomotives, Passenger Trains and Cars,* (Modesto, Calif.), p. 324.

At Ogden, the Overland Route's great intermountain transfer point, the *City of St. Louis* pauses on its way to Los Angeles.

31
Colorado

Colorado railroading, like the topography of the state, was of two different worlds. And in the higher one—the realm of mountain railroading—the dimension of elevation was seemingly defied with impunity. The state's railroad history was rich, and its themes course like whitewater rushing through Royal Gorge. Chapters included the warfare between rival track gangs, their frenetic construction activity to reach mining boom towns, and then the emergence of Colorado as the world power of narrow-gauge railroading.

Once into the standard-gauge era, the story included the laying of the highest and bleakest mountain crossings in America, one of which helped produce the country's largest railroad abandonment up to its time. All the while, the metropolis of Denver remained a mile high and dry, lacking any direct rail connection to the West until 1934.

Six main lines of five railroads entered Colorado from the east, four from Kansas, two from Nebraska. Their targets were, variously, Denver, Colorado Springs, and Pueblo, to the west of all of which lay the formidable stone wall of the Rocky Mountains.

Three of the railroads, the Burlington, Rock Island, and Missouri Pacific went no further. The mountains deflected two other thrusts: the Union Pacific to the north, the Santa Fe to the south. Running through the uplands of eastern Colorado was one thing; crossing the Rockies was another matter. Only one of the railroads that dared scale the Colorado peaks, the Denver & Rio Grande Western, survived. It faced its first major struggle as a narrow-gauge infant headed, under terms of its charter, for El Paso.

The violence of the Royal Gorge track war of the late 1870's has been exaggerated, but the armistice made a major imprint on Southwestern railroad history. If Hollywood were to film this little drama, the premise would be easily stated: "Hombre, there's room in this gorge for just one railroad"—and it was true: Over millions of years, the Arkansas River had sawed through the granite a canyon 1,100 feet deep and, in places, a mere thirty feet wide at its base, where the roiling waters cascaded.

The two contenders for this meager, sometimes nearly vertical, riparian acreage were the Santa Fe and the Denver & Rio Grande, and the construction crews of each raced west from Pueblo to build through the slim passage. The two had met before, at Raton Pass.

At first, the railroads waged economic warfare by trying to hire away the enemy's manpower. Soon enough, the struggle got ugly. Gunmen started taking potshots, and both railroads were prompted to build stone forts to protect their workers. Vandals burned bridges and dismantled or blocked rival trackage. Lawyers worked overtime.

In the final court settlement, the D&RG gave up its RG ambitions and agreed not to strike for El Paso,

Denver Union Station circa 1906 (top) hosted a mixture of standard-gauge trains and "little fellows." The three-rail track permitted handling of both regular and narrow-gauge trains at the same platforms. The Royal Gorge was the Rio Grande's great scenic attraction, and its trains commonly stopped (bottom) for ten minutes so passengers could view the roiling Arkansas River. As most of them view the spectacle, however, the inevitable railfan (foreground) checks out the D&RGW 4-8-4 on this July 16, 1938.

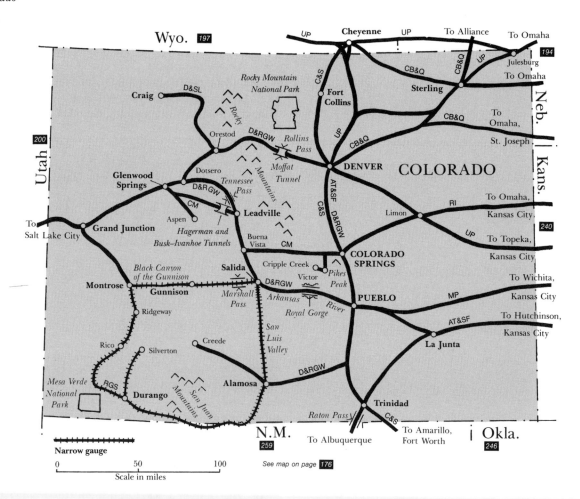

See map on page 176

Narrow gauge

0   50   100

Scale in miles

and the Santa Fe promised to turn its back on the Rocky Mountain Empire and forge on through New Mexico.[1]

The D&RG bought some of the Santa Fe trackage and continued on to Salida. By 1883 it had extended its narrow-gauge main line via Gunnison over the Continental Divide, using Marshall Pass, to Ogden. But now a new competitor was on the scene, one that threatened to upset the narrow-gauge ecology of Colorado's mountain region. The Colorado Midland was planning a standard-gauge line over the Rockies.

The Midland route proceeded west from Colorado Springs, skirting Pikes Peak on the way to Buena Vista, providing a much shorter route than the D&RG's through Pueblo and the Royal Gorge. The CM pushed on to booming Leadville, America's highest city, then crossed the Divide with a series of tortuous loops leading to the Hagerman Tunnel at more than two miles elevation above sea level, the highest standard-gauge crossing of the Rockies at that time.

The D&RG, meanwhile, seeing that it would have to go standard-gauge or lose valuable traffic, constructed a new line to Leadville and on over Tennessee Pass to Glenwood Springs, then set about widening the whole route. The two competitors were able to cooperate in the use of a common line into Grand Junction, where the tracks of the Rio Grande Western came in from Ogden.

For several decades, in a period of uneasy truce and sometimes Byzantine ownership schemes, sleeping cars rode over the Rockies on both lines, offering transcontinental passengers some of the most scenic travel in America. The D&RG's principal train was the *Scenic Special;* the CM operated the generic *Pacific / Atlantic Expresses,* the westbound *Ute,* eastbound *Coyote,* and the overnight *11 Come 7.* The Midland, in 1902, had the distinction of operating the first observation cars in the Rockies, the 10-section *Starlight* and *San Benito.*[2]

The life of the Colorado Midland, however, was short and never easy. Since it violated the mountain railroad convention of "build below the timber line," its passes were subject to the full fury of winter storms. Sometimes its trains were marooned in snow for days. During one winter the entire line was shut down for months, with a stock train and its frozen cargo completely buried. The newer Busk-Ivanhoe Tunnel helped somewhat, but it didn't cure the Midland's ills.

After the CM fell victim to several corporate squeeze-plays by other lines, principally the Denver & Rio Grande, the United States Railroad Administration dealt the death blow. This agency, which ran American railroads during World War I, diverted much of the Midland's traffic to other lines, making its financial outlook hopeless. Abandonment was the only answer, and so the costliest line ever built in the West operated its last train on August 4, 1918, and the Colorado Midland became, at that time, the largest railroad in history to be torn up and junked.

The D&RG, meanwhile, began having troubles of its own after overextending itself in the financing of the Western Pacific. This was an alternate line from Salt Lake City to San Francisco Bay, through California's Feather River Canyon, that was intended to break the SP's monopoly in that region. The D&RG followed the WP into bankruptcy in 1920 and, having previously been consolidated with the RGW, finally emerged as the Denver & Rio Grande Western.

Despite the railroads' financial difficulties, the Rio Grande/WP route proved to be a popular way west, offering first the accommodations of the *Scenic Limited*, starting in 1915, then the *Royal Gorge* after World War II.

Ten years after the death of the Colorado Midland, a tunnel was constructed that would once again give Colorado a second transcontinental route. The Denver & Salt Lake had been built early in this century by David Moffat, who planned to reach Utah but could finance his line no further than Craig. Traffic was sparse and the mountain crossing arduous. The D&SL went over desolate Rollins Pass, which was higher than even the Midland's Hagerman tunnel.

With the help of the city of Denver, which used a portion of the facility to supply its water, the 6.2-mile Moffat Tunnel—the Western Hempisphere's second longest—was completed in 1928, reducing the road's maximum elevation by 2,400 feet and substantially lessening the grades. Six years later, the Rio Grande, which by then controlled the Moffat Road, built the Dotsero Cutoff, linking its main line with the tunnel route, and for the first time Denver had a direct rail line to the West.

The Moffat Tunnel route was 175 miles shorter than the one going through Royal Gorge, and it became the Rio Grande's main line west. The *Panoramic* was inaugurated on the Moffat line in 1934, the *Exposition Flyer* in 1939, and the famed *California Zephyr* in 1949.

Some insist that the essence of Colorado's railroad charm was its system of narrow-gauge railroads. The southwestern portion of the state was the homeland of these pint-sized conveyances, and they were built in this primarily mountainous terrain for one main reason: they were more economical. Their trains could negotiate tighter curves and required smaller tunnels, substantially reducing construction costs.

General William Jackson Palmer, impressed by the economy of a Lilliputian railway in the mountains of Wales, ordered his Denver & Rio Grande built in the narrow, three-foot gauge. Pullman obliged by building a line of small sleeping cars to accommodate overnight passengers on the general's trains and others.

It was not long, however, before standard gauge became the norm because interchanging freight between trains of different gauges was both cumbersome and costly, and even mountain railroads had to conform to majority design.

For many years, however, a narrow gauge empire survived in southwestern Colorado, conducting its business, though not always profitably, in the manner in which it had been born. As late as 1949, a loop of nearly 600 miles, plus branches, was still in service, incorporating tracks of the D&RGW and the Rio Grande Southern, which the D&RGW controlled. The old Marshall Pass route from Salida to Montrose formed the northern segment. From Salida south, tracks went through the San Luis Valley to Alamosa. From there, the line wended its way over the San Juan Mountains to Durango, from which the loop continued west, skirting Mesa Verde National Park before heading north to Montrose. The railroad actively promoted "around the circle" tours, with stopovers at Mesa Verde.

The San Luis Valley route had the distinction of having a fifty-three-mile segment of straight track, the longest on the whole Rio Grande system. The Alamosa-Durango leg offered America's last regularly scheduled name train on a narrow-gauge line, the *San Juan,* which even carried parlor and dining accommodations. The Rio Grande discontinued operation of the *San Juan* in 1951, just when tourist interest in its branch line from Durango to Silverton was rapidly increasing. Happily, the open-car trains of the Silverton operation thrived and it eventually was granted landmark status.

Although Denver lacked a direct route across the Rockies until 1934, it was still the metropolis of the mountains and had by far the greatest choice of Pullman service of any city in the region.

The Rio Grande, while offering the most scenic route to Salt Lake City, Ogden, and California, had the schedule disadvantage of longer operation through Pueblo and the Royal Gorge, at least until the Moffat Tunnel route was opened. As a result, the Santa Fe and Union Pacific both gathered liberal shares of the western business. The Rio Grande, however, covered Colorado like powder snow at Vail. At one time or another, the D&RGW scheduled sleepers—whether narrow gauge, standard, or first one and then the other—from Denver to fourteen communities in the state. The selection included Alamosa, Aspen (1895), Craig, Creede, Durango (narrow), Glenwood Springs, Grand Junction, Gunnison (narrow), Leadville, Montrose, Rico (Rio Grand Southern, narrow, 1895), Salida, Silverton (narrow), and

Trinidad (1890).

After World War II, the Rio Grande scheduled Pullmans to Ogden on the *Royal Gorge* (via its namesake route) and to Salt Lake City on the *Prospector,* which traveled the Moffat Tunnel route. (A pre-war version of the *Prospector,* a Budd-built, self-propelled streamliner with sleeper, coach, and dining facilities was discontinued within months of its inauguration for lack of sufficent power to handle the mountain grades. A standard train was put in service at the end of the war.)

Although the Rio Grande carried sleepers from St. Louis and Chicago for the West Coast under numerous interline arrangements, its dedicated service from Denver was sparse. Early in the century it scheduled cars for Los Angeles (Union Pacific), San Francisco (Southern Pacific), and Portland (UP), but its later offerings were confined to San Francisco, such as the Pullmans that rode the *Scenic Limited* (Western Pacific).

One of the Rio Grande's memorable attractions were the sojourns in Royal Gorge, where its trains paused a short time at the famous Hanging Bridge, while passengers marveled at the soaring canyon walls above and the roaring water below. Earlier, the road operated open-top observation cars through Royal Gorge, as well as on its Marshall Pass line through the Black Canyon of the Gunnison. Schedules for the summer of 1917 showed the cars operating on the principal trains, with a fare of "25 cents per capita."

To the east, the D&RGW favored interline connections with the Missouri Pacific, both of which roads had once been part of George Jay Gould's transcontinental ambitions. The most handsome collaboration of the two systems, which met at Pueblo, was the streamlined *Colorado Eagle* to St. Louis, introduced in

1942. Rio Grande/MoPac Pullman service included Wichita, St. Louis (via Kansas City and early in the century also via Wichita), and Atlanta (Frisco/Southern).

The Rio Grande's eastern partner on the *California Zephyr* was universally acclaimed for its silvery fleet that carried the name of the legendary west wind, and the user-friendly Burlington had good reason to honor Denver with its own *Zephyr.* It was in the Mile High City, at 5:05 on the morning of May 26, 1934, that the modern legend began.

The Burlington had served the Denver-Chicago market well since the late 1880's. In 1930 it inaugurated the luxurious *Aristocrat,* one of three new eightieth anniversary trains, as its principal flyer on this run. As the Depression worsened and the performance of the fledgling airline industry improved, however, it became clear to the American railroads that they had to provide better and faster service to attract passengers. The CB&Q's answer was a three-car, articulated, stainless steel streamliner built by the E.G. Budd company. It was capable of speeds up to 120 miles an hour.

The sumptuous but relatively sluggish *Aristocrat* was then making the run in about twenty-six hours. Burlington President Ralph Budd decided to cap the highly successful promotional tour of the little shovel-nosed train he had named *Zephyr* with a dawn-to-dusk, non-stop run from Denver to Chicago that would essentially cut the time the *Aristocrat* consumed almost in half. To give the operation a little dramatic flair, Budd arranged to have the train glide onto a stage before a live audience at Chicago's Century of Progress Exposition.

After leaving Denver on that spring morning, the Burlington *Zephyr* whizzed through town and country,

Otto C. Perry/Denver Public Library

A decade after the Colorado Midland became the nation's largest railroad abandonment, in 1918, the Busk-Ivanhoe Tunnel, west of Leadville, finds use serving a motor road.

Richard Kindig

D&RGW's *Exposition Flyer* emerges from Tunnel 8 near Scenic, Colo., with a seven-car consist on Jan. 12, 1941.

passing countless witnesses stationed almost like side-boys along the CB&Q main line east. Then a short-circuit occurred, and the diesel engine had to be shut down—but the train was luckily on a downgrade. A courageous engineer from Electro-Motive Division, however, saved the day before the train had coasted to a stop by splicing two bare cables together, burning his hands in the process, but enabling the crew to restart the engine. The *Zephyr* again was under way, and news bulletins continued to report its progress.

Thirteen hours after it left Denver, the train was within the city limits of Chicago, and at 8:09 p.m. local time it rolled onstage, to the thunderous acclaim of the audience at the fair. The age of the streamliner had begun.[3]

Within two years of this dazzling performance, Denver had its own *Zephyr*, the first long-distance train in the CB&Q fleet and which, in 1956, would be America's last train of its kind to be completely re-equipped. The Chicago market was competitive (Union Pacific, Rock Island, and Santa Fe were all contenders), but the *Denver Zephyr* remained a redoubtable train of quality, carrying sleepers, slumbercoaches, Vista-Dome parlor cars, and diners, to the very end.

In addition to Chicago, Pullman service on the Burlington to the east included Omaha, Kansas City, St. Louis, and Peoria.

The Burlington's control of the Colorado & Southern Railway gave the Q an important north-south line that started on its own property from Billings and extended all the way to Galveston. Part of the southern segment of the line had the corporate identity of the Fort Worth and Denver Railway. The various elements of the C&S, which also had some Colorado narrow-gauge trackage, were consolidated under auspices of the Union Pacific, but in 1908 the Burlington acquired control.[4]

In the summer of 1940, the *Texas Zephyr* was put into service, carrying Pullmans from Denver to Fort Worth and Dallas. Over the years, the C&S/FW&D also carried sleepers to Trinidad, Amarillo, and on an interline basis to San Antonio (Missouri-Kansas-Texas), Galveston (SP), and New Orleans (either SP or Texas & Pacific).

Over the C&S/CB&Q line to Billings, the Burlington scheduled Pullmans to Casper, Cody, Billings, Butte (Northern Pacific), Gardiner (NP), and Seattle (either NP or Great Northern). Sleepers to Torrington, Wyo., Edgemont, S.D., and Alliance began riding the main line east before branching off to the north.

Inasmuch as the ill-fated Colorado Midland originated in Colorado Springs, the railroad required an interline arrangement to serve Denver. For this service, the CM initially relied on the Santa Fe and, later, the Colorado & Southern, which owned a half-interest in the CM during the first decade of the Twentieth Century. Pullmans departed from Denver via C&S/CM for Cripple Creek (Midland Terminal), Leadville, Grand Junction, and Ogden (RGW).

After snatching Raton Pass from Rio Grande track crews, the Santa Fe pushed westward and became a transcontinental railroad with its own line through to San Diego in 1885. It was not content, however, merely to crease the southeastern uplands of Colorado. Two years later it built a line into Denver, and in 1890 it leaped the formidable Rockies—thus accomplishing what it had been denied at Royal Gorge—by acquiring the Colorado Midland. The adoption unraveled after the financial chaos of 1893, but the Santa Fe continued handling some of the CM's interline Pullmans from Denver, sleepers to Cripple Creek (MT, 1896), Victor (MT, 1895) Aspen (1887-90), Salt Lake City (RGW, 1892), and, with RGW/UP, Boise (1898), Huntington, Ore. (1898), and Portland (1897).

Although the Santa Fe served Denver with a stub line from La Junta, the railroad's passenger service was of the usual high quality. It scheduled Pullmans to various destinations, ranging from Chicago, to Texas, to Los Angeles. Service included sleepers to Trinidad, Wichita, Kansas City, Oklahoma City, Tulsa, Fort Worth/Dallas, New Orleans (T&P), Galveston, San Antonio (MP), El Paso, and Phoenix.

Like the Santa Fe main line, the principal stem of the Union Pacific's Overland Route also bypassed Denver, much to the disappointment of prominent early settlers, who tried to coax it south. It was therefore not surprising that Colorado's first railroad was the Denver Pacific's feeder to the UP at Cheyenne. In the end, it hardly mattered that Denver was off the main line, considering the level of passenger service the UP provided. The original Kansas Pacific line to Kansas City and the connection to Julesburg, where the Overland Route briefly dipped into Colorado, provided good service to the East.

During the steam era, the Chicago-Denver trains included the *Colorado Express,* summertime *Denver Special,* and the *Columbine,* which in its 1927 inaugural edition carried an elegant heavyweight observation car in blue, gold, and silver livery adorned with the flag and crest of Colorado.

With the summer of 1936, however, came the crack *City of Denver,* the yellow, twelve-car streamliners that averaged 65.5 miles an hour on their sixteen-hour overnight journey. The *City* acclimated eastern cosmopolitans to the wild west with its rustic bar car decorated in the manner of a frontier shack. The *National Parks Special* was a summertime companion.

The UP line from Kansas City brought through trains for the West Coast, some of which originated in St. Louis: the older *Pacific Coast Limited* and the streamline-era *City of St. Louis* and *Pony Express.* The *Portland Rose,* originally a Chicago train that the streamlined *City of Portland* supplanted, originated in Denver in later years.

To the west, Union Pacific trains carried Denver Pullmans to Salt Lake City, Los Angeles, San Francis-

Union Pacific Railroad

When the Chicago-Denver race heated up in the mid-1930's, the *City of Denver* became the UP/C&NW's flashy entrant.

co (SP), and Portland; to the east to Omaha, Chicago (either Chicago & North Western or Milwaukee Road), Kansas City, St. Louis (Wabash), San Antonio (MKT), Shreveport (Kansas City Southern), and New Orleans (Frisco/Illinois Central).

Union Pacific tracks also accommodated Rock Island trains serving Denver. When the Rock Island was building its line to Colorado, the only promise of a standard gauge line over the Rockies in the vicinity was the Colorado Midland's. The Rock therefore headed for Colorado Springs, which it reached in 1888. To get access to Denver, it arranged trackage rights with the UP.

Although the Rock Island trains never matched the speed of the competing UP and Burlington flyers, the railroad nonetheless offered exemplary passenger service from Chicago, benefiting from the economy of serving both Denver and Colorado Springs with one train. The *Rocky Mountain Limited* entered service in 1898, operating until November 1939, when the silver-sided *Rocky Mountain Rocket,* under tow of red and maroon diesels, took its place. At Limon, the sections of the train were separated for their respective destinations.

In addition to Chicago Pullmans, the Rock Island scheduled sleepers from Denver to its junction at McFarland, Kans., and to Omaha, Kansas City, and St. Louis. For many years, the Rock Island served as the western leg of the Frisco/Southern Railway's Kansas City-Florida line. Under this interline arrangement, it carried Pullmans to Kansas City for Birmingham, Atlanta, Jacksonville, and (with Seaboard Air Line) St. Petersburg and Tampa. In two variations on this theme at Memphis, a sleeper returned to the Rock Island for delivery to Hot Springs, and a New Orleans Pullman went over to the IC.

Denver Union Station was born of a remodeling of its historic 1880's-vintage depot just before World

War I. A large, central waiting room, with high arched windows looking out over the platforms, was built between existing portions of the station. The through tracks and their platforms were raised to make underground access possible. In its early days, the rebuilt terminal was like a beehive, when 200 trains a day steamed in and out with their loads of summertime tourists.

Until early in this century, all the terminal tracks were three-rail to accommodate the trains of both standard and narrow gauges, and the little fellows continued serving the facility until the eve of World War II.

South of Denver, what often appeared to be one main line running to Trinidad along the morning side of the mountains actually involved three railroads, Santa Fe, Rio Grande, and C&S. Various arrangements—paired track, trackage rights, and joint trackage—governed the use of the rails by each of the three along certain segments, and the variety of trains north of Pueblo, as a result, was substantial.

Between the bustling commercial center of Denver to the north, and the coal, steel, and manufacturing communities to the south lay Colorado Springs, a settlement that prided itself for its salubrious atmosphere and mineral waters and genteel manner of living. Millionaires from the gold fields of Cripple Creek and eastern nabobs mixed in perfect comity in the posh precincts of this noted resort. The Broadmoor Hotel anticipated the whims of the jet set by forty years.

The Chicago Pullman trade from Colorado Springs was long the specialty of the Rock Island with its *Rocky Mountain Limited* and *Rocket*. The Santa Fe, however, briefly tried its hand in the late 1920's, and with the opening of the Air Force Academy, the Burlington added Colorado Springs sleepers to its newly equipped *Denver Zephyr* in 1956. The Rio Grande's *Royal Gorge* carried them south of Denver. Chicago, however, was the extent of Pullman service from Colorado Springs in the twilight of private passenger railroading.

The Rock Island earlier had carried sleepers to St. Louis and Kansas City, where the Frisco picked up other cars, for Brunswick, Ga., and Jacksonville, that it delivered to the Southern.

Before the turn of the century, on its overnight *11 Come 7*, the Colorado Midland carried a sleeper on the short journey to Cripple Creek (MT).[5] The Santa Fe scheduled Pullmans from Colorado Springs to Oklahoma City and Tulsa, the C&S to Fort Worth.

Pueblo, once an important sleeper terminus, faded as Denver's stature grew. Santa Fe provided pre-1900 service to the junction of Kinsley, Kans., and thriving Las Vegas, N.M. Through a trackage agreement with the Rio Grande, the Rock Island scheduled Pullmans to Chicago, Kansas City, St. Louis, and Hot Springs (Frisco/RI). Sleepers traveled from Pueblo via MoPac to Wichita, Kansas City, St. Louis, and Memphis. The

Burlington's *Aristocrat*, here east of Derby in 1935, was luxurious but slow. The *Denver Zephyr* solved the speed problem.

The *California Zephyrs* meet in Glenwood Canyon, where the idea for the dome car was born, for a spectacular setup shot.

Rio Grande carried a Pullman to Alamosa on its *Colorado & New Mexico Express.*

Before the demise of the Colorado Midland, D&RGW scheduled a sleeper from Glenwood Springs to Ogden, and on the Rio Grande Southern, a narrow gauge sleeping car tottered through the night from Durango to Ridgway.

Most of it, like the wildflower specials the Midland used to operate to the high country, is gone, but not all. Like the mountain columbines, the tourists still come back every summer to ride the wonderful living legacy of Colorado's early railroad history, the little train to the once-great boom town of Silverton.

1. James Marshall, *Santa Fe: The Railroad That Built an Empire* (New York: Random House, 1945), p. 156.
2. Letter of July 22, 1986, received from Tom Greco, Colorado Midland scholar. Mr. Greco reports that recent research by the railroad's historical society indicates that the Midland's own passenger equipment was painted a maroon/tuscan color.
3. Richard C. Overton, *Burlington Route: A History of the Burlington Lines* (New York: Knopf, 1968), pp. 395-7.
4. George H. Drury, *The Historical Guide to North American Railroads* (Milwaukee: Kalmbach, 1985), p. 100.
5. Mr. Greco speculates that the Colorado Springs-Cripple Creek run of fifty-seven miles might have been the shortest sleeping car line in America.

32
Nevada

America's first transcontinental railway was Nevada's only trunk line until early in the Twentieth Century. After the Chinese construction gangs brought the rails of the Central Pacific down from the Donner Pass in the Sierra Nevada, they raced ahead across the comparatively easier terrain of northern Nevada, following the course of the Humboldt River for much of the way. In May 1869, they met the Union Pacific crews at Promontory, Utah, and for almost four decades, this was the route of all the Overland trains serving California, north and south. (The Southern Pacific, which leased the Central Pacific, for some reason chose to avoid using the Overland name for much of this period.)

Shortly after the turn of the century, the San Pedro, Los Angeles & Salt Lake (which became the LA&SL) was built across southern Nevada, giving the UP, which later controlled the line, direct access to southern California. The new route obviated the need to operate through cars for Los Angeles by way of Sacramento or Oakland. In 1909, the Western Pacific line from Salt Lake City to San Francisco was completed, paralleling the SP across north-central Nevada, and crossing the Sierra through Feather River Canyon, north of Donner Pass.

The historic SP line carried such trains as the *Overland, San Francisco* and *Pacific Limiteds,* and the

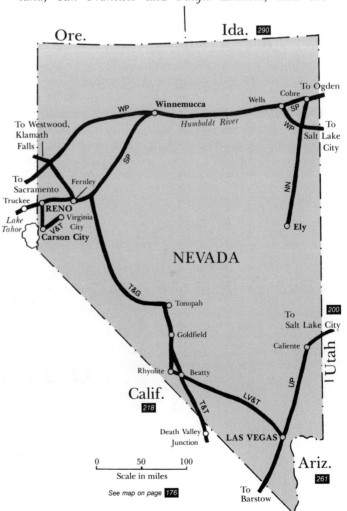

See map on page 176

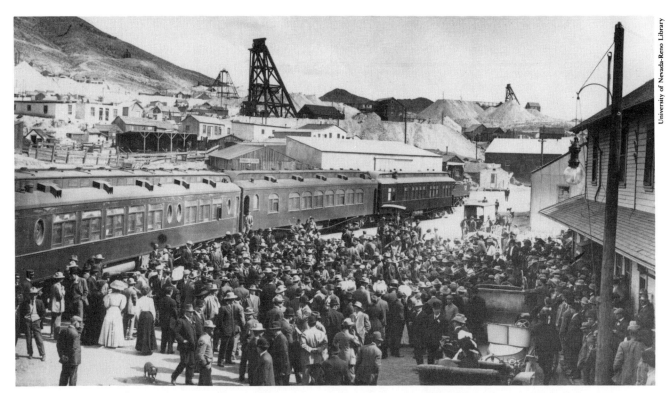

Tonopah hails a dignitary ca. 1906. Tonopah & Goldfield train carries a rarity for the time, all-room (7c-2dr) Pullman *Omena*.

streamlined *City of San Francisco*. From Reno, the SP scheduled sleepers to both Sacramento and San Francisco.

In an arrangement that originated during World War I, the Southern Pacific and Western Pacific operated under a paired-track agreement between Winnemucca and Wells, bringing such WP trains as the *Scenic Limited, Exposition Flyer,* and *California Zephyr* into close proximity with their Overland competitors.

The UP's *Los Angeles Limited* became the premier train through southern Nevada after opening of the LA&SL line, which later became the raceway of the

streamlined *City of Los Angeles* and *Challenger*. The UP scheduled Pullman service over this route to the gambling mecca of Las Vegas from both Los Angeles and Salt Lake City, and to Caliente from Salt Lake City.

Both the UP and SP, as well as the Santa Fe, which entered Southern California from Arizona, joined in interline sleeping car operations with some of Nevada's colorful, small railroads that served the state's booming mining communities.

In 1859, gold and silver were discovered in the eastern foothills of the Sierra Nevada near Lake Tahoe, and the legendary Comstock Lode fueled mining

Splashed in the sunshine of a July morning in 1938, WP's nine-car *Scenic Limited* rolls west near Portola, Calif.

enterprises there for the next several decades. During the height of the silver extraction, Virginia City was reputedly the most important urban center in the entire West. Within three years of the transcontinental railway completion, the Virginia & Truckee Railroad had connected with it at Reno, and soon sleeping cars were scheduled between Carson City and the San Francisco Bay railhead at Vallejo.

The fortunes of Virginia City waned with the depletion of the silver ore, but Carson City maintained its own importance as capital of Nevada. One of the big events in heavyweight boxing took place there on March 17, 1897, when Robert Fitzsimmons knocked out James Corbett in fourteen rounds. Dozens of special trains, many all-sleeper, brought in sports fans from around the country to witness the championship match.

Early in the Twentieth Century, two more Nevada boom towns burgeoned with the discovery of silver and gold, Tonopah and Goldfield. The tiny railroad of that name connected with the Southern Pacific's Mina branch, and the two provided interline Pullman service from Tonopah to San Francisco.

Another short-line railroad, the Tonopah & Tidewater, came up from Southern California to Goldfield, skirting Death Valley on the way. The T&T connected with the Santa Fe at Ludlow, Calif., and the two roads participated in joint Pullman service from Los Angeles to both Beatty and Goldfield. The T&T later worked with the UP on the Los Angeles-Beatty sleeper.

Shortly after the turn of the century, Rhyolite had a feverish surge of mining activity, and a Los Angeles Pullman was inaugurated. The Las Vegas & Tonopah Railroad carried the sleeper from Rhyolite, just west of Beatty, to Las Vegas, where the LA&SL took it over for the remainder of the trip. Later in the same year, 1907, another sleeper was put into service from Goldfield, traveling to Beatty on the colorfully named Bull-

An oil-burning Union Pacific 4-6-6-4 hauls 14 cars of the eastbound *Pony Express* near Caliente on June 9, 1947.

frog Goldfield Railroad, thence LV&T/LA&SL to Los Angeles.[1]

Two possible trunk lines that might have been built, but never were, involved the Nevada Northern, a railroad serving the great copper mines near Ely. Extensions of the line to both Tonopah and the LA&SL tracks in the south were proposed, but that was all. Meanwhile, the NN, connecting with the SP at Cobre, originated Pullmans for both San Francisco and Salt Lake City (UP).

1. David F. Myrick, *Railroads of Nevada and Eastern California* (Burbank, Calif.: Howell-North Books, 1963), pp. 486-7. UP (LA&SL) schedule information courtesy library of Nevada State Railroad Museum.

In the spring of 1950, SP No. 26 winds through the Truckee River canyon near Verdi with two Oakland-Reno Pullmans, one more than the schedule advertises.

33
California

Neither San Francisco nor Los Angeles figured prominently in the plans for the first transcontinental railroads that eventually came to their doorsteps. Investors in San Francisco, accustomed to maritime trade, showed little interest in backing the Central Pacific Railway, and so the four principals of the company—all Sacramento businessmen—raised the necessary capital themselves to lay the iron to Promontory, Utah.

When the Central Pacific began laying tracks south through the San Joaquin Valley, the plan was to head for the Colorado River near Needles and connect with lines building west from Texas. Its detour through Los Angeles was a subsequent accommodation. And when the Santa Fe became a transcontinental in 1885, San Diego, not Los Angeles, was its Pacific railhead.

The Sacramento merchants soon came to be known as the Big Four—Leland Stanford, Collis P. Huntington, Mark Hopkins, and Charles Crocker—and they built the Central Pacific across the Sierra Nevada to its historic rendezvous with the Union Pacific. The four entrepreneurs, of course, had the help of more than 15,000 track layers, mostly Chinese, who carved and filled a roadbed through the same pass where the Donner party had met disaster.

In the home state, after the golden spike ceremony in 1869, the Big Four lost no time constructing and taking over other railroads, doing business as both the CP and the Southern Pacific, over which they had common control.

Gradually the SP spread its tentacles throughout California, north and south, acquiring a virtual monopoly on rail transportation—but for the Santa Fe's incursion in the far south, which the Big Four had bitterly oppposed. The SP used its substantial political clout (Stanford was a former California governor) to advance its interests, and in time the tentacles became the metaphor for Frank Norris's famous novel *The Octopus.*

The line through the San Joaquin Valley and across the Mojave Desert to Needles, however, provided the ultimate basis of an accommodation that both the SP and Santa Fe saw to be in their own best interests.

Responding to overtures from Los Angeles, the SP laid a line south from Mojave into the City of Angels and continued it eastward, past the area now occupied by the Salton Sea, to Yuma. There it joined with the rails of other chartered projects that together would form the Sunset Route to El Paso.

The Santa Fe continued pushing its tracks toward the Pacific through New Mexico and Arizona, but with a charter good only to the Colorado River and a California legislature swayed by SP influence, it had to use the SP line west of Needles. Meanwhile, a group of San Diego investors were building the Southern California Railroad through Temecula Can-

On a summer morning in the late 1960's, the *City of San Francisco* snakes down the Sierra Nevada past Crystal Lake. It's a long train for the waning railroad passenger industry, and an EMD helper is on the point.

Richard Steinheimer/DeGolyer Library

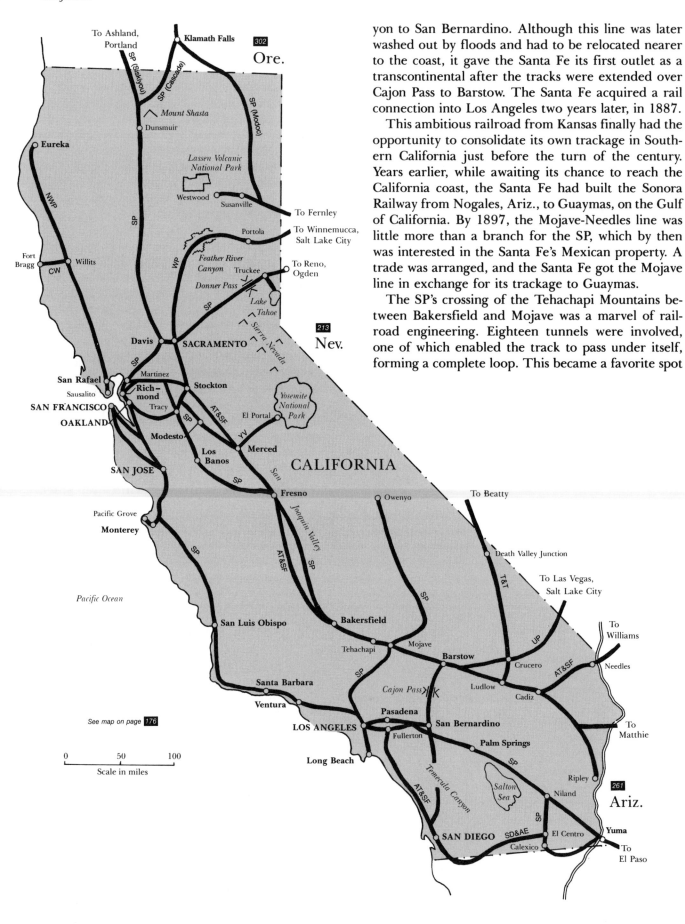

yon to San Bernardino. Although this line was later washed out by floods and had to be relocated nearer to the coast, it gave the Santa Fe its first outlet as a transcontinental after the tracks were extended over Cajon Pass to Barstow. The Santa Fe acquired a rail connection into Los Angeles two years later, in 1887.

This ambitious railroad from Kansas finally had the opportunity to consolidate its own trackage in Southern California just before the turn of the century. Years earlier, while awaiting its chance to reach the California coast, the Santa Fe had built the Sonora Railway from Nogales, Ariz., to Guaymas, on the Gulf of California. By 1897, the Mojave-Needles line was little more than a branch for the SP, which by then was interested in the Santa Fe's Mexican property. A trade was arranged, and the Santa Fe got the Mojave line in exchange for its trackage to Guaymas.

The SP's crossing of the Tehachapi Mountains between Bakersfield and Mojave was a marvel of railroad engineering. Eighteen tunnels were involved, one of which enabled the track to pass under itself, forming a complete loop. This became a favorite spot

Silver Crescent's porter awaits the ferries bringing *California Zephyr* passengers to the train at Oakland Mole, summer 1955.

to photograph one end of a long train crossing the other. The Santa Fe had trackage rights on this busy portion, but laid its own line north of Bakersfield to Oakland.

Mountains were always part of the mix in California's rail construction, especially on the SP's Shasta Route north to Oregon. The original line crossed the rugged Siskiyou Range, providing the first service to Portland from the south, in 1887. Forty years later, SP completed a more favorable crossing of the Cascade Mountains. Like this line, all the other major rail construction projects in California awaited the Twentieth Century.

From San Francisco, the city's only trunk line went south to San Jose, from which the SP had a charter to continue building down the coast. The Coast Line to Los Angeles was finally completed in 1901 and slightly shortened three years later. This was the most scenic seaside line in America, incorporating more than a hundred miles of running along sheer ocean cliffs and beaches north of Ventura.

The San Pedro, Los Angeles & Salt Lake gave its namesake California terminals much more direct connection with the Overland Route, at Ogden. This line through southern Nevada joined the Santa Fe just east of Barstow and, under a 1901 agreement, used the railroad's tracks over Cajon Pass to San Bernardino. Reorganized without the "San Pedro" in its name, the LA&SL was leased by and became part of the Union Pacific.

America's last route to the West Coast was opened in 1909, with the completion of the Western Pacific Railroad from Salt Lake City to Oakland. The SP monopoly of direct service to the east was of concern both to California business interests and to George Gould, the eastern rail mogul who found himself excluded from the West Coast markets by Edward Harriman's control of the SP/UP. Construction of the WP, with the financial backing of Gould's Denver & Rio Grande, was the result.

Builders of the Western Pacific availed themselves of a scenic, though much lower crossing of the Sierra Nevada than the SP used. By using Feather River Canyon and Beckwourth Pass, WP trains experienced a maximum of one per cent grades.

The Northwestern Pacific, a consolidation of roads extending from San Francisco Bay north to Eureka, was a cooperative enterprise of both the SP and Santa Fe for about two decades until 1929, when the latter sold its interest. Until the Golden Gate Bridge took away its traffic, this line through the redwood empire ran a substantial electric suburban service to Sausalito, from which its ferries carried commuters to San Francisco.

After its early glory as a trancontinental terminus, San Diego remained basically a branch—and monopoly—operation of the Santa Fe from its major railhead at Los Angeles. The San Diego & Arizona Railway, however, was designed to give the hometown better connections to the east by way of a sinuous line that dipped into Mexico, traversed the stark Carriso Gorge, and joined the SP's alternate route through the Imperial Valley at El Centro. The SD&A, completed in 1919, exchanged through cars with some of the secondary trains on the Sunset and Golden State Route. Like the NWP, the reorganized San Diego & Arizona Eastern became an SP subsidiary.

Another Twentieth Century construction project was comparatively short, just over a mile in length, but it continued a tradition of superlatives. Originally the Central Pacific trains entered Oakland from the south, via Tracy and the Altamont Pass. The grades and the mileage were excessive, however, and so the railroad instituted ferry service to carry trains from its Vallejo Line across the Carquinez Strait to Port Costa, from which they continued on to Oakland over a new line skirting San Francisco Bay.

The paddle-wheeled *Solano* began service in 1879 and was joined in 1915 by the *Contra Costa*, both of which were the largest ferries ever built up to their time. Passenger trains of the Shasta and Overland Routes were broken into as many as four sections for the trip across the water, after which they were reassembled and continued on their way.

Ferry service, however, gave way to more expeditious handling of the trains after a half-century of colorful operation. In 1930, the SP opened its heavy, double-tracked bridge, the longest such span west of the Mississippi, across the strait between Benicia and Martinez.

San Francisco was always the nominal destination of transcontinental rail service to Northern California, a tradition reinforced with such train names as the *City of San Francisco* and *San Francisco Chief*. Geography of the Bay Area, however, dictated otherwise, and with the exception of certain trains from the south, San Francisco was the terminus in name only. All other trains discharged their passengers in the East Bay, mostly in Oakland, and travelers going to San Francisco had to complete their journey on what was advertised as the world's largest ferry system.

Befitting its early image as a monstrous octopus, the Southern Pacific had gained control of the entire Oakland waterfront through a deal a subsidiary had cut with the city's mayor in the late 1860's.[1] Despite subsequent challenges, the SP monopolized the East Bay terminal scene until early in the Twentieth Century, when the Western Pacific, under cover of darkness and armed guards, laid tracks along the north quay protecting Oakland's inner harbor and won court approval of their bold coup. The Santa Fe, meanwhile, had to content itself with a more distant boat terminal called Ferry Point, in Richmond.

But even with the contenders in residence, the SP's dominance was never in dispute. From its ferry terminal at Alameda and the huge Oakland Mole, the railroad operated as many as forty-three double-ended

Passengers check in at the pier for the *Oakland Lark*, which in later years combined with the San Francisco train in San Jose.

Little more than a week old, the *San Francisco Chief* heads for the Bay Area through the Tehachapi Pass in June 1954.

boats carrying passengers and vehicles across San Francisco Bay.[2] On the eve of World War I, 1,000 trains a day served the Oakland pier, most of them the red, multiple-unit electric trains that plied SP's large commutation network in the East Bay.[3] In addition, all the steam trains of the Shasta and Overland routes as well as other California lines called at Oakland.

Before the bridges opened, the only way to cross the bay was by ferry: on the orange boats of the Key System, which also operated an extensive East Bay rail network funneling into its own tidewater terminal; the Western Pacific's *Edward T. Jeffery*, serving the railroad's own mole just to the south of the SP's; the Santa Fe's boat service from Ferry Point; and of course the SP's great white fleet serving both Oakland and Alameda, and its subsidiary Northwestern Pacific ferries, which extended the network across the Golden Gate to Marin County. (In 1933, on the eve of the bridge openings and the twilight of the ferry era, the Southern Pacific agreed to let Santa Fe and WP trains use Oakland Mole and their passengers use SP ferries.)

The destination for all this activity in San Francisco was the Ferry Building, with its soaring clock tower, at the foot of Market Street, where the thoroughfare's famed four-track, street car right-of-way terminated.

The peak year for ferry traffic was 1930—just as the Depression was beginning to take its toll—when the SP boats alone carried more than 40 million passengers. The coup de grace, of course, was delivered with the opening of the bridges. Auto ferry service ended almost immediatly, and the SP made its last commuter run in early 1939.

The San Francisco Bay Bridge was originally built with two train tracks on its lower level. These accommodated the commuter trains of the SP's subsidiary Interurban Electric, the Key System, and the Sacramento Northern. The trains proceeded over a rail right of way to an elevated terminal south of Market Street in downtown San Francisco. Only the trains of the Key System survived World War II, and that service ended in early 1958 after motor vehicle interests demanded more bridge lanes, and the tracks were ripped out and paved over.[4]

The remnants of trans-Bay ferry service ended that summer, awaiting rediscovery a number of years later. Train passengers, themselves an endangered species, thereafter departed San Francisco by bus from the SP's mission-style terminal at Third and Townsend.

The *Golden Gate Special, Overland Limited, San Francisco Limited, Pacific Limited, Treasure Island Special, For-*

Caption text below image:

Northwestern Pacific's No. 3, with its Pullman from Eureka, begins its final climb to the tunnel north of San Rafael, home.

ty-Niner, City of San Francisco—over the years the fleet of SP trains leaving Oakland for the east was substantial. The railroad's principal partner was the Union Pacific, but it also ran interline sleeping car service with the Rio Grande, at least when the Goulds and Harrimans were not snubbing one another.

Regionally on the Donner Pass line, the SP carried Pullmans to Sacramento, Truckee, Lake Tahoe, Susanville, and Westwood; in Nevada to Reno, Carson City (Virginia & Truckee, 1876, from Vallejo), Tonopah (Tonopah & Goldfield), and Ely (Nevada Northern), and to Ogden and Salt Lake City. The Union Pacific forwarded sleepers to Denver and Kansas City, and an interline tourist car operated via D&RGW to Denver. The main eastern destinations were Chicago and, secondarily, St. Louis. Post-war Pullmans operated through to New York and Washington.

To the north, the SP ran its Shasta service, with sleepers to Dunsmuir, Klamath Falls, Portland, Spokane (UP), and Seattle (Pool service). This was the line of the all-Pullman *Shasta Limited* and successor *Cascade*, the *Klamath*, *Oregonian*, and *Shasta Daylight*.

After opening of the Cascade Line in 1926, passenger service was gradually shifted to it from the older Siskiyou route.

Service to the south was more complicated in its routing. The SP had two main lines between Northern and Southern California, the Coast Line and the San Joaquin Valley line(s). The latter had alternate routes, intersecting at Fresno, from Tracy to a point north of Bakersfield.

The Coast Line forked at San Jose, with service from Oakland but principally from San Francisco. Trains for the Valley line left Oakland going north by way of Martinez before curving around to the southeast and traveling one of the two lines through Fresno.

After completion of the Coast Line, by far the more scenic, most of the long-distance Pullman traffic to San Francisco traveled that route, although there were exceptions. San Francisco, until World War II, was the terminus of the Golden State/Sunset routes, and through cars from Chicago were routed on both lines at different times.

The Coast Line was the route of the *Lark*, an all-Pullman train from San Francisco to Los Angeles that began running in May 1910. Although the *Lark* was inaugurated before Hollywood's emergence as motion picture capital, the train became a favorite of movie people. As a prewar streamliner, the Lark carried the SP's nighttime two-tone gray paint and included in its consist one of the triple unit lounge/diner/kitchen cars.

A companion all-Pullman train, the *Padre*, also ran on the Coast Line, but from Oakland, until replaced early in the Depression by the *Oakland Lark*. Two more overnight trains were still working the Coast route with sleepers at this time, the *Sunset Limited* and the leisurely *Coaster*.

This almost beach-combing trackage of the SP, however, was probably most widely remembered for what many thought was the most beautiful train of all, the *Daylight*, in its striking red, orange, and black livery, hauled by streamlined, silver nosed GS 4-8-4 locomotives.

The *Daylight* began service in 1922 with a heavyweight consist but stepped-up schedule that made it the SP's fastest train. A 1930 trainset was done in gray, but the 1937 streamlined version came on the scene like a sunburst, to instant popularity. A second train was introduced in 1940, making morning and noon editions that together carried peak loads of more than 1,500 passengers from Third and Townsend streets. A *San Joaquin Daylight* went into service in 1941 and *Sacramento* and *Shasta Daylights* after World War II. (The *Noon Daylight* was suspended during the war, then reinstated for several years before becoming the overnight, all-coach *Starlight*.)

In addition to Los Angeles service on the Coast

Fred Matthews/California State Railroad Museum

Line, San Francisco had a choice of Pullmans to Monterey, Santa Barbara, San Antonio, New Orleans, and Chicago, and tourist cars to St. Louis (Missouri-Kansas-Texas), Cincinnati (Illinois Central/Baltimore & Ohio), and Washington (Louisville & Nashville/West Point Route/Southern).

Sleepers departed Oakland on the Valley route for Merced, El Portal (Yosemite Valley), Fresno, Bakersfield, and Los Angeles. The long-time champion on this line—and of course antedating the Coast Line trains—was the overnight *Owl*, a onetime all-Pullman train inaugurated in 1898.

Unlike its two competitors connecting San Francisco and the east, the Santa Fe had longer, more southerly routings for its two lines to Kansas City. Just like its other passenger accommodations, service was first-rate.

The Santa Fe's line from Oakland all but paralleled the SP's San Joaquin operation in addition to sharing its tracks over the Tehachapi Mountains. The Santa Fe's long-distance service involved cutting San Francisco cars in and out of the Chicago-Los Angeles trains at Barstow. Over the years, the *California Limited*, *Navajo*, and *Missionary* all carried San Francisco Pullmans. The *Grand Canyon*, introduced in 1929, operated in several sections, including one going through to the Bay area. The *San Francisco Chief* replaced this in 1954. In addition to Chicago service, the Santa Fe scheduled Pullmans to Burrton, Kans. (for pre-1900 interchange with the Frisco), Kansas City (tourist), and New Orleans (Missouri Pacific).

For intrastate service, the Santa Fe was handicapped by the rather indirect route to Los Angeles over Cajon Pass. Nonetheless, in 1912 it went into direct competition with the SP's *Lark* with its own

southbound *Angel* and northbound *Saint*. The trains, which offered compartment/drawing room cars for both Los Angeles and San Diego, were discontinued during World War I. The Santa Fe also scheduled Pullmans from Oakland for El Portal (YV), Fresno, Bakersfield, Barstow (tourist), and Needles.

In 1938, the Santa Fe devised more competitive service to Los Angeles with its streamlined, twin *Golden Gate* trains. These operated from Oakland to Bakersfield, where passengers boarded buses for the remainder of the trip. The *Valley Flyer*, a steam-powered train, supplemented the *Golden Gates*.

San Francisco's most publicized service, at least in modern times, was provided by the silvery *California Zephyr*, which after going on display on the city's Embarcadero in 1949, began a remarkable twenty-one year career of traversing the Feather River Canyon and Colorado Rockies en route to Chicago. In its prime, it carried through Pullmans for New York, but the stainless steel cars for Chicago, including five Vista-Domes, were always the main and most memorable part of the consist.

The Western Pacific invariably interchanged at Salt Lake City with the Rio Grande, which had sponsored the WP's construction. Earlier collaborations of the two roads included the *Exposition Flyer* via the Moffat Route and the *Scenic Limited* via the Royal Gorge, the latter carrying Denver and St. Louis (MP) Pullmans. The Western Pacific scheduled sleepers on its own trains to Portola and Salt Lake City.

Sausalito, a busy railhead where the Northwestern Pacific's electric commuter lines terminated before abandonment in 1941, had once been considered as a possible Bay area terminus of the Overland Route. A

An SP cab-forward guides the southbound *West Coast* through Soledad Canyon. The train, which bypassed the Bay Area, once carried Los Angeles-Seattle sleepers.

Donald Sims

predecessor line of the NWP was briefly owned by the Central Pacific when this subject was broached in the early 1870's, but of course the decision was otherwise.[5]

But even after the third-rail interurbans were gone, and the NWP shifted from Sausalito to San Rafael, the railroad was still carrying a Pullman on its flood-prone route through thousand-year-old redwoods to Eureka. In the 1920's service also included sleepers to Willits and Fort Bragg, the latter by way of the tiny California Western, renowned for its colorful series of railcars dubbed "Skunk."[6]

Los Angeles did not have direct service to the east until twelve years after the Overland Route was opened. In 1881 the SP joined tracks with the Texas & Pacific, and two years later the modern Sunset Route was opened through to New Orleans. The Santa Fe arrived in Los Angeles in 1887. By then the boom was on, fueled by a rate war between the SP and Santa Fe, and Los Angeles's slow start as a rail terminus was quickly forgotten.

The Southern Pacific's line through Palm Springs to Yuma was part of the western leg of three different routes, the Sunset to New Orleans, the SP/Rock Island line through Tucumcari, known as the Golden State, and the Sunshine Route through Texas to St. Louis. The SP had alternates to this basic route, in California, Arizona, and New Mexico.

The *Sunset Limited,* first operated in 1894, was always the premier train to New Orleans, the *Argonaut* its secondary companion. Over this route the SP carried Pullmans to Phoenix, Tucson, San Antonio, Houston, New Orleans, and Jacksonville (L&N/Seaboard Air Line), and tourist sleepers to El Paso, Cincinnati (IC/B&O) and Washington (L&N/WPR-/Southern).

Los Angeles also had sleeping car service to Mexico by way of this line and a branch to Nogales, Ariz., where the SP's onetime subsidiary Sud-Pacifico de Mexico took over. Pullmans traveled south to Guaymas, Mazatlán, Tepic, Guadalajara, and Mexico City (Nacionales de Mexico).

The *Golden State Limited,* born in 1902, was the flagship of the SP/Rock Island interline service to Chicago, over the years running in the company of the *Californian, Apache,* and *Imperial.* Pullmans from Los Angeles to Chicago, Minneapolis/St. Paul, St. Louis, and Memphis were commonly part of the consists, and after World War II the *Golden State Limited* offered through service to New York. Additionally, the SP originated sleepers to Calexico, Little Rock (RI), and Kansas City (RI).

With the inauguration of the *Sunshine Special* in 1915, the Missouri Pacific/T&P gave travelers from California to St. Louis and Memphis greatly improved service. The MKT also interchanged St. Louis Pullmans with the SP at San Antonio. The Southern Pacific scheduled sleepers to Fort Worth/Dallas (T&P) and on its own lines just to Dallas.

Los Angeles had Pullman service to both San Francisco and Oakland by way of the Coast Line and to Oakland through the San Joaquin Valley. Most of the intrastate cars traveled the Valley route, sleepers to Bakersfield, Fresno, Merced, El Portal (YV), and Sacramento. On its line through the Owens Valley from Mojave, the SP scheduled a sleeper to Owenyo. The *West Coast,* a train that bypassed Oakland, carried Pullmans to Portland and Seattle (Pool).

Until the LA&SL connection to Salt Lake City, the Overland Route was not so convenient for Los Angeles travel as other lines. Nonetheless, the SP previously carried sleepers that were cut into Overland trains, either at Oakland or Sacramento, and long after the opening of the LA&SL, tourist cars for Chicago and St. Louis were still traveling the old route.

With such trains as the *Los Angeles Limited* and the crack *City of Los Angeles,* the Union Pacific captured a fair share of the market, offering Pullman service to Chicago, the Twin Cities, St. Louis, and, after World War II, New York and Washington. Among the intermediate destinations were Las Vegas, Cedar City, Utah (summers), Salt Lake City, Butte, Sun Valley,

Robert O. Hale

Santa Fe's **Grand Canyon** pops off as it tops the grade of Cajon Pass. The mountain crossing also saw Overland trains.

Ida. (winter), Denver (UP alone and also with D&RGW), Omaha, and Kansas City. At Crucero the UP interchanged sleepers for Death Valley Junction and Beatty, Nev., with the Tonopah & Tidewater Railroad. Winter 1952 schedules list a once-weekly sleeper from Spokane. Both Long Beach and Pasadena had dedicated Pullman service from Chicago. For a period of several years starting in 1907, sleepers traveled the UP to Las Vegas, whence the Las Vegas & Tonopah Railroad carried them on to both Rhyolite and Goldfield, Nev. (Bullfrog Goldfield Railroad).

Despite the SP and UP's best efforts, however, the Santa Fe dominated service to the east with an array of trains including the *De Luxe, California Limited, Grand Canyon, Chief,* and the monarch of them all, the *Super Chief.* Just as its competitors did, the Santa Fe carried Pullmans to Chicago, the Twin Cities (Chicago Great Western), St. Louis, Washington (B&O), and New York (either Pennsylvania Railroad or New York Central), in addition to sleepers to Phoenix, Grand Canyon, Denver, Kansas City, Dallas/Fort Worth, and New Orleans (MP) and from Galveston. Within California, the railroad offered sleepers to San Diego, Barstow, Merced, Oakland, and Ripley. Interline service with the Tonopah & Tidewater included sleeping cars to Beatty and Goldfield, Nev.

Los Angeles Union Passenger Terminal was the last architecturally notable American train station built on a grand scale. The California Mission style structure, shared by the SP, UP, and Santa Fe, was opened in May 1939 just a few blocks from the downtown Civic Center. From the large waiting room with ornately tiled floor, pitched-beam ceiling, and heavy, colonial chandeliers, passengers entered an underground concourse from which the eight train platforms were accessible by ramps.

Another unique feature of the Los Angeles passenger scene was the Pacific Electric Railway, which

Southern Pacific's eastbound *Sunset Limited* pauses at Colton.

billed itself as the largest interurban in the world. The huge network, over which 2,300 daily trains traveled on the eve of World War I, extended from the San Fernando Valley to Orange County, from San Bernardino and Redlands to the ocean. This SP subsidiary offered interline connections in its big red cars to the parent's long-distance passengers, scheduling, for instance, no fewer than 542 daily trains to Hollywood at the start of the Depression. Passenger service on the once-mighty system ended in 1961.

San Diego, with a short-lived exception starting in the 1920's, was Santa Fe country, connected to Los Angeles by the Surf Line, another scenic route that gave passengers the sensation of running barefoot

Before consolidation of passenger stations, the *Los Angeles Limited* departs over an L.A. street at 10 mph in April 1934.

The California Mission-style Los Angeles Union Passenger Terminal opened in 1939, with UP, Santa Fe, and SP sharing its tracks.

down the beach. From its attractive station, designed in a Spanish Colonial motif, the Santa Fe operated the *San Diegan* streamliners, stainless steel consists hauled by diesels in their classic yellow and red warbonnets.[7] Despite the crack daytime trains, however, Pullman service from San Diego was relatively modest.

Aside from the sleepers to Los Angeles and Oakland once carried by the *Saint* and *Angel,* the Santa Fe's other Pullman service was by way of connecting trains at Los Angeles, such as the *Chief.* In addition to the eastern destinations of Chicago, Minneapolis/St. Paul (Burlington/Illinois Central/Minneapolis & St. Louis), St. Louis (Frisco), and Washington (B&O), the railroad also scheduled Pullmans from San Diego to San Bernardino (1895).

After completion of the San Diego & Arizona in 1919, the city had a connection for several decades with the SP's southern transcontinental route, which offered Pullman service to both New Orleans and Chicago (RI).

Sacramento, the state capital, was a major junction of the Overland Route, the Western Pacific, and several other SP lines. Sacramento had Pullman service on the SP to Los Angeles, San Francisco, Portland, and Reno.

When San Francisco was the western terminus of the Sunset Route, Santa Barbara had its own sleepers to Chicago on the *Golden State Limited* and to New Orleans on the *Sunset Limited.* Pacific Grove, on the Monterey peninsula, also had a dedicated car on the *Sunset* to New Orleans.

Yosemite National Park is one of the nation's most grandly scenic preserves, containing, among other attractions, the country's highest waterfall. Opened in 1890, the park was accessible at first only by means of a two-day stage trip from Merced, but in 1907, the

Yosemite Valley Railroad began service and soon was carrying through Pullmans from both Los Angeles (SP) and Oakland (either SP or Santa Fe). This husky little carrier ran from Merced through the Merced River Canyon to a station called El Portal, at the park's entrance, where it also built a rustic though luxurious lodge. A fifteen-mile stage or motor coach shuttle completed the trip into the Yosemite Valley.

By 1916, however, more people were entering the park by automobile than by train, and the decline continued. After continuing financial and operating difficulties, the Yosemite Valley Railroad ceased operation in 1945 and was torn up for scrap.[8]

Another tourist passenger line to the north met a similar fate. In 1900, a narrow-gauge railroad began operation between Truckee and Lake Tahoe, carrying passengers from Southern Pacific main line trains to the Tahoe Tavern, built on the shore of the pristine lake. Both the resort, which had accommodations for 400 tourists, and railway were enterprises of the Bliss family, as was the steamer *Tahoe,* which made excursions around the lake's 73-mile circumference. (The Blisses pioneered the lumber industry across the lake during the Comstock heyday.)

In 1925, the Southern Pacific acquired the lease of the Lake Tahoe Railway and converted the line to standard gauge. The following spring, the SP began Pullman service from Oakland, with the sleeper arriving as the sun was rising over the opposite shore of the lake. Although a day train with parlor observation car also was scheduled from Oakland, as well as two shut-

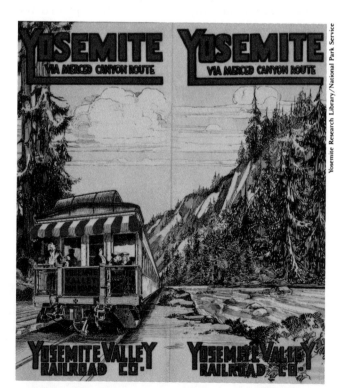

The little trains of the Yosemite Valley Railroad carried Los Angeles and Oakland sleepers to El Portal, the park station.

It's the summer of 1927, the YV train has arrived at El Portal, and this stage full of happy sightseers is off to Yosemite Park.

tle trains to Truckee, it was not long before Greyhound buses were providing most of the connecting service for the few tourists who were not streaming into the area in their own automobiles.[9]

1. Gilbert H. Kneiss, *Trains & Travel*, Vol. 13, No. 9 (July 1953), p. 54.
2. Neill C. Wilson and Frank J. Taylor, *Southern Pacific: The Roaring Story of a Fighting Railroad* (New York: McGraw-Hill, 1952), p. 198.
3. John A. Droege, *Passenger Terminals and Trains* (Milwaukee: Kalmbach, 1969. Reprint of original, New York: McGraw-Hill, 1916), p. 104.
4. About this time, local transportation authorities were still seeking to replace San Francisco's historic cable cars with "more efficient" diesel buses. Several years later, the Bay Area Rapid Transit system planners went to work on a trans-bay tunnel to replace what had just been ripped off the Bay Bridge and paved over.
5. George H. Drury, *The Historical Guide to North American Railroads* (Milwaukee: Kalmbach, 1985), p. 240.
6. Although special ski trains are not within the purview of this book, it is interesting to look at the blueprints of a weekend trip that took place in the relatively primitive days—at least for skiing—of 1941. The Northwestern Pacific and Southern Pacific jointly operated a *Snowball Special* on February 9 from Santa Rosa to Norden, where the SP ran its own Ski Hut, near the Sugar Bowl ski area in the Donner Pass. Area sporting goods shops and hardware stores (they commonly sold and installed bindings) took reservations for the train, which departed late Saturday night and arrived in time for about six hours of skiing Sunday before loading for the return trip home. Intermediate-class, round-trip fare was $4.55, plus $2.05 for an upper berth ($2.35 for two persons—yes, skiing was a fast-track recreation even in those days) in the tourist Pullmans. The consist also included a dance car, diner, and coaches.—From an advertisement submitted by Ed Nervo to *The Headlight*, newsletter of the Northwestern Pacific Railroad Historical Society, IV, No. 6 (June 1988).
7. Under Amtrak, traffic on the Surf Line between Los Angeles and San Diego became almost embarrassingly successful, resulting in more train service than the line had ever seen under Santa Fe operation.
8. Alfred Runte, "Yosemite Valley Railroad," *National Railway Historical Society Bulletin*, Vol. 39-40, p. 10. Reprinted from *National Parks & Conservation Magazine*, (December 1974). The author, an expert on the role of railroads in the development of national parks, presents an intelligent case for reconstructing the YV Railroad to protect the park from the increasing congestion of private automobiles and the ecological harm they cause.
9. SP schedules and other information courtesy of the library of the Nevada State Railroad Museum, Richard C. Datin, curator.

Pennsylvania Railroad/courtesy Conrail

J. Parker Lamb

Santa Fe Southern Pacific Corp.

Otto C. Perry/Denver Public Library

34
Missouri

It had the powerful and imposing appearance of a medieval fortress, and its gray stonework seemingly was built for the ages. When St. Louis Union Station opened in 1894, it was superlative in every respect; it was the largest railroad terminal in the world.

Union Station's turreted, Romanesque facade, which was longer than two football fields, was accented by a massive clock tower that soared above the chateau-like roof-line of the complex. Opposite the entrance, an open square featured an elaborate display of fountain and statuary. Within the terminal's walls was a hotel with hundreds of rooms.

A voluminous trainshed covered the thirty-two stub tracks that entered the station by way of overlapping, double wyes. Altogether, more than six miles of trackage lay under this shelter. In the 1920's, ten new tracks were added, further increasing the capacity of a station that was known for accommodating more railroads than any other.

Eighteen major railways scheduled sleeping cars from St. Louis Union Station to more than 130 destinations over nearly twice that number of lines. There were just two American cities from which passengers could travel directly to any other part of the country, and St. Louis was one of those cities. St. Louis was second only to Chicago as a railroad center, but what Chicago dispersed among six terminals, St. Louis brought together into one. In its prime, St. Louis Union Station accommodated about 300 trains a day, virtually all of them intercity. During the streamline era, the station's tenants produced a color palette unequaled anywhere.

St. Louis was the natural rail gateway to the Southwest for travelers from the East. For years, passengers rode the crack limiteds of the New York Central, Pennsylvania, Baltimore & Ohio, and Nickel Plate to St. Louis, then, along Union Station's boulevard-like Midway, sought their connections to Texas or other destinations from among the premier flyers of the Missouri Pacific, St. Louis-San Francisco, Missouri-Kansas-Texas, and St. Louis Southwestern. For a few years after World War II, entire sections of trains bringing through Pullmans obviated the need to transfer. These sleepers, mostly from New York and bound for Oklahoma, Texas, and Mexico City, traveled principally on two trains, the Pennsylvania's *Penn Texas* and the New York Central's *Southwestern Limited*.

The Pennsylvania scheduled the greatest flow of trains to the East, led by the all-Pullman *"Spirit of St. Louis"* and *American*. As late as the 1950's, the Pennsy was offering name trains accounting for a half dozen round trips, including the *St. Louisan*, all-coach *Jeffersonian*, *Penn Texas*, *Metropolitan*, *Statesman*, and *Allegheny*. St. Louis had Pullman service on the Pennsylvania to Indianapolis, Cincinnati (B&O, 1890), Columbus, Cleveland, Pittsburgh, Washington, New York, and

St. Louis Union Station (top left) was the greatest single gathering point of different railroads and their passenger trains in America. Kansas City Union Station (left center) was second in terms of the number of railroads it served. A suburban St. Louis departure (top right), the *Colorado Eagle* heads west out of Kirkwood in June 1959. With a Pennsy shark-nosed 4-4-4-4 on the point, the *"Spirit of St. Louis"* speeds through the Illinois countryside in June 1946 with 14 cars at 80 miles an hour.

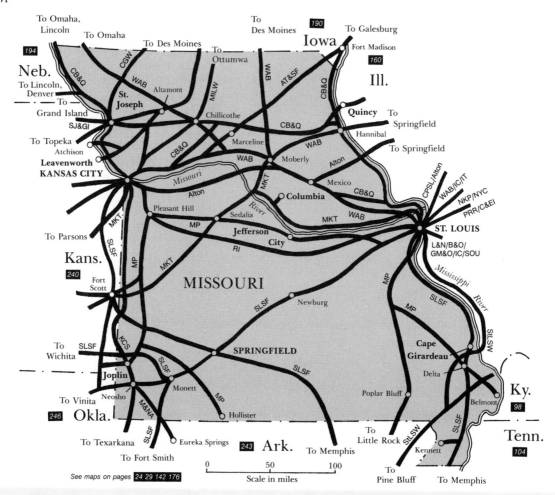

To Omaha, Lincoln
To Omaha
To Des Moines
To Ottumwa
To Des Moines
190
To Galesburg
Iowa
Fort Madison
160
194
Neb.
Ill.
CGW
CB&Q
WAB
MILW
WAB
AT&SF
CB&Q
To Lincoln, Denver
To Grand Island
St. Joseph
Altamont
Chillicothe
Quincy
To Springfield
SJ&GI
CB&Q
Marceline
Hannibal
To Topeka
Atchison
WAB
Moberly
WAB
To Springfield
Leavenworth
KANSAS CITY
CB&Q
Missouri River
Alton
Alton
Mexico
CB&Q
CPSL/Alton
WAB/IC/IT
NKP/NYC
Columbia
MKT
WAB
PRR/C&EI
Pleasant Hill
Sedalia
MKT
WAB
ST. LOUIS
To Parsons
MP
Jefferson City
L&N/B&O/
GM&O/IC/SOU
Kans.
MKT
SLSF
RI
Mississippi River
240
Fort Scott
MP
MISSOURI
MP
MKT
SLSF
Newburg
MP
SLSF
To Wichita
SLSF
KCS
SLSF
SPRINGFIELD
Cape Girardeau
StLSW
Joplin
SLSF
SLSF
Delta
Ky.
To Vinita
Neosho
Monett
MP
Poplar Bluff
Belmont
98
246
Okla.
M&NA
Hollister
Tenn.
To Texarkana
SLSF
Eureka Springs
243
Ark.
To Memphis
To Little Rock
StLSW
SLSF
Kennett
104
To Fort Smith
0   50   100
Scale in miles
To Pine Bluff
To Memphis

See maps on pages 24 29 142 176

Boston (New Haven).

The *Southwestern Limited* led New York Central's Big Four fleet, which included the *Knickerbocker* and *Missourian*. The Central scheduled fewer trains but about the same number of Pullman destinations, including sleepers to Indianapolis, Cincinnati, Columbus, Cleveland, Detroit, Buffalo, New York, and Boston. In company with the Chesapeake & Ohio, the Central offered Pullmans to Washington; Richmond and Newport News, Va.; and, via Erie/Pittsburgh & Lake Erie to Pittsburgh.

In 1857, a predecessor of the B&O became the first railroad to reach St. Louis from the East. The Mississippi River had already been bridged, but not at St. Louis, where passengers arriving from the East had to be ferried from the Illinois side. St. Louis had been a river town all along, and East St. Louis, Ill., was becoming a massive railhead. The two sides were finally connected in 1874 by the Eads Bridge, which at once made St. Louis a true rail gateway and paved the way for the construction of Union Station two decades later.

Inaugurated in 1925, the *National Limited* was the B&O's premier train on its St. Louis-New York route, also served by the *Diplomat* and *Metropolitan Special*. The B&O carried Pullmans to French Lick, Ind. (Monon), Cincinnati, Louisville, Pittsburgh, Washing-

ton, Norfolk (Norfolk & Western), and New York. Early service (1890) included cars to Baltimore, Philadelphia, and New York (Erie). In the 1960's, the C&O assisted on a sleeper to Washington.

The Nickel Plate carried Pullmans from St. Louis to Toledo, Cleveland, and New York (Lackawanna).

Before World War I, during the control of Jay Gould and his son George, the Wabash extended its reach east, offering Pullman service from St. Louis to Buffalo, New York, and Boston.[1] Historically, the Wabash also provided St. Louis a connection with cities in the Far West, making it the only midcontinent railroad originating sleepers to both coasts.

In more recent times, the Wabash concentrated on serving just mid-America, but it always was unique in straddling both east and west, with one foot in Detroit and the other in Omaha or Kansas City. Within this territory, the Wabash carried sleepers to Chicago (see Illinois chapter), Toledo, Detroit, Des Moines, Kansas City (this service, like Chicago's, included setouts at St. Louis's West End Delmar Station), Minneapolis/St. Paul (Minneapolis & St. Louis[2]), Omaha, and Ottumwa, Iowa (1890). Its daylight trains included the *Banner Blue* and *Blue Bird* to Chicago, *Wabash Cannonball* to Detroit, and the *City of Kansas City* to the western Missouri metropolis. Pullman trains included the *Midnight* to Chicago, *Midnight Limited* to

Kansas City, and *City of St. Louis* to the West Coast.

The Wabash was St. Louis's link to the Union Pacific's Overland Route at Kansas City. The railroad carried Pullmans for Denver, Salt Lake City, West Yellowstone, Los Angeles, San Francisco (Southern Pacific), Portland, and Seattle. (See chapter "Going to California.")

St. Louis had four rail connections to the South, but unlike the pattern in Washington, Cincinnati, and Louisville, very little passenger traffic moved from the north to these lines. St. Louis was primarily a gateway to the Southwest.

Although the famous Florida-bound *Dixie Flyer* had its Nineteenth Century origins in St. Louis, in more recent times only sections of the *Dixie* trains served the city, via Evansville, Ind. The Louisville & Nashville inaugurated its post-war *Georgian* between St. Louis and Atlanta, but when the expected traffic didn't materialize, that too was rerouted and St. Louis, for the L&N, remained somewhat of a secondary operation.

From Union Station, the L&N carried Pullmans to Evansville, Louisville, Nashville, Montgomery, and Pensacola. The railroad cooperated with the Nashville, Chattanooga & St. Louis on sleeping car lines to Atlanta and Charleston, S.C. (Georgia/Southern, 1898), and originated numerous Florida Pullmans that traveled over several routes (see chapter "Going to Florida").

Southern Railway service from St. Louis also lacked name-train glamor. The railroad scheduled Pullman service to French Lick; Louisville; Danville, Ky.; Knoxville; Asheville and Goldsboro, N.C.; and Jacksonville.

The Illinois Central tapped the St. Louis market with a branch from its main line at Carbondale, Ill. There, many of the major trains from Chicago combined with sections serving St. Louis: the all-Pullman *Panama Limited* and coach streamliner *City of New Or-*leans to the Crescent City, and the *Floridan*, *Seminole*, and *City of Miami* to Florida. The *Chickasaw* was a St. Louis-Memphis train.

Over IC rails, St. Louis had Pullman service to Fulton and Paducah, Ky., Memphis, Gulfport, Miss., and New Orleans, and from Cairo. Around the turn of the century, the railroad used an interchange with the NC&StL at Martin, Tenn. Sleepers to Nashville, Chattanooga, and Augusta, Ga. (Georgia Railroad) traveled this route, as did cars for Florida. In the 1890's, the IC delivered a sleeper for Memphis to the L&N at Milan, Tenn.

Through Illinois, the IC carried Pullmans to Chicago. In 1904, to accommodate Canadian visitors to the World's Fair in St. Louis, the IC picked up Montreal sleepers from the Grand Trunk Western in suburban Chicago.[3]

The Gulf, Mobile & Ohio was formed by the merger in 1940 of two parallel roads that ran north from Mobile. One of them, the Mobile & Ohio, carried sleepers on its *Gulf Coast Special* (later *Gulf Coast Rebel*) to Mobile. Early in the century, the M&O had provided New Orleans service jointly with the Southern, and for a few years after the GM&O merger, the railroad operated its mini-streamliner *Rebel*, with company-operated sleeper, from St. Louis to New Orleans and a regular Pullman to Montgomery. In one of the many 1890-vintage routes to San Francisco, the M&O turned over a sleeper in New Orleans to the Southern Pacific.

The Alton, a railroad that once had generated far more California traffic, was merged into the GM&O in 1947, making a new system, like the Wabash, that fanned out from, rather than terminating in, St. Louis. In its midwestern territory, the Alton scheduled sleepers from St. Louis to Chicago and Kansas City.

Two major roads serving Texas, the Missouri Pacific and the St. Louis-San Francisco, commonly known as the Frisco, originated the largest selection of Pullmans from St. Louis Union Station. The MoPac offered a slightly longer menu of Pullman destinations than the Frisco's, though not by much, considering its larger territory. The MP's former Iron Mountain line, which ran southwest out of St. Louis, was the route of the *Sunshine Special* and post-war *Texas Eagles*. These trains and others carried sleepers from St. Louis to Poplar Bluff; Little Rock, Pine Bluff, El Dorado, and Hot Springs, Ark.; Monroe, La., and (with Kansas City Southern) Shreveport, La., and Port Arthur, Tex.

In association with the Texas & Pacific, the Missouri Pacific scheduled Pullmans to Texas for El Paso, Dallas/Fort Worth, San Antonio, Laredo, Houston, Galveston, Brownsville, Mission, Big Spring, Sweetwater, and Taylor (1890), and to Mexico City (Nacionales de

New York Central's top St. Louis train, the **Southwestern Limited** heads east through Mitchell, Ill., in June 1951.

Mexico).

On its line to Louisiana, the Missouri Pacific carried sleepers to Memphis, Helena, Ark., Lake Charles, La., and New Orleans.

The *Scenic Limited* and *Colorado Eagle*, which replaced it in 1942, commanded MP rails from St. Louis to the west. The *Missourian, Westerner, Sunflower,* and *Missouri River Eagle* also plied this route. The railroad scheduled Pullman service to Denver (via either Kansas City, or Wichita, Kans., both with Denver & Rio Grande Western) and Pueblo, Colo.; Wichita, Arkansas City, and Hutchinson, Kans.; Omaha, Lincoln, Kansas City, St. Joseph, and Joplin.

Predecessor roads of the Frisco were twice involved in attempts to lay their rails to the Pacific Coast. The first was abandoned at the onset of the Civil War. The second involved an association with the Santa Fe that resulted in difficulties culminating in the Santa Fe's acquisition of the new line and reorganization of the Frisco into a purely regional system.

The Frisco, nonetheless, continued a working relationship with the Santa Fe (which briefly owned it) for several decades, exchanging through passenger cars at both Burrton, Kans., and Paris, Tex.

In the early 1890's, the Frisco delivered St. Louis sleepers for El Paso and California to the Santa Fe at Burrton, which also had its own car. This arrangement lasted only briefly, and St. Louis-Wichita Pullmans were the main modern survivors on this line. For a time in the late 1920s, however, the Frisco carried a St. Louis-Blackwell, Okla., sleeper by way of Beaumont, Kans.

At Paris, the Frisco fed traffic to the Santa Fe from its line through Fort Smith, Ark., until about World War I. This included Pullmans from St. Louis to the Texas cities of Beaumont, Dallas, Galveston, Houston, and San Antonio (MP). The Southern Pacific later took over a Houston sleeper at Paris, which also had its own car. In addition, the Frisco carried Pullmans for Eureka Springs, Ark. (Missouri & North Arkansas) and Fort Smith.

The *Texas Special* became a joint Frisco-Katy train in 1917, using the SLSF line to Vinita, Okla., and the MKT south of there, and both railroads promoted it as their premier train. During the life of the agreement, all the Frisco's Pullmans to Dallas and Fort Worth, Waco, San Antonio, and Los Angeles (SP) traveled the joint line. A decade later, the two railroads inaugurated on this route the *Bluebonnet*, named in honor of the Texas state flower, but the train later became MKT-only, operating from Kansas City.

Although it had to compete with several lines for traffic between St. Louis and Texas, the Frisco had but one competitor into Oklahoma, the Katy, and it pressed its advantage with the *Meteor*, a crack train to Tulsa introduced in 1902. The Frisco extended *Meteor* service to Texas briefly before the joint *Texas Special* began operation, but after that the *Meteor* became an Oklahoma City train. As a steam-powered heavyweight, the *Meteor* carried a blue and silver livery, but after World War II it was reequipped with stainless steel and red coaches similar to the *Texas Special's*. The *Oklahoma Special, Southwest Limited,* and *Will Rogers* also carried through sleepers over this route.

On its main line from St. Louis, the Frisco scheduled Pullmans to Springfield, Joplin, and Monett; Sapulpa, Tulsa, Enid, Lawton, Oklahoma City, and Okmulgee, Okla.; Quanah and Amarillo, Tex. (Rock Island).

In 1901, the Frisco got a toehold in the Deep South with its new line to Birmingham, opening the way for through sleepers from St. Louis to Kennett, Memphis, Birmingham, and several Florida cities. This was the route of the *Memphian* and a section of the *Sunnyland*.

By terms of its charter, the MKT was oriented to Kansas City rather than St. Louis, into which it was a relative latecomer. The Katy commonly ran sections of some of its name trains from both cities, putting them together in Parsons, Kans., for the remainder of the journey into Texas, where its route structure was stronger than the Frisco's. The *Katy Limited, Bluebon-*

Otto C. Perry/Denver Public Library

Wabash No. 3, the overnight *Detroit-St. Louis Special*, brings its ten-car consist into the Gateway City in September 1935.

Donald Sims

The *Katy Flyer* awaits late evening departure time from St. Louis Union Station in the early 1950's.

net, and *Katy Flyer* all had roots in St. Louis, but near the end, only the *Flyer* maintained its ties, basically as a connecting train at Parsons.

The MKT, of course, handled the *Texas Special* south of Vinita. From St. Louis, the Katy carried Pullmans to Columbia, Sedalia, Oklahoma City, Parsons, and Dallas, Austin, San Antonio, Wichita Falls, Houston, Galveston, and Brownsville, Tex. (SP). At San Antonio, the MKT delivered Pullmans from St. Louis to the SP, including cars for California and Mexico City (NdeM, via Eagle Pass).

The Cotton Belt was the fourth carrier operating between St. Louis and the Southwest and the most marginal from a passenger standpoint. The St. Louis Southwestern actually approached St. Louis from the southeast, coming up the Illinois side of the Mississippi River on Missouri Pacific tracks. Overshadowed by more powerful competitors, the Cotton Belt's *Lone Star* trains nonetheless carried a fair selection of sleepers over the years from St. Louis, to Pine Bluff, Texarkana, Shreveport, and Dallas, Houston (SP), Tyler, and Waco, Tex. In later years, SSW cars wore the red and orange of parent Southern Pacific.

The Burlington provided St. Louis with service to the upper Midwest and the plains states over a main line that went north along the Mississippi River. Northwest of the city, the Kansas City "short line" branched off to Mexico, Mo., from which the Burlington had trackage rights on the Alton. The main line continued north to Hannibal, and Burlington, Iowa; another line went west from Hannibal to St. Joseph. Trains to the west traveled over both routes: the *Nighthawk* via Mexico, the *Overland Express* and the *Colorado/St. Louis Limiteds* by way of Hannibal. The daytime streamliner *General Pershing Zephyr* traveled the former route, and the *Mark Twain Zephyr* the latter.

The Burlington carried Pullmans from St. Louis to Kansas City, St. Joseph, Omaha, Edgemont, S.D., Denver, Billings, and Butte, Mont. (Northern Pacific, 1899); Seattle (NP), Los Angeles (D&RGW/SP, Tourist), and San Francisco (D&RGW/SP).

Sleepers on the CB&Q traveled north through Hannibal to Burlington, Cedar Rapids (Rock Island), and Quincy, Galesburg (1890), Savanna, and Rock Island, Ill. (1890). Before the turn of the century, the railroad carried sleeping cars to Minneapolis/St. Paul up the Illinois/Wisconsin side of the Mississippi River and also by way of Burlington in conjunction with Rock Island/M&StL. The latter service evolved into the joint CB&Q/RI line, over which the streamlined *Zephyr Rockets* operated after their introduction in 1941.

The Rock Island gave St. Louis Pullman connections to its Golden State Route to California at Kansas City and also scheduled sleepers to Denver, Pueblo, and Colorado Springs.

In addition to the colorful trains of the IC, GM&O, and Wabash, St. Louis had Pullman service to Chicago via Chicago & Eastern Illinois. In the 1890's, predecessor lines of the Chicago & Illinois Midland also carried sleeping cars to Chicago and Peoria.

Three of the railroads serving the Chicago market also provided accommodations for summertime travel from hot, humid St. Louis to the cool woods of northern Michigan. The Alton and, later, C&EI carried Pullmans for Bay View that were given over to the Pere Marquette in Chicago, and the IC scheduled

Union Pacific PR, St. Louis

Headlights from the approaching second section of MP's *Texas Eagle* illuminate the first at the Poplar Bluff station.

sleepers for Harbor Springs (NYC/PRR). The more enduring service, however, proved to be that of the Pennsylvania, which handled the cars on its own tracks all the way. The Pennsy once offered service from St. Louis to Mackinaw City and, in its final days of resort runs, Harbor Springs.

St. Louis had one further tie to the cities of Illinois, though not involving trains from huge Union Station. The Illinois Terminal, an electric interurban system, provided sleepers to Springfield, Bloomington, Champaign, Decatur, and Peoria. IT trains crossed the Mississippi on their own bridge after departing their own station in downtown St. Louis via private underground entrance. The last sleeping car service, to Peoria, ended in 1940.

After the railroads left St. Louis Union Station, the great terminal underwent a benign restoration in the mid-1980's and reopened as a hotel and convention center. Under the refurbished train shed, where multiple sections of the *Sunshine Special* had once awaited their passengers, an eleven-acre shopping and entertainment area was built.

Missouri's great railroad terminus on the Mississippi River had a pronounced eastern accent with the cosmopolitan grays, tuscan red, and royal blue of the roads coming in from New York and Washington. The state's great rail terminus on the Missouri River, however, had a thoroughly western flavor, especially given by the silvery Santa Fe trains and their throbbing EMD and Alco diesels done up in yellow and red warbonnets.

Kansas City Union Station was second only to its older and larger St. Louis counterpart in the number of railroads using it. Twelve roads shared the cost of constructing the Bedford limestone, modified Renaissance structure shortly before World War I, and, like most of its contemporaries, it had the appearance of classic durability. The main waiting room, extending 352 feet over its sixteen through tracks, served as a concourse for the terminal.[4] Unlike St. Louis, not all of Kansas City's railroads terminated there. Through trains of the Santa Fe, Rock Island, Burlington, and Missouri Pacific paused briefly for servicing and crew changes before continuing on their way east or west.

The Santa Fe was born in Kansas, but Kansas City, Mo., became the eastern terminus of the road in 1875 and continued to originate passenger trains west, even after the line to Chicago was opened thirteen years later. Although six different railroads (including the Burlington, Alton, Milwaukee Road, Rock Island, and Wabash) competed for Pullman traffic to Chicago, the Santa Fe's tracks were the fastest, and the railroad always provided the widest selection of service with some of the finest trains in the country.

In the summer of 1951, for instance, the eastbound *Super Chief, Chief,* and *Texas Chief* rolled into town early in the morning; the *Grand Canyon* and streamlined *Chicagoan* offered daylight departures for Chicago, and the late evening lineup included the *Kansas City Chief, California Limited,* and all-coach *El Capitan.* West of Kansas City, the older trains to California mostly traveled via Topeka, the newer streamliners via Ottawa. The southern section of the *Grand Canyon* and the later *San Francisco Chief* worked the Santa Fe's southern main line, via Wichita and Amarillo.

Kansas City had Pullman service on the Santa Fe over the full length and breadth of the system, to Chicago, Galveston, San Antonio (MP), Los Angeles, and San Francisco, as well as the intermediate destinations of Albuquerque and Carlsbad, N.M.; Denver, El Paso, Fort Worth, Oklahoma City, Tulsa, and Sweetwater, Tex. In addition, the railroad carried sleepers to the Kansas communities of Wichita, Galena (Frisco, 1890), Hutchinson, Independence, and Wellington (1890).

The Rock Island was another major presence in Kansas City, which it used as a hub for its Chicago-California, Twin Cities-Texas, and St. Louis-Colorado trains. The *Golden State Limited, Imperial,* and earlier *Apache* and *Californian* (which the Alton carried from Chicago) plied the Golden State Route to California; the *Twin Star Rocket* and *Mid Continent Special* connected Minneapolis and St. Paul with the cities of Texas; the *Colorado Flyer* carried Pullmans to Denver and Colorado Springs.

The Rock Island scheduled sleepers from Kansas City east to Rock Island and Chicago; south to Wichita and Enid, Oklahoma City, and Shawnee, Okla., and Dallas/Fort Worth and Houston; west to Hutchinson, Los Angeles (SP), Denver, Colorado Springs, and Pueblo; and north to Des Moines and St. Paul/Minneapolis.

Kansas City was also an important hub for the western lines of the Missouri Pacific to St. Louis, Omaha, Pueblo, and Little Rock, the latter of which tapped the railroad's main routes to the Gulf Coast. The *Scenic Limited* and later *Colorado Eagle* traveled the route from St. Louis to Denver (D&RGW). The *Missourian, Missouri River Eagle,* and *Sunflower* (once a St. Louis-southern Kansas train) maintained the schedule to Omaha, and the *Southerner* and *Rainbow Special* carried sleepers to Little Rock for southern destinations. The MoPac dominated the Kansas City-St. Louis trade, with five trains a day for many years.

Its Pullman service in Missouri included sleepers from Kansas City to St. Louis, Joplin, and Hollister; in Kansas to Coffeyville, Downs, Wichita, and Winfield (1890). The MoPac carried Pullmans north to Lincoln and Omaha and, in conjunction with the Chicago & North Western, Sioux City and Minneapolis/St. Paul. The railroad scheduled sleepers west to Pueblo and south to Little Rock, Hot Springs, Memphis (1890), and New Orleans.

The Burlington provided a fair amount of Pullman service from Kansas City, most of it to the west. Its trackage to Chicago was less direct than some of its competitors', although construction in the early 1950's of the Kansas City short cut improved its schedule considerably. This seventy-one mile segment in northern Missouri was the longest new line built since the 1920's. The Burlington inaugurated the *Kansas City Zephyr* over the new route as a daytime companion to the overnight *American Royal Zephyr.* The Chicago cars were the Burlington's sole Pullman service to the east apart from the earlier operation with the Alton to St. Louis.

For many years, the Burlington's No. 43 departed Kansas City every night at 6:30 for Billings, Mont. This unnamed train and another traveled by way of St. Joseph and Lincoln, providing Kansas City with sleepers all the way to Seattle (via either Great Northern or Northern Pacific). This was the route of the original *Zephyr,* North America's first regularly scheduled, streamlined diesel train, which the Burlington inaugurated between Kansas City and Lincoln (via Omaha) in November 1934. The *Silver Streak Zephyr* entered service as a replacement on the route in April 1941.

The Burlington scheduled Pullmans to Lincoln, Omaha, and Alliance, Nebr.; Minneapolis/St. Paul (C&NW), Denver, Billings, and, via Northern Pacific in summers, to Butte and Gardiner, Mont., and, via Great Northern, to Glacier Park and Great Falls, Mont., and Spokane.

Four other railroads served the Upper Midwest from Kansas City: the Chicago Great Western, Alton, Wabash, and Milwaukee Road.

The Milwaukee Road's premier train, the *Southwest Limited,* had separate sections serving Chicago, Milwaukee, and Cedar Rapids, each of which city had its own sleepers, as did Ottumwa and Davenport.

The Alton also scheduled Pullmans to Chicago on its line that was served by one of the nation's earliest name trains, the *Hummer,* starting in 1880. In addition, the road scheduled sleepers to St. Louis (over its own route and jointly with the Burlington), Peoria, and Springfield, Ill.

Except for Pullmans to St. Louis and Detroit, Kansas City's service on the Wabash was early and short-lived: Chicago and Decatur (1890) and Des Moines. In 1964, Missouri got what to it was a new name in railroading, when the Norfolk & Western leased (and later merged) the Wabash.

The Chicago Great Western's *Tri-State* and *Mill Cities Limiteds* carried sleepers north to Des Moines, Rochester, and St. Paul/Minneapolis.

Missouri's connection to the Overland Route was at Kansas City, by way of the former Kansas Pacific line to Denver. From St. Louis, sleepers came by way of the Wabash to go aboard such Union Pacific trains as the *Pacific Limited* and *Pony Express.* The post-war *City of St. Louis,* which carried a complement of Pullmans for the West Coast, operated over Wabash and UP rails to Los Angeles. The Union Pacific scheduled sleepers from Kansas City to Salina, Kans., Denver,

The Santa Fe's *Hopi*, a Los Angeles-Chicago train, departs Kansas City Union Station with 11 cars on Sept. 13, 1931.

Salt Lake City, Los Angeles, San Francisco (SP), Portland, and Seattle.

Before completion of the Overland Route, it was thought that the Union Pacific would benefit by drawing traffic from the south. Accordingly, the Union Pacific Railway, Southern Branch was chartered for that purpose in 1865, with the provision of a land grant. This line became the Missouri-Kansas-Texas Railroad,[5] which grew into a north-south system extending from Kansas City and St. Louis to San Antonio and Galveston.

Earlier, the Katy's primary passenger thrust was from St. Louis, although it operated sections from Kansas City carrying Pullmans for its own *Katy Limited* and the *Texas Special*. In later years, it shifted emphasis to Kansas City and originated the *Bluebonnet* there.

The MKT carried Pullmans from Kansas City to Bartlesville, Okla., Oklahoma City, Dallas, Fort Worth, Wichita Falls, Waco, Houston, Galveston, San Antonio, and Corpus Christi (SP).

Although the rival Frisco scheduled more Pullman service from Kansas City, it tended to leave the Texas traffic to the Katy and focus on its unique link to the Southeast. Just after the turn of the century, the Frisco acquired its line to Birmingham, where it exchanged cars with the Southern. This was the route of the *Kansas City-Florida Special* and the *Sunnyland*, over which the Frisco carried Pullmans from Kansas City to Memphis, New Orleans (IC), and Birmingham, and with the Southern to Atlanta, Jacksonville, and Miami (Florida East Coast).

The Frisco once operated a section of the *Meteor* from Kansas City to Tulsa as well as the daytime *Firefly*, inaugurated in 1939 as a blue-and-silver, heavyweight, steam-powered streamliner, and later the *Oklahoman*. Pullman service to the Sooner State included cars to Tulsa, Sapulpa, Oklahoma City, and Ada; and to Fort Worth. The Frisco also carried sleepers to Springfield, Joplin, Monett and Fort Smith. On the eve of World War II, the small hamlet of Newburg, the railroad station of choice for the Army's Fort Leonard Wood, also had scheduled Pullman service, a modest warmup for the endless flow of troop trains that would roll into and out of town over the next five years.

The third, and smallest railroad connecting Kansas City with the Gulf of Mexico was the Kansas City Southern, which threaded its way in and out of six states en route to Port Arthur, Tex. For the grain-growing area the KCS served, the Gulf was the closest access to sea-going freighters. Although its namesake city was the only major metropolitan area on the KCS, at least until the merger with the Louisiana & Arkansas took it into New Orleans, the KCS prided itself in running good passenger service.

The *Flying Crow*, inaugurated in 1928, was one of two through trains working the line and carried Pullmans from Kansas City to Port Arthur and Houston (T&P/MP). Intermediate service included cars to Joplin and Fort Smith. With the L&A merger and introduction of the streamlined *Southern Belle*, the KCS extended Pullman service to New Orleans.

The St. Joseph & Grand Island Railway carried a sleeper for many years over the length of its system, offering passengers connections at Grand Island, Neb., to the westward flow of Overland Route trains. Schedules from 1909 show a Kansas City-Grand Island sleeping car also going into the consist of the night train from St. Joseph.

Aside from the SJ&GI, five other railroads serving St. Joseph, the eastern terminus of the Pony Express,

A name train dating back to 1880, the historic Alton *Hummer* arrives in Kansas City after its overnight trek from Chicago.

originated sleepers there. The Rock Island put a Chicago Pullman aboard the *Apache* at its main line junction of Altamont and also scheduled a St. Joseph sleeper to Caldwell, Kans. The *American Royal* carried a Chicago Pullman for the Burlington, which also scheduled sleepers from St. Joseph to Omaha and St. Louis. The Santa Fe carried sleeping cars to Wichita and the junction of Newton, Kans. (1890). The Chicago Great Western operated an early line to Minneapolis/St. Paul (1890), as did the Missouri Pacific to St. Louis.

Springfield was the main crossroads of the Frisco system, which gave the city its sole sleeping car connections, including St. Louis service on the *Will Rogers*. The red and bronze diesels brought Frisco name trains through town in goodly numbers—the *Texas Special, Meteor, Kansas City-Florida Special, Sunnyland*—mostly at night. The railroad scheduled sleepers to Springfield from Kansas City, Birmingham, Memphis (1890), Oklahoma City, and Wichita.

The lead- and zinc-mining center of Joplin had a variety of service to Kansas City; KCS, Frisco, and MoPac all scheduled sleepers, and the latter two provided accommodations to St. Louis. The Frisco carried a car from Joplin to Oklahoma City, as did the Missouri, Oklahoma & Gulf/Fort Smith & Western. The Missouri & North Arkansas offered service to Helena, Ark.

Monett, a junction on the Frisco where its line to Paris, Tex., branched off from the main line to Oklahoma, had its own sleepers to Dallas, Little Rock (MP), and Burrton, Kans. (1890).

Two other Missouri communities had their own Pullman service: The Wabash scheduled a Moberly-Chicago car, and the MKT briefly carried a sleeper between Sedalia and Muskogee.

In a brief, pre-turn of the century arrangement, Chicago-Houston sleepers, an interline operation involving the Burlington and Katy, traveled over Wabash trackage between Hannibal and Moberly, bypassing the St. Louis gateway.

1. Before the Wabash got trackage rights to Buffalo over the Grand Trunk in 1898, its sleepers traveled there via Michigan Central from Detroit. The West Shore and Lackawanna forwarded cars to New York City, and West Shore and Boston & Maine handled the Boston sleepers.
2. Before the turn of the century, this also involved the Milwaukee Road in at least two different routings into the Twin Cities.
3. Arthur D. Dubin, *Some Classic Trains* (Milwaukee: Kalmbach, 1964), p. 335.
4. John A. Droege, *Passenger Terminals and Trains* (Milwaukee: Kalmbach, 1969. Reprint of original, New York: McGraw-Hill, 1916), p. 95.
5. George H. Drury, *The Train-Watcher's Guide to North American Railroads* (Milwaukee: Kalmbach, second printing, 1985), p. 125.

Preston George

Santa Fe Southern Pacific Corp.

Otto C. Perry/Denver Public Library

Richard Kindig

35
Kansas

KCS' northbound *Southern Belle* (top) winds along the foot of Oklahoma's Black Fork Mountain in August 1946. The *Scout* (left center) was inaugurated in 1916 and, though never the Santa Fe's premier flyer, for many years was designated Santa Fe train No. 1. It ran from Chicago to Los Angeles via the southern route. The RI/SP's extra-fare *Arizona Limited* (right center) operated on the eve of World War II, alternate days, with all-room sleepers, Chicago to Phoenix. The Rock Island's *Rocky Mountain Limited* (bottom) departs Denver in May 1939.

Construction of the Atchison, Topeka & Santa Fe Railway began in Kansas shortly after the Civil War, with the intention of tapping the trade of the historic Santa Fe Trail. By that time, huge herds of cattle were being driven north from Texas and Oklahoma to Dodge City and Wichita, and the new railroad was ideally situated to haul the livestock onward to mid-western markets. Inasmuch as the Santa Fe had land grants to dispense, it soon found itself carrying wheat grown by the farmers it had induced to settle the area. Attracted by the rich coal and mineral deposits, the railroad pushed on into Colorado, and this lucrative freight business gave it a momentum that propelled it into transcontinental ranks by 1885.

Atchison remained the eastern terminus only in the railroad's marvelously melodic name. Kansas City, Mo., briefly assumed that role before the Santa Fe, out of competitive necessity, decided to lay tracks to Chicago, thereby becoming the first and, eventually, sole-surviving, transcontinental passenger carrier under single management.

The eastern half of Kansas was laced with branch lines of the Santa Fe and other railroads sidling up to the grain elevators that periodically erupted from the vast expanse of farmland. Santa Fe main lines, in their several bifurcations, also contributed significantly to the state's rail coverage.

One line went west from Kansas City through Topeka, and a cutoff ran southwest through Ottawa, the two joining at Emporia. Further down the line at Newton, the main passenger route to the west continued through Hutchinson and Dodge City to Raton Pass in Colorado, and the alternate, more southerly Belen (N.M.) Cutoff branched off through Wichita. The two lines came together south of Albuquerque.

Kansas's most populous city, Wichita, also had the state's largest selection of originating Pullman service, and the Santa Fe was for many years the major provider. The *Scout, Grand Canyon* (southern section), and the streamlined *San Francisco Chief* were among the railroad's western trains serving Wichita, and the *Ranger, Antelope, Texas Chief,* and *Kansas Cityan/Chicagoan* went through town en route to Oklahoma and Texas. The Santa Fe scheduled dedicated sleepers to Wichita from Chicago, Kansas City, St. Joseph, Mo., Topeka, Denver, Galveston, Amarillo, and Tulsa.

The Missouri Pacific, too, scheduled a goodly amount of Pullman service to Wichita, although its main line running west across the state passed to the north of the city, as did the Santa Fe's. Connecting trains carried sleepers between Wichita and Geneseo for such main line flyers as the heavyweight *Scenic Limited* and streamlined *Colorado Eagle*. The *Sunflower,* however, gave the city direct service to St. Louis. On these and other trains, the MoPac scheduled Pullmans from Wichita to Pueblo, Denver (Denver & Rio Grande Western), Kansas City, St. Louis, and Hot

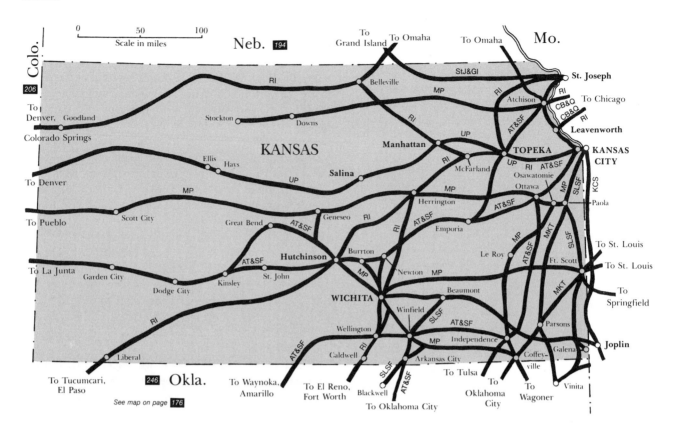

KANSAS

Neb. 194   Mo.

Colo.   206

See map on page 176

Springs.

Wichita was also on the Rock Island's main line from Kansas City to Texas and served by such trains as the *MidContinent Special, Firefly,* and streamlined *Twin Star Rocket,* which carried through sleepers from the Twin Cities to Houston. The Rock Island scheduled dedicated Pullmans from Wichita to Oklahoma City, Kansas City, and Omaha.

Well-to-do Kansans had an endless choice of accomodations to book for a winter journey to sunny California, but the Frisco gamely countered by wooing travelers the other way, to Florida. In 1926, the railroad was briefly scheduling thrice-weekly Pullman service to Jacksonville (Southern) from Wichita. Beyond this, Wichita had dedicated sleepers by way of the Frisco to Tulsa, and St. Louis and Springfield, Mo.

The state capital, Topeka, being near to and well connected with Kansas City, had little dedicated Pullman service—a Santa Fe sleeper to Wichita was one exception. Topeka, however, did have the through service of the Santa Fe, Union Pacific, and Rock Island, including the latter's trains to Texas and California.

Topeka, moreover, had a distinction all its own in the realm of railway cuisine, for it was there in 1875 that Fred Harvey opened his first restaurant for travelers on the Santa Fe. From this beginning, the Harvey House—and the creator's cookery—went on to

Santa Fe's answer to streamlined steam was its "Blue Goose" locomotives, this shown getting the *Chief* under way.

become celebrated amenities of the railroad's passenger service throughout all its history.

Hutchinson, which witnessed the main flow of the Santa Fe's crack West Coast trains, such as the *California Limited, Chief,* and *Super Chief,* had varied but limited Pullman service of its own. The Santa Fe brought in a sleeper from Kansas City, as did the Rock Island. The Missouri Pacific carried a Pullman to St. Louis.

Atchison had similarly diverse service: Chicago sleepers via the Burlington (1890) and Rock Island, and a car on the MoPac to Stockton.

Burrton was the terminus of several unusual sleeping car lines, when the Santa Fe briefly owned the Frisco in the early 1890's. The two railroads interchanged through cars from Chicago to the West Coast at Burrton and also used the town as a terminus. The Santa Fe brought in sleepers from San Francisco, and the Frisco scheduled cars from Monett, Mo., and St. Louis to this onetime thriving passenger junction.

Several Kansas communities that grew up where lines of the same or different railroads crossed or branched out also were on the Pullman route map. Coffeyville (MP) had sleepers to both Kansas City and Little Rock; McFarland (RI) to Denver and Fort Worth/Dallas; Newton (Santa Fe) to St. Joseph (1890) and Dallas/Fort Worth; and Caldwell (RI) to St. Joseph and Omaha/Council Bluffs. Parsons, the nerve center of the Katy, had sleepers on the MKT to St. Louis and Hillsboro, Tex.

Leavenworth, site of the federal penitentiary, had dedicated Pullmans to Chicago on the Rock Island and Burlington.

Much of the remainder of Kansas's Pullman service

came from Kansas City: cars to Downs (MP), Galena (Frisco/Santa Fe, 1890), Independence (Santa Fe), Wellington (Santa Fe, 1890), and Winfield (MP, 1890).

Residents of Kansas City, Kans., a sizable city in its own right, used the huge, classic rail terminal in Kansas City, Mo., for virtually all their Pullman connections after the station was built just before World War I.

Elsewhere in Kansas, sleepers operated between Kinsley and Pueblo, Colo. (Santa Fe, 1895), and Arkansas City and St. Louis (MP).

The tracks of the Rock Island's main line to Colorado crossed the uplands in the northwestern part of the state. This was the route of the *Rocky Mountain Limited* and streamline-era *Rocket,* which carried through Pullmans from Chicago to Denver and Colorado Springs.

Salina, which also had its own Pullman to Kansas City, over the Union Pacific's former Kansas Pacific line to Denver, was the namesake of one of the country's historic trains. This three-car, Pullman-built, lightweight flyer was born in February 1934 as the M-10000—the world's first streamliner. The UP proudly sent it on tour to whet the appetite of rail travelers and hoped that it would help bring the passenger business out of the doldrums of the Depression. The M-10000, which was the progenitor of the UP's famous fleet of *City* streamliners, went into service between Kansas City and Salina, via Topeka, in early 1935, thereby becoming the *City of Salina.*[1]

---

1. Karl Zimmermann, "From Salina to Everywhere," *Passenger Train Journal,* (April/May 1976), pp. 16-17.

MoPac's eastbound *Colorado Eagle* pauses at Jefferson City, Mo., in 1944.

Union Pacific PR, St. Louis

J. Parker Lamb

Three Alco PA units lead the Missouri Pacific's northbound *Southerner* out of Little Rock for St. Louis in June 1960.

36
Arkansas

The main street of Arkansas passenger railroading was the Missouri Pacific's line that divided the state diagonally on its way from St. Louis to Texarkana. This often double-tracked right-of-way was the route of the multi-section *Sunshine Special* and *Texas Eagle*. At the state capital and rail hub of Little Rock, during the hours between midnight and dawn, these crack trains pulled into the station, one section after another, sometimes as many as ten altogether.

The Missouri Pacific inaugurated the *Sunshine Special* in December 1915, immediately upgrading service between St. Louis and Texas. During the 1920's the *'Shine* developed into a train of three separate sections that carried Pullmans for the West Coast, Mexico City, South and West Texas, and Louisiana (in Texas, the routing involved the Texas & Pacific, for many years controlled by and later merged into Mo-Pac). Some of the sections originated and much of the switching occurred in Little Rock, where connecting trains to Memphis also added to the mix.

For a time after World War II, the *Sunshine Special* operated from New York on the Pennsylvania Railroad, which carried through Pullmans from the East to Texas and Mexico City. The Pennsy's contribution later became the *Penn Texas.*

Originally, the Missouri Pacific planned to modernize the train with streamlined equipment and append the successful *Eagle* name that its premier Colorado

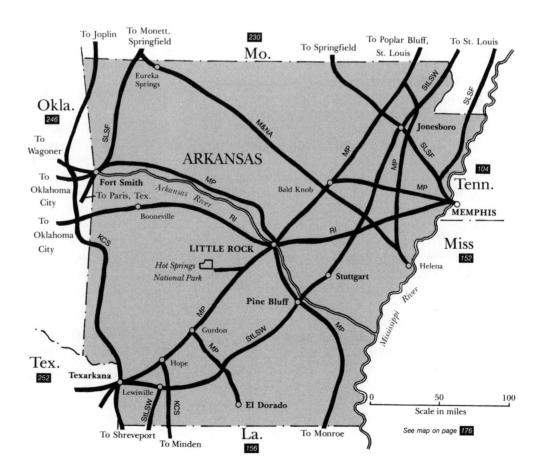

See map on page 176

trains bore. The *Sunshine Eagle*, however, was instead introduced as the *Texas Eagle*, which made its first appearance in August 1948 in a new livery of blue, gray, white, and gold. Moving out of St. Louis, the West and South Texas sections of this crack new streamliner were carded just three-minutes apart.

Central Arkansas had remarkably cosmopolitan Pullman accommodations to both coasts and Mexico City. Little Rock had considerable sleeping car service of its own, in addition to the heavy through traffic of both the MP and Rock Island. Nearby Hot Springs, of course, attracted substantial resort trade.

The line of the *Sunshine Special* and *Eagles* gave Little Rock Pullman service to El Dorado, Dallas-/Fort Worth, El Paso, Houston, San Antonio, Mexico City (Nacionales de Mexico), St. Louis, and Memphis, and from Shreveport (Kansas City Southern).

Another line of the Missouri Pacific went up the Arkansas River valley into Oklahoma and Kansas, bringing scheduled sleepers to Little Rock from Fort Smith, Coffeyville, Kans., Monett, Mo. (Frisco), Kansas City, and Omaha. Pullmans to Lake Charles, La., and New Orleans traveled the line through Pine Bluff.

Around the turn of the century, the Rock Island acquired a line that ran west from Memphis, through Little Rock to the interchange with Southern Pacific at Tucumcari, N.M. The *Memphis Californian, Choctaw Limited,* and later *Choctaw Rocket* worked this route, over which Little Rock had Pullman service to Oklahoma City and Los Angeles, as well as Memphis.

The same two railroads that served Little Rock with sleepers, Missouri Pacific and Rock Island, also monopolized the traffic to Hot Springs, whose curative waters began attracting national attention in the 1830's. For nearly six decades, starting in 1890, the MP's *Hot Springs Special* operated from St. Louis with sleepers, including cars from Chicago (Wabash; later, the Gulf, Mobile and Ohio made the connection).

From Memphis, the Rock Island operated its *Hot Springs Limited* as well as what it called the *Hot Springs-Panama Limited,* both of which carried Pullmans delivered from Chicago by the Illinois Central, the latter train having the better pedigree. In addition to Memphis-Hot Springs service, the Rock Island also scheduled Pullmans from Denver and Pueblo. The Frisco's *Kansas City-Florida Special* carried the cars across Missouri and northeastern Arkansas, between Kansas City and Memphis, whence they resumed their journey to the Springs under Rock auspices.

MoPac scheduled sleepers to Hot Springs from Kansas City, Wichita, Memphis, and New Orleans and cooperated on a Pullman line from Minneapolis/St. Paul (Minneapolis & St. Louis/Wabash).

Four railroads gave Fort Smith its relatively limited Pullman service. The Kansas City Southern's main line to the Gulf, which ducked in and out of every state it traversed on its way south, happened to have ducked into Oklahoma where it passed Fort Smith, but by way of a branch the railroad delivered Pullmans from Kansas City and Shreveport. The Frisco also scheduled sleepers to Kansas City as well as St. Louis. Fort Smith had Pullman service east via MP to Little Rock, and the Fort Smith & Western carried a former Wagner car to Oklahoma City.

The Missouri & North Arkansas cut across the state diagonally from Eureka Springs to Helena, perpendicular to the Missouri Pacific main line, but the M&NA never enjoyed the sort of prosperity the giant MoPac did. Its dreams of substantial bridge-route traffic never were fulfilled, and it was plagued by labor unrest and financial troubles stemming from a tragic rail car accident.

In happier days, however, it cooperated with the Frisco on a sleeper from St. Louis to Eureka Springs, the resort that happened to be the site of the last temperance speech of Carrie Nation, whose devoted followers erected Hatchet Hall within its precincts for the training of future prohibitionists. The M&NA also carried Pullmans from Helena for both Joplin, Mo., and Kansas City shortly before World War I. Helena, an ante bellum center of cultured society and Arkansas's major river port, also had its own sleeper to St. Louis on the Missouri Pacific.

Arkansas had another major railroad paralleling the Missouri Pacific main line, the St. Louis Southwestern. This road, more commonly called the Cotton Belt, was a subsidiary of giant Southern Pacific, bringing SP's reach up to the Memphis and St. Louis gateways. The *Lone Star Limited* was the Cotton Belt's premier train on the Memphis-Dallas/Fort Worth route. The railroad carried Pullmans from Texarkana and Pine Bluff to St. Louis, and Texarkana to Memphis.

The Kansas City Southern main line ran through Texarkana, giving the town the service of the *Southern Belle* and *Flying Crow*.

The MP also scheduled sleepers to St. Louis from both Pine Bluff and the oil capital of El Dorado.

37
Oklahoma

Most of what is now Oklahoma was called the Indian Territory, wherein lived members of the "five civilized tribes"—Choctaw, Chickasaw, Cherokee, Creek, and Seminole. Under terms of the original treaties, the land was to be for the Indians; no white settlers were allowed.

When the Pacific Railroad surveys were made in the 1850's, however, one of the potential routes surveyed led west from Fort Smith, Ark., across the Indian Territory, and in a treaty signed in 1855, the United States gained for itself, or a corporation, the right to acquire rights-of-way on the land for telegraph and rail lines.

After the Civil War, these provisions were reinforced with new treaties that were more punitive, inasmuch as some of the Indians had sided with the Confederacy. These treaties laid the groundwork for the construction of the north-south Missouri-Kansas-Texas line, starting in 1871, and the presumably east-west St. Louis-San Francisco the same year.[1]

After a failed attempt to become a transcontinental line and then reorganization as a southwestern regional system, the SLSF pushed its line, which had terminated at Sapulpa, on to Oklahoma City and then into Texas in 1887.

Oklahoma City didn't begin to achieve its current importance until the territory achieved statehood in 1907, but the newly chosen capital city soon became the key hub of rail travel within Oklahoma. Five railroads provided Pullman service to Oklahoma City, and until the later years, the SLSF, better known as the Frisco, accounted for nearly half the total.

The Frisco's main competitor for access to the St. Louis gateway was the MKT—more familiarly known as the Katy, the first railroad to provide sleeping car service in Oklahoma, in the early 1870's. The Frisco, nonetheless, had the advantage of a main line that was not only fifty miles shorter, it also directly served Tulsa as well.

In 1902, the Frisco inaugurated the *Meteor* between St. Louis and Oklahoma. To meet the competition of the Katy's new *Texas Special*, *Meteor* service was extended to Texas in 1915. Two years later, however, when the *Texas Special* became a jointly operated train of both the Frisco and the Katy,[2] the *Meteor* was shifted to St. Louis-Oklahoma City service.

As a heavyweight in the early 1940's, some of the *Meteor's* train sets wore a blue and silver livery with its own logo on the tender. For a time, service was extended to Lawton, and its consists also included Pullmans from Kansas City. During the post-war extension of eastern sleeping car service to the Southwest, it carried accommodations for both New York and Washington. In the spring of 1948, the train was re-equipped with streamlined stainless steel and red coaches from Pullman Standard pulled by EMD diesels. Although popular at the start, patronage on the new flyer inevitably dwindled, and the *Meteor* ended service as the Frisco's premier train in 1965.

The *Oklahoma Special,* whose numbers were later assumed by the *Will Rogers* and the *Southwest Limited,* also worked the line to St. Louis. The *Governor* was an overnight train from Oklahoma City to Tulsa and Muskogee.

The Frisco worked its home territory north through the Ozarks to St. Louis, scheduling Pullmans from Oklahoma City to Sapulpa, Tulsa, Okmulgee, and Muskogee; Joplin, Springfield, St. Louis, and Kansas City, Mo.; Memphis; and Chicago (Gulf, Mobile & Ohio or Illinois Central). Post-war, the radius of through Pullman travel was greatly extended with sleeping cars to New York (New York Central or Baltimore & Ohio), and Washington (B&O).

For a short time in the 1920's, Oklahoma City had dedicated Pullman service via the Frisco to Birmingham and, by extension, three times weekly, to Jacksonville (Southern), where passengers could transfer to any number of flyers serving the East or West Coast of Florida. Accommodations from Oklahoma City west included cars to Wichita Falls (MKT) and Floydada, Tex., (Quanah, Acme & Pacific).

Although the Katy was the first railroad in Oklahoma, its tracks didn't reach Oklahoma City until 1904.

Despite the superior Frisco competition, the MKT nonetheless inaugurated the *Sooner* from Kansas City and added a section to the *Katy Limited* to serve Oklahoma City. Pullman service on these and other trains included cars from Oklahoma City to Muskogee, Tulsa, St. Louis, and Kansas City.

Before statehood, Guthrie was the territorial capital, and at least two railroads were seemingly indifferent to running their lines through Oklahoma City. The Fort Smith & Western opted for Guthrie as its western terminus, which time proved to be the wrong choice, and the Rock Island's main line from Wichita to Texas passed twenty-five miles to the west of Oklahoma City, through El Reno.

The Rock Island, nonetheless, provided Oklahoma with Pullman service to the four points of the compass by way of its Memphis-Tucumcari, N.M., line, which crossed its tracks leading to Texas at El Reno. The east-west line was the route of the *Memphis-Californian* and later *Cherokee,* which linked the Tennessee city with the Golden State Route to the West Coast, and of

Santa Fe's northbound *Texas Chief* (opposite top) departs Oklahoma City with 11 cars on July 8, 1948. A similar-length *Memphis Californian* (center), eastbound under Rock Island power, rounds a curve near Elk City, Sept. 15, 1945. The *Texas Special* (bottom), here under Katy steam, was a joint Frisco/MKT operation for more than four decades.

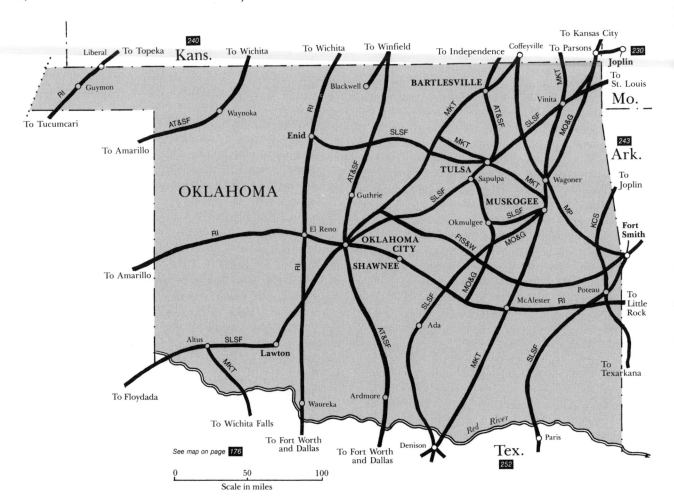

See map on page 176

0   50   100

Scale in miles

the *Choctaw Limited* and daytime streamliner *Choctaw Rocket.* Through trains on the Texas line included the *Firefly, Mid-Continent Special,* and streamlined *Twin Star Rocket.*

Oklahoma City had Pullman service on the Rock Island east to McAlester, Muskogee (Missouri, Oklahoma & Gulf), Little Rock, and Memphis; north to Wichita and Kansas City, west to Amarillo, and south to Fort Worth/Dallas. The Rock Island and Frisco shared a handsome mission-style station, opened in 1932.

Santa Fe trains serving Oklahoma City included the heavyweight *Ranger, Antelope,* and *Texas/Chicago Expresses* and the streamlined *Kansas Cityan, Chicagoan,* and *Texas Chief.* By the 1950's, the Santa Fe was providing most of Oklahoma City's dedicated Pullman service, including sleepers to Dallas/Fort Worth, Kansas City, Denver, and Chicago. Cars to Colorado Springs, however, had long since disappeared, as had the intrastate sleeper to Ardmore.

The little Fort Smith & Western, which ended its rather short and not very prosperous life in 1939, carried sleepers from Oklahoma City to its namesake terminus in Arkansas. The FS&W, which had trackage rights from the Katy into Oklahoma City, briefly scheduled an interline sleeping car with the MO&G to Joplin.

What was true of the Santa Fe in Oklahoma City applied to the Frisco in Tulsa: It was providing most of the city's Pullman service at mid-century. The main-line *Meteor, Southwest Limited,* and *Will Rogers* came through town en route to Oklahoma City, and several Frisco trains terminated in Tulsa, including the *Oil Fields Special* and *Black Gold* from Fort Worth and Dallas, and the daytime *Firefly* from Kansas City.[3]

Tulsa had Pullman service on the Frisco to Oklahoma City, Dallas, Fort Worth, Wichita, Kansas City, St. Louis, Chicago (Chicago & Eastern Illinois), and New York (Pennsylvania or NYC).

Although Tulsa was on a stub-end line of the Santa Fe, the road of the *Chiefs* was the principal passenger carrier to Chicago. During the heyday of rail travel, usually two trains a day made fast time on the Chicago-Kansas City run, then cut back the trottle slightly on the final lap through Kansas and Oklahoma. Not even the crack *Meteor* with good connections at St. Louis, however, could match the Santa Fe's Chicago-Tulsa card.

In the streamline era, the *Oil Flyer* was the overnight extension from Kansas City to Tulsa, carrying the Chicago sleepers. The *Tulsan,* which carried through chair cars, was a daytime link with the *Kansas Cityan/Chicagoan.* The Santa Fe also scheduled Pullmans over the years from Tulsa to Amarillo, Wichita, Kansas City, Denver, and Colorado Springs.

Tulsa was not a main line operation for the Katy, which, by the early 1950's had relegated the city to a single daily mixed-train. The *Bluebonnet* and *Katy Limited,* however, once carried dedicated Pullmans for Tulsa from Waco, Dallas, and Fort Worth.

After opening of Tulsa's art deco style Union Depot in 1931, all three railroads used the facility.

Numerous other passenger trains made brief transits of Oklahoma.

The Kansas City Southern's *Flying Crow* and *Southern Belle* passed through Poteau, where the Frisco's line from Fort Smith to Paris, Tex., intersected. Once owned by the Santa Fe, the Frisco had interchanged considerable Pullman traffic with its parent at Paris, but with the Santa Fe's preference for doing things its own way and the joint Frisco/Katy line to St. Louis, this evaporated.

In the northeastern corner of Oklahoma, the *Southerner* and *Rainbow Special* plied the tracks of the Missouri Pacific between Kansas City and Little Rock.

Waynoka was a short-lived air/rail terminal of the historic coast-to-coast "Lindbergh Line," where passengers arrived on the Ford trimotors of Transcontinental Air Transport to await the arrival of the Santa Fe's westbound *Missionary.* Patrons of this pioneering

In early 1946, a 14-car, northbound Santa Fe *Ranger* follows the Washita River through Arbuckle Canyon.

Preston George

Frisco's southbound *Meteor* winds through an S-curve north of Oklahoma City early on Columbus Day, 1946.

service slept en route in Pullman cars and resumed flying the next morning from Clovis, N.M. (See chapter The Transcontinentals.)

This southern line of the Santa Fe, generally used for freight traffic because of its favorable grades, also carried such passenger trains as the *Scout, Grand Canyon's* southern section, and *San Francisco Chief*. Waynoka itself had dedicated sleepers to Clovis and Albuquerque; in summers, Albuquerque cars detoured via Carlsbad Caverns, where they laid over during the day.

To the west, the Rock Island's line from Chicago to Tucumcari made a diagonal cut through the Panhandle. This was the route of the *Golden State Limited, Apache, Californian,* and *Imperial Limited*.

Several other Oklahoma communities had Pullman service, mostly to the large cities of Missouri: The Frisco carried sleepers to St. Louis from Enid, Blackwell, Lawton, Okmulgee, and Sapulpa. For Kansas City, the Frisco scheduled Pullmans from Ada and Sapulpa, the Rock Island from Shawnee and Enid, and the Katy from Bartlesville.

Muskogee, in addition to having three different lines to Oklahoma City (MKT, RI/MO&G, or SLSF), once was served with a sleeper to Sedalia, Mo., via the Katy.

1. Letter of August 18, 1988 from Preston George, the Oklahoma rail historian.
2. The Frisco operated the *Texas Special* over its shorter main line from St. Louis to Vinita, Okla., where the train went over to the MKT tracks for the remainder of the trip to Texas. The joint agreement remained in force until 1959.
3. The *Firefly*, not to be confused with the similarly named Rock Island train, was introduced in 1939 as a blue and silver consist of rebuilt heavyweights under tow of a streamlined steam locomotive. Originally scheduled to Oklahoma City, it was later cut back to Tulsa.

The Frisco's *Will Rogers* worked the St. Louis-Oklahoma City main line, carrying several local Pullmans.

38
Texas

Three of the Lone Star State's premier trains: MoPac's *Texas Eagle* (top left) passes a namesake sign in Austin while heading south in October 1963. Fort Worth & Denver's *Texas Zephyr* (top right) runs through the yards at Fort Worth in December 1964. New Orleans-bound, the *Sunset Limited* (bottom) takes a sharp curve after leaving San Antonio station in March 1963.

One doesn't have to be a Texan to appreciate the immensity of the state. In the heyday of rail travel it was possible to retire in the same Pullman sleeper for two consecutive nights and still remain under Lone Star sovereignty. Passengers on the Sunset Route between Orange on the east and El Paso on the west notched more than 925 miles of travel on Texas soil, and several other railroads scheduled runs of almost equal dimension elsewhere in the state.

Despite its enormous size, however, Texas was similar to several other mid-American states in that it was the eastern staging area of a transcontinental railway and with few exceptions its passenger service was mostly directed west of the Mississippi River. After World War II, through Pullmans were instituted between several Texas cities and New York and Washington, but destinations traditionally ranged from New Orleans, Atlanta, the Midwest, Denver, to California.

The country's second transcontinental railway resulted from the meeting of the Texas & Pacific and Southern Pacific at Sierra Blanca in 1881. The T&P, then under control of Jay Gould, extended across the breadth of Texas, through Dallas and Fort Worth, and by the following year had a line through to New Orleans, via Shreveport. By agreement, it used SP tracks from Sierra Blanca to El Paso.

In the eastern part of the state, the Texas & Pacific served as a bridge between the northern and southern components of the Missouri Pacific (which controlled the T&P for many years and finally merged it). In west Texas it played a similar role in the MP/T&P/SP Sunshine Route from St. Louis to California.

What the T&P was to the central part of the state, the SP's Sunset line through Houston and San Antonio was to southern Texas. Completed in 1883, the Sunset Route linked New Orleans and California with the country's southernmost and lowest crossing of the formidable western mountain barriers.

Within ten years of the SP's completion, six modern systems or their predecessors had entered Texas from the north: Missouri-Kansas-Texas (the first), Missouri Pacific, Santa Fe, Frisco, Burlington, and Rock Island, and the predecessor of the modern Cotton Belt, meanwhile, had pushed narrow-gauge rails north to a connection with St. Louis and the east.[1]

Not surprisingly, Texas had more large cities with substantial Pullman service than any other state. Dallas and Fort Worth, although distinct from and highly competitive with one another, shared considerable service because of their proximity, and the two together had the largest selection in the state.

The Texas & Pacific and Rock Island served the two cities sequentially, while the Katy, Frisco, SP, and Cotton Belt generally scheduled Pullmans to each city separately, and the Santa Fe and Burlington offered

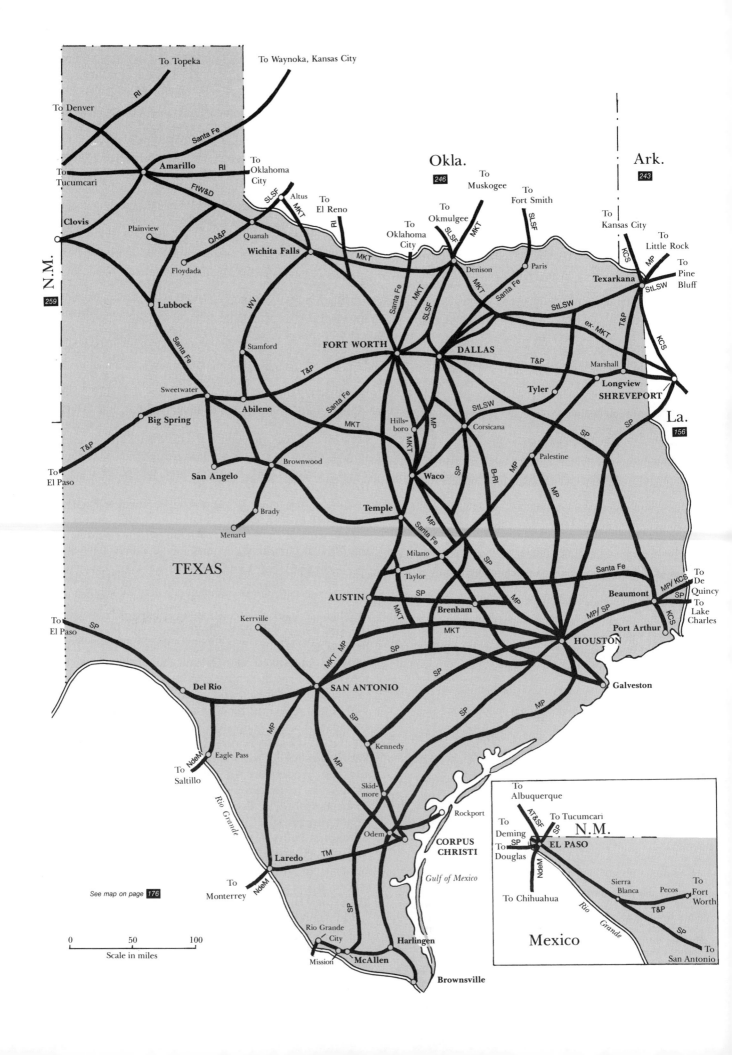

To Topeka

To Waynoka, Kansas City

To Denver

RI

Santa Fe

Okla.
246

To Muskogee

To Fort Smith

Ark.
243

To Amarillo

RI

To Oklahoma City

To El Reno

To Okmulgee

To Oklahoma City

To Kansas City

To Little Rock

To Pine Bluff

To Tucumcari

FtW&D

SLSF

Altus

MKT

RI

SLSF

MKT

SLSF

KCS

MP

Texarkana

StLSW

Clovis

Plainview

QA&P

Quanah

Wichita Falls

MKT

Denison

Paris

Santa Fe

StLSW

T&P

KCS

N.M.
259

Floydada

WV

Santa Fe

MKT

SLSF

Santa Fe

MKT

ex-MKT

StLSW

Lubbock

FORT WORTH

DALLAS

T&P

Marshall

Stamford

T&P

Tyler

Longview
SHREVEPORT

La.
156

Sweetwater

Santa Fe

Santa Fe

MKT

Hills-boro

MP

StLSW

Corsicana

SP

Palestine

SP

Abilene

Big Spring

MKT

MP

B-RI

MP

T&P

Brownwood

Waco

SP

To El Paso

San Angelo

Temple

MP

MKT

San Antonio

Brady

Santa Fe

Milano

MP

SP

Menard

TEXAS

Taylor

Santa Fe

Beaumont

MP/KCS

To De Quincy

To El Paso

SP

Kerrville

AUSTIN

SP

Brenham

MKT

MP

MP/SP

KCS

To Lake Charles

Port Arthur

MKT

HOUSTON

Del Rio

SP

SAN ANTONIO

SP

MKT MP

SP

SP

MP

Galveston

Eagle Pass

MP

Kennedy

To NdeM

To Saltillo

MP

Skid-more

Rockport

To Albuquerque

Laredo

TM

Odem

CORPUS CHRISTI

AT&SF

To Tucumcari

N.M.

To Deming

SP

EL PASO

To Monterrey

NdeM

Gulf of Mexico

To Douglas

NdeM

Sierra Blanca

Pecos

To Fort Worth

See map on page 176

SP

Rio Grande City

To Chihuahua

Rio Grande

T&P

SP

Mission

McAllen

Harlingen

Mexico

To San Antonio

Brownsville

0   50   100
Scale in miles

Rio Grande

Dallas and Fort Worth both separate and sequential service. The MP's International-Great Northern line served Fort Worth only.

The T&P main line carried such trains as the northern section of the *Sunshine Special, Texan, Texas Ranger,* the streamlined *Texas Eagle,* and the *Westerner.* The T&P connected with the Missouri Pacific at Texarkana providing interline Pullman service from Fort Worth and Dallas to Little Rock, Memphis, St. Louis, New York (New York Central), Washington (Pennsylvania or Baltimore & Ohio), and Chicago (Gulf, Mobile & Ohio). SP picked up sleepers for Los Angeles at El Paso.

On its main line east, the T&P carried Pullmans for Shreveport, Atlanta (Illinois Central/Southern), and New Orleans. To the west it scheduled intrastate sleepers to Abilene, Sweetwater, Big Spring, Pecos, El Paso, and, jointly with the Santa Fe, cars for Amarillo, Lubbock, and San Angelo.

The Southern Pacific operated two crack daytime trains between Fort Worth/Dallas and Houston, the streamliners *Hustler* and *Sunbeam*—wearing SP's Daylight livery of red, orange, and black—plus the overnight *Owl.* Both Fort Worth and Dallas had dedicated Pullmans from Galveston, Houston, New Orleans, and San Antonio by way of the SP, which also scheduled sleepers from Dallas to Austin, Beaumont, Port Arthur, and Los Angeles.

The Santa Fe offered wide-ranging service over its lines extending from Chicago to the Gulf Coast at Galveston, then in a broad arc to California. The Chicago-Galveston line through Fort Worth carried the *Texas Express* until the train was discontinued early in the Depression. The *Ranger* was put into service on the eve of World War I, and the *Antelope* was of 1920's vintage. The streamlined, flagship *Texas Chief,* inaugurated in 1948, grew to acquire a Dallas section several years later.

Santa Fe scheduled Pullmans to Dallas/Fort Worth from Chicago, Newton, Kans., Oklahoma City, Clovis, N.M., Denver, Los Angeles, and San Antonio (MP), in addition to sleepers from each city to Galveston, San Angelo, and St. Louis (Frisco).

Santa Fe ownership of the Frisco in the early 1890's resulted in some interline sleeper service—such as the St. Louis accommodations—that outlived the corporate relationship. In Texas itself, the Frisco scheduled Pullmans from Fort Worth to Brownwood, Brady, and Menard jointly with its Fort Worth & Rio Grande line, which it sold to the Santa Fe in 1937. To the north, the Frisco carried sleepers from Dallas to Monett, Mo., Tulsa, and Chicago (Chicago & Eastern Illinois), and from Fort Worth to Tulsa and Kansas City.

Dallas and Fort Worth saw less purely Frisco service than otherwise might have been expected because the railroad shared operation of the *Texas Special* and *Bluebonnet* with the MKT. In 1917, the Katy's *Texas Special* began operating from St. Louis on the Frisco's tracks to Vinita, Okla., where it switched to the MKT for the remainder of the trip. The *Bluebonnet* followed the same routing, which offered faster schedules over the Frisco's shorter, northern portion of the route. Until the arrangement was terminated in 1959, the Pullmans carried on these trains from St. Louis and the post-war sleepers from New York (NYC) were of joint Frisco/MKT sponsorship.

The Katy had separate lines through both Dallas and Fort Worth (the one through Fort Worth involved trackage rights over the T&P), and the railroad scheduled Pullmans to each city from Houston, Austin, San Antonio, Tulsa, Kansas City, and Minneapolis/St. Paul.

In addition to joint operation with the Frisco, the MKT also scheduled trains on its own line from St.

Named for the Texas state flower, the *Bluebonnet* was a joint Frisco-Katy endeavor upon its introduction in 1927. Here it's on MKT tracks in Fort Worth, June 1938.

Harold K. Vollrath collection

Otto C. Perry/Denver Public Library

T&P's *Texas Ranger*, which operated the breadth of the state and carried many Dallas Pullmans, departs El Paso July 28, 1939.

Louis, some of which included sections from Kansas City. Among these were the *Katy Flyer* and onetime all-Pullman *Katy Limited*. The *11 O'Clock Katy* was an overnight train between Fort Worth, Dallas, and Houston named for its departure times from each terminus. The MKT also scheduled sleepers from Dallas to Galveston, Wichita Falls, St. Louis, and Shreveport. (It sold the Shreveport line to a predecessor of Kansas City Southern in 1923).

Before the Rock Island's streamlined *Twin Star Rocket* began carrying Pullmans from Minneapolis/St. Paul through to Houston, the railroad terminated its *Mid-Continent Special* and *Firefly* in Dallas. The Rock Island's sleeper schedules from Dallas/Fort Worth included cars to Oklahoma City, McFarland, Kans., Kansas City, Chicago, and Minneapolis.

The Burlington served the Dallas/Fort Worth passenger market by way of the Colorado & Southern/Fort Worth & Denver lines. This became the route of the crack Denver-to-Dallas *Texas Zephyr* in the summer of 1940. FW&D Pullman destinations from Dallas/Fort Worth included Wichita Falls, Lubbock, Abilene (Wichita Valley), Plainview, Amarillo, and Denver, in addition to a Fort Worth-Colorado Springs sleeper. The daytime streamliner *Sam Houston Zephyr* was the Burlington's challenge to the SP's *Sunbeams* and *Hustlers*.[2]

The St. Louis Southwestern Railway, better known as the Cotton Belt, in honor of the principal cargo it was built to haul, inaugurated its *Lone Star* between Dallas, Fort Worth and Memphis in 1916. This and a few other solid, comfortable trains provided Cotton

Belt passenger service until the end came in 1959. The SSW scheduled Pullmans from both Dallas and Fort Worth to Texarkana and Memphis, and from Dallas to St. Louis.

The MP lines served Fort Worth with Pullmans from Mission and Houston.

Dallas's Union Station served all eight of the city's passenger carriers. Fort Worth had two main terminals, the tenants of which, at mid-century, were: Rock Island, Frisco, Santa Fe, and Southern Pacific in Union Station; and Burlington, MKT, SSW, and Texas & Pacific at the T&P's facility. Before service ended, SSW used the latter, and MP used Union.

The Missouri Pacific and Southern Pacific provided the largest portion of Houston's Pullman service, much of which was intrastate. SP, the sole occupant of the grandiously named Grand Central Station, operated the overnight *Owl*, with Pullmans to both Dallas and Fort Worth, and the daytime *Sunbeam* and *Hustler* to those cities. The *Sunset Limited* and *Argonaut* covered the Sunset Route between New Orleans and the West Coast, and several other trains operated from New Orleans into Texas: the *Sunset Mail* and more recent *Alamo* to San Antonio, and *Texas Limited* and *Acadian* to Houston.

Pullman service on the transcontinental line included cars to New Orleans, Chicago (IC), San Antonio (both main line and via Kennedy), and Los Angeles. The SP also scheduled through sleepers to Mexico in the 1920's by way of Corpus Christi and Laredo (Texas Mexican/National Railways of Mexico) to Monterrey, Tampico, and Mexico City. Within Texas, the railroad carried sleeping cars from Houston to McAllen, Brownsville, Corpus Christi, Austin, and Waco. SP service also included Pullmans to Shreveport and St. Louis (Frisco or Cotton Belt). Summer 1928 schedules show a Pullman for Kerrville in the scenic uplands northwest of San Antonio.

The MoPac's International-Great Northern and Gulf Coast Lines in eastern Texas offered stiff competition to the SP, whose service it often duplicated. The MP's main distinction were the accommodations it provided on the *Sunshine Special* and streamlined *Texas Eagle* to St. Louis and the north. Over this route, involving the T&P as a bridge between other MP components, Pullmans departed from Houston to Little Rock, Memphis, Kansas City (KCS), St. Louis, Chicago (Alton), and, after World War II, New York (PRR or NYC) and Washington (B&O or PRR).

The Missouri Pacific, an old hand at through service to Mexico, carried the Monterrey and Tampico Pullmans (TM/NdeM) to Corpus Christi for several years before the SP's participation and also scheduled sleepers to New Orleans. Within Texas, Pullmans moved over MP lines from Houston to Austin, Fort Worth, Corpus Christi, Brownsville, and Mission. The overnight *Pioneer* and daytime *Valley Eagle* worked the line to Brownsville.

In addition to the Missouri Pacific, two other railroads scheduled Pullman service from Houston's Union Station. The Rock Island's *Twin Star Rocket*, from Minneapolis/St. Paul to Houston, traveled the longest distance of any single-railroad, north-south train in the country. The *Rocket* carried sleepers from Houston to Kansas City and the Twin Cities.

Santa Fe scheduled Pullmans from Houston to San Angelo and Clovis and in an early interline operation turned over a St. Louis sleeper to the Frisco at Paris.

The Katy, operating from its own station in Houston, had a fairly roundabout route north; consequently MKT service from Houston was slower and not so glamorous as its competitors', and the railroad itself considered San Antonio its principal southern terminus. Nonetheless, sections of such trains as the *Katy Limited* and *Bluebonnet* served Houston, connecting with the San Antonio trains at Waco.

In addition to Dallas and Fort Worth cars, the railroad scheduled sleepers to San Antonio, Wichita Falls, Kansas City, St. Louis, Minneapolis/St. Paul (Chicago Great Western), and Chicago (Burlington, 1895, at Hannibal, Mo.).

In addition to being a terminus for Katy trains, San Antonio also stood at the crossing of the SP's Sunset Route and the principal modern line to Mexico, the Missouri Pacific.

The *Sunset Limited* and *Argonaut* were the SP's through trains, and the *Sunset Mail* and *Alamo* offered accommodations to New Orleans. San Antonio had Pullman service east to Houston, Galveston, New Orleans, Atlanta (Louisville & Nashville/West Point Route), and Chicago (IC), and west to Los Angeles and San Francisco. Service to other Texas communities included Dallas—where also a Memphis sleeper was turned over to the Cotton Belt—Fort Worth, Brownsville, Corpus Christi, McAllen, and the Gulf resort community of Rockport.

In San Antonio the MKT and Southern Pacific transferred interline Pullmans operating between the Midwest and California, cars delivered by such trains as the *Texas Special, Bluebonnet,* and *Katy Flyer,* which terminated in the city of the Alamo.

The Katy carried Pullmans to Waco, Dallas, Fort Worth, Houston, Galveston, Kansas City, and St. Louis. Its other service was interline, principally with the Frisco, and this included sleepers to St. Louis, Chicago (Wabash), Washington (B&O), and New York (PRR or NYC). An alternate line to Chicago involved the Alton, and a sleeper to Minneapolis/St. Paul went via CGW. San Antonio once had two Pullman routes to Denver on the Katy, one via the Burlington at Fort Worth, and an earlier and longer line via Union Pacific at Kansas City.

In the Missouri Pacific's massive shuffling of Pullmans, San Antonio was served by the South Texas sections of the *Sunshine Special* and streamlined *Texas Eagle* as well as the *Southerner* and *Texan.* Accommodations included sleepers to Little Rock, St. Louis, and Washington (NYC/Chesapeake & Ohio). The MP also transfered to the Santa Fe at Milano some interline Pullmans for Fort Worth, Denver, Kansas City, and St. Louis (Frisco).

Within Texas, the MP scheduled sleepers from San Antonio to Corpus Christi, Brownsville, Mission, Rio Grande City, Laredo, Waco, and Brownwood (Santa Fe).

The ancestral line of the National Railways of Mexico to Nuevo Laredo originally was narrow gauge when it opened in 1888, and passengers going to Mexico City were transported in small Pullman cars aquired from two Colorado mountain railroads. Conversion of the line to standard gauge fifteen years later made possible the scheduling of through cars from the United States. Both San Antonio and the Missouri Pacific played key roles in the international operation.

Through Pullmans for Mexico City originated in Little Rock, St. Louis, Chicago, and, after World War II, even New York City, and they were carried principally in the MP's *Sunshine Special* and post-war, streamlined *Texas/Aztec Eagles.* From 1937-40, the MP/NdeM offered a once-weekly luxury train from St. Louis, called the *City of Mexico,* with a bilingual staff. When through cars were not scheduled, Pullmans originated in San Antonio, and this arrangement included sleepers for Tampico and Mexico City.[3]

A second line into Mexico, by way of a branch of the Southern Pacific to Eagle Pass/Piedras Negras, had the advantage of being standard gauge from the start, and upon completion in 1889, the SP and predecessor lines of the NdeM put into service the deluxe *Montezuma Special*. This train originated in New Orleans and operated about every tenth day. San Antonio later had dedicated sleepers over this route to both Saltillo and Mexico City.

El Paso was yet another rail gateway to Mexico and the first to have through service to Mexico City, in 1884. The line, however, never carried the traffic from north of the Rio Grande that the other two did, and the El Paso sleepers to Mexico City, carried by NdeM and predecessor Mexican Central, were mostly an accommodation for passengers transferring from the three major American railroads calling at El Paso's Union Depot.

At El Paso the Golden State and Sunset Routes of the Southern Pacific came together from Chicago and New Orleans, respectively, bringing to this West Texas outpost such trains as the *Golden State Limited, Californian, Apache, Imperial, Sunset Limited*, and *Argonaut*. The SP scheduled Pullmans on these and other trains from El Paso to Douglas, Tucson, Globe, and Phoenix, Ariz.; Los Angeles (tourist); and Chicago (RI).

El Paso was the Texas & Pacific's western terminus, which put such trains as the *Sunshine Special, Texas Ranger, Westerner*, and *Texas Eagle* on the arrival board. The T&P carried sleepers to El Paso from New Orleans, Texarkana, and Dallas/Fort Worth, and in company with the Missouri Pacific from Little Rock, St. Louis, Chicago (GM&O), and New York (PRR).

The Santa Fe's line to the north provided El Paso with Pullman service to Albuquerque, Denver, Kansas City, and St. Louis (Frisco, 1890, at Burrton, Kans.).

Galveston reputedly was once the richest city in Texas from the cotton shipments it handled, and this business helped give it an unusually high level of Pullman service. In the last years of private railroading, Santa Fe carried Chicago Pullmans on its crack *Texas Chief;* earlier it scheduled sleeping cars from Galveston to Temple, Sweetwater, Dallas, Fort Worth, Wichita, St. Louis (Frisco), Kansas City, Denver, and Los Angeles.

Three other railroads once carried sleepers from Galveston to St. Louis: The SP made an interline transfer with the MKT at Denison (1890); shortly after the turn of the century, Galveston cars made the entire trip on the *Katy Flyer*. The MoPac's Pullmans to St. Louis, however, outlived all the others.

The SP also exchanged Chicago sleepers with the MKT (1890, Burlington via Hannibal) at Denison as well as serving this key rail junction with its own car. Galveston had dedicated Pullman service on the SP to San Antonio, Dallas, Fort Worth, Denver (Burlington) and New Orleans.

MP offered competing accommodations to New Orleans as well as a post-war sleeper to Washington (PRR). The Katy scheduled Pullmans from Galveston to San Antonio, Dallas, and Kansas City.

Waco was always busy with MKT activity, and it was there that train sections and cars were sorted according to routes or destinations, Dallas or Fort Worth, San Antonio or Houston, etc. The Katy scheduled Pullmans from Waco to Stamford, Tex., Tulsa, Kansas City, and jointly with the Frisco to St. Louis and New York City (NYC).

Both the MKT and MP carried sleepers from Waco to San Antonio.

Over the years, the Cotton Belt was also prominent on the scene at Waco, one of several communities where the tracks of the SSW met those of its parent Southern Pacific, which brought Pullmans into town from Houston.

The Cotton Belt's first sleepers to depart Waco, in the mid-1880's, were narrow-gauge cars, some of which traveled all the way to Cairo, Ill. After the line was converted to standard gauge, the IC took over Waco-Chicago Pullmans at Cairo. Subsequent sleeping car service included Texarkana, Memphis (1895), Nashville (1897), and St. Louis.

Three regional passenger routes intersected in the Texas Panhandle at Amarillo: the Santa Fe's Belen Cutoff, served by such trains as the southern section of the *Grand Canyon, Scout, Missionary*, and *San Francisco Chief;* the Rock Island's Choctaw line from Tucumcari to Memphis, which linked the Tennessee city with the Golden State Route; and the Fort Worth & Denver, route of the *Texas Zephyr*.

The Santa Fe scheduled Pullmans from Amarillo to Albuquerque, Tulsa, and Wichita; the Rock Island to Oklahoma City, St. Louis (Frisco), and Memphis; and the FW&D to Wichita Falls and Denver. Both the FW&D and Santa Fe/T&P carried sleepers to Fort Worth/Dallas.

Pullman service to Wichita Falls by way of the Katy included cars from Houston, Dallas, Kansas City, Oklahoma City (Frisco), and St. Louis (Frisco). In addition to the accommodations from Amarillo, the FW&D scheduled sleepers from both Dallas and Fort Worth.

Texas' southernmost Pullman destination, Brownsville, had service on both the MP and SP from both Houston and San Antonio. A sleeper arrived from New Orleans on the MP, as did a Chicago car delivered by IC to the Crescent City. The SP once brought another sleeping car from Chicago to Brownsville by

Donald Sims

The colorful *Southern Belle*, with a Kansas City-Port Arthur sleeper, stops at the KCS/MP station in Beaumont, about 1951.

way of Fort Worth (Rock Island). Pullmans from St. Louis came by way of the MP's *Sunshine Special* and on the *Texas Special* (Frisco/MKT/SP).

Laredo had dedicated accommodations by way of Missouri Pacific lines to San Antonio, St. Louis, and Chicago (Wabash, 1895), and via NdeM to Mexico City.

Port Arthur had Pullman service on the Kansas City Southern to Kansas City and St. Louis (MP). The *Flying Crow* was the principal KCS train serving the community after its inauguration in June 1928. The SP scheduled sleepers from Port Arthur to Dallas.

Texans with business in the state capital at Austin could book overnight accommodations from Houston on either MP or SP, or from Dallas on SP or MKT. The Katy also scheduled Pullmans from Austin to Fort Worth and St. Louis.

Both SP and MP served Corpus Christi, and each of them carried Pullmans to San Antonio and Houston. The SP also originated a sleeper for Kansas City (MKT).

The stateline railroad junction of Texarkana was the terminus of the T&P's 863-mile Pullman line from El Paso as well as a busy center of Cotton Belt operation. The SSW scheduled sleepers from Texarkana to Dallas, Fort Worth, Waco, Memphis, and St. Louis.

Pullman service to Mission was a MoPac endeavor and included sleepers to Houston, San Antonio, Fort Worth, and St. Louis.

Abilene residents had a choice of Pullman service to Fort Worth/Dallas, either directly via the T&P, or roundabout by way of the Wichita Valley/FW&D through Wichita Falls. The T&P once offered through sleepers from Abilene to New Orleans.

The SP scheduled Pullmans from Beaumont to Dallas. By way of the Santa Fe/Frisco interchange at Paris, Beaumont also had a sleeper to St. Louis.

Several other Texas communities had sleeping car service to St. Louis: Big Spring, Sweetwater, and Taylor (1890) via T&P/MP lines; Paris and Quanah on the Frisco, and Tyler via the Cotton Belt. Other sleeper lines included Hillsboro-Parsons, Kans. (MKT), Floydada-Oklahoma City (Quanah, Acme & Pacific/Frisco), and Sweetwater-Kansas City (SF).

---

1. Texas law once required railroads operating in the state to be owned by Texas corporations. As a result, the Santa Fe did business under the Texas corporate name of Gulf, Colorado & Santa Fe Railway, and so forth down the line. Mostly I use the general system name, unless the Texas designation is particularly relevant.

2. Both the Rock Island and Burlington Lines shared, under several ownership arrangements, the use of the Trinity & Brazos Valley Railway. This line, plus some trackage agreements, extended their operation from Fort Worth/Dallas to Houston/Galveston. Thus the *Sam Houston Zephyrs* and *Twin Star Rockets* shared the same iron and appeared in both the parents' Texas schedules. The *Sam Houston* inaugurated the streamliner competition in 1936 with a trainset of one of the original *Twin Zephyrs;* the SP responded to the challenge with its splashy *Sunbeams* the following year.

3. Arthur D. Dubin, "Mexicano de Lujo!" *Trains,* Vol. 27, No. 4 (February 1967), pp. 34–38.

39
New Mexico

Possession is nine points in the law, and Santa Fe track gangs beat their Denver & Rio Grande rivals to occupancy of Raton Pass by all of thirty minutes one morning, closing the case. The Atchison, Topeka & Santa Fe Railroad, as it was formally known then, had been incorporated in Kansas and was generally following the Santa Fe trail toward its namesake destination in New Mexico. The gradual slope of the high plains of Colorado was now behind it; the wall of the Rocky Mountains lay ahead.

The climb through a spur of the Rockies to Raton Pass begins at Trinidad, just as abruptly as the Great Plains end. While its elevation of just over 7,500 feet is not so high as other major modern crossings, Raton's 3.5 per cent grades are by far the most challenging. Nor was it all downhill beyond the summit. After flanking the Turkey Mountains and the Sangre de Cristo Range, Santa Fe tracks turned west, crossed the Pecos River, and began their climb over Glorieta Pass.

When the Santa Fe built this line and continued west from Albuquerque to California in the early 1880's, the Raton/Glorieta route was all but dictated by the exclusive reservation, for Indian tribes, of what is now Oklahoma. The subsequent land rushes and railroad building, however, changed that, and by early in this century, the Santa Fe had completed a low-grade line through Clovis, called the Belen Cutoff. Although most of the passenger trains continued using the line over Raton Pass, the Santa Fe put its heavy freight traffic on the southern line.

Not so glamorous as other Santa Fe flyers, the *California Limited* (top) nonetheless was the amply luxurious workhorse of the Chicago-Los Angeles fleet. Here it's gathering speed near Romero, N.M., on Oct. 4, 1947. A 12-car *Californian*, with SP's streamlined *Daylight* locomotive No. 4454, pauses in Deming on Sept. 25, 1947.

Santa Fe Southern Pacific Corp.

Santa Fe's famed, train-stop complex at Albuquerque included the Alvarado Hotel and Fred Harvey Indian Museum.

The city of Santa Fe, New Mexico's capital, was connected with Southern Colorado by the Rio Grande's narrow-gauge Chili Line, but only by a branch from the main Santa Fe line at Lamy. Albuquerque, meanwhile, became the state's major metropolis, well served by the ambitious transcontinental railway from Kansas.

While such trains as the *De Luxe, California Limited, Chief,* and *Super Chief* paused for brief servicing, passengers engaged in the railroad's ritual of wandering the platforms and station at Albuquerque to buy native American arts and crafts.

The Indian-detours were a favorite promotion of the Santa Fe in New Mexico. "Harveycars" and buses met the trains at Albuquerque, Lamy and other stations and took passengers on two- or three-day guided tours of "hidden primitive Indian pueblos, Spanish missions, pre-historic cliff-dwellings and buried cities," according to the ads.

Albuquerque accounted for most of New Mexico's dedicated Pullman service, which was Santa Fe all the way. The railroad brought sleepers to Albuquerque from Chicago, Kansas City, and Denver. Over the Belen Cutoff, the Santa Fe carried Pullmans to Clovis, Roswell, Amarillo, and Waynoka, Okla. On the line south from Belen it scheduled sleepers to Deming, Silver City, and El Paso.

The southern main line through Clovis did not ac-commodate the premier passenger trains, but it was the main line of the Santa Fe's California-Texas service, and carried a fair portion of the Chicago-Kansas City-California traffic. Over the years such stalwarts as the *Scout, Missionary, Grand Canyon's* southern section, and the *San Francisco Chief* plied the route. The Santa Fe scheduled Pullman service to Clovis from Waynoka, Dallas, and Houston.

A Santa Fe line ran south from Clovis through Roswell and Carlsbad, site of the national park containing the world's largest known natural cavern. Much like its Grand Canyon operation, the railroad scheduled sleepers on the *Scout* between Waynoka and Albuquerque and routed them in the middle of the night to Carlsbad. There they laid over during the day while their occupants toured the caves. That night the cars departed, to be cut back into the *Scout* at Clovis and resume their east-west journey early in the morning. The Santa Fe also scheduled sleepers to Carlsbad from Kansas City.

In part to accommodate travelers to and from Mesa Verde National Park, to the north, Gallup had its own Pullman to Phoenix. And before the turn of the century, a dedicated sleeper connected the bustling commercial center of Las Vegas with the key rail junction of Pueblo, Colo., over Raton Pass.

Two lines of the Southern Pacific fanned out west of El Paso through the portion of New Mexico acquired in the Gadsden Purchase of 1853. The original Sunset Route ran through Deming, and the former El Paso & Southwestern line through Columbus. The latter formed part of the Golden State Route, which connected with the Rock Island at Tucumcari.

The *Golden State Limited* was the flagship of its namesake route from Chicago to San Francisco, which also hosted the *Californian, Apache,* and *Imperial Limited*. The Rock Island's Choctaw line from Memphis also terminated in Tucumcari.

The SP's venerable *Sunset Limited* began sporadic service over its route in 1894 and finally became a daily train in 1913. Its secondary companion on the New Orleans-California haul through southern New Mexico was the *Argonaut*.

The tracks of the Burlington's Colorado & Southern cut across the northeast corner of the state through Clayton. This line between Denver and Dallas/Fort Worth was the route of the unnamed, though hard-working express trains Nos. 1 and 2, and the *Texas Zephyr* of 1940 vintage.

40 Arizona

Two transcontinental lines, the Southern Pacific and Santa Fe, gave Arizona an enviable assortment of luxurious passenger trains. The SP crossed the southern half of the state, through the land acquired in the Gadsden Purchase of 1853, and the Santa Fe bisected the northern half. The two of them provided through service from the Great Lakes and Mississippi River Valley to the Pacific Coast.

The Southern Pacific and Texas & Pacific completed a new transcontinental line in 1881, when their track gangs met at Sierra Blanca, Tex., and the same year the Santa Fe became a link in a third, but temporary, transcontinental line after connecting with the SP at Deming, N.M.

The SP/T&P tracks formed a route from San Francisco, by way of Los Angeles, through Yuma, Maricopa, Tucson, and Bowie, to El Paso, Fort Worth/Dallas and the east. Two years later, in 1883, the SP opened its Sunset Route to New Orleans.

In the same year, the Santa Fe made its connection with the SP at Needles, Calif. (this SP line, through Barstow, Calif., later became the Santa Fe's), and in 1885 the future road of the *Chiefs* dispatched its first transcontinental train east from San Diego.

The Santa Fe became justly famous for its crack passenger fleet, which over the years included the *California Limited, De Luxe, Chief,* all-coach *El Capitan,* and the incomparable *Super Chief.*

Shortly after the turn of the century, the railroad began promoting the great natural wonder of the

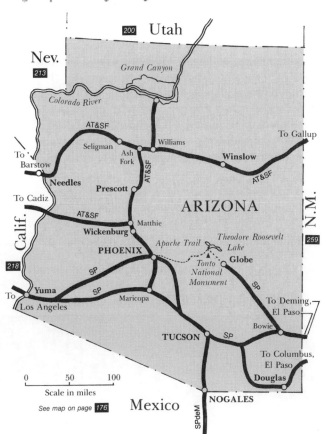

See map on page 176

Santa Fe Southern Pacific Corp.

A dry snow glistens as the eastbound, all-room, all-Pullman *Super Chief* glides over the highlands near Flagstaff.

Grand Canyon. The Santa Fe acquired and extended a small freight line from Williams to the south rim of the canyon and in 1901 began operating through Pullmans during the summer travel season. Several years later it built the spectacular El Tovar Hotel on the edge of the canyon.

In 1929, the Santa Fe introduced its *Grand Canyon* trains, which commonly operated in two or more sections. From both Chicago and Los Angeles they carried Pullmans that detoured to the park at night so passengers could enjoy a day of sight-seeing before resuming their journey. That evening, the Santa Fe hauled the cars back to the main line at Williams and cut them into the appropriate east- or westbound trains.

On the eve of World War I, the Southern Pacific began promoting a detour of its own to spice up travel across the desert country. One line of the Arizona Eastern, an SP subsidiary, connected with the Sunset Route at Bowie, and the SP began running Pullmans from New Orleans and El Paso over this route to the copper-mining town of Globe. There, passengers boarded autos or stages and traveled over what was called the Apache Trail to Phoenix, whence they resumed their rail trip to Los Angeles.

The Apache Trail offered scenic travel on mountain roads and access to Theodore Roosevelt Lake and Dam and the Tonto National Monument. The service involved such trains as the *Sunset Limited* and *Californian* and was scheduled during the winter sea-

Otto C. Perry/Denver Public Library

The Santa Fe served Phoenix with accommodation trains that tied into the main line at Ash Fork. No. 42, shown here in 1933, carried Chicago sleepers.

The *Golden State Limited*, here eastbound at Alhambra, Calif., was a joint operation of Southern Pacific and Rock Island.

sons.

A second main line in Arizona went through Douglas, the former El Paso & Southwestern from Tucson to Tucumcari, N.M., and this had become part of the Golden State Route upon inauguration of the namesake train in November 1902. The SP took over the El Paso road in 1924.

The *Golden State Limited* was faithful to this route and for some years was the only first-class train on the line. By mid-century most of the SP trains, including the *Sunset Limited*, were traveling through Douglas, but near the end of the road, the combined *Sunset-/Golden State Limited* was back on the Bowie tracks. Regionally, Douglas had Pullman service to El Paso.

Phoenix, the capital of Arizona, was originally a branch line operation for both the Santa Fe and Southern Pacific. The SP shuttled Phoenix cars to its main line at Maricopa, and the Santa Fe conducted a similar maneuver on its line through Prescott to Ash Fork.

In the early 1920's, civic pride and political pressure built to the point where the two railroads agreed to erect a union station for the rapidly growing city. Completed in late 1923, the long, colonnaded structure incorporated an appropriate accommodation to the city's desert climate, an open-air waiting room. Three years later, the SP completed its new main line through Phoenix, whereupon the eagerly awaited

trains—eventually including the *Sunset, Golden State, Californian, Imperial,* and at times the *Argonaut*—began calling.

Briefly before World War II, Arizona had its own winter-season train in addition to the parade of California varnish flowing through the state. The *Arizona Limited,* inaugurated in 1940, was a joint SP/Rock Island venture, an all-Pullman, extra-fare streamliner operating on alternate days between Phoenix and Chicago. Hundreds of people crowded the platform to greet the premier train upon its arrival from the Midwest on December 17 of that year.

Phoenix had Pullman service on the SP to Tucson, Douglas, Globe, El Paso, Chicago (Rock Island), and Los Angeles. The Santa Fe scheduled sleepers to Winslow, Gallup, Denver, Chicago, and Los Angeles.

Tucson became a one-railroad town after the SP acquisition of the El Paso & Southwestern. The consolidation, however, had little effect on Pullman service. The Southern Pacific scheduled sleepers from Tucson to El Paso, Chicago (RI), Phoenix, Los Angeles, and Guaymas (Southern Pacific of Mexico).

The Pullman to Guaymas was one of many operating between the United States and Mexican destinations as far south as Mexico City. Through sleepers from both Los Angeles and El Paso came through Tucson, from which they were routed south to Nogales for their SPdeM connection.

41
West by Northwest

The Union Pacific's last *Portland Rose* (top) crosses the Blue Mountains westbound near Kamela, Ore., on Amtrak day, May 1, 1971. Running combined (bottom), the Soo Line's eastbound *Winnipeger* and *Soo-Dominion* arrive in Minneapolis on May 4, 1958. In summertime, the *Soo-Dominion* gave way to the *Mountaineer*, offering through sleepers to Vancouver, B.C.

When President Abraham Lincoln signed the charter for what was to become the first of the northern transcontinental railroads, the plan was to link the Pacific Northwest with rails running from the Great Lakes at Duluth. Minneapolis and St. Paul were not yet connected with a through route from Chicago and were still growing into their roles as the railroad metropolises of the upper Midwest. Two of the roads did build lines from Lake Superior, but the Twin Cities quickly established their supremacy and eventually became the eastern terminus for all four of the northern routes to the Pacific Coast.

The Northern Pacific, a land-grant road whose charter Lincoln signed, was the first railroad from the Midwest to reach Puget Sound, in 1883. Ten years later, without the array of subsidies commonly available to the western roads,[1] James J. Hill finished driving his Great Northern to the same goal. Also in 1893, another northern route was formed, when the Soo Line connected with the Canadian Pacific, at Portal, N.D. The last of the four was the Pacific extension of the Milwaukee Road, completed in 1909.

The Milwaukee, of course, had its own tracks to Chicago, but the other railroads had to negotiate this vital link. In 1901, the Great Northern and Northern Pacific solved the problem by buying control of the Chicago, Burlington & Quincy, and in 1909 the Soo assured itself of a Chicago connection by leasing the Wisconsin Central. Meanwhile in Washington state, the GN and NP were backing construction of the Spokane, Portland & Seattle Railway to give them both access to Portland via the Columbia River valley.

Fittingly, the senior northern transcontinental also had the oldest name train: The NP's *North Coast Limited*, which began service in 1900. Following part of the trail blazed by Lewis and Clark and offering views of 28 mountain ranges on its way west, the *North Coast Limited* was a pioneering provider of such luxuries as open observation cars, private baths for men and women, a barbershop, a soda fountain, and the electronic marvel of radio. In 1911, when the train began running to and from Chicago, rather than originating in St. Paul, it was operated east of the Twin Cities by the Chicago & North Western, through Milwaukee. The Burlington took over in 1918. After World War II, the train was streamlined in a handsome color scheme of apple and dark green designed by Raymond Loewy, and dome cars were added to the five sets of equipment in 1954. The *North Coast Limited* survived as a train of fine breeding until the advent of Amtrak in 1971. The *Mainstreeter* was the Northern Pacific's secondary train during much of the streamline era.

In what was also a fitting touch, the premier train of the rival Great Northern was named for the railroad's forceful builder, James J. Hill. The classic *Oriental Limited*, which was renowned for tea service in

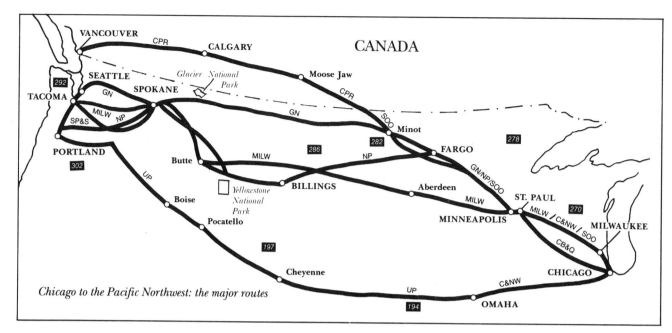

Chicago to the Pacific Northwest: the major routes

the observation lounge every afternoon at four, had been the road's flagship since 1905. But in the entrepreneurially minded 1920s, management decided to honor the old builder himself, and upon arrival of its luxurious new equipment in 1929, the train was renamed the *Empire Builder*. It was just in time to use the newly opened, second Cascade Tunnel in Washington, at 7.79 miles the longest underground bore in the Americas. The old *Oriental Limited* was operated until 1931, then dropped.

The *Empire Builder* became the first postwar transcontinental streamliner when it began service in February 1947 with five sets of equipment operating on a much faster schedule than had been offered before. Now Seattle had 45-hour, two-night service from Chicago, thanks to this striking train dressed in deep orange and olive green and trimmed in gold. In 1951, the *Empire Builder* again was reequipped, and in 1955 twenty-two new dome cars arrived from the manufac-

turers, including six full-length Great Domes from Budd for use of Pullman passengers. The *Oriental Limited,* which had a brief post-war resurrection, vanished forever in 1951, and from the *Builder's* four-year-old equipment, the *Western Star* was born. At various times, during the Depression and near the end of the road, the Burlington combined the *Empire Builder* and the *North Coast Limited* between Chicago and the Twin Cities.

Since hundreds of miles of track that once carried Chicago, Milwaukee, St. Paul & Pacific trains to Puget Sound have been ripped out, the superlative service the railroad once offered was what Lucius Beebe would have called a fragrant memory. Twenty sets of train cars, painted in the characteristic orange and maroon, were built to cover the new service inaugurated by the *Olympian* and *Columbian* in May 1911. At that time the railroad operated all its own sleeping cars, but when a new *Olympian* came along in 1927,

Great Northern's premier *Empire Builder* against a backdrop of the "Shining Mountains" of Glacier National Park, Mont.

Around the bend, the eastbound *North Coast Limited* will enter Stampede Tunnel through Washington's Cascade Mountains.

Pullmans came with it. The most striking version of the train, however, was its third and last: the streamlined *Olympian Hiawatha* that went into service in June 1947—not quite fully dressed, although within two years it was—and which in 1952 was given full-length Super Domes to go with the Milwaukee's unique Sky Top Lounges. The view that these cars afforded was reminiscent of the ride the railroad offered in its wooden-bench, open-air observation cars that ran under wire over the western mountains before World War II. The secondary *Columbian* was dropped during the Depression but brought back for a decade of postwar service to small communities on the main line. The *Olympian Hiawatha* made its final run in 1961.

The fourth northern route was an international operation controlled by the Canadian Pacific. In June 1900, the Soo began running a section of the *Imperial Limited* from St. Paul, turning it over to the CPR at Portal for the rest of its journey to Vancouver.[2] In July 1907, the *Soo-Pacific Train De-Luxe* was inaugurated, carrying a St. Paul-Portland sleeper (via Kingsgate, B.C., and the Spokane International and Union Pacific), other sleeping cars for both Vancouver and Seat-

tle, and open-air observation cars through the mountains. The *Soo-Pacific* evolved into the *Mountaineer*.

For a time the *Mountaineer* was the longest international run in North America. This summertime heavyweight carried sleepers from Chicago to Banff and Vancouver over Soo and Canadian Pacific tracks. Originally, as an all-Pullman train in the 1920's, it operated from Chicago to St. Paul on Soo tracks, but in the early '30s it was combined with the Chicago & North Western's *Viking* and ran by way of Milwaukee. After Chicago service ended in 1949, the train originated in St. Paul. The *Soo-Dominion* was the off-season version. The end for the *Mountaineer* came in 1960.

As the tourist industry grew, the national parks in the west became leading attractions and the railroads played a major role in their development, underwriting construction of hotels and laying special trackage to accommodate the trainloads of visitors. The Northern Pacific, the first railroad to become involved with a national park, started promoting Yellowstone in the 1880's, and by the following decade it was calling itself the "Yellowstone Park Line." For years it ran the summertime *Yellowstone Comet* with Pullmans from

Richard Steinheimer/DeGolyer Library

The westbound *Olympian Hiawatha* is ready to roll out of Butte, Mont., in March 1953. At left, a Butte, Anaconda & Pacific local.

Chicago and the Twin Cities. The Great Northern played a similar role in developing and serving Glacier Park in Montana, and the two railroads and the Milwaukee Road actively promoted tourism to Mount Rainier in Washington.

Despite the long-standing agreements among the railroads serving the Pacific Northwest, sleepers and entire trains historically have moved over a variety of routes from Chicago. Northern Pacific trains have rolled into Chicago over not just the Burlington and North Western, but the Soo and the Milwaukee Road as well.

Seattle and Portland, of course, were not entirely dependent on the northern transcontinentals for their service to Chicago. Pullmans that had traveled the Union Pacific's Overland and Oregon Short Line routes arrived in both cities, and Portland had a further Overland connection via the briefly used Southern Pacific "Modoc" route through Fernley, Nev.

Long before the SP&S was built, Northern Pacific sleepers entered Portland over the UP from Spokane and later over the road's own rails through Tacoma.

After the opening of the Panama Canal, which drew away much of the Far Eastern traffic the railroads had expected to carry, it became clear that the northern systems were overbuilt and some pruning would be necessary. Eventually the Great Northern and Northern Pacific combined with the Burlington, in 1970, and the Pacific extension of the Milwaukee Road was amputated. Nonetheless, the great passenger trains of the Big Sky country will continue to resound in the memory, crossing the Bitterroot Mountains and Marias Pass, fulfilling the sort of dreams that once built empires.

1. A predecessor road got land grants for building sixty-odd miles of track in Minnesota.
2. Arthur D. Dubin, *More Classic Trains* (Milwaukee: Kalmbach, 1974), p. 388.

42
Wisconsin

On a map of Wisconsin, many of the main passenger train lines rose out of the southeast like a volley of roman candles that arced in varying trajectories before exiting the northwestern part of the state. Some fed into lines that crossed plains and mountains to reach the Pacific Ocean, and others continued on but half that distance into grange country and Badlands. The rest headed north toward Lake Superior to traverse the great iron and copper mining regions. The location of Chicago and the settlement of Minneapolis and St. Paul were the key factors in this grand design: The Illinois and Minnesota metropolises were like magnetic poles that exerted considerable pull on the rails of the Badger state.

Always in Wisconsin passenger train history, the finest trains were the swift runners between Chicago and the Twin Cities. Over the years, eight different railroads competed for this sleeping car traffic on at least eleven different routes, five of which were still in operation after World War II.

The Milwaukee Road, which accounted for three of the choices, began through service on its line via Portage and La Crosse after acquiring trackage to the Twin Cities in 1872. Its overnight train was a success from the start, and after a quarter century of carrying thousands of new settlers to Minnesota, the train was fittingly honored with the name *Pioneer Limited* when its fine new livery was introduced in May 1898. At times, the *Pioneer* ran in two sections, one a solid train of sleeping cars.

F. Scott Fitzgerald, recalled this route of his youth through the voice of Nick Carraway in *The Great Gatsby*: Students, going home for the holidays, greeted one another in Chicago's Union Station before boarding "the murky yellow cars of the Chicago, Milwaukee & St. Paul railroad looking cheerful as Christmas itself on the tracks beside the gate.

"When we pulled out into the winter night and the real snow, our snow, began to stretch out beside us and twinkle against the windows, and the dim lights of small Wisconsin stations moved by, a sharp wild brace came suddenly into the air. We drew in deep breaths of it as we walked back from dinner through the cold vestibules, unutterably aware of our identity with this country for one strange hour, before we melted indistinguishably into it again.

"That's my Middle West—not the wheat or the prairies or the lost Swede towns, but the thrilling returning trains of my youth. . . ."[1]

The route of the *Pioneer Limited* became a speedway when the Milwaukee Road introduced the first steam-powered streamliner, the *Hiawatha*, in May 1935. This new daytime flyer in the railroad's customary orange and maroon frequently exceeded 100 miles an hour as it traveled from Chicago to Milwaukee before rushing across the farmlands to the Wisconsin River valley and the scenic formations of the Dells, tunnel-

ing through the divide to the Mississippi River at La Crosse, then gliding up the west bank to St. Paul and Minneapolis.

To meet the rapidly heating competition, the Milwaukee introduced a new version of the *Hiawatha* in October 1936 and a third one less than two years later. In early 1939, twice-daily service began. Post war, with a fourth generation of equipment, the Milwaukee Road's streamliner family came into full fruition on this route, with *Morning Hiawatha, Afternoon Hiawatha,* and *Olympian Hiawatha* racing in both directions across Wisconsin with their glass-palace Sky Top Lounge cars and, by 1952, Superdomes.

In addition to this high-speed main line, which the *Columbian* also worked overnight, the Milwaukee Road operated sleepers on two other routes to the Twin Cities: an even earlier one by way of Madison,

Prairie du Chien, and Austin, Minn.,[2] and the other by way of Savanna, Ill., Dubuque, Iowa, and the west bank of the Mississippi River. (The Milwaukee Road had an on-and-off relationship with Pullman, which finally took over operation of the road's through sleeper service in 1927, but even in later years the railroad continued to run its own sleeping cars on local runs.) The *Pioneer Limited* offered the road's last Chicago-Twin Cities sleeper accommodations before the historic overnighter was discontinued in 1970—it was the Milwaukee Road's oldest train—after completing nearly a century of operation.

On the Wisconsin side of the Mississippi River lay the superlative main line of the Chicago, Burlington & Quincy, another Chicago/Twin Cities raceway. From the palisades of Illinois north, the Burlington's 300-mile run along the Father of Water was grandly scenic. There the two fastest trains in the world for much of the postwar era, the gentle winds of the west known as the *Twin Zephyrs*, made their daily circuits with several stainless steel, glass-topped Vista-Dome coaches, a Burlington innovation, in tow.

This route along the Mississippi was a relatively late addition (1886) to the Chicago-Twin Cities competition as well as to the railroad system itself. But the Burlington's fast, water-level run quickly made the line an important contender for traffic to and from Minnesota and the northwest.

In 1897, the Burlington introduced its luxurious overnight *Burlington Limited*, which carried one of Pullman's early all-compartment cars. Among the heirs to this service were the *Minnesota Limited, Commercial Traveler* and finally the *Blackhawk* (later *Black Hawk*), one of the CB&Q's all-new trains of 1930.

In 1901, the Northern Pacific and Great Northern acquired joint ownership of the Chicago, Burlington & Quincy Railroad to assure themselves of access to Chicago from St. Paul; thereafter Burlington locomotives began hauling the great trains of both northern roads—*Empire Builder, Oriental Limited, North Coast Limited, Western Star*—to and from America's rail crossroads.

In 1935, the Burlington introduced the stainless steel daytime *Twin Zephyrs*, with 44 sets of real twins riding the inaugural trains running in parallel. The *Zephyrs* were reequipped in 1936 and again in 1947, by then with the new Vista-Domes.

The flagship of the Chicago & North Western Railway for many years was its overnight train to the Twin Cities, the *North Western Limited*. This service, like the *Pioneer Limited's*, evolved before the train acquired its formal name, which probably happened when it was reequipped with a elegant consist of Wagner cars in the late 1890's.[3] Before completion of its new main line early in this century, the railroad operated the train by way of Madison rather than Milwaukee. Visually, the train made an exotic departure from the prevailing Pullman drab of the time. For about fifteen years, until the late 1920's, the *Limited* was painted yellow and dark green with black stripes, presaging the color scheme of the famous *400* streamliners more than a decade later.

On January 2, 1935, the original *400* inaugurated high-speed daylight travel to St. Paul and Minneapolis on the North Western's main line through Milwaukee. For a brief time, the steam-powered, heavyweight *400* was the fastest long-distance train in the world. (Speed was a hallmark of passenger railroading in Wisconsin. In 1953, the Burlington accounted for the six fastest diesel runs in the United States over its trackage north and south of Prairie du Chien, and the CB&Q, Milwaukee Road, and North Western all together were credited with half of the country's fifty fastest runs, all twenty-five of which were in Wisconsin.)

In 1939, a new streamlined *400* made its debut, and its great popularity prompted the railroad eventually to inaugurate an entire fleet of *400's*—*Twin Cities, Flambeau, Peninsula, Shoreland*—from Chicago

William D. Middleton

On its overnight run from Chicago in September 1959, the Milwaukee Road's historic *Pioneer Limited*, will soon be in St. Paul.

to the upper Midwest, all but one of which traveled through Wisconsin. But the good years were few, and by 1959 the road was in for a drastic pruning. As part of the remedy, the venerable *North Western Limited* was discontinued, leaving the Twin Cities Pullman runs to just the Milwaukee and Burlington.

When the Soo Line dropped its Chicago to Minneapolis sleeping car in 1953, it ended a service that had been both homely and cosmopolitan. By leasing the Wisconsin Central in 1909, the Minneapolis, St. Paul & Sault Ste. Marie Railroad, or Soo, as it was nicknamed and finally officially called, acquired an entrance to Chicago. Over the years the road provided sleeper service not only to the Twin Cities, but also from Chicago to Vancouver (with Canadian Pacific) via the year-round *Soo-Dominion* and the summertime *Mountaineer* (the North Western moved the cars to the Twin Cities for about fifteen years, until Chicago service ended in 1949).

It seems by modern standards that four railroads providing seven different sleeper routes between Chicago and the Twin Cities would be ample, yet four other lines were at one time involved in this remarkable competition. None of these, however, traveled through Wisconsin, and no doubt their longer journeys through Iowa contributed to their earlier demise. In the 1880's, the Minneapolis & St. Louis Railway and the Rock Island were carrying sleepers in their crack collaboration known as the *Cannon Ball*. After the Rock laid tracks into St. Paul in 1902 and established its own through service, the M&StL teamed with the Illinois Central, sending sleepers to the Windy City on the new *North Star Limited*.[4] The

Chicago Great Western also offered overnight service, which for a time in the 1920's attained a rather luxurious level with the introduction of the *Legionnaire*.

As an adjunct to their Twin Cities service, both the North Western and the Soo operated sleepers to Duluth-Superior from Chicago and Milwaukee, and the North Western also offered accommodations through Madison. On this run the *Duluth-Superior Limited* also carried Chicago Pullmans for Vancouver and, in summers, for Jasper Park; these moved west on the Canadian National's *Continental Limited*.

The Soo stopped running cars into Milwaukee (they had used Milwaukee Road tracks) in 1938; thereafter passengers rode the interurban to and from Waukesha. The Soo's Ashland sleeper also moved in the Duluth train (in recent years known as the *Laker*). For a time the Soo carried a car from Neenah to Chicago that it returned to Stevens Point. The Wisconsin Central also had run Milwaukee-Stevens Point service in the 1890's. (Like the Milwaukee Road, the Soo operated many of its own sleeping cars.)

The Milwaukee Road operated sleeping car service on its north woods line to the Minoqua area from both Milwaukee and Chicago. Wausau, the town with the station pictured in the insurance company logo, once had sleepers running to Chicago and the Twin Cities on this route as well as to Milwaukee and Chicago on the North Western. Another unusual Pullman once passed through Wausau in the middle of the night: the North Western's Fond du Lac-Minneapolis car that ran by way of Green Bay. During the 1920s, the Milwaukee Road offered summertime sleeping car service between Minoqua and the Twin Cities.

In the 1880's, the North Western and Milwaukee Road began advertising in magazines, touting their wondrous fishing holes, and until the automobile took away the last vestige of such service, the railroads commonly scheduled seasonal sleepers to the resorts of Wisconsin and Michigan's Upper Peninsula. The Milwaukee's *North Woods Fisherman*, once a summertime daily, bolstered the regular Minoqua service with sleepers to Star Lake and Tomahawk from both Chicago and Milwaukee. The road ran sleeping cars from Milwaukee to Champion, Channing, Iron River, Menominee (1890), and Ontonagon, Mich. During several post-war winters, it ran a weekend skiers sleeper from Chicago, variously to Wausau or La Crosse, or to Houghton or Iron Mountain, Mich.[5]

The North Western offered Pullmans from Milwaukee to Florence (1890) and the Michigan communities of Iron Mountain, Ironwood, and Watersmeet, as well as service to Milwaukee from Escanaba. Accommodations to Chicago included cars from Phelps, Drummond, Eagle River, Mercer, and Sturgeon Bay (this a late-1920's operation involving the tiny Ahnapee and Western Railway and Kewaunee, Green Bay and Western Railroad). Rhinelander had a sleeper

Chicago, Burlington & Quincy's 4-6-4 *Aeolus* brings the *Black Hawk* into Chicago with nine cars on Aug. 10, 1939.

Otto C. Perry/Denver Public Library

It's a cold New Year's Day 1959, and the Soo *Laker* is about to begin its run through central Wisconsin from Duluth to Chicago.

from Chicago. On its Ashland line, the Soo carried fishermen and campers between Chicago and Park Falls.

The modern history of Green Bay began in 1634, when the French explorer Jean Nicolet stepped ashore to the astonishment of a group of native Americans—and probably himself as well. From that time on, the city's importance as a transportation center—water and eventually rail—grew. The original main line of the North Western went to Green Bay by way of Janesville. Shortly after the Civil War, the railroad scheduled sleepers from Chicago to both Fond du Lac and Green Bay over that route. In modern times the Green Bay Pullmans traveled via Milwaukee, one through Fond du Lac and the other through Manitowoc. The Milwaukee Road also offered a Chicago-Green Bay sleeper.

The tracks of both railroads continued north into Michigan—where they accommodated the early mining and lumbering enterprises—and a North Western line crossed the state to Ashland. The road served Ashland with Pullmans from Chicago and Milwaukee.

Many of the North Western's *400* streamliners served Green Bay by one route or the other; these trains, some of them bilevel, were the road's last hurrah in interstate passenger service. Though briefly augmented by the sprightly *Chippewa Hiawatha,* the reigning flyer on the Milwaukee Road's line north was the *Copper Country Limited,* a train that began operating in the 19th Century and provided Michigan's Up-per Peninsula with its last sleeping car service.

As Wisconsin's largest city, Milwaukee originated the most sleeping car traffic in the state. Over the years Milwaukee had dedicated service to more than two dozen communities, including five lines to the Twin Cities. When the *North Western Limited* operated by way of Madison, sleepers from Milwaukee joined the train there; with the new main line (which also brought Pullman service from Rochester), the cars became hometown setouts. The North Western also had a third, roundabout Twin Cities Pullman line by way of Fond du Lac, Green Bay, Eland, and Merrillan. In addition to the North Western lines, the Milwaukee Road and Soo also scheduled Minneapolis sleepers. On its Twin Cities run, the Milwaukee Road offered a setout car for La Crosse, as well.

Trains operating on the Milwaukee Road's charter route, which ran from Milwaukee to Madison and along the Wisconsin River to the Mississippi, carried sleepers for Prairie du Chien (1890), Mason City, Iowa, and Mitchell, S.D. At Prairie du Chien, which also had accommodations to Chicago (1890), the trains crossed into Iowa on pontoon bridges originally built in 1874. The bridges, which replaced ferries and wintertime sleigh crossings, rested on barge hulls, and the single track could be raised or lowered as the water level dictated. The bridges could pivot open to allow river traffic or ice floes to pass through.

The *Southwest Limited* was one of the Milwaukee Road's crack trains from Chicago that also had a Wisconsin

William D. Middleton

Wisconsin's capitol gleams in the background as the C&NW *Duluth-Superior Limited* pauses at Madison in April 1957.

section. This departed Milwaukee with sleepers for Omaha and Kansas City traveling via Beloit to Savanna, Ill., where it merged with the Illinois section. Service once included cars for Rock Island/Davenport.

Until 1965, when the new Union Station opened in Milwaukee, Wis., the city's two railroads used their own 1880's vintage depots: the Milwaukee Road's turreted modern-gothic station just north of the new facility, and the North Western's classic lakefront structure near the present War Memorial and Art Museum.

Although Chicago and Milwaukee were but eighty-five miles apart, sleeper accommodations between the two cities were once offered, according to 1876 schedules of the North Western. In modern times, however, with nearly 100 daily trains operating between the two cities, the fastest carded at a mere seventy-five minutes, there was little need for overnight Pullmans.

Even though it grew to become Wisconsin's second city, Madison, saw limited sleeping car service of its own. Both the Milwaukee Road and North Western

offered accommodations to Chicago, and the latter provided Pullman service to Minneapolis. Long before the turn of the century, the Milwaukee Road operated a car to Sanborn, Iowa, over what became the route of the *Sioux,* a Chicago-Rapid City train.

In modern times, however, the capital city was served by through sleepers to Duluth and South Dakota. Among the North Western's trains were the *Minnesota & Black Hills Express* and the *Dakota 400,* the only member of that famous fleet to include a Pullman sleeper (carried just on the western end of its run) in its consist. In Madison, trains of the Milwaukee Road and North Western had the unique distinction of crossing each other's paths in Lake Monona, which intersection afforded passengers a striking view of the state Capitol and the city's skyline.

La Crosse, where the *Zephyrs* and *Hiawathas* crossed paths before the latter bridged the Mississippi River, was the major rail center in scenic western Wisconsin. Day and night, the crack trains of the Burlington and Milwaukee Road stopped for brief servicing before continuing on their way. Most of the North Western's

William D. Middleton

Milwaukee Road's *Sioux* stands in Madison before moving out under Fairbanks Morse power for its trip west through the Wisconsin River Valley.

trains to southern Minnesota and the Black Hills also served La Crosse. But in spite of all the through traffic, La Crosse had little dedicated Pullman service. Before the turn of the century the North Western scheduled a sleeper to Chicago (1876), and the Milwaukee Road offered one to Jackson, Minn. (1890). Surviving into modern times was the Milwaukee's set-out car from its namesake city.

Several other Wisconsin communities had Pullman service: The North Western scheduled sleepers from Eau Claire to Minneapolis and from Oshkosh to Chicago, and the Northern Pacific, on its overnight Twin Cities-Duluth train, once assigned Superior its own sleeping car.

Northern Wisconsin had through service on two unusual Pullman lines, one of which provided accommodations, via the Sault Ste. Marie, all the way to the New England coast—as improbable a trek as could be imagined by empire builders of today. The Soo Line and Canadian Pacific, which controlled the Soo, scheduled Pullmans from the Twin Cities to both Montreal and Boston (Boston & Maine). And the Duluth, South Shore & Atlantic, which carried sleepers mostly between Duluth and Michigan's U.P., also offered Pullman service for a time between its namesake city and Montreal (CPR).

In the end, although it was intriguing to consider a train trip to the Atlantic Coast from northern Wisconsin, it was far more memorable to recall the flyers that blazed their way in the opposite direction, to the Pacific Northwest.

1. F. Scott Fitzgerald, *The Great Gatsby* (New York: Charles Scribner's Sons, 1953), pp. 117-8.
2. This line was in place west of Milwaukee before the newer main line through Portage and La Crosse. In 1869, the railroad was offering sleepers, which operated as far as Prairie du Chien, on its overnight expresses between Milwaukee and the Twin Cities.
3. Arthur D. Dubin, *Some Classic Trains* (Milwaukee: Kalmbach, 1964), p. 168.
4. Frank P. Donovan Jr., *Mileposts On the Prairie* (New York: Simmons-Boardman Publishing Corp., 1950), p. 141.
5. Patrick C. Dorin, *The Milwaukee Road East* (Seattle: Superior Publishing Co., 1978), p. 22.

43

Minnesota

Minneapolis and St. Paul, which grew up alongside a meandering Mississippi River in the home state of its headwaters, were like other early river towns west of the Appalachians, dependent upon the land's natural waterways for transportation before the railroads came. Although Spanish moss and magnolias were not in evidence at their Minnesota landfalls, Mississippi River steamboats were every bit as important to the trade and development of the Upper Midwest as they were in the Deep South.

Before mid-Nineteenth Century, when Galena, not Chicago, was the metropolis of Illinois, steamboats departed from the lead-mining center for upriver destinations with great regularity. Upon completion of its charter railway line into Galena, the Illinois Central set up coordinated rail/steamboat service for settlers and other travelers heading up the Mississippi.[1]

Boats that came from far downriver even helped establish Minnesota's reputation as a cool, summer retreat long before the territory achieved statehood. Wealthy planters brought their families and retinues of servants north to escape the oppressive heat of the Deep South. But just as it happened everywhere else, the railroads were quick to take over the steamboats' business.

The Milwaukee Road was the first railway to offer the Twin Cities connections to the south, over a line through Austin to North McGregor, Iowa, across from Prairie du Chien, Wis., where the Milwaukee's lake-to-river charter line terminated. Passengers were ferried across the Mississippi until a pontoon bridge was built in 1874. Timetables for 1869 show two "express trains" a day operating on roughly twenty-three-hour schedules, and sleepers were offered between Milwaukee and Prairie du Chien. The Milwaukee Road's .modern main line through Portage and La Crosse opened two years later, and the forebear of the *Pioneer Limited* began through service from Chicago in 1872.

After the Milwaukee Road became the fourth northern transcontinental line, in 1909, it instituted service from Chicago to Seattle/Tacoma on the *Olympian* and *Columbian*. Cars for western destinations originated in Chicago; the Milwaukee concentrated its sleeping car service from the Twin Cities in the eastern portion of the system.

The Milwaukee carried sleepers for Wausau, Minocqua (summers), and Milwaukee, Wis., and Chicago on its main line, and to North McGregor (1890), Dubuque (1886), and Chicago down the west bank of the Mississippi. The line through Austin carried yet another Chicago sleeping car[2] (via Madison), as well as pre-1900 sleepers to Ottumwa, Iowa (Minneapolis & St. Louis), and St. Louis (M&StL/Wabash). To the west, the Milwaukee scheduled a car to Aberdeen, S.D.

Like the Milwaukee Road, the Chicago, Burlington

The dour assemblage portrayed upper right was photographed May 28, 1907 on the observation platform of the Great Northern's principal train. Four decades later, open-air observations were fading fast, but the Milwaukee's *Hiawatha* fleet carried superlative successors (upper left). Duluth, Winnipeg & Pacific No. 19 (bottom), which in the early 1950's was still offering a Winnipeg sleeper three nights a week, is about ready to depart from Duluth.

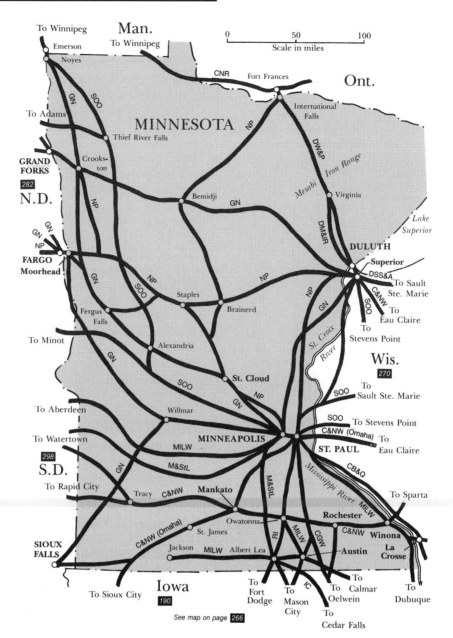

See map on page 266

& Quincy brought some superlative transcontinental trains to the Twin Cities: its parent Northern Pacific's *North Coast Limited* and *Mainstreeter* and the Great Northern's *Empire Builder, Oriental Limited,* and *Western Star.* The CB&Q's own *Black Hawk* and *Zephyrs* plied the fast tracks down the Wisconsin side of the Mississippi to Chicago. The Burlington carried Pullmans from Minneapolis and St. Paul to Chicago, Galesburg, Peoria, Rock Island, and St. Louis (1890).

Both the Northern Pacific and Great Northern are best remembered for their fine trains to the West Coast, yet the two railroads served numerous intermediate destinations with sleeping car accommodations. Although the NP regularly scheduled Pullmans from the Twin Cities to Tacoma, Portland, and Seattle, the GN preferred to originate most of its West Coast sleepers in Chicago.

The Great Northern used two main routes east of Minot, N.D. Within Minnesota, the original line to

Fargo went through St. Cloud, but after the early 1930's, most of the transcontinental traffic used the line via Willmar. In North Dakota, some trains traveled via Grand Forks, others by way of New Rockford over a cutoff opened in 1912 that shortened the transcontinental line by sixty miles. The *Red River Limited, Dakotan,* and *Winnipeg Limited* were among the GN trains serving this region.

The sleeping car lineup from St. Paul and Minneapolis on the GN included Crookston; Fargo, Grand Forks, Devils Lake (summers), and Minot, N.D.; Sioux City (1890); Yankton and Aberdeen, S.D.; Butte, Helena, and Great Falls, Mont. (1890), and, to a limited degree, Tacoma and Seattle. The Great Northern also carried sleepers to Winnipeg, where, in summers, it turned over cars for Vancouver to Canadian National (it had a similar arrangement with Canadian Pacific in the early 1890's, before the Soo came into the picture). The Great Northern operated

The C&NW/Omaha Road's *Victory,* a Chicago to Twin Cities train via Madison, steams into St. Paul with a consist of a dozen cars.

its own sleeping cars until 1922.

The Northern Pacific scheduled Pullmans from the Twin Cities to Fergus Falls (1890), Bemidji, and International Falls; Fargo, Jamestown, and Mandan, N.D.; Gardiner, Mont. (the NP's Yellowstone Park entrance, served in summers by the *Comet*), and Winnipeg (1890).

Both NP and GN, as well as the Soo, carried sleeping cars to Duluth, and each railroad originated cars in both Minneapolis and St. Paul. In the late 1920's, the three railroads pooled their service between the Twin Cities and Duluth, and the Northern Pacific operated the overnight train, which for a time included a Pullman for the Twin Port city of Superior.

The Soo Line, once more formally known as the Minneapolis, St. Paul & Sault Ste. Marie, was built by local interests seeking to beat the high cost of shipping Minnesota grains by way of Chicago. The line connected with the Canadian Pacific at Sault Ste. Marie and offered an alternate route east. After consolidating several lines in the Upper Midwest, the Soo made another connection with the CPR, at Portal, N.D., in 1893, and came under control of the Canadian road the same year. As a result, the Soo's sleeping car service (much of which was railroad-operated) took on a distinctly international character.

The *Mountaineer, Soo-Dominion,* and their luxurious predecessors that carried sleepers to Banff, Vancouver, Seattle, and Portland fit into an established and highly competitive pattern of traffic. But the Soo's line to the east was another matter and led to one of the continent's more improbable sleeping car treks. After the completion of the international bridge at Sault Ste. Marie in 1888, the Soo and Canadian Pacific inaugurated through sleepers from Minneapolis/St. Paul to Montreal and Boston (Boston & Maine). This line gave the Twin Cities service to both coasts until the Boston car was dropped about the time of World War I. Service eventually terminated at Sault

Ste. Marie.

Apart from its accommodations to the West Coast, the Soo carried a number of sleepers to intermediate destinations: Oakes, Enderlin, Minot, and Adams, N.D., and (with CPR) Moose Jaw and Calgary, in Canada. Its *Winnipeger* and predecessor carried sleeping cars to Thief River Falls and Winnipeg (CPR).

The Soo got a leg into Chicago by leasing the Wisconsin Central in 1909. Its overnight express from the Twin Cities, which was combined with the Duluth-Chicago train in the early 1930's, carried sleepers to Milwaukee and Chicago.

The Chicago & North Western was a strong contender for Chicago traffic with its venerable *North Western Limited* and daytime *400.* The railroad also furnished one of the Twin Cities' connections to the West Coast over its line to Omaha, where through cars were put aboard Overland Route trains. This was the main line of the Chicago, St. Paul, Minneapolis & Omaha, the "Omaha Road," which the North Western gained control of in 1882. The passenger operations of both railroads were integrated.

Over its many—no doubt too many—lines through Wisconsin, the North Western carried Pullmans to the Twin Cities from Chicago (via both Milwaukee and Madison), Milwaukee (via either Clyman Junction, Madison, or Green Bay), Madison, Fond du Lac, and Eau Claire. Over the Omaha Road, sleepers traveled to Sioux City, Omaha, Kansas City (either Burlington or Missouri Pacific), and Los Angeles (Union Pacific), and also to the South Dakota communities of Watertown, Redfield, Huron, Sioux Falls, and Mitchell.

Two railroads that eventually became part of the North Western system, the Minneapolis & St. Louis and Chicago Great Western, also provided Minneapolis and St. Paul with Pullman service.

The M&StL, which made an early withdrawal from first-class passenger service, had an ambitious sleeping car operation during the heyday of rail travel, with a

Otto C. Perry/Denver Public Library

The Great Western's Chicago-Twin Cities *Minnesotan*, shown here in St. Paul in 1931, replaced the more opulent *Legionnaire*.

choice of destinations ranging from Columbus, Ohio, to Hot Springs, Ark., to San Diego.[3] The system consisted of lines running south through Iowa, then east to Peoria—hence "the Peoria Gateway" slogan—and a western route to South Dakota.

Before the turn of the century, the M&StL was the Rock Island's access to the Twin Cities, and the two railroads conducted interline sleeper service to both Chicago and St. Louis, exchanging cars at Albert Lea. The M&StL then worked with the IC on yet another Chicago run, as well as Pullman lines to Omaha and Kansas City (MP) and tourist sleepers to Los Angeles (RI/Denver & Rio Grande Western/Southern Pacific) and San Diego (CB&Q/Santa Fe). Still later, the railroad cooperated with the Wabash on Pullmans to St. Louis and Hot Springs (MP).

The M&StL scheduled sleepers on its own to Aberdeen and Watertown, S.D.; Fort Dodge, Des Moines, and Oskaloosa, Iowa; and Peoria, at which central Illinois gateway it turned over the Columbus Pullman to the New York Central.

The Chicago Great Western was another home-grown effort that started contruction in the mid-1880's, when many thought its chosen territory to be already overbuilt. The CGW, nonetheless, grew into a system extending to Chicago, Kansas City, and Omaha and bravely engaged in passenger competition with both luxurious and imaginative hardware that reached its high point in the 1920's. The road's marketing efforts even included promotion of joint air/rail service to St. Louis and Cleveland.

The *Great Western Limited* became the elegant *Legionnaire*, which served the Chicago market with brand new train sets. The railroad put in service to Rochester the *Blue Bird*, an all-blue "deluxe motor train" rebuilt from old McKeen cars, and the steam-powered *Red Bird*, wearing a startling red livery. The

Mill Cities and *Tri-State Limiteds* served Des Moines and Kansas City. During this expansive era, the Corn Belt Route worked with the Missouri-Kansas-Texas to send Pullmans to Dallas/Fort Worth, San Antonio, and Houston and with the Santa Fe on a line to Los Angeles. On its own system, the Great Western carried sleepers to Chicago, Des Moines, Kansas City, Omaha, and St. Joseph (1890).

After the Rock Island built its line into St. Paul and Minneapolis in 1902, it expanded the cities' Pullman selection beyond the few offerings it had shared with the M&StL. The Rock continued Pullman service to Chicago and St. Louis (CB&Q) and added cars to Rock Island and Peoria. It scheduled sleepers to and through Des Moines to Omaha, Kansas City, Dallas-/Fort Worth, and Houston and tapped into its Golden State Route at Kansas City with cars for Los Angeles.

The heavyweight *Mid-Continent Special* carried an assortment of Pullmans for various destinations. During the streamline era, the *Twin Star Rocket* operated to Houston, the longest north-south run made by an American passenger train on a single railroad. The *Zephyr Rocket* service to St. Louis was an interline effort of the Rock Island and Burlington.

The Twin Cities enjoyed well planned, centralized terminal facilities. The Great Northern station in Minneapolis opened just before World War I, St. Paul Union Depot just after, and one or the other accommodated virtually all the passenger traffic in and out of the Twin Cities. The only exception was the Milwaukee Road Station in Minneapolis, which handled the railroad's own trains and those of the Rock Island and Soo in that city. In its heyday in the 1920's, St. Paul Union Depot accommodated more than 280 trains a day,[4] Minneapolis' GN station handled about half that number.

For passenger convenience, setout sleepers were

scheduled to Duluth, Chicago, and Des Moines from whichever Twin City the night train didn't originate it. For Duluth, GN setouts were in Minneapolis, Soo and NP cars in St. Paul. Both the *North Western Limited* and the Milwaukee's *Pioneer Limited* took on Chicago sleepers in St. Paul, where the Rock Island did the same with a Des Moines Pullman.

Probably the most substantial monument to rail travel to and through the Twin Cities was the double-tracked, Stone Arch Bridge built across the Mississippi at Minneapolis by the Empire Builder James J. Hill. Opened in 1883, the granite and limestone span composed of twenty-three arches facilitated Hill's terminal plans and for nearly a century carried all the passenger trains between the Twin Cities, except for those using the Milwaukee Road Station.

At the height of iron ore mining in the Mesabi Range, Duluth ranked second only to the port of New York in total tonnage, achieving this with just the eight-month shipping season dictated by the ice on frigid Lake Superior. Built long and narrow on lava-rock bluffs rising six hundred feet or more from the lake, Duluth was the chartered eastern terminus of the Northern Pacific Railway.

Over the years, seven railroads originated sleeping cars in Duluth, and apart from service to Milwaukee, Chicago, and the Twin Cities, all the lines served destinations in the north.

For the pool service to the Twin Cities, the Great Northern scheduled the *Gopher Limited* and *Badger Express;* the Northern Pacific had the *Twin City-Twin Ports Express* and *Lake Superior Limited;* the Soo operated nameless No. 62 and 63. All three railroads at one time offered overnight accommodations to Minneapolis and St. Paul. On the lines connecting with their northern transcontinental routes, the GN scheduled sleepers to Grand Forks, the Northern Pacific to Fargo and Jamestown.

Both the Soo and the Omaha Road provided sleepers to Milwaukee and Chicago, and for a time before and after World War II, they pooled their service. The C&NW carried Pullmans via both Milwaukee and Madison, the latter the route of the *Duluth-Superior Limited.* The Soo overnighter acquired the name of *Laker.*

A railroad built to serve logging and mining enterprises scheduled the largest selection of sleeping car service from Duluth, its western terminus. The Duluth, South Shore & Atlantic worked the iron and copper country of Michigan's Upper Peninsula with its main line to Sault Ste. Marie and St. Ignace.

The railroad's most unusual sleeping car run took place from its namesake city aboard the grandiously named *Boston Express*. The Canadian Pacific, which had acquired a controlling interest in both the Soo

and South Shore before the turn of the century, planned to operate through sleepers between Duluth and Boston.[5] While that may have been overly ambitious, the South Shore and CPR did schedule a Duluth-Montreal sleeping car.

In its own territory, the South Shore operated sleepers between Duluth and the Michigan communities of Calumet, Mackinaw City, Marquette, and Sault Ste. Marie. Like those of the Soo, with which it merged in 1961, South Shore cars were adorned in the wine-red of the Canadian Pacific.

The Duluth, Winnipeg & Pacific, which became part of the Canadian National system, carried sleepers to Winnipeg, including Chicago-Vancouver cars the North Western had originated. Before the "Peg" laid tracks into Duluth in 1912, however, Winnipeg sleepers traveled to Virginia, Minn., over the Duluth, Missabe & Iron Range.

Duluth's Union Depot, a castlelike, French Norman structure, served the trains of the northern transcontinentals, DM&IR, and South Shore. The Omaha Road had its own station, also used by the Peg, and the Soo operated its own facility, to which the DSS&A later moved. The three stations were virtually adjacent to one another.

Rochester, with its famed Mayo Clinic, saw a steady flow of prospective patients arriving in Pullman accommodations on both the Great Western and North Western railroads, each of which provided service from Chicago. The C&NW even carried special cars with side hatches that facilitated the loading of passengers being carried on stretchers. The North Western also scheduled sleepers to Milwaukee, Huron, and Omaha, the CGW to Kansas City.

The North Western gave several Minnesota communities their sole Pullman service. It scheduled sleepers from Winona to Chicago (1876) and Omaha, and from both Tracy (1890) and Mankato to Chicago. The Milwaukee Road provided berths from Jackson to La Crosse (1890) and Chicago, and from Austin to Chicago.

According to summer of 1909 schedules, the Great Northern carried sleeping cars from Willmar to Yankton.

1. Carlton J. Corliss, *Main Line of Mid-America* (New York: Creative Age Press, 1950), pp. 74-8.
2. All nine Pullman-carrying railroads serving Minneapolis and St. Paul originated sleepers for Chicago, though just a mere seven of them from the Twin Cities themselves. Both the Great Northern and Northern Pacific were spared this close-quarter competition because their tracks ran the wrong way.
3. Frank P. Donovan Jr., *Mileposts On the Prairie* (New York: Simmons-Boardman, 1950), pp. 114, 143.
4. Steven Glischinski, "A Tale of Twin Cities," *Trains,* Vol. 46, No. 12 (October 1986), p. 36.
5. Aurele A. Durocher, "The Duluth, South Shore, and Atlantic Railway Company," *Bulletin of the Railway and Locomotive Historical Society, Inc.,* No. 111 (1964), p. 57.

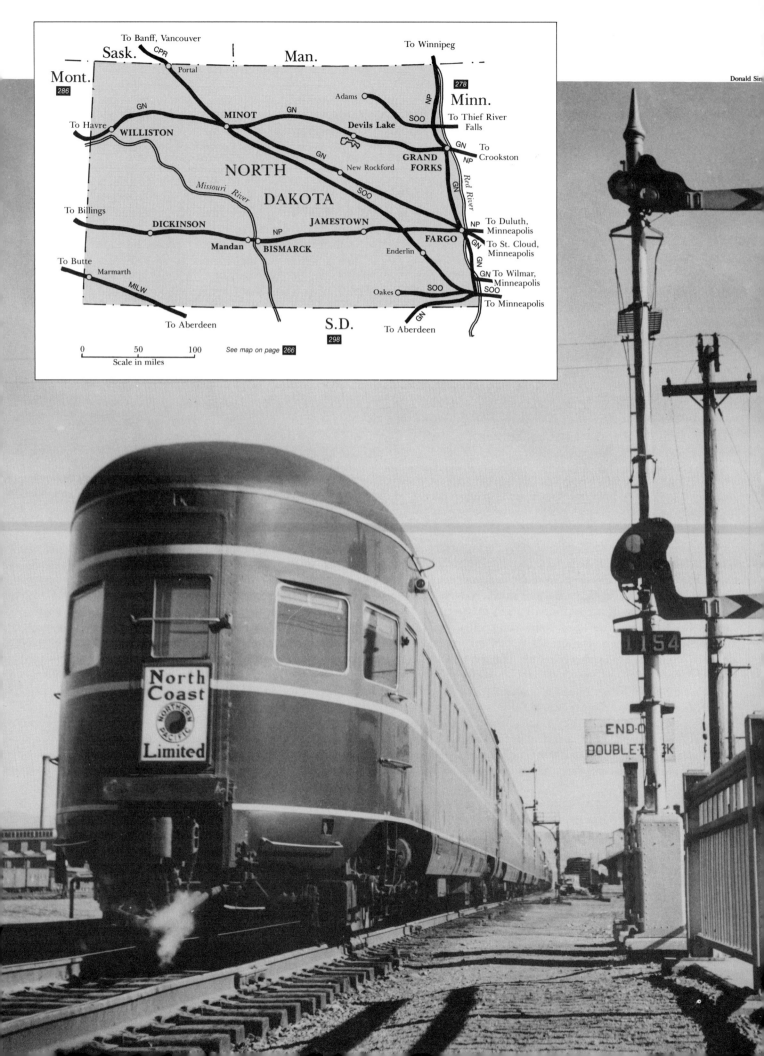

To Banff, Vancouver
CPR
Sask.   Man.
Mont.   To Winnipeg
286   278
Portal   NP
GN   Adams   SOO   Minn.
MINOT   GN   Devils Lake   To Thief River
To Havre   WILLISTON   Falls
GN   GRAND   GN   To Crookston
NP   New Rockford   FORKS   NP
NORTH
Missouri River   GN   SOO   Red River
DAKOTA
To Billings   JAMESTOWN   To Duluth,
NP   NP   Minneapolis
DICKINSON   NP   FARGO   To St. Cloud,
Mandan   BISMARCK   Enderlin   GN   Minneapolis
To Butte   GN   To Wilmar,
Marmarth   GN   Minneapolis
MILW   Oakes   SOO   SOO   To Minneapolis
To Aberdeen   S.D.   To Aberdeen
298   GN
0   50   100   See map on page 266
Scale in miles

Donald Sin

1154

END OF
DOUBLE TRACK

North
Coast
NORTHERN PACIFIC
Limited

44
North Dakota

The four northern transcontinental rail routes were the main lines of North Dakota. The trails of both the Northern Pacific and Great Northern ran nearly parallel to each other across the state from the more populous Red River Valley communities west to Montana: the NP from Fargo, through Jamestown, Bismarck, and Dickinson, and the GN from Grand Forks through Devils Lake, Minot and Williston. Most communities in North Dakota were within sixty miles of

Northern Pacific's *North Coast Limited* was the oldest name train to the Pacific Northwest, dating to 1900.

one line or the other.

The Great Northern's direct line ran northwest from Fargo, through New Rockford, joining the line from Grand Forks just east of Minot. Running on the same bias was the Soo's line to Portal, where it connected with the rails of the Canadian Pacific. The Milwaukee Road's main line only briefly creased the southwest corner of the state.

Fargo, the largest city in North Dakota, was served by trains of both the Northern lines: the NP's venerable *North Coast Limited, Mainstreeter, Comet, Atlantic-/Pacific Expresses,* and *Alaskan,* and the GN's *Empire Builder, Oriental Limited, Western Star, Dakotan, Winnipeg Limited,* and *Red River.*

At Fargo, the alternate main lines of the Great Northern converged. Passenger trains from the east came to town via either St. Cloud or Willmar, Minn., in more recent times mostly the latter. Westward, the *Empire Builder* traveled the direct route through New Rockford, and some of the secondary trains looped north via Grand Forks.

Although the principal trains of all the railroads carried numerous through sleepers between Chicago and the Pacific Northwest, virtually all of North Dakota's dedicated service was directed to the three largest cities of Minnesota. In one of the few exceptions, the Great Northern scheduled Pullmans from Fargo to Aberdeen, S.D., but the city's other options fit the pattern: sleepers on the Northern Pacific to Duluth, and both GN and NP to the Twin Cities (in later years the two roads alternated with each other for four-month periods).

The Great Northern and the Soo's main lines crossed at Minot, and both railroads brought dedicated sleepers from the Twin Cities to town. Summertime traffic on this joint Soo/CPR line to Banff and Vancouver was heavy through most of the 1950's, with tour business swelling the consist of the seasonal *Mountaineer.* The *Soo-Dominion* worked the line year-round. In addition to the Minot cars, the Soo scheduled sleepers from the Twin Cities to Enderlin, Adams, and Oakes.

The Northern Pacific was the first of the northern transcontinentals, and by 1881 its tracks had reached Montana. Both Mandan and Jamestown had their own sleepers to Minneapolis/St. Paul, and the NP operated a car from Jamestown to Duluth early in the century.

The Great Northern carried Pullmans from Grand Forks to both Duluth and the Twin Cities and from Devils Lake to Minneapolis/St. Paul. The GN, according to July 1909 schedules, offered sleepers from Williston to both Butte and Havre, Mont.

Marmath was the principal stop in North Dakota for the Milwaukee Road's *Olympian Hiawatha* and *Columbian,* both of which offered through sleeping accommodations to Chicago and Seattle/Tacoma.

45
Montana

The Great Northern's *Western Star*, the principal train serving Glacier National Park in the early 1950's, winds along the preserve's scenic boundaries.

Over the vast rangelands of eastern Montana, the three northern transcontinental railroads slowly climbed for their appointment with the Rocky Mountains in the west. Two of the lines, the first and last to be built, chose routes through the southern part of the state. The northern route was true to its name and the declamation of its enterprising builder: "No man on whom the snow does not fall ever amounts to a tinker's damn." The three railroads together accounted for most of Montana's passenger service, which was almost without exception superlative.

The land-grant Northern Pacific completed its route to the West Coast in 1883 and brought Montana's first sizeable influx of settlers. Within a decade, the state was crossed by a second transcontinental, James J. Hill's Great Northern, whose chosen route proved to be the shortest of all and the one with the lowest grades (it was also 150 miles closer to the Arctic Circle). The Milwaukee Road capped the era of Montana railroad building with its Pacific extension, opened in 1909.

The Northern Pacific followed the course of the Yellowstone River through the rolling plains of eastern Montana to Billings and on to Livingston. The Bozeman tunnel took the tracks through the natural divide between the Yellowstone River and the Gallatin Valley, where, at Three Forks, the headwaters of the Missouri River are formed. The original line went through Helena, the capital city, then crossed the Continental Divide at Mullan Pass through another tunnel. Several years later, the NP built a second line that served Butte, making a somewhat higher crossing of the Divide by way of Homestake Tunnel. The railroad's principal passenger trains generally traveled by way of Butte.

After entering Montana, Hill's Great Northern followed the profile of the Missouri and Milk rivers. Upon reaching Havre, he momentarily diverted from his northern route, running a line southwest to Great Falls. The Montana Central, whose construction he had supported, had a line from there to Helena and Butte, giving him both a connection with the Union Pacific and access to communities of growing importance. Still wanting to push west to Puget Sound, however, he pursued the search for a pass through the Continental Divide that had been rumored to exist since the Lewis and Clark explorations in 1806.

The existence of the pass was no secret to Blackfoot Chief Little Dog, who had conveyed a description to one of the government's Pacific Railway surveyors as early as 1853. The pass, nonetheless, remained lost to the white man until Hill's reconnaissance engineer, John F. Stevens, "found" it just as winter closed in on the mountains in late 1889. This was the Marias Pass, which gave Hill the most favorable crossing of the Divide anywhere between Canada and New Mexico. He completed his line to the coast in 1893.[1]

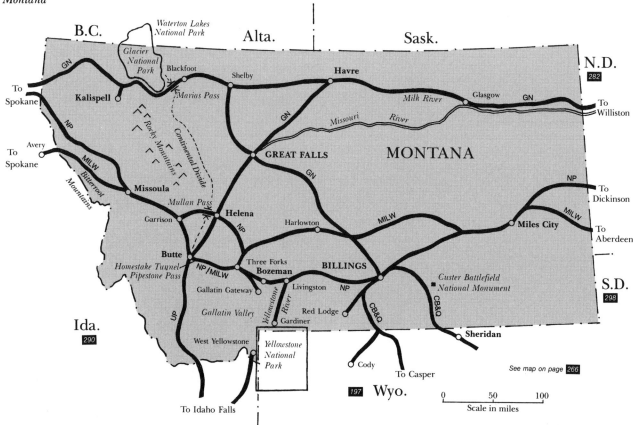

Eager to tap the growing traffic from the Pacific Northwest, the Chicago, Milwaukee & St. Paul Railway made a move that it hoped would put it on a more competitive footing with the family of lines that would one day make up the Burlington Northern. It decided to become much more than a midwestern, granger system, ending at the Missouri River, by building an extension to the Pacific Coast, not a financially easy task for a latecomer. By 1908, however, the new line had been built to Butte, and a year later it was completed through to Puget Sound.

Except for a portion of its tracks in central Montana, the Milwaukee Road paralleled the Northern Pacific for much of the way. North of Three Forks,

the tracks of each railroad lay on opposite sides of the Missouri River, with the curious result that trains bound for the same destination passed each other in opposite directions. Milwaukee Road trains crossed the Continental Divide in a tunnel under Pipestone Pass, near Butte.

The Milwaukee's main line through western Montana had one main difference from the other two northern routes. By late 1916, more than 400 miles of it had been electrified, from Harlowton to Avery, Ida. Using the region's plentiful hydroelectric power, this first of two Milwaukee Road electrification projects made possible substantial economies in operation. Crossing the several mountain ranges, the powerful

Northern Pacific No. 2, the *Main-streeter*, pauses at Missoula in May 1953. The eclectic consist, including a Milwaukee Road diner, was used while awaiting delivery of new cars.

bi-polar electric locomotives made good use of regenerative braking, whereby on the downgrades the force of the descending trains drove the motors, actually generating electricity. This cinder-free locomotion also enhanced the memorable summertime journeys passengers made while riding open observation gondolas through the Bitterroot Mountains.

Billings was one of two Montana cities with direct rail connections to the south, the Chicago, Burlington & Quincy's line to Lincoln, Neb., and the Burlington-/Colorado & Southern line to Denver. By way of the first, Billings had Pullman service to Alliance, Lincoln, and Omaha, Neb.; Kansas City and St. Louis; and, via the other, to Cody and Cheyenne, Wyo., and Denver. The city was an important transfer point for travelers continuing on to the Northwest on trains of the Northern Pacific and Great Northern, especially in summertime.

The Northern Pacific's main line, of course, served Billings, bringing the premier *North Coast Limited,* as well as the *Mainstreeter, Alaskan,* and summertime *Yellowstone Comet.* Billings had its own Pullmans on the NP to Chicago (CB&Q), Helena, and Spokane.

Butte was Montana's other gateway city to the south, with the Union Pacific line to Pocatello, Ida., and Ogden, Utah. The *Butte Special* carried sleepers to both Salt Lake City and Los Angeles. This great copper capital was also served by all three northern transcontinentals.

The Northern Pacific, apart from scheduling a

Pullman to Spokane, worked with the Burlington on interline sleepers from Butte to Denver, Kansas City, and St. Louis (1899). Early in this century, the Great Northern carried sleeping cars to Havre, Glasgow, Williston, N.D., and Minneapolis/St. Paul. The Milwaukee Road's *Olympian* and *Columbian* paused in Butte with through sleepers for Chicago and the Pacific Northwest.

Northern Montana was entirely the province of the Great Northern, whose flagship *Empire Builder, Oriental Limited* and its post-war successor *Western Star* worked the line between St. Paul and Puget Sound, doing heavy summertime business at Glacier National Park. Most of the Great Northern's dedicated Pullman service from Montana's urban areas originated not on this main trunk, however, but on the line from Havre to Butte. A Havre-Williston sleeper was one exception.

At Great Falls, this Butte line intersected another one running diagonally between Shelby and Billings. The CB&Q delivered Pullmans from Kansas City and Omaha through this gateway, and the GN forwarded them to Great Falls. This picturesque city on the upper Missouri River had sleeping car service on the Great Northern to Havre, Minneapolis/St. Paul (1890), and Chicago (CB&Q) to the east, and Blackfoot, Kalispell, and Spokane to the west. In later years, the *Western Star* operated between Shelby and Havre through Great Falls.

Service to Helena, apart from the through sleepers

Donald Sims

A late winter sun bathes the sides of the Milwaukee's *Olympian Hiawatha,* ready to depart Butte in the early 1950's.

America was still in the horse and buggy era in 1905, when this NP train (above) pulled into the loop at Gardiner, the railroad's entrance to Yellowstone Park. A generation later, a special help train (left) arrives June 15, 1930 with mostly college youths who will spend their summer working at the park.

carried on the secondary NP trains and the Billings car, was in the hands of the Great Northern, which scheduled Pullmans to Havre, Minneapolis/St. Paul, and Chicago (CB&Q).

Vacation traffic accounted for a large number of sleeping car movements in Montana, and Glacier National Park was the main scenic attraction on the Great Northern. The railroad actively promoted cre-

ation of the park, and Congress passed the enabling legislation in 1910. The main line crossing Marias Pass was the southern boundary of the new park, giving the Great Northern monopoly rail access. Its stewardship, however, was benign, and the GN financed construction of numerous lodges and chalets in the park to accommodate the tourists that its trains delivered. In company with the Burlington, the Great

It's 1934, in the middle of the Depression, but people are still riding the posh *Empire Builder* to Glacier National Park.

Northern scheduled Pullmans to Glacier from Chicago, Kansas City, Omaha, and Cody. Another Omaha sleeper terminated at Shelby.

Yellowstone National Park, created in 1872, was the first federal preserve of its kind, and the Northern Pacific had taken a similarly active interest in bringing it into being. The major land area of the park lies within Wyoming, but portions extend into both Idaho and Montana. The NP, which soon began calling itself the "Yellowstone Park Line," monopolized the early rail transportation by way of a branch that eventually reached Gardiner, on the northern boundary of the park.

Just before the park opened in the summer, special trains stopped on the great turning loop at Gardiner and disgorged legions of college students hired for the season. A few days later, the Northern Pacific started bringing the paying customers to Gardiner in sleepers from Chicago (CB&Q), Minneapolis/St. Paul, Seattle, and Portland (Spokane, Portland & Seattle), as well as interline cars via Billings from Denver (C&S/CB&Q) and Kansas City (CB&Q). In later years, the NP used Red Lodge as a gateway, with Pullmans from Chicago, and also a third entrance, Cody, accessible by way of Burlington tracks from Billings. During the park season, the NP scheduled a Chicago-Bozeman sleeper (CB&Q).

Shortly after the turn of the century, the UP built its own line from Idaho Falls to the western boundary of the park and tapped the lucrative market with its *Yellowstone Express* and *Yellowstone Special,* which picked up Pullmans at Pocatello from elsewhere on the Overland system. UP service to West Yellowstone included sleepers from Pocatello, Salt Lake City, Portland, Chicago (Chicago & North Western), St. Louis (Wabash), and even Jacksonville, Fla. This car, which operated in the summer of 1925, traveled via Atlantic Coast Line, Central of Georgia, Nashville, Chattanooga & St. Louis, Louisville & Nashville, and Wabash, before the UP finally took over at Kansas City.

The Milwaukee Road, handicapped by the greatest distance of any of the railroads from Yellowstone, scheduled Pullmans from Chicago and Seattle to a spur terminating in Gallatin Gateway, south of Bozeman.

1. Ralph Budd, "The Conquest of the Rockies," *Trains,* Vol. 7, No. 12 (October 1947), p. 52.

John Aquirre collection

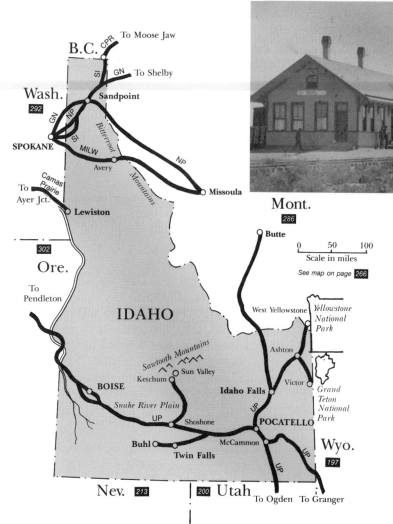

John Aquirre collection

The UP's Oregon Short Line originally snubbed Boise, but by the time of this late 1920's visit of the *Portland Limited* (top), OSL trains were looping through the capital city. On a more rustic scale, an OSL local (lower) delivers the Salt Lake City sleeper to Buhl, circa 1912.

46
Idaho

America's three northern transcontinental railways—Northern Pacific, Great Northern, and Milwaukee Road—crossed Idaho far up in the Panhandle, where scouts and surveyors had been able to find only moderately forbidding topography. As a result, the dense pine forests in the central part of the state never echoed with the steam whistles or diesel horns of crack limiteds working their way to Puget Sound. The major rail activity in the broad southern base of the Gem State took place on the Oregon Short Line of the Union Pacific, which made its way from Pocatello across the Snake River Plain to Boise, and eventually Portland. For dedicated Pullman service, Idaho looked entirely to the UP.

At McCammon, the tracks from Granger, Wyo., merged with the Ogden-Butte line and headed into Pocatello, the major rail marshaling center in southern Idaho. In Pocatello the Overland trains dropped and added Pullmans, some to be routed north to Yellowstone National Park, others to continue on the OSL main line to the Pacific Northwest. This was the route of the *City of Portland, Portland Rose, Idahoan,*

and the older *Portland Limited.* From Salt Lake City came such trains as the *Butte Special, Butte Express, Yellowstone Special,* and *Yellowstone Express.*

Salt Lake City was the destination of about half the Pullman movements originating in Idaho, and this gave the northern mountain area access to both the Los Angeles and San Francisco lines of the Overland Route. From the Utah capital city, the Union Pacific carried sleepers to Idaho Falls (tourist), Ashton, Victor, and West Yellowstone (just over the Montana line and right at the western entrance to the park). Tourists bound for Grand Teton National Park left their Pullmans at Victor, where they transferred to a bus for a trip across an 11,500-foot mountain pass to Jackson, Wyo.[1] Sleepers also departed from Pocatello to West Yellowstone as an accommodation to passengers from Overland or OSL trains.

The Union Pacific scheduled sleeping cars by way of Pocatello from Salt Lake City to Boise and Buhl. Wintertime service included through sleepers from Los Angeles to Ketchum, a mile from the ski resort the UP had built at Sun Valley, in the Sawtooth Mountains, in 1936.

For many years, Boise did without main line passenger service. When the OSL was laying track in the vicinity in 1883, it bypassed Boise by twelve miles after the community failed to post the required bond. Merchants eventually built a spur line, but not until 1926 did the capital city see the through service of the UP, when the railroad opened its newly constructed loop through town and thousands of joyful citizens welcomed the first arrival.[2]

Boise had its own Pullman accommodations to Spokane and Lewiston, the latter with the help of the Camas Prairie Railroad, a joint operation of the UP and Northern Pacific in Washington and western Idaho. Briefly before the turn of the century, Boise also had sleeping car service to Denver by way of the UP, Rio Grande Western and ill-fated Colorado Midland.

Lewiston's sleeper to Portland also began its journey on the Camas Prairie before parent UP took it over for the remainder of the trip down the south bank of the Columbia River.

The tiny Spokane International Railway, in Idaho's Panhandle, at one time was a key link in yet another northern transcontinental line that saw sleeper service between the Twin Cities of Minnesota, and Portland. The *Soo-Spokane-Portland Limited* traveled the Soo-Canadian Pacific interline route to Kingsgate/Eastport, whence the SI forwarded the Portland sleepers on to Spokane and the UP. After this early-20th Century service ceased, the SI continued to carry a Spokane-Calgary sleeper.

---

1. Letter dated June 15, 1988, from Idaho railroad historian Jim Witherell.
2. Witherell, letter dated July 11, 1988.

Otto C. Perry/Denver Public Library

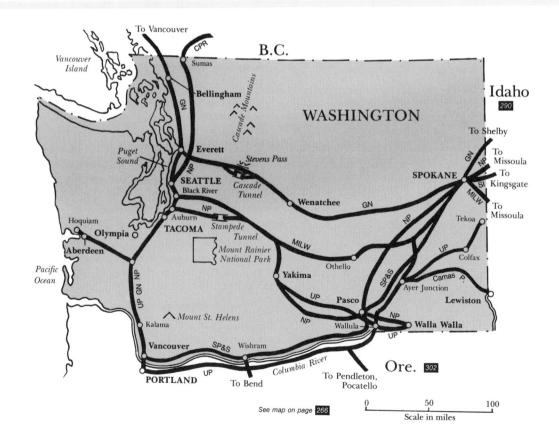

See map on page 266

47
Washington

A crewmember's lantern illuminates this time-exposure of the Milwaukee Road's eastbound *Olympian*, shortly before its departure from Tacoma in October 1931. The GE bipolar locomotive takes it through the western part of the railroad's electrified territory.

Not until 1870, a year after the great central route was completed, did construction of America's first northern transcontinental railroad begin. The first segment of the Northern Pacific to be built in Washington started at Kalama, on the Columbia River, where freight boats unloaded rails and other materials for the crew laying the line north to Tacoma. Financing of the system was accomplished by Jay Cooke.

In the next three years, the builders reached Tacoma and those in the east got to the Missouri River, but financial disorder and Indian warfare stalled further construction for much of the rest of the decade. Cooke's empire crumbled in the Panic of 1873, and the job of finishing construction of the NP fell to the German-born financier Henry Villard.

Owner of the Oregon Railway & Navigation Company, a railroad running east from Portland along the Columbia River, Villard took steps to tie the NP into this property rather than building a competing line across the Cascade Mountains to Tacoma, as originally planned. And so, when the east and west lines were finally joined in Montana in 1883, the NP was essentially a Duluth-to-Portland railway, with branches to Minneapolis/St. Paul at one end, and to Tacoma on the other.

Villard, however, soon encountered financial difficulties of his own, and the new management pushed construction across the Cascades from Pasco. They completed a line with numerous switchbacks in 1887, and the following year opened the 1.8-mile Stampede Tunnel (the bore so named for a "stampede" of striking track layers down the mountain).

Five years later, in 1893, James J. Hill was about to crack the Northern Pacific monopoly by running his own Great Northern line from Minnesota across the same mountains. Hill had triumphed in his crossing of the Rockies in Montana with the relatively lowlevel Marias Pass, discovered by his savvy engineer, John F. Stevens. In the Cascades, where snowfall from the moisture-laden maritime air often measured a foot an hour, Stevens found another suitable pass and hurdled it with a temporary line stitched with switchbacks.

Although Hill now had his railroad to Puget Sound, the Stevens Pass, with its horrendous four per cent grades, was only marginally suitable, and work began immediately on what would be the first of two tunnels. The early one, opened in 1900, eliminated the switchbacks, cut nine miles off the route, and reduced grades to two per cent. The Great Northern, however, was not yet out of the weather.

The 2.6-mile tunnel was a hellhole for locomotive crews, who faced the hazards of torrid steam exhaust and the noxious fumes of stack gas. Snowsheds were necessary protection against the frequent avalanches on the steep slopes. In one instance, passengers and trainmen were trapped for ten days after an avalanche sealed them in one of the sheds—only fast action by the engine crew in moving the train under cover had saved it from being swept away in the snow slide.[1]

Electrification in 1909 eliminated the main hazards to the engine crews, but the problem of snow was always great until the second Cascade Tunnel was opened in 1929, just south of Stevens Pass. This longest tunnel in the Americas, running 7.79 miles underground, reduced the elevation of the crossing by almost 1,200 feet from the original switchback line. Until diesels were the rule, the Great Northern operated under seventy miles of electric wire west from Wenatchee through the Cascade Tunnel.

The third and last northern transcontinental, the Chicago, Milwaukee, St. Paul & Pacific, completed its line to the coast in 1909, and eventually boasted the longest main-line electrification in America. In addition to 438 miles in Montana and Idaho, the Milwaukee also electrified its main line west of Othello.

All three railroads came from Idaho into Spokane before fanning out over the Columbia River Basin and beginning their climb into the Cascades through the apple-growing areas on the eastern side of the mountains.

The Great Northern ended its westward thrust at Everett, and the Northern Pacific and Milwaukee lines forked between Seattle and Tacoma. The Great Northern originally scheduled through sleepers from the east to both Seattle and Tacoma, but the other lines came to dominate the Tacoma market with more directly routed, dedicated sleepers.

Serving both cities, the Northern Pacific cut the Tacoma cars of its trains in or out at East Auburn. The Milwaukee Road had the unusual practice of running its through trains into Seattle, then backing them to Tacoma. Eastbound trains were hauled backwards from Tacoma to Seattle. In general, the schedules favored Seattle over Tacoma.

After the Northern Pacific's early alliance with the Oregon Railway & Navigation Company, the OR&N became part of the Union Pacific system, and the NP and GN, which had jointly bought control of the Burlington in 1901 to gain access to Chicago, pooled their efforts in building the Spokane, Portland & Seattle Railway several years later. From Portland east, the SP&S line occupied the north bank of the Columbia River to Pasco, where it connected with the NP. It interchanged with the GN at Spokane. SP&S trains carried the through cars of both its parent roads to Portland.

Most of Seattle's sleeping car service moved in and out of King Street Station, opened in 1906 and

The eastbound *North Coast Limited* slips under Milwaukee Road tracks west of Easton, Wash., on April 12, 1963.

SP&S No. 1 awaits filling out of its consist (background) in Spokane for the Portland section of the *Empire Builder*.

shared by both the Northern Pacific and Great Northern. Here such trains as the NP's two-tone green *North Coast Limited* and *Mainstreeter* and GN's green and orange *Empire Builder*, *Oriental Limited*, and *Western Star* began and ended their journeys across the northern states. Both roads expanded their Pullman menu with interline service over the Burlington.

The Northern Pacific scheduled sleepers from Seattle to Yakima, Walla Walla, and Spokane; Gardiner, Mont., one of the entrances to Yellowstone Park (sum-

Spokane International No. 2, near Sandpoint, Ida., carried a Calgary-Spokane Pullman in cooperation with Canadian Pacific.

mers); Minneapolis/St. Paul, and Chicago (CB&Q or Chicago & North Western. The *North Coast Limited* was operated to Chicago by the C&NW for several years before World War I). At Billings, Mont., the NP picked up Pullmans delivered by the Burlington from Denver (Colorado & Southern/CB&Q), St. Louis, Kansas City, and Lincoln.

Great Northern trains plied the same territory, bringing sleepers to Seattle from Wenatchee, Spokane, Minneapolis/St. Paul, and Chicago (CB&Q), and, via the Burlington at Billings, from Denver (C&S/CB&Q), Lincoln, and Kansas City. Over its scenic line north, the GN carried Pullmans to Bellingham, and Vancouver, B.C.

The Milwaukee Road, operating from Seattle's Union Station, focused most of its efforts on through sleepers to Chicago, plus summertime service to Gallatin Gateway—its Yellowstone station—and cars to Spokane.

Union Station was also the Union Pacific's terminal. By way of Portland and the Overland Route, the UP provided Seattle with through Pullmans to Portland, Chicago (C&NW), Kansas City, and St. Louis (Wabash), and south to San Francisco and Los Angeles on such Southern Pacific trains as the *Shasta Limited*, *Cascade*, and *West Coast*. The UP also carried sleepers locally to Aberdeen and Hoquiam.

Early in this century, sleepers operated over a fourth route to the Twin Cities. The Northern Pacific carried cars to Sumas, on the Canadian border, whence they traveled the Canadian Pacific/Soo interline east.

Tacoma's classic domed Union Station accommodated all the pool trains on its through tracks, and this joint UP/GN/NP service between Seattle and Portland also included Pullmans from Tacoma to Portland. Apart from this activity, however, most of Tacoma's dedicated sleeping car service was courtesy of the Northern Pacific, whose monad symbol adorned the huge, arched entrance to the station. Before shuttle trains and finally buses took over the link to Auburn, the NP scheduled sleepers from Tacoma to Yakima, Minneapolis/St. Paul, Chicago (CB&Q), and Omaha (CB&Q). Before it shifted focus to Seattle, the Great Northern also scheduled a sleeper to the Twin Cities.

From its own facility in Tacoma, the Milwaukee Road carried sleeping cars to both Spokane and Chicago until fairly late in its transcontinental operation.

Also until well in the 1950's, Spokane had Pullman service to Seattle on all three of the northern roads. As the urban hub of the Inland Empire, Spokane was the first major city on the three railroad lines west of Minneapolis, as well as the first point at which all their crack passenger trains converged. The city also was a terminus of three other roads, the Spokane Interna-

Jim Fredrickson

On a foggy night in October 1962, Northern Pacific's eastbound *Mainstreeter* stops at East Auburn for the Tacoma traffic.

tional, SP&S, and UP.

Spokane's dedicated sleeping car service on the Northern Pacific was mostly regional, with accommodations to Billings and Butte, Mont., to the east, and from Walla Walla, Yakima, and Tacoma in Washington state. The Great Northern scheduled Pullmans to Wenatchee, Great Falls, Mont., and Kansas City (CB&Q). The Milwaukee's *Olympian* once carried a Spokane car to Chicago.

Aside from transporting the through cars for Portland from such GN trains as the *Oriental Limited, Empire Builder,* and *Western Star,* the SP&S also flaunted its own colors, yellow and Pullman green, and carried a Spokane-Portland Pullman.

The UP's *Spokane* provided competing accommodations to Portland, where the SP took over a San Francisco sleeper. Spokane also had Pullman service on the UP to Walla Walla, Boise, Salt Lake City, and Los Angeles.

Although in recent times a part of the UP system, the Spokane International once served as a bridge route to a connection with the Canadian Pacific at Kingsgate, B.C. While under CPR control, it cooperated with its parent in an interline sleeper from Spokane to Calgary.

The Union Pacific scheduled dedicated sleepers from other Washington communities, all directed to Portland: Colfax, Hoquiam, Tekoa, Walla Walla, and Yakima.

In Washington, just as in several other western states, the railroads played an important part in promoting the founding and use of national parks. Under terms of the Northern Pacific's land grant of 1864, the railroad had been given possession of alternate sections of land from which Mount Rainier National Park was created in 1899. The railroad willingly exchanged its land for acreage elsewhere and agreed to promote the park as well.

In 1980, Mount St. Helens became the world's newest active volcano, belching ash and lava from . . . former Northern Pacific land. Two years later, the Burlington Northern completed the circle by deeding its NP holdings on the summit back to the federal government.[2]

1. D.W. Mc Laughlin, "How Great Northern Conquered the Cascades," *Trains,* Vol. 22, No. 1 (November 1961), p. 28.
2. Alfred Runte, *Trains of Discovery: Western Railroads and the National Parks* (Flagstaff, Ariz.: Northland Press, 1984), p. 77.

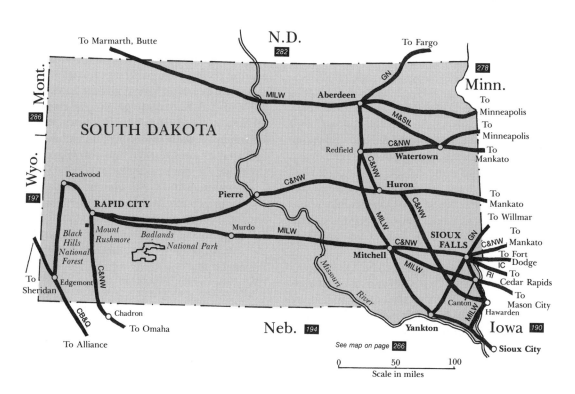

48
South Dakota

The eastern half of South Dakota was covered by a network of railroads in typical granger pattern, but in the western part the lines were sparse. The plains west of the Missouri River, which winds its way down the middle of the state, were reserved for the Sioux, and as a result railroad building was temporarily halted at the river.

The Black Hills, in the southwestern portion of the state, were sacred to the Indians, and under the terms of a treaty, the sanctity of them was to be maintained. In 1873, however, a report of gold in the western slopes sent an expedition led by General George Armstrong Custer to investigate, and the later confirmation triggered a rush. Although Custer was supposed to chase out the miners, it was the Sioux who were eventually driven out. So much for the treaty; the sacred Black Hills would be forevermore open to mining and tourism.[1]

Rapid City grew up to be South Dakota's second largest city and gateway to this area of uncommon scenic attraction. Both the Chicago & North Western and Milwaukee Road rushed to build lines there, arriving in the summer of 1907. Tourism got a big boost in the 1920's with the sculpting of Mount Rushmore by Gutzon Borglum and the visit of President Calvin Coolidge, who in 1927 established his summer White House in Rapid City. Both of the main line railroads terminating there enjoyed the prosperity.

The Chicago & North Western's *Minnesota & Black Hills Express,* via Elroy and Madison, Wis., carried Pullmans to Chicago. In the streamline era, the Rapid City sleepers rode the *Dakota 400* between Huron and Mankato, Minn., giving the train the distinction of being the only *400* ever to carry a sleeping car. The *Dakota 400* originally terminated in Huron, later in Rapid City, before deserting South Dakota completely in 1960.[2]

The Milwaukee's *Sioux* also carried a Rapid City Chicago sleeper (in later years the eastbound car terminated and laid over in Sioux Falls during the day). Early in the century, the North Western operated a Pullman between Rapid City and Deadwood.

This was a reasonably quiet period in the tumultuous history of Deadwood, South Dakota's most colorful and notorious community. The Black Hills' first great gold rush brought an instant population of 25,000, including Wild Bill Hickok and Calamity Jane, to the stark gulch in which the town took root. While miners labored to reclaim the yellow dust, a service force of saloon keepers, prostitutes, card dealers, and other enterprising souls labored for their share of the pot, too. Dwellings and shops were built helter-skelter up the steep slopes, creating an archetypal western mining town. The boom inevitably subsided but reignited in the 1930's after the repeal of Prohibition and revaluation of gold, which once again made the mines profitable. By then, Deadwood's no-

General Custer, the Burlington's No. 41, crosses the Cheyenne River at Edgemont on June 10, 1938.

C&NW's *Minnesota & Black Hills Express* approaches its destination of Rapid City, June 21, 1936.

toriety was secure, and tourists came from afar to see the wicked place for themselves.

The North Western carried sleepers to Deadwood from Omaha and, in summers, Chicago. The Chicago, Burlington & Quincy once served Deadwood with Pullmans from Omaha and Lincoln, but after buses began to offer tours through the area, the sleepers terminated in Edgemont, on the CB&Q's main line to Billings, Mont. Service on the Q included cars from Chicago, St. Louis, Omaha, Lincoln, Denver, and Cody, Wyo.

Sioux Falls, South Dakota's first white settlement (1859) and ultimately largest city, had four different sleeping car lines to Chicago, two of which were on the Milwaukee Road. One sleeper went aboard the *Sioux*, which traveled via Mason City, Iowa, and Madison, Wis.; the other was carried on the *Arrow*, which traveled the Omaha main line east of Manilla, Iowa. This was also the route of the Sioux Falls section of the *Midwest Hiawatha*. The Milwaukee Road also scheduled sleeping cars to Aberdeen (through Canton and Mitchell) and Dubuque (1890).

The Illinois Central, which operated a section of the *Hawkeye Limited* to Sioux Falls, also carried sleepers to Dubuque (1890) and Chicago. In an almost similar vein, the Rock Island scheduled cars to Peoria (1890) and Chicago. The Chicago, St. Paul, Minneapolis & Omaha provided Pullman service to the Twin Cities and Omaha.

Four railroads scheduled sleeping cars to Aberdeen, which was on the Milwaukee Road's main line to the Pacific Northwest. Both the *Olympian* and *Columbian* offered convenient through service, but the faster, postwar schedules put the *Olympian Hiawatha* in Aberdeen in the early hours of the morning. Aberdeen became the western terminus of the *Columbian* in 1955 shortly before its discontinuance.[3]

Three railroads offered service from Aberdeen to Minneapolis/St. Paul, including the Milwaukee, Great Northern, and Minneapolis & St. Louis. Sleepers on the Milwaukee also went to Sioux City, those on the GN to Fargo, and the North Western scheduled service to Omaha.

Mitchell, home of an ornate Corn Palace with towers and minarets, had sleeping cars on the Milwaukee Road to Chicago and Mason City and from Milwaukee. The North Western scheduled Pullmans to the Twin Cities and Omaha.

Pierre, the state capital, was on the main line of the North Western to the Black Hills, but its Pullman service consisted of through cars from Rapid City. The C&NW scheduled sleepers to Huron from Minneapolis/St. Paul, Rochester, and Omaha. Huron's reputation as the capital of pheasant hunting brought peak loads of well-armed Pullman passengers to town in autumn.

Both Watertown and Redfield had sleeping car service to the Twin Cities by way of the North Western, and the M&StL also scheduled a car from Watertown.

A line of the Great Northern extending to Yankton gave the town sleeping car connections to Willmar and Minneapolis/St. Paul.

The Milwaukee Road scheduled sleepers from Murdo to Sioux City, and, after service to Rapid City was cut back, a Canton-Chicago car.

1. Robert J. Casey and W.A.S. Douglas, *Pioneer Railroad: The Story of the Chicago and North Western System* (New York: Whittlesey House [McGraw-Hill], 1948), p. 166.
2. Jim Scribbins, *The 400 Story* (Park Forest, Ill.: PTJ Publishing, 1982), p. 102.
3. Scribbins, *The Hiawatha Story* (Milwaukee: Kalmbach, 1978), pp. 178-81.

49 Oregon

The two western roads of the Overland Route—Southern Pacific and Union Pacific—also were Oregon's original rail connections to California and the east. Their routes intersected perpendicularly at Portland, the top, east-west leg formed by the UP line along the Columbia River, and the supporting leg to the south being the SP. This was the state's basic, arterial rail design, little changed by subsequent modifications.

Portland's window to the east was opened several years before it had rail connection to California. The Oregon Railway & Navigation Company, the ancestral UP line along the Columbia, was the temporary western leg of the transcontinental Northern Pacific upon completion of the NP in 1883. The following year, another UP member, the Oregon Short Line, connected with the OR&N, giving Portland its link to the Overland Route at Granger, Wyo. The original SP line south was not completed until 1887.

This Siskiyou Line, as it came to be known, had smooth sailing through the Willamette River Valley in Oregon and in the Sacramento Valley of California. The hardship was in the rugged mountains in between, where the line went up and down over economically burdensome grades of as much as 3.3 per cent. (Interstate Highway 5, through Grants Pass, Medford and Ashland follows the Siskiyou route.) The SP set about improving operating conditions and, in 1926, opened the new Cascade Line through the great forest areas to the east.

The new line ran from Eugene and curved around northeast of Crater Lake National Park, named for America's deepest body of water (1,932 feet), and rejoined the old line in California, near Mount Shasta. The Cascade Line had the greater elevation, but smoother grades and curves once it attained plateau altitude, and in the spring of 1927, the SP began shifting its main passenger trains to the new route.

Although traffic up and down the West Coast was important and SP had a monopoly on it, it offered fewer potential Pullman lines than was the case to the east. The UP, as a result, scheduled sleepers to more destinations than the SP did from Portland. All the major railroads used Union Station, which dated from the early 1890's, and the UP accounted for about half the terminal's Pullman listings.

The *Portland Limited* made the trek on the historic OSL/OR&N line, with an assortment of Pullmans from as far east as Chicago, until the *Portland Rose* supplanted it in the early 1930's. The *Oregon Trail Express* joined the flow from Salt Lake City. But while the UP service during the steam era was both solid and scenic, it was the streamlined *City of Portland* that put on the speed, cutting three-quarters of a day off the schedule to Chicago. The *Idahoan* was a secondary train during the *City's* years.

The UP carried Pullmans from Portland to Pendle-

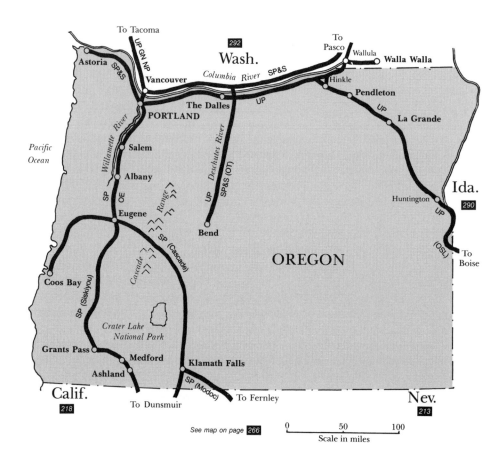

See map on page 266

Scale in miles

ton and La Grande; West Yellowstone, Mont. (summers); Salt Lake City, Kansas City, St. Louis (Wabash), Chicago (Chicago & North Western or Milwaukee Road), and Denver. Two other lines to Denver around the turn-of-the-century involved UP to Salt Lake City, thence Denver & Rio Grande Western or D&RGW/Colorado Midland/Santa Fe to Denver.

Over the UP line running northeast to Spokane, Portland had Pullman service to Colfax, Tekoa, Walla Walla, Yakima, and Spokane, Wash., and Lewiston, Idaho (Camas Prairie).

Before the Northern Pacific built its own line to Tacoma across the Cascade Mountains, its Minneapolis/St. Paul sleepers operated to Portland over the OR&N. And in the first decade of this century, another Portland/Twin Cities sleeping car line rode UP tracks to Spokane, then Spokane International/Canadian Pacific/Soo tracks east.

On the line north through Vancouver, Wash., the Union Pacific carried a Pullman to Hoquiam.

In 1925, the Union Pacific, Great Northern, and Northern Pacific began pooling their service between Portland and Seattle. Under this arrangement, Portland continued to have dedicated Pullman service to both Tacoma and Seattle.

With the OR&N line along the Columbia River firmly controlled by the Union Pacific of Edward H. Harriman, his archrival, James J. Hill, decided to meet the competition head-on. Hill, who had control of both the NP and GN early in this century, built the competing Spokane, Portland & Seattle Railway on the north bank of the river, connecting the first two cities of its name in 1909. The SP&S thereby became the GN and NP's link to Portland, and the trains of the "Northwest's Own Railroad" were often a colorful mixture of sections for both the parent roads. The *North Coast Limited* once was spliced onto the *Empire Builder,* later the *Western Star.* Similarly, the *Oriental Limited-Comet* was a temporarily hyphenated train.

In addition to carrying Portland-Chicago sleepers for both northern roads, the SP&S moved cars for the NP destined for Gardiner, Mont. (summer), and the Twin Cities, as well as its own Pullman to Spokane. The SP&S had one more standard sleeping car line, through its subsidiary Oregon Trunk, from Portland to Bend, in central Oregon.

The saga of the OT's construction to Bend is reminiscent of the great track war that took place in Colorado's Royal Gorge, three decades earlier. Bend lies on a high plateau several thousand feet above the Columbia River level, and it was accessible to rail builders only through the rugged canyon of the Deschutes River, whose sheer cliffs rise hundreds of feet above the roiling water. Building one line would have been challenging enough, but this route to the untapped timber and mineral riches of central Oregon was coveted by both Hill and Harriman, adding the

At Wishram, Wash., across the Columbia River from Oregon, the SP&S eastbound *Empire Builder* takes orders on the fly.

spice of often unfriendly competition.

Hill sent his Oregon Trunk crews into the canyon, and Harriman countered with his own. The two teams laid tracks and brawled their way up 150 miles of river, and in October 1911 they miraculously reached Bend. Despite the combat, much of the resulting line was a patchwork of cooperative effort, which the trains of both Oregon Trunk and Union Pacific shared. Twenty years later, Hill's Great Northern fulfilled another dream of the late Empire Builder by pushing south from Bend and forging a route through to California.

The SP&S had yet another subsidiary that contributed to the lore of overnight train travel, the Oregon Electric, which ran south from Portland in the Willamette River Valley through Salem and Albany to Eugene. This classy interurban system with green and black coaches was one of three electric lines in the United States that carried sleepers. For somewhat more than a decade, ending in the mid 1920's, the OE ran sleeping cars from Portland to Eugene. During the heyday of interurbans, there was talk of an all-electric Portland-San Francisco line, with the Oregon Electric possibly connecting with the Sacramento Northern, but the OE never went south of Eugene. Another Hill enterprise, however, did get him into

California shortly before World War I, when he took to the high seas.

Hill's Great Northern Pacific Steamship Company operated what the ads billed as the "twin palaces of the Pacific," steamships that made twenty-six-hour trips between Astoria and San Francisco. They were the *SS Great Northern* and *SS Northern Pacific*. The boats made thrice-weekly sailings with accommodations ranging in price from eight to twenty dollars. An SP&S boat train obligingly made a connection with Portland and also brought to dockside a sleeper the GN had dispatched from Vancouver, B.C. In 1917, the boats were taken over by the federal government for the war effort, ending Hill's maritime extension.[1]

Both the SP&S parents once had limited sleeper operations of their own north of Portland. The GN carried sleeping cars to Vancouver, B.C., from Portland, and before the turn of the century the NP scheduled Chicago sleepers by way of Tacoma; east of the Twin Cities, they traveled either Milwaukee Road or Soo/Wisconsin Central.

The pride of the Southern Pacific's new Cascade Line, upon its opening to passenger traffic in April 1927, was the namesake train itself. The *Cascade* was an all-Pullman, extra-fare train from Seattle and Port-

The first section of the *Shasta*, once the SP's premier Portland-Oakland train, rolls into Williams, Calif., July 25, 1938.

Otto C. Perry/Denver Public Library

land to San Francisco. South of Portland it carried an open-air observation car for the scenic trip over the new line. The Depression, however, brought the temporary addition of coaches and an end to the extra fare. The streamlined post-war version, wearing the SP's nighttime livery of two-tone gray, was introduced in August 1950 with consists of eight all-room sleepers and the articulated, three-unit diner/club/galley cars. Despite the mountainous running, the *Cascade* was put on a sixteen-and-a-half-hour schedule, giving Portland afternoon departures and morning arrivals, but losing much of the choice scenery to the darkness.

Before the opening of the Cascade Line, the SP's premier train to California was the luxurious *Shasta Limited*, which had been born as the *Shasta Express* in 1899. Offering such amenities as a library/observation car, barbershop, and shower, and the services of stenographers, ladies maids, and valets, the *Shasta* (the *Limited* was later dropped) also advertised a rather uncommon attraction. California's Mount Lassen, forty-some miles east of the SP right of way, had erupted in 1915, and for a time the railroad advised travelers that "when crater is in eruption, passengers have the novel experience of a car window view of a volcano in action." Extra fare, but—the gods willing—worth it.

During the steam era, when San Francisco was roughly a day's trip from Portland, all through trains commonly carried sleepers. Three years after the Cascade Line was opened, the *Klamath* and the *West Coast* were backing up the *Cascade* on the new line, and the *Shasta* and *Oregonian* were serving the Siskiyou trade.

The *West Coast*, which carried the SP's last open observation car, was additionally notable in carrying through Pullmans from Seattle and Portland to Los Angeles, a service that perished shortly after World War II. The *West Coast*, which bypassed the San Francisco Bay area, was also the SP's link on a brief Portland-Chicago sleeping car line. In the late 1920's, the railroad constructed a freight cutoff called the Modoc Line, which ran from Klamath Falls to a connection with the Overland Route at Fernley, Nev. In the winter of 1929-30 it offered through Pullmans fom Portland that finished their journey on the UP/C&NW *Continental Limited*.

In the summer of 1949, the *Shasta Daylight*, in blazing SP red and orange, went into Cascade Line service between Portland and San Francisco. In the twilight of SP passenger service, the *Rogue River* was the surviving Pullman train on the Siskiyou Line, carrying a sleeper to Ashland. Previously, the Southern Pacific had scheduled sleeping cars from Portland to Eugene, Coos Bay (Marshfield), Klamath Falls, and Sacramento.

In Oregon, most of the Pullman activity was Portland's, but Klamath Falls had its own sleeper to San Francisco, and little Huntington, on the eastern edge of the state, where the OR&N and OSL originally had met, was served by Pullman connections on the UP to Salt Lake City and, before the turn of the century, Denver (D&RGW/CM).

1. Donald Sims, "The SP&S Story," *Trains*, Vol. 20, No. 5 (March 1960), p. 40.

Four Union Pacific E9 units power the eastbound Domeliner *City of Portland* through the Columbia River Gorge.

Brian Downes

Donald Sims

"*A hog can cross America without changing trains—but YOU can't.*"

**—Robert R. Young, chairman
of the Chesapeake & Ohio Railway,
in post-World War II advertisements.**

"*I do not feel that the airplane will measurably compete with the train, the steamship or the automobile . . .*"

**—General W. W. Atterbury, president
of the Pennsylvania Railroad, in 1928.**

50
The
Transcontinentals

The *Sunset Limited* (top right) carried the Washington-Sunset tourist Pullmans at times, a Jacksonville-Los Angeles sleeper in the 1920's, and a regular Washington-Los Angeles Pullman in the final days of transcontinental rail travel. Before the post-war coast-to-coast service, changing trains in Chicago was routine. Fred and Phyllis Astaire (top left) make the switch between *Super Chief* and *Twentieth Century Limited* in 1940. The *California Zephyr* (bottom) carried New York-San Francisco sleepers.

In America's first great age of coast-to-coast transportation, the wagon and the sailing ship around Cape Horn were the enabling conveyances. The journey in either case, however, was long, difficult, costly, and dangerous. This all changed in 1869, when the locomotives of the Union Pacific and Central Pacific finally met, pilot to pilot, at Promontory, Utah. The golden spike ceremony held on that May day commemorated completion of America's first transcontinental railroad and ushered in the beginning of the second great age of cross-country travel.

Other lines reached the West Coast over the next four decades, and trains steamed off to California and the Pacific Northwest in ever-increasing numbers, carrying evermore luxurious accommodations. Individual railroads, which were rapidly consolidating into regional systems, cooperated in exchanging through sleeping cars. Transcontinental systems, however, were still the lusty ambitions of a few incipient rail barons, and so passengers could not travel coast-to-coast in a completely seamless manner. Always there was the necessity to change trains, if not in Chicago, at least somewhere in the vicinity of Meridian 90 West. There was one early exception.

Just seven years after the ceremony at Promontory, the same year that General Custer succumbed to superior forces, a train called the *Jarrett & Palmer Special* made the first totally transcontinental trip and electrified the nation in the process.

In the spring of 1876, a journey from New York to San Francisco required a full week, including changing of trains. Now, the Jarrett & Palmer theatrical partnership hired a train to carry three famous actors from New York to San Francisco in time to appear in a production of Shakespeare's *Henry V.* Their engagement was set for June 5; the train was scheduled to leave June 1. The New York *Herald* compared this travel package to the American counterpart in *Around the World in Eighty Days,* but the paper found Jules Verne's version far less imaginative.[1]

The *Special,* carrying the ornate Pullman buffet/sleeper *Marlborough,* departed on its historic run across the country via the Pennsylvania Railroad and the Chicago & North Western/Union Pacific/Central

Pacific Overland Route. Eighty-four hours and seventeen minutes after leaving Jersey City, it arrived at Oakland pier to immense fanfare on June 4.

Despite this record-breaking run, however, the railroads continued to emphasize comfort rather than speed. As the trackage and motive power improved, so did the schedules. Nonetheless, until the beginning of the streamline era in the 1930's, a traveler riding the best of trains would have found it impossible to improve upon the speed record of the *Jarrett & Palmer Special.* And despite the train's pioneering coast-to-coast sleeping car, the railroads offered very little similar service until after World War II.

Nonetheless, at least ten transcontinental sleeper lines operated in this seventy-year period, three of them being tourist cars of relatively spartan furnishings. New York, Washington, and Jacksonville all originated lines, but the majority, six, emanated from the hub of the universe itself, Boston.

During the autumn of 1891, weekly Pullman service was offered on three lines between Boston and Los Angeles, via Oakland. The Boston & Maine/Central Vermont/Grand Trunk handled the eastern segment, and west of Chicago one line involved the Santa Fe, another went via Alton/Missouri Pacific/Denver & Rio Grande Western/Southern Pacific, and a third traveled over Rock Island/D&RGW/SP. The latter line was resurrected the following spring for a two-year-period, and for the spring of 1895, the Burlington stood in for the Rock Island.[2]

Boston had two other transcontinental sleeper lines, also originating on the B&M, tourist cars in each case. For a number of years, until World War I, one of these Pullmans—billed as the longest such run in North America—operated weekly via Montreal (Boston & Maine/Canadian Pacific) to Vancouver, B.C. The other car went via West Shore and Nickel Plate to Chicago, where it made its connection to the West.[3]

From January to April 1892, a "Palace Buffet Sleeping Car" left New York's Grand Central Station once a week and ran through to San Francisco on a five-day schedule. The car traveled on the New York Central's *North Shore Limited* to Chicago, then the Milwaukee Road/Union Pacific/Southern Pacific to California.[4]

In the winter of 1897-8, a weekly New York-El Paso Pullman line was inaugurated on the Central of New Jersey, a commuter carrier whose sleeper schedulings were true rarities. These Pullmans traveled B&O to St. Louis, then Missouri Pacific/Texas and Pacific. The line offered an extension "to Oakland when needed," presumably via SP, according to an internal Pullman Company compilation.[5]

No doubt the most famous transcontinental car was the tourist Pullman of the Washington-Sunset Route, which operated daily from the nation's capital all the way to San Francisco, by way of the Southern Railway, West Point Route, Louisville & Nashville, and Southern Pacific. West of New Orleans, the sleeper sometimes rode the SP's *Sunset Limited.* Passengers leaving Washington at 4:35 p.m. on a Sunday, according to sample 1917 schedules, would arrive in San Francisco at 1 p.m. Friday. A berth cost nine dollars.

The *Sunset Limited* carried another coast-to-coast Pullman for about four years during and after the great Florida land boom of the Roaring Twenties. The service, a Jacksonville-Los Angeles sleeper, started in late 1924, with the introduction of the Seaboard Air Line/L&N *New Orleans Florida Limited.* The SAL proudly, though inaccurately, billed the Pullman as the "only transcontinental sleeping car service between the Atlantic and Pacific."[6] Soon after this line was terminated, a new development completely changed the rules of transcontinental travel.

On the evening of July 7, 1929, an unusual partnership began, a brief and colorful association that paired "the Standard Railroad of the World" with the airline that would one day become Howard Hughes's own— but that's getting ahead of the story.

In the summer of 1929, with the nation's business activity running at full steam and the superheated stock market several notches on the throttle beyond that, there was the numbing realization that coast-to-coast travel by rail still took a minimum of 88 hours,[7] a lackluster standard in an age that had thrilled to the achievements of Col. Charles A. Lindbergh and other aviation pioneers. This was all the more intolerable inasmuch as Jarrett & Palmer had done better a half century earlier.

Although World War I had established the usefulness of the airplane and supplied the labor force with a cadre of trained aviators who, when they weren't cracking up and killing themselves, were flying the mails and barnstorming during the Roaring Twenties, the air transport industry was just getting off the ground when the decade ended. By that time, however, there were several aircraft available, notably the Ford and Fokker trimotors, that could fly a dozen passengers in moderate comfort and safety. And so with these brave new birds, scheduled air service began, but it was confined to daylight hours until the necessary safeguards for nighttime passenger travel could be developed.

Several people in both the railroad and the airline camps, meanwhile, saw that each had needs that the other could help fulfill. The railroads, already losing passengers to the highways, had to put on some speed, and any airline with transcontinental ambitions had to do something with its potential passengers at night. Clearly, the airplane could help the railroads, and the old-hat, though ever-reliable Pullman sleeper could help the fledgling airline industry. A marriage of con-

Pennsylvania Railroad/courtesy Conrail

This detail from a Grif Teller mural depicts the Pennsylvania Railroad's *Airway Limited* speeding passengers to Port Columbus, Ohio, where they boarded waiting trimotor aircraft in the nation's most elaborate air/rail operation.

venience was arranged.

Northwest Airways pioneered joint air-rail service in the summer of 1928 between Chicago and St. Paul-/Minneapolis, offering a daytime aerial bridge between the trains of the eastern lines—the New York Central, Pennsylvania, and Baltimore & Ohio—terminating in Chicago and those linking the Twin Cities with the Pacific Coast: Northern Pacific, Great Northern, and Milwaukee Road. The eastbound morning flight and westbound afternoon flight were scheduled so as to save an entire business day for travelers going between cities in the East and the Northwest.

Giving details of a far more ambitious joint air-rail plan the following December, General W. W. Atterbury, president of the Pennsylvania Railroad, pointed out that the Northwest line had been planned by none other than Colonel Lindbergh, who would have the same responsibilities in bringing a new coast-to-coast service, of which the mighty Pennsy was a part, into operation.

Transcontinental Air Transport was organized for this purpose, and the Pennsylvania bought a 20 per cent interest in the company, which was promoted as "the Lindbergh Line." Passengers would leave Penn Station in New York on the new *Airway Limited* at 6:05 in the evening, and after a special ceremonial meal in the diner, they could go to sleep in their Pullman accommodations as the train speeded overnight to Columbus, Ohio.

Next morning at Port Columbus, the airport, they transferred from train to Ford Trimotor and flew all through the day—except for landings at Indianapolis, St. Louis, Kansas City, and Wichita—to Waynoka, Okla. There they spent four hours or so, in what for some metropolitans must have been rather appalling-

ly rural surroundings, waiting for the Santa Fe's westbound *Missionary*. Again, they slept the night away in Pullman comfort, transferring to another TAT "tin goose" the next morning in Clovis, N.M. By 5 o'clock, they were in Los Angeles. The joint transcontinental service was scheduled to shave travel time by a day and half, to about 50 hours.

Colonel Lindbergh, in California, inaugurated the service on that July night by pressing a button that gave the *Airway Limited* the highball from Pennsylvania Station, while a band played *California, Here I Come*. At Port Columbus the next morning, five thousand spectators were standing in the rain to watch the transfer of passengers (about a thousand requests for bookings on the first trip had been recorded). Among the lucky few making the inaugural flight was Amelia Earhart, who was briefly employed as a TAT executive. The Pacific Coast ceremonies were no less impressive: After Gloria Swanson and Mary Pickford christened his trimotor, Colonel Lindbergh himself flew the first leg eastbound.[8]

The initial service proved to be successful, despite the sting of seeing competitors preempt their way into the record book. Because of delays in the Pennsy/TAT operation, the rival New York Central and Universal Air Lines got the jump on the Lindbergh people and started their own joint service several weeks earlier. They made their train-plane transfer at Cleveland and put passengers on Santa Fe rails at Garden City, Kans. This routing, however, was much slower and involved three nights of Pullman travel.

Two air crashes, one involving TAT (which stood for "take a train," the joke went), put a chill on marketing efforts, but 30,000 passengers, nonetheless, made the trip during the first year of operation.[9]

Management held out lower fares to help overcome the public's understandable fear of flying in the primitive airliners, but this did not help earnings. The operation lost $2.75 million in its first 18 months.[10]

But even if the partnership had been profitable, it was not destined to last. The Depression was putting a crimp into travel of all kinds, and the airlines were benefiting from rapidly improving technology. By late 1932, night flying had made possible coast-to-coast travel in just 24 hours, and it was clear that the airlines did not need the railroads anymore. The Pennsy continued the New York-Port Columbus service a while longer, but eventually quit and finally disposed of its stock in TAT, in 1936.[11]

"I recognized that aviation was certainly destined to become a most valuable ally of commercial railroading if it were properly handled," General Atterbury had said in announcing the joint air/rail service, no doubt totally unaware of the irony in his remarks. In any case, so much for the handling: The bird was out of the cage—or was it the genie out of the bottle?—and within two decades, in 1949, the air transport industry would be carrying more passengers than Pullman. TAT, the little airline of the tin trimotors, grew up to become Trans World Airlines.

Despite all the rhetoric in Robert Young's "hog" ads promoting coast-to-coast Pullman travel, proposals for joint operations were discussed before World War II. The war naturally delayed the service, which eventually began on March 31, 1946 and provided through sleepers from New York and Washington to San Francisco, Los Angeles, and, later, San Diego. A line to Seattle, involving the New York Central's *Iroquois* and *Fifth Avenue* in the East and *Empire Builder* in the northwest, was proposed but never started.[12]

Chicago, the nation's rail hub, had been a major impediment to through traffic from the very beginning. All its six major stations were essentially stub terminals, and interline switching of cars was in some cases rather complicated. And so, with the few prewar exceptions mentioned, through service historically was not offered. As a result, travelers customarily had the better part of a day between a morning arrival and late afternoon departure. If you found yourself between trains in Chicago, you simply engaged in the sort of diversions that were determined by who you were.

If you were Lucius Beebe, you probably had an elaborate lunch at Ernie Byfield's Pump Room. Passengers of moderate means often spent the time shopping on State Street. For a minor sum, the budget-class traveler could see the city by riding the red-and-cream-colored streetcars of the world's largest trolley system. In any case, this forced layover ended in the first postwar spring. (Technically, the layover was still

there, but if you chose to do so, you could loaf on your Pullman couch while your car was being switched and never be exposed to the Windy City's forceful breath.)

New York and Los Angeles had the most service and, with six different routes, the greatest variety. East of Chicago, the New York Central and the Pennsylvania Railroad carried the cars. West of Chicago the Pullmans traveled to Los Angeles on either the Santa Fe, the Chicago & North Western/Union Pacific Overland Route, or the Rock Island/Southern Pacific Golden State Route. San Francisco sleepers traveled via either the C&NW/UP/SP Overland Route or the *California Zephyr* route of the Burlington/Denver & Rio Grande Western/Western Pacific. On some of the lines, cars moved on alternate days.

The Central and Pennsy paired their prime paladins—the *Twentieth Century Limited* and the *Broadway Limited*—with the Santa Fe's crack *Chief*, each of the eastern trains often turning over two Pullmans each.[13] Although the all-sleeper, extra-fare *Chief* had lost its premier status to the *Super Chief*, it was no second-rate train. It merely provided more manageable connections in Chicago, allowing for a four- or five-hour layover.

Beyond these first-string pairings, the initial assignments in 1946 went to the Central's westbound *Iroquois* and eastbound *Fifth Avenue* and the Pennsy's westbound *Golden Arrow* and eastbound *Manhattan Limited*. Their western partners were both the *Los Angeles Limited* and *Overland*.

By the early 1950's, the historic *Los Angeles Limited*, which still carried the Union Pacific's train numbers 1 and 2 after losing top billing to the *City of Los Angeles*, generally took on a Pullman each from the Central's *Advance Commodore Vanderbilt* and the *Pennsylvania Limited*, still No. 1 and 2 on the PRR card. And until the service disappeared in the summer of 1951, the *Golden State* (RI/SP) received sleepers from both the eastern roads on alternate days.

San Francisco had service on the *California Zephyr* one day and the *San Francisco Overland* the next (the *Exposition Flyer* handled the cars before the *CZ* was born in 1949). The eastern connections were the *Lake Shore Limited* (NYC) and *Pennsylvania Limited/Admiral* (PRR) (the Central's *Commodore Vanderbilt* and PRR's *General* were the original carriers). Travelers on the Pennsy/Overland Route retraced the movement of the *Jarrett & Palmer Special* three quarters of a century earlier.

Both the Pennsylvania and Baltimore & Ohio provided coast-to-coast Pullman service from Washington, but most of it was short-lived. Sleepers from the Pennsylvania went via the Overland Route to both San Francisco and Los Angeles. The B&O's *Capitol Limited* carried San Francisco sleepers for an Overland connection and Los Angeles Pullmans for the Santa Fe's *Chief.* The latter run, extended to San Diego, was the sole survivor of Washington's service by the early 1950's, although in the summer of 1956, the Washington-Sunset service was enjoying a brief revival. This time it was a regular Pullman, not a tourist car, making the trip, and it rode the Southern Railway's *Crescent* and SP's *Sunset Limited.* It was not long, however, before the disenchantment of the SP and other railroads with their passenger service deepened considerably and all the transcontinental service was finally put to rest.

The Pennsylvania ended its participation in late 1957, and the following April the Central quit, ending all transcontinental service (which at the close involved both the *Super Chief* and *City of Los Angeles*). The year 1958 was an ominous one on the Central in more ways than one. In January, the railroad announced it was severing its relationship with the Pullman company (which roughly had the effect of severing Pullman's femoral artery—the Central accounted for 15 per cent of the company's business) and thereafter would operate its own sleepers. Mid-year the railroad downgraded the *Twentieth Century Limited* by putting coaches aboard, and in December it dealt another company a death blow. Although the Central was a major owner of the Railway Express Agency, it announced it was terminating its contract with the company. It was not a good year for either the New York Central or the passenger train.

The transcontinental journey was more than ever in vogue, however, and now it was entering its third great age, one in which the railroads would play only a bit part. By this time there were airborne conveyances, requiring crews of fewer than ten persons, that could carry across the country in hours the same number of passengers that took an entire train, with its staffing of several hundred crew members, days to move.

The traffic flow never abated, not even pausing for a moment to mourn the passing of a wonderful tradition. It just kept growing to levels that General Atterbury, sitting in his somber offices in Philadelphia's old Broad Street Station, would have found incomprehensible. But the vapor trails in the sky spoke volumes, and the conclusion was incontrovertible: The third great age of transcontinental travel belonged to the jet airliner.

Symbolically standing in for all conductors for all time, NKP's George Zeigler might well be saluting the end of America's great age of railroad passenger transportation.

1. "The Jarrett & Palmer Special," *Trains,* Vol. 3, No. 2 (December 1942). p. 10. The article credits the Pennsylvania Railroad's *Mutual Magazine* for its information.
2. "Coast to Coast Service, Years 1887 to 1899 Inclusive," Pullman Company internal report, Nov. 28, 1945. From the collection of Arthur D. Dubin.
3. Details are sketchy east of Chicago and lacking on the western end of the operation. See the New England chapter, endnote No. 9.
4. Charles E. Fisher, "Through Car Service From New England," *RLHS Bulletin,* No. 87 (October 1952), p. 48.
5. Dubin collection.
6. The Washington-Sunset service was in its heyday at this point, although it was a tourist sleeper, not a regular Pullman like the Jacksonville-Los Angeles car. And purists may insist that Washington is not truly "East Coast." The SAL, however, erred in calling their new service "the first and only coast-to-coast sleeping car," considering the Pullmans that went from New York and Boston to California in the early 1890's.
7. Carl W. Condit, *The Port of New York, Volume II* (Chicago: University of Chicago Press, 1980), p. 157.
8. Robert Serling, *Howard Hughes' Airline* (New York: St. Martin's/Marek, 1983), pp. 89.
9. Condit, p. 157.
10. E. Paul Kutta, editor's note: "The Airway Limited." *NRHS Bulletin.* XL, No. 1 (1975), p. 20.
11. Kutta, p. 20.
12. An internal Pullman Company circular (194621), written three days before the start of the service, listed the Seattle line, but a later draft written just hours later omitted it. An appended note explained that the service was "temporarily" out. (Arthur D. Dubin collection.)
13. A typical *Twentieth Century Limited* consist: (4c-4dbr-2dr) and (10r-6dbr). For the *Broadway:* two (4c-4dbr-2dr).

Donald Sims

Key to "Midnight Sleepers" listings

Train name and number ————————————————→ **Kansas City-Florida Special, No. 7**
Train origin and destination ————————————————→ Kansas City to Jacksonville
Routing or scheduling information ————————————→ Via Jesup
Operating railroad at midnight ————————————→ Southern Railway
Location of train at midnight ————————————————→ 35 miles northwest of Macon

PULLMAN ORIGIN and DESTINATION ————————→ *KANSAS CITY to JACKSONVILLE*
    Car number/s and accommodations ——————————→   *F67: 8s-5dbr*   *F69: 10s-c-2dbr*
Routing of Pullman cars: Railroad, train number, ori-   SLSF 105, Kansas City (11:30p Sat)
gin, (departure time), destination, (arrival time).   to Birmingham (4:25p)
                                                         SOU 8, Birmingham (4:40p Sun) to Atlanta (10:25p)
[When Pullmans travel more than one night, day   SOU 7, Atlanta (10:45p) to Jacksonville (7:30a Mon)
names are used for example. The trains are assumed
to be within the state under which they are listed at   *ATLANTA to JACKSONVILLE*
midnight on a Sunday.]   *63: 8s-c-diner-lounge*
                                                         SOU 7, Atlanta (10:45p) to Jacksonville (7:30a)
(Further information about the above consist or the
train as it existed in 1951-2 is given in italics, within *ATLANTA to BRUNSWICK*
parentheses.)   *51: 10s-dr-2c*
                                                         SOU 7, Atlanta (10:45p) to Jesup (5:10a)
Information of historical interest on the train or Pull- SOU motor train, Jesup (5:40a) to Brunswick (7:10a)
man lines is presented in this manner. Railroads in
parentheses, such as the Rock Island (RI), also con-   *(Dining cars: Kansas City to Birmingham, Birmingham*
tributed to the Pullman movements.   *to Atlanta; lounge car Kansas City to Birmingham. An-*
                                                         other Kansas City-Florida Special, 24 hours behind this
                                                         one in Georgia, is now in Kansas.)

Abbreviations of accommodations:

section (s)	roomette (r)
bedroom (br)	duplex roomette (dup/r)
single bedroom (sbr)	Standard (Std)
double bedroom (dbr)	master bedroom (mbr)
compartment (c)	observation (obs)
drawing room (dr)	

The Frisco's connection with Southern Railway at Birmingham after the turn of the century opened a Florida route that brought sleepers to the Sunshine State from Oklahoma City, Wichita, Colorado Springs, and Denver. Atlanta had dedicated service from Kansas City and Denver (RI or MP/D&RGW), and Brunswick sleepers from Colorado Springs (RI) and Birmingham were scheduled.

51
Midnight Sleepers

Midnight Sleepers is an appendix that examines the Pullman operation as a total system, showing the individual parts and how they all fit together. For purpose of example, it "stops" all Pullman-carrying trains on a hypothetical midnight in March 1952 and describes their location, consists, routings, and schedules to show where the cars had come from and where they were going. Setouts are similarly described. *Midnight Sleepers* is organized alphabetically by states.

In 1952, America's first-class rail system was in its last great days. Much of the post-war streamliner equipment was in use and more dome cars were on the way. Schedules were fast, all-Pullman trains were still reasonably abundant, transcontinental service had settled into a well-run operation, and the best of the trains were truly superlative. In 1951, the airlines had succeeded in out-scoring the railroads in passenger miles for the first time, and the passenger train ecology would soon become increasingly fragile—five years later the story would be completely different. But for the moment, the railroads were still competing for business, offering Pullman service over nearly 100,000 route miles, about a quarter less than had existed during the golden years of the mid-1920's.

The "night" portrayed in *Midnight Sleepers* is actually a composite, rather than an actual one, so as to permit inclusion of Pullman lines that did not operate daily or year-round. Most lines operated at least six nights a week (some rested on Saturdays, when business traffic shriveled); those with less frequent scheduling are so designated.

Trains are assumed to be within the state under which they are listed at midnight on a typical Sunday, even those that did not actually operate on a Sunday. Using one day as a base time permits showing all trains in their geographic relationship to one another. Schedules of long-distance trains that were under way more than one night show other days of the week (Sat., Sun., Mon., etc.) to help clarify the times. The days given, therefore, are used only for example and should not be read as actual schedule information.

All schedule and consist information comes from the March 1952 issue of the *Official Guide of the Railways,* except for summer-only trains, for which I used the July 1951 issue. In certain cases winter and summer schedules differ, but the March 1952 information should be considered bench mark data, and all summer schedules, whether for added trains or added cars, are specifically marked. I interpolated public timetable figures to estimate each train's location at midnight.

There's no doubt that Pullman service had undergone a great pruning by 1951-2, even though the major, high-volume routes were still intact and competitive. There were still four ways from Chicago to the Twin Cities, Omaha and Seattle, five to New York, and so forth. Smaller cities and towns, however, had lost much of their dedicated service: Of fifteen lines that had

A flagman protects the eastbound *Imperial* on a stop at Colton, Calif., in the mid-1950's. The SP/RI train served its namesake Imperial Valley en route from Los Angeles to Chicago.

served Peoria over the years, only one remained. Now, even though most of the historic route structure was still intact, many more passengers in less-populous communities had to make do with through service, as opposed to dedicated cars that loaded at the passengers' convenience. Riding a through sleeper, for instance, might mean boarding or leaving a train in the middle of the night. So, despite the fact that a Pullman car still rolled down the same tracks, the service, for many passengers, was losing much of its convenience. In a few years, most of the business travelers would attest to this and defect to the airlines and highways.

But even though the time was getting short, 1952 was still a wonderful year for Pullman travel. Picture a land shedding its bleak winter garb, under a midnight sky, and in the distance a white shaft of light pierces the darkness. Chances are good there will be a whiff of coal smoke in the air, the mournful chime of a steam whistle, and the darkened passage of one or more heavyweight sleepers in their characteristic Pullman green. Or maybe the countryside is alerted to the clarion call of an exuberantly painted brace of EMD or Alco diesels, pulling a string of stainless steel cars with window shades pulled low.

In these last great days, in liveries old and new, there were still more than 750 night trains carrying Pullmans across America, doing the work that only this remarkable, bygone system could do.

Alabama

See map on page 146

South Wind, No. 6-11
Streamliner
Miami to Chicago
Two days out of three
Atlantic Coast Line
15 miles northwest of Dothan

MIAMI to CHICAGO
SW41: 10r-6dbr   SW42: 6dbr-lounge
SW43: 4c-2dr-4dbr   SW44: 13dbr
SW45: 6s-4r-4dbr
FEC 6, Miami (12:01p) to Jacksonville (7p)
ACL 6-11, Jacksonville (7:20p)
          to Montgomery (2:15a)
L&N 16, Montgomery (2:25a) to Louisville (11:35a)
PRR 303, Louisville (11:40a) to Chicago (5:45p)

JACKSONVILLE to CHICAGO
SW40: 10r-6dbr
As above, Jacksonville (7:20p) to Chicago (5:45p)

(Dining cars and observation-lounge. Off-season, the train reverts to a one-in-three-day schedule.)

The tuscan red *South Wind* was one of the three streamlined, coach trains introduced in 1940 to offer faster and better service to Florida from Chicago. Nine railroads participated in the package, which also resulted in creation of the *City of Miami* and the *Dixie Flagler*. Eventually all three trains added Pullman cars, and the *City* and *South Wind* increased their frequency of operation.

No. 12
Birmingham to Atlanta
Southern Railway
20 miles west of Anniston

BIRMINGHAM to NEW YORK
S45: 10s-1dr-2c
SOU 12, Birmingham (10:30p Sun)
          to Atlanta (6a Mon)
SOU 34, Atlanta (9a) to Washington (1:25a Tue)
PRR 108, Washington (2:10a)
          to New York (6:25a Tue)

(No. 34 is the Piedmont Limited, carrying a diner to Monroe, Va., and lounge to Washington. This is one of four Birmingham-New York Pullman lines still operating in 1952.)

The Southern scheduled sleepers from Birmingham to Brunswick, Ga., Jacksonville, Charlotte, Richmond, and Washington.

South Wind, No. 15
Streamliner
Chicago to Miami
Two out of three days
Louisville & Nashville
16 miles north of Montgomery

CHICAGO to MIAMI
SW41: 10r-6dr   SW42: 6dbr-lounge
SW43: 4dbr-4c-2dr   SW44: 13dbr
SW45: 6s-4r-4dbr
PRR 304, Chicago (9a) to Louisville (2:50p)
L&N 15, Louisville (3p) to Montgomery (12:20a)
ACL 12-5, Montgomery (12:30a)
          to Jacksonville (9:25a)
FEC 5, Jacksonville (9:45a) to Miami (4:45p)

CHICAGO to JACKSONVILLE
SW40: 10r-6dbr
As above, Chicago (9a) to Jacksonville (9:25a)

(Observation-lounge and dining cars. The train reverts to one-in-three-day operation off-season)

Piedmont Limited, No. 33
New York to New Orleans
Louisville & Nashville
38 miles south of Montgomery

NEW YORK to MOBILE
SR10: 10s-2c-dr
PRR 141, New York (10:15p Sat)
          to Washington (2:10a)
SOU 33, Washington (2:40a Sun)
          to Atlanta (6:30p)
WPR 33, Atlanta (7:15p) to Montgomery (10:40p)
L&N 33, Montgomery (11:10p)
          to Mobile (4:10a Mon)

NEW YORK to NEW ORLEANS
SR11: 8s-5dbr   SR12: 10s-2c-dr
As above, New York (10:15p Sat)
          to Montgomery 10:40p Sun)
L&N 33, Montgomery (11:10p)
          to New Orleans (8:05a Mon)

WASHINGTON to NEW ORLEANS
S56: 10s-lounge
As above, Washington (2:40a Sun)
          to New Orleans (8:05a Mon)

ATLANTA to NEW ORLEANS
A5: 12s-1dr
As above, Atlanta (7:15p) to New Orleans (8:05a)

(Diners: Monroe, Va., to Atlanta, Atlanta to Montgomery, Montgomery to New Orleans)

The *Piedmont Limited*, which played second fiddle to the more glamorous *Crescent*, at one time served Mobile with sleepers to and from Montgomery and to New Orleans and also counted Boston-New Orleans Pullmans in its consist.

Piedmont Limited, No. 34
New Orleans to New York
Louisville & Nashville
48 miles north of Flomaton

NEW ORLEANS to ATLANTA
L39: 12s-1dr
L&N 34, New Orleans (5p) to Montgomery (2a)
WPR 34, Montgomery (2:35a) to Atlanta (8:20a)

NEW ORLEANS to WASHINGTON
L40: 10s-lounge
As above, New Orleans (5p Sun)
          to Atlanta (8:20a Mon)
SOU 34, Atlanta (9a) to Washington (1:25a Tue)

NEW ORLEANS to NEW YORK
L42: 10s-2c-dr   L43: 8s-5dbr
As above, New Orleans (5p Sun)
          to Washington (1:25a Tue)
PRR 108, Washington (2:10a)
          to New York (6:25a Tue)

MOBILE to NEW YORK
L44: 10s-2c-dr
L&N 34, Mobile (8:45p Sun) to Montgomery (2a Mon)
As above, Montgomery (2:35a)
          to New York (6:25a Tue)

(Dining cars: New Orleans to Mobile, Montgomery to Atlanta, Atlanta to Monroe, Va.)

Pullmans from Montgomery traveled the West Point Route to Atlanta, Washington (SOU), and New York (SOU/PRR).

Pelican, No. 41
Washington to New Orleans
Southern Railway
4 miles east of Livingston

NEW YORK to NEW ORLEANS
N80: 10r-6dbr
PRR 139, New York (6:50p Sat)
    to Washington (11:10p)
SOU 41, Washington (11:55p)
    to Lynchburg, Va. (4:12a Sun)
N&W 41, Lynchburg (4:20a) to Bristol, Va. (10a)
SOU 41, Bristol (10:10a)
    to New Orleans (7:30a Mon)

NEW YORK to SHREVEPORT
N81: 10s-2c-dr
As above, New York (6:50p Sat) to Bristol (10a Sun)
SOU 41, Bristol (10:10a)
    to Meridian, Miss. (1:15a Mon)
IC 205, Meridian (2:50a)
    to Shreveport (1:30p Mon)

WASHINGTON to NEW ORLEANS
S64: 10s-lounge
As above, Washington (11:55p Sat)
    to New Orleans (7:30a Mon)

ATLANTA to JACKSON, MISS.
S17: 12s-dr
SOU 29, Atlanta (4:45p) to Birmingham (8:30p)
SOU 41, Birmingham (8:50p) to Meridian (1:15a)
IC 205, Meridian (2:50a) to Jackson (5:40a)

(Dining cars: New York to Washington, Roanoke to Birmingham. IC 205, the Southwestern Limited, carries a cafe-lounge, Jackson to Shreveport.)

The Southern scheduled sleepers from Birmingham to New Orleans, Mobile (M&O), Jackson (IC), and Shreveport (IC); in the opposite direction to Cincinnati and Asheville.

Tennessean, No. 46
Streamliner
Memphis to Washington
Southern Railway
Approaching Decatur

MEMPHIS to NEW YORK
S73: 14r-4dbr
SOU 46, Memphis (7:40p Sun)
    to Bristol (10:40a Mon)
N&W 46, Bristol (10:50a) to Lynchburg (3:50p)
SOU 46, Lynchburg (3:55p) to Washington (8:10p)
PRR 108, Washington (2:10a Tue)
    to New York (6:25a Tue)

MEMPHIS to WASHINGTON
S66: 14r-4dbr
As above, Memphis (7:40p) to Washington (8:10p)

MEMPHIS to KNOXVILLE
S20: 10s-2dbr-c
SOU 46, Memphis (7:40p) to Knoxville (7:10a)

MEMPHIS to CHATTANOOGA
62: 8s-c-diner-lounge
SOU 46, Memphis (7:40p) to Chattanooga (4:10a)

(Memphis to Washington, lounge; Knoxville to Washington, diner)

The streamlined *Tennessean* made its debut in May 1941. Historically, Memphians rode Pullmans to Washington over five different routes, with gateways ranging from Cincinnati to Birmingham. This historic line of the Southern Railway System operates via Chattanooga.

City of Miami, No. 52
Streamliner
Chicago to Miami
Two out of three days
Central of Georgia
28 miles northwest of Opelika

CHICAGO to JACKSONVILLE
    CM30: 6s-6r-4dbr
IC 53, Chicago (8:10a) to Birmingham (9:35p)
CofG 52, Birmingham (9:50p)
    to Albany, Ga. (4:25a)
ACL 15, Albany (4:35a) to Jacksonville (8:15a)

CHICAGO to MIAMI
    CM31: 6s-6r-4dbr   CM32: 10r-6dbr
    CM33: 10r-6dbr   CM35: 10r-6dbr
    CM34: 3dbr-dr-c-lounge
As above, Chicago (8:10a) to Jacksonville (8:15a)
FEC 3, Jacksonville (8:40a) to Miami (3:40p)

ST. LOUIS to MIAMI
    CM36: 10r-6dbr
IC 201, St. Louis (9:22a) to Carbondale (12:15p)
IC 53, Carbondale (1:02p) to Birmingham (9:35p)
As above, Birmingham (9:50p)
    to Miami (3:40p)

(Operates every third day off-season. Diners and observation-lounge, Chicago to Miami.)

The *City of Miami* was inaugurated in December 1940, running every third day in rotation with the *South Wind* and *Dixie Flagler*. A second *City* train was added in 1951-2, increasing operation to two out of three days. During winters of the late 1960s, the Illinois Central leased domed sleepers from the Northern Pacific for the train.

During the heavyweight era, the IC's premier, seasonal train to Florida was the *Floridan*, with Pullmans to Palm Beach, Miami, Sarasota, and St. Petersburg.

Birmingham had sleeping car service to both Savannah and Jacksonville (ACL) via the Central of Georgia.

Pan-American, No. 98
New Orleans to Cincinnati
Louisville & Nashville
34 miles west of Mobile

NEW ORLEANS to CINCINNATI
78: 10s-lounge
L&N 98, New Orleans (9p) to Cincinnati (9:15p)

NEW ORLEANS to PITTSBURGH
L62: 10s-2dbr-dr
L&N 98, New Orleans (9p Sun)
    to Cincinnati (9:15p Mon)
PRR 202, Cincinnati (11:20p)
    to Pittsburgh (7a Tue)

(Montgomery to Cincinnati, dining car)

Inaugurated in December 1921, the *Pan-American* became the flagship of the Louisville & Nashville Railroad. The train was upgraded to deluxe, all-Pullman status in 1925, though during the Depression coaches were once again added. About that time, radio station WSM in Nashville began broadcasting the sounds of the *Pan-American* passing the station's tower south of the city, a practice that compounded its already substantial renown over the next twelve years. The *Pan-American* ceased operation, just short of a half-century of service, at the beginning of Amtrak.

Pan-American, No. 98
Louisville & Nashville
Mobile setout
Standing in Mobile

MOBILE to BIRMINGHAM
72: 12s-dr
L&N 98, Mobile (12:55a) to Birmingham (7:55a)

The *Pan-American* carried a Pullman from Mobile to Cincinnati; the Alabama seaport also had its own sleeper to New Orleans.

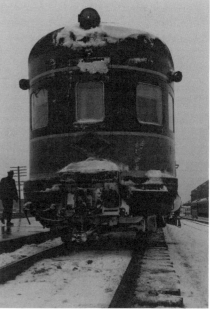

A *City of Miami* pauses on its way south at Champaign, Ill., in December 1960.

Pan-American, No. 99
Cincinnati to New Orleans
Louisville & Nashville
59 miles south of Montgomery

CINCINNATI to NEW ORLEANS
77: 10s-lounge
L&N 99, Cincinnati (9a) to New Orleans (7:10a)

PITTSBURGH to NEW ORLEANS
205: 10s-2dbr-dr
PRR 203, Pittsburgh (10:40p Sat)
    to Cincinnati (6:45a Sun)
L&N 99, Cincinnati (9a)
    to New Orleans (7:10a Mon)

BIRMINGHAM to MOBILE
71: 12s-dr
L&N 99, Birmingham (8:20p) to Mobile (3:25a)

(Dining cars: Pittsburgh to Cincinnati, Cincinnati to Birmingham)

Birmingham had a dedicated sleeper on the *Pan-American* to New Orleans and, elsewhere on the L&N, to Nashville. Montgomery had Pullman service to Pensacola and Mobile, and Nashville, Louisville, and Cincinnati.

Sunnyland, No. 108
Birmingham to Memphis
Frisco
48 miles west of Jasper

ATLANTA to MEMPHIS
37: 10s-2c-dr
SOU 29, Atlanta (4:45p) to Birmingham (8:30p)
SLSF 108, Birmingham (9p) to Memphis (6:30a)

BIRMINGHAM to MEMPHIS
S17: 12s-dr
SLSF 108 Birmingham (9p) to Memphis (6:30a)

(Atlanta to Birmingham, diner)

In addition to serving the Florida vacation trade, the *Sunnyland* carried some wide-ranging sleepers: New Orleans-Denver and Memphis to Portsmouth, Va., and Pensacola. This Frisco line provided Birmingham with dedicated Pullman service to Springfield, St. Louis, Kansas City, Oklahoma City, and Denver (RI).

J. Parker Lamb

Arizona

See map on page 261

Southern Pacific trains had two sets of alternate routes in Arizona, one of which extended the choice all the way to El Paso. The El Paso & Southwestern, which ran from Tucson to El Paso, via Douglas, then on to a connection with the Rock Island in New Mexico, became the middle link in the Golden State Route in 1902, providing, in the process, an alternative to the historic SP line through Bowie. The EP&SW eventually came into the SP system, and Golden State and Sunset Route trains used both lines between Tucson and El Paso. As of early 1952, all the major trains—*Sunset Limited, Golden State,* and *Imperial*—were routed via Douglas, except the eastbound *Argonaut,* which operated via Bowie.

As for the other Arizona alternative, the original SP line went through Maricopa, bypassing Phoenix. After the captial city began its rapid, modern growth, a new line through Phoenix was constructed in the late 1920s, bringing most of the major Sunset and Golden State trains to town. At this scheduling, the *Argonaut,* both east- and westbound, was the only cross-country train serving Maricopa.

Although three Southern Pacific trains are plying their assigned routes through the Arizona desert on this night (two *Golden States* and a westbound *Argonaut*), far more action is to be seen on a 125-mile stretch of the Santa Fe main line to the north. Six West Coast trains, including all four *Grand Canyons,* the eastbound *Chief,* and westbound *Super Chief,* are working that segment, from Williams west.

Golden State, No. 3

Chicago to Los Angeles
Via Douglas, Phoenix
Southern Pacific
Leaving Phoenix

CHICAGO to LOS ANGELES
　35: 10r-6dbr　　　36: 4dbr-4c-2dr
　Obs.-lounge-dr-2dbr, or　30: 6dbr-10r
RI 3, Chicago (1:20p Sat)
　　　to Tucumcari, N.M. (10:16a CST Sun)
SP 3, Tucumcari (9:30a MST)
　　　to Los Angeles (7:35a Mon)

MINNEAPOLIS/ST. PAUL to LOS ANGELES
　575: 6s-6r-4dbr
RI 507, Minneapolis (12:01p Sat) and
　　　St. Paul (12:25p) to Kansas City (9:15p)
RI 3, Kansas City (11p)
　　　to Tucumcari (10:16a CST Sun)
SP 3, Tucumcari (9:30a MST)
　　　to Los Angeles (7:35a Mon)

ST. LOUIS to LOS ANGELES
　115: 6s-6r-4dbr
MP 11, St. Louis (4p Sat) to Kansas City (9p)
RI 3, Kansas City (11p)
　　　to Tucumcari (10:16a CST Sun)
SP 3, Tucumcari (9:30a MST)
　　　to Los Angeles (7:35a Mon)

(The Golden State has a dining car, a coffee shop-lounge-diner, and, when not carrying the observation sleeper, another first-class lounge in its consist. It is an extra-fare train. MP No. 11 is the Colorado Eagle, which carries dining-lounge, grill-coach, and Planetarium coach. Rock Island No. 507 is the Twin Star Rocket, which carries a dining car and observation lounge.)

Golden State, No. 3

Southern Pacific
Phoenix setout
Standing in Phoenix

CHICAGO to PHOENIX　　　*Winter season only*
　34: 4dbr-2dr-4c
RI 3, Chicago (1:20p Sat)
　　　to Tucumcari (10:16a CST Sun)
SP 3, Tucumcari (9:30a MST)
　　　to Phoenix (11:45p Sun)
　Can be occupied until 8 a.m. (Mon)

Golden State, No. 4

Los Angeles to Chicago
Via Phoenix, Douglas
Southern Pacific
39 miles west of Tucson

LOS ANGELES to CHICAGO
　45: 104-6dbr　　　　46: 4dbr-4c-2dr
　Obs.-lounge-dr-2dbr, or　40: 6dbr-10r
SP 4, Los Angeles (1:30p Sun)
　　　to Tucumcari (1:43p MST Mon)
RI 4, Tucumcari (2:49p CST)
　　　to Chicago (11:45a Tue)

PHOENIX to CHICAGO　　　*Winter season only*
　44: 4dbr-2dr-4c
SP 4, Phoenix (10:35p Sun)
　　　to Tucumcari (1:43p MST Mon)
RI 4, Tucumcari (2:49 CST)
　　　to Chicago (11:45a Tue)

LOS ANGELES to ST. PAUL/MINNEAPOLIS
　4508: 6s-6r-4dbr
SP 4, Los Angeles (1:30p Sun)
　　　to Tucumcari (1:43p MST Mon)
RI 4, Tucumcari (2:49p CST)
　　　to Kansas City (1:25p Tue)
RI 508, Kansas City (9:10a) to St. Paul (6p)
　　　and Minneapolis (6:30p Tue)

LOS ANGELES to ST. LOUIS
　41: 6s-6r-4dbr
SP 4, Los Angeles (1:30p Sun)
　　　to Tucumcari (1:43p MST Mon)
RI 4, Tucumcari (2:49p CST)
　　　to Kansas City (1:25a Tue)
MP 20, Kansas City (1:45a) to St. Louis (8:05a Tue)

(A dining car and coffee shop-lounge-diner are in the consist of the extra-fare Golden State, as is a first-class lounge car when the observation sleeper is not included. Rock Island No. 508, the Twin Star Rocket, carries a dining car and observation lounge to Minneapolis. MoPac No. 20, the Sunflower, has a grill-coach on the St. Louis leg. East- and westbound editions of the Golden State are also now in Kansas.)

Golden State, No. 4

Southern Pacific
Tucson setout
Standing in Tucson

TUCSON to CHICAGO　　　*Winter season only*
　43: 4dbr-4c-2dr, or　　12dbr
SP 4, Tucson (12:55p Mon)
　　　to Tucumcari (1:43p MST)
RI 4, Tucumcari (2:49p CST)
　　　to Chicago (11:45a Tue)

Tucson formerly had its own sleeping cars to El Paso, Los Angeles, and Guaymas, Mexico. Other through sleepers to Mexico left the main line at Tucson for the branch to Nogales.

Argonaut, No. 5

New Orleans to Los Angeles
Via Douglas, Maricopa
Southern Pacific
3 miles east of Douglas

NEW ORLEANS to LOS ANGELES
　53: 8s-5dbr　　　54: Std: s-dr-c
SP 5, New Orleans (11a Sat)
　　　to Los Angeles (4p Mon)

SAN ANTONIO to LOS ANGELES
　55: Std: s-dr
SP 5, San Antonio (2:05a Sun)
　　　to Los Angeles (4p Mon)

(Dining car and first-class lounge, New Orleans to Los Angeles. Another westbound Argonaut is now in Texas.)

No. 14

Williams to Grand Canyon
Atchison, Topeka & Santa Fe
Standing in Williams

CHICAGO to LOS ANGELES
　8s-2c-2dbr
Santa Fe 23, Chicago (12:01p Sat)
　　　to Williams (10:30p Sun)
Santa Fe 14, Williams (4:15a Mon)
　　　to Grand Canyon (7a)
Santa Fe 15, Grand Canyon (8p)
　　　to Williams (10:20p)
Santa Fe 123, Williams (10:40p Mon)
　　　to Los Angeles (10:25a Tue)

(No. 14 and its counterpart, No. 15, shuttle sleepers between Williams and the Grand Canyon for the benefit of tourists who want to spend the day sightseeing in the spectacular national park. The Santa Fe's Grand Canyon trains—Nos. 23 and 24 are the Southern Section, Nos. 123 and 124 the Northern one—have the main line duty. No. 14 awaits the arrival of the Los Angeles-Chicago Pullman on No. 124.)

Donald Sims

Super Chief, No. 17
All-Pullman streamliner
Chicago to Los Angeles
Via Raton Pass
Atchison, Topeka & Santa Fe
Four miles east of Ash Fork

CHICAGO to LOS ANGELES
>   2dr-4c-4dbr   2dr-4c-4dbr
>   2dr-4c-4dbr   10r-6dbr
>   10r-6dbr   4dr-dbr-observation
>   Santa Fe 17, Chicago (7p Sat)
>   to Los Angeles (8:45a Mon)

KANSAS CITY to LOS ANGELES
>   10r-6dbr
>   Santa Fe 17, Kansas City (2:45a Sun)
>   to Los Angeles (8:45a Mon)

(The Super Chief *carries one of the Pleasure Dome lounges with the Turquoise Room for private dining parties as well as a full diner and additional lounge. Extra fare on the luxury train is $15.)*

Chief, No. 20
All-Pullman streamliner
Los Angeles to Chicago
Via Raton Pass
Atchison, Topeka & Santa Fe
10 miles west of Ash Fork

LOS ANGELES to CHICAGO
>   17r   2dr-4c-4dbr
>   Observation-4dr-1dbr
>   Santa Fe 20, Los Angeles (12:30p Sun)
>   to Chicago (10:30a Tue)

LOS ANGELES to NEW YORK
>   2dr-4c-4dbr   10r-6dbr
>   Santa Fe 20, Los Angeles (12:30p Sun)
>   to Chicago (10:30a Tue)
>   NYC 26, Chicago (4:30p) to New York (9:30a Wed)

LOS ANGELES to NEW YORK
>   2dr-4c-4dbr
>   Santa Fe 10, Los Angeles (12:30p Sun)
>   to Chicago (10:30a Tue)
>   PRR 28, Chicago (4:30p) to New York (9:30a Wed)

SAN DIEGO to WASHINGTON
>   10r-6dbr
>   Santa Fe 71, San Diego (7:45a Sun)
>   to Los Angeles (10:30a)
>   Santa Fe 20, Los Angeles (12:30p)
>   to Chicago (10:30a Tue)
>   B&O 6, Chicago 4:30p)
>   to Washington (8:55a Wed)

(Like the Chief, *the three eastern trains, NYC's Twentieth Century Limited,* Pennsy's *Broadway Limited, and the B&O's* Capitol Limited, *are all-Pullman and carry lounges and diners. The San Diego train has a parlor-lounge.)*

The westbound *Golden State* kicks up dust at Beaumont, Calif., in June 1950.

Chief, No. 20
Atchison, Topeka & Santa Fe
Phoenix setout
Standing in Ash Fork

PHOENIX to CHICAGO
>   2dr-4c-4dbr
>   Santa Fe 42, Phoenix (5p Sun)
>   to Ash Fork (11:15p)
>   Santa Fe 20, Ash Fork (12:30a Mon)
>   to Chicago (10:30a Tue)

(Phoenix to Ash Fork, cafe-parlor car)

Grand Canyon, Southern Section, No. 23
Chicago to Los Angeles
Via Amarillo, Clovis, N.M.
Atchison, Topeka & Santa Fe
9 miles east of Seligman

CHICAGO to LOS ANGELES
>   10r-3dbr-2c   14s
>   Santa Fe 23, Chicago (12:01p Sat)
>   to Los Angeles (10:40a Mon)

CHICAGO to SAN FRANCISCO
>   6r-6s-4dbr
>   Santa Fe 23, Chicago (12:01p Sat)
>   to Barstow (5:55a Mon)
>   Santa Fe 23, Barstow (6:20a) to Oakland (8p)
>   Bus to San Francisco (8:30p Mon)

DALLAS to LOS ANGELES
>   10r-5dbr
>   Santa Fe 112, Dallas (8:15p Sat) to Fort Worth (9p)
>   Santa Fe 77, Fort Worth (9:20p)
>   to Brownwood, Tex. (1:40a Sun)
>   Santa Fe 75, Brownwood (2:05a)
>   to Sweetwater, Tex. (4:45a)
>   Santa Fe 94, Sweetwater (5:05a)
>   to Texico, Tex. (ntg)
>   Santa Fe 97, Texico (11:05a)
>   to Clovis (11:30a CST)
>   Santa Fe 23, Clovis (11a MST)
>   to Los Angeles (10:40a Mon)

NEW ORLEANS to SAN FRANCISCO
>   6s-6r-4dbr
>   MP 3, New Orleans (8:35a Sat) to Houston (6:10p)
>   Santa Fe 66, Houston (6:45p) to Temple (10:40p)
>   Santa Fe 75, Temple (10:50p)
>   to Sweetwater (4:45a Sun)
>   As above, Sweetwater (5:05a)
>   to Clovis (11:30a CST)
>   Santa Fe 23, Clovis (11a MST)
>   to Barstow (5:55a Mon)
>   Santa Fe 23, Barstow (6:20a)
>   to Bakersfield (10:50a)
>   Santa Fe 61, Bakersfield (11:20a) to Oakland (5p)
>   Ferry to San Francisco (5:25p Mon)

(Dining and lounge service between New Orleans, Chicago and Los Angeles. No. 61, a Golden Gate, *carries a chair-lounge and lunch counter dining car. The Southern Section of the Santa Fe's* Grand Canyon *works the railroad's lower-grade, principally freight route through the Texas Panhandle.)*

Grand Canyon, Southern Section, No. 24
Grand Canyon sleeper
Atchison, Topeka & Santa Fe
Standing in Williams

LOS ANGELES to CHICAGO
>   8s-2c-2dbr
>   Santa Fe 124, Los Angeles (1:30p Sat)
>   to Williams (3:25a Sun)
>   Santa Fe 14, Williams (4:15a)
>   to Grand Canyon (7a)
>   Santa Fe 15, Grand Canyon (8p)
>   to Williams (10:20p)
>   Santa Fe 24, Williams (3:40a Mon)
>   to Chicago (3:45p Tue)

Grand Canyon, Southern Section, No. 24
Los Angeles to Chicago
Via Clovis, Amarillo
Atchison, Topeka & Santa Fe
75 miles west of Seligman

LOS ANGELES to CHICAGO
>   10r-3dbr-2c   14s
>   Santa Fe 24, Los Angeles (1:40p Sun)
>   to Chicago (3:45p Tue)

LOS ANGELES to DALLAS
>   10r-5dbr
>   Santa Fe 24, Los Angeles (1:40p Sun)
>   to Clovis (2:30p MST Mon)
>   Santa Fe 98, Clovis (4p CST)
>   to Farwell-Texico (ntg)
>   Santa Fe 95, Texico (4:15p) to Sweetwater (9:50p)
>   Santa Fe 76, Sweetwater (9:50p)
>   to Brownwood (12:45a Tue)
>   Santa Fe 78, Brownwood (1:10a))
>   to Fort Worth (6:25a)
>   Santa Fe 111, Fort Worth (7:15a) to Dallas (8a Tue)

SAN FRANCISCO to NEW ORLEANS
>   6s-6r-4dbr
>   Bus from San Francisco (7:45a Sun)
>   Santa Fe 60, Oakland (8:07a)
>   to Bakersfield (2:05p)
>   Santa Fe 24, Bakersfield (2:15p) to Barstow (6p)
>   Santa Fe 24, Barstow (6:25p)
>   to Clovis (2:30p MST Mon)
>   Santa Fe 98 Clovis (4p CST) to Texico (ntg)
>   Santa Fe 95, Texico (4:15p) to Sweetwater (9:50p)
>   Santa Fe 76, Sweetwater (9:50p)
>   to Temple (3:55a Tue)
>   Santa Fe 65, Temple (4:07a) to Houston (8:15a)
>   MP 58, Houston (9:05a)
>   to New Orleans (7:15p Tue)

SAN FRANCISCO to CHICAGO
>   6s-6r-4dbr
>   As above, San Francisco (7:45a Sun)
>   to Clovis (2:30p MST Mon)
>   Santa Fe 24, Clovis (3:40p CST)
>   to Chicago (3:45p Tue)

(Dining and lounge service Oakland to New Orleans, Los Angeles to Chicago. New Orleans passengers must travel on at least nine different conveyances, including seven Santa Fe trains to reach their destination; it's the most complex Pullman line in the country at this scheduling. Other east- and westbound editions of the Grand Canyon *are in Kansas.)*

No. 47
Ash Fork to Phoenix
Atchison, Topeka & Santa Fe
Standing in Ash Fork

DENVER to PHOENIX
>   8s-dr-2c
>   Santa Fe 101, Denver (2:10p Sat) to Pueblo (4:45p)
>   Santa Fe 2, Pueblo (4:50p) to La Junta (6:10p)
>   Santa Fe 3, La Junta (9:50p)
>   to Ash Fork (7:15p Sun)
>   Santa Fe 47, Ash Fork (2a Mon)
>   to Phoenix (8:30a Mon)

CHICAGO to PHOENIX
>   2dr-4c-4dbr
>   Santa Fe 19, Chicago (1:30p Sat)
>   to Ash Fork (10:05p Sun)
>   Santa Fe 47, Ash Fork (2a Mon)
>   to Phoenix (8:30a Mon)

(These two Pullmans await departure of No. 47, the Phoenix accommodation train, at 2 a.m. The Denver sleeper arrived in Ash Fork on the California Limited, and the Chicago car on much swifter schedule aboard the Chief. *Having no diner west of La Junta, the* California Limited *made meal stops at Fred Harvey dining rooms in Albuquerque, Gallup, N.M., and Williams. Dining and lounge cars carried Chicago to Ash Fork on the* Chief, *and Ash Fork to Phoenix.)*

The first section of No. 23, the Santa Fe's *Grand Canyon*, near Taiban, N.M., Oct. 2, 1947.

Arizona   *(continued)*

Grand Canyon, Northern Section, No. 123

Kansas City to Los Angeles
Via Topeka, Raton Pass
Atchison, Topeka & Santa Fe
12 miles west of Ash Fork

CHICAGO to LOS ANGELES
    24dup/r   8s-2c-2dbr
    8s-dr-2c
    Santa Fe 23, Chicago (12:01p Sat)
        to Kansas City (9p)
    Santa Fe 123, Kansas City (9:30p)
        to Los Angeles (10:25a Mon)

DENVER to LOS ANGELES
    6s-6r-4dbr
    Santa Fe 141, Denver (11:40p Sat)
        to Pueblo (2:15a Sat)
    Santa Fe 14, Pueblo (2:20a) to La Junta (5a)
    Santa Fe 123, La Junta (7:30a)
        to Los Angeles (10:25a Mon)

CHICAGO to LOS ANGELES
    8s-2c-2dbr
    Santa Fe 23, Chicago (12:01p Fri)
        to Williams (10:30p Sat)
    Santa Fe 14, Williams (4:15a Sun)
        to Grand Canyon (7a)
    Santa Fe 15, Grand Canyon (8p)
        to Williams (10:20p)
    Santa Fe 123, Williams (10:40p Sun)
        to Los Angeles (10:25a Mon)

(Dining and lounge cars on No. 23, the Grand Canyon, Southern Section, and on No. 123. At this scheduling, the Grand Canyon operates as a single unit, Nos. 23 and 24, between Chicago and Kansas City, where the Northern Sections, Nos. 123 and 124, are made up and terminate, respectively. West of Kansas City, two separate trains operate all the way to Los Angeles, the northern one via the Raton Pass, the other by way of Amarillo in the Texas Panhandle. In California, both trains enter Los Angeles via Fullerton, rather than Pasadena.)

Grand Canyon, Northern Section, No. 124

Los Angeles to Kansas City
Via Raton Pass, Topeka
Atchison, Topeka & Santa Fe
60 miles west of Seligman

LOS ANGELES to CHICAGO
    24dup/r   8s-2c-2dbr
    8s-dr-2c
    Santa Fe 124, Los Angeles (1:30p Sun)
        to Kansas City (6:30a Tue)
    Santa Fe 24, Kansas City (7a)
        to Chicago (3:45p Tue)

LOS ANGELES to CHICAGO
    8s-2c-2dbr
    Santa Fe 124, Los Angeles (1:30p Sun)
        to Williams (3:25a Mon)
    Santa Fe 14, Williams (4:15a)
        to Grand Canyon (7a)
    Santa Fe 15, Grand Canyon (8p)
        to Williams (10:20p)
    Santa Fe 24, Williams (3:40a Tue)
        to Chicago (3:45p Wed)

LOS ANGELES to DENVER
    6s-6r-4dbr
    Santa Fe 124, Los Angeles (1:30p Sun)
        to La Junta (6:05p Mon)
    Santa Fe 13, La Junta (6:40p) to Pueblo (7:45p)
    Santa Fe 130, Pueblo (7:55p)
        to Denver (10:40p Mon)

(Dining and lounge cars, Los Angeles to Kansas City and Chicago, on Nos. 124 and 24. Other eastbound and westbound editions of the Grand Canyon Northern Section are in Kansas.)

Grand Canyon, Northern Section, No. 124

Atchison, Topeka & Santa Fe
Phoenix setout
Standing in Ash Fork

PHOENIX to DENVER
    8s-dr-2c
    Santa Fe 42, Phoenix (5p Sun)
        to Ash Fork (11:15p)
    Santa Fe 124, Ash Fork (2:10a Mon)
        to Albuquerque (10:25a)
    Santa Fe 4, Albuquerque (8p) to La Junta (6a)
    Santa Fe 1, La Junta (8a) to Pueblo (9:45a)
    Santa Fe 102, Pueblo (9:55a) to Denver (1p Tue)

(This inordinately slow schedule did not necessarily require a passenger to spend the day in Albuquerque or a second night on the train: One could remain on No. 124, in parlor car or other first-class space, and by connecting train at La Junta be in Denver late the same evening. The main advantage of this two-night schedule was that it gave Albuquerque passengers convenient overnight service to Denver. Dining and lounge cars between Phoenix and Ash Fork and Albuquerque; cafe-observation car La Junta to Denver.)

Arkansas

See map on page 243

The two and a half hours starting at 11:15 p.m. were memorable with sight and sound for those who happened to be at Missouri Pacific trackside in Little Rock during the late winter of 1952. Within that time span, all six *Texas Eagles* and *Sunshines* arrived and departed after busy terminal switching activity.

The *Sunshine Special* was introduced in 1915, improving Texas's service to the east and St. Louis's to the west. Pullmans from St. Louis and Memphis bound for California were an important part of the *'Shine's* business, most of which it conducted while operating in three sections. By these schedules, the two, 1948-vintage, streamlined *Texas Eagles* have supplanted the West and South Texas sections of the *Sunshine Special*, but Nos. 31 and 32 survive in good health, providing St. Louis with a half dozen Pullmans to San Antonio and points in Louisiana and Arkansas.

Southern Belle, No. 1

Kansas City to New Orleans
Streamliner, via Coushatta, La.
Kansas City Southern
79 miles north of Texarkana

KANSAS CITY to NEW ORLEANS
    1: 14r-4dbr
    KCS 1, Kansas City (4p) to New Orleans (10:15a)

KANSAS CITY to PORT ARTHUR
    3: 14r-4dbr
    KCS 1, Kansas City (4p) to Shreveport (3a)
    KCS 101, Shreveport (3:30a) to Port Arthur (9:30a)

(Kansas City to New Orleans, diner and observation-lounge; Shreveport to Port Arthur, dining car)

The merger of the Louisiana & Arkansas and the Kansas City Southern in 1939 provided the impetus to create the new streamlined *Southern Belle*. She wore a striking livery of blackish-green accented with yellow and red striping, with roofs painted silver. Although the KCS main line went to the west of Fort Smith, the railroad did carry dedicated sleepers to the community from both Kansas City and Shreveport.

Texas Eagle, No. 1

Streamliner
St. Louis to El Paso
Missouri Pacific
20 miles northeast of Little Rock

ST. LOUIS to FORT WORTH
    19: 14r-dr-2dbr   17: 5br-lounge
    MP 1, St. Louis (5:30p Sun)
        to Texarkana (3:05a Mon)
    T&P 1, Texarkana (3:25a) to Dallas (7:30a)
        and Fort Worth (8:30a Mon)

NEW YORK to EL PASO
    MP2: 14r-4dbr
    PRR 3, New York (7:35p Sat) to St. Louis (3p Sun)
    MP 1, St. Louis (5:30p Sun)
        to Texarkana (3:05a Mon)
    T&P 1, Texarkana (3:25a) to El Paso (10:15p Mon)

WASHINGTON to FORT WORTH
    MP30: 14r-4dbr
    B&O 1, Washington (6:30p Sat)
        to St. Louis (1p Sun)
    As above, St. Louis (5:30p) to Dallas (7:30a)
        and Fort Worth (8:30a Mon)

(Dining car, St. Louis to Fort Worth; diner-lounge, Fort Worth to El Paso)

Texas Eagle, No. 1
Memphis Pullman
Missouri Pacific
Standing in Little Rock

MEMPHIS to FORT WORTH
   18: *14r-4dbr*
MP 201, Memphis (8:30p) to Little Rock (11:45p)
MP 1, Little Rock (12:40a) to Texarkana (3:05a)
T&P 1, Texarkana (3:25a) to Dallas (7:30a)
     and Fort Worth (8:30a)

Lone Star, No. 1
Memphis to Dallas
St. Louis Southwestern
5 miles southwest of Stuttgart

MEMPHIS to DALLAS
   17: *12s-dr*
StLSW 1, Memphis (9:10p) to Dallas (11:20a)

MEMPHIS to SHREVEPORT
   15: *10s-observation*
StLSW 1, Memphis (9:10p)
    to Lewisville, Ark. (4:25a)
StLSW 201, Lewisville (4:30a)
    to Shreveport (6:30a)

The *Lone Star Limited* was inaugurated in March 1916. Over the years it and the other Memphis trains handled sleepers to Fort Worth, San Antonio, Texarkana, and Waco, in addition to the above. The Cotton Belt, as the railroad was more popularly known, ended all passenger service in 1959.

Texas Eagle, No. 2
Streamliner
El Paso to St. Louis
Missouri Pacific
13 miles north of Gurdon

FORT WORTH to ST. LOUIS
   23: *14r-dr-2dbr*   29: *5br-lounge*
T&P 2, Fort Worth (5:15p) and
    Dallas (6:15p) to Texarkana (10:30p)
MP 2, Texarkana (10:40p) to St. Louis (8:10a)

FORT WORTH to WASHINGTON
   BO32: *14r-4dbr*
As above, Fort Worth (5:15p Sun)
    to St. Louis (8:10a Mon)
B&O 2, St. Louis (10:05a)
    to Washington (7:25a Tue)

EL PASO to NEW YORK
   PR9: *14r-4dbr*
T&P 2, El Paso (12:50a Sun) to Texarkana (10:30p)
MP 2, Texarkana (10:40p) to St. Louis (8:10a Mon)
PRR 4, St. Louis (10a) to New York (7:25a Tue)

FORT WORTH to MEMPHIS
   266: *14r-4dbr*
T&P 2, Fort Worth (5:15p Sun)
    to Texarkana (10:30p)
MP 2, Texarkana (10:40p) to Little Rock (1:10a)
MP 202, Little Rock (2:30a) to Memphis (6:30a)

(El Paso to Fort Worth, dining-lounge car; Fort Worth to St. Louis, dining car. Lounges and diners on B&O 2, the National Limited, and PRR 4, the Penn Texas.)

Lone Star, No. 2
Dallas to Memphis
St. Louis Southwestern
39 miles east of Texarkana

DALLAS to MEMPHIS
   16: *12s-dr*
StLSW 2, Dallas (4:45p) to Memphis (7:15a)

SHREVEPORT to MEMPHIS
   14: *10s-observation*
StLSW 202, Shreveport (9:30p)
    to Lewisville (11:25p)
StLSW 2, Lewisville (11:44p) to Memphis (7:15a)

No. 6
Texarkana to St. Louis
St. Louis Southwestern
7 miles north of Jonesboro

PINE BLUFF to ST. LOUIS
   2: *10s-2c-dr*
StLSW 6, Pine Bluff (7p) to St. Louis (7:43a)

Despite the heavy competion of Missouri Pacific, Frisco, and Katy, the Cotton Belt at times went head-to-head, offering sleepers on its St. Louis trains to Dallas, Houston, Shreveport, Texarkana, Tyler, and Waco. Even Pine Bluff for a time was a contested market, with both the Cotton Belt and Missouri Pacific offering sleepers to St. Louis.

Texas Eagle, No. 21
Streamliner
St. Louis to southern Texas
Missouri Pacific
10 miles northwest of Little Rock

ST. LOUIS to GALVESTON
   212: *14r-dr-2dbr*
MP 21, St. Louis (5:30p) to Texarkana (3a)
T&P 221, Texarkana (3:10a)
    to Longview, Tex. (4:45a)
MP 21, Longview (4:50a) to Palestine, Tex. (6:30a)
MP 25, Palestine (6:50a) to Houston (10a)
    and Galveston (12:01p)

WASHINGTON to HOUSTON
   MP5-6dbr
PRR 575, Washington (7:45p Sat)
    to Harrisburg, Pa. (10:44p)
PRR 3, Harrisburg (10:53p) to St. Louis (3p Sun)
As above, St. Louis (5:30p) to Houston (10a Mon)

NEW YORK to HOUSTON
   MP4-10r-6dbr
PRR 3, New York (7:35p Sat) to St. Louis (3p Sun)
As above, St. Louis (5:30p) to Houston (10a Mon)

NEW YORK to SAN ANTONIO
   MP3: *10r-6dbr*
PRR 3, New York (7:35p Sat) to St. Louis (3p Sun)
As above, St. Louis (5:30p)
    to Palestine (6:30a Mon)
MP 21, Palestine (6:45a)
    to San Antonio (11:40a Mon)

ST. LOUIS to SAN ANTONIO
   215: *14r-4dbr*   216: *14r-4dbr*
As above, St. Louis (5:30p)
    to San Antonio (11:40a)

(Dining-lounges, St. Louis to Houston and San Antonio. PRR 3, the Penn Texas, carries dining and lounge cars; PRR 575, a buffet-lounge.)

Although St. Louis-Houston Pullmans no longer nominally existed in 1952, these sleepers from the east more than sufficed. In addition to the MoPac, the Katy, Cotton Belt (SP), and Frisco (SP or Santa Fe) all scheduled Pullmans between the two cities.

Texas Eagle, No. 21
Memphis Pullman
Missouri Pacific
Standing in Little Rock

MEMPHIS to HOUSTON
   217: *14r-4dbr*
MP 201, Memphis (8:30p) to Little Rock (11:45p)
MP 21, Little Rock (12:20a) to Texarkana (3a)
As above, Texarkana (3:10a) to Houston (10a)

Texas Eagle, No. 22
Southern Texas to St. Louis
Missouri Pacific
3 miles southwest of Gurdon

GALVESTON to ST. LOUIS
   267: *14r-dr-2dbr*
MP 26, Galveston (2p) to Palestine (6:55p)
MP 22, Palestine (7:20p) to Longview (9p)
T&P 222, Longview (9:05p) to Texarkana (10:40p)
MP 22, Texarkana (10:50p) to St. Louis (8:20a)

HOUSTON to MEMPHIS
   265: *14r-4dbr*
MP 26, Houston (4p) to Palestine (6:55p)
As above, Palestine (7:20p) to Texarkana (10:40p)
MP 22, Texarkana (10:50p) to Little Rock (1:20a)
MP 202, Little Rock (2:30a) to Memphis (6:30a)

HOUSTON to NEW YORK
   PR12, *10r-6dbr*
MP 26, Houston (4p Sun) to Palestine (6:55p)
As above, Palestine (7:20p)
    to St. Louis (8:20a Mon)
PRR 4, St. Louis (10a) to New York (7:25a Tue)

HOUSTON to WASHINGTON
   PR11, *10r-6dbr*
As above, Houston (4p Sun)
    to St. Louis (8:20a Mon)
PRR 4, St. Louis (10a) to Harrisburg (3:52a Tue)
PRR 504, Harrisburg (4:15a)
    to Washington (7:35a Tue)

SAN ANTONIO to NEW YORK
   PR 15: *10r-6dbr*
MP 22, San Antonio (2:10p Sun) to Palestine (7p)
As above, Palestine (7:20p)
    to New York (7:25a Tue)

SAN ANTONIO to ST. LOUIS
   227: *14r-4dbr*   228: *14r-4dbr*
MP 22, San Antonio (2:10p) to Palestine (7p)
As above, Palestine (7:20p) to St. Louis (8:20a)

(Dining-lounges, Houston and San Antonio to St. Louis. PRR 4, the Penn Texas, carries diner and lounge cars.

Sunshine Special, No. 31
St. Louis to San Antonio
Missouri Pacific
6 miles southwest of Bald Knob

ST. LOUIS to HOT SPRINGS
   311: *14r-4dbr*
MP 31, St. Louis (5:35p) to Little Rock (1:05a)
MP 831, Little Rock (2:30a) to Hot Springs (4a)

ST. LOUIS to LAKE CHARLES
   312: *8s-5dbr*
MP 31, St. Louis (5:35p) to Little Rock (1:05a)
MP 131, Little Rock (1:30a)
    to Lake Charles (12:15p)

ST. LOUIS to EL DORADO
   315: *8s-dr-2c*
MP 31, St. Louis (5:35p) to Gurdon (2:55a)
MP 821, Gurdon (6:05a) to El Dorado (8:20a)

ST. LOUIS to SHREVEPORT
   314: *14r-4dbr*
MP 31, St. Louis (5:35p) to Hope (3:35a)
KCS 3, Hope (4:30a) to Shreveport (7:30a)

ST. LOUIS to SAN ANTONIO
   317: *12s-dr*   M31: *14s*
MP 31, St. Louis (5:35p) to Texarkana (4:50a)
T&P 231, Texarkana (5:15a) to Longview (7a)
MP 31, Longview (7:15a) to San Antonio (4:25p)

(St. Louis to Poplar Bluff, Mo., diner-lounge; Texarkana to San Antonio, cafe-parlor car. No. 131 is the Louisiana Sunshine Special, which carries a grill-coach from Little Rock to Alexandria.)

Little Rock once had its own sleeper to Lake Charles.

Arkansas   *(continued)*

Sunshine Special, No. 32
San Antonio to St. Louis
Missouri Pacific
Departing Little Rock

SAN ANTONIO to ST. LOUIS
320: 12s-dr   M100: 14s
MP 32, San Antonio (8:15a) to Longview (5:15p)
T&P 232, Longview (5:30p) to Texarkana (7:35p)
MP 32, Texarkana (7:50p) to St. Louis (7:50a)

SHREVEPORT to ST. LOUIS
323: 14r-4dbr
KCS 16, Shreveport (5:10p) to Texarkana (6:50p)
MP 32, Texarkana (7:50p) to St. Louis (7:50a)

EL DORADO to ST. LOUIS
324: 8s-dr-2c
MP 822, El Dorado (6:30p) to Gurdon (8:45p)
MP 32, Gurdon (9:15p) to St. Louis (7:50a)

LAKE CHARLES to ST. LOUIS
1021: 8s-5dbr
MP 132, Lake Charles (1:05p)
        to Little Rock (11:35p)
MP 32, Little Rock (11:59p) to St. Louis (7:50a)

HOT SPRINGS to ST. LOUIS
325: 14r-4dbr
MP 832, Hot Springs (8p) to Little Rock (9:45p)
MP 32, Little Rock (11:59p) to St. Louis (7:50a)

LITTLE ROCK to ST. LOUIS
327: 10s-dr-2c
MP 32, Little Rock (11:59p) to St. Louis (7:50a)

(San Antonio to Texarkana, cafe-parlor car; Poplar Bluff to St. Louis, diner-lounge. No. 132, the Louisiana Sunshine Special, carries a grill-coach from Alexandria to Little Rock.)

The *Sunshine Special's* involvement with Hot Springs traffic was long-standing. In the heyday of rail travel, both the Missouri Pacific and Rock Island aggressively courted visitors to Hot Springs National Park. In addition to the *Sunshine Special* and the *Southerner*, the MP scheduled from St. Louis its *Hot Springs Special*, which also carried a Chicago sleeper (Wabash), and other trains came in from Kansas City, Memphis, and New Orleans, all of which cities had dedicated service to the resort. Wichita had its own sleeper, as Minneapolis/St. Paul once did (via the Minneapolis & St. Louis and Wabash).

The Rock Island used Memphis as its gateway, tying in with the Illinois Central's *Panama Limited*, which was part of a third Hot Springs-Chicago Pullman line. The Rock's other partner was the Frisco, which forwarded sleepers for Denver and Pueblo back into the Rock Island's own hands at Kansas City. Memphis, too, had dedicated service to the Springs via Rock Island.

Harold K. Vollrath collection

The St. Louis Southwestern's *Lone Star Limited* near Greenville, Tex., in 1948.

Texan, No. 25
St. Louis to Fort Worth
Missouri Pacific
15 miles northeast of Hope

LITTLE ROCK to FORT WORTH
254: 10s-dr-2c
MP 25, Little Rock (9:50p) to Texarkana (1:30a)
T&P 15, Texarkana (2a) to Fort Worth (8:15a)

On this route of the *Sunshine Specials*, Little Rock had dedicated Pullman service to El Paso, San Antonio, Mexico City (NdeM), Houston, Memphis, and St. Louis.

Cherokee, No. 111
Memphis to Tucumcari, N.M.
Rock Island
15 miles west of Little Rock

MEMPHIS to LOS ANGELES
1115: 10s-dr-2c
RI 111, Memphis (8p Sun)
        to Tucumcari (8p CST Mon)
SP 39, Tucumcari (8:10p MST)
        to Los Angeles (11:30p Tue)

LITTLE ROCK to LOS ANGELES
1117: 8s-dr-2c
RI 111, Little Rock (11:35p Sun)
        to Tucumcari (8p Mon)
SP 39, Tucumcari (8:10p)
        to Los Angeles (11:30p Tue)

(Little Rock to Tucumcari, diner-lounge. SP 39, the Imperial, carries dining and lounge cars.)

Cherokee, No. 112
Tucumcari to Memphis
Rock Island
Standing in Booneville

LOS ANGELES to MEMPHIS
404: 10s-dr-2c
SP 40, Los Angeles (11p Fri)
        to Tucumcari (5:50a Sun)
RI 112, Tucumcari 7:20a CST)
        to Memphis (6:50a Mon)

LOS ANGELES to LITTLE ROCK
403: 8s-dr-2c
SP 40, Los Angeles (11p Fri)
        to Tucumcari (5:50a Sun)
RI 112, Tucumcari (7:20a CST)
        to Little Rock (3a Mon)

(Tucumcari to Little Rock, diner-lounge. No. 40, the Imperial, carries dining and lounge cars.)

Nos. 111 and 112 previously carried Little Rock-Memphis Pullmans, and the *Choctaw Limited* offered sleepers in the other direction, from Little Rock to Oklahoma City.

Memphian, No. 806
Memphis to St. Louis
Frisco
30 miles north of Memphis

FRISCO

MEMPHIS to ST. LOUIS
86: 8s-5br
SLSF 806, Memphis (11:15p) to St. Louis (7:40a)

(Buffet car)

The *Memphian* formally carried a Jacksonville Pullman from St. Louis, and over this route the Frisco also scheduled sleepers to Birmingham, Pensacola, Jacksonville, and St. Petersburg (SOU/SAL).

California

See map on page **218**

All Santa Fe trains from the east enter and leave Los Angeles by way of Pasadena, except for the two sections of the *Grand Canyon*, which operate via Fullerton.

Sunset Limited, No. 2
Los Angeles to New Orleans
Via Phoenix and Douglas, Ariz.
Southern Pacific
Entering Niland

LOS ANGELES to NEW ORLEANS
26: 10r-6dbr   27: 10r-6dbr
29: 10r-6dbr
SP 2, Los Angeles (8p Sun)
        to New Orleans (4p Tue)

LOS ANGELES to SAN ANTONIO
25: 10r-6dbr
SP 2, Los Angeles (8p Sun)
        to San Antonio (4:30a Tue)

LOS ANGELES to DALLAS
24: 10r-6dbr
SP 2, Los Angeles (8p Sun) to El Paso (3p Mon)
T&P 6, El Paso (3:30p) to Dallas (8:45a Tue)

(Los Angeles to New Orleans, dining car and lounge car with shower. T&P 6, the Westerner, has a diner from El Paso to Big Spring and a grill coach going to Dallas. This edition of the Sunset is one of the five red and orange, streamlined trainsets placed in service in August 1950, when the train went to its new forty-two-hour schedule, five hours faster than the previous carding. An eastbound Sunset Limited is now in Texas.)

The Sunset Route, itself, had notable Pullman service: the long-running Washington-Sunset tourist car that was inaugurated in 1898, and the Los Angeles-Jacksonville sleeper of the late 1920's that was the longest regular Pullman line, and the only transcontinental one, of its time. Briefly after World War II, the *Golden State Limited* carried New York-Los Angeles sleepers. Besides the destinations of Phoenix and Tucson, the SP moved Mexico-bound sleeping cars to the gateway of Nogales, including Pullmans from Los Angeles to Guaymas, Mazatlan, Tepic, Guadalajara, and Mexico City.

No. 3
Eureka to San Francisco
Northwestern Pacific
50 miles north of Willits

EUREKA to SAN FRANCISCO
Std: s-c-dr
NWP 3, Eureka (7:30p) to San Rafael (8:10a)
Bus to San Francisco (9a)

(Eureka to San Rafael, lounge)

No. 4
San Francisco to Eureka
Northwestern Pacific
37 miles north of Willits

SAN FRANCISCO to EUREKA
Std: s-c-dr
Bus from San Francisco (7:40p)
NWP 4, San Rafael (8:30p) to Eureka (9a)

(San Rafael to Eureka, lounge)

Northwestern Pacific trains departed from Sausalito before the Golden Gate Bridge stifled the railroad's suburban electric trains and ferry service, which ended in 1941. NWP Pullmans also traveled to Willits and Fort Bragg (CW)

Argonaut, No. 6

Los Angeles to New Orleans
Via Maricopa, Ariz., and Deming, N.M.
Southern Pacific
12 miles east of Palm Springs

LOS ANGELES to NEW ORLEANS
64: 8s-5dbr   65: Std: s-dr-c
66: Std: s-dr
SP 6, Los Angeles (8:30p Sun)
    to New Orleans (7a Wed)

(Los Angeles to New Orleans, dining car and lounge. Two more editions of the eastbound Argonaut are under way, both in Texas.)

The *Argonaut*, the hardworking understudy of the more prestigious *Sunset Limited*, was introduced in 1926 and carried Washington-Sunset tourist Pullmans to Los Angeles and San Francisco. The *Argonaut* also took on West Coast sleepers from St. Louis and Memphis at El Paso and carried San Diego Pullmans from New Orleans.

Cascade, No. 12

Streamliner
San Francisco to Portland
Southern Pacific
7 miles north of Dunsmuir

SAN FRANCISCO to PORTLAND
Four all-room sleeping cars
Ferry from San Francisco (5p)
SP 12, Oakland pier (5:32p) to Portland (9:30a)

SAN FRANCISCO to SEATTLE
122: 22r   123: 6dbr-10r
124: 6dbr-10r   125: 4c-4dbr-2dr
As above, San Francisco (5p) to Portland (9:30a)
Pool 407, Portland (10a) to Seattle (2p)

(Oakland to Portland, articulated dining car/full lounge; Portland to Seattle, diner)

Although a "coastal" train, the *Cascade* climbed to a nearly mile-high summit on the route through its namesake mountains between California's Sacramento River Canyon and the Willamette River Valley in Oregon. The eight sleepers on this 1950 streamlined version contain all private rooms.

Super Chief, No. 18

All-Pullman streamliner
Los Angeles to Chicago
Via Raton Pass
Atchison, Topeka & Santa Fe
25 miles west of Ludlow

LOS ANGELES to CHICAGO
Observation: 4dr-dbr   2dr-4c-4dbr
2dr-4c-4dbr   2dr-4c-4dbr
10r-6dbr   10r-6dbr
Santa Fe 18, Los Angeles (8p Sun)
    to Chicago (1:45p Tue)

LOS ANGELES to KANSAS CITY
10r-6dbr
Santa Fe 18, Los Angeles (8p Sun)
    to Kansas City (5:35a Tue)

(Los Angeles to Chicago: extra fare $15; dining car, dormitory/lounge, Pleasure Dome lounge and Turquoise private dining room; barber, shower, radio, and recorded music. Although some of the stainless steel trainsets were just three years old, the Santa Fe re-equipped the Super Chief in 1951, and these immaculately maintained cars are in the above consist. The maitre d'hotel will greet you in formal attire, and your waiter will gladly bring you the dining car specialty of freshly grilled, Colorado mountain trout. The Super Chief that left Los Angeles the previous night is now in Kansas.)

A Santa Fe helper on the point of the streamlined *Chief* digs into the grade at Cajon Pass.

California Limited, No. 3
Chicago to Los Angeles
Via Raton Pass, Topeka
Atchison, Topeka & Santa Fe
3 miles west of Cadiz

CHICAGO to LOS ANGELES
12s-dr   8s-dr-2c
Santa Fe 3, Chicago (8:45p Fri)
    to Los Angeles (7a Mon)

CHICAGO to LOS ANGELES
10s-dr-2c
Santa Fe 3, Chicago (8:45p Fri)
    to Kansas City (7:15a Sat)
Santa Fe 5, Kansas City (8:50a)
    to Newton, Kans. (12:10p)
Santa Fe 105, Newton (12:55p)
    to Belen, N.M. (6:35a Sun)
Santa Fe 14, Belen (6:55a) to Albuquerque (7:45a)
Santa Fe 3, Albuquerque (8:30a)
    to Los Angeles (7a Mon)

CHICAGO to SAN FRANCISCO
10s-dr-2c
Santa Fe 3, Chicago (8:45p Fri)
    to Barstow (2:15a Mon)
Santa Fe 23, Barstow (6:20a) to Oakland (8p)
Bus to San Francisco (8:30p Mon)

PHOENIX to LOS ANGELES
10s-2dr
Santa Fe 42, Phoenix (5p Sun)
    to Wickenburg, Ariz. (6:40p)
Santa Fe 170, Wickenburg (7p)
    to Matthie, Ariz. (7:10p)
Santa Fe 117, Matthie (7:10p) to Cadiz (11:20p)
Santa Fe 3, Cadiz (11:55p)
    to Los Angeles (7a Mon)

(On the California Limited*: dining car, Chicago to La Junta; Fred Harvey station dining room service west of La Junta; lounge, Chicago to Los Angeles. No. 23, dining car, Barstow to Oakland. One Los Angeles sleeper travels the southern route on 5-105, the Scout, which carries diners from Kansas City to Amarillo, Tex. The Scout serves Carlsbad Caverns. Other westbound editions of the* California Limited *are in Colorado and Illinois.)*

Inaugurated in 1892, the *California Limited* was the Santa Fe's heavy-volume luxury train. During summertime peaks of the 1920's, seven different sections, each with as many as eleven Pullmans, were dispatched, one after another. The *California Limited* set a movement record of forty-five sections simultaneously under way, twenty-two westbound and twenty-three eastbound. The train, of course, was superseded by the *Chief* in the '20s and the *Super Chief* in the '30s, but as these 1952 schedules indicate, the *California Limited* was still carrying its weight as a prime Santa Fe flyer long after its more glamorous brethren came along.

California Limited, No. 4
Los Angeles to Chicago
Via Raton Pass, Topeka
Atchison, Topeka & Santa Fe
2 miles east of Ludlow

LOS ANGELES to CHICAGO
12s-dr   8s-dr-2c
Santa Fe 4, Los Angeles (6:15p Sun)
    to Chicago (8:30a Wed)

LOS ANGELES to CHICAGO
10s-dr-2c
Santa Fe 4, Los Angeles (6:15p Sun)
    to Albuquerque (6:45p)
Santa Fe 13, Albuquerque (9p Mon)
    to Belen (9:40p)
Santa Fe 106 Belen (9:55p) to Newton (5:25p Tue)
Santa Fe 6, Newton (6:10p)
    to Kansas City (10:15p)
Santa Fe 4, Kansas City (11p)
    to Chicago (8:30a Wed)

LOS ANGELES to PHOENIX
10s-2dr
Santa Fe 4, Los Angeles (6:15p Sun)
    to Cadiz (12:54a Mon)
Santa Fe 118, Cadiz (1a) to Matthie (ntg)
Santa Fe 181, Matthie (6:30a)
    to Wickenburg (6:40a)
Santa Fe 47, Wickenburg (6:50a)
    to Phoenix (8:30a Mon)

SAN FRANCISCO to CHICAGO
10s-dr-2c
Bus from San Francisco (9a)
Santa Fe 4, Oakland (9:25a Sun)
    to Barstow (10:10p)
Santa Fe 4, Barstow (10:45p)
    to Chicago (8:30a Wed)

(No. 4, diner, Oakland to Barstow. California Limited: Fred Harvey station dining room service east to La Junta; dining car, La Junta to Chicago; lounge, Los Angeles to Chicago, No. 106-6, the Scout, dining cars Amarillo (7a) to Kansas City. Eastbound editions of the California Limited *are also now in New Mexico and Missouri.)*

Trains like the *California Limited* gave the Santa Fe a prestigious image of being primarily a cross-country passenger carrier that dominated the Chicago-Los Angeles market. Before Californians had their freeways, however, the Santa Fe provided a good deal of close-in Pullman support, with sleepers from Los Angeles to Barstow, Merced, San Francisco, Blythe, Ripley, and San Diego, as well as Goldfield and Beatty, Nev., in cooperation with the Tonopah & Tidewater.

San Francisco, too, had considerable intrastate Pullman service on the Santa Fe: to El Portal, Fresno, Bakersfield, Barstow, Needles, and San Diego.

California   (continued)

Chief, No. 19

All-Pullman streamliner
Chicago to Los Angeles
Via Raton Pass
Atchison, Topeka & Santa Fe
Arriving in Needles

CHICAGO to LOS ANGELES
17r   2dr-4c-4dbr
   Observation-4dr-dbr
Santa Fe 19, Chicago (1:30p Mon)
   to Los Angeles (8:30a Mon)

NEW YORK to LOS ANGELES
2501: 2dr-4c-4dbr   2502: 10r-6dbr
NYC 25, New York (6p Fri) to Chicago (9a Sat)
Santa Fe 19, Chicago (1:30p)
   to Los Angeles (8:30a Mon)

NEW YORK to LOS ANGELES
2dr-4c-4dbr
PRR 29, New York (6p Fri) to Chicago (9a Sat)
Santa Fe 19, Chicago (1:30p)
   to Los Angeles (8:30a Mon)

WASHINGTON to SAN DIEGO
10r-6dbr
B&O 5, Washington (5:30p Fri) to Chicago (8a Sat)
Santa Fe 19, Chicago (1:30p)
   to Los Angeles (8:30a Mon)
Santa Fe 74, Los Angeles (11:30a)
   to San Diego (2:15p Mon)

(Chicago to Los Angeles: $10 extra fare; baggage/lounge, lounge, dining car; valet, barber, shower bath, radio. The Chief's premier transcontinental sleeper lines involved four all-Pullman trains, including the Twentieth Century Limited, NYC No. 25; the Broadway Limited, PRR No. 29; and the Capitol Limited, B&O No. 5. All the eastern trains carried dining cars and various lounges. The other westbound Chief is now in Kansas.)

By 1952, San Diego's Pullman service had dwindled to just this Washington sleeper. Never extensive, the schedule included cars on the Santa Fe to Chicago, a tourist car to Minneapolis/St. Paul, and pre-1900 lines to San Bernardino and St. Louis. The late-coming and short-lived San Diego & Arizona Eastern interchanged Pullmans for New Orleans and Chicago with parent SP in the Imperial Valley, and that was pretty much the story. By the time San Diego merited major league passenger service, the railroads were throwing in the towel.

Klamath, No. 20

San Francisco to Portland
Southern Pacific
62 miles north of Davis

SAN FRANCISCO to PORTLAND
Std: s-dr-c
Ferry from San Francisco (7:30p)
SP 20, Oakland pier (8p) to Portland (8:45p)

(Oakland to Portland, diner/coffee shop)

The *Klamath* carried a Pullman from Oakland for Spokane, and one of its sister trains, the *Oregonian*, handled the Dunsmuir sleeper.

Gold Coast, No. 23

Chicago to San Francisco
Southern Pacific
7 miles west of Truckee

CHICAGO to SAN FRANCISCO
12s-dr   12s-dr
12s-dr
C&NW 23, Chicago (8p Fri) to Omaha (7:30a Sat)
UP 23, Omaha (8:20a) to Ogden (6:15a Sun)
SP 23, Ogden (8:15a) to Oakland pier (7a Mon)
Ferry to San Francisco (7:35a Mon)

RENO to SAN FRANCISCO
Std: s-dr-c
SP 23, Reno (10:40p) to Oakland pier (7a)
Ferry to San Francisco (7:35a)

DENVER to SAN FRANCISCO
12s-dr
UP 37, Denver (5:35p Sat) to Ogden (6:35a Sun)
As above, Ogden (8:15a)
   to San Francisco (7:35a Mon)

OGDEN to SAN FRANCISCO
Std: s-dr-c
As above, Ogden (8:15a) to San Francisco (7:35a)

(Dining car and lounge on No. 23, Chicago to Oakland, and on No. 37, the Pony Express, Denver to Ogden. In addition to California, the Gold Coast has east- and westbound editions on the road in both Iowa and Wyoming.)

The *Gold Coast* was introduced in 1926 as one of three daily trains to San Francisco, a level of service the Overland Route hadn't seen since World War I. The train also carried sleepers for Los Angeles during its early years, but it was withdrawn in 1933, a Depression cutback. The *Gold Coast*, however, had a postwar reprise, when it replaced the *Challenger* in 1947.

Gold Coast, No. 24

San Francisco to Chicago
Southern Pacific
40 miles west of Truckee

SAN FRANCISCO to CHICAGO
12s-dr   12s-dr
12s-dr
Ferry from San Francisco (6p Sun)
SP 24, Oakland pier (6:30p) to Ogden (6:30p Mon)
UP 24, Ogden (7:15p) to Omaha (7:50p Tue)
C&NW 24, Omaha (8:30p) to Chicago (8:15a Wed)

SAN FRANCISCO to DENVER
12s-dr
As above, San Francisco (6p Sun)
   to Ogden (6:30p Mon)
UP 38, Ogden (7:10p) to Denver (8a Tue)

SAN FRANCISCO to OGDEN
Std: s-dr-c
As above, San Francisco (6p Sun)
   to Ogden (6:30p Mon)

(Diner and lounge on No. 24, Oakland to Chicago, and on No. 38, the Pony Express, Ogden to Denver.)

Not all the cars in the consist of a *Gold Coast* or other crack flyers were bound for Chicago or the major cities en route. The *Gold Coast* carried Westwood Pullmans, and other trains crossing the Sierra Nevada hauled sleepers for Sacramento, Truckee, and Susanville, and the Nevada communities of Tonopah (T&G), Ely (NN) and, dating to the antiquity of 1876, Carson City (V&T).

No. 26

San Francisco to Sparks, Nev.
Southern Pacific
3 miles west of Davis

SAN FRANCISCO to RENO
Std: s-dr-c
Ferry from San Francisco (9p)
SP 26, Oakland pier (9:35p) to Reno (7:50a)

Pony Express, No. 37

Kansas City to Los Angeles
Union Pacific
15 miles east of Crucero

KANSAS CITY to LOS ANGELES
12s-dr
UP 37, Kansas City (11:59p Fri)
   to Los Angeles (7a Mon)

LAS VEGAS to LOS ANGELES
8s-dr-3dbr
UP 37, Las Vegas (9:15p) to Los Angeles (7a)

SUN VALLEY, IDA., to LOS ANGELES   Winter only
10r-6dbr
UP 56, Sun Valley (10:30p Sat))
   to Shoshone, Ida. (12:45a)
UP 12, Shoshone (1:30a Sun) to Pocatello (3:40a)
UP 32, Pocatello (4:15a) to Salt Lake City (8:50a)
UP 37, Salt Lake City (9:40a)
   to Los Angeles (7a Mon)

BUTTE to LOS ANGELES
10r-6dbr
UP 30, Butte (7p Sat) to Salt Lake City (7:35a Sun)
UP 37, Salt Lake City (9:40a)
   to Los Angeles (7a Mon)

CHICAGO to LOS ANGELES
10s-3dbr   14s
C&NW 23, Chicago (8p Fri) to Omaha (7:30a Sat)
UP 23, Omaha (8:20a) to Ogden (6:15a Sun)
UP 37, Ogden (7a) to Los Angeles (7a Mon)

(No. 37, the Pony Express, dining car, Kansas City to Las Vegas; first-class lounge, Denver to Los Angeles. No. 32, Pocatello to Salt Lake City, cafe-lounge car. UP 30, Butte Special, cafe-lounge car, Butte to Pocatello; first-class lounge, Butte to Salt Lake City. C&NW/UP 23, Gold Coast, Chicago to Ogden, dining car and first-class lounge car. Westbound editions of the Pony Express are now in Missouri and Wyoming.)

Imperial, No. 40

Los Angeles to Chicago
Via Calexico and Douglas
Southern Pacific
33 miles east of Los Angeles

LOS ANGELES to CHICAGO
405:10s-dr-2c   406: 6s-6dbr
SP 40, Los Angeles (11p Sun)
   to Tucumcari, N.M. (5:50a MST)
RI 40, Tucumcari (7:20a CST Tue)
   to Chicago (8:40a Wed)

LOS ANGELES to MEMPHIS
404: 10s-dr-2c
SP 40, Los Angeles (11p Sun)
   to Tucumcari (5:50a MST Tue)
RI 112, Tucumcari (7:20a CST)
   to Memphis 6:50a Wed)

LOS ANGELES to LITTLE ROCK
403: 8s-dr-2c
SP 40, Los Angeles (11p Sun)
   to Tucumcari (5:50a MST Tue)
RI 112, Tucumcari (7:20a CST)
   to Little Rock (3a Wed)

LOS ANGELES to CALEXICO
400: 10s-dr-2c
SP 40, Los Angeles (11p) to Calexico (5:15a)

(Dining cars: Los Angeles to El Paso, Tucumcari to Chicago; lounge car, Los Angeles to Chicago. No. 112 is the Cherokee, which carries a diner from Tucumcari to Little Rock and a lounge car all the way to Memphis. The train was formerly known as the Memphis Californian. Other eastbound editions of the Imperial are now in New Mexico and Iowa.)

Pony Express, No. 38
Los Angeles to Kansas City
Union Pacific
8 miles west of Barstow

LOS ANGELES to KANSAS CITY
12s-dr
UP 38, Los Angeles (7:30p Sun)
   to Kansas City (10:30p Tue)

LOS ANGELES to LAS VEGAS
8s-dr-3dbr
UP 38, Los Angeles (7:30p) to Las Vegas (5:10a)

LOS ANGELES to BUTTE
10r-6dbr
UP 38, Los Angeles (7:30p Sun)
   to Salt Lake City (5:15p)
UP 29, Salt Lake City (8p Mon) to Butte (9:30a Tue)

LOS ANGELES to SUN VALLEY *Winter only*
10r-6dbr
UP 38, Los Angeles (7:30p Sun)
   to Salt Lake City (5:15p)
UP 29, Salt Lake City (8p Mon)
   to Pocatello (12:45a Tue)
UP 25, Pocatello (1:10a) to Shoshone (3:15a)
UP 55, Shoshone (4:30a) to Sun Valley (7:30a Tue)

LOS ANGELES to CHICAGO
10s-3dbr   14s
UP 38, Los Angeles (7:30p Sun)
   to Ogden (6:40p Mon)
UP 24, Ogden (7:15p) to Omaha (7:50p Tue)
C&NW 24, Omaha (8:30p) to Chicago (8a Wed)

(No. 38, the Pony Express, *Los Angeles to Denver, first-class lounge; Las Vegas to Kansas City, dining car. No. 29, Butte Special, Salt Lake City to Butte, first-class and cafe-lounge cars. No. 24, Gold Coast, Ogden to Chicago, dining and first-class lounge cars. The* Pony Express *that departed Los Angeles the previous day is now in Wyoming.)*

The sleek, yellow streamliners are the lasting memory of the great Union Pacific passenger service, but in earlier times, when gold and silver brought fortune seekers back to Nevada after the turn of the century, the UP shared in the transport of Pullmans in and out of the state. Hauling sleepers up from Los Angeles, over the Cajon pass, and into the desert, the UP cut out cars for Death Valley Junction and Beatty at Crucero (T&T), and delivered those bound for Goldfield and Rhyolite to the Las Vegas & Tonopah at Las Vegas.

No. 43
Tucumcari to Los Angeles
Via Deming
Southern Pacific
5 miles west of Palm Springs

CALEXICO to LOS ANGELES
431:10s-dr-2c
SP 347, Calexico (7:30p) to Niland (9p)
SP 43, Niland (9:30p) to Los Angeles (3:45a)

Owl No. 57
Los Angeles to San Francisco
Via San Joaquin Valley, Los Banos
Southern Pacific
27 miles north of Bakersfield

LOS ANGELES to SAN FRANCISCO
Std: sbr-dbr-lounge Std: s-dbr-c-dr
SP 57, Los Angeles (5:40p) to Oakland pier (7:40a)
Ferry to San Francisco (8:15a)

(Dining Car)

One of several *Owl* overnight trains in the country, this venerable example began serving the San Joaquin Valley in 1898. The SP scheduled sleepers on this route to Fresno, Merced, El Portal at Yosemite Park, and on to Portland and Seattle.

Owl No. 58
San Francisco to Los Angeles
Via San Joaquin Valley, Los Banos
Southern Pacific
31 miles south of Tracy

SAN FRANCISCO to LOS ANGELES
Std: sbr-dbr-lounge Std: s-dbr-c-dr
Ferry from San Francisco (8:05p)
SP 58, Oakland pier (8:35p)
   to Los Angeles (10:45a)

(Diner, serving breakfast)

The *Owl* once carried a San Francisco-Bakersfield sleeper, and its companion *Sequoia* moved Pullmans to Merced, El Portal, and Fresno.

West Coast, No. 59
Los Angeles to Sacramento
Southern Pacific
15 miles north of Tehachapi

LOS ANGELES to SACRAMENTO
Std: s-br-r
SP 59, Los Angeles (7:30p) to Sacramento (8:30a)

(Los Angeles to Sacramento, diner-lounge)

Lark, No 75
All-Pullman streamliner
Los Angeles to San Francisco, Oakland
Southern Pacific, Coast Line
14 miles west of Santa Barbara

LOS ANGELES to SAN FRANCISCO
Std: dbr-dr-c-r
SP 75, Los Angeles (9p) to San Francisco (9a)

LOS ANGELES to OAKLAND
Std: dbr-dr-c-r
SP 75, Los Angeles (9p) to San Jose (7:20a)
SP 73, San Jose (7:47a) to Oakland (9:05a)

(Los Angeles to San Francisco, articulated lounge and dining cars; Los Angeles to Oakland, buffet-lounge sleeper)

The Southern Pacific's Coast Line was San Francisco's only source of Pullmans directly serving the city; all others were accessible only by boat or motor vehicle. Over the years, several trains, including the *Sunset Limited*, brought sleepers to the station at Third and Townsend, but by 1952 only the *Lark* was still performing this service.

West Coast, No. 60
Sacramento to Los Angeles
Southern Pacific
24 miles south of Merced

SACRAMENTO to LOS ANGELES
Std: s-br-r
SP 60, Sacramento (8:45p) to Los Angeles (8:50a)

(Sacramento to Los Angeles, diner-lounge)

By this schedule the *West Coast* has been cut back from its onetime operation through to Portland, with cars for Seattle. Though it once carried Pullmans for San Francisco, the train itself bypassed the Bay Area. Sacramento had dedicated sleepers both to Portland and, over the Sierra Nevada, to Reno.

Lark, No. 76
All-Pullman streamliner
San Francisco, Oakland to Los Angeles
Southern Pacific, Coast Line
Leaving Salinas

SAN FRANCISCO to LOS ANGELES
Std: dbr-dr-c-r
SP 76, San Francisco (9p) to Los Angeles (9a)

OAKLAND to LOS ANGELES
Std: dbr-dr-c-r
SP 74, Oakland (8:43p) to San Jose (9:50p)
SP 76, San Jose (10:15p) to Los Angeles (9a)

(San Francisco to Los Angeles, articulated lounge and dining cars; Oakland to Los Angeles, buffet-lounge sleeper)

The *Lark* began operating between California's two largest cities in 1910, and within a decade was the train of choice for making the overnight run. The *Oakland Lark* was added in 1931. Though still listed as a separate section in 1952, the Oakland train and the regular *Lark* were combined in San Jose. The two-tone gray streamliner, nonetheless, was still all-Pullman, the only purely western train able to make that claim.

On other Coast Line trains, San Francisco had Pullman service to Monterey and Santa Barbara. Santa Barbara, in turn, had its own sleepers to Chicago on the *Golden State Limited* and to New Orleans on the *Sunset Limited*.

Fred Matthews/California State Railroad Museum

A pair of cab-forwards take water at Mojave, Calif., with the Oakland-bound *Owl* in 1952.

Colorado

See map on page 206

California Limited, No. 3

Chicago to Los Angeles
Via Topeka, Raton Pass
Atchison, Topeka & Santa Fe
2 miles west of Trinidad

CHICAGO to LOS ANGELES
12s-dr   8s-dr-2c
Santa Fe 3, Chicago (8:45p Sat)
        to Los Angeles (7a Tue)

CHICAGO to SAN FRANCISCO
10s-dr-2c
Santa Fe 3, Chicago (8:45p Sat)
        to Barstow (2:15a Tue)
Santa Fe 23, Barstow (6:20a) to Oakland (8p)
Bus to San Francisco (8:30p Tue)

DENVER to PHOENIX
8s-dr-2c
Santa Fe 101, Denver (2:10p Sun)
        to Pueblo (4:45p)
Santa Fe 2, Pueblo (4:50p) to La Junta (6:10p)
Santa Fe 3, La Junta (9:50p)
        to Ash Fork (7:15p Mon)
Santa Fe 47, Ash Fork (2a Tue)
        to Phoenix (8:30a Tue)

(Lounge, Chicago to Los Angeles; Diners, Chicago to La Junta and Barstow to Oakland; Fred Harvey station dining room service La Junta to Los Angeles. At this point, the train is beginning its climb to the summit of Raton Pass. Other westbound editions of the California Limited are now in Illinois and California.)

No. 7

Denver to Dallas
Colorado & Southern
38 miles north of Trinidad

DENVER to AMARILLO
72: Std: s-dr
C&S 7, Denver (7:10p) to Amarillo (9a)

(Diner, Amarillo to Fort Worth)

No. 7 once carried a Denver sleeper to Trinidad, and other C&S service to the southeast included Pullmans to San Antonio (MKT), Galveston (SP), and New Orleans (SP or T&P)

Prospector, No. 7

Denver to Grand Junction, Ogden
Streamliner, via Moffat Tunnel
Denver & Rio Grande Western
46 miles east of Grand Junction

DENVER to SALT LAKE CITY
D5: 5s-5r-6dbr   D7: 5s-5r-6dbr
D&RGW 7, Denver (5:30p)
        to Grand Junction (12:53a)
D&RGW 7-1, Grand Junction (1:10a)
        to Salt Lake City (8:15a)

(Denver to Salt Lake City, diner-lounge)

The *Prospector* was inaugurated three weeks before Pearl Harbor as a stainless steel, articulated streamliner. The little train, however, proved to be inadequately powered for the mountain grades, and it was discontinued in 1942. Reborn as a regular train in 1945, it handled the Rio Grande's overnight traffic between Denver and Ogden.

No. 7-1

Grand Junction to Ogden
Denver & Rio Grande Western
Awaiting makeup in Grand Junction

DENVER to OGDEN
D22: 10s-2c-dr
D&RGW 1, Denver (8:50a)
        to Grand Junction (10:50p)
D&RGW 7-1, Grand Junction (1:10a)
        to Ogden (9:50a)

(Denver to Grand Junction, grill-lounge; Grand Junction to Salt Lake City, diner-lounge; Denver to Ogden, Vista Dome coach. No. 1 is the Royal Gorge, traveling its namesake route. No. 7 is the Prospector, operating via the Moffat Tunnel. The two trains combine in Grand Junction.)

The *Royal Gorge* was the post-war heir to the Rio Grande's *Scenic Limited*, which dated back to 1915 and Panama-Pacific Exposition in San Francisco. The *Scenic* carried through sleepers from the Midwest for California, with the Missouri Pacific's version of the train interchanging St. Louis Pullmans at Pueblo and the Burlington bringing cars from Chicago through Denver. The much shorter Moffat Tunnel route, however, soon diverted most of the Chicago traffic from the Royal Gorge, and the MoPac's successor to the *Scenic*, the *Colorado Eagle*, as of this scheduling, was turning over its sole Los Angeles sleeper to the *Golden State Limited* in Kansas City. Earlier intrastate service on the Royal Gorge line had included sleepers from Denver to Salida, Leadville, Glenwood Springs, and Grand Junction.

Prospector/Royal Gorge, No. 8-2

Ogden to Grand Junction, Denver
Denver & Rio Grande Western
11 miles west of Grand Junction

SALT LAKE CITY to DENVER
D6: 5s-5r-6dbr   D8: 5s-5r-6dbr
D&RGW 8-2, Salt Lake City (5:30p)
        to Grand Junction (12:15a)
D&RGW 8, Grand Junction (12:32a)
        to Denver (8:15a)

OGDEN to DENVER
D10: 10s-2c-dr
D&RGW 8-2, Ogden (3:50p)
        to Grand Junction (12:15a)
D&RGW 2, Grand Junction (2a) to Denver (3:30p)

(In Grand Junction, No. 8-2 divides into two trains for Denver: the Prospector, No. 8, carries the two Salt Lake City Pullmans and diner-lounge via the Moffat Tunnel route; the Royal Gorge, No. 2, takes the sleeper and the Vista Dome chair car that originated in Ogden and cuts in a grill-lounge for the spectacular journey through the Grand Canyon of the Arkansas River, the Royal Gorge. There the train will pause for the customary ten minutes, while passengers get out to view the river rushing through the narrow base of the towering canyon walls.)

No. 10

Craig to Denver
Denver & Rio Grande Western
14 miles west of Orestod

CRAIG to DENVER
D4: 12s-dr
D&RGW 10, Craig (9p) to Orestod (12:50a)
D&RGW 20, Orestod (1:58a) to Denver (7:20a)

(At Orestod, the Craig sleeper goes aboard No. 20, the Mountaineer, which handles the Montrose-Denver Pullman)

The Rio Grande's No. 10 is the local night train serving the communities on what had been David Moffat's original Denver & Salt Lake Railway.

It's Oct. 23, 1936, and the new *Denver Zephyr* is about to leave on its record 12-hour, 17-minute run from Chicago.

Mountaineer, No. 19

Denver to Montrose
Via Moffat Tunnel
Denver & Rio Grande Western
12 miles east of Orestod

DENVER to MONTROSE
D1: 14s
D&RGW 19, Denver (7:30p)
        to Grand Junction (6:40a)
D&RGW 320, Grand Junction (7:10a)
        to Montrose (9:50a)

DENVER to CRAIG
D3: 12s-dr
D&RGW 19, Denver (7:30p) to Orestod (12:40a)
D&RGW 9, Orestod (3:12a) to Craig (7:45a)

Mountaineer, No. 20

Montrose to Denver
Via Moffat Tunnel
Denver & Rio Grande Western
7 miles west of Dotsero

MONTROSE to DENVER
D2: 14s
D&RGW 319, Montrose (5:30p)
        to Grand Junction (8:05p)
D&RGW 20, Grand Junction (8:20p)
        to Denver (7:20a)

No. 29

Denver to Billings, Mont.
Colorado & Southern
4 miles south of Fort Collins

DENVER to BILLINGS
C291: Std: s-dr
C&S 29, Denver (9:20p) to Wendover, Wyo. (5:25a)
CB&Q 29, Wendover (5:27a) to Billings (5:25p)

(Diner, Casper to Lovell, Wyo.)

No. 141

Denver to Pueblo
Atchison, Topeka & Santa Fe
11 miles south of Denver

DENVER to LOS ANGELES
6s-6r-4dbr
Santa Fe 141, Denver (11:40p Sun)
        to Pueblo (2:15a Mon)
Santa Fe 14, Pueblo (2:20a) to La Junta (5a)
Santa Fe 123, La Junta (7:30a)
        to Los Angeles (10:25a Tue)

(No. 123, the Grand Canyon's northern section, carries a diner and lounge.)

Over the years the Santa Fe fed considerable Pullman traffic from Denver into the main line at La Junta. Among the bygone sleepers were Denver to Dallas/Fort Worth, El Paso, Galveston, New Orleans (T&P), San Antonio (MP), Trinidad, Tulsa, and Wichita.

Connecticut

See map on page 14

Delaware

See map on page 58

District of Columbia

State of Maine, No. 124

New York to Portland, Me.
New Haven Railroad
Departing New London

NEW YORK to PORTLAND
12s-dr   14r-4dbr
2c-dr-3sbr-buffet-lounge
NH 124, New York (9p) to Worcester (2:24a)
B&M 81, Worcester (2:55a) to Portland (6:40a)

NEW YORK to CONCORD, N.H.
8s-5sbr
NH 124, New York (9p) to Worcester (2:24a)
B&M 81, Worcester (2:55a) to Lowell (3:58a)
B&M 303, Lowell (7:11a) to Concord (8:40a)

NEW YORK to FARMINGTON, ME.
Summertime, Friday only
1247: 10s-dr-2c
As above, New York (9p) to Portland (6:40a)
MEC 7, Portland (7:40a) to Farmington (10:45a)

(Saturday schedules different.)

Night White Mountains, No. 166

New York to New Hampshire
Summertime, Fridays only
New Haven Railroad
Departing New Haven

NEW YORK to BRETTON WOODS-FABYAN
1661: 10s-dr-2c   1662: 6c-3dr
1663: 6c-buffet-lounge   1664: 14sbr
NH 166, New York (10:15p) to Springfield (ntg)
B&M 71, Springfield (ntg) to Bretton Woods (8a)

WASHINGTON to BRETTON WOODS-FABYAN
680: 6s-6dbr
PRR 168, Washington (4:10p) to New York (8p)
NH 168, New York (8:25p) to New Haven (10:09p)
NH 166, New Haven (12m) to Springfield (ntg)
B&M 71, Springfield (ntg) to Bretton Woods (8a)

NEW YORK to WOODSVILLE
NH 166, New York (10:15p) to Springfield (ntg)
B&M 71, Springfield (ntg) to Woodsville (6:10a)

(Diner, Washington to New York. Because the Night White Mountains *departs from New York's Grand Central Terminal, the Washington sleeper travels on No. 168, the* Montrealer, *via Penn Station and the Hell Gate route and joins the train in New Haven.)*

Federal, No. 173

New Haven Railroad
Springfield setouts
Standing in New Haven

SPRINGFIELD to WASHINGTON
8s-dr-3dbr   14r-4dbr (Mon-Fri)
NH 97, Springfield (8:55p) to New Haven (10:55p)
NH 173, New Haven (2:20a) to New York (3:50a)
PRR 173, New York (4:02a) to Washington (8:20a)

(The Federal *is now arriving in Providence)*

Shoreliner, No. 179

Boston to New York
New Haven Railroad
26 miles east of New Haven

PROVIDENCE to WASHINGTON
1791: 14r-4dbr
NH 179, Providence (10:05p)
        to New Haven (12:30a)
NH 173, New Haven (2:20a) to New York (3:50a)
PRR 173, New York (4:02a) to Washington (8:20a)

(Weekend schedules different. The Providence sleeper finishes its journey to Washington on the Federal.)

Cavalier, No. 469

New York to Cape Charles, Va.
Pennsylvania Railroad
Leaving Wilmington

NEW YORK to CAPE CHARLES
18r   8s-4dbr
12s-dr-c
PRR 469, New York (9p) to Cape Charles (6a)
Ferry, Cape Charles (6:55a) to Norfolk (9:45a)

PHILADELPHIA to CAPE CHARLES
10s-3dbr
PRR 469, Philadelphia (11:10p)
          to Cape Charles (6a)
Ferry, Cape Charles (6:55a) to Norfolk (9:45a)

(The Philadelphia Pullman departs Broad Street Station at 10 p.m. for 30th Street)

Palmland, No. 9

New York to Florida
Richmond, Fredericksburg & Potomac
Standing in Washington

WASHINGTON to HAMLET, N.C.
10s-lounge
RF&P 9, Washington (1:10a) to Richmond (4:05a)
SAL 9, Richmond (4:05a) to Hamlet (9:40a)

(Diner, south of Raleigh)

Apart from Florida destinations, RF&P/SAL scheduled dedicated sleepers from Washington to Columbia, S.C., and Columbus, Ga.

Metropolitan Special, No. 12

St. Louis to New York
Baltimore & Ohio
Standing in Washington

ST. LOUIS to NEW YORK
12s-dr
B&O 12, St. Louis (11:30p Sat)
        to Jersey City (6:50a Mon)

WASHINGTON to NEW YORK
12s-dr   8s-5dbr
B&O 12, Washington (1a) to Jersey City (6:50a)

(Dining-lounge car, Cincinnati to Washington; lunch counter: St. Louis to Cincinnati, Washington to New York. B&O trains terminate in Jersey City. Passengers take buses or ferry to New York.)

The B&O provided Washington with dedicated Pullman service to at least 35 cities and towns. Over the main line from St. Louis came sleepers from Houston (MP), Oklahoma City (SLSF), Cincinnati, Louisville, Fairmont, Grafton, and Cumberland.

The Southern Railway's *Asheville Special* arriving in Biltmore, N.C., Sept. 15, 1938.

District of Columbia (continued)

Piedmont Limited, No. 33
Southern Railway
Washington setout
Standing in Washington

WASHINGTON to NEW ORLEANS
S56: 10s-lounge
SOU 33, Washington (2:40a) to Atlanta (6:30p)
WPR 33, Atlanta (7:15p) to Montgomery (10:40p)
L&N 33, Montgomery (11:10p)
to New Orleans (8:05a)

(Diner, Monroe, Va., to New Orleans. The train that will pick up this sleeper-lounge is now leaving Philadelphia with Pullmans from New York for New Orleans, Mobile, and Charlotte—in addition to the consist of the Havana Special.)

The *Piedmont Limited* carried the tourist Pullmans that ran from Washington to Los Angeles and San Francisco (SP).

Pelican, No. 41
New York to New Orleans
Southern Railway
Departing Washington

NEW YORK to NEW ORLEANS
N80: 10r-6dbr
PRR 139, New York (6:50p Sun)
to Washington (11:10p)
SOU 41, Washington (11:55p)
to Lynchburg, Va. (4:12a Mon)
N&W 41, Lynchburg (4:20a) to Bristol, Va. (10a)
SOU 41, Bristol (10:10a)
to New Orleans (7:30a Tue)

NEW YORK to SHREVEPORT
N81: 10s-dr-2c
As above, New York (6:50p Sun)
to Bristol (10a Mon)
SOU 41, Bristol (10:10a)
to Meridian, Miss. (1:15a Tue)
IC 205, Meridian (2:50a) to Shreveport (1:30p Tue)

NEW YORK to KNOXVILLE
N82: 10r-6dbr
As above, New York (6:50p) to Bristol (10a)
SOU 41, Bristol (10:10a) to Knoxville (1:35p)

NEW YORK to BRISTOL
N83: 10r-6dbr
As above, New York (6:50p) to Bristol (10a)

NEW YORK to WILLIAMSON, W.VA.
N84: 10r-6dbr
As above, New York (6:50p) to Lynchburg (4:12a)
N&W 41, Lynchburg (4:20a) to Roanoke (5:30a)
N&W 15, Roanoke (6:25a) to Williamson (1:35p)

WASHINGTON to NEW ORLEANS
S64: 10s-lounge
As above, Washington (11:55p Sun)
to New Orleans (7:30a Tue)

WASHINGTON to ROANOKE
S15: 10r-6dbr
As above, Washington (11:55p)
to Roanoke (5:30a)

(Dining cars: New York to Washington, Roanoke to Birmingham. IC 205, the Southwestern Limited, carries cafe-lounge from Jackson, Miss., to Shreveport. N&W 15, the Cavalier, has a diner-lounge. The Pullman-laden Pelican, which is about to emerge from the tunnel under Capitol Hill, is the third of four Southern Railway trains to depart Washington at night for New Orleans. The all-sleeper Crescent pulls out at 7:15, the mostly coach Southerner at 8:50, and the Piedmont Limited ends the night's activity at 2:40 a.m.)

This joint Southern/N&W line to New Orleans brought Pullmans to Washington from Williamson and Bluefield, W.Va., Bristol, Knoxville, Chattanooga, Nashville, and Shreveport (IC).

Havana Special, No. 75
Richmond, Fredericksburg & Potomac
Washington setout
Standing in Washington

WASHINGTON to ORLANDO
8s-2c-dr
RF&P 75, Washington (3:05a) to Richmond (5:50a)
ACL 75, Richmond (6:30a) to Orlando (3:55a)

(Dining car and tavern-lounge, Richmond to Jacksonville. The Havana Special is now departing Philadelphia with New York sleepers for Miami, Tampa, and Fort Myers. The Pennsylvania operates the Havana Special and Piedmont Limited as one train.)

Washington had its own dedicated Pullmans to Key West when the *Havana Special* was the flagship of the Oversea Extension of Henry Flagler's Florida East Coast Railway. The RF&P/ACL also carried sleepers to New Bern, N.C. (NS) and Charleston, S.C.

Edison, No. 102
Washington to New York
Pennsylvania Railroad
Standing in Washington

WASHINGTON to NEW YORK
Std: s-r-dr-c-sbr-dbr-dup-/r
PRR 102, Washington (1a) to New York (6:05a)

No. 108
Washington to New York
Pennsylvania Railroad
Standing in Washington

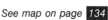

WASHINGTON to NEW YORK
Std: r-dbr
PRR 108, Washington (2:10a) to New York (6:25a)

WASHINGTON to PHILADELPHIA Mon.-Fri.
Std: s-dr
PRR 108, Washington (2:10a)
to Philadelphia (5:30a)

MEMPHIS to NEW YORK
S73: 14r-4dbr
SOU 46, Memphis (7:40p Sat)
to Bristol (10:40a Sun)
N&W 46, Bristol (10:50a) to Lynchburg (3:50p)
SOU 46, Lynchburg (3:55p) to Washington (8:10p)
PRR 108, Washington (2:10a Mon)
to New York (6:25a Mon)

(No. 108, an all-Pullman train, has an eclectic assortment of sleepers: the Memphis car is from the Tennessean; Pullmans from Richmond and Norfolk will soon arrive on the Havana Special, and the consist will be completed when sleepers come in on the Piedmont Limited from New Orleans, Mobile, and Salisbury, N.C.)

Florida

See map on page 134

At least forty-three cities and towns had dedicated Pullman service to Florida at one time or another, including: Asheville, Atlanta, Augusta, Birmingham, Boston, Buffalo, Charlotte, Chattanooga, Chicago, Cincinnati, Cleveland, Colorado Springs, Columbus (Ohio), Denver, Detroit, Grand Rapids, Greensboro, Indianapolis, Kansas City, Los Angeles, Louisville, Macon, Memphis, Montgomery, Montreal, Nashville, New Orleans, New York, Norfolk, Oklahoma City, Philadelphia, Pittsburgh, Portsmouth (Va.), Quebec, Roanoke, Savannah, Springfield (Mass.), St. Louis, Toledo, Washington, West Yellowstone, Wichita, and Winston-Salem.

Among the Florida Pullman destinations were: Boca Grande, Clewiston, Fort Myers, Jacksonville, Key West, Knight's Key, Miami, Naples, Orlando, Palatka, Palm Beach, Panama City, Pensacola, Sarasota, Sebring, St. Augustine, St. Petersburg, Tallahassee, Tampa, Venice, and West Palm Beach.

The *Havana Special*, once the principal train operating down the Atlantic Seaboard and onto the Florida East Coast's Oversea Extension to Key West, was still important in 1952, now for its diversity of destinations rather than maritime involvement. The *Havana Special* was particularly illustrative of how some trains were broken apart like seed pods to serve resorts on both coasts of Florida. Note the various trains numbered 35, 75, 76, 275, and 276 and how they interrelated to form the venerable *Havana Special* to and from New York.

Palmland, No. 1
New York to Florida
Seaboard Air Line
57 miles south of Jacksonville

CLEVELAND to TAMPA
4332: 10s-1dr-2c
NYC 433, Cleveland (12:30p Sat)
to Cincinnati (6:45p)
SOU 1, Cincinnati (9:15p)
to Jacksonville (7:55p Sun)
SAL 1, Jacksonville (10:30p)
to Tampa (5:10a Mon)

(Though now aboard a New York train, this sleeper from Cleveland came south on Southern No. 1, the Ponce de Leon. Diner-lounge, Cleveland to Cincinnati; restaurant-lounge, Cincinnati to Jacksonville.)

Palmland, No. 2
Florida to New York
Seaboard Air Line
Departing Plant City

TAMPA to CLEVELAND
B123: 10s-1dr-2c
SAL 2, Tampa (11:10p Sun)
to Jacksonville (6a Mon)
SOU 2, Jacksonville (9a) to Cincinnati (7:55a Tue)
NYC 426, Cincinnati (9:15a) to Cleveland (3:45p)

(This is the northbound version of No. 1, above, with the same accommodations in reverse)

Dixie Flagler, No. 4
Streamliner
Miami to Chicago
Every third day
Florida East Coast
3 miles north of St. Augustine

MIAMI to NASHVILLE
   N11: 8s-5dbr
   FEC 4, Miami (5:40p) to Jacksonville (12:40a)
   ACL 14-6, Jacksonville (1a) to Atlanta (8:45a)
   NC&StL 12, Atlanta (8:55a) to Nashville (2:05p)

MIAMI to CHICAGO
   DF91: 10r-6dbr   DF92: 6s-6dbr
   DF93: 6s-6dbr   DF94: 6c-3dbr
   DF90: 5dbr-lounge
   As above, Miami (5:40p) to Nashville (2:05p)
   L&N 12, Nashville (2:15p) to Evansville (5:15p)
   C&EI 10, Evansville (5:25p) to Chicago 10:55p)

(Dining car, lounge)

   One of the three streamliners created in 1940 to improve Chicago-Florida train service, the *Dixie Flagler* was made up of stainless-steel coaches that had been in FEC service. Pullmans came later. The train joined the famous "Dixie" fleet that had been pioneered more than a half-century earlier by the *Dixie Flyer*.

City of Miami, No. 4
Streamliner
Miami to Chicago
Two days out of three
Florida East Coast
3 miles north of St. Augustine

MIAMI to CHICAGO
   CM31: 6s-6r-4dbr   CM32: 10r-6dbr
   CM33: 10r-6dbr   CM34: 3dbr-dr-c-lounge
   CM35: 10r-6dbr
   FEC 4, Miami (5:40p) to Jacksonville (12:40a)
   ACL 14, Jacksonville (1a) to Albany, Ga. (4:40a)
   CofG 53, Albany (4:45a) to Birmingham (9:20a)
   IC 52, Birmingham (9:30a) to Chicago (10:55p)

MIAMI to ST. LOUIS
   CM36: 10r-6dbr
   As above, Miami (5:40p) to Birmingham (9:20a)
   IC 52, Birmingham (9:30a)
      to Carbondale, Ill. (5:45p)
   IC 226, Carbondale (5:55p) to St. Louis (8:45p)

(Diners and lounge, Miami to Chicago. Operates every third day off season.)

City of Miami, No. 4
Atlantic Coast Line
Jacksonville setout
Standing in Jacksonville

JACKSONVILLE to CHICAGO
   CM30: 6s-6r-4dbr
   As above, Jacksonville (1a) to Chicago (10:55p)

Miamian, No. 8
Miami to New York
Florida East Coast
16 miles north of St. Augustine

MIAMI to NEW YORK
   F173: 7c-lounge   F174: 10s-2c-dr
   F175: 6c-3dbr   F176: 10s-2c-dr
   F177: 12r-sb-4dbr
   FEC 8, Miami (5:05p) to Jacksonville (12:25a)
   ACL 8, Jacksonville (12:40a)
      to Richmond (11:30a)
   RF&P 8, Richmond (11:40a) to Washington (2p)
   PRR 106, Washington (2:25p) to New York (6:25p)

MIAMI to WASHINGTON
   F507: 6s-6dbr
   As above, Miami (5:05p) to Washington (2p)

(Miami to New York: dining car, lounge; Miami to Washington, dining car)

   The *Miamian* was a deluxe, all-Pullman train introduced in the late 1920's, principally serving New York and Washington, though also carrying a thrice-weekly sleeper to and from Quebec.

Seven cars make up a summer 1947 consist of the southbound *Dixie Flagler* at Danville, Ill.

New Royal Palm, No. 10
Streamliner
Miami to the Midwest
Florida East Coast
Leaving Ormond Beach

MIAMI to CINCINNATI
   F706: 12r-2sb-3dbr
   FEX 10, Miami (6:30p Sun)
      to Jacksonville (1:30a Mon)
   SOU 6, Jacksonville (1:50a) to Cincinnati (10:15p)

MIAMI to DETROIT
   F705: 13dbr   F704: 10r-5dbr
   F703: 10r-6dbr   F702: 4c-2dr-4dbr
   F700: 5dbr-obs.-lounge
   As above, Miami (6:30p Sun)
      to Cincinnati (10:15p Mon)
   NYC 302, Cincinnati (11:15p) to Detroit (7a Tue)

MIAMI to CLEVELAND
   F707: 10r-6dbr
   As above, Miami (6:30p Sun)
      to Cincinnati (10:15p Mon)
   NYC 442, Cincinnati (11:30p)
      to Cleveland (6:20a Tue)

MIAMI to BUFFALO
   F708: 14r-4dbr
   As above, Miami (6:30p Sun)
      to Cleveland (6:20a Tue)
   NYC 90, Cleveland (6:35a) to Buffalo (10:20a)

MIAMI to CHICAGO
   F710: 10r-6dbr   F709: 10r-6dbr
   As above, Miami (6:30p Sun)
      to Cincinnati (10:15p Mon)
   NYC 437, Cincinnati (11:55p)
      to Chicago (7:30a Tue)

(Miami to Cincinnati, diner-lounge and dining car. NYC 90 is the Chicagoan, *which carries a dining car and sleeper-lounge. NYC 437 also has a sleeper-lounge in its consist.)*

Palmland, No. 9
New York to Florida
Seaboard Air Line
23 miles north of Wildwood

NEW YORK to MIAMI
   S117: 8s-5dbr
   PRR 143, New York (8:30p Sat)
      to Washington (12:25a Sun)
   RF&P 9, Washington (1:10a) to Richmond (4:05a)
   SAL 9, Richmond (4:05a) to Miami (7:55a Mon)

(Raleigh, N.C., to Jacksonville, diner)

Palmland, No. 10
Miami to New York
Seaboard Air Line
41 Miles south of Sebring

MIAMI to NEW YORK
   B122: 8s-5dbr
   SAL 10, Miami (9p Sun) to Richmond (1:10a Tue)
   RF&P 110, Richmond (1:10a) to Washington (4a)
   PRR 112, Washington (5:05a) to New York (9:20a)

MIAMI to TAMPA
   CS2: 10s-dr-2c
   SAL 10, Miami (9p) to West Lake Wales (1:40a)
   SAL 428-27, West Lake Wales (2:30a)
      to Tampa (4:40a)

(Diner-lounges: Jacksonville to Raleigh, Washington to New York)

No. 28-427
Tampa to Miami
Seaboard Air Line
Entering Plant City

TAMPA to MIAMI
   CS2: 10s-dr-2c
   SAL 28-427, Tampa (11:20p)
      to West Lake Wales (1:15a)
   SAL 9, West Lake Wales (2:50a) to Miami (7:55a)

(This cross-Florida sleeper is now entering what the Indians called Ichepucksassa, but renamed Plant City, not for the surrounding strawberry fields, but for the rail magnate Henry B. Plant. The Pullman will complete its trip to Miami on No. 9, the Palmland.)

   In addition to Tampa-Miami, the SAL's night service across the state also once included sleepers from St. Petersburg.

Florida   (continued)

Southland, No. 32

Florida to the Midwest
Atlantic Coast Line
20 miles south of Perry

ST. PETERSBURG to CHICAGO
    R303: 8s-2c-dr   R304: 13dbr or 4c-4dr
ACL 132, St. Petersburg (6:20p Sun)
        to Trilby (8:35p)
ACL 32, Trilby (9:10p) to Albany (3:35a Mon)
CofG 32-33, Albany (3:45a) to Atlanta (8:40a)
L&N 32, Atlanta (9:25a) to Cincinnati (9:25p)
PRR 201, Cincinnati (11:35p)
        to Chicago (6:40a Tue)

ST. PETERSBURG to DETROIT
    B100: 10s-dr-2c
As above, St. Petersburg (6:20p Sun)
        to Cincinnati (9:25p)
B&O 58, Cincinnati (11:45p Mon)
        to Detroit (7:20a Tue)

SARASOTA/TAMPA to CHICAGO
    R302: 8s-dr-3dbr
ACL 76, Sarasota (4:45p Sun) to Tampa (7:10p)
ACL, Tampa (7:35p) to Trilby (8:40p)
As above, Trilby (9:10p) to Chicago (6:40a Tue)

SARASOTA/TAMPA to ATLANTA
    11: 8s-buffet-lounge
As above, Sarasota (4:45p) and
        Tampa (7:35p) to Atlanta (8:40a)

(Dining cars: St. Petersburg to Thomasville, Ga., Macon to Cincinnati. Sleeper-buffet-lounge, Atlanta to Chicago.)

The *Southland* made its bow in 1915. With the opening of the Perry cutoff in 1928, the train concentrated on serving the Gulf Coast resorts, with dedicated Pullmans to such northern cities as Chicago, Detroit, Cleveland, and Grand Rapids.

Havana Special, No. 75

New York to Florida
Atlantic Coast Line
9 miles north of Palatka

NEW YORK to TAMPA
    A162: 10s-c-dr
PRR 141, New York (10:15p Sat)
        to Washington (2:10a Sun)
RF&P 75, Washington (3:05a) to Richmond (5:50a)
ACL 75, Richmond (6:30a) to Jacksonville (8:30p)
ACL 75, Jacksonville (11p) to Tampa (7a Mon)

WASHINGTON to ORLANDO
    R508: 8s-2c-dr
As above, Washington (3:05a)
        to Jacksonville (8:30p)
ACL 75, Jacksonville (11p) to Orlando (3:55a)

CHICAGO to TAMPA/SARASOTA   alternate days
    903: 8s-2c-dr
IC 9, Chicago (4:20p Sat)
        to Birmingham (8:20a Sun)
CofG 10, Birmingham (8:40a) to Albany (4:25p)
ACL 17, Albany (4:30p) to Jacksonville (9p)
ACL 75, Jacksonville (11p) to Tampa (7a Mon)
        and Sarasota (11:30a Mon)

ST. LOUIS to TAMPA/SARASOTA   alternate days
    905: 8s-2c-1dr
IC 205, St. Louis (6:45p Sat) to Carbondale (9:15p)
IC 9, Carbondale (10:28p)
        to Birmingham (8:20a Sun)
As above, Birmingham (8:40a) to Tampa (7a Mon)
        and Sarasota (11:30a Mon)

(Washington to Jacksonville, diner; Richmond to Jacksonville, lounge. The Tampa/Sarasota Pullmans came to Jacksonville not on the Havana Special, but on the Seminole from the Midwest.)

Advance Havana Special, No. 35

New York to Miami
Florida East Coast
Approaching Ormond Beach

NEW YORK to MIAMI
    A161: 10-2c-2dbr
PRR 141, New York (10:15p Sat)
        to Washington (2:10a Sun)
RF&P 75, Washington (3:05a) to Richmond (5:50a)
ACL 75, Richmond (6:30a) to Jacksonville (8:30p)
FEC 35, Jacksonville (10:30p) to Miami (7a Mon)

JACKSONVILLE to MIAMI
    S: 10s-dr-2c   G: 10s-observation
FEC 35, Jacksonville (10:30p) to Miami (7a)

(Washington to Jacksonville, diner; Richmond to Jacksonville, lounge. The FEC's regular Havana Special, offering coaches only, follows the Advance Havana a half-hour later.)

When the Oversea Extension of Henry Flagler's FEC was completed to Key West in 1912, through service from New York began, and the *Havana Special* was the flagship of the operation until the disastrous hurricane of Labor Day 1935 closed it forever. Thereafter, the train terminated in Miami.

Seminole, No. 37

Chicago to Florida Gulf Coast
Atlantic Coast Line
24 miles north of Gainesville

CHICAGO to ST. PETERSBURG   alternate days
    907: 8s-2c-dr
IC 9, Chicago (4:20p Sat)
        to Birmingham (8:20a Sun)
CofG 10, Birmingham (8:40a) to Albany (4:25p)
ACL 17, Albany (4:30p) to Jacksonville (9p)
ACL 37, Jacksonville (10:30p)
        to St. Petersburg (9a Mon)

ST. LOUIS to ST. PETERSBURG   alternate days
    909: 8s-2c-dr
IC 205, St. Louis (6:45p Sat) to Carbondale (9:15p)
IC 9, Carbondale (10:28p)
        to Birmingham (8:20a Sun)
As above, Birmingham (8:40a)
        to St. Petersburg (9a Mon)

(Chicago to Jacksonville, diner-lounge)

The *Seminole* became the first year-round train to Jacksonville from Chicago, in 1909. For a number of years it also carried Pullmans to Savannah. By the time of this scheduling, the *Seminole* worked the Gulf Coast, the *City of Miami* the ocean side.

Seminole, No. 38

Florida West Coast to Chicago
Atlantic Coast Line
60 miles south of Gainesville

ST. PETERSBURG to CHICAGO   alternate days
    907: 8s-2c-dr
ACL 38, St. Petersburg (6:30p Sun)
        to Jacksonville (6a)
ACL 18, Jacksonville (7:45a Mon)
        to Albany (12:25p)
CofG 9, Albany (12:30p) to Birmingham (6:10p)
IC 10, Birmingham (6:30p)
        to Chicago (10:45a Tue)

ST. PETERSBURG to ST. LOUIS   alternate days
    909: 8s-2c-dr
As above, St. Petersburg (6:30p Sun)
        to Birmingham (6:10p)
IC 10, Birmingham (6:30p Mon)
        to Carbondale (4:35a Tue)
IC 16, Carbondale (4:50a) to St. Louis (7:47a)

(Jacksonville to Chicago: diner-lounge)

Gulf Wind, No. 60

Jacksonville to Flomaton, Ala.
Louisville & Nashville
13 miles east of Pensacola

JACKSONVILLE to NEW ORLEANS
    B1: 6s-6dbr   B2: 10s-buffet-lounge
    B5: 6s-4r-4dbr
SAL 39, Jacksonville (5p)
        to Chattahoochee (9:35p)
L&N 60, Chattahoochee (9:45p)
        to Flomaton (1:40a)
L&N 99, Flomaton (1:50a) to New Orleans (7:10a)

(Jacksonville to Tallahassee, diner. At Flomaton, the Gulf Wind becomes part of No. 99, the Pan-American.)

The post-war *Gulf Wind* continued service that once had included sleepers from Jacksonville to both Tallahassee and Pensacola, as well as an unusual, transcontinental Pullman line that operated briefly during the 1920's to Los Angeles.

Gulf Wind, No. 61

Flomaton to Jacksonville
Louisville & Nashville
13 miles east of Pensacola

NEW ORLEANS to JACKSONVILLE
    L10: 6s-4r-4dbr   L11: 10s-buffet-lounge
    L12: 6s-6dbr
L&N 34, New Orleans (5p) to Flomaton (10:15p)
L&N 61, Flomaton (10:30p)
        to Chattahoochee (4:25a)
SAL 38, Chattahoochee (4:35a)
        to Jacksonville (9a)

(This Jacksonville train leaves New Orleans as part of the L&N's Piedmont Limited, which carries a diner to Mobile. The Seaboard cuts another diner into the Gulf Wind at Tallahassee, serving breakfast into Jacksonville.)

Havana Special, No. 76

Florida Gulf Coast section
Atlantic Coast Line
25 miles east of Plant City

SARASOTA/TAMPA to CHICAGO   alternate days
    903: 8s-2c-dr
ACL 76, Sarasota (4:45p Sun) and
        Tampa (10:30p Sun)
        to Jacksonville (6:30a Mon)
ACL 18, Jacksonville (7:45a) to Albany (12:25p)
CofG 9, Albany (12:30p) to Birmingham (6:10p)
IC 10, Birmingham (6:30p)
        to Chicago (10:45a Tue)

SARASOTA/TAMPA to ST. LOUIS   alternate days
    905: 8s-2c-dr
As above, Sarasota (4:45p Sun) and
        Tampa (10:30p Sun)
        to Birmingham (6:10p Mon)
IC 10, Birmingham (6:30p)
        to Carbondale (4:35a Tue)
IC 16, Carbondale (4:50a) to St. Louis (7:47a)

TAMPA to NEW YORK
    A162: 10s-c-dr
ACL 76, Tampa (10:30p Sun)
        to Jacksonville (6:30a Mon)
ACL 76, Jacksonville (8:10a)
        to Richmond (10:05p)
RF&P 76, Richmond (10:30p)
        to Washington (1:10a Tue)
PRR 110, Washington (2:15a) to New York (7:05a)

(In Jacksonville, the Sarasota/Tampa Pullmans will become part of the Seminole, which carries a diner-lounge to Chicago. The Tampa-New York sleeper will become part of the regular Havana Special, with dining car to Washington, lounge to Richmond.)

Havana Special, No. 76
Miami to New York
Florida East Coast
Standing in West Palm Beach

MIAMI to JACKSONVILLE
  S: 10s-2c-dr   G: 10s-observation
FEC 76, Miami (10p) to Jacksonville (7a)

MIAMI to NEW YORK
  F161: 10s-2dbr-c
FEC 76, Miami (10p Sun) to Jacksonville (7a Mon)
ACL 76, Jacksonville (8:10a)
      to Richmond (10:05p)
RF&P 76, Richmond (10:30p)
      to Washington (1:10a Tue)
PRR 110, Washington (2:15a)
      to New York (7:05a Tue)

(Jacksonville to Washington, dining car; Jacksonville to Richmond, lounge)

Havana Special, No. 76
Atlantic Coast Line
Orlando setout
Standing in Orlando

ORLANDO to WASHINGTON
  R508: 8s-2c-dr
ACL 76, Orlando (1:35a Mon)
      to Jacksonville (6:30a)
As above, Jacksonville (8:10a)
      to Washington (1:10a Tue)

No. 275
Jacksonville to Fort Myers
Atlantic Coast Line   Logo 1
12 miles south of Palatka

NEW YORK to FORT MYERS
  A163: 10s-2c-dr
PRR 141, New York (10:15p Sat)
      to Washington (2:10a Sun)
RF&P 75, Washington (3:05a) to Richmond (5:50a)
ACL 75, Richmond (6:30a) to Jacksonville (8:30p)
ACL 275, Jacksonville (10:30p)
      to Fort Myers (8:30a Mon)

(This sleeper travels to Jacksonville on the Havana Special.*)*

No. 276
Fort Myers to Jacksonville
Atlantic Coast Line
8 miles south of Orlando

FORT MYERS to NEW YORK
  A163: 10s-2c-dr
ACL 276, Fort Myers (6:45p Sun)
      to Jacksonville (5:45a)
ACL 76, Jacksonville (8:10a Mon)
      to Richmond (10:05p)
RF&P 76, Richmond (10:30p)
      to Washington (1:10a Tue)
PRR 110, Washington (2:15a)
      to New York (7:05a Tue)

(This is the northbound version of No. 275)

Everglades, No. 376
Jacksonville to Washington
Atlantic Coast Line
25 miles north of Jacksonville

JACKSONVILLE to WASHINGTON
  R503: 8s-2c-dr
ACL 376, Jacksonville (11:35p)
      to Richmond (1:15p)
RF&P 376, Richmond (1:25p)
      to Washington (3:45p)

(Florence, S.C., to Washington, dining car)

Introduced as an all-Pullman train at the end of the Roaring '20s, the *Everglades* carried sleepers from both coasts of Florida, principally to New York and Boston (including a car from Palm Beach), though a Miami-Springfield, Mass., Pullman also made the trip four times a week.

See map on page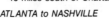

Royal Palm, No. 3
Midwest to Jacksonville
Via Valdosta
Southern Railway
20 miles north of Macon

CINCINNATI to JACKSONVILLE
  80: 10s-lounge
SOU 3, Cincinnati (8:45a) to Jacksonville (8:15a)

CHICAGO TO JACKSONVILLE
  4101: 10s-dr-2c
NYC 410, Chicago (11:30p Sat)
      to Cincinnati (8a Sun)
SOU 3, Cincinnati (8:45a)
      to Jacksonville (8:15a Mon)

DETROIT to JACKSONVILLE
  3093: 8s-dr-2c
NYC 309, Detroit (11:30p Sat)
      to Cincinnati (6:50a Sun)
SOU 3, Cincinnati (8:45a)
      to Jacksonville (8:15a Mon)

(Chicago to Cincinnati, buffet-lounge; Cincinnati to Atlanta, dining car)

The *Royal Palm* was introduced just before World War I to the Chicago-Jacksonville trade, and service expanded to include other Great Lakes cities and resorts on both coasts of Florida. In the late 1920's, a seasonal all-Pullman section was operated as the *Royal Palm De Luxe,* and a post-World War II, seasonal streamliner carried the name *New Royal Palm.*

Over this line the Southern carried dedicated sleepers from Atlanta to Chattanooga, Louisville, Cincinnati, Chicago and Detroit (NYC), and to Macon and Miami (SAL).

No. 3
Augusta to Atlanta
Georgia Railroad
Standing in Augusta

AUGUSTA to ATLANTA
  12s-dr
Georgia 3, Augusta (2:20a) to Atlanta (7a)

The Georgia Railroad and NC&StL cooperated on Pullman service from Augusta to Chicago (L&N/C&EI) and St. Louis (IC)

No. 3
Savannah to Atlanta and Columbus
Central of Georgia
4 miles west of Midville

SAVANNAH to ATLANTA
  A: 10s-dr-2c
CofG 3, Savannah (9:30p) to Macon (2:30a)
CofG 3, Macon (3:30a) to Atlanta (6:30a)

SAVANNAH to COLUMBUS
  C: 10s-dr-2c
CofG 3, Savannah (9:30p) to Macon (2:30a)
CofG 3, Macon (3:40a) to Columbus (6:45a)

Savannah had dedicated sleeping cars to Augusta, Birmingham, and Chicago (IC).

No. 4
Atlanta to Nashville
Nashville, Chattanooga & St. Louis
40 miles south of Chattanooga

ATLANTA to NASHVILLE
  1: 14s
NC&StL 4, Atlanta (9p) to Nashville (6:30a)

Pullmans from Atlanta traveled on the NC&StL to Chattanooga and Memphis (L&N).

No. 4
Atlanta to Augusta
Georgia Railroad
15 miles west of Camak

ATLANTA to AUGUSTA
  12s-dr
Georgia 4, Atlanta (9p) to Augusta (1:40a)

ATLANTA to FLORENCE, S.C.
  10s-dr-c
Georgia 4, Atlanta (9p) to Augusta (1:40a)
ACL 54, Augusta (2:10a) to Florence (7:05a)

ATLANTA to WILMINGTON
  10s-dr-2c
Georgia 4, Atlanta (9p) to Augusta (1:40a)
ACL 54, Augusta (2:10a) to Wilmington (11:35a)

No. 4
Atlanta to Savannah
Central of Georgia
43 miles northwest of Macon

ATLANTA to SAVANNAH
  A: 10s-dr-2c
CofG 4, Atlanta (10:30p) to Macon (1:45a)
CofG 4, Macon (2:30a) to Savannah (7:45a)

ATLANTA to PANAMA CITY
  40: 10s-dr-2c
CofG 4, Atlanta (10:30p) to Macon 1:45a)
CofG 7, Macon (2:40a) to Dothan (8:45a)
A&StAB 1, Dothan (8:50a) to Panama City (11:05a)

The Central of Georgia routed many sleepers through Macon, where they were switched in the middle of the night and dispatched to their final destinations. Atlanta service included cars to Albany, Montgomery, Tallahassee and Tampa (SAL), Thomasville and Jacksonville (ACL), and Valdosta and Palatka (SOU). Macon had its own sleepers scheduled to Columbus, Savannah, and Cincinnati (L&N).

No. 4
Columbus to Savannah
Central of Georgia
48 miles east of Columbus

COLUMBUS to SAVANNAH
  C: 10s-dr-2c
CofG 4, Columbus (10:45p) to Macon (1:40a)
CofG 4, Macon (2:30a) to Savannah (7:45a)

Royal Palm, No. 4
Jacksonville to Midwest
Via Valdosta
Southern Railway
24 miles east of Valdosta

JACKSONVILLE to CINCINNATI
  80: 10s-lounge
SOU 4, Jacksonville (10p) to Cincinnati (9:55p)

JACKSONVILLE to CHICAGO
  S200: 10s-dr-2c
SOU 4, Jacksonville (10p Sun)
      to Cincinnati (9:55p Mon)
NYC 437, Cincinnati (11:55p)
      to Chicago (7:30a Tue)

JACKSONVILLE to DETROIT
  S-690: 8s-dr-2c
SOU 4, Jacksonville (10p Sun)
      to Cincinnati (9:55p Mon)
NYC 302, Cincinnati (11:15p) to Detroit (7a Tue)

(Atlanta to Cincinnati, dining car; Cincinnati to Chicago, buffet-lounge)

At Macon, the Southern's two lines from Jacksonville came together. Macon had Pullman service from Jacksonville as well as to Atlanta, Asheville, Louisville, and Cincinnati, and to Washington and New York (PRR).

Georgia   (continued)

New Royal Palm, No. 5
Midwest to Miami, winter
Via Valdosta
Southern Railway
18 miles north of Tifton

CINCINNATI to MIAMI
706: 12r-2sbr-3dbr
SOU 5, Cincinnati (8:30a) to Jacksonville (4:05a)
FEC 9, Jacksonville (4:25a) to Miami (11:25a)

DETROIT to MIAMI
700: Obs. buffet-5dbr   702: 4c-4dbr-2dr
703: 10r-6dbr   704: 10r-5dbr
705: 13dbr
NYC 309, Detroit (11:30p Sat)
to Cincinnati (6:50a Sun)
As above, Cincinnati (8:30a)
to Miami (11:25a Mon)

CLEVELAND to MIAMI
707: 10r-6dbr
NYC 417, Cleveland (12:10a Sun)
to Cincinnati (7:10a)
As above, Cincinnati (8:30a)
to Miami (11:25a Mon)

BUFFALO to MIAMI
708: 14r-4dbr
NYC 5, Buffalo (8p Sat) to Cleveland (11:55p)
As above, Cleveland (12:10a Sun)
to Miami (11:25a Mon)

CHICAGO to MIAMI
709: 10r-6dbr   710: 10r-6dbr
NYC 410, Chicago (11:30p Sat)
to Cincinnati (8a Sun)
As above, Cincinnati (8:30a)
to Miami (11:25a Mon)

(Chicago to Cincinnati, buffet-lounge; Cincinnati to Miami, diner and lounge car)

Like tree branches, the network of the New York Central flowed into a main trunk at Cincinnati, where Florida trains for several routes were consolidated. A generation earlier, the consist of the *Royal Palm De Luxe* would also have included Pullmans from Grand Rapids and Indianapolis.

Sunland, No. 8
Florida to Washington
Seaboard Air Line
10 miles north of Savannah

JACKSONVILLE to WASHINGTON
B110: 10s-dr-2c
SAL 8, Jacksonville (9p) to Richmond (11a)
RF&P 108, Richmond (11:25a)
to Washington (2:20p)
TAMPA to PORTSMOUTH, Va.
B18: 10s-dr-2c
SAL 108, Tampa (2:35p) to Wildwood, Fla. (4:45p)
SAL 8, Wildwood (5:10p) to Norlina, N.C. (9a)
SAL 18, Norlina (9a) to Portsmouth (12:25p)

(Wildwood to Jacksonville, SAL parlor-dining car; Hamlet, N.C. to Washington, SAL diner)

No. 8
Dothan to Macon
Central of Georgia
10 miles north of Americus

PANAMA CITY to ATLANTA
41: 10s-2c-dr
A&StAB 4, Panama City (5:10p) to Dothan (7:15p)
CofG 8, Dothan (7:25p) to Macon (1:55a)
CofG 3, Macon (3:30a) to Atlanta (6:30a)

Kansas City-Florida Special, No. 7
Kansas City to Jacksonville
Via Jesup
Southern Railway
35 miles northwest of Macon

KANSAS CITY to JACKSONVILLE
F67: 8s-5dbr   F69: 10s-c-2dbr
SLSF 105, Kansas City (11:30p Sat)
to Birmingham (4:25p)
SOU 8, Birmingham (4:40p Sun)
to Atlanta (10:25p)
SOU 7, Atlanta (10:45p)
to Jacksonville (7:30a Mon)

KANSAS CITY to MIAMI
F68: 14r-4dbr
As above, Kansas City (11:30p Sat)
to Jacksonville (7:30a)
FEC 73, Jacksonville (9:50a Mon)
to Miami (5:15p Mon)

ATLANTA to JACKSONVILLE
63: 8s-c-diner-lounge
SOU 7, Atlanta (10:45p) to Jacksonville (7:30a)

ATLANTA to BRUNSWICK
51: 10s-dr-2c
SOU 7, Atlanta (10:45p) to Jesup (5:10a)
SOU motor train, Jesup (5:40a)
to Brunswick (7:10a)

(Dining cars: Kansas City to Birmingham, Birmingham to Atlanta, and two diners Jacksonville to Miami on the Vacationer; lounge car Kansas City to Birmingham. Another Kansas City-Florida Special, 24 hours behind this one, is now in Kansas.)

The Frisco's connection with Southern Railway at Birmingham after the turn of the century opened a Florida route that brought sleepers to the Sunshine State from Oklahoma City, Wichita, Colorado Springs, and Denver. Atlanta had dedicated service from Kansas City and Denver (RI or MP/D&RGW), and Brunswick sleepers from Colorado Springs (RI) and Birmingham were scheduled.

Kansas City-Florida Special, No. 8
Jacksonville to Kansas City
Via Jesup
Southern Railway
18 miles northwest of Jesup

MIAMI to KANSAS CITY
F68: 14r-4dbr
FEC 74, Miami (1p Sun) to Jacksonville (8:20p)
SOU 8, Jacksonville (9:20p) to Atlanta (6:30a Mon)
SOU 7, Atlanta (7:30a) to Birmingham (11:30a)
SLSF 106, Birmingham (12:05p)
to Kansas City (7:30a Tue)

JACKSONVILLE to KANSAS CITY
S67: 8s-5dbr   S69: 10s-c-2dbr
As above, Jacksonville (9:20p Sun)
to Kansas City (7:30a)

JACKSONVILLE to ATLANTA
63: 8s-c-diner-lounge
SOU 8, Jacksonville (9:20p) to Atlanta (6:30a)

BRUNSWICK to ATLANTA
51: 10s-dr-2c
SOU motor train, Brunswick (9:35p)
to Jesup (11p)
SOU 8, Jesup (11:25p) to Atlanta (6:30a)

(Diners: two cars Miami to Jacksonville aboard the Vacationer, and one each Atlanta to Birmingham, and Birmingham to Kansas City; lounge Birmingham to Kansas City)

Dixie Flagler, No. 11
Chicago to Miami streamliner
Every third day
Nashville, Chattanooga & St. Louis
On the outskirts of Atlanta

CHICAGO to JACKSONVILLE
95: 6s-4r-4dbr
C&EI 9, Chicago (8:10a) to Evansville (1:40p)
L&N 11, Evansville (1:50p) to Nashville (4:50p)
NC&StL 11, Nashville (5p) to Atlanta (12:15a)
ACL 5-15, Atlanta (12:25a) to Jacksonville (8:15a)

CHICAGO to MIAMI
90: 5dbr-lounge   91: 10r-6dbr
92: 6s-6dbr   93: 6s-6dbr
94: 6c-3dr

As above, Chicago (8:10a) to Jacksonville (8:15a)
FEC 3, Jacksonville (8:40a) to Miami (3:40p)

NASHVILLE to MIAMI
N11: 8s-5dbr
As above, Nashville (5p) to Miami (3:40p)

(Chicago to Miami: dining car, coffee shop-lounge, observation lounge. The Dixie Flagler alternated service with the City of Miami and the South Wind.)

These three streamliners, originally all coach, were inaugurated in 1940 by nine participating railroads to perk up Florida service from the Midwest. The *Dixie Flagler* served Atlanta; the other operated through southern Georgia.

No. 11
Atlanta to Birmingham
Southern Railway
On the outskirts of Atlanta

ATLANTA to BIRMINGHAM
54: 10s-dr-2c
SOU 11, Atlanta (11:45p) to Birmingham (5a)

Beyond Birmingham, the Southern and Illinois Central carried interline sleeping cars from Atlanta to Shreveport and Fort Worth (T&P).

Silver Star, No. 22
Florida to New York
Seaboard Air Line
18 miles south of Thalmann

MIAMI to NEW YORK
B45: 10r-6dbr   B46: 8s-5dbr
B47: 6c-buffet-lounge
SAL 22, Miami (4:40p) to Richmond (11:10a)
RF&P 22, Richmond (11:10a)
to Washington (1:50p)
PRR 190, Washington (2:10p) to New York (6:15p)

PORT BOCA GRANDE to NEW YORK
B55: 6c-3dr
SAL 322-29, Port Boca Grande (3p)
to Tampa (6:50p)
SAL 122, Tampa (7:15p) to Wildwood (9p)
SAL 22, Wildwood (9:25p) to Richmond (11:10a)
As above, Richmond (11:10a) to New York (6:15p)

ST. PETERSBURG to NEW YORK
B56: 8s-5dbr
SAL 122, St. Petersburg (5:30p) to Wildwood (9p)
As above, Wildwood (9:25p) to New York (6:15p)

(Miami to New York: SAL diner, tavern coach, observation coach; St. Petersburg to New York, SAL dining car)

No. 57
Savannah to Montgomery
Atlantic Coast Line
26 miles east of Valdosta

SAVANNAH to MONTGOMERY
5: 12s-dr
ACL 57, Savannah (7:30p) to Montgomery (7:25a)

The ACL carried a Savannah-Jacksonville sleeper and a Thomasville car from New York (PRR/RF&P).

Skyland Special, No. 24
Jacksonville to Charlotte
Southern Railway
21 miles south of Savannah

JACKSONVILLE to ASHEVILLE
 S124: 8s-dr-3dbr
 SOU 24, Jacksonville (9:20p) to Columbia (4:25a)
 SOU 9, Columbia (4:50a) to Asheville (11:30a)

JACKSONVILLE to CHARLOTTE
 S126: 12s-dr
 SOU 24, Jacksonville (9:20p) to Charlotte (7:55a)

(Dinette-coach, Jacksonville to Asheville)

The *Skyland Special* was a dual-origin train whose elements came together at Columbia and continued on to Jacksonville. At one time the Asheville sleeper went all the way to Cincinnati, and the Charlotte service extended to Greensboro.

Savannah had Pullman service to Columbia, Asheville, and Charlotte.

No. 36
New Orleans to Washington
Southern Railway
Standing in Atlanta

NEW ORLEANS to WASHINGTON
 L17: 12s-dr
 L&N 4, New Orleans (8:20a)
  to Montgomery (5:25p)
 WPR 36, Montgomery (6p) to Atlanta (11:40p)
 SOU 36, Atlanta (12:30a) to Washington (5:45p)

ATLANTA to WASHINGTON
 58: 10s-c-2dbr
 SOU 36, Atlanta (12:30a) to Washington (5:45p)

ATLANTA to COLUMBIA
 31: 12s-dr
 SOU 36, Atlanta (12:30a) to Greenville (4:45a)
 SOU 18, Greenville (5:15a) to Columbia (10a)

(Diners: New Orleans-Montgomery, Greenville to Monroe, Va.)

Atlanta had Pullman service on this line from Montgomery and San Antonio (SP/L&N).

No. 58
Montgomery to Savannah
Atlantic Coast Line
At the Alabama state line.

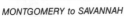

MONTGOMERY to SAVANNAH
 5: 12s-dr
 ACL 58, Montgomery (6:45p) to Savannah (9:10a)

Southland No. 32-33
Midwest to Florida Gulf Coast
Central of Georgia
Entering Americus

CHICAGO to ST. PETERSBURG
 R303: 8s-2c-dr
 R304: 13dbr, or   4c-4dr
 PRR 200, Chicago (11:35p Sat)
  to Cincinnati (7:40a Sun)
 L&N 33, Cincinnati (8a) to Atlanta (7:10p)
 CofG 32-33, Atlanta (8p) to Albany (1a Mon)
 ACL 33, Albany (1:15a) to Trilby, Fla. (7:30a)
 ACL 133, Trilby (8:05a)
  to St. Petersburg (10:10a Mon)

CHICAGO to TAMPA/SARASOTA
 R302: 8s-dr-3dbr
 As above, Chicago (11:35p Sat)
  to Trilby (7:30a Mon)
 ACL 33-37, Trilby (8a) to Tampa (9:15a) and
  Sarasota (11:30a)

DETROIT to ST. PETERSBURG
 B100: 10s-2c-dr
 B&O 57, Detroit (11:40p Sat)
  to Cincinnati (6:55a Sun)
 As above, Cincinnati (8a)
  to St. Petersburg (10:10a Mon)

ATLANTA to TAMPA/SARASOTA
 11: 8s-buffet-lounge
 As above, Atlanta (8p) to Trilby (7:30a)
 ACL 33-37, Trilby (8a) to Tampa (9:15a)
  and Sarasota (11:30a)

(Dining cars: Cincinnati to Macon, Thomasville to St. Petersburg; lounge-sleeper, Chicago to Atlanta)

The Pennsylvania Railroad set out to meet the competition of the *Royal Palm* with the advent of the *Southland* in 1915. In addition to its own network, the Pennsy availed itself of other northern connections, such as here with the B&O and earlier with the Wabash, which delivered Detroit Pullmans to Fort Wayne. After the opening of the Perry Cutoff in Florida in 1928, the *Southland* concentrated on serving the Gulf Coast.

Dixie Flyer, No. 94
Jacksonville to Chicago
Atlantic Coast Line
26 miles east of Tifton

JACKSONVILLE to CHICAGO
 51: 10s-2c-dr
 ACL 94, Jacksonville (9p Sun) to Albany (2a Mon)
 CofG 94-5, Albany (2:10a) to Atlanta (8a)
 NC&StL 94, Atlanta (9:35a) to Nashville (5:10p)
 L&N 94, Nashville (5:30p) to Evansville (10p)
 C&EI 6, Evansville (10:20p) to Chicago (5:20a Tue)

JACKSONVILLE to CHICAGO
 R301: 8s-buffet-lounge
 As above, Jacksonville (9p Sun)
  to Atlanta (8a Mon)
 L&N 32, Atlanta (9:25a) to Cincinnati (9:25p)
 PRR 201, Cincinnati (11:35p)
  to Chicago (6:40a Tue)

JACKSONVILLE to ATLANTA
 950: 10s-2c-dr
 As above, Jacksonville (9p) to Atlanta (8a)

(Atlanta to Evansville, dining car; the Jacksonville-Chicago sleeper/buffet-lounge R301 becomes part of the Southland north of Atlanta.)

The *Dixie Flyer*, the oldest name in midwestern train travel to Florida, could trace the origins of its service, between St. Louis and Jacksonville, to the 1880's. The onetime all-Pullman train eventually became the namesake for an entire Dixie fleet.

Dixie Flyer, No. 94-95
Chicago to Jacksonville
Central of Georgia
10 miles south of Macon

CHICAGO to JACKSONVILLE
 51: 10s-2c-dr
 C&EI 5, Chicago (10:45p Sat)
  to Evansville (5:20a Sun)
 L&N 95, Evansville (5:40a) to Nashville (10:10a)
 NC&StL 95, Nashville (10:30a) to Atlanta (7:15p)
 CofG 94-94, Atlanta (9:05p) to Albany (2:35a Mon)
 ACL 95, Albany (2:50a)
  to Jacksonville (7:45a Mon)

CHICAGO to JACKSONVILLE
 R301: 8s-buffet-lounge
 PRR 200, Chicago (11:35p Sat)
  to Cincinnati (7:40a Sun)
 L&N 33, Cincinnati (8a) to Atlanta (7:10p)
 As above, Atlanta (9:05p)
  to Jacksonville (7:45a Mon)

ATLANTA to JACKSONVILLE
 950: 10s-2c-dr
 As above, Atlanta (9:05p) to Jacksonville (7:45a)

(Dining car, Evansville, to Atlanta. The Chicago-Jacksonville buffet-lounge sleeper R301 makes the trip to Atlanta in the Southland.)

Everglades, No. 375
Washington to Jacksonville
Atlantic Coast Line
8 miles north of Savannah

WASHINGTON to JACKSONVILLE
 R503: 8s-2c-dr
 RF&P 375, Washington (10:50a)
  to Richmond (1:15p)
 ACL 375, Richmond (1:25p)
  to Jacksonville (3:30a)

(Dining car, Washington to Florence, S.C.)

In 1929, the all-Pullman *Everglades* began service to both coasts of Florida from New York and Boston and entered the Havana competition by offering connections at Miami to the flying boats of the young and enterprising Pan-American Airways.

Frank Ardrey

A Central of Georgia K-1 hefts the *Seminole* near Weems, Ala., in April 1948.

Idaho

See map on page 290

Alaskan, No. 4

Seattle to St. Paul
via Helena, Mont.
Northern Pacific
3 miles east of Sandpoint

SPOKANE to BILLINGS, MONT.
47: 10s-c-dr
NP 4, Spokane (9:45p) to Billings (6:35p)

(Seattle to Spokane, cafe car; station and meal stops east of Spokane. Other eastbound editions of the Alaskan, a three-night train, are now in Montana and Minnesota.)

The *Alaskan* was the NP's principal secondary train until the introduction of the streamlined *Mainstreeter* in the early 1950's.

Idahoan, No. 11

Omaha to Portland
Union Pacific
20 miles east of Shoshone

ST. LOUIS to SEATTLE
6s-6r-4dbr
Wabash 9, St. Louis (4p Sat) to Kansas City (9p)
UP 9, Kansas City (9:30p)
    to Green River, Wyo. (3:35p Sun)
UP 11, Green River (4:30p) to Portland (3:30p Mon)
Pool 459, Portland (5p) to Seattle (9:15p Mon)

SALT LAKE CITY to PORTLAND
10s-dr-c
UP 31, Salt Lake City (5:25p Sun)
    to Pocatello (9:55p)
UP 11, Pocatello (10:25p) to Portland (3:30p Mon)

(First-class lounge, diner: St. Louis to Green River, and North Platte, Neb., to Portland on No. 11; Cafe-lounge Salt Lake City to Pocatello; diner Portland to Seattle)

Idahoan, No. 12

Portland to Omaha
Union Pacific
75 miles east of Boise

SEATTLE to ST. LOUIS
6s-6r-4dbr
Pool 402, Seattle (11:45p Sat)
    to Portland (6:45a Sun)
UP 12, Portland (8:10a) to Green River (9:50a Mon)
UP 10, Green River (10:45a)
    to Kansas City (6:45a Tue)
Wabash 10, Kansas City (7a)
    to St. Louis (11:59a Tue)

PORTLAND to SALT LAKE CITY
10s-dr-c
UP 12, Portland (8:10a) to Pocatello (3:40a)
UP 32, Pocatello (4:15a) to Salt Lake City (8:50a)

BOISE to SALT LAKE CITY
10s-3dbr
UP 12, Boise (10:15p) to Pocatello (3:40a)
UP 32, Pocatello (4:15a) to Salt Lake City (8:50a)

(Diner, first-class lounge: Portland to North Platte on No. 12, Green River to St. Louis; cafe-lounge, Pocatello to Salt Lake City)

Aside from the Pullmans to West Yellowstone and Victor for the benefit of national park visitors, the UP carried regular sleepers from Salt Lake City to Ashton and Buhl and a tourist car to Idaho Falls. Service from Boise also included Pullmans for Lewiston and Spokane.

Olympian Hiawatha, No. 15

Streamliner
Chicago to Seattle/Tacoma
Milwaukee Road
35 miles west of Avery

CHICAGO to SEATTLE/TACOMA
151: 10r-6dbr   152: 8dbr-Skytop lounge
A15: 14s Touralux   B15: 14s Touralux
MILW 15, Chicago (3:30p Sat)
    to Seattle (10:30a Mon)
    and Tacoma (11:45a)

(Diner, Tip Top grill car. The Olympian Hiawatha *is one of the two-night trains between Chicago and Seattle. The other westbound Hi is now in Minnesota.)*

Olympian Hiawatha, No. 16
Streamliner
Tacoma/Seattle to Chicago
Milwaukee Road
35 miles west of Avery

TACOMA/SEATTLE to CHICAGO
161: 10r-6dbr   162: 8dbr-Skytop lounge
A16: 14s Touralux   B16: 14s Touralux
MILW 16 Tacoma (1:30p Sun) and
    Seattle (2:45p) to Chicago (1:45p Tue)

(Diner, Tip Top grill car. Another eastbound Hiawatha is now in South Dakota.)

Mail Express, No. 25
Pocatello to Portland
Union Pacific
Standing in Pocatello

SALT LAKE CITY to BOISE
10s-3dbr
UP 31, Salt Lake City (5:25p) to Pocatello (9:55p)
UP 25, Pocatello (1:10a) to Boise (6:45a)

(Salt Lake City to Pocatello, cafe-lounge)

Butte Special, No. 29
Salt Lake City to Butte
Union Pacific
6 miles south of McCammon

LOS ANGELES to BUTTE
10r-6dbr
UP 38, Los Angeles (7:30p Sat)
    to Salt Lake City (5:15p Sun)
UP 29, Salt Lake City (8p Sun)
    to Butte (9:30a Mon)

SALT LAKE CITY to BUTTE
10s-dr-c
UP 29, Salt Lake City (8p) to Butte (9:30a)

LOS ANGELES to SUN VALLEY   *Winter only*
10r-6dbr
UP 38, Los Angeles (7:30p Sat)
    to Salt Lake City (5:15p)
UP 29 Salt Lake City (8p Sun)
    to Pocatello (12:45a Mon)
UP 25, Pocatello (1:10a) to Shoshone (3:15a)
UP 55, Shoshone (4:30a) to Sun Valley (7:30a Mon)

SALT LAKE CITY to WEST YELLOWSTONE
Summer only
6c-3dr   14s
UP 29, Salt Lake City (8p) to Pocatello (12:45a)
UP 35, Pocatello (1:35a)
    to West Yellowstone (7:30a)

SALT LAKE CITY to VICTOR   *Summer only*
8s-dr-2c
UP 29, Salt Lake City (8p) to Pocatello (12:45a)
UP 35, Pocatello (1:35a) to Ashton (5:10a)
UP 47, Ashton (5:20a) to Victor (7a)

(Los Angeles to Salt Lake City, first-class lounge; Las Vegas to Salt Lake City, diner; Salt Lake City to Butte: first-class and cafe-lounges)

Butte Special No. 30

Butte to Salt Lake City
Union Pacific
50 miles north of Idaho Falls

BUTTE to LOS ANGELES
10r-6dbr
UP 30, Butte (7p Sun)
    to Salt Lake City (7:35a Mon)
UP 37, Salt Lake City (9:40a)
    to Los Angeles (7a Tue)

BUTTE to SALT LAKE CITY
10s-dr-c
UP 30 Butte (7p) to Salt Lake City (7:35a)

(Butte to Pocatello, first-class and cafe-lounges; Salt Lake City to Las Vegas, diner, and to Los Angeles, first-class lounge)

Yellowstone Special, No. 35
Pocatello to West Yellowstone
Summers only
Union Pacific
Standing in Pocatello

CHICAGO to WEST YELLOWSTONE
Std: s-dr-c
C&NW 27, Chicago (3:30p Sat)
    to Omaha (12:30a Sun)
UP 27, Omaha (12:55a) to Green River (4:20p)
UP 11, Green River (4:30p) to Pocatello (9:45p)
UP 35, Pocatello (1:35a)
    to West Yellowstone (7:30a Mon)

POCATELLO to WEST YELLOWSTONE
Std: 10-dr-c
UP 35, Pocatello (1:35a)
    to West Yellowstone (7:30a)

(First-class lounge)

Santa Fe Southern Pacific Corp.

Illinois

See map on page [160]

The *Chief*, about to depart Chicago, obliges the official Santa Fe camera.

Pioneer Limited, No. 1
Chicago to Minneapolis
Milwaukee Road
39 miles north of Chicago

CHICAGO to ST. PAUL/MINNEAPOLIS
11: 16dup/r-4dbr   12: 16dup/r-4dbr
14: 8dup/r-6r-4dbr   15: 8dup/r-6r-4dbr
16: 8dup/r-6r-4dbr
MILW 1, Chicago (11:15p) to St. Paul (8a)
     and Minneapolis (8:40a)

CHICAGO to MINOCQUA
18: 10s-c-dr
MILW 1, Chicago (11:15p) to New Lisbon (3:34a)
MILW 217, New Lisbon (4:15a)
     to Minocqua (10:40a)

(Tip top tap diner. No. 217 is the Tomahawk.*)*

No. 2
Chicago to Grand Rapids
Chesapeake & Ohio
20 miles south of Chicago

CHICAGO to GRAND RAPIDS
21: 10r-6dbr
C&O 2, Chicago (11:15p) to Grand Rapids (5a)

CHICAGO to MUSKEGON
202: 10r-6dbr
C&O 2, Chicago (11:15p) to Holland (4:10a)
C&O 201, Holland (5a) to Muskegon (6:20a)

California Limited, No. 3
Chicago to Los Angeles
Atchison, Topeka & Santa Fe
40 miles east of Galesburg

CHICAGO to LOS ANGELES
12s-dr   8s-dr-2c
Santa Fe 3, Chicago (8:45p Sun)
     to Los Angeles (7a Wed)

CHICAGO to LOS ANGELES
10s-dr-2c
Santa Fe 3, Chicago (8:45p Sun)
     to Kansas City (7:15a Mon)
Santa Fe 5, Kansas City (8:50a)
     Newton, Kans. (12:10p)
Santa Fe 105, Newton (12:55p)
     to Belen, N.M. (6:35a Tue)
Santa Fe 14, Belen (6:55a) to Albuquerque (7:45a)
Santa Fe 3, Albuquerque (8:30a)
     to Los Angeles (7a Wed)

CHICAGO to SAN FRANCISCO
10s-dr-2c
Santa Fe 3, Chicago (8:45p Sun)
     to Barstow (2:15a Wed)
Santa Fe 23, Barstow (6:20a) to Oakland (8p)
Bus to San Francisco (8:30p Wed)

CHICAGO to GALVESTON
10s-2dbr-c
Santa Fe 3, Chicago (8:45p Sun)
     to Kansas City (7:15a Mon)
Santa Fe 5, Kansas City (8:50a)
     to Galveston (9:50a Tue)

(Lounge, Chicago to Los Angeles; diners, Chicago to La Junta, Kansas City to Fort Worth, and Barstow to Oakland; Fred Harvey station dining room service La Junta to Los Angeles.)

Ak-Sar-Ben, No. 3
Chicago to Lincoln
Burlington
45 miles east of Galesburg

CHICAGO to QUINCY
Std: s-dr
CB&Q 3, Chicago (10p) to Galesburg (12:45a)
CB&Q 3, Galesburg (3:20a) to Quincy (6:30a)

CHICAGO to OMAHA
Std: s-r-dup/r-dbr-dr-c
CB&Q 3, Chicago (10p) to Omaha (8a)

CHICAGO to LINCOLN
Std: as for Omaha, plus lounge
CB&Q 3, Chicago (10p) to Lincoln (9:40a)

(Diner, Creston, Iowa, to Lincoln)

Louisiane, No. 3
Chicago to New Orleans
Illinois Central
14 miles south of Effingham

CHICAGO to MEMPHIS
301: 8s-5dbr   303: 18r
IC 3, Chicago (7:05p) to Memphis (8:35a)

(Dining cars: Chicago to Champaign, Fulton to Memphis; parlor-lounge, Chicago to Memphis.)

Louisiane, No. 4
Memphis to Chicago
Illinois Central
Entering North Cairo

MEMPHIS to CHICAGO
402: 8s-5dbr   404: 18r
IC 4, Memphis (8p) to Chicago (8:15a)

(Diner, Champaign to Chicago; parlor-lounge, Memphis to Chicago. The northbound Louisiane *originates in Memphis, not New Orleans.)*

No. 4
Kansas City to Galesburg
Burlington
35 miles south of Galesburg

QUINCY to CHICAGO
Std: s-dr
CB&Q 4, Quincy (10:05p) to Galesburg (1a)
CB&Q 30, Galesburg (5:15a) to Chicago (8:10p)

(No. 30, the Ak-Sar-Ben, *carries a diner from Galesburg to Chicago)*

Dixie Flyer, No. 5
Chicago to Jacksonville
Chicago & Eastern Illinois
68 miles south of Chicago

CHICAGO to JACKSONVILLE
51: 10s-2c-dr
C&EI 5, Chicago (10:45p Sun)
     to Evansville (5:20a Mon)
L&N 95, Evansville (5:40a) to Nashville (10:10a)
NC&StL 95, Nashville (10:30a) to Atlanta (7:45p)
CofG 94-5, Atlanta (9:05p)
     to Albany, Ga. (2:35a Tue)
ACL 95, Albany (2:50a) to Jacksonville (7:45a Tue)

CHICAGO to EVANSVILLE
57: 8s-2c-dr
C&EI 5, Chicago (10:45p) to Evansville (5:20a)

(Diner, Evansville to Atlanta)

Des Moines-Omaha Limited, No. 5
Chicago to Omaha
Rock Island Railroad
20 miles west of Joliet

CHICAGO to DES MOINES
52: 16dup/r   53: 8s-5dbr
RI 5, Chicago (10:45p) to Des Moines (7:45a)

(Cafe-lounge, Chicago to Des Moines)

Yellowstone Special, No. 36
West Yellowstone to Pocatello
Summers only
Union Pacific
Standing in Idaho Falls

WEST YELLOWSTONE to POCATELLO
Std: s-dr-c
UP 36, West Yellowstone (7:30p)
     to Pocatello (1:35a)

WEST YELLOWSTONE to SALT LAKE CITY
6c-3dr   14s
UP 36, West Yellowstone (7:30p)
     to Pocatello (1:35a)
UP 30, Pocatello (3:15a) to Salt Lake City (7:30a)

WEST YELLOWSTONE to CHICAGO
Std: s-dr-c
UP 36, West Yellowstone (7:30p Sun)
     to Pocatello (1:35a)
UP 12, Pocatello (4:10a Mon)
     to Green River (9:50a)
UP 28, Green River (10:15a) to Omaha (2:50a Tue)
C&NW 28, Omaha (3:15a) to Chicago (12:30p Tue)

VICTOR to SALT LAKE CITY
8s-dr-2c
UP 48, Victor (8:15p) to Ashton (9:55p)
UP 36, Ashton (10:10p) to Pocatello (1:35a)
UP 30, Pocatello (3:15a) to Salt Lake City (7:30a)

(First-class lounge: West Yellowstone to Pocatello, Pocatello to Salt Lake City and to North Platte on No. 12 and Green River to Chicago; diner Pocatello to North Platte and Green River to Chicago)

No. 56
Sun Valley to Shoshone
Union Pacific
20 miles north of Shoshone

SUN VALLEY to LOS ANGELES   Winter only
10r-6dbr
UP 56, Sun Valley (10:30p Sun)
     to Shoshone (12:45a Mon)
UP 12, Shoshone (1:30a) to Pocatello (3:40a)
UP 32, Pocatello (4:15a) to Salt Lake City (8:50a)
UP 37, Salt Lake City (9:40a)
     to Los Angeles (7a Tue)

(Shoshone to Salt Lake City, cafe-lounge; Salt Lake City to Los Angeles, first-class lounge; Salt Lake City to Las Vegas, dining car)

Illinois   (continued)

Allegheny, No. 6
St. Louis to Pittsburgh
Pennsylvania Railroad
15 miles west of Vandalia

ST. LOUIS to PITTSBURGH
    12s-dr
PRR 6, St. Louis (11p) to Pittsburgh (11:59a)

(Sleeper-lounge, Indianapolis to Pittsburgh)

Midnight Special, No. 7
Chicago to St. Louis
Gulf Mobile & Ohio
25 miles southwest of Chicago

CHICAGO to ST. LOUIS
    90: 10s-dr-2c   91: 12s-dr
    92: 13dbr   93: 4s-8r-c-3dbr
GM&O 7, Chicago (11:25p) to St. Louis (7:08a)

CHICAGO to SPRINGFIELD
    Alternates monthly with Illinois Central
    200: 12s-dr
GM&O 7, Chicago (11:25p) to Springfield (4:05a)

Midnight Special, No. 8
St. Louis to Chicago
Gulf Mobile & Ohio
Departing East St. Louis

ST. LOUIS to CHICAGO
    100: 10s-dr-2c   101: 12s-dr
    102: 13dbr   103: 4s-8r-c-3dbr
GM&O 8, St. Louis (11:45p) to Chicago (7a)

HOT SPRINGS to CHICAGO
    83: 10r-6dbr
MP 220, Hot Springs (1:45p) to Little Rock (3p)
MP 8, Little Rock (3:20p) to St. Louis (11p)
GM&O 8, St. Louis (11:45p) to Chicago (7a)

(MP 8, the Southerner, carries a diner-lounge)

**Midnight Special, No. 8
or Night Diamond, No. 18**
GM&O and IC alternate monthly
Springfield setout
Standing in Springfield

SPRINGFIELD to CHICAGO
    200: 12s-dr
GM&O 8, Springfield (2:20a) to Chicago (7a)

SPRINGFIELD to CHICAGO
    182: 12s-dr
IC 18, Springfield (2:25a) to Chicago (7a)

No. 9
Chicago to Des Moines
Rock Island
Standing in Chicago

CHICAGO to ROCK ISLAND
    95: 14s
RI 9, Chicago (1a) to Rock Island (6:25a)

CHICAGO to PEORIA
    91: 14s
RI 9, Chicago (1a) to Bureau (4:25a)
RI 209, Bureau (4:30a) to Peoria (5:45a)

Kansas City Chief, No. 9
Chicago to Kansas City
Atchison, Topeka & Santa Fe
60 miles southwest of Joliet

CHICAGO to KANSAS CITY
    Std: s-r-dbr-dup/r
Santa Fe 9, Chicago (10p) to Kansas City (7:45a)

(Diner-lounge)

Hawkeye, No. 11
Chicago to Sioux City
Illinois Central
13 miles east of Rockford

CHICAGO to WATERLOO
    113: 12s-dr
IC 11, Chicago (10:40p) to Waterloo (5:30a)

CHICAGO to SIOUX CITY
    111: 8s-5dbr
IC 11, Chicago (10:40p) to Sioux City (1:20p)

(Waterloo to Sioux City, buffet-lounge)

Metropolitan Special, No. 12
St. Louis to Washington
Balitmore & Ohio
12 miles east of St. Louis

ST. LOUIS to NEW YORK
    12s-dr
B&O 12, St. Louis (11:30p Sun)
          to Jersey City (6:50a Tue)
Ferry or bus to New York

ST. LOUIS to CINCINNATI
    8s-4dbr
B&O 12, St. Louis (11:30p) to Cincinnati (7:50a)

ST. LOUIS to LOUISVILLE
    12s-dr
B&O 12, St. Louis (11:30p)
          to North Vernon, Ind. (4:47a)
B&O 59, North Vernon (5:55a) to Louisville (7:40a)

(St. Louis to Cincinnati, lunch counter; Cincinnati to Washington, dining-lounge car)

La Salle Street Limited, No. 14
Rock Island (Ill.) setout
Rock Island Railroad
Standing in Rock Island

ROCK ISLAND to CHICAGO
    145: 14s
RI 14, Rock Island (3:30a) to Chicago (7:30a)

(The Omaha-to-Chicago La Salle Street Limited is now in Iowa, 27 miles east of Des Moines. It carries a cafe-lounge car.)

Chickasaw, No. 15
St. Louis to Memphis
Illinois Central
41 miles southeast of St. Louis

ST. LOUIS to MEMPHIS
    151: 8s-dr-3dbr
IC 15, St. Louis (10:35p) to Memphis (7:05a)

No. 15
Chicago to Boone, Iowa
Chicago & North Western
42 miles west of Chicago

CHICAGO to CEDAR RAPIDS
    12s-dr
C&NW 15, Chicago (11:01p) to Cedar Rapids (7a)

The Midnight, No. 17
Chicago to St. Louis
Wabash
Arriving Englewood, Chicago

CHICAGO to ST. LOUIS
    65: 12r-4dbr   67: 12r-4dbr
    69: 8s-4dbr   71: 12s-dr
Wabash 17, Chicago (11:50p) to St. Louis (7:30a)

CHICAGO to DECATUR
    63: 8s-dr-3dbr
Wabash 17, Chicago (11:50p) to Decatur (3:44a)

Night Diamond, No. 17
Chicago to St. Louis
Illinois Central
Approaching Woodlawn, Chicago

CHICAGO to ST. LOUIS
    175: 10s-c-2dbr   173: 12r-dr-dbr-2sbr
IC 17, Chicago (11:50) to St. Louis (7:15a)

CHICAGO to SPRINGFIELD
    Alternates monthly with Gulf, Mobile & Ohio
    171: 12s-dr
IC 17, Chicago (11:50) to Springfield (4:35a)

Night Diamond, No. 18
St. Louis to Chicago
Illinois Central
Arriving East St. Louis

ST. LOUIS to CHICAGO
    184: 12r-dr-dbr-2sbr   186: 10s-c-2dbr
IC 18, St Louis (11:50p) to Chicago (7a)

The Midnight, No. 18
Wabash
Decatur setout
Standing in Decatur

DECATUR to CHICAGO
    64: 8s-dr-3dbr
Wabash 18, Decatur (3a) to Chicago (7:40a)

(The Midnight is now departing St. Louis's Delmar Boulevard station)

No. 47
St. Louis to Savanna
Burlington
6 miles south of Savanna

ROCK ISLAND to ST. PAUL/MINNEAPOLIS
    10s-c-dr
CB&Q 47, Rock Island (9:30p) to Savanna (12:15a)
CB&Q 53, Savanna (2:31a) to St. Paul (8:15a)
              and Minneapolis (8:50a)

(A dining-lounge car will be cut into No. 53, the combined Black Hawk and Western Star, at La Crosse, Wis.)

North Coast Limited, No. 51
Chicago to Seattle
Burlington
10 miles west of Aurora

CHICAGO to SEATTLE
    252: 3br-c-6r-8dup/r   BN: 14s (tourist)
    254: 4br-c-obs.-lounge
CB&Q 51, Chicago (11p Sun) to St. Paul (8a Mon)
NP 1, St. Paul (9a) to Seattle (7:30a Wed)

CHICAGO to PORTLAND
    251: 3br-c-6r-8dup/r   BR: 14s (tourist)
CB&Q 51, Chicago (11p Sun) to St. Paul (8a Mon)
NP 1, St. Paul (9a) to Pasco, Wash. (11:40p Tue)
SP&S 3, Pasco (1:35a) to Portland (7a Wed)

CHICAGO to BILLINGS
    253: 3br-c-6r-8dup/r
CB&Q 51, Chicago (11p Sun) to St. Paul (8a Mon)
NP 1, St. Paul (9a) to Billings (4:55a Tue)

CHICAGO to CODY   *Summertime only*
    B38: 10s-c-dr
As above, Chicago (11p Sun)
              to Billings (4:55a Tue)
CB&Q 24-27, Billings (8a) to Cody (11:15a Tue)

(Dining car)

Western Star/Black Hawk, No. 53

Chicago to Seattle/Portland
Burlington
Arriving in Aurora

CHICAGO to SEATTLE
  34: 4s-8dup/r-4dbr   35: 16 dup/r-4dbr
  39: dr-2dbr-obs.-buffet-lounge
  CB&Q 53, Chicago (11:15p Sun)
      to St. Paul (8:15a Mon)
  GN 3, St. Paul (8:45a) to Seattle (7:30a Wed)

CHICAGO to PORTLAND
  32: 4s-8dup/r-4dbr
  CB&Q 53, Chicago (11:15p Sun)
      to St. Paul (8:15a Mon)
  GN 3, St. Paul (8:45a) to Spokane (9:30p Tue)
  SP&S 3, Spokane (9:45p) to Portland (7a Wed)

CHICAGO to GREAT FALLS, MONT.
  38: 16dup/r-4dbr
  CB&Q 53, Chicago (11:15p Sun)
      to St. Paul (8:15a Mon)
  GN 3, St. Paul (8:45a) to Great Falls (7:50a Tue)

CHICAGO to ST. PAUL/MINNEAPOLIS
  Std: s-dbr
  CB&Q 53, Chicago (11:15p) to St. Paul (8:15a)
      and Minneapolis (8:50a)

(Chicago to Seattle, diner, coffee shop car; Spokane to Portland, observation-lounge. The Black Hawk, which picks up a sleeper from Rock Island and a dining-lounge car, is the Burlington's over-nighter to the Twin Cities. The Western Star is the Great Northern's secondary train to the Pacific Northwest.)

American Royal, No. 55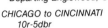

Chicago to Kansas City/St. Joseph
Burlington
32 miles northeast of Quincy

CHICAGO to KANSAS CITY
  8s:5dbr
  CB&Q 55, Chicago (6:30p)
      to Cameron Junction, Mo. (5:33a)
  CB&Q 55, Cameron Junction (5:47a)
      to Kansas City (7:30a)

CHICAGO to ST. JOSEPH
  12s-2dbr
  CB&Q 55, Chicago (6:30p)
      to Cameron Junction, Mo. (5:33a)
  CB&Q 55, Cameron Junction (5:55a)
      to St. Joseph (7:10a)

(Dining car, Chicago to Galesburg)

No. 55

St. Louis to Evansville
Louisville & Nashville
16 miles west of Mount Vernon

ST. LOUIS to NASHVILLE
  49: 12s-dr
  L&N 55, St. Louis (10:30p) to Evansville (2:20a)
  L&N 53, Evansville (2:35a) to Nashville (7:35a)

ST. LOUIS to LOUISVILLE
  29: 12s-dr
  L&N 55, St. Louis (10:30p) to Evansville (2:20a)
  L&N 155, Evansville (2:40a) to Louisville (7:10a)

Motor City Special, No. 316

Chicago to Detroit
New York Central
Departing Chicago

CHICAGO to DETROIT
  10r-6dbr   10r-6dbr
  22r   22r
  22r   13dbr
  8s-dr-2c   6dbr-buffet-lounge
  NYC 316, Chicago (11:59p) to Detroit (7:35a)

Southland, No. 200

Chicago to Florida
Pennsylvania Railroad
Departing Englewood, Chicago

CHICAGO to CINCINNATI
  10r-5dbr
  PRR 200, Chicago (11:35p) to Cincinnati (7:40a)

CHICAGO to DAYTON
  6s-6dbr   18r
  PRR 200, Chicago (11:35p)
      to Richmond, Ind. (4:20a)
  PRR 100, Richmond (4:30a) to Dayton (6:18a)

CHICAGO to SPRINGFIELD, OHIO
  12s
  PRR 200, Chicago (11:35p) to Richmond (4:20a)
  PRR 100, Richmond (4:30a) to Springfield (8a)

CHICAGO to JACKSONVILLE
  R301: 8s-buffet-lounge
  PRR 200, Chicago (11:35p Sun)
      to Cincinnati (7:40a Mon)
  L&N 33, Cincinnati (8a) to Atlanta (7:10p)
  CofG 94-95, Atlanta (9:05p) to Albany (2:35a Tue)
  ACL 95, Albany (2:50a) to Jacksonville (7:45a Tue)

CHICAGO to ST. PETERSBURG
  R303: 8s-2c-dr   Alternate days:
  R304: 13dbr, or   4c-4dr
  As above, Chicago (11:35p Sun)
      to Atlanta (7:10p Mon)
  CofG 32-33, Atlanta (8p) to Albany (1a Tue)
  ACL 33, Albany (1:15a) to Trilby, Fla. (7:30a)
  ACL 133, Trilby (8:05a)
      to St. Petersburg (10:10a Tue)

CHICAGO to TAMPA/SARASOTA
  R302: 8s-dr-3 dbr
  As above, Chicago (11:35p Sun)
      to Trilby (7:30a Tue)
  ACL 33-37, Trilby (8a) to Tampa (9:15a)
      and Sarasota (11:30a Tue)

(Dining cars: Cincinnati to Macon, Thomasville to St. Petersburg; Sleeper-buffet-lounge, Atlanta to Tampa/ Sarasota. The Jacksonville sleeper-lounge travels aboard the Dixie Flyer south of Atlanta.)

No. 217

Chicago to Green Bay
Via Fond du Lac
Chicago & North Western
Standing in Chicago

CHICAGO to GREEN BAY
  8s-dr-4dbr
  C&NW 217, Chicago (1a) to Green Bay (9a)

CHICAGO to GREEN BAY
  10s-dr-c
  C&NW 217, Chicago (1a) to Milwaukee (3:05a)
  C&NW 317, Milwaukee (4:15a) to Green Bay (8a)

(The first Pullman has a relatively late arrival in Green Bay, but this enables it to give through service to the Fox River valley cities from Fond du Lac, where it arrives at 5:45 a.m., north. The second car, on No. 317, travels via Manitowoc. Sunday schedules vary slightly.)

No. 230

Peoria to Bureau
Rock Island Railroad
Standing in Peoria

PEORIA to CHICAGO
  61: 14s
  RI 230, Peoria (1:30a) to Bureau (2:45a)
  RI 6, Bureau (3:25a) to Chicago (5:55a)

North Western Limited, No. 405

Chicago to Minneapolis
Chicago & North Western
7 miles south of Kenosha

CHICAGO to ST. PAUL/MINNEAPOLIS
  6sbr-2dbr-lounge
  16dup/r-3dbr-c cars as necessary
  C&NW 405, Chicago (11p) to St. Paul (7:40a)
      and Minneapolis (8:15a)

CHICAGO to DULUTH
  10s-dr-c
  C&NW 405, Chicago (11p) to Altoona, Wis. (5:21a)
  C&NW 513, Altoona (ntg) to Duluth (12:15p)

(Diner)

A NYC Hudson with the *Twentieth Century Limited* on the outskirts of Chicago, 1934.

No. 410

Chicago to Cincinnati
New York Central
19 miles south of Chicago

CHICAGO to INDIANAPOLIS
  12s-2dbr
  NYC 410, Chicago (11:30p) to Indianapolis (4a)

CHICAGO to CINCINNATI
  6dbr-buffet-lounge
  NYC 410, Chicago (11:30p) to Cincinnati (8a)

CHICAGO to JACKSONVILLE
  10s-dr-2c
  NYC 410, Chicago (11:30p Sun)
      to Cincinnati (8a Mon)
  SOU 3, Cincinnati (8:45a)
      to Jacksonville (8:15a Tue)

CHICAGO to MIAMI
  10r-6dbr   10r-6dbr
  NYC 410, Chicago (11:30p Sun)
      to Cincinnati (8a Mon)
  SOU 5, Cincinnati (8:30a)
      to Jacksonville (4:05a Tue)
  FEC 9, Jacksonville (4:25a) to Miami (11:25a Tue)

(Southern No. 3 is the Royal Palm, carrying from Cincinnati a diner to Atlanta and sleeper-lounge to Jacksonville. No. 5 is the streamlined New Royal Palm, with lounge and dining car to Miami.)

Cleveland-Cincinnati Special, No. 446

St. Louis to Cleveland
New York Central
38 miles east of St. Louis

ST. LOUIS to CLEVELAND
  8s-buffet-lounge
  NYC 446, St. Louis (11:10p) to Cleveland (11:40a)

ST. LOUIS to CINCINNATI
  10r-5dbr
  NYC 446, St. Louis (11:10p) to Indianapolis (3:55a)
  NYC 410, Indianapolis (4:20a) to Cincinnati (8a)

Indiana

See map on page  80

Detroit Limited, No. 2
St. Louis to Detroit
Wabash
4 miles west of Lafayette

ST. LOUIS to DETROIT
50: 8s-dr-2c   52: 12r-4dbr
54: 8s-5dbr
Wabash 2, St. Louis (6:15p) to Detroit (7:50a)

ST. LOUIS to TOLEDO
46: 8s-dr-c's
Wabash 2, St. Louis (6:15p) to Fort Wayne (3:20a)
Wabash 12, Fort Wayne (3:50a) to Toledo (7:50a)

(St. Louis to Detroit, diner-lounge)

No. 3
Chicago to Evansville
Chicago & Eastern Illinois
12 miles south of Terre Haute

CHICAGO to NASHVILLE
30: 10s-c-dr
C&EI 3, Chicago (8:05p) to Evansville (2a)
L&N 53, Evansville (2:35a) to Nashville (7:35a)

(Chicago to Evansville, club coach. A St. Louis-Nashville sleeper will also join the consist of No. 53 in Evansville.)

St. Louis Limited, No. 3
Detroit to St. Louis
Wabash
5 miles east of Logansport

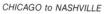

DETROIT to ST. LOUIS
23: 8s-5dbr   25: 8s-dr-2c
27: 12r-4dbr
Wabash 3, Detroit (8:10p) to St. Louis (7:40a)

TOLEDO to ST. LOUIS
45: 8s-dr-c's
Wabash 13, Toledo (7:40p) to Fort Wayne (9:25p)
Wabash 3, Fort Wayne (10:08p)
    to St. Louis (7:40a)

(Dining-lounge car, Detroit to St. Louis)

Inter-City Limited, No. 6
Chicago to Montreal
Grand Trunk Western
45 miles southeast of Chicago

CHICAGO to LANSING
58: 12s-dr
GTW 6, Chicago (11p) to Lansing (5:20a)

CHICAGO to PORT HURON
248: 12s-2dbr
GTW 6, Chicago (11p) to Port Huron (9a)

CHICAGO to DETROIT
56: 8s-dr-2c
GTW 6, Chicago (11p) to Durand, Mich. (6a)
GTW 28, Durand (6:10a) to Detroit (7:50a)

(Durand to London, Ont., buffet-parlor)

Nickel Plate Limited, No. 6
Nickel Plate
Fort Wayne setout
Standing in Fort Wayne

FORT WAYNE to CLEVELAND
601: 10r-6dbr
NKP 6, Fort Wayne (2:55a) to Cleveland (8a)

Dixie Flyer, No. 6
Jacksonville to Chicago
Chicago & Eastern Illinois
37 miles south of Terre Haute

JACKSONVILLE to CHICAGO
A44: 10s-2c-dr
ACL 94, Jacksonville (9p Sat) to Albany (2a Sun)
CofG 94-5, Albany (2:10a) to Atlanta (8a)
NC&StL 94, Atlanta (9:35a) to Nashville (5:10p)
L&N 94, Nashville (5:30p) to Evansville (10p)
C&EI 6, Evansville (10:20p)
    to Chicago (5:20a Mon)

EVANSVILLE to CHICAGO
67: 8s-2c-dr
C&EI 6, Evansville (10:20p) to Chicago (5:20a)

(Atlanta to Evansville, dining car)

The C&EI also offered Terre Haute-Chicago Pullman service on this line.

Nickel Plate Limited, No. 6
Chicago to Buffalo
Nickel Plate
Approaching Hammond

CHICAGO to CLEVELAND
61: 10r-6dbr   62: 10r-6dbr
63: dbr's   64: 10r-6dbr
NKP 6, Chicago (11:30p) to Cleveland (8a)

CHICAGO to NEW YORK
60: 10r-6dbr
NKP 6, Chicago (11:30p Sun) to Buffalo (12:40p)
DL&W 10, Buffalo (5:45p)
    to Hoboken, N.J. (5:30a Tue)
Ferry to New York

(Chicago to Buffalo, diner-lounge. DL&W 10, the New York Mail, carries a buffet-lounge to Elmira, N.Y.)

The *Nickel Plate Limited* made its inaugural run in February 1928.

Shenandoah, No. 8
Chicago to New York
Baltimore & Ohio
Arriving in Gary

CHICAGO to AKRON
8s-4dbr
B&O 8, Chicago (11p) to Akron (6:20a)

CHICAGO to PITTSBURGH
8s-2c-dr   10r-5dbr
B&O 8, Chicago (11p) to Pittsburgh (8:50a)

CHICAGO to WASHINGTON
10s-2c-dr
5r-sbr-3dr Strata Dome, alternate days
B&O 8, Chicago (11p) to Washington (4:30p)

CHICAGO to NEW YORK
8s-buffet-lounge   10s-2c-dr
B&O 8, Chicago (11p) to Jersey City (9:50p)
Bus or ferry to New York

(Diner, Youngstown to Jersey City)

Atlantic Express, No. 8
Chicago to New York
Erie Railroad
31 miles east of North Judson

CHICAGO to MEADVILLE, PA.
12r-2sbr-3dbr
Erie 8, Chicago (10p) to Meadville (9:35a)

CHICAGO to AKRON
10r-6dbr
Erie 8, Chicago (10p) to Akron (6:40a)

(Diner-lounge, Akron to Jersey City. By cutting out the Chicago sleeper at the division point of Meadville, the car could be turned around for a Youngstown departure the same night on the Erie Limited. Sunday schedules slightly different.)

No. 9
Cleveland to St. Louis
Nickel Plate
16 miles west of Muncie

CLEVELAND to ST. LOUIS
91: 10r-6dbr
NKP 9, Cleveland (6:30p) to St. Louis (7:40a)

(Diner-lounge)

No. 10
St. Louis to Cleveland
Nickel Plate
Standing in Frankfort

ST. LOUIS to CLEVELAND
101: 10r-6dbr
NKP 10, St. Louis (6:25p) to Cleveland (8:55a)

(Diner-lounge)

St. Louisan, No. 32
St. Louis to New York
Pennsylvania Railroad
Departing Richmond

ST. LOUIS to NEW YORK
6dbr-lounge   21r
6s-6dbr
PRR 32, St. Louis (6:30p) to New York (4:30p)

ST. LOUIS to WASHINGTON
12s-dr
PRR 32, St. Louis (6:30p) to Harrisburg, Pa. (1p)
PRR 554, Harrisburg (2:03p)
    to Washington (5:10p)

LOUISVILLE to NEW YORK
6s-6dbr
PRR 327, Louisville (8p) to Indianapolis (10:25p)
PRR 32, Indianapolis (10:50p) to New York (4:30p)

(St. Louis to New York, diner; Louisville to Indianapolis, parlor-cafe. The St. Louisan will pick up a Cincinnati-New York Pullman at Pittsburgh in the morning.)

Missourian, No. 40
St. Louis to New York
New York Central
9 miles east of Muncie

ST. LOUIS to NEW YORK
6dbr-buffet-lounge   22r
NYC 40, St. Louis (6p) to New York (5:15p)

ST. LOUIS to BUFFALO
8s-5dbr
NYC 40, St. Louis (6p) to Buffalo (8:55a)

ST. LOUIS to CLEVELAND
10r-5dbr
NYC 40, St. Louis (6p) to Cleveland (5:17a)

ST. LOUIS to DETROIT
10r-5dbr
NYC 40, St. Louis (6p)
    to Bellefontaine, Ohio (2:20a)
NYC 304, Bellefontaine (2:50a) to Detroit (7:45a)

INDIANAPOLIS to CLEVELAND
6s-6dbr
NYC 40, Indianapolis (10:35p) to Cleveland (5:17a)

(Dining cars: St. Louis to Indianapolis, Cleveland to New York; parlor-observation, Buffalo to New York)

No. 44
Chicago to Pittsburgh
Pennsylvania Railroad
20 miles southeast of Chicago

CHICAGO to CANTON, OHIO
12s-dr
PRR 44, Chicago (11:30p) to Canton (9:21a)

Gotham Limited, No. 54
Chicago to New York
Pennsylvania Railroad
33 miles southeast of Chicago

CHICAGO to NEW YORK
5dbr-lounge	21r
12s-dr	12s-dr

PRR 54, Chicago (11:15p) to New York (5:40p)

CHICAGO to WASHINGTON
8s-2c-dr
PRR 54, Chicago (11:15p) to Harrisburg (1:54p)
PRR 554, Harrisburg (2:03p)
    to Washington (5:10p)

CHICAGO to PITTSBURGH
6s-6dbr	21r (Sat. only)
12dup/r-4dbr (Sat. only)	

PRR 54, Chicago (11:15p) to Pittsburgh (8:50a)

(Chicago to New York, diner)

Once an all-Pullman train, the *Gotham Limited* carried a diverse assortment of cars, from St. Louis, Louisville, Detroit, Toledo, Columbus, as well as Chicago.

No. 56
Louisville to North Vernon
Baltimore & Ohio
22 miles north of New Albany

LOUISVILLE to ST. LOUIS
12s-dr
B&O 56, Louisville (11:25p)
    to North Vernon (12:37a)
B&O 11, North Vernon (1:01a) to St. Louis (7:40a)

(No. 11 is the Metropolitan Special, which carries a lunch counter-coach. The train is now departing Cincinnati.)

Golden Triangle, No. 62
Chicago to Pittsburgh
Pennsylvania Railroad
44 miles west of Fort Wayne

CHICAGO to PITTSBURGH
12s-dr	3dbr-dr-buffet
12dup/r-5dbr	10r-6dbr
21r	12dup/r-4dbr
4c-2dr-4dbr	

PRR 62, Chicago (10:15p) to Pittsburgh (8a)

Chicagoan, No. 90
Chicago to New York
New York Central
12 miles east of Gary

CHICAGO to NEW YORK
6dbr-buffet-lounge	10r-6dbr
8s-dr-2c	

NYC 90, Chicago (11:15p) to New York (7p)

CHICAGO to BUFFALO
10r-6dbr
NYC 90, Chicago (11:15p) to Buffalo (10:20a)

(The Chicagoan will depart Cleveland at 6:35 a.m. with a diner for New York and a Pullman for Buffalo that came from Miami on the New Royal Palm)

Ohioan, No. 110
Chicago to Pittsburgh
Pennsylvania Railroad
7 miles north of North Judson

CHICAGO to COLUMBUS
10s-lounge	12dup/r-5dbr
18r	

PRR 110, Chicago (10:45p) to Columbus (7:20a)

Southland, No. 201
Pennsylvania Railroad
Springfield, Dayton Pullmans
Standing in Richmond

SPRINGFIELD, OHIO to CHICAGO
12s-dr
PRR 101, Springfield (10p)
    to Richmond (11:30p CST)
PRR 201, Richmond (12:55a) to Chicago (6:40a)

DAYTON to CHICAGO
18r	6s-6dbr

PRR 101, Dayton (11:35p)
    to Richmond (11:30p CST)
PRR 201, Richmond (12:55a) to Chicago (6:40a)

West Virginia Night Express, No. 246
Chicago to Wheeling, W.Va.
Baltimore & Ohio
6 miles west of Avilla

CHICAGO to WHEELING
10s-2c-dr
B&O 246, Chicago (8:55p) to Wheeling (10:45a)

(Newark, Ohio, to Wheeling, diner-lounge)

Forest City, No. 290
Chicago to Cleveland
New York Central
Departing Gary

CHICAGO to CLEVELAND
6dbr-buffet-lounge	10r-6dbr
10r-6dbr	10r-6dbr
22r	22r

NYC 290, Chicago (11:25p) to Cleveland (7:30a)

CHICAGO to TOLEDO
22r
NYC 290, Chicago (11:25p) to Toledo (4:30a)

The *Forest City* once was all-Pullman as well as all-Cleveland.

Kentuckian, No. 305
Louisville to Chicago
Pennsylvania Railroad
23 miles north of Louisville

LOUISVILLE to CHICAGO
10s-lounge	10r-6dbr
12s-dr	

PRR 305, Louisville (11:30p) to Chicago (7a)

LOUISVILLE to PITTSBURGH
8s-buffet-lounge
PRR 305, Louisville (11:30p)
    to Indianapolis (1:55a)
PRR 6, Indianapolis (3:05a) to Pittsburgh (11:59a)

(No. 6 is the Allegheny, now 15 miles west of Vandalia, Ill.)

Kentuckian, No. 305
Pennsylvania Railroad
Indianapolis setout
Standing in Indianapolis

INDIANAPOLIS to CHICAGO
10s-3dbr
PRR 305, Indianapolis (2:10a) to Chicago (7a)

Kentuckian, No. 306
Chicago to Louisville
Pennsylvania Railroad
30 miles northwest of North Judson

CHICAGO to INDIANAPOLIS
10s-3dbr
PRR 306, Chicago (11p) to Indianapolis (3:20a)

CHICAGO to LOUISVILLE
10s-lounge	10r-6dbr
12s-dr	

PRR 306, Chicago (11p) to Louisville (6:10a)

No. 437
New York Central
Indianapolis setout
Standing in Indianapolis

INDIANAPOLIS to CHICAGO
12s-2dbr
NYC 437, Indianapolis (1:55a)
    to Chicago (7:30a)

(No. 437 is now departing Cincinnati)

The Monon also scheduled sleepers from Indianapolis to Chicago, but beyond that the capital city's Pullman service was dominated by the New York Central and Pennsylvania. In addition to the Chicago cars, they both carried sleepers to Columbus, Detroit, Evansville, Grand Rapids, Jacksonville, New York, St. Louis, and Washington.

No. 501
Richmond to Fort Wayne
Pennsylvania Railroad
Standing in Richmond

CINCINNATI to GRAND RAPIDS
12s-dr
PRR 237, Cincinnati (11p) to Richmond (11:45p)
PRR 501, Richmond (12:35a)
    to Fort Wayne (2:55a)
PRR 509, Fort Wayne (3:20a)
    to Grand Rapids (8:30a)

(Sunday schedule slighly different. Additional summertime sleepers scheduled.)

Indianapolis sleepers for Traverse City, Northport, and Mackinaw City were transported to Richmond, where they were cut into the summertime trains to the Michigan resorts. Richmond also had dedicated service to Mackinaw City early in this century as an accommodation to passengers arriving on trains from St. Louis, Cincinnati, and Pittsburgh.

The Monon's *Bluegrass*, at Bedford, Ind., May 7, 1949, was running as one section of three Kentucky Derby specials.

Charles Herley/Monon Railroad HTS

Posed against a backdrop of the Rockies, the *City of Denver* flaunts its UP/C&NW logos.

Iowa

See map on page 190

Denver Zephyr, No. 1
Streamliner
Chicago to Denver
Burlington Route
7 miles south of Council Bluffs

CHICAGO to DENVER
Accommodations: 36s-10dbr-4c-2dr-4chamb-4r
CB&Q 1, Chicago (5p) to Denver (8:30a)

(Chicago to Denver: dining car, lounge, parlor-buffet-lounge, valet and shower facilities)

Fast Mail, No. 6
Cedar Rapids to Chicago
Sleeping car passengers only
Chicago & North Western
Standing in Cedar Rapids

CEDAR RAPIDS to CHICAGO
12s-dr
C&NW 6, Cedar Rapids (1:30a) to Chicago (6:30a)

Hawkeye, No. 12
Sioux City to Chicago
Illinois Central
Entering Waterloo

SIOUX CITY to CHICAGO
122: 8s-5dbr
IC 12, Sioux City (5:15p) to Chicago (7:40a)

(Sioux City to Waterloo, buffet-lounge; Freeport, Ill., to Chicago, diner)

The *Hawkeye* was the major night train on the IC's western lines, at one time carrying Pullmans to Chicago from Sioux Falls, Omaha, and Dubuque as well as the cities in the present consist. In more leisurely days one of the Sioux City Pullmans was an open observation car.

Hawkeye, No. 12
Illinois Central
Waterloo setout
Standing in Waterloo

WATERLOO to CHICAGO
120: 12s-dr
IC 12, Waterloo (12:20a) to Chicago (7:40a)

(Freeport to Chicago, diner)

The Chicago Great Western also offered a Waterloo-Chicago Pullman and the IC scheduled one from neighboring Cedar Falls.

La Salle Street Limited, No. 14
Omaha to Chicago
Rock Island
27 miles east of Des Moines

DES MOINES to CHICAGO
142: 16dup/single rooms   143: 8s-5dbr
RI 14, Des Moines (11:30p) to Chicago (7:30a)

(Des Moines to Chicago, cafe-lounge. Shortly after 3 a.m., the Limited will cut in a setout sleeper at Rock Island, Ill.)

Once known as the *Iowa-Nebraska Limited*, No. 14 was still, as of 1952, a Nebraska train, albeit a local west of Des Moines. The Omaha-Chicago Pullman now rode the *Rocky Mountain Limited*, and the sleepers from Omaha to both Des Moines and Rock Island were gone. So too, were the Chicago Pullmans that had departed Des Moines on the Chicago Great Western, North Western, and Milwaukee Road.

St. Louis Limited, No. 14-18
Des Moines to Moberly/St. Louis
Wabash
2 miles north of Missouri line

DES MOINES to ST. LOUIS
32: 12s-dr
Wabash 14-18, Des Moines (8:50p)
to St. Louis (7:35a)

(At Moberly, the Des Moines sleeper goes aboard the Omaha-St. Louis night train, also No. 14-18, departing 3:45 a.m.)

Mid-Continent Special, No. 15
Minneapolis to Kansas City
Rock Island
Leaving Iowa Falls

MINNEAPOLIS/ST. PAUL to KANSAS CITY
151: 8s-dr-2c   152: 8s-5dbr
RI 15, Minneapolis (6:30p) and
St. Paul (7:10p) to Kansas City (7:30a)

(Minneapolis to Kansas City, snack-lounge car)

The *Mid-Continent Special*, the Rock Island's historic workhorse on its north-south line from the Twin Cities, once carried Pullmans to Dallas, Los Angeles, and even Chicago, when the railroad was still competing for that traffic. Now Kansas City suffices as a destination, and the *Twin Star Rockets* haul the Minneapolis sleepers to Texas. Three other railroads or combinations competed for the Minneapolis-Kansas City overnight trade: Great Western, North Western (CB&Q or MP), and M&StL-IC-MP.

Mid-Continent Special, No. 15
Rock Island
Des Moines setout
Standing in Des Moines

DES MOINES to KANSAS CITY
156: 10s-dr-c
RI 15, Des Moines (2:05a) to Kansas City (7:30a)

(Minneapolis to Kansas City, snack-lounge car)

The Rock Island's prior competitors for the Des Moines-Kansas City Pullman market were the Great Western and Wabash.

Mid-Continent Special, No. 18
Kansas City to Minneapolis
Rock Island
5 miles south of Chariton

KANSAS CITY to ST. PAUL/MINNEAPOLIS
181: 8s-dr-2c   182: 8s-5dbr
RI 18, Kansas City (8:15p) to St. Paul (8:30a)
and Minneapolis (9:15a)

(Kansas City to Minneapolis, snack-lounge)

Mid-Continent Special, No. 18
Rock Island
Des Moines setout
Standing in Des Moines

DES MOINES to ST. PAUL/MINNEAPOLIS
185: 10s-3dbr
RI 18, Des Moines (1:40a) to St. Paul (8:30a)
and Minneapolis (9:15a)

(Kansas City to Minneapolis, snack-lounge)

Once again, the Rock island was the sole survivor, in this case in the sleeping car competition between Des Moines and the Twin Cities. The field had included the Minneapolis & St. Louis and the Great Western.

Sioux, No. 22
Canton, S.D., to Chicago
Milwaukee Road
35 miles east of Mason City

CANTON to CHICAGO
221: 10s-c-dr
MILW 22, Canton (5:30p) to Chicago (10a)

(Madison, Wis., to Chicago, dining car)

The *Sioux* by this carding only touches South Dakota, with taxi connections from Canton to Sioux Falls, but formerly it originated in Rapid City and carried sleepers from there and other upper midwestern communities to Chicago. Among the cars was one from Minneapolis, cut into the train at Calmar, yet another of the many lines linking the Twin Cities and Chicago.

Corn Belt Rocket, No. 23
Chicago to Omaha
Rock Island
Leaving Davenport

CHICAGO to OMAHA
233: 8dup/r-6r-4dbr
RI 23, Chicago (8:05p) to Omaha (7:25a)

(Chicago to Davenport, observation-dining-parlor car)

Chicago to Omaha was one of American railroading's most competitive corridors. The Rock Island, when these schedules were current, was just one of four railroads providing Pullman service, including the Burlington, Milwaukee Road, and North Western, and the list once included both the Illinois Central and Great Western

As of 1952, the Rock Island was still carrying sleepers from Chicago to the railroad's namesake city across the Mississippi River, but the separate Davenport and Moline cars were by then gone.

Gold Coast, No. 23
Chicago to San Francisco
Chicago & North Western
30 miles east of Cedar Rapids

CHICAGO to SAN FRANCISCO
    12s-dr cars as needed
C&NW 23, Chicago (8p Sun)
        to Omaha (7:30a Mon)
UP 23, Omaha (8:20a) to Ogden (6:15a Tue)
SP 23, Ogden (8:15a) to Oakland pier (7a Wed)
Ferry to San Francisco (7:35a Wed)

CHICAGO to OMAHA
    10s-3dbr
C&NW 23, Chicago (8p) to Omaha (7:30a)

CHICAGO to SIOUX CITY
    12s-2dbr
C&NW 23, Chicago (8p) to Carroll (4:56a)
C&NW 215, Carroll (5:10a) to Sioux City (9a)

CHICAGO to PORTLAND
    10s-dr-c
C&NW 23, Chicago (8p Sun)
        to Omaha (7:30a Mon)
UP 23, Omaha (8:20a)
        to Green River, Wyo. (1:35a Tue)
UP 17, Green River (3:25a) to Portland (6a Wed)

CHICAGO to LOS ANGELES
    10s-3dbr   *14s*
As above, Chicago (8p Sun) to Ogden (6:15a Tue)
UP 37, Omaha (7a) to Los Angeles (7a Wed)

(Chicago to Oakland, Portland, and Los Angeles: dining and lounge service. Breakfast not served into Portland on No. 17, the Portland Rose, or into Los Angeles on No. 37, the Pony Express.)

Apart from the heavy work of dispatching Pullmans to the West Coast, the *Gold Coast* was now performing the discontinued *Corn King's* overnight delivery of sleepers from Chicago to Sioux City and Omaha.

Gold Coast, No. 24
San Francisco to Chicago
Chicago & North Western
8 miles west of Ames

PORTLAND to CHICAGO
    10s-dr-c
UP 18, Portland (9:45p Fri)
        to Green River (11:05p Sat)
UP 24, Green River (12:15a Sun) to Omaha (7:50p)
C&NW 24, Omaha (8:30p) to Chicago (8:15a Mon)

LOS ANGELES to CHICAGO
    10s-3dbr   *14s*
UP 38, Los Angeles (7:30p Fri)
        to Odgen (6:40p Sat)
UP 24, Ogden (7:15p) to Omaha (7:50p Sun)
C&NW 24, Omaha (8:30p) to Chicago (8:15a Mon)

SAN FRANCISCO to CHICAGO
    12s-dr cars as needed
Ferry from San Francisco (6p Fri)
SP 24, Oakland pier (6:30p) to Ogden (6:30p Sat)
As above, Ogden (7:15p) to Chicago (8:15a Mon)

OMAHA to CHICAGO
    10s-3dbr
C&NW 24, Omaha (8:30p) to Chicago (8:15a)

SIOUX CITY to CHICAGO
    12s-2dbr
C&NW 216, Sioux City (6:15p) to Carroll (10:20p)
C&NW 24, Carroll (10:40p) to Chicago (8:15a)

(First-class lounge service to Chicago from Los Angeles, Portland, and San Francisco; dining service east of Las Vegas on No. 38, the Pony Express, and east of Pendleton on No. 18, the Portland Rose. East- and westbound editions of the Gold Coast are now in both Wyoming and California.)

Southwest Limited, No. 25
Milwaukee to Kansas City
Milwaukee Road
16 miles west of Davenport

CHICAGO to KANSAS CITY
    1074: 10s-c-dr
MILW 107, Chicago (6:10p) to Savanna (9:25p)
MILW 25, Savanna (9:50p) to Kansas City (7:25a)

(Chicago to Savanna, dining car. In the last year, the Southwest Limited has suffered severe pruning. Now gone: the Milwaukee-Kansas City sleeper, 1074: 8s-2c, and the parlor and dining cars in Milwaukee-Savanna service, leaving No. 25 coach-only over that portion.)

Southwest Limited, No. 26
Kansas City to Milwaukee
Milwaukee Road
5 miles west of Ottumwa

KANSAS CITY to CHICAGO
    260: 10s-c-dr
MILW 26, Kansas City (7:30p) to Savanna (4:45a)
MILW 108, Savanna (5:20a) to Chicago (8:50a)

(Savanna to Chicago, diner. Both east-and westbound, the remaining sleeper on the Southwest Limited and the diner are part of the Arrow, Nos. 107 and 108, between Savanna and Chicago.)

In its heyday in the late 1920's, the *Southwest Limited* from Kansas City operated three distinct sections. At Ottumwa the Cedar Rapids sleeper was cut out to form the nucleus of the Iowa section. At Davenport, which had its own Pullman from Kansas City and another to Milwaukee, the Wisconsin section was formed This carried Milwaukee sleepers from Kansas City and one from Omaha the train later picked up from the *Arrow*. The Illinois section accommodated the remaining Chicago Pullmans.

San Francisco Overland, No. 27
Chicago to San Francisco
Chicago & North Western
13 miles northeast of Council Bluffs

CHICAGO to SAN FRANCISCO
    4dbr-4c-2dr   *6s-6r-4dbr*
    10r-6dbr
C&NW 27, Chicago (3:30p Sun)
        to Omaha (12:30a Mon)
UP 27, Omaha (12:55a) to Ogden (8:33p)
SP 27, Ogden (9p) to Oakland pier (1:45p Tue)
Ferry to San Francisco (2:20p Tue)

NEW YORK to SAN FRANCISCO
    10r-6dbr, alternating days New York to Chicago
    between:
NYC 17, New York (6:15p Sat)
        to Chicago (11:15a Sun)
PRR 1, New York (6:45p Sat)
        to Chicago (10:25a Sun)
As above, Chicago (3:30p)
        to San Francisco (2:20p Tue)

CHICAGO to WEST YELLOWSTONE, MONT.
    10s-dr-2c   *Summer only*
C&NW 27, Chicago (3:30p Sun)
        to Omaha (12:40a Mon)
UP 27, Omaha (12:55a) to Green River (4:20p)
UP 11, Green River (4:30p) to Pocatello (9:45p)
UP 35, Pocatello (1:35a)
        to West Yellowstone (7:30a Tue)

(Dining cars and lounges: Chicago to Oakland; New York to Chicago on NYC No. 17, the Wolverine, and PRR No. 1, the Pennsylvania Limited, and Green River to Pocatello on UP 11, the Idahoan. Another westbound Overland is in Nevada.)

Ak-Sar-Ben, No. 30
Lincoln to Chicago
Burlington Route
18 miles east of Red Oak

LINCOLN to CHICAGO
    Std: s-r-dup/r-sbr-dbr-c-dr-lounge
CB&Q 30, Lincoln (9p) to Chicago (8:10a)

OMAHA to CHICAGO
    Std: dr-c-dbr-sbr-dub/r-r-s
CB&Q 30, Omaha (10:30p) to Chicago (8:10a)

(Diner serving breakfast)

The *Ak-Sar-Ben* bore the name of the large Omaha civic organization. It was one of the 80th-anniversary trains the Burlington introduced in 1930.

Imperial, No. 39
Chicago to Los Angeles
Via Douglas, Phoenix, El Centro
Rock Island
Entering Davenport

CHICAGO to LOS ANGELES
    398: 6s-6dbr   *391:10s-dr-2c*
RI 39, Chicago (8:15p Sun)
        to Tucumcari, N.M. (8:26p CST Mon)
SP 39, Tucumcari (8:10p MST)
        to Los Angeles (11:30p Tue)

CHICAGO to PHOENIX   *Winter only*
    394: 12r-sbr-4dbr
RI 39, Chicago (8:15p Sun)
        to Tucumcari (8:26p CST Mon)
SP 39, Tucumcari (8:10p MST)
        to Phoenix (12:25p Tue)

CHICAGO to KANSAS CITY
    390: 10s-c-2dbr
RI 39, Chicago (8:15p) to Kansas City (7:20a)

(Chicago to Los Angeles, dining car and first-class lounge)

The *Imperial*, which went into service in 1938, was the secondary train on the Rock Island/Southern Pacific's Golden State Route. It traveled through California's Imperial Valley, hence the name. The *Imperial* train took over the function of the *Apache* and *Californian*.

Imperial, No. 40
Los Angeles to Chicago
Via El Centro, Phoenix, Douglas
Rock Island
6 miles east of Centerville

LOS ANGELES to CHICAGO
    406: 6s-6dbr   *405:10s-dr-2c*
SP 40, Los Angeles (11p Fri)
        to Tucumcari (5:50a MST Sun)
RI 40, Tucumcari (7:20a CST)
        to Chicago (8:40a Mon)

PHOENIX to CHICAGO   *Winter only*
    401: 12r-sbr-4dbr
SP 40, Phoenix (12:25p Sat)
        to Tucumcari (5:50a MST Sun)
RI 40, Tucumcari (7:20a CST)
        to Chicago (8:40a Mon)

KANSAS CITY to CHICAGO
    402: 10s-c-2dbr
RI 40, Kansas City (7p) to Chicago (8:40a)

(Los Angeles to Chicago, diner and first-class lounge. Both east- and westbound editions of the Imperial are in New Mexico at this time, and one bound for Chicago is still in California, soon to make a run through its namesake valley.)

Iowa   (continued)

Zephyr Rocket, No. 61-562
St Louis to Minneapolis
Rock Island
14 miles south of West Liberty

ST. LOUIS to ST. PAUL/MINNEAPOLIS
   B155: 10s-3dbr   B156: 8s-dr-3dbr
CB&Q 15, St. Louis (5p) to Burlington (10:30p)
RI 61-562, Burlington (10:55p) to St. Paul (7:10a)
    and Minneapolis (8a)

(St. Louis to Minneapolis, observation-parlor-dining car)

The *Zephyr Rockets* were inaugurated in January 1941 as an interline venture of the Burlington and Rock Island in a marriage of the two fleet names. The silvery trains plied Mark Twain country, connecting the two great metropolitan areas of the upper Mississippi River. By long tradition, the two roads had made these night-time interline transfers in Burlington. The Burlington Railroad formerly provided the Iowa community with Pullman service to both St. Louis and Chicago, the latter route seeing much greater passenger activity as the Q's main line to Omaha and Denver.

City of San Francisco, No. 101
Extra-fare streamliner
Chicago to San Francisco
Chicago & North Western
3 miles east of Ames

CHICAGO to SAN FRANCISCO
   Consist varies among:
   4c-3dr   12s
   10r-6dbr   5dbr-12 duplex sb
   4dbr-4c-2dr   6s-6r-4dbr
C&NW 101, Chicago (7p Sun)
    to Omaha (2:45a Mon)
UP 101, Omaha (2:55a) to Ogden (6:05p)
SP 101, Ogden (6:22p) to Oakland pier (8:40a Tue)
Ferry to San Fancisco (9:15a Tue)

(Dining car and first-class lounge, Chicago to Oakland. The previous day's City of San Francisco *is now in Nevada.)*

In 1952, the bright yellow *City* streamliners were crossing Iowa at night on the main line of the Chicago & North Western, mostly the way it had been for Overland trains since the beginning. Arrangements dramatically changed in 1955, however, when the entire streamlined fleet moved to Milwaukee Road tracks. Apart from Council Bluffs, in all of Iowa only the little town of Tama was visited by the *City* streamliners both before and after the switch, for it was there, in midstate, the North Western and Milwaukee Road tracks crossed.

City of Los Angeles, No. 103
Extra-fare streamliner
Chicago to Los Angeles
Chicago & North Western
16 miles east of Ames

CHICAGO to LOS ANGELES
   Consist varies among:
   11dbr   12s
   10r-6dbr   18r
   13r   4dbr-4c-2dr
   4c-3dr   6s-6r-4br
   2dr-c-dbr-obs.-lounge 4dbr-obs.-lounge
   Obs.-buffet-lounge
C&NW 103, Chicago (7:15p Sun)
    to Omaha (3a Mon)
UP 103, Omaha (3:10a to Los Angeles (9a Tue)

(Dining car and first-class lounge with bath, Chicago to Los Angeles. The City of Los Angeles *that passed here 24 hours ago, is now in Nevada.)*

City of Portland, No. 105
Streamliner
Chicago to Portland
Chicago & North Western
17 miles west of Carroll

CHICAGO to PORTLAND
   2dr-4c-4dbr   6s-6r-4dbr
   6s-6r-4dbr
C&NW 105, Chicago (5:30p Sun)
    to Omaha (1:30a Mon)
UP 105, Omaha (1:40a) to Portland (7:30a Tue)

CHICAGO to SEATTLE
   12r-4dbr
As above, Chicago (5:30p Sun)
    to Portland (7:30a Tue)
Pool 457, Portland (8a) to Seattle (11:59a Tue)

(Chicago to Portland, dining car and first-class lounge with bath. Portland to Seattle, Astra Dome observation-lounge and Astra Dome dining car. Another westbound City of Portland *is now in Oregon.)*

Arrow, No. 107
Chicago to Omaha
Milwaukee Road
Leaving Marion

CHICAGO to SIOUX CITY/SIOUX FALLS
   1070: 6sbr-2dbr-lounge   1072: 12s-dr
MILW 107, Chicago (6:10p) to Manilla (5a)
MILW 217, Manilla (5:30a) to Sioux City (8a)
    and Sioux City (10:55a)

CHICAGO to OMAHA
   1071: 8s-2c-dr
MILW 107, Chicago (6:10p) to Omaha (7:30a)

(Dining-buffet service: Chicago to Savanna, Manilla to Sioux Falls)

The Milwaukee Road introduced the *Arrow* in 1920 as its prime overnight train to Omaha. By way of Madrid, Des Moines once had dedicated sleepers from Chicago and to Sioux City.

Arrow, No. 108
Omaha to Chicago
Milwaukee Road
Standing in Madrid

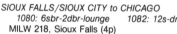

SIOUX FALLS/SIOUX CITY to CHICAGO
   1080: 6sbr-2dbr-lounge   1082: 12s-dr
MILW 218, Sioux Falls (4p)
    and Sioux City (6:45p) to Manilla (9:20p)
MILW 108, Manilla (9:50p) to Chicago (8:50a)

OMAHA to CHICAGO
   1081: 8s-2c-dr
MILW 108, Omaha (7:40p) to Chicago (8:50a)

(Dining-buffet service: Sioux Falls to Manilla, Savanna to Chicago)

Sioux City-Chicago service was still operating at a high level in 1952, with dedicated sleepers on the IC and North Western and the *Arrow's* through cars from Sioux Falls. The Milwaukee Road formerly offered a Sioux City-Aberdeen sleeper.

City of Denver, No. 111
Streamliner
Chicago to Denver
Chicago & North Western
30 miles east of Council Bluffs

CHICAGO to DENVER
   8s-3dbr   4r-4dbr-3c-dr
   Observation-6dbr   12s cars as necessary
C&NW 111, Chicago (5p) to Omaha (12:40a)
UP 111, Omaha (12:05a) to Denver (8:30a)

(Chicago to Denver: "Frontier Shack" lounge and dining car)

The *City of Denver* made its debut in 1936, the same year the Burlington's *Denver Zephyr* went into service. Both cut travel time from Chicago to a fast overnight schedule.

Sioux, Austin section, No. 122
Austin, Minn., to Calmar
Milwaukee Road
13 miles northwest of Calmar

AUSTIN to CHICAGO
   1220: 8s-2c-dr
MILW 122, Austin (10:30p) to Calmar (12:25a)
MILW 22, Calmar (1:40a) to Chicago (10a)

(Dining car, Madison, Wis., to Chicago)

Nightingale, No. 202
Omaha to Minneapolis
Chicago & North Western
42 miles north of Council Bluffs

OMAHA to ST. PAUL/MINNEAPOLIS
   10s-3dbr   8s-5dbr
C&NW 202, Omaha (10:25p) to St. Paul (7:45a)
    and Minneapolis (8:20a)

St. James, Minn., to Minneapolis, cafe-lounge)

Nightingale, No. 202
Chicago & North Western
Sioux City setout
Standing in Sioux City

SIOUX CITY to ST. PAUL/MINNEAPOLIS
   12s-dr
C&NW 202, Sioux City (1:20a) to St. Paul (7:45a)
    and Minneapolis (8:20a)

St. James, Minn., to Minneapolis, cafe-lounge)

In the other direction, Sioux City had Pullman service on the Omaha Road to Omaha and Kansas City (Missouri Pacific).

Zephyr Rocket, No. 561-62
Minneapolis to St. Louis
Rock Island
14 miles south of West Liberty

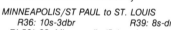

MINNEAPOLIS/ST PAUL to ST. LOUIS
   R36: 10s-3dbr   R39: 8s-dr
RI 561-62, Minneapolis (5p) and
    St. Paul (5:30p) to Burlington (1:10a)
CB&Q 8, Burlington (1:35a) to St. Louis (7:58a)

(Minneapolis to St. Louis, observation-parlor-dining car)

A forerunner of the *Zephyr Rocket* carried a Cedar Rapids-St. Louis Pullman. The Rock Island-Burlington interline Pullman operation proved to be the strongest Twin Cities-St. Louis routing, but earlier competitors had included various combinations involving the M&StL, Wabash, Rock Island, and Milwaukee Road, and before the turn of the century, the Burlington did it all itself using its lines along or paralleling the Mississippi River.

Kansas

See map on page 240

In Kansas, the home of the Atchison, Topeka & Santa Fe, the railroad had two pairs of alternate routes that its passenger trains used when traveling west via the Raton Pass. The alternatives involved either Topeka or Ottawa and St. John or Great Bend. The only through train in 1952 using the Great Bend route was the *California Limited*, which also served Topeka. The other main trains through the capital were the *Kansas Cityan/Chicagoan*, and *Grand Canyon*, Northern Section. The southern route trains traveling through Wichita and the Texas Panhandle branched off at Newton.

Golden State, No. 3

Chicago to Los Angeles
Rock Island
9 miles east of Topeka

CHICAGO to LOS ANGELES
    35: 10r-6dbr   36: 4dbr-4c-2dr
    2dbr-dr-obs.-lounge or   30: 6dbr-10r
    RI 3, Chicago (1:20p Sun)
        to Tucumcari, N.M. (10:16a CST Mon)
    SP 3, Tucumcari (9:30a MST)
        to Los Angeles (7:35a Tue)

CHICAGO to PHOENIX   Winter season only
    34: 4dbr-2dr-4c
    RI 3, Chicago (1:20p Sun)
        to Tucumcari (10:16a CST Mon)
    SP 3, Tucumcari (9:30a MST)
        to Phoenix (11:45p Mon)

CHICAGO to TUCSON   Winter season only
    301: 4dbr-4c-2dr, or   12dbr
    RI 3, Chicago (1:20p Sun)
        to Tucumcari (10:16a CST Mon)
    SP 3, Tucumcari (9:30a MST)
        to Tucson (9:17p Mon)

MINNEAPOLIS/ST. PAUL to LOS ANGELES
    575: 6s-6r-4dbr
    RI 507, Minneapolis (12:01p Sun) and
        St. Paul (12:25p) to Kansas City (9:15p)
    RI 3, Kansas City (11p)
        to Tucumcari (10:16a CST Mon)
    SP 3, Tucumcari (9:30a MST)
        to Los Angeles (7:35a Tue)

ST. LOUIS to LOS ANGELES
    115: 6s-6r-4dbr
    MP 11, St. Louis (4p Sun) to Kansas City (9p)
    RI 3, Kansas City (11p)
        to Tucumcari (10:16a CST Mon)
    SP 3, Tucumcari (9:30a MST)
        to Los Angeles (7:35a Tue)

(MP No. 11 is the Colorado Eagle, *which carries dining-lounge, grill-coach, and Planetarium coach. Rock Island No. 507 is the* Twin Star Rocket, *which carries a dining car and observation lounge. The* Golden State *has a dining car, a coffee shop-lounge-diner, and, when not carrying the observation sleeper, another first-class lounge in its consist. It is an extra-fare train. East- and westbound editions of the train are also now in Arizona.)*

The *Golden State Limited* was inaugurated in November 1902 as an all-Pullman seasonal train. Originally, the Alton handled the train between Chicago and Kansas City until the Rock Island consolidated its line. Though not so fast as the crack Santa Fe and Union Pacific competition, the *Golden State* made the most southerly and lowest-altitude crossing of the Rockies, a fact that was commonly stressed in its advertising. In years past, the Pullman consist included Chicago cars for both Santa Barbara and San Diego (SD&AE)

This publicity shot caught the semi-streamlined *City of St. Louis* in its namesake city.

Golden State, No. 4

Los Angeles to Chicago
Rock Island
Entering Topeka

LOS ANGELES to CHICAGO
    45: 10r-6dbr   46: 4dbr-4c-2dr
    2dbr-dr-obs.-lounge, or   40: 6dbr-10r
    SP 4, Los Angeles (1:30p Sat)
        to Tucumcari (1:43p MST Sun)
    RI 4, Tucumcari (2:49p CST)
        to Chicago (11:45a Mon)

PHOENIX to CHICAGO   Winter season only
    44: 4dbr-2dr-4c
    SP 4, Phoenix (10:35p Sat)
        to Tucumcari (1:43p MST Sun)
    RI 4, Tucumcari (2:49 CST)
        to Chicago (11:45a Mon)

TUCSON to CHICAGO   Winter season only
    43: 4dbr-4c-2dr, or   12dbr
    SP 4, Tucson (12:55a Sun)
        to Tucumcari (1:43p MST)
    RI 4, Tucumcari (2:49p CST)
        to Chicago (11:45a Mon)

LOS ANGELES to ST. PAUL/MINNEAPOLIS
    4508: 6s-6r-4dbr
    SP 4, Los Angeles (1:30p Sat)
        to Tucumcari (1:43p MST Sun)
    RI 4, Tucumcari (2:49p CST)
        to Kansas City (1:25a Mon)
    RI 508, Kansas City (9:10a Mon) to St. Paul (6p)
        and Minneapolis (6:30p Mon)

LOS ANGELES to ST. LOUIS
    41: 6s-6r-4dbr
    SP 4, Los Angeles (1:30p Sat)
        to Tucumcari (1:43p MST Sun)
    RI 4, Tucumcari (2:49p CST)
        to Kansas City (1:25a Mon)
    MP 20, Kansas City (1:45a)
        to St. Louis (8:05a Mon)

(Rock Island No. 508, the Twin Star Rocket, *carries a dining car and observation lounge to Minneapolis. MoPac No. 20, the* Sunflower, *has a grill-coach on the St. Louis leg. A dining car and coffee shop-lounge-diner are in the consist of the extra-fare* Golden State, *as is a first-class lounge car when the observation sleeper is not included.)*

City of St. Louis, No. 9

Streamliner
St. Louis to Los Angeles
Union Pacific
25 miles west of Manhattan

KANSAS CITY to DENVER
    6s-6r-4dbr
    UP 9, Kansas City (9:30p) to Denver (7:40a)

ST. LOUIS to LOS ANGELES
    01: 10r-1dbr
    Wabash 9, St. Louis (4p Sun) to Kansas City (9p)
    UP 9, Kansas City (9:30p) to Los Angeles (3p Tue)

ST. LOUIS to SAN FRANCISCO
    92: 6s-6r-4dbr
    Wabash 9, St. Louis (4p Sun) to Kansas City (9p)
    UP 9, Kansas City (9:30p)
        to Green River, Wyo. (3:35p)
    UP 27, Green River (4:40p Mon) to Ogden (8:33p)
    SP 27, Ogden (9p) to Oakland (1:45p Tue)
    Ferry to San Francisco (2:20p Tue)

ST. LOUIS to SEATTLE
    93: 6s-6r-4dbr
    As above, St. Louis (4p Sun)
        to Green River (3:35p Mon)
    UP 11, Green River (4:30p) to Portland (3:30p Tue)
    UP 459, Portland (5p) to Seattle (9:15p Tue)

(Dining cars and first-class lounges carried on all segments. The Oakland Pullman reaches its destination via the San Francisco Overland, No. 27; *the Seattle sleeper travels to Portland aboard the* Idahoan, No. 11.)*

The *City of St. Louis* was inaugurated in 1946, continuing the UP-Wabash partnership represented by the old *Pacific Coast Limited*.

Kansas   *(continued)*

City of St. Louis, No. 10
Streamliner
Los Angeles to St. Louis
Union Pacific
Standing in Ellis

LOS ANGELES to ST. LOUIS
   104: 12r-4dbr
UP 10, Los Angeles (10:30a Sat)
      to Kansas City (6:45a Mon)
Wabash 10, Kansas City (7a)
      to St. Louis (11:59a Mon)

SAN FRANCISCO to ST. LOUIS
   288: 6s-6r-4dbr
Ferry from San Francisco (11a Sat)
SP 28, Oakland pier (11:30a) to Ogden (6:25a Sun)
UP 10, Ogden (6:50a) to Kansas City (6:45a Mon)
Wabash 10, Kansas City (7a)
      to St. Louis (11:59a Mon)

SEATTLE to ST. LOUIS
   4020: 6s-6r-4dbr
UP 402, Seattle (11:45p Fri) to Portland (6:45a Sat)
UP 12, Portland (8:10a) to Green River (9:50a Sun)
UP 10, Green River (10:45a)
      to Kansas City (6:45a Mon)
Wabash 10, Kansas City (7a)
      to St. Louis (11:59a Mon)

DENVER to KANSAS CITY
   6s-6r-4dbr
UP 10, Denver (6:45p) to Kansas City (6:45a)

(Dining cars and lounges on all segments. SP No. 28 is the San Francisco Overland; UP 12 the Idahoan. East- and westbound editions of the City of St. Louis *are also now in Utah.)*

Colorado Eagle, No. 11
Streamliner
St. Louis to Denver
Missouri Pacific
14 miles east of Herrington

ST. LOUIS to DENVER
   111: 14r-dr-2dbr   112: 6s-6r-6dbr
MP 11, St. Louis (4p) to Pueblo (7a)
D&RGW 4, Pueblo (7:10a) to Denver (9:40a)

(Grill coach, dining-lounge, and Planetarium coach)

The *Colorado Eagle* made its debut early in World War II, replacing the venerable *Scenic Limited*, which had operated since 1915 and given St. Louis through service to the West Coast via the Rio Grande/Western Pacific route.

Colorado Eagle, No. 12
Streamliner
Denver to St. Louis
Missouri Pacific
35 miles west of Geneseo

DENVER to ST. LOUIS
   31: 14r-dr-2dbr   32: 6s-6r-6dbr
D&RGW 3, Denver (4:10p) to Pueblo (6:40p)
MP 12, Pueblo (6:50p) to St. Louis (12:01p)

DENVER to WICHITA
   34: 8s-dr-3dbr
D&RGW 3, Denver (4:10p) to Pueblo (6:40p)
MP 12, Pueblo (6:50p) to Geneseo (1:51a)
MP 412, Geneseo (4a) to Wichita (7a)

(Denver to St. Louis: Grill coach, dining-lounge, and Planetarium coach)

The *Eagle's* predecessor, the *Scenic Limited*, carried a Pueblo-Wichita Pullman. To the east, Wichita had dedicated sleepers to Kansas City and Hot Springs.

Texas Chief, No. 16
Galveston to Chicago
Streamliner, via Wichita, Ottawa
Atchison, Topeka & Santa Fe
6 miles east of Ottawa

GALVESTON to CHICAGO
   4c-4dbr-2dr   10r-3dbr-2c
   24dup/r
Santa Fe 16, Galveston (6:45a) to Chicago (9a)

OKLAHOMA CITY to CHICAGO
   10r-6dbr
Santa Fe 16, Oklahoma City (5:55p)
      to Chicago (9a)

WICHITA to CHICAGO
   4c-4dbr-2dr   10r-3dbr-2c
Santa Fe 16, Wichita (9:15p) to Chicago (9a)

(Dining car, lounge)

The *Texas Chief* was a post-war addition to the Santa Fe's fleet of streamliners, going into operation in April 1948. Dallas got its own service in 1955.

Super Chief, No. 18
All-Pullman streamliner
Los Angeles to Chicago
Via Raton Pass, Ottawa
Atchison, Topeka & Santa Fe
Standing in Dodge City

LOS ANGELES to CHICAGO
   Observation: 4dr-dbr   2dr-4c-4dbr
   2dr-4c-4dbr   2dr-4c-4dbr
   10r-6dbr   10r-6dbr
Santa Fe 18, Los Angeles (8p Sat)
      to Chicago (1:45p Mon)

LOS ANGELES to KANSAS CITY
   10r-6dbr
Santa Fe 18, Los Angeles (8p Sat)
      to Kansas City (5:35a Mon)

(Los Angeles to Chicago: extra fare $15; dining car, dormitory/lounge, Pleasure Dome Lounge and Turquoise private dining room; barber, shower, radio, and recorded music. Another Super Chief *heading east is now in California.)*

Chief, No. 19
All-Pullman streamliner
Chicago to Los Angeles
Via Ottawa, Raton Pass
Atchison, Topeka & Santa Fe
14 miles east of Emporia

CHICAGO to LOS ANGELES
   17r   2dr-4c-4dbr
   Observation-4dr-dbr
Santa Fe 19, Chicago (1:30p Sun)
      to Los Angeles (8:30a Tue)

NEW YORK to LOS ANGELES
   2501: 2dr-4c-4dbr   2502: 10r-6dbr
NYC 25, New York (6p Sat) to Chicago (9a Sun)
Santa Fe 19, Chicago (1:30p)
      to Los Angeles (8:30a Tue)

NEW YORK to LOS ANGELES
   2dr-4c-4dbr
PRR 29, New York (6p Sat) to Chicago (9a Sun)
Santa Fe 19, Chicago (1:30p)
      to Los Angeles (8:30a Tue)

WASHINGTON to SAN DIEGO
   10r-6dbr
B&O 5, Washington (5:30p Sat)
      to Chicago (8a Sun)
Santa Fe 19, Chicago (1:30p)
      to Los Angeles (8:30a Tue)
Santa Fe 74, Los Angeles (11:30a)
      to San Diego (2:15p Tue)

(Chicago to Los Angeles: $10 extra fare; baggage/ lounge, lounge, dining car; valet, barber, shower bath, radio. No. 74 carries a diner. Three of the supporting trains here are all-Pullman: The Twentieth Century Limited, *NYC No. 25; the* Broadway Limited, *PRR No. 29; and the* Capital Limited, *B&O No. 5. All the eastern trains carry dining cars and various lounges, including observation cars, and the* Capital Limited *offers facilities in a Strata Dome sleeper every other day. The other westbound* Chief *is just entering Needles, Calif.)*

The Santa Fe introduced the *Chief* as its standard-bearer in November 1926, and for a time it was the fastest and only extra-fare train ($10) to Southern California. Hollywood movie people took to the *Chief* with such affection that *Variety* published daily passenger lists. A Chicago-San Diego Pullman was included in the otherwise all-Los Angeles consist.

In Winfield, Ala., two *Kansas City-Florida Specials* meet in May 1955. Frisco diesels were named for famous racehorses; *Gallant Fox* (right) is southbound.

Chief, No. 20

All-Pullman streamliner
Los Angeles to Chicago
Via Raton Pass, Ottawa
Atchison, Topeka & Santa Fe
Standing in Emporia

LOS ANGELES to CHICAGO
   17r   2dr-4c-4dbr
   Observation: 4dr-1dbr
   Santa Fe 20, Los Angeles (12:30p Sat)
      to Chicago (10:30a Mon)

LOS ANGELES to NEW YORK
   2dr-4c-4dbr   10r-6dbr
   Santa Fe 20, Los Angeles (12:30p Sat)
      to Chicago (10:30a Mon)
   NYC 26, Chicago (4:30p) to New York (9:30a Tue)

LOS ANGELES to NEW YORK
   2dr-4c-4dbr
   Santa Fe 20, Los Angeles (12:30p Sat)
      to Chicago (10:30a Mon)
   PRR 28, Chicago (4:30p) to New York (9:30a Tue)

SAN DIEGO to WASHINGTON
   10r-6dbr
   Santa Fe 71, San Diego (7:45a Sat)
      to Los Angeles (10:30a)
   Santa Fe 20, Los Angeles (12:30p)
      to Chicago (10:30a Mon)
   B&O 6, Chicago (4:30p) to Washington (8:55a Tue)

PHOENIX to CHICAGO
   2dr-4c-4dbr
   Santa Fe 42, Phoenix (5p Sat)
      to Ash Fork, Ariz. (11:15p)
   Santa Fe 20, Ash Fork (12:30a Sun)
      to Chicago (10:30a Mon)

(Like the Chief, *the three eastern trains, NYC's Twentieth Century Limited,* Pennsy's Broadway Limited, *and the B&O's* Capitol Limited, *are all-Pullman and carry lounges and diners. The San Diego train has a parlor-lounge, the Phoenix connection a cafe-parlor.)*

Grand Canyon, Southern Section, No. 23

Chicago to Los Angeles
Via Ottawa, Wichita, Amarillo
Atchison, Topeka & Santa Fe
4 miles west of Emporia

CHICAGO to LOS ANGELES
   10r-3dbr-2c   14s
   Santa Fe 23, Chicago (12:01p Sun)
      to Los Angeles (10:40a Tue)

CHICAGO to LOS ANGELES
   8s-2c-2dbr
   Santa Fe 23, Chicago (12:01p Sun)
      to Williams, Ariz. (10:30p Mon)
   Santa Fe 14, Williams (4:15a Tue)
      to Grand Canyon (7a)
   Santa Fe 15, Grand Canyon (8p)
      to Williams (10:20p)
   Santa Fe 123, Williams (10:40p)
      to Los Angeles (10:30a Wed)

CHICAGO to SAN FRANCISCO
   6r-6s-4dbr
   Santa Fe 23, Chicago (12:01p Sun)
      to Barstow (5:55a Tue)
   Santa Fe 23, Barstow (6:20a) to Oakland (8p)
   Bus to San Francisco (8:20p Tue)

(Dining and lounge service over all segments except Grand Canyon-Williams. The Southern Section of the Santa Fe's Grand Canyon *works the railroad's lower-grade, principally freight route through the Texas Panhandle. At Williams, one Los Angeles Pullman will be cut out for a visit to the Grand Canyon, where it stands all day while the passengers sight-see. After returning to Williams, it will be taken aboard the next night's edition of the* Grand Canyon, Northern Section, No. 123, *for the remainder of the trip west.)*

Grand Canyon, Southern Section, No. 24

Los Angeles to Chicago
Via Amarillo, Wichita, Ottawa
Atchison, Topeka & Santa Fe
14 miles west of Wellington

LOS ANGELES to CHICAGO
   10r-3dbr-2c   14s
   Santa Fe 24, Los Angeles (1:40p Sat)
      to Chicago (3:45p Mon)

SAN FRANCISCO to CHICAGO
   6s-6r-4dbr
   Bus from San Francisco (7:45a Sat)
   Santa Fe 60, Oakland (8:07a)
      to Bakersfield (2:05p)
   Santa Fe 24, Bakersfield (2:15p) to Barstow (6p)
   Santa Fe 24, Barstow (6:25p)
      to Chicago (3:45p Mon)

LOS ANGELES to CHICAGO
   8s-2c-2dbr
   Santa Fe 124, Los Angeles (1:30p Fri)
      to Williams (3:25a)
   Santa Fe 14, Williams (4:15a Sat)
      to Grand Canyon (7a)
   Santa Fe 15, Grand Canyon (8p)
      to Williams (10:20p)
   Santa Fe 24, Williams (3:40a Sun)
      to Chicago (3:45p Mon)

(Los Angeles to Chicago, dining car and lounge; Oakland to Bakersfield, lunch-counter dining car. Just as the westbound operation works, the Pullman that travels to the rim of the Grand Canyon left Los Angeles on the Grand Canyon's *Northern Section the day before the rest of this train's consist was assembled. Sleepers for Dallas and New Orleans were cut out at Clovis. Other east- and westbound editions of the* Grand Canyon Southern Section *are in Arizona.)*

Bluebonnet, No. 27

Kansas City to San Antonio, Houston
Via Dallas/Fort Worth
Missouri-Kansas-Texas
22 miles north of Parsons

KANSAS CITY to SAN ANTONIO   *via Dallas*
   71: 8s-5dbr
   MKT 27, Kansas City (9:40p) to Parsons (12:25a)
   MKT 7, Parsons (12:35a) to Denison (6:05a)
   MKT 27-7, Denison (6:20a) to Dallas (8:40a) and
      to Waco (12:01p)
   MKT 7, Waco (12:15p) to San Antonio (5:45p)

KANSAS CITY to WACO   *via Fort Worth (8:45a)*
   72: 10s-dr-2c
   As above, Kansas City (9:40p) to Denison (6:05a)
   MKT 27-7, Denison (6:20a) to Waco (11:40a)

KANSAS CITY to DALLAS
   73: 8s-6r-dbr-sbr
   As above, Kansas City (9:40p) to Dallas (8:40a)

(The Bluebonnet *demonstrates the Katy's classic way of serving the Big Texas cities: the train divides at Denison, and one section serves Fort Worth, the other Dallas. Although the trains meet again in Hillsboro, Waco is where the next switching takes place, and train No. 7 for San Antonio and No. 27 for Houston are assembled. In this consist, sleepers 71 and 73 go to or through Dallas, as do a lounge car, diner—the latter cut in at Denison—and coach for San Antonio. The Fort Worth section consists of sleeping car 72, the dining car from Kansas City, and Denison-Houston and Fort Worth-San Antonio coaches. This routing permits in-train transfer of passengers traveling to and from the two pairs of large cities. The same sort of flexibility was also available to Kansas City and St. Louis passengers, with the division of some of the Katy's Texas trains in the junction of Parsons, but as of this scheduling, service to St. Louis has become minimal.)*

No. 28

Oklahoma City to Newton
Atchison, Topeka & Santa Fe
22 miles south of Wichita

OKLAHOMA CITY to KANSAS CITY
   12s-dr
   Santa Fe 28, Oklahoma City (8:10p)
      to Newton (1:25a)
   Santa Fe 124, Newton (2a) to Kansas City (6:30a)

OKLAHOMA CITY to DENVER
   8s-dr-2c
   Santa Fe 28, Oklahoma City (8:10p)
      to Newton (1:25a)
   Santa Fe 123, Newton (1:50a) to La Junta (7:05a)
   Santa Fe 1, La Junta (8a) to Pueblo (9:45a)
   Santa Fe 102, Pueblo (9:55a) to Denver (1p)

(Nos. 123 and 124 are the Grand Canyon's *Northern Section, carrying lounges and diners. La Junta to Denver, cafe-observation car.)*

Kansas City-Florida Special, No. 105

Kansas City to Jacksonville
Frisco
14 miles north of Paola

KANSAS CITY to JACKSONVILLE
   F69: 10s-2br-c   F67: 8s-5dbr
   SLSF 105, Kansas City (11:30p Sun)
      to Birmingham (4:25p Mon)
   SOU 8, Birmingham (4:40p) to Atlanta (10:25p)
   SOU 7, Atlanta (10:45p)
      to Jacksonville (7:30a Tue)

KANSAS CITY to MIAMI
   F68: 14r-4dbr
   As above, Kansas City (11:30p Sun)
      to Jacksonville (7:30a Tue)
   FEC 73, Jacksonville (9:50a) to Miami (5:15p Tue)

(Dining cars, all segments; lounge cars, all segments except Birmingham-Atlanta. FEC 73 is the Vacationer.)

Rainbow Special, No. 117

Osawatomie to Little Rock
Missouri Pacific
4 miles south of Le Roy

KANSAS CITY to LITTLE ROCK
   1171: 10s-dr-2c
   MP 15, Kansas City (8:30p)
      to Osawatomie (10:25p)
   MP 117, Osawatomie (10:40p)
      to Little Rock (11:59a)

(Coffeyville to Little Rock, grill coach)

Grand Canyon, Northern Section, No. 123

Kansas City to Los Angeles
Via Topeka, Raton Pass
Atchison, Topeka & Santa Fe
8 miles east of Emporia

CHICAGO to LOS ANGELES
   24dup/r   8s-2c-2dbr
   8s-dr-2c
   Santa Fe 23, Chicago (12:01p Sun)
      to Kansas City (9p)
   Santa Fe 123, Kansas City (9:30p)
      to Los Angeles (10:25a Tue)

KANSAS CITY to DENVER
   6s-6r-4dbr
   Santa Fe 123, Kansas City (9:30p)
      to La Junta (7:05a)
   Santa Fe 1, La Junta (8a) to Pueblo (9:45a)
   Santa Fe 102, Pueblo (9:55a) to Denver (1p)

KANSAS CITY to OKLAHOMA CITY
   12s-dr
   Santa Fe 123, Kansas City (9:30p)
      to Newton (1:30a)
   Santa Fe 27, Newton (2:45a)
      to Oklahoma City (8:50a)

(Chicago to Los Angeles, dining car, lounge; La Junta to Denver, cafe-observation car.)

Kansas

(continued)

Oklahoman, No. 111
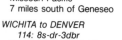
Kansas City to Tulsa
Frisco
13 miles south of Kansas City

KANSAS CITY to TULSA
    61: 102-dr-2c
    SLSF 111, Kansas City (11:35p) to Tulsa (7:35a)

Grand Canyon, Northern Section, No. 124
Los Angeles to Kansas City
Via Raton Pass, Topeka
Atchison, Topeka & Santa Fe
25 miles east of Kinsley

LOS ANGELES to CHICAGO
    24dup/r   8s-2c-2dbr
    8s-dr-2c
    Santa Fe 124, Los Angeles (1:30p Sat)
        to Kansas City (6:30a Mon)
    Santa Fe 24, Kansas City (7a)
        to Chicago (3:45p Mon)

DENVER to KANSAS CITY
    6s-6r-4dbr
    Santa Fe 101, Denver (2:10p) to Pueblo (4:45p)
    Santa Fe 2, Pueblo (4:50p) to La Junta (6:10p)
    Santa Fe 124, La Junta (6:40p)
        to Kansas City (6:30a)

DENVER to OKLAHOMA CITY
    8s-dr-2c
    As above, Denver (2:10p) to La Junta (6:10p)
    Santa Fe 124, La Junta (6:40p) to Newton (1:50a)
    Santa Fe 27, Newton (2:45a)
        to Oklahoma City (8:50a)

(Dining car and lounge, Los Angeles to Chicago.)

No. 411
Wichita to Geneseo
Missouri Pacific
7 miles south of Geneseo

WICHITA to DENVER
    114: 8s-dr-3dbr
    MP 411, Wichita (9:30p) to Geneseo (12:15a)
    MP 11, Geneseo (1:32a) to Pueblo (7a)
    D&RGW 4, Pueblo (7:10a) to Denver (9:40a)

(Dining-lounge, grill coach, and Planetarium coach on No. 11, the streamlined Colorado Eagle)

No. 420
Witchita to Pleasant Hill, Mo.
Missouri Pacific
Standing in Fort Scott

WICHITA to ST. LOUIS
    202: 10r-5dbr
    MP 420, Wichita (7:45p) to Pleasant Hill (2:25a)
    MP 20, Pleasant Hill (2:45a) to St. Louis (8:05a)

(Wichita to Pleasant Hill, grill coach)

Twin Star Rocket, No. 507
Streamliner
Minneapolis to Houston
Rock Island
22 miles northeast of Herrington

MINNEAPOLIS/ST. PAUL to HOUSTON
    573: 8dup/r-6r-4dbr
    RI 507, Minneapolis (12:01p) and
        St. Paul (12:25p) to Houston (1:35p)

KANSAS CITY to HOUSTON
    571: 6s-6r-4dbr
    RI 507, Kansas City (9:45p) to Houston (1:35p)

(Minneapolis to Houston, diner and parlor-observation-lounge car)

C&O's photog caught the *Sportsman* (or *GW?*—he didn't note which) at Ashland in 1948.

Kentucky

See map on page 98

George Washington, No. 1

Chesapeake & Ohio
Washington Pullman
Standing in Ashland

WASHINGTON to CINCINNATI
    474: 10r-6dbr
    C&O 5, Washington (10:40a)
        to Charlottesville, Va. (1:20p)
    C&O 47, Charlottesville (1:30p)
        to Ashland (11:31p)
    C&O 1, Ashland (5:40a) to Cincinnati (8:45a)

(Coach-lunch, Washington to Charlottesville; dining cars: Charlottesville to Charleston, Ashland to Cincinnati. This Washington-Cincinnati Pullman, which arrived in Ashland on the Sportsman, *waits overnight for the* George Washington *to complete the trip.)*

Ashland formerly had Pullman service on the C&O to Cincinnati, Richmond, Newport News (Phoebus), and Washington.

Ponce de Leon, No. 1
Cincinnati to Jacksonville
Southern Railway
32 miles south of Danville

CLEVELAND to TAMPA
    4332: 10s-dr-2c
    NYC 433, Cleveland (12:30p Sun)
        to Cincinnati (6:45p)
    SOU 1, Cincinnati (9:15p)
        to Jacksonville, Fla. (7:55p)
    SAL 9-1, Jacksonville (10:30p Mon)
        to Tampa (5:10a Tue)

CINCINNATI to JACKSONVILLE
    170: 10s-diner-lounge
    SOU 1, Cincinnati (9:15p) to Jacksonville (7:55p)

(Cleveland to Cincinnati, dining and lounge cars)

The *Ponce de Leon* once carried a Louisville-Jacksonville sleeper.

Azalean, No. 1
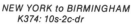
Cincinnati to New Orleans
Louisville & Nashville
Standing in Bowling Green

NEW YORK to BIRMINGHAM
    K374: 10s-2c-dr
    PRR 37, New York (11:20p Sat)
        to Pittsburgh (7:50a Sun)
    PRR 205, Pittsburgh (10:15a) to Cincinnati (5:40p)
    L&N 1, Cincinnati (7:05p)
        to Birmingham (7:20a Mon)

(Dining cars: Pittsburgh to Columbus, Cincinnati to Louisville, Birmingham to New Orleans. PRR 37 is the Iron City Express.*)*

Louisville had Pullman service on this line from Cincinnati and to Memphis, Nashville, and Montgomery.

No. 3

Cincinnati to Montgomery
Louisville & Nashville
41 miles east of Louisville

CINCINNATI to NASHVILLE
    3: 12s-dr
    L&N 3, Cincinnati (10:50p) to Nashville (8:05a)

Sportsman, No. 4-46

Detroit to Washington/Newport News
Chesapeake & Ohio
67 miles west of Ashland

CHICAGO to NEWPORT NEWS (PHOEBUS)
    4068: 10r-6dbr
    NYC 406, Chicago (12:50p) to Cincinnati (9:05p)
    C&O 4-46, Cincinnati (10:10p) to Phoebus (5:30p)

CHICAGO to CLIFTON FORGE, VA.
    4069: 10r-6dbr
    NYC 406, Chicago (12:50p) to Cincinnati (9:05p)
    C&O 4-46, Cincinnati (10:10p)
        to Clifton Forge (8:35a)

ST. LOUIS to RICHMOND
    C247: 10r-6dbr
    NYC 24, St. Louis (12:45p) to Indianapolis (5:02p)
    NYC 406, Indianapolis (5:15p)
        to Cincinnati (9:05p)
    C&O 4-46, Cincinnati (10:10p)
        to Richmond (2:30p)

(NYC 406 is the Carolina Special, *with dining service and coach-lounge. NYC 24 is the* Knicker-bocker, *offering dining service and a sleeper-lounge. Ashland to Richmond, diner.)*

Humming Bird, No. 5
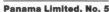
Streamliner
Cincinnati to New Orleans
Louisville & Nashville
5 miles north of Bowling Green

CINCINNATI to NEW ORLEANS
    5: 12s-dr   *19: 6s-6dbr*
L&N 5, Cincinnati (8p) to New Orleans (3:10p)

CINCINNATI to MEMPHIS
    73: 10s-2c-dr   *79: 10s-lounge*
L&N 5, Cincinnati (8p) to Bowling Green (12:05a)
L&N 101, Bowling Green (12:35a)
          to Memphis (7:50a)

(Cincinnati to New Orleans, diner and lounge)

Seminole, No 9
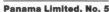
Chicago to Jacksonville
Illinois Central
32 miles north of Fulton

CHICAGO to ST. PETERSBURG   *alternate days*
    907: 8s-2c-dr
IC 9, Chicago (4:20p Sun)
        to Birmingham (8:20a Mon)
CofG 10, Birmingham (8:40a) to Albany (4:25p)
ACL 17, Albany (4:30p) to Jacksonville (9p)
ACL 37, Jacksonville (10:30p)
        to St. Petersburg (9a Tue)

CHICAGO to TAMPA/SARASOTA alternate days
    903: 8s-2c-dr
As above, Chicago (4:20p Sun)
        to Jacksonville (9p Mon)
ACL 75, Jacksonville (11p) to Tampa (7a Tue)
        and Sarasota (11:30a Tue)

CHICAGO to BIRMINGHAM
    901: 8s-2c
IC 9, Chicago (4:20p) to Birmingham (8:20a)

ST. LOUIS to TAMPA/SARASOTA   *alternate days*
    905: 8s-2c-dr
IC 205, St. Louis (6:45p Sun) to Carbondale (9:15p)
IC 9, Carbondale (10:28p)
        to Birmingham (8:20a Mon)
As above, Birmingham (8:40a)
        to Jacksonville (9p)
ACL 75, Jacksonville (11p) to Tampa (7a Tue)
        and Sarasota (11:30a Tue)

ST. LOUIS to ST. PETERSBURG   *alternate days*
    909: 8s-2c-dr
As above, St. Louis (6:45p Sun)
        to Jacksonville (9p Mon)
ACL 37, Jacksonville (10:30p)
        to St. Petersburg (9a Tue)

(The Florida sleepers are scheduled so that one departs every day from Chicago and St. Louis, with one going to St. Petersburg and the other to Sarasota. On any given night, only half the cars will be in the Seminole's consist. The northbound operation alternates similarly. Chicago to Jacksonville, dining car.)

The Illinois Central inaugurated the *Seminole Limited* in 1909, upon establishment of its interline route to Florida with the Central of Georgia, which it had just bought. The *Seminole* was the first year-round, Chicago-to-Jacksonville train. Its consist over the years included Chicago sleepers for Paducah, Savannah, Ga., Fort Myers, and Miami.

Panama Limited, No. 5

All-Pullman streamliner
Chicago to New Orleans
Illinois Central
Departing Fulton

CHICAGO to NEW ORLEANS
    509: 10r-6dbr   *511: 2dr-4c-4dbr*
    513: 10r-6dbr   *515: 10r-5dbr*
    517: 6s-6r-4dbr   *519: 6s-6r-4dbr*
    501: 2dbr-dr-2c-observation
IC 5, Chicago (5p) to New Orleans (9:30a)

ST. LOUIS to NEW ORLEANS
    505: 6s-6r-4dbr   *507: 10r-6dbr*
IC 205, St. Louis (6:45p) to Carbondale, Ill. (9:15p)
IC 5, Carbondale (9:41p) to New Orleans (9:30a)

CHICAGO to JACKSON, MISS.
    503: 6s-6r-4dbr
IC 5, Chicago (5p) to Jackson (6a)

(Chicago to New Orleans, twin diners and lounge car; St. Louis to Carbondale, parlor-buffet-lounge. This crack streamliner averages in excess of 55 mph over its entire trip, despite as many as 19 intermediate stops.)

Flamingo, No. 17
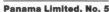
Cincinnati to Jacksonville
Louisville & Nashville
Switching in Corbin

CINCINNATI to ATLANTA
    17: 8s-dr-3dbr
L&N 17, Cincinnati (7:50p) to Atlanta (8:10a)

CINCINNATI to KNOXVILLE
    9: 10s-2c-dr
L&N 17, Cincinnati (7:50p) to Knoxville (3:50a)

LOUISVILLE to ATLANTA
    21: 8s-dr-3dbr, or   *10s-dr-2dbr*
L&N 21, Louisville (6:45p) to Corbin (11:35p CST)
L&N 17, Corbin (1:05a EST) to Atlanta (8:10a)

(Dining car cut in at Etowah, Tenn., 5:23 a.m., to Atlanta)

When the *Flamingo* carried through cars to Miami and Fort Myers during the 1920s, Detroit and Cleveland sleepers came to Cincinnati via New York Central, and the Chicago section traveled to Louisville on the Pennsylvania. The L&N joined the two trains at Corbin.

At Corbin, cars from Louisville were put aboard trains on the Cincinnati-Atlanta line of the L&N, including sleepers for Knoxville and Jacksonville, and others were dispatched for Lynch and Norton, Va.

No. 24

Louisville to Ashland
Chesapeake & Ohio
Entering Ashland

LOUISVILLE to WASHINGTON
    466: 10r-6dbr
C&O 24, Louisville (5:45p) to Ashland (12:01a)
C&O 4-46 Ashland (2:16a)
        to Charlottesville (11:05a)
C&O 4-46-4, Charlottesville (11:45a)
        to Washington (2:30p)

(Dining cars on all segments. No. 4-46 is the Sportsman.)

Separate sections of C&O main line trains served Louisville and Lexington from Ashland. Pullmans arrived in Louisville from Ashland, Hinton, Clifton Forge, Charlottesville, Richmond, and Newport News (Phoebus). Lexington had its own service from Washington.

Carolina Special, No. 27

Chicago to the Carolinas
Southern Railway
7 miles south of Danville

CHICAGO to CHARLESTON, S.C.
    4062: 8s-5dbr
NYC 406, Chicago (12:50p) to Cincinnati (9:05p)
SOU 27, Chicago (9:50p) to Asheville, N.C. (11a)
SOU 28, Asheville (11:20a) to Charleston (8:30p)

CHICAGO to GREENSBORO, N.C.
    4061: 10s-dr-2c
As above, Chicago (12:50p) to Asheville (11a)
SOU 22, Asheville (11:25a) to Greensboro (6:15p)

CINCINNATI to ASHEVILLE
    47: 10s-diner-lounge
SOU 27, Cincinnati (9:50p) to Asheville (11a)

LOUISVILLE TO COLUMBIA, S.C.
    21: 12s-dr
SOU 23, Louisville (6:45p) to Danville (10p)
SOU 27-28, Danville (11:50p) to Columbia (4:45p)

(Diner, Chicago to Cincinnati, Knoxville to Columbia; dinette coach, Asheville to Goldsboro. In Asheville, the Carolina Special divides into two sections, the South Carolina portion going to Columbia and Charleston, and the North Carolina section bound for Greensboro and Goldsboro.)

The *Carolina Special* began service in 1911 as a Charleston train and later acquired the North Carolina section. Goldsboro and Spartanburg sleepers were once included.

Sportsman, No. 47

Newport News/Washington to Detroit
Chesapeake & Ohio
8 miles west of Ashland

NEWPORT NEWS (PHOEBUS) to DETROIT
    471: 10r-6dbr
C&O 47, Phoebus (8:20a) to Detroit (8:20a)

RICHMOND to CHICAGO
    C475: 10r-6dbr
C&O 47, Richmond (10:55a) to Toledo (6:05p)
NYC 19, Toledo (8:40a) to Chicago (12:30p)

CLIFTON FORGE to COLUMBUS
    470: 10r-6dbr
C&O 47, Clifton Forge (4:40p)
        to Columbus (2:25a)

CLIFTON FORGE to CLEVELAND
    C472: 10r-6dbr
C&O 47, Clifton Forge (4:40p)
        to Columbus (2:25a)
NYC 442, Columbus (3:10a) to Cleveland (6:20a)

(Dining cars: Richmond to Charleston, Columbus to Detroit. Coach-lunch, Washington to Charlottesville. NYC 19 is the Lake Shore Limited, with dining and lounge facilities. NYC 442 is the Midnight Special. The Clifton Forge sleepers served the resorts at Hot Springs, Va., and White Sulphur Springs, W. Va.)

Irvin S. Cobb, No. 103
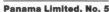
Louisville to Fulton
Illinois Central
40 miles east of Nortonville

CINCINNATI to MEMPHIS
    1031: 10s-c-dr
B&O 63, Cincinnati (6:30p) to Louisville (8:50p)
IC 103, Louisville (9:10p) to Fulton (4:20a)
IC 15, Fulton (4:35a) to Memphis (7:05a)

LOUISVILLE to PADUCAH
    1033: 8s-dr-2c
IC 103, Louisville (9:10p) to Paducah (2:59a)

(Snack car, Cincinnati to Louisville. No. 15 is the St. Louis-to-Memphis Chickasaw.)

Kentucky   *(continued)*

Humming Bird, No. 93
Chicago/St. Louis to New Orleans
Louisville & Nashville
20 miles north of Guthrie

CHICAGO to NEW ORLEANS
175: 8s-2c-dr
C&EI 17, Chicago (4:15p) to Evansville (10p)
L&N 93, Evansville (10:20p) to Nashville (1:40a)
L&N 5, Nashville (2:05a) to New Orleans (3:10p)

ST. LOUIS to MONTGOMERY
31: 10s-c-dr
L&N 93, St. Louis (4:35p) to Nashville (1:40a)
L&N 5, Nashville (2:05a) to Montgomery (7:55a)

CHICAGO to BIRMINGHAM
171: 6s-4r-4dbr
As above, Chicago (4:15p) to Nashville (1:40a)
L&N 1, Nashville (2:30a) to Birmingham (7:20a)

(Dining car and lounge: Chicago to Evansville, Nashville to New Orleans; diner, St. Louis to Evansville. The southbound Humming Bird draws from three sources: Chicago, St. Louis, and Cincinnati, where L&N 5 originates. It all comes together in Nashville. The Birmingham sleeper, however, is cut into the Azalean, No. 1.)

Irvin S. Cobb, No. 104
Fulton to Louisville
Illinois Central
21 miles north of Fulton

MEMPHIS to CINCINNATI
410: 10s-c-dr
IC 4, Memphis (8p) to Fulton (10:50p)
IC 104, Fulton (11:25p) to Louisville (7:30a)
B&O 64, Louisville (8:20a) to Cincinnati (12:30p)

(Louisville to Cincinnati, snack car. IC 4 is the Louisane.)

Irvin S. Cobb, No. 104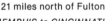
Illinois Central
Paducah setout
Standing in Paducah

PADUCAH to LOUISVILLE
1040: 8s-dr-2c
IC 104, Paducah (12:50a) to Louisville (7:30a)

Paducah formerly had Pullman service on the IC to both St. Louis and, on the *Seminole*, Chicago.

No. 156
Louisville to Evansville
Louisville & Nashville
55 miles west of Louisville

LOUISVILLE to ST. LOUIS
10: 12s-dr
L&N 156, Louisville (10:45p) to Evansville (3:05a)
L&N 56, Evansville (3:50a) to St. Louis (7:55a)

(No. 56, the St. Louis section of the Humming Bird, carries a dining car.)

No. 156-56 once included a Louisville-Evansville sleeper.

Louisiana

See map on page 156

Sunset Limited, No. 1
Streamliner, extra fare
New Orleans to Los Angeles
Southern Pacific
Standing in New Orleans

NEW ORLEANS to LOS ANGELES
15: 10r-6dbr   16: 10r-6dbr
17: 10r-6dbr   19: 10r-6dbr
SP 1, New Orleans (12:30a Mon)
          to Los Angeles (4:30p Tue)

(Dining car, first-class lounge. Of the two dozen trains operating between the Mississippi River valley and the West Coast, the westbound Sunset Limited is the only one spanning the distance within two calendar days. Although it is two nights under way, the Sunset departs New Orleans post-midnight, making this schedule exception possible. Passengers, however, could board the Pullmans as early as 10 p.m. The edition of the Sunset Limited that departed 24 hours earlier is now in New Mexico.)

Acadian, No. 3
New Orleans to Houston
Southern Pacific
43 miles east of Lafayette

NEW ORLEANS to HOUSTON
31: Std: s-dr-c   33: 8s-5dbr
SP 3, New Orleans (9p) to Houston (7:20a)

NEW ORLEANS to LAKE CHARLES
34: 10s-2dr
SP 3, New Orleans (9p) to Lake Charles (3:30a)

(First-class lounge)

The *Acadian* plies the easternmost leg of the Southern Pacific's Sunset Route, host to the namesake *Sunset Limited* and *Argonaut*. Night trains on this southern transcontinental route carried Pullmans from New Orleans to Lafayette, as well as Galveston, Dallas, Fort Worth, Denver, Globe (Ariz.), Mexico City, San Diego, Santa Barbara, Pacific Grove, and San Francisco.

Acadian, No. 4
Houston to New Orleans
Southern Pacific
Switching at Lake Charles

HOUSTON to NEW ORLEANS
42: Std: s-dr-c
SP 4, Houston (8:30p) to New Orleans (6:45a)

LAKE CHARLES to NEW ORLEANS
43: s-dr
SP 4, Lake Charles (12:04a)
          to New Orleans (6:45a)

(First-class lounge)

Houstonian, No. 9
New Orleans to Houston
Missouri Pacific
8 miles east of Baton Rouge

NEW ORLEANS to HOUSTON
95: 8s-5dbr
MP 9, New Orleans (10:10p) to Houston (7:50a)

Though it never achieved the dominance that the SP's Sunset Route did, this Missouri Pacific line accommodated dedicated sleepers from New Orleans to Galveston, Brownsville, and Los Angeles (Santa Fe).

No. 9
Kansas City to New Orleans
Streamliner, via Minden
Kansas City Southern
17 miles east of Shreveport

KANSAS CITY to NEW ORLEANS
9: 10s-3dbr
KCS 9, Kansas City (9:30a) to New Orleans (7:45a)

SHREVEPORT to NEW ORLEANS
109: 14r-4dbr
KCS 9, Shreveport (11:25p)
          to New Orleans (7:45a)

(Diner-lounge)

The KCS also scheduled sleepers from Shreveport to Lake Charles and, in the other direction, Little Rock (MP) and Fort Smith, Ark., and Denver (UP). The MKT offered service to Dallas, over its line that the KCS later bought.

Texas & Pacific's *Louisiana Eagle* on the outskirts of Shreveport, La., May 1954.

No. 10
New Orleans to Kansas City
Streamliner, via Minden
Kansas City Southern
14 miles north of Baton Rouge

NEW ORLEANS to KANSAS CITY
10: 10s-3dbr
KCS 10, New Orleans (9:30p)
to Kansas City (7:45p)

NEW ORLEANS to SHREVEPORT
110: 14r-4dbr
KCS 10, New Orleans (9:30p)
to Shreveport (6:25a)

(Diner-lounge)

Louisiana Eagle, No. 21
New Orleans to Fort Worth
Texas & Pacific
21 miles south of Alexandria

NEW ORLEANS to LITTLE ROCK
1161: 10s-dr-2c
T&P 21, New Orleans (7:50p)
to Alexandria (12:30a)
MP 116, Alexandria (1:25a) to Little Rock (9:10a)

NEW ORLEANS to MONROE
1162: 10s-dr-2c
T&P 21, New Orleans (7:50p)
to Alexandria (12:30a)
MP 116, Alexandria (1:25a) to Monroe (3:55a)

NEW ORLEANS to FORT WORTH
210: Std: s-buffet-lounge 213: Std: r-br
T&P 21, New Orleans (7:50p) to Fort Worth (9:05a)

(Dining-lounge car, Marshall to Fort Worth. No. 116, the Southerner, carries a grill-coach, Monroe to Little Rock.)

The *Louisiana Limited*, predecessor of the *Eagle*, carried Pullmans for Shreveport, Dallas/Fort Worth, and Denver. Other T&P night trains working this line offered sleepers from New Orleans to Alexandria, Abilene, and El Paso.

Louisiana Eagle, No. 22
Fort Worth to New Orleans
Texas & Pacific
53 miles south of Shreveport

FORT WORTH to NEW ORLEANS
220: Std: s-buffet-lounge 222: Std: r-br
T&P 22, Fort Worth (4:45p) to New Orleans (7a)

(Fort Worth to Marshall, dining-lounge car)

No. 27
Shreveport to Houston
Southern Pacific
15 miles south of Shreveport

SHREVEPORT to HOUSTON
270: Std: s-dr
SP 27, Shreveport (11:30p) to Houston (6:45a)

No. 103
Little Rock to Alexandria
Missouri Pacific
30 miles south of Monroe

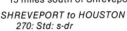

LITTLE ROCK to NEW ORLEANS
1031: 10s-dr-2c
MP 103, Little Rock (6p) to Alexandria (1:40a)
T&P 22, Alexandria (2:10a) to New Orleans (7a)

MONROE to NEW ORLEANS
1032: 10s-dr-2c
MP 103, Monroe (11:25p) to Alexandria (1:40a)
T&P 22, Alexandria (2:10a) to New Orleans (7a)

(T&P 22 is the Louisiana Eagle)

This route of the *Louisiana Sunshine Special* through Monroe brought to that city Pullmans from St. Louis and Memphis.

Maine

See map on page 14

The Gull, No. 8
Halifax, N.S., to Boston
Via Brunswick and Dover
Maine Central
36 miles west of Vanceboro

HALIFAX to BOSTON
10s-dr-2c
CNR 59, Halifax (8:15a) to Moncton, N.B. (1:55p)
CNR 13, Moncton (2:55p)
to Saint John, N.B. (5:45p)
CPR 113, Saint John (8:30p)
to Vanceboro (10:40p)
MEC 8, Vanceboro (11p) to Portland (5:50a)
B&M 8, Portland (6:25a) to Boston (8:45a)

SAINT JOHN to BOSTON
435: 10s-dr-2c
As Above, Saint John (8:30p) to Boston (8:45a)

(Diner on CNR 59, the Scotian; cafe-parlor Moncton to Saint John)

The *Gull* was introduced in the summer of 1928 to connect Boston with the Maritime Provinces and supplanting the *Yankee*. It also carried a Moncton sleeper.

The Gull, No. 23
Boston to Halifax
Via Dover, Lewiston
Maine Central
Standing in Portland

BOSTON to BANGOR
230: 12s-dr
B&M 23, Boston (9:30p) to Portland (11:50p)
MEC 23, Portland (12.10a) to Bangor (3:05a)

BOSTON to VAN BUREN
235: 10s-dr-2c
As above, Boston (9:30p) to Bangor (3:35a)
BAR 1, Bangor (4:40a) to Van Buren (11:45a)

BOSTON to CALAIS
231: 12s-dr
As above, Boston (9:30p) to Bangor (3:35a)
MEC 123, Bangor (5a) to Calais (10a)

BOSTON to SAINT JOHN
232: 10s-dr-2c
As above, Boston (9:30p) to Bangor (3:35a)
MEC 23, Bangor (4:35a) to Vanceboro (7:50a)
CPR 114, Vanceboro (9:15a)
to Saint John (12:20p)

BOSTON to HALIFAX
234: 10s-dr-2c
As above, Boston (9:30p) to Saint John (12:20p)
CNR 14, Saint John (1p) to Moncton (3:35p)
CNR 60, Moncton (4:30p) to Halifax (10:30p)

(Diners: On Bangor & Aroostook's No. 1, the Potatoland Special, from Bangor to Van Buren, and on the Scotian from Moncton to Halifax; cafe-parlor, Saint John to Moncton; McAdam, N.B., breakfast stop.)

Especially in summertime, when numerous sections of the *Bar Harbor Express* arrived, Portland was a busy interchange between the Boston & Maine and Maine Central railroads. Portland, however, once had important ties of its own, with through service to Montreal and Chicago via both Canadian National and Canadian Pacific routings.

Penobscot, No. 22
Bangor to Boston
Via Brunswick and Portsmouth
Maine Central
Standing in Waterville

BANGOR to BOSTON
220: 12s-dr
MEC 22, Bangor (10:30p) to Portland (2:30a)
B&M 22, Portland (3:20a) to Boston (6:20a)

CALAIS to BOSTON
221: 12s-dr
MEC 116, Calais (4:45p) to Bangor (9:35p)
As above, Bangor (10:30p) to Boston (6:20a)

VAN BUREN to BOSTON   Mon-Fri
70: 10s-dr-2c
BAR 8, Van Buren (3p) to Bangor (9:40p)
As above, Bangor (10:30p) to Boston (6:20a)

(Van Buren to Bangor, dining car)

Boston had considerable Pullman service to and from Maine: Caribou, Farmington, Greenville, Kineo Station, Mount Desert Ferry, Portland, Rockland, and Waterville.

No. 39
Saint John to Montreal
Canadian Pacific
11 miles west of Mattawamkeag

SAINT JOHN to MONTREAL
4c-dr-buffet-lounge   14s
CPR 39, Saint John (8p) to Montreal (11:25p)

(Just as New York Central trains cross Ontario between Buffalo and Detroit, Canadian trains break the night stillness of northern Maine, carrying sleepers between Montreal and Saint John. No. 39 is one of three such Canadian Pacific trains on this stretch of trackage at midnight. The level of traffic to and from Saint John indicates the importance of this seaport on the Bay of Fundy to Montreal commercial interests.)

No. 40
Montreal to Saint John
Canadian Pacific
16 miles east of Greenville

MONTREAL to SAINT JOHN
4c-dr-buffet-lounge
CPR 40, Montreal (3:15p) to Saint John (6:40p)

No. 41
Saint John to Montreal
Canadian Pacific
15 miles west of Greenville

SAINT JOHN to MONTREAL
10r-5dbr   14s
4c-dr-buffet-lounge
CPR 41, Saint John (5p) to Montreal (7:55a)

(Sunday schedule different)

Maryland

See map on page **58**

Diplomat, No. 3
New York to St. Louis
Baltimore & Ohio
30 miles west of Cumberland

NEW YORK to ST. LOUIS
   8s-buffet-lounge   14r-4dbr
   8s-4dbr
Ferry or bus from New York
B&O 3, Jersey City (3:55p) to St. Louis (3:15p)

NEW YORK to LOUISVILLE
   12s-dr
B&O 3, Jersey City (3:55p)
   to North Vernon, Ind. (9:38a)
B&O 51, North Vernon (10a) to Louisville (11:15a)

(Dining car)

Shenandoah, No. 7
New York to Chicago
Baltimore & Ohio
12 miles north of Washington

NEW YORK to CHICAGO
   8s-buffet-lounge   10s-2c-dr
Ferry or bus from New York
B&O 7, Jersey City (6:50p) to Chicago (3p)

WASHINGTON to CHICAGO
   5r-sbr-3dr Strata-Dome on alternate days
   10s-2c-dr
B&O 7, Washington (11:40p) to Chicago (3p)

NEW YORK to PITTSBURGH
   8s-dr-3dbr
B&O 7, Jersey City (6:50p) to Pittsburgh (6:35a)

WASHINGTON to PITTSBURGH
   12r-sbr-4dr, or 17r, alternate days
   10s-3dbr
B&O 7, Washington (11:40p) to Pittsburgh (6:35a)

(Dining cars: New York to Washington, Pittsburgh to Chicago)

Metropolitan Special, No. 12
Baltimore & Ohio
Baltimore setout
Standing in Baltimore

BALTIMORE to NEW YORK
   8s-5dbr
B&O 12, Baltimore (2:03a) to Jersey City (6:50a)
Ferry or bus to New York

The Baltimore & Ohio gave its namesake city dedicated sleeping car service also to Fairmont, Grafton, Parkersburg, Pittsburgh, Cincinnati, Louisville, Toledo, and Chicago.

Statesman, No. 51
Washington to Pittsburgh
Pennsylvania Railroad
15 miles south of Baltimore

WASHINGTON to PITTSBURGH
   8s-buffet-lounge   10r-5dbr
   8s-2c-dr
PRR 51, Washington (11:35p)
   to Pittsburgh (8:25a)

WASHINGTON to CHICAGO
   8s-2c-dr
PRR 51, Washington (11:35p)
   to Harrisburg (2:45a)
PRR 55, Harrisburg (2:54a) to Chicago (3:40p)

(No. 55 is the Gotham Limited, *which carries a sleeper-lounge and diner)*

Statesman, No. 51
Pennsylvania Railroad
Baltimore setout
Standing in Baltimore

BALTIMORE to PITTSBURGH
   8s-5dbr
PRR 51, Baltimore (12:32a) to Pittsburgh (8:25a)

(Sleeper-lounge)

Edison, No. 102
Pennsylvania Railroad
Baltimore setout
Standing in Baltimore

BALTIMORE to NEW YORK
   Std: s-dr-c-r
PRR 102, Baltimore (1:55a) to New York (6:05a)

(The Edison is now standing in Washington)

Though the distance was short, Baltimore had its own Pullman to Philadelphia.

Palmland, No. 143
New York to Florida
Pennsylvania Railroad
16 miles south of Baltimore

NEW YORK to MIAMI
   8s-5dbr
PRR 143, New York (8:30p Sun)
   to Washington (12:25a)
RF&P 9, Washington (1:10a Mon)
   to Richmond (4:05a)
SAL 9, Richmond (4:05a) to Miami (7:55a Tue)

NEW YORK to HAMLET, N.C.
   8s-dr-3dbr   12r-sbr-4dbr
   6c-3dr, Thu and Fri only
As above, New York (8:30p) Richmond (4:05a)
SAL 9, Richmond (4:05a) to Hamlet (9:40a)

(Diner, Raleigh to Jacksonville. The Hamlet cars serve the resort area of Southern Pines.)

The Federal, No. 172
Washington to Boston
Pennsylvania Railroad
18 miles north of Baltimore

WASHINGTON to BOSTON
   6sbr-buffet-lounge   2c-dr-2dbr-2sbr-lounge
   14s   12s-dr
   13dbr   14r-4dbr (3 days/week)
   18r   18r
PRR 172, Washington (11p) to New York (2:58a)
NH 172, New York (3:10a) to Boston (8:10a)

WASHINGTON to PROVIDENCE
   14r-4dbr
PRR 172, Washington (11p) to New York (2:58a)
NH 172, New York (3:10a) to Providence (7:05a)

WASHINGTON to SPRINGFIELD
   14r-4dbr
PRR 172, Washington (11p) to New York (2:58a)
NH 172, New York (3:10a) to New Haven (4:40a)
NH 412, New Haven (6:55a) to Springfield (8:45a)

(The Federal *will pick up yet another Boston sleeper in New York's Pennsylvania station.)*

In addition to its usual complement, the *Federal* carried a number of Florida-to-Boston Pullmans. For about five years before the opening of the Hell Gate Bridge in 1917, the *Federal Express* traveled a route that bypassed New York City, crossing the Hudson River at Poughkeepsie, N.Y. Before that Washington-Boston sleepers were barged across New York Harbor. The Hell Gate Bridge permitted a continuous flow of passenger traffic between the Pennsylvania and New Haven railroads.

Massachusetts

See map on page **14**

The Owl, No. 3
Boston to New York
New Haven Railroad
Standing in Boston

BOSTON to NEW YORK
   10 Pullmans, Cars 30-39: Std-s-dr-c-r-sbr-dbr
NH 3, Boston (12:30a) to New York (6:20a)

(The consist of the Owl *will swell to an even dozen sleepers when the two Providence cars are cut in. The* Owl, *like other New Haven trains terminating in New York, uses Grand Central Terminal.)*

The New Haven scheduled sleepers from Boston to New Haven. From New York, the Pennsylvania forwarded other cars on to Chicago, Cincinnati, St. Louis, New Orleans, Jacksonville, and Palm Beach.

South Shore Express, No. 43
Boston to Buffalo
Boston & Albany
Departing Worcester

BOSTON to BUFFALO
   12s-dr
B&A 43, Boston (10:30p) to Albany (4:35a)
NYC 43, Albany (5:20a) to Buffalo (1p)

BOSTON to ALBANY
   8s-4dbr
B&A 43, Boston (10:30p) to Albany (4:35a)

WORCESTER to NEW YORK
   8s-6r-2sbr
B&A 43, Worcester (11:57p) to Springfield (1:10a)
NH 55, Springfield (1:55a) to New York (6:55a)

(Albany to Buffalo, diner-lounge. No. 43 will pick up a New York-Syracuse sleeper in Albany.)

No. 55
Springfield to New York
New Haven Railroad
Standing in Springfield

SPRINGFIELD to NEW YORK
   551: Std-s-dr
NH 55, Springfield (1:55a) to New York (6:55a)

WHITE RIVER JUNCTION, VT. to NEW YORK
   Std: s-dr-c
B&M 728, White River Jct. (7:05p)
   to Springfield (11:25p)
NH 55, Springfield (1:55a) to New York (6:55a)

(Mon-Fri only; Sunday schedule different)

The New Haven originated Pullmans from Springfield to Philadelphia, Miami, and St. Petersburg. To the north, the B&M scheduled sleepers to Montreal and Sherbrooke, Que.

State of Maine, No. 82
Portland to New York
Boston & Maine
West of Lowell Junction

PORTLAND to NEW YORK
   12s-dr   14r-4dbr
   2c-dr-3sbr-buffet-lounge
B&M 82, Portland (9:30p) to Worcester (1:35a)
NH 125, Worcester (2:10a) to New York (7:30a)

FARMINGTON, ME., to NEW YORK
   245: 10s-dr-2c   Summer, Sun only
MEC 722, Farmington (5:15p) to Portland (8p)
As above, Portland (9:30p) to New York (7:30a)

(The State of Maine *carries three additional Portland sleepers in its summer consist)*

State of Maine, No. 82
Boston & Maine
Concord Pullman
Standing in Lowell

CONCORD, N.H. to NEW YORK
  8s-5sbr
B&M 24, Concord (8:20p) to Lowell (9:50p)
B&M 82, Lowell (12:20a) to Worcester (1:35a)
NH 125, Worcester (2:10a) to New York (7:30a)

(No. 24 is the Winnipesaukee*)*

Bar Harbor Express, No. 84
All-Pullman, Maine to Washington
Summertime only, thrice weekly
Boston & Maine
ca. Lawrence

ELLSWORTH to WASHINGTON
  849: 10s-dr-2c
MEC 840, Ellsworth (5p) to Portland (9:40p)
B&M 84, Portland (10p) to Worcester (ntg)
NH 185, Worcester (ntg) to New York (7:05a)
PRR 111, New York (7:30a)
      to Washington (11:45a)

ELLSWORTH to PHILADELPHIA
  842: 6c-3dr   843: 8s-buffet-lounge
  844: 13dbr
As above, Ellsworth (5p) to New York (7:05a)
PRR 185, New York (7:05a) to Philadelphia (9:05a)

ELLSWORTH to NEW YORK
  845: 12r-4dbr-sbr   846: 10s-dr-2c
  847: 12r-4dbr-sbr
As above, Ellsworth (5p) to New York (7:05a)

ROCKLAND to PHILADELPHIA
  560: 10s-dr-2c
MEC 56, Rockland (5:10p) to Portland (8p)
As above, Portland (10p) to Philadelphia (9:05a)

ROCKLAND to NEW YORK
  561: 6s-4r-4dbr   562: 10s-dr-2c
As above, Rockland (5:10p) to New York (7:05a)

PORTLAND to PHILADELPHIA
  848: 12r-3dbr-2sbr
As above, Portland (10p) to Philadelphia (9:05a)

PLYMOUTH, N.H. to NEW YORK Sunday only
  265: 10s-dr-2c
B&M 6056, Plymouth (6:50p) to Concord (8:10p)
B&M 26, Concord (8:20p) to Lowell (9:28p)
B&M 84, Lowell (ntg) to Worcester (ntg)
NH 185, Worcester (ntg) to New York (7:05a)

(Diners, Ellsworth and Rockland to Portland; New Haven to Philadelphia on No. 185; New York to Washington on No. 111. Philadelphia arrivals at 30th Street Station. Tuesday, Thursday, and Sunday operation. Schedules and consists differ on Sundays.)

The *Bar Harbor Express* was the patrician among the summer vacation trains, carrying well-to-do easterners and their families to and from Maine between late spring flower time and the start of fall color, for more than a half-century. Inaugurated as a New York train in 1902, it began carrying through Washington and Philadelphia accommodations after the opening of the Hell Gate Bridge route. By the early 1920s, separate sections were operated to and from the three cities daily. Passengers bound for Bar Harbor rode Pullmans to Mount Desert Ferry until a causeway was built connecting the island to the mainland in 1931 and the cars terminated at Ellsworth. When service resumed after World War II, thrice-weekly operation was the norm. The end for the storied *Bar Harbor* came in 1960.

Streamlined New Haven Baldwins powered such trains as the *Pittsburgh Express*.

Night Cape Codder, No. 111
Cape Cod to New York
Summertime Sundays only
New Haven Railroad
13 miles west of Buzzards Bay

HYANNIS to NEW YORK
  3 Pullmans: s-dr-c-sbr
NH 840, Hyannis (10:35p)
      to Buzzards Bay (11:23p)
NH 111, Buzzards Bay (11:30p)
      to New York (5:55a)

WOODS HOLE to NEW YORK
  4 Pullmans: s-dr-c-sbr
NH 838, Woods Hole (10:40p)
      to Buzzards Bay (ntg)
NH 111, Buzzards Bay (11:30p)
      to New York (5:55a)

HYANNIS to WASHINGTON
  8400: s-dr-c
NH 840, Hyannis (10:35p)
      to Buzzards Bay (11:23p)
NH 111, Buzzards Bay (11:30p) to New Haven (4a)
NH 185, New Haven (4:05a) to New York (7:05a)
PRR 111, New York (7:30a)
      to Washington (11:45a)

(Diner, New York to Washington. The Washington sleeper is given over to No. 185, the Bar Harbor, in New Haven and goes through New York's Pennsylvania Station. The New York Pullmans arrive in Grand Central Terminal.)

Montrealer, No. 733
Washington to Montreal
Boston & Maine
Standing in Springfield

WASHINGTON to MONTREAL
  687: 8s-5dbr   688: 5c-lounge
PRR 112, Washington (4:10p) to New York (8p)
NH 168, New York (8:25p) to Springfield (11:50p)
B&M 733, Springfield (12:07a)
      to White River Jct. (3:10a)
CV 325, White River Jct. (3:22a)
      to East Alburgh, Vt. (ntg)
CNR 21, East Alburgh (ntg) to Montreal (8:15a)

NEW YORK to ST. ALBANS, VT.
  684: 12s-3dbr
As above, New York (8:25p)
      to White River Jct. (3:10a)
CV 303, White River Jct. (4:55a) to St. Albans (9a)

NEW YORK to MONTREAL
  685: 14sbr   686: 8s-5dbr
As above, New York (8:25p) to Montreal (8:15a)

(Washington to New York, dining car. The Montrealer combines with the Boston-Montreal New Englander at White River Junction. The southbound train is the Washingtonian.)

Bar Harbor Express, No. 184
All-Pullman, Washington to Maine
Summertime only, thrice weekly
New Haven Railroad
22 miles south of Worcester

WASHINGTON to ELLSWORTH
  W2: 10s-dr-2c
PRR 184, Washington (2p) to New York (7:15p)
NH 184, New York (7:30p) to Worcester (ntg)
B&M 85, Worcester (ntg) to Portland (4:15a)
MEC 85, Portland (4:50a) to Ellsworth (9:35a)
Bus to Bar Harbor (10:25a)

PHILADELPHIA to ELLSWORTH
  P8: 6c-3dr   P10: 8s-buffet-lounge
  P12: 13dbr
PRR 184, Philadelphia (5:30p) to New York (7:15p)
As above, New York (7:30p) to Ellsworth (9:35a)

PHILADELPHIA to PORTLAND
  P20: 12r-3dbr-2sbr
As above, Philadelphia (5:30p) to Portland (4:15a)

PHILADELPHIA to ROCKLAND
  P22: 10s-dr-2c
As above, Philadelphia (5:30p) to Portland (4:15a)
MEC 55, Portland (7:10a) to Rockland (10:35a)

NEW YORK to ELLSWORTH
  N14: 12r-4dbr-sbr   N16: 10s-dr-2c
  N18: 12r-4dbr-sbr
As above, New York (7:30p) to Ellsworth (9:35a)

NEW YORK to ROCKLAND
  N24: 6s-4r-4dbr   N26: 10s-dr-2c
As above, New York (7:30p) to Rockland (10:35a)

NEW YORK to PLYMOUTH, N.H. Fri only
  N30: 10s-dr-2c
NH 184, New York (7:30p) to Worcester (ntg)
B&M 85, Worcester (ntg) to Lowell (ntg)
B&M 1, Lowell (3:45a) to Concord (5:10a)
B&M 4301, Concord (5:20a) to Plymouth (6:57a)

(Dining cars: Philadelphia to New Haven, Portland to Ellsworth and Rockland. Departures Monday, Wednesday, and Friday. Additional sleepers scheduled on Friday. Philadelphia departures from 30th Street Station.)

Quaker, No. 187
Boston to Philadelphia
New Haven Railroad
Departing Boston Back Bay Station

BOSTON to PHILADELPHIA
  14r-4dbr   14sbr
  8s-dr-3dbr
NH 187, Boston (11:50p) to New York (5:15a)
PRR 187, New York (6a) to Philadelphia (7:45a)

BOSTON to NEW YORK
  12s-2dbr
NH 187, Boston (11:50p) to New York (5:15a)

Michigan

See map on page 86, 270

La Salle, No. 5
Montreal to Chicago
Grand Trunk Western
30 miles east of Flint

PORT HURON to CHICAGO
  183: 12s-2dbr
GTW 5, Port Huron (11:20p) to Chicago (7a)

(The La Salle will pick up a sleeper from Detroit at Durand and the Lansing setout, listed below)

The Grand Trunk also provided Flint and Pontiac with dedicated service to Chicago.

La Salle, No. 5
Grand Trunk Western
Lansing setout
Standing in Lansing

LANSING to CHICAGO
  57: 12s-dr
GTW 5, Lansing (2:30a) to Chicago (7a)

No. 7
Grand Rapids to Chicago
Chesapeake & Ohio
Departing Grand Rapids

GRAND RAPIDS to CHICAGO
  71: 10r-6dbr
C&O 7, Grand Rapids (11:59p) to Chicago (5a)

No. 7
Marquette to Duluth
Duluth, South Shore & Atlantic
92 miles west of Champion

MARQUETTE to DULUTH   Mon, Wed, Fri only
  Std.
DSS&A 7, Marquette (6:10p) to Duluth (5:40a)

No. 14
Chicago to Montreal
Grand Trunk Western
Standing in Battle Creek

CHICAGO to TORONTO
  2: 24dup/r   18: 8s-dr-2c
  4: 2c-dr-3sbr-buffet-lounge
GTW/CNR 14, Chicago (8p) to Toronto (8:55a)

(Diner: London, Ont., to Toronto)

Canadian, No. 20
Chicago to Montreal
Canadian Pacific
Standing in Detroit

DETROIT to TORONTO
  8s-4dbr
CPR 20, Detroit (3:15a) to Toronto (8:50a)

(London to Toronto, cafe-parlor car. Two Chicago-Toronto sleepers will become part of the Canadian after their arrival in Detroit on NYC 358, now in Battle Creek.)

No. 22
Muskegon to Detroit
Grand Trunk Western
Standing in Grand Rapids

MUSKEGON to DETROIT
  64: 12s-dr
GTW 22, Muskegon (10p) to Detroit (7:35a)

Pennsylvania Railroad's *Detroit Express* trundles into town on a bright morning in 1941.

No. 55
Detroit to Durand
Grand Trunk Western
Arriving in Pontiac

DETROIT to MUSKEGON
  63:12s-dr
GTW 55, Detroit (10:45p) to Durand (12:50a)
GTW 57, Durand (2:25a) to Muskegon (8:05a)

With this scheduling, the Muskegon sleeper remained parked in Grand Rapids until 7 a.m., thus providing the furniture capital with Pullman service from Detroit that the Michigan Central and Pere Marquette no longer offered.

No. 57
Detroit to Muskegon
Grand Trunk Western
11 miles south of Pontiac

DETROIT to CHICAGO
  55: 8s-dr-2c
GTW 57, Detroit (11:30p) to Durand (1:15a)
GTW 5, Durand (1:40a) to Chicago (7a)

(No 5 is the La Salle)

No. 57
Detroit to Louisville
Baltimore & Ohio
10 miles south of Detroit

DETROIT to DAYTON
  8s-4dbr
B&O 57, Detroit (11:40p) to Dayton (5:09a)

DETROIT to CINCINNATI
  8s-4dbr
B&O 57, Detroit (11:40p) to Cincinnati (6:55a)

DETROIT to LOUISVILLE
  8s-4dbr
B&O 57, Detroit (11:40p) to Louisville (10:05a)

DETROIT to ST. PETERSBURG
  10s-2c-dr
B&O, Detroit (11:40p Sun)
      to Cincinnati (6:55a Mon)
L&N 33, Cincinnati (8a) to Atlanta (7:10p)
CofG 32-33, Atlanta (8p) to Albany, Ga. (1a Tue)
ACL 33, Albany (1:15a) to Trilby, Fla. (7:30a)
ACL 133, Trilby (8:05a)
      to St. Petersburg (10:10a Tue)

(Cincinnati to Louisville, dining car. L&N 33 is the Southland which carries dining cars from Cincinnati to Macon, and Thomasville, Ga., to St. Petersburg, as well as sleeper-lounges.)

The Baltimore & Ohio served Detroit with dedicated sleepers to Akron, Pittsburgh, Wheeling, Washington, Philadelphia, and, via Southern, Jacksonville and St. Augustine (FEC).

Pittsburgh Express, No. 106
Detroit to Pittsburgh
Pennsylvania Railroad
37 miles south of Detroit

DETROIT to PITTSBURGH
  6s-6dbr   10r-5dbr
PRR 106, Detroit (11:10p) to Pittsburgh (7:20a)

No. 208
Muskegon to Holland
Chesapeake & Ohio
Departing Grand Haven

MUSKEGON to CHICAGO
  281: 10r-6dbr
C&O 208, Muskegon (11:30p) to Holland (12:30a)
C&O 7, Holland (1:13a) to Chicago (5a)

Summertime service on this line included a Chicago-Pentwater sleeper.

Ohio Special, No. 309
Detroit to Cincinnati
New York Central
19 miles south of Detroit

DETROIT to CINCINNATI
  10r-6dbr
NYC 309, Detroit (11:30p) to Cincinnati (6:50a)

DETROIT to JACKSONVILLE
  8s-dr-2c
NYC 309, Detroit (11:30p Sun)
      to Cincinnati (6:50a Mon)
SOU 3, Cincinnati (8:45a)
      to Jacksonville (8:15a Tue)

DETROIT to MIAMI
  4c-4dbr-2dr   10r-6dbr
  10r-5dbr   13dbr
  5dbr-buffet-observation
NYC 309, Detroit (11:30p Sun)
      to Cincinnati (6:50a Mon)
SOU 5, Cincinnati (8:30a)
      to Jacksonville (4:05a Tue)
FEC 9, Jacksonville (4:25a) to Miami (11:25a Tue)

(SOU 3 is the Royal Palm, *carrying diner and sleeper-lounge from Cincinnati to Atlanta. SOU 5 is the streamlined* New Royal Palm, *with dining and lounge cars from Cincinnati to Miami.)*

Detroit Pullmans traveled south through the Cincinnati gateway, departing on the Southern for Atlanta, New Orleans, St. Augustine, Tampa, St. Petersburg, and Miami; and on the L&N for Fort Myers and Orlando. The NYC also scheduled sleepers to Dayton, Columbus, and Norfolk (N&W).

No. 311
Detroit to Toledo
New York Central
Departing Detroit

DETROIT to PITTSBURGH
  10r-6dbr
NYC 311, Detroit (11:45p) to Toledo (1:20a)
NYC 288, Toledo (1:50a) to Cleveland (3:40a)
Erie 688, Cleveland (3:55a) to Youngstown (5:32a)
P&LE 88, Youngstown (5:40a)
        to Pittsburgh (7:10a)

DETROIT to CLEVELAND
  22r
As above, Detroit (11:45p) to Cleveland (3:40a)

This interline operation also included a Detroit-Youngstown sleeper.

Motor City Special, No. 315
Detroit to Chicago
New York Central
22 miles west of Detroit

DETROIT to CHICAGO
  13dbr   *22r*
  22r   *22r*
  8s-dr-2c   *12s-dr*
  6dbr-buffet-lounge
NYC 315, Detroit (11:30p) to Chicago (7:20a)

Once all-Pullman, the *Motor City Special* at this scheduling still carried a heavy first-class consist, in addition to coaches. Kalamazoo and Lansing formerly had their own sleepers to Chicago.

Northerner, No. 337
Detroit to Mackinaw City
New York Central
48 miles north of Detroit

DETROIT to MACKINAW CITY
  10s-3dbr
NYC 337, Detroit (10:45p)
        to Mackinaw City (8:10a)

(A 6dbr-buffet-lounge sleeper is added to the summer consist of the Northerner)

The *Northerner* offered the last year-round Pullman service to the northern tip of Michigan's lower peninsula, although summertime service lasted until the early 1960's. The Michigan Central carried dedicated sleepers from Detroit to Bay City, Alpena (D&M), Grayling, and by ferry over the Straits of Mackinac and DSS&A, Calumet and Sault Ste. Marie (Soo). Summertime Pullmans to Mackinaw City from Toledo, Cincinnati, Columbus, and Pittsburgh also traveled this route.

Northerner, No. 338
Mackinaw City to Detroit
New York Central
Standing in Grayling

MACKINAW CITY to DETROIT
  10s-3dbr
NYC 338, Mackinaw City (9:15p) to Detroit (7:05a)

No. 352
Bay City to Jackson
New York Central
Departing Lansing

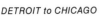

BAY CITY to CHICAGO
  8s-dr-2c
NYC 352, Bay City (9:30p) to Jackson (12:55a)
NYC 315, Jackson (1:55a) to Chicago (7:20a)

SAGINAW to CHICAGO
  6s-6dbr
NYC 352, Saginaw (10:10p) to Jackson (12:55a)
NYC 315, Jackson (1:55a) to Chicago (7:20a)

(No. 315 is the Motor City Special, which carries a sleeper-buffet-lounge)

No. 510
Grand Rapids to Fort Wayne
Pennsylvania Railroad
16 miles south of Kalamazoo

GRAND RAPIDS to CINCINNATI
  12s-dr
PRR 510, Grand Rapids (10p)
        to Fort Wayne (1:15a)
PRR 500, Fort Wayne (1:50a)
        to Richmond, Ind. (4a)
PRR 200, Richmond (4:40a) to Cincinnati (7:40a)

(No. 200 is the Florida-bound Southland*)*

In addition to the heavy summertime Pullman traffic that moved over the old Grand Rapids & Indiana line to and from the northern resorts, the Pennsylvania also scheduled sleepers from Grand Rapids to Chicago (NYC), Indianapolis, Pittsburgh, and, via the *Southland,* St. Petersburg.

Northern Arrow, No. 519
Cincinnati to Mackinaw City
Summertime only, thrice weekly
Pennsylvania Railroad
7 miles south of Kalamazoo

CINCINNATI to MACKINAW CITY
  6dbr-lounge   *4c-2dr-4dbr*
  10s-2dr
PRR 519, Cincinnati (5:40p)
        to Mackinaw City (8:45a)

CINCINNATI to HARBOR SPRINGS
  10s-2c-dr
PRR 519, Cincinnati (5:40p) to Harbor Springs (8a)

CINCINNATI to TRAVERSE CITY
  10s-2c-dr
PRR 519, Cincinnati (5:40p)
        to Traverse City (6:30a)

ST. LOUIS to HARBOR SPRINGS
  6s-6dbr
PRR 30, St. Louis (12:30p)
        to Richmond, Ind. (6:01p)
PRR 519, Richmond (6:40p) to Harbor Springs (8a)

CHICAGO to HARBOR SPRINGS
  6s-4r-4dbr
PRR 2, Chicago (6p) to Fort Wayne (8:21p)
PRR 519, Fort Wayne (9p) to Harbor Springs (8a)

CHICAGO to MACKINAW CITY
  10s-2c-dr   *12r-2sbr-3dbr*
PRR 2, Chicago (6p) to Fort Wayne (8:21p)
PRR 519, Fort Wayne (9p)
        to Mackinaw City (8:45a)

(Dining cars all segments except on branches to Traverse City and Harbor Springs. No. 30 is the "Spirit of St. Louis." No. 2 is the Pennsylvania Limited.)

The *Northern Arrow* continued the vacation travel tradition started by the Grand Rapids & Indiana, which the Pennsy originally leased before taking it over. The GR&I billed itself as "The Fishing Line" in a logo showing a jumping fish. Such trains as the *Northland Limited* and *North Michigan Resorter* carried sleepers to the north from Louisville, Indianapolis, Columbus, Dayton, and Richmond, in addition to the cities carded in these schedules. The *Northern Arrow,* which replaced the *Limited* in 1935, made the last Michigan resort run in 1961.

Canadian-Niagara, No. 358
Chicago to Buffalo
Via Niagara Falls
New York Central
Arriving in Battle Creek

CHICAGO to BUFFALO
  8s-buffet-lounge
NYC 358, Chicago (7:35p) to Buffalo (8:50a)

CHICAGO to TORONTO
  12s-dr   *10r-6dbr*
NYC 358, Chicago (7:35p) to Detroit (2:50a)
CPR 20, Detroit (3:15a) to Toronto (8:50a)

CHICAGO to SAGINAW
  6s-6dbr
NYC 358, Chicago (7:35p) to Jackson (12:57a)
NYC 351, Jackson (2a) to Saginaw (5:30a)

CHICAGO to BAY CITY
  8s-dr-2c
NYC 358, Chicago (7:35p) to Jackson (12:57a)
NYC 351, Jackson (2a) to Bay City (6a)

(CPR 20 carries a cafe-parlor car from London to Toronto)

No. 358
New York Central
Detroit setout
Standing in Detroit

DETROIT to BUFFALO   *via Niagara Falls*
  22r
NYC 358, Detroit (3:10a) to Buffalo (8:50a)

Northern Arrow, No. 520
Mackinaw City to Cincinnati
Summertime only, thrice weekly
Pennsylvania Railroad
20 miles north of Grand Rapids

MACKINAW CITY to CINCINNATI
  6dbr-lounge   *4c-2dr-4dbr*
  10s-2dr
PRR 520, Mackinaw City (6p) to Cincinnati (8:40a)

HARBOR SPRINGS to CINCINNATI
  10s-2c-dr
PRR 520, Harbor Springs (6:30p)
        to Cincinnati (8:40a)

TRAVERSE CITY to CINCINNATI
  10s-2c-dr
PRR 520, Traverse City (7:55p)
        to Cincinnati (8:40a)

HARBOR SPRINGS to ST. LOUIS
  6s-6dbr
PRR 520, Harbor Springs (6:30p)
        to Richmond (5:35a)
PRR 31, Richmond (6:31a) to St. Louis (12:10p)

HARBOR SPRINGS to CHICAGO
  6s-4r-4dbr
PRR 520, Harbor Springs (6:30p)
        to Fort Wayne (3:15a)
PRR 23, Fort Wayne (3:43a) to Chicago (6:20a)

MACKINAW CITY to CHICAGO
  10s-2c-dr   *12r-2sbr-3dbr*
PRR 520, Mackinaw City (6p)
        to Fort Wayne (3:15a)
PRR 23, Fort Wayne (3:43a) to Chicago (6:20a)

(Dining cars all segments except on branches from Traverse City and Harbor Springs. No. 31 is the "Spirit of St. Louis." No. 23 is the Manhattan Limited.)

Minnesota

See map on page **278**

Empire Builder, No. 1

Streamliner
Chicago to Seattle
via Wilmar, New Rockford
Great Northern
52 miles south of Fargo

CHICAGO to SEATTLE
Cars 14-17, Std: s-r-dbr-c-dr-dup/r
10: Obs.-buffet-lounge sleeper
CB&Q 49, Chicago (1p Sun) to St. Paul (7:45p)
GN 1, St. Paul (8:15p) to Seattle (8a Tue)

CHICAGO to PORTLAND
Cars 11-12, Std: s-r-dbr-c-dr-dup/r
CB&Q 49, Chicago (1p Sun) to St. Paul (7:45p)
GN 1, St. Paul (8:15p) to Spokane (11:30p Mon)
SP&S 1, Spokane (12:06a Tue)
to Portland (7:30a Tue)

(Dining car, Ranch coffee shop-lounge. Only two trains, Empire Builder and the Milwaukee's Olympian Hiawatha, make the Chicago-Puget Sound trip in two nights. The other westbound Builder has just been divided in Spokane; one portion leaving for Seattle, and the other departing in six minutes for Portland.)

Secondary and regional trains of the GN carried sleepers west from the Twin Cities to: Crookston; Devils Lake and Minot, N.D.; Aberdeen, S.D.; Butte, Helena, and Great Falls, Mont.; and Tacoma and Seattle, Wash.

Soo-Dominion, No. 3

St. Paul to Moose Jaw, Sask.
Except summers; see No. 3-13
Soo Line
63 miles west of Minneapolis

ST. PAUL to CALGARY, ALTA.
8s-dr-2c
SOO 3, St. Paul (9:35p Sun) and
Minneapolis (10:25p)
to Portal, N.D. (2p CST Mon)
CPR 13, N. Portal, Sask. (2:20p MST)
to Moose Jaw (7:15p)
CPR 7, Moose Jaw (8:50p) to Calgary (8:30a Tue)

ST. PAUL to MOOSE JAW
16s (tourist)
As above, St. Paul (9:35p) to Moose Jaw (7:15p)

(Dining car, Enderlin, N.D., to Portal; buffet parlor car North Portal to Moose Jaw. Canadian Pacific 7 is the Dominion, a Montreal-Vancouver train that interchanges through sleepers with the Soo-Dominion at Moose Jaw.)

Mountaineer, No. 3-13

St. Paul to Vancouver, B.C.
Summers only
Soo Line
61 miles west of Minneapolis

ST. PAUL to VANCOUVER

6s-4dbr-4r	2c-dr-lounge
6c-3dr	8s-dr-2c
14s tourist	

Soo 3-13, St. Paul (9:35p Sun) and
Minneapolis (10:25p)
to Portal (2:30p Mon)
CPR 3-13, N. Portal (2:20p MT)
to Vancouver (7:30a Wed)

ST. PAUL to BANFF
6s-6dbr
As above, St. Paul (9:35p Sun)
to Portal (2:30p Mon)
CPR 3-13, N. Portal (2:20p MT)
to Banff (10:10a Tue)

(Dining car, Enderlin to Vancouver; open observation car, Calgary to Vancouver)

Alaskan, No. 3

St. Paul to Spokane
Northern Pacific
17 miles north of St. Cloud

ST. PAUL to MANDAN, N.D.
35: 10s-c-dr
NP 3, St. Paul (9:30p) and
Minneapolis (10:15p) to Mandan (12:40p)

ST. PAUL to FARGO
30: 10s-c-dr
NP 3, St. Paul (9:30p) and
Minneapolis (10:15p) to Fargo (5:30a)
(Alternates service with GN every four months)

Pioneer Limited, No. 4

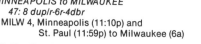

Minneapolis to Chicago
Milwaukee Road
Leaving St. Paul

MINNEAPOLIS to CHICAGO

41: 16dup/r-4dbr	42: 16dup/r-4dbr
44: 8dup/r-6r-4dbr	45: 8dup/r-6r-4dbr

MILW 4, Minneapolis (11:10p) and
St. Paul (11:59) to Chicago (8a)

ST. PAUL to CHICAGO
46: 8dup/r-6r-4dbr
MILW 4, St. Paul (11:59p) to Chicago (8a)

MINNEAPOLIS to MILWAUKEE
47: 8 dup/r-6r-4dbr
MILW 4, Minneapolis (11:10p) and
St. Paul (11:59p) to Milwaukee (6a)

(Tip Top Tap diner)

Alaskan, No. 4

Seattle to St. Paul
Northern Pacific
4 miles south of Fargo

MANDAN, N.D. to ST. PAUL
45: 10s-c-dr
NP 4, Mandan (4:20p) to Minneapolis (6:55a)
and St. Paul (7:30a)

FARGO to ST. PAUL
40: 10s-c-dr
NP 4, Fargo (11:40p) to Minneapolis (6:55a)
and St. Paul (7:30a)
(Alternates service with GN every four months)

(Other eastbound editions of the Alaskan, a three-night train, are in Idaho and Montana)

Winnipeg Limited, No. 7

St. Paul to Winnipeg
Great Northern
9 miles south of Fergus Falls

ST. PAUL to WINNIPEG

72: 6s-6dbr	73: 6s-6dbr
74: 8s-diner-lounge	

GN 7, St. Paul (8p) and
Minneapolis (8:30p) to Winnipeg (8a)

ST. PAUL to VANCOUVER　　Summertime only
75: 6c-3dr
As above, St. Paul (8p Sun) to Winnipeg (8a Mon)
CNR 3, Winnipeg (9:50a)
to Vancouver (7:45a Wed)

(CNR 3 is the Continental Limited. This crack transcontinental offers a diner and lounge and cuts in a mountain observation car for the Jasper-Kamloops Jct. leg through the Canadian Rockies.)

Winnipeg Limited, No. 8

Winnipeg to St. Paul
Great Northern
Entering Crookston

WINNIPEG to ST. PAUL

82: 6s-6dbr	83: 6s-6dbr
84: 8s-diner-lounge	

GN 8, Winnipeg (7:30p) to Minneapolis (7:25a)
and St. Paul (8a)

VANCOUVER to ST. PAUL　　Summertime only
6c-3dr
CNR 2, Vancouver (7:30p Fri) to Winnipeg (6:10p)
GN 8, Winnipeg (7:30p Sun)
to Minneapolis (7:25a Mon)
and St. Paul (8a Mon)

(CNR 2 is the Continental Limited, offering dining and lounge service, and mountain observation car through the Canadian Rockies between Kamloops Jct. and Jasper.)

Frank C. Hoffman

Second No. 1, the GN's *Empire Builder*, arrives with nine cars in Minneapolis, July 29, 1937. Three sections ran that day.

Dakotan, No. 10
Williston N.D., to St. Paul Via Wilmar
Great Northern
40 miles south of Fargo

GRAND FORKS to ST. PAUL
106: 10s-2c-dr
GN 10, Grand Forks (7:55p) to Minneapolis (7:05a)
and St. Paul (7:40a)

(Coach-lunch, Minot, N.D., to St. Paul. A Fargo-St. Paul Pullman alternates every four months between the Dakotan *and the NP's* Alaskan.*)*

No. 11
St. Paul to International Falls
Northern Pacific
4 miles south of Brainerd

ST. PAUL to INTERNATIONAL FALLS
111: 6s-6dbr   *Sun, Tue, Thu*
NP 11, St. Paul (8:15p) and Minneapolis (8:55p)
to International Falls (7:45a)

No. 12
International Falls to St. Paul
Northern Pacific
49 miles north of Brainerd

INTERNATIONAL FALLS to ST. PAUL
121: 6s-6dbr   *Mon, Wed, and Fri*
NP 12, International Falls (7p)
to Minneapolis (6:40a)
and St. Paul (7:15a)

Olympian Hiawatha, No. 15
Streamliner
Chicago to Seattle/Tacoma
Milwaukee Road
85 miles west of Minneapolis

CHICAGO to SEATTLE/TACOMA
151: 10r-6dbr   152: 8dbr-Skytop lounge
A15: 14s tourist   B15: 14s tourist
MILW 15, Chicago (3:30p Sun)
to Seattle (10:30a Tue)
and Tacoma (11:45a)

(Diner and Tip Top Grill car. The other westbound Olympian Hiawatha, *a two-night train, is in Idaho.)*

Short Line Express, No. 17
Minneapolis to Kansas City
Rock Island
43 miles south of St. Paul

MINNEAPOLIS to DES MOINES
175: 10s-3dbr
RI 17, Minneapolis (9:45p) and
St. Paul (10:30p) to Des Moines (8a)

The *Short Line Express* formerly handled the Kansas City sleeper now moved on the *Mid Continent Special.* The Rock Island once moved Pullmans from the Twin Cities to Rock Island, Ill., Peoria, Chicago, and Dallas/Fort Worth. In the late 1950s it instituted brief Omaha service.

Columbian, No. 18
Tacoma/Seattle to Chicago
Milwaukee Road
Standing in St. Paul

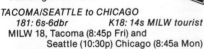

TACOMA/SEATTLE to CHICAGO
181: 6s-6dbr   K18: 14s MILW tourist
MILW 18, Tacoma (8:45p Fri) and
Seattle (10:30p) Chicago (8:45a Mon)

(Dining-lounge car. Eastbound editions of the Columbian, *a three-night train, are also in Washington and Montana.)*

No. 19
Duluth to Winnipeg
Duluth, Winnipeg & Pacific
35 miles south of International Falls

DW&P

DULUTH to WINNIPEG
1: 10s-dr-buffet   *Tue, Thu, Sat*
DW&P 19, Duluth (7:15p) to Fort Frances (1:35a)
CNR 19, Fort Frances (2a) to Winnipeg (9:45a)

(Cafe parlor, Fort Frances to Winnipeg)

The DW&P once moved a Chicago-Vancouver sleeper over this route. The car rode the Chicago & North Western south of Duluth and the Canadian National west of Winnipeg.

No. 20
Winnipeg to Duluth
Canadian National
57 miles west of Fort Frances

WINNIPEG to DULUTH
2: 10s-dr-buffet   *Mon, Wed, Fri*
CNR 20, Winnipeg (6:30p) to Fort Frances (2:25a)
DW&P 20, Fort Frances (3a) to Duluth (9:05a)

(Cafe parlor, Winnipeg to Fort Frances)

No. 29
St. Paul to Grand Forks
Great Northern
12 miles south of St. Cloud

ST. PAUL to GRAND FORKS
296: 10s-2c-dr
GN 29, St. Paul (9:40p) and
Minneapolis (10:30p)
to Grand Forks (8:20a)

Twin Ports-Twin City Express, No. 66
Minneapolis to Duluth
Northern Pacific
On the outskirts of St. Paul

MINNEAPOLIS to DULUTH
660: 10s-c-dr
NP 66, Minneapolis (11p) and
St. Paul (11:40p) to Duluth (6:10a)

Competition for passenger traffic between the Twin Cities and Twin Ports (Duluth, and Superior, Wis.) was heated until the 1920's, when a pool arrangement was approved. Until that time, the GN, NP and SOO all had scheduled dedicated sleepers from each of the Twin Cities to Duluth and in at least one case to Superior; thereafter, the NP got the overnight assignment, for which by 1952 a single Pullman sufficed.

Winnipeger, No. 109
St. Paul to Winnipeg
Soo Line
Entering Alexandria

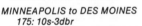

ST. PAUL to WINNIPEG
10s-dr-c   8s-5dbr
SOO 109, St. Paul (8p) and
Minneapolis (8:40p)
to Noyes, Minn. (6:15a)
CPR 109, Emerson, Man. (6:45a)
to Winnipeg (8:30a)

(Dining-lounge car)

Winnipeger, No. 110
Winnipeg to St. Paul
Soo Line
26 miles south of Thief River Falls

WINNIPEG to ST. PAUL
10s-dr-c   8s-5dbr
CPR 110, Winnipeg (7p) to Emerson, Man. (8:55p)
SOO 110, Noyes, Minn. (9:10p)
to Minneapolis (7:05a)
and St. Paul (7:45a)

(Dining-lounge car)

Thief River Falls once had its own sleeper to the Twin Cities

Nightingale, No. 201
Minneapolis to Omaha
Chicago & North Western
2 miles north of Mankato

MINNEAPOLIS to OMAHA
10s-3dbr   8s-5dbr
C&NW 201, Minneapolis (9:25p) and
St. Paul (10:10p) to Omaha (7:45a)

MINNEAPOLIS to SIOUX CITY
12s-dr
C&NW 201, Minneapolis (9:25p) and
St. Paul (10:10p) to Sioux City (4:50a)

(Cafe-lounge, Minneapolis to St. James. The Nightingale *traveled a main trunk of the Chicago, St. Paul, Minneapolis & Omaha, part of the North Western system and popularly known as the "Omaha Road.")*

This line to Omaha was tapped by granger branches running into South Dakota, over which Twin Cities Pullmans to Mitchell, Redfield, Sioux Falls, and Watertown, S.D., were operated. Both the Missouri Pacific and Burlington interchanged Kansas City sleeper with the Omaha Road. Former competitors for Minneapolis-Omaha business included the Great Western and M&StL-IC.

North Western Limited, No. 406
Streamliner
Minneapolis to Chicago
Chicago & North Western
12 miles east of St. Paul

MINNEAPOLIS to CHICAGO
6sbr-2dbr-lounge
16dup/r-3dbr-c cars as necessary
C&NW 406, Minneapolis (11p) and
St. Paul (11:40p) to Chicago (8a)

MINNEAPOLIS to MILWAUKEE
16dup/r-3dbr-c
C&NW 406, Minneapolis (11p) and
St. Paul (11:40p) to Milwaukee (5:45a)

ST. PAUL to CHICAGO
16dup/r-3dbr-c
C&NW 406, St. Paul (11:40p) to Chicago (8a)

(Diner)

The *City* streamliners may have been more press-worthy, but the true flagship of the C&NW was its *North Western Limited,* whose predecessors went back to the 1880's and included luxurious, vestibuled, Wagner consists. Now in its last decade, this classic yellow-and-green flyer finally succumbed in June 1959. The North Western had two basic routes to Chicago, including this and the original line through Madison, which the *North Western Limited* followed until the Milwaukee one opened for traffic.

Mississippi

See map on page 152

Rebel, No. 1
Streamliner
Jackson, Tenn., to New Orleans
Gulf Mobile & Ohio
44 miles south of New Albany

JACKSON, TENN., to NEW ORLEANS
   5: 6s-dr-obs.-lounge (GM&O owned)
GM&O 1, Jackson (8p) to New Orleans (11a)

(Jackson, Miss., to New Orleans, economy buffet)

The *Rebels* entered service on the Gulf Mobile &
Northern in the summer of 1935, the country's first
nonarticulated streamliners. The red and silver
flyers additionally were the first trains to carry
hostesses and the first diesel streamliners in the
South.

Rebel, No. 2
Streamliner
New Orleans to Jackson, Tenn.
Gulf Mobile & Ohio
50 miles east of Jackson, Miss.

NEW ORLEANS to JACKSON, TENN.
   6: 6s-dr-obs.-lounge (GM&O owned)
GM&O 2, New Orleans (5:30p) to Jackson (7:45a)

(Economy buffet, New Orleans to Jackson, Miss.)

Jackson, Miss., had Pullman service on the
GM&N to Mobile and Jackson, Tenn.

Northern Express, No. 26
New Orleans to Chicago
Illinois Central
28 miles south of Jackson

NEW ORLEANS to MEMPHIS
   264: 6s-4r-4dbr
IC 26, New Orleans (8:30p) to Memphis (6:35a)

The Illinois Central served Jackson with sleepers to New Orleans and Memphis.

Crescent, No. 38
Streamliner
New Orleans to New York
Louisville & Nashville
26 miles west of Gulfport

NEW ORLEANS to NEW YORK
   L25: 10r-6dbr   *L26: 10r-6dbr*
   L27: 10r-6dbr   *L28: 5dbr-obs.-lounge*
   L24: 2dr-master room-lounge
L&N 38, New Orleans (11p Sun)
         to Montgomery (7:20a Mon)
WPR 38, Montgomery (7:40a) to Atlanta (1:15p)
SOU 38, Atlanta (1:45p) to Washington (4:15a Tue)
PRR 118, Washington (4:55a)
         to New York (9:15a Tue)

NEW ORLEANS to WASHINGTON
   L19: 10r-6dbr
As above, New Orleans (11p Sun)
         to Washington (4:15a Tue)

*(Diner. The Crescent is all-Pullman north of At-
lanta. The master room in sleeper L24 has its own
shower and radio.)*

No. 208
Pensacola to Amory
Frisco
2 miles south of Amory

PENSACOLA to MEMPHIS
   38: 6s-2dbr (Frisco-owned)
SLSF 208, Pensacola (3:30p) to Amory (12:05a)
SLSF 108, Amory (1:30a) to Memphis (6:30a)

(Pensacola to Memphis, buffet car)

Pelican, No. 42
New Orleans to New York
Southern Railway
6 miles south of Hattiesburg

NEW ORLEANS to NEW YORK
   N9: 10r-6dbr
SOU 42, New Orleans (9p Sun)
         to Bristol, Va. (7:40p Mon)
N&W 42, Bristol (7:55p) to Lynchburg, Va. (2a Tue)
SOU 42, Lynchburg (2:10a) to Washington (6:45a)
PRR 122, Washington (7:20a)
         to New York (11:30a Tue)

NEW ORLEANS to WASHINGTON
   S44: 10s-lounge
As above, New Orleans (9p Sun)
         to Washington (6:45a Tue)

*(Dining cars: Birmingham to Roanoke, Washington
to New York. As the Crescent makes it way along
the Gulf Coast, the less-exclusive Pelican heads
for the same destination, though by way of the joint
Southern/Norfolk & Western route across the Appalachians. The consist of the Pelican will swell considerably, with Pullmans from Shreveport, Jackson, Knoxville, Bristol, Williamson, W.Va., and Roanoke.)*

Sunnyland, No. 107
Memphis to Birmingham
Frisco
2 miles northwest of New Albany

MEMPHIS to BIRMINGHAM
   36: 10s-dr-2c
SLSF 107, Memphis (9:30p) to Birmingham (7a)

MEMPHIS to ATLANTA
   37: 10s-dr-2c
SLSF 107, Memphis (9:30p) to Birmingham (7a)
SOU 30, Birmingham (7:25a) to Atlanta (1:25p)

MEMPHIS to PENSACOLA
   38: 6s-2dbr (Frisco-owned)
SLSF 107, Memphis (9:30p) to Amory (2a)
SLSF 207, Amory (3a) to Pensacola (11:45a)

*(Buffet car, Memphis to Pensacola; diner, Bir-
mingham to Atlanta)*

Northeastern Limited, No. 226
Shreveport to Meridian
Illinois Central
28 miles east of Jackson

SHREVEPORT to NEW YORK
   2264: 10s-dr-2c
IC 226, Shreveport (4:30p Sun)
         to Meridian (2a Mon)
SOU 42, Meridian (2:55a) to Bristol (7:40p)
N&W 42, Bristol (7:55p) to Lynchburg (2a Tue)
SOU 42, Lynchburg (2:10a) to Washington (6:45a)
PRR 122, Washington (7:20a)
         to New York (11:30a Tue)

JACKSON to ATLANTA
   2262: 12s-dr
IC 226, Jackson (11:15p) to Meridian (2a)
SOU 42, Meridian (2:55a) to Birmingham (7:05a)
SOU 30, Birmingham (7:25a) to Atlanta (1:25p)

*(The Northeastern Limited carries a cafe-lounge
from Shreveport to Jackson. No. 42, the Pelican,
has diners from Birmingham to Roanoke, and
Washington to New York. On No. 30, the
Sunnyland, a dining car operates through to
Atlanta.)*

Trains working the IC's east-west Yazoo main
line carried sleepers from Shreveport and Jackson
to Birmingham, and between Shreveport and Chicago, and Atlanta and Fort Worth (T&P).

Missouri

See map on page 230

Texas Special, No. 1
Streamliner
St. Louis to San Antonio
Frisco
8 miles east of Neosho

ST. LOUIS to FORT WORTH
   18: 14r-4dbr
SLSF 1, St. Louis (5:30p) to Vinita, Okla. (ntg)
MKT 1-11, Vinita (ntg) to Denison, Tex. (5:05a)
MKT 11, Denison (5:20a) to Fort Worth (7:35a)

ST. LOUIS to DALLAS
   14: 14r-4dbr   *16: 14r-4dbr*
As above, St. Louis (5:30p) to Dennison (5:05a)
MKT 1, Denison (5:20a) to Dallas (7:30a)

ST. LOUIS to SAN ANTONIO
   10: 12s-dr
   11: 2br-dr-obs-lounge (2 days out of 3)
As above, St. Louis (5:30p) to Denison (5:05a)
MKT 1, Denison (5:20a) to San Antonio (1:55p)

NEW YORK to SAN ANTONIO
   15: 14r-4dbr, or 10r-6dbr
PRR 3, New York (7:35p Sat) to St. Louis (3p Sun)
As above, St. Louis (5:30p)
         to San Antonio (1:55p Mon)

WASHINGTON to SAN ANTONIO
   12: 14r-4dbr
B&O 1, Washington (6:30p Sat)
         to St. Louis (1p Sun)
As above, St. Louis (5:30p)
         to San Antonio (1:55 Mon)

*(Diners: St. Louis to Springfield, Muskogee to San
Antonio. Full lounge carried when observation
sleeper 11 is not in consist. PRR 3, the Penn Texas,
carries dining and recreation cars, and B&O 1, the
National Limited, has a diner and sun room-observa-
tion-lounge for Pullman passengers.)*

Will Rogers, No. 3
St. Louis to Oklahoma City
Frisco
28 miles west of St. Louis

CHICAGO to OKLAHOMA CITY
   35: 14r-4dbr
GM&O 3, Chicago (4:50p) to St. Louis (10p)
SLSF 3, St. Louis (11:20p)
         to Oklahoma City (1:30p)

ST. LOUIS to SPRINGFIELD
   33: 12s-dr
SLSF 3, St. Louis (11:20p) to Springfield (4:50a)

ST. LOUIS to JOPLIN
   39: 12s-2dbr
SLSF 3, St. Louis (11:20p) to Monett (6:25a)
SLSF 303, Monett (6:50a) to Joplin (8:25a)

(Springfield-Oklahoma City, diner-parlor car)

Important as a junction, Monett had dedicated
sleepers from both Little Rock and Dallas.

Will Rogers, No. 4
Oklahoma City to St. Louis
Frisco
13 miles east of Monett

*(Oklahoma City to Springfield, dining car. No
Pullmans are in the consist at midnight, but in a
little more than an hour, the Springfield sleeper
will be cut in.)*

No. 4 once carried an Amarillo-St. Louis Pullman
by interline arrangement with the Rock Island.

Will Rogers, No. 4
Frisco
Springfield setout
Standing in Springfield

SPRINGFIELD to ST. LOUIS
43: 12sdr
SLSF 4, Springfield (1:15a) to St. Louis (7:25a)

Springfield, the nerve center of the Frisco system, also had its own setouts for Memphis, Birmingham, and Oklahoma City.

California Limited, No. 4
Los Angeles to Chicago
Via Raton Pass, Topeka
Atchison, Topeka & Santa Fe
50 miles east of Kansas City

LOS ANGELES to CHICAGO
12s-dr   8s-dr-2c
Santa Fe 4, Los Angeles (6:15p Fri)
to Chicago (8:30a Mon)

LOS ANGELES to CHICAGO
10s-dr-2c
Santa Fe 4, Los Angeles (6:15p Fri)
to Albuquerque (6:45)
Santa Fe 13, Albuquerque (9p Sat)
to Belen, N.M. (9:40p)
Santa Fe 106, Belen (9:55p)
to Newton, Kans. (5:25p Sun)
Santa Fe 6, Newton (6:10p)
to Kansas City (10:15p)
Santa Fe 4, Kansas City (11p)
to Chicago (8:30a Mon)

SAN FRANCISCO to CHICAGO
10s-dr-2c
Bus from San Francisco (9a)
Santa Fe 4, Oakland (9:25a Fri)
to Barstow (10:10p)
Santa Fe 4, Barstow (10:45p)
to Chicago (8:30a Mon)

GALVESTON to CHICAGO
10s-2dbr-c
Santa Fe 6, Galveston (7:30p Sun)
to Kansas City (10:15p)
Santa Fe 4, Kansas City (11p Sun)
to Chicago (8:30a Mon)

(No. 4, diner, Oakland to Barstow. California Limited: Fred Harvey station dining room service west of La Junta, Colo.; dining car, La Junta to Chicago; lounge, Los Angeles to Chicago. No. 106, the Scout, dining car starting at 7 a.m. Amarillo to Newton. No 6, diner Fort Worth to Kansas City. Eastbound editions of the California Limited are now in California and New Mexico.)

No Santa Fe train gave Missouri more continuing service than the California Limited, which made its first appearance as a daily train in 1892. When the railroad decided to build its line from Kansas City to Chicago, it purposely avoided populous areas and headed straight for the metropolis on Lake Michigan's shore. This route gave Santa Fe passenger trains unparalleled speed credentials.

No. 5
St. Louis to Texarkana
Cotton Belt (StLSW)
Leaving Delta

ST. LOUIS to PINE BLUFF, ARK.
1: 10s-2c-dr
StLSW 5, St. Louis (8p) to Pine Bluff (8a)

(The sleeper is cut out at Pine Bluff "Shops," where the train makes a 25-minute meal stop before moving on to Pine Bluff station, one mile south, and Texarkana)

The Cotton Belt's Pullman service from St. Louis included cars for Texarkana, Shreveport, Dallas, Houston (SP), Tyler, and Waco.

Southerner, No. 7
St. Louis to Texarkana
Missouri Pacific
Standing in St. Louis

ST. LOUIS to SAN ANTONIO
75: 10s-dr-2c
MP 7, St. Louis (12:45a Mon) to Texarkana (2:30p)
T&P 237, Texarkana (4:15p)
to Longview, Tex. (6:50p)
MP 37, Longview (7:10p)
to San Antonio (7:10a Tue)

CHICAGO to HOT SPRINGS
71: 10r-6dbr
GM&O 3, Chicago (4:50p Sun) to St. Louis (10p)
MP 7, St. Louis (12:45a Mon)
to Little Rock (10:30a)
MP 219, Little Rock (11:20a)
to Hot Springs (12:45p)

(St. Louis to Longview, dining-lounge car)

Meteor, No. 9
Streamliner
St. Louis to Oklahoma City
Frisco
19 miles northeast of Springfield

ST. LOUIS to OKLAHOMA CITY
93: 8s-dr-2c   94: 14r-4dbr
SLSF 9, St. Louis (7p) to Oklahoma City (8:25a)

ST. LOUIS to TULSA
91: 8s-dr-3dbr   92: 14r-4dbr
SLSF 9, St. Louis (7p) to Tulsa (5:30a)

ST. LOUIS to FORT SMITH, ARK.
97: 10s-3dbr
SLSF 9, St. Louis (7p) to Monett (1:40a)
SLSF 709, Monett (2:15a) to Fort Smith (7:10a)

(St. Louis to Oklahoma City, observation-diner-lounge)

The Frisco inaugurated the Meteor in 1902. Its service was extended to Texas to compete with the Katy's Texas Special, but when that train went under joint operation in 1917, the Meteor reverted to its old territory of Oklahoma. In 1948, both the Meteor and Texas Special were requipped as streamliners with Pullman Standard cars of stainless steel sides and red letter boards.

Missourian, No. 9
St. Louis to Kansas City
Missouri Pacific
Leaving St. Louis (Tower Grove)

ST. LOUIS to KANSAS CITY
91: 8s-5dbr   92: 10s-dr-2c
MP 9, St. Louis (11:50p) to Kansas City (7:20a)

Missouri Pacific and Wabash were still serving the St. Louis-Kansas City market at this scheduling; other competitors had been Alton, Burlington/Alton, and Santa Fe/Wabash.

Missourian, No. 10
Kansas City to St. Louis
Missouri Pacific
18 miles southeast of Kansas City

KANSAS CITY to ST. LOUIS
105: 10s-dr-2c   106: 8s-5dbr
MP 10, Kansas City (11:30p) to St. Louis (7:28a)

The Missourian formerly carried St. Louis sleepers from both Lincoln and Omaha. Kansas City had similar service on another train.

Kansas City Chief, No. 10
Kansas City to Chicago
Atchison, Topeka & Santa Fe
7 miles west of Marceline

KANSAS CITY to CHICAGO
10r-6dbr   6r-6s-4dbr
24dup/r   17r
Santa Fe 10, Kansas City (10p) to Chicago (7:30a)

TULSA to CHICAGO
6r-6s-4dbr
Santa Fe 48, Tulsa (2p) to Kansas City (8:40p)
Santa Fe 10, Kansas City (10p) to Chicago (7:30a)

(Tulsa to Kansas City, diner and cafe-observation; Kansas City to Chicago, lounge and diner-lounge. No. 48 is the Oil Flyer.)

The Santa Fe offered sleeping cars to Kansas City from Galena and Independence, Kans., and Tulsa. The Kansas City-Chicago market was dominated by the Santa Fe, but three other railroads were still offering overnight accommodations in 1952: Burlington, Milwaukee Road, and Rock Island. The Alton and Wabash once made the total six.

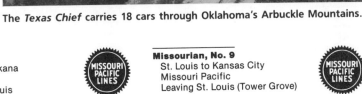
The *Texas Chief* carries 18 cars through Oklahoma's Arbuckle Mountains.

Missouri   *(continued)*

Meteor, No. 10
Streamliner
Oklahoma City to St. Louis
Frisco
Entering Neosho

OKLAHOMA CITY to CHICAGO
105: 14r-4dbr
SLSF 10, Oklahoma City (7p) to St. Louis (7:45a)
GM&O 2, St. Louis (8:58a) to Chicago (2:08p)

OKLAHOMA CITY to ST. LOUIS
103: 8s-dr-2c   104: 14r-4dbr
SLSF 10, Oklahoma City (7p) to St. Louis (7:45a)

TULSA to ST. LOUIS
101: 8s-dr-3dbr   102: 14r-4dbr
SLSF 10, Tulsa (9:45p) to St. Louis (7:45a)

(Oklahoma City to St. Louis, observation-diner-lounge)

The *Meteor* once carried sleepers for St. Louis from Enid, Okmulgee, and Quanah. Other trains operating over the Frisco main line through Springfield brought Pullmans from Monett, Eureka Springs, Ark. (Missouri & Northern Arkansas), and Sapulpa, Blackwell, and Lawton, Okla.

Meteor, No. 10
Frisco
Joplin and Fort Smith setouts
Standing in Monett

JOPLIN to ST. LOUIS
109: 12s-2dbr
SLSF 304, Joplin (9p) to Monett (10:50p)
SLSF 10, Monett (1a) to St. Louis (7:45a)

FORT SMITH to ST. LOUIS
107: 10s-3dbr
SLSF 704, Fort Smith (6p) to Monett (10:55p)
SLSF 10, Monett (1a) to St. Louis (7:45a)

Omaha Limited, No. 11
St. Louis to Omaha
Wabash
11 miles east of Chillicothe

ST. LOUIS to OMAHA
39: 10s-dr-c
Wabash 11, St. Louis (7:15p) to Omaha (7:45a)

(Combined with Des Moines Limited, also No. 11, St. Louis to Moberly, where separate trains are created)

Des Moines Limited, No. 11
St. Louis to Des Moines
Wabash
51 miles north of Moberly

ST. LOUIS to DES MOINES
87: 12s-dr
Wabash 11, St. Louis (7:15p) to Des Moines (6a)

Before the turn of the century, the Des Moines night train carried a sleeper for Ottumwa.

St. Louis Limited, No. 14
Omaha to St. Louis
Wabash
44 miles northwest of Chillicothe

OMAHA to ST. LOUIS
30: 10s-dr-c
Wabash 14, Omaha (7:15p) to St. Louis (7:35a)

(No. 14 will combine with No. 18, the Midnight Limited, *the Wabash's overnight train from Kansas City, in Moberly)*

Texas Chief, No. 15
Streamliner
Chicago to Galveston
Atchison, Topeka & Santa Fe
11 miles west of Marceline

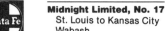

CHICAGO to GALVESTON
24dup/r   4c-4dbr-2dr
10r-3dbr-2c
Santa Fe 15, Chicago (6p) to Galveston (8:15p)

CHICAGO to WICHITA
10r-3dbr-2c
Santa Fe 15, Chicago (6p) to Wichita (5:35a)

CHICAGO to OKLAHOMA CITY
10r-6dbr
Santa Fe 15, Chicago (6p) to Oklahoma City (9a)

CHICAGO to TULSA
6r-6s-4dbr
Santa Fe 15, Chicago (6p) to Kansas City (1:30a)
Santa Fe 47, Kansas City (2a) to Tulsa (9:35a)

(Chicago to Galveston, lounge and dining car. No. 47 is the Oil Flyer, *which carries a diner and cafe-observation.)*

The Santa Fe scheduled Kansas City Pullmans to Wichita, Hutchinson, and Wellington, Kans., and Fort Worth, Galveston, San Antonio (MP), and Sweetwater, Tex.

Short Line Express, No. 16
Kansas City to Minneapolis
Rock Island
Leaving Kansas City

KANSAS CITY to DES MOINES
166: 10s-dr-c
RI 16, Kansas City (11:59p) to Des Moines (6:30a)

The Wabash and Great Western also once scheduled Kansas City-Des Moines sleepers.

Super Chief, No. 17
Atchison, Topeka & Santa Fe
Kansas City setout
Standing in Kansas City

KANSAS CITY to LOS ANGELES
10r-6dbr
Santa Fe 17, Kansas City (2:45a Mon)
to Los Angeles (8:45a Tue)

Super Chief, No. 17
All-Pullman streamliner
Chicago to Los Angeles
Via Ottawa, Kans., and Raton Pass
Atchison, Topeka & Santa Fe
57 miles northeast of Marceline

CHICAGO to LOS ANGELES
2dr-4c-4dbr   2dr-4c-4dbr
2dr-4c-4dbr   10r-6dbr
10r-6dbr   4dr-dbr-observation
Santa Fe 17, Chicago (7p Sun)
to Los Angeles (8:45a Tue)

(The Super Chief *carries one of the Pleasure Dome lounges with the Turquoise Room for private dining parties as well as a full diner and additional lounge; valet, barber, and shower bath facilities available. Extra fare on the luxury train is $15. Another westbound edition of the train is now in Arizona.)*

Midnight Limited, No. 17
St. Louis to Kansas City
Wabash
Northwest to Delmar Station

ST. LOUIS to KANSAS CITY
111: 12s-2dbr
Wabash 17, St. Louis (11:40p)
to Kansas City (7:25a)

ST. LOUIS (DELMAR) to KANSAS CITY
113: 8s-5dbr
Wabash 17, Delmar Station (11:54p)
to Kansas City (7:25a)

(From its Delmar Boulevard Station, serving the West-End business and residential district of St. Louis, the Wabash offers setout sleepers to both Kansas City and Chicago. The trains pull in just before midnight, five minutes apart.)

The Wabash, with the *City of St. Louis* and earlier trains, gave St. Louis its connection to the Overland Route. Pullmans for Denver, West Yellowstone, Salt Lake City, and Portland were among those scheduled.

The Midnight, No. 18
St. Louis to Chicago
Wabash
Delmar Station

ST. LOUIS to CHICAGO
68: 12r-4dbr   72: 12s-dr
Wabash 18, St. Louis (11:50p) to Chicago (7:40a)

ST. LOUIS (DELMAR) to CHICAGO
66: 12r-4dbr   70: 8s-4dbr
Wabash 18, Delmar Station (11:59p)
to Chicago (7:40a)

The term Rainbow Race was applied to the passenger train service between St. Louis and Chicago, referring to IC green, Alton reds, and Wabash blue (the C&EI dropped out before its streamlined, blue and orange livery could figure in). This spirited overnight competition was memorably portrayed by artist Gil Reid, whose painting showed a "rainbow" trio, their locomotives steaming impatiently just before midnight against a backdrop of the huge St. Louis train shed.

Midnight Limited, No. 18
Kansas City to St. Louis
Wabash
Leaving Kansas City

KANSAS CITY to ST. LOUIS
112: 12s-2dbr   118: 8s-5dbr
Wabash 18, Kansas City (11:59p)
to St. Louis (7:35a)

Sunflower, No. 19
St. Louis to Kansas City
Missouri Pacific
Outskirts of Kansas City

ST. LOUIS to OMAHA
191: 10s-dr-2c
MP 19, St. Louis (5p) to Kansas City (12:15a)
MP 119, Kansas City (12:40a) to Omaha (7:45a)

(Grill coaches, St. Louis until arrival at Pleasant Hill at 11:10 p.m. and Kansas City to Omaha. The Sunflower *carries the numbers 19-119-419; the Wichita section is handled by No. 419 west of Pleasant Hill.)*

The two-stemmed *Sunflower,* in addition to its Omaha and Wichita service, once carried Pullmans from St. Louis to Joplin, Lincoln, and Hutchinson and Arkansas City, Kans. MP's western trains also carried Pullmans for St. Joseph, Pueblo, and Los Angeles (D&RG/SP, tourist) and San Francisco (D&RGW/WP).

Sunflower, No. 20
Kansas City to St. Louis
Missouri Pacific
Standing in Kansas City

OMAHA to ST. LOUIS
204: 10s-dr-2c
MP 110, Omaha (4p) to Kansas City (9:20p)
MP 20, Kansas City (1:45a) to St. Louis (8:05a)

(The eastbound Sunflower *name also applies to No. 110 from Omaha and No. 420 from Wichita, whose sleeper will be cut in at Pleasant Hill. No. 20 also awaits the arrival in Kansas City at 1:25 a.m. of the Los Angeles-St. Louis Pullman on the Rock Island's Golden State. The* Sunflower *carries grill coaches Omaha-Kansas City, and Wichita-St. Louis.)*

No. 23
Kansas City to Omaha
Burlington
Leaving Kansas City

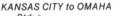

KANSAS CITY to OMAHA
Std: s
CB&Q 23, Kansas City (11:58p) to Omaha (7a)

The Burlington's line north of Kansas City, which paralleled the Missouri River, accommodated traffic headed for three general destinations: Denver, Billings, and Omaha, all of which had dedicated Pullman service. The Billings traffic was the greatest, inasmuch as it involved interline transfers with parents NP and, in some cases, GN of sleepers for Gardiner, Butte, Glacier Park, Spokane, and Seattle. The Lincoln Pullman was cut out en route; another, for the Twin Cities, traveled to Omaha, where the North Western took over.

Pony Express, No. 37
Kansas City to Los Angeles
Union Pacific
Leaving Kansas City

KANSAS CITY to LOS ANGELES
12s-dr
UP 37, Kansas City (11:59p Sun)
    to Los Angeles (7a Wed)

KANSAS CITY to PORTLAND
10s-dr-2c
UP 37, Kansas City (11:59p Sun)
    to Denver (3:30p Mon)
UP 17, Denver (5:40p) to Portland (6a Wed)

(Kansas City to Las Vegas, Nev., diner; Denver to Los Angeles and Portland, first-class lounges; Denver to Hinkle, Ore., diner. No. 17 is the Portland Rose. *The earlier editions of the westbound* Pony Express *are now in Wyoming and California.)*

American Royal, No. 56
Kansas City/St. Joseph to Chicago
Burlington
37 miles east of Chillicothe

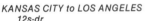

KANSAS CITY to CHICAGO
8s-5dbr
CB&Q 56, Kansas City (8p)
    to Cameron Junction (9:22p)
CB&Q 56, Cameron Junction (9:39p)
    to Chicago (8:25a)

ST. JOSEPH to CHICAGO
12s-2dbr
CB&Q 56, St. Joseph (8:15p)
    to Cameron Junction (9:10p)
CB&Q 56, Cameron Junction (9:39p)
    to Chicago (8:25a)

(The two sections of the American Royal, *soon to be accorded* Zephyr *status with streamlined equipment, are combined in Cameron Junction. Dining car, Galesburg to Chicago.)*

Kansas City-Florida Special, No. 106
Jacksonville to Kansas City
Frisco
97 miles southeast of Springfield

MIAMI to KANSAS CITY
F68: 14r-4dbr
FEC 74, Miami (1p Sat) to Jacksonville (8:20p)
SOU 8, Jacksonville (9:20p) to Atlanta (6:30a Sun)
SOU 7, Atlanta (7:30a) to Birmingham (11:30a)
SLSF 116, Birmingham (12:05p)
    to Kansas City (7:30a Mon)

JACKSONVILLE to KANSAS CITY
S67: 8s-5dbr   S69: 10s-2br-c
As Above, Jacksonville (9:20p Sat)
    to Kansas City (7:30a Mon)

(Diners, all segments; lounges, all segments but Atlanta-Birmingham. No. 74 is the all-Pullman Vacationer. *Another edition of the* Kansas City-Florida Special *is now in Georgia, 24 hours behind this one.)*

The *Kansas City-Florida Special* and *Sunnyland*, working the Frisco's line to Birmingham, carried Pullmans for Springfield, Memphis, New Orleans (IC), Birmingham, and Atlanta (Southern).

Sunflower, No. 419
Pleasant Hill to Wichita
Missouri Pacific
13 miles south of Pleasant Hill

ST. LOUIS to WICHITA
192: 10r-5dbr
MP 19, St. Louis (5p) to Pleasant Hill (11:10p)
MP 419, Pleasant Hill (11:40p) to Wichita (7:15a)

(St. Louis to Wichita, grill coach)

Memphian, No. 805
St. Louis to Memphis
Frisco
36 miles south of St. Louis

ST. LOUIS to MEMPHIS
85: 8s-5dbr
SLSF 805, St. Louis (11p) to Memphis (7:10a)

(Buffet and lounge cars)

The *Memphian* and other Frisco trains fed into the railroad's Kansas City-Birmingham line at Memphis, bringing with them Pullmans from St. Louis for Kennett, Birmingham, Pensacola, Jacksonville (Southern), and St. Petersburg (Southern/Seaboard Air Line). St. Louis also had service to Jacksonville via the Southern, L&N, and IC.

Montana
See map on page 286

North Coast Limited, No. 1
Chicago to Seattle
Northern Pacific
64 miles east of Miles City

CHICAGO to SEATTLE
252: 3br-c-6r-8dup/r   BN: 14s (tourist)
254: 4br-c-obs.-lounge
CB&Q 51, Chicago (11p Sat) to St. Paul (8a Sun)
NP 1, St. Paul (9a) to Seattle (7:30a Tue)

CHICAGO to PORTLAND
251: 3br-c-6r-8dup/r   BR: 14s (tourist)
CB&Q 51, Chicago (11p Sat) to St. Paul (8a Sun)
NP 1, St. Paul (9a) to Pasco, Wash. (11:40p Mon)
SP&S 3, Pasco (1:35a Tue) to Portland (7a Tue)

CHICAGO to BILLINGS
253: 3br-c-6r-8dup/r
CB&Q 51, Chicago (11p Sat) to St. Paul (8a Sun)
NP 1, St. Paul (9a) to Billings (4:55a Mon)

CHICAGO to CODY   Summertime only
B38: 10s-c-dr
As above, Chicago (11p Sat)
    to Billings (4:55a Mon)
CB&Q 24-27, Billings (8a) to Cody (11:15a Mon)

(Dining car. During summers, two additional Chicago sleepers join the consist: 253: 4dbr-sbr-12r for Seattle, and the Pullman for Cody, Wyo., eastern gateway to Yellowstone National Park on the Burlington's line. Other westbound editions of the North Coast Limited *are now beginning their journey in Illinois and on the final leg of it in Washington.)*

North Coast Limited, No. 2
Seattle to Chicago
Northern Pacific
40 miles west of Billings

SEATTLE to CHICAGO
262: 3br-c-6r-8dup/r   NB: 14s (tourist)
264: 4br-c-obs.-lounge
NP 2, Seattle (10p Sat) to St. Paul (10p Mon)
CB&Q 50, St. Paul (10:45p) to Chicago (7:45a Tue)

PORTLAND to CHICAGO
261: 3br-c-6r-8dup/r   PN 14s (tourist)
SP&S 4, Portland (9:15p Sat)
    to Pasco, Wash. (2:30a Sun)
NP 2, Pasco (5:25a) to St. Paul (10p Mon)
CB&Q 50, St. Paul (10:45p) to Chicago (7:45a Tue)

(Diner. The summertime consist will also include additional Seattle-Chicago sleeper 263: 4dbr-sbr-12r. Other eastbound editions of the three-night North Coast Limited *are now in Washington and Wisconsin.)*

Great Northern *Western Stars* in a photo-opportunity meeting at Whitefish Lake, Mont.

Montana

(continued)

North Coast Limited, No. 2
Northern Pacific
Billings and Cody setouts
Standing in Billings

BILLINGS to CHICAGO
263: 3br-c-6r-8dup/r
NP 2, Billings (1a Mon) to St. Paul (10p)
CB&Q 50, St. Paul (10:45p) to Chicago (7:45a Tue)

CODY to CHICAGO   *Summertime only*
B38: 10s-c-dr
CB&Q 28, Cody (7:45p Sun) to Billings (11p)
As above, Billings (1a Mon)
       to Chicago (7:45a Tue)

Alaskan, No. 3
St. Paul to Spokane
Northern Pacific
13 miles west of Billings

BILLINGS to SPOKANE
37: 10s-c-dr
NP 3, Billings (11:40p) to Spokane (6p)

   A predecessor of the *Alaskan* carried a sleeper
from Billings to Helena.

Western Star, No. 3
Chicago to Seattle
Streamliner via Great Falls
Great Northern
85 miles east of Glasgow

CHICAGO to SEATTLE
34: 4s-8dup/r-4dbr   35: 16 dup/r-4dbr
39: dr-2dbr-obs.-buffet-lounge
CB&Q 53, Chicago (11:15p Sat)
       to St. Paul (8:15a Sun)
GN 3, St. Paul (8:45a) to Seattle (7:30a Tue)

CHICAGO to PORTLAND
32: 4s-8dup/r-4dbr
CB&Q 53, Chicago (11:15p Sat)
       to St. Paul (8:15a Sun)
GN 3, St. Paul (8:45a) to Spokane (9:30p Mon)
SP&S 3, Spokane (9:45p) to Portland (7a Tue)

CHICAGO to GREAT FALLS
38: 16dup/r-4dbr
CB&Q 53, Chicago (11:15p Sat)
       to St. Paul (8:15a Sun)
GN 3, St. Paul (8:45a) to Great Falls (7:50a Mon)

*(Chicago to Seattle: diner, coffee shop car;
Spokane to Portland, observation-lounge. The
Western Star, the main GN train serving Glacier
National Park, takes three nights between Chicago
and Seattle. The first-night edition of the west-
bound train is now in Illinois, the third-nighter in
Washington.)*

   When the *Empire Builder* got its second streamlined train-sets in 1951, the GN created the
Western Star with the old equipment and retired
the venerable *Oriental Limited,* until that time its
secondary train.

   Great Falls was Great Northern all the way. The
railroad scheduled dedicated sleepers from there
to Blackfoot, Havre, Kalispell, St. Paul, and
Spokane.

Western Star, No. 4
Streamliner
Seattle to Chicago
Great Northern
22 miles southwest of Havre

SEATTLE to CHICAGO
44: 4s-8dup/r-4dbr   45: 16dup/r-4dbr
49: dr-2dbr-obs.-buffet-lounge
GN 4, Seattle (10:15p Sat) to St. Paul (10:30p Mon)
CB&Q 54, St. Paul (11p) to Chicago (8a Tue)

PORTLAND to CHICAGO
42: 4s-8dup/r-4dbr
SP&S 4, Portland (9:15p Sat)
       to Spokane (6:35a Sun)
GN 4, Spokane (7:20a) to St. Paul (10:30p Mon)
CB&Q 54, St. Paul (11p) to Chicago (8a Tue)

GREAT FALLS to CHICAGO
48: 16dup/r-4dbr
GN 4, Great Falls (9:35p Sun)
       to St. Paul (10:30p Mon)
CB&Q 54, St. Paul (11p) to Chicago (8a Tue)

*(Diner, coffee shop. Other editions of the east-
bound train are in Washington and Wisconsin.)*

Columbian, No. 17
Chicago to Seattle/Tacoma
Milwaukee Road
24 miles east of Miles City

CHICAGO to SEATTLE/TACOMA
171: 6s-6dbr   K17: 14s MILW tourist
MILW 17, Chicago (10p Sat) to Seattle (7:30a Tue)
       and Tacoma (9:05a)

*(Diner-lounge. Westbound Columbians are also
now in Wisconsin and Washington.)*

Columbian, No. 18
Tacoma/Seattle to Chicago
Milwaukee Road
25 miles west of Harlowton

TACOMA/SEATTLE to CHICAGO
181: 6s-6dbr   K18: 14s MILW tourist
MILW 18, Tacoma (8:45p Sat) and
       Seattle (10:30p) to Chicago (8:45a Tue)

*(Diner-lounge. Other eastbound editions of the
train are now in Washington and Minnesota.)*

No. 42
Great Falls to Billings
Great Northern
Standing in Great Falls

GREAT FALLS to OMAHA
G420: 12s-dr
GN 42, Great Falls (12:15a Mon) to Billings (7a)
CB&Q 42, Billings (9a) to Lincoln (6:40a Tue)
CB&Q 20, Lincoln (7:30a) to Omaha (8:30a Tue)

*(Dining cars: Billings to Lincoln, Lincoln to Omaha.
No. 20 is the Burlington's Silver Streak Zephyr.)*

No. 43
Billings to Great Falls
Great Northern
18 miles north of Billings

KANSAS CITY to GREAT FALLS
B430: 12s-dr
CB&Q 21, Kansas City (4p Sat) to Lincoln (10:15p)
CB&Q 43, Lincoln (12:05a Sun) to Billings (8p)
GN 43, Billings (11:30p) to Great Falls (6:35a Mon)

*(Diners: Kansas City to Lincoln, Lincoln to Billings.
CB&Q 21 is the northbound Silver Streak Zephyr.
GN 43 connects at Great Falls with the Western
Star and in summer carries Pullmans for Glacier
National Park.)*

   Some of the cities having dedicated Pullman
service to Glacier Park included Chicago, Omaha,
Kansas City, and Cody.

Nebraska

See map on page 194

Los Angeles Limited, No. 1
Chicago to Los Angeles
Union Pacific
Standing in Grand Island

CHICAGO to LOS ANGELES
12r-4dbr   14s
2dr-4c-4dbr
C&NW 1, Chicago (12:01p Sun) to Omaha (9p)
UP 1, Omaha (9:25p) to Los Angeles (10:40a Tue)

MINNEAPOLIS/ST. PAUL to LOS ANGELES
6s-6r-4dbr
C&NW 203, Minneapolis (11:40a Sun)
       and St. Paul (12:10p) to Omaha (8:40p)
UP 1, Omaha (9:25p) to Los Angeles (10:40p Tue)

OMAHA to LOS ANGELES
14s
UP 1, Omaha (9:25p Sun)
       to Los Angeles (10:40p Tue)

NEW YORK to LOS ANGELES
10r-6dbr
PRR 49, New York (5p Sat) to Chicago (8:20a Sun)
As above, Chicago (12:01p)
       to Los Angeles (10:40a Tue)

NEW YORK to LOS ANGELES
10r-6dbr
NYC 65, New York (3:30p Sat)
       to Chicago (7:30a Sun)
As above, Chicago (12:01p Sun)
       to Los Angeles (10:40a Tue)

*(Diner and first-class lounge, New York to Chicago
and Chicago to Los Angeles; cafe lounge Minneapolis to Omaha. Both PRR 49, the General, and
NYC 65, the Advance Commodore Vanderbilt, are
all-Pullman trains. Another westbound Los Angeles Limited is in Nevada.)*

Los Angeles Limited, No. 2
Los Angeles to Chicago
Union Pacific
66 miles west of Grand Island

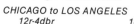

LOS ANGELES to CHICAGO
2dr-4c-4dbr   12r-4dbr
14s
UP 2, Los Angeles (12:01p Sat)
       to Omaha (3:50a Mon)
C&NW 2, Omaha (4:30a) to Chicago (2p Mon)

LOS ANGELES to NEW YORK
10r-6dbr
As above, Los Angeles (12:01p Sat)
       to Chicago (2p Mon)
NYC 22, Chicago (6:30p) to New York (12:30p Tue)

LOS ANGELES to NEW YORK
10r-6dbr
As above, Los Angeles (12:01p Sat)
       to Chicago (2p Mon)
PRR 2, Chicago (6:30p) to New York (11:59a Tue)

LOS ANGELES to ST. PAUL/MINNEAPOLIS
6s-6r-4dbr
UP 2, Los Angeles (12:01p Sat)
       to Omaha (3:50a Mon)
C&NW 204, Omaha (9:25a) to St. Paul (6:15p Mon)
       and Minneapolis (6:50p)

LOS ANGELES to OMAHA
14s
UP 2, Los Angeles (12:01p Sat)
       to Omaha (3:50a Mon)

*(Dining cars and lounges, Los Angeles to Chicago
and Chicago to New York; cafe lounge Omaha to
Minneapolis. NYC 22 is the Lake Shore Limited,
and PRR 2 is the Pennsylvania Limited. Another
Los Angeles Limited is in Utah.)*

Coloradoan, No. 6
Denver to Chicago
Burlington
47 miles west of McCook

DENVER to OMAHA
12s-dr
CB&Q 6, Denver (8:45p) to Omaha (8:30a)

(Lincoln to Chicago, dining-parlor car)

The Burlington formerly carried a Peoria sleeper from Omaha over this line.

Rocky Mountain Rocket, No. 7
Streamliner
Chicago to Colorado
Rock Island
16 miles east of Lincoln

CHICAGO to DENVER
70: 5dbr-obs-buffet-lounge   72:10s-4r
RI 7, Chicago (1:55p) to Denver (8:25a)

CHICAGO to COLORADO SPRINGS
73: 8s-2br-2c
RI 7, Chicago (1:55p) to Limon, Colo. (6:31a)
RI 7, Limon (6:51a) to Colorado Springs (8:35a)

(Dining car, Chicago to Denver, club diner to Colorado Springs. At Limon, a convertible "B" diesel unit became the locomotive for the Colorado Springs section, and the remainder of the train darts for Denver on UP rails.)

Although it didn't come into being until 1939, three years after the *Denver Zephyr* and *City of Denver*, the silver, red, and maroon *Rocky Mountain Rocket* had its own loyal clientele, thanks partly to its Colorado Springs service. Despite the *Rocket's* later start, the train itself was a seasoned traveler on the route, dating to its birth in 1898 as the *Rocky Mountain Limited*.

Rocky Mountain Rocket, No. 8
Colorado to Chicago
Rock Island
Leaving Omaha

DENVER to CHICAGO
80: 5dbr-obs.-buffet-lounge   82: 10s-4r
RI 8, Denver (1p) to Chicago (8:50a)

COLORADO SPRINGS to CHICAGO
83: 8s-2br-2c
RI 8, Colorado Springs (12:45p) to Limon (2:25p)
RI 8, Limon (2:40p) to Chicago (8:50a)

OMAHA to CHICAGO
81: 4dbr-6r-8dup/r
RI 8, Omaha (11:59p) to Chicago (8:50a)

(At Limon, the procedure for the westbound train is reversed: The Colorado Springs locomotive becomes a B unit for the combined train. Denver to Chicago, dining car; Colorado Springs to Chicago, club diner.)

The Rock Island scheduled a dedicated sleeper to Chicago from Lincoln. The predecessor *Rocky Mountain Limited* carried Pullmans from Omaha to Des Moines; other trains had cars to Rock Island, Ill., and Denver.

Denver Zephyr, No. 10
Streamliner
Denver to Chicago
Burlington
Entering Lincoln

DENVER to CHICAGO
Total accommodations:
36s-10dbr-4c-2dr-4 chambrettes-4r
CB&Q 10, Denver (4p) to Chicago (9:05a)

(Denver to Chicago: dining car, parlor-lounge, lounge)

Laden with head-end cars, the *Utahn* labors up Sherman Hill, Wyo., in August 1947.

Idahoan, No. 11
Omaha to Portland
Union Pacific
49 miles east of Grand Island

OMAHA to DENVER
12s-2dbr
UP 11, Omaha (9:45p) to North Platte (4:10a CST)
UP 85, North Platte (3:35a MST) to Denver (9:15a)

(North Platte to Denver, cafe-lounge)

Idahoan, No. 12
Portland to Omaha
Union Pacific
in North Platte

DENVER to OMAHA
12s-2dbr
UP 86, Denver (6p) to North Platte (11:15p MST)
UP 12, North Platte (12:45a CST) To Omaha (7a)

(Denver to North Platte, cafe-lounge)

California Zephyr, No. 17
Vista Dome streamliner
Chicago to San Francisco
Burlington
Leaving Omaha

CHICAGO to SAN FRANCISCO
CZ10: Obs-dr-3dbr   CZ12: 16s
CZ14: 6dbr-10r   CZ15: 6dbr-10r
CB&Q 17, Chicago (3:30p Sun)
　　to Denver (8:20a Mon)
D&RGW 17, Denver (8:40a)
　　to Salt Lake City (10:20p)
WP 17, Salt Lake City (10:40p)
　　to Oakland (4:15p Tue)
Ferry to San Francisco (4:50p)

NEW YORK to SAN FRANCISCO
CZ11: 6dbr-10r, alternating days between:
NYC 19, New York (6:30p Sat)
　　to Chicago (12:30p Sun)
PRR 1, New York (6:45p Sat)
　　to Chicago (10:25a Sun)
As above, Chicago (3:30p)
　　to San Francisco (4:50p Tue)

(Chicago to Oakland, via Moffat Tunnel route: Vista Dome observation-lounge with shower, Vista Dome buffet-lounge, dining car; New York to Chicago, dining cars and lounges. NYC 19 is the Lake Shore Limited, and PRR 1 is the Pennsylvania Limited.)

No. 13
Omaha to Rapid City
Chicago & North Western
14 miles west of Omaha

OMAHA to CHADRON
14s
C&NW 13, Omaha (11:30p) to Chadron (1:50p)

Omaha had dedicated Pullman service on this line through Chadron to Casper and Lander, Wyo., and Deadwood, S.D.

No. 14
Rapid City to Omaha
Chicago & North Western
6 miles west of O'Neill

CHADRON to OMAHA
14s
C&NW 14, Chadron (3:55p) to Omaha (6:30a)

California Zephyr, No. 18
Vista Dome streamliner
San Francisco to Chicago
Burlington
Standing in McCook

SAN FRANCISCO to CHICAGO
CZ10: Observation-dr-3dbr   CZ12: 16s
CZ14: 6dbr-10r   CZ15: 6dbr-10r
Ferry from San Francisco (9a Sat)
WP 18, Oakland pier (9:26a)
　　to Salt Lake City (5:20a Sun)
D&RGW 18, Salt Lake City (5:40a) to Denver (7p)
CB&Q 18, Denver (7:15p) to Chicago (1:30p Mon)

SAN FRANCISCO to NEW YORK
CZ11: 6dbr-10r
As above, San Francisco (9a Sat)
　　to Chicago (1:30p Mon)
Alternating days from Chicago between:
PRR 70, Chicago (5:30p) to New York (11:45a Tue)
NYC 22, Chicago (6:30p) to New York (12:30p Tue)

(Oakland to Chicago, via Moffat Tunnel route: Vista Dome observation-lounge with shower, Vista Dome buffet-lounge, dining car; Chicago to New York, dining cars and lounges. PRR 70 is the Admiral and NYC 22 the Lake Shore Limited. The other east- and westbound Zephyrs are now in Nevada.)

Nebraska   (continued)

Coloradoan, No. 19
Chicago to Denver
Burlington
Standing in Lincoln

OMAHA to DENVER
12s-dr
CB&Q 19, Omaha (10:20p) to Denver (8:10a)

No. 22
Omaha to Kansas City
Burlington
Leaving Omaha

OMAHA to KANSAS CITY
Std: s
CB&Q 22, Omaha (11:45p) to Kansas City (6:55a)

San Francisco Overland, No. 28
San Francisco to Chicago
Union Pacific
15 miles west of Grand Island

SAN FRANCISCO to CHICAGO
10r-6dbr   2dr-4c-4dbr
6s-6r-4dbr
Ferry from San Francisco (11a Sat)
SP 28, Oakland pier (11:30a) to Ogden (6:25a Sun)
UP 28, Ogden (6:55a) to Omaha (3:20a Mon)
C&NW 28, Omaha (3:45a) to Chicago (1p Mon)

SAN FRANCISCO to NEW YORK
10r-6dbr
As above, San Francisco (11a Sat)
          to Chicago (1p Mon)
Alternating days from Chicago between:
PRR 70, Chicago (5:30p) to New York (11:45a Tue)
NYC 22, Chicago (6:30p) to New York (12:30p Tue)

WEST YELLOWSTONE, MONT. to CHICAGO
10s-dr-2c   summer only
UP 36, West Yellowstone (7:30p Sat)
          to Pocatello (1:35a)
UP 12, Pocatello (4:10a Sun)
          to Green River, Wyo. (9:50a)
UP 28, Green River (10:15a) to Omaha (2:50a Mon)
C&NW 28, Omaha (3:15a)
          to Chicago (12:30p Mon)

(Oakland to Chicago, Pocatello to Green River: dining cars and first-class lounges; Chicago to New York: diners and lounges on both the Pennsylvania's Admiral, *No. 70, and New York Central's* Lake Shore Limited, *No. 22. The other eastbound* Overland *is in Nevada.)*

No. 42
Billings, Mont., to Lincoln
Burlington
96 miles east of Alliance

GREAT FALLS, MONT., to OMAHA
12s-dr
GN 42, Great Falls (12:15a Sun) to Billings (7a)
CB&Q 42, Billings (9a) to Lincoln (6:40a Mon)
CB&Q 20, Lincoln (7:30a) to Omaha (8:30a Mon)

CASPER, WYO. to OMAHA
10s-2dbr-c
CB&Q 32, Casper (3:30p) to Alliance (9:35p)
CB&Q 42, Alliance (10:15p) to Lincoln (6:40a)
CB&Q 20, Lincoln (7:30a) to Omaha (8:30a)

SHERIDAN, WYO., to CHICAGO Summertime only
Std. s-c-dr
CB&Q 42, Sheridan (12:50p) to Lincoln (7a)
CB&Q 6, Lincoln (7:45a) to Chicago (9:30p)

(Diner, Billings to Omaha; cafe coach, Casper to Alliance. No. 20 is the Burlington's Silver Streak Zephyr.)*

No. 43
Lincoln to Billings
Burlington
Standing in Lincoln

KANSAS CITY to GREAT FALLS
B430: 12s-dr
CB&Q 21, Kansas City (4p Sun)
          to Lincoln (10:15p)
CB&Q 43, Lincoln (12:05a Mon) to Billings (8p)
GN 43, Billings (11:30p) to Great Falls (6:35a Tue)

OMAHA to GREAT FALLS
12s-dr
CB&Q 19, Omaha (10:20p Sun) to Lincoln (11:20p)
As above, Lincoln (12:05a Mon)
          to Great Falls (6:35a Tue)

(Inasmuch as GN schedules don't acknowledge this and the Burlington's are ambiguous, its seems likely this car terminated in Alliance, becoming an eastbound sleeper to Kansas City on No. 44)

OMAHA to CASPER, WYO.
10s-2dbr-c
CB&Q 10, Omaha (10:20p) to Lincoln (11:20p)
CB&Q 43, Lincoln (12:05a) to Alliance (6:55a)
CB&Q 31, Alliance (7:20a) to Casper (1:40p)

CHICAGO to SHERIDAN   Summertime only
Std. s-c-dr
CB&Q 19, Chicago (11a) to Lincoln (11:20p)
CB&Q 43, Lincoln (12:05a) to Sheridan (4:15p)

(Lincoln to Billings, diner. The Silver Streak Zephyr, *CB&Q 21, also offered dining service on the Kansas City-Lincoln leg. CB&Q 19 is the* Coloradoan.)*

Burlington's Nos. 42 and 43 over the years provided the railroad's northwest connection to its parental roads from the cities of the lower Missouri River valley. Early in the century, through cars moved to Seattle and Tacoma; summertime sleepers to Yellowstone and Glacier National Parks were common quite late in the game. Always the connections to through trains were there in Montana, at Billings for the NP and Great Falls or Shelby for the GN.

The Burlington dominated the passenger rail scene in Lincoln, which formerly had its own Pullman service west to Alliance, Deadwood, Edgemont, Cody, Billings, and Seattle (GN or NP); and east to Kansas City and Peoria.

No. 44
Alliance-Lincoln-Kansas City
Burlington
Standing in Lincoln

ALLIANCE to KANSAS CITY
12s-dr
CB&Q 44, Alliance (10:30a) to Lincoln (8p)
CB&Q 44, Lincoln (12:15a) to Kansas City (6:55a)

The Burlington formerly scheduled overnight accommodations from Alliance to Billings, Denver, Lincoln, Omaha, Chicago, and St. Louis.

City of San Francisco, No. 102
Extra-fare streamliner
San Francisco to Chicago
Union Pacific
48 miles west of Grand Island

SAN FRANCISCO to CHICAGO Typical consist:
4c-3dr   5dbr-12dup/sbr
12s   4dbr-4c-2dr
10r-6dbr   6s-6r-4br
Ferry from San Francisco (5p Sat)
SP 102, Oakland pier (5:27p) to Ogden (9:40a Sun)
UP 102, Ogden (9:55a) to Omaha (2:50a Mon)
C&NW 102, Omaha (3a) to Chicago (10:45a Mon)

(Oakland to Chicago, diner and first-class lounge. The other eastbound City of San Francisco *is in Nevada.)*

City of Los Angeles, No. 104
Extra-fare streamliner
Los Angeles to Chicago
Union Pacific
37 miles west of Grand Island

LOS ANGELES to CHICAGO   Range of consist:
11dbr   12s
10r-6dbr   18r
13r   4dbr-4c-2dr
4c-3dr   6s-6r-4br
2dr-dbr-c-obs.-lounge   4dbr-obs.-lounge
UP 104, Los Angeles (11p Sat)
          to Omaha (2:40a Mon)
C&NW 104, Omaha (2:50a)
          to Chicago (10:45a Mon)

(Los Angeles to Chicago: dining car, first-class lounge with bath, and possible observation-buffet-lounge, depending on sleeper consist. Another City of Los Angeles, *24 hours behind this edition, is now in Nevada.)*

City of Portland, No. 106
Streamliner
Portland to Chicago
Union Pacific
60 miles west of Grand Island

PORTLAND to CHICAGO
2dr-4c-4dbr   6s-6r-4dbr
6s-6r-4dbr
UP 106, Portland (5:30p Sat) to Omaha (3a Mon)
C&NW 106 Omaha (3:10a)
          to Chicago (11:30a Mon)

SEATTLE to CHICAGO
12r-4dbr
UP 408, Seattle (12:30p Sat) to Portland (4:30p)
As above, Portland (5:30p)
          to Chicago (11:30a Mon)

(Seattle to Portland, diner; Portland to Chicago, diner and first-class lounge with bath. Another eastbound City of Portland *is passing its first midnight in Oregon.)*

City of Denver, No. 112
Streamliner
Denver to Chicago
Union Pacific
39 miles east of Grand Island

DENVER to CHICAGO
Observation-6dbr   12s
8s-3dbr   dr-3c-4dbr-4r
UP 112, Denver (4:30p) to Omaha (1:35a)
C&NW 112, Omaha (1:45a) to Chicago (9:35a)

(Denver to Chicago: dining car, first-class lounge)

The *City of Denver* became the UP's challenger in the competitive Chicago-Denver market when inaugurated in 1936. Its rustic Frontier Shack lounge was one of the West's more popular mile-and-a-quarter-a-minute saloons.

Nevada

See map on page 213

Los Angeles Limited, No. 1
Chicago to Los Angeles
Union Pacific
South of Caliente

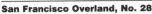

CHICAGO to LOS ANGELES
2dr-4c-4dbr   12r-4dbr
14s
C&NW 1, Chicago (12:01p Sat) to Omaha (9p)
UP 1, Omaha (9:25p) to Los Angeles (10:40a Mon)

MINNEAPOLIS to LOS ANGELES
6s-6r-4dbr
C&NW 203, Minneapolis (11:40a Sat) and
    St. Paul (12:10p) to Omaha (8:40p)
UP 1, Omaha (9:25p) to Los Angeles (10:40a Mon)

OMAHA to LOS ANGELES
14s
UP 1, Omaha (9:25p Sat)
    to Los Angeles (10:40a Mon)

NEW YORK to LOS ANGELES
10r-6dbr
PRR 49, New York (5p Fri) to Chicago (8:20a Sat)
As above, Chicago (12:01p)
    to Los Angeles (10:40a Mon)

NEW YORK to LOS ANGELES
10r-6dbr
NYC 65, New York (3:30p Fri)
    to Chicago (7:30a Sat)
As above, Chicago (12:01p)
    to Los Angeles (10:40a Mon)

(New York to Chicago: dining car, lounge. Minneapolis to Omaha: cafe-lounge. Chicago to Los Angeles: first-class lounge with bath and radio, dining car. Both of the eastern trains are all-Pullman: the New York Central's Advance Commodore Vanderbilt and the Pennsylvania's General. Another Los Angeles Limited is 24 hours behind this, in Nebraska.)

Until the coming of the *City* streamliners, the *Los Angeles Limited* was the premier train on the UP's Los Angeles & Salt Lake line.

California Zephyr, No. 17
Vista Dome streamliner
Chicago to San Francisco
Western Pacific
At Utah state line

CHICAGO to SAN FRANCISCO
CZ10: 3dbr-dr-Vista Dome-obs.-lounge
CZ12: 16s   CZ14: 6dbr-10r
CZ15: 6dbr-10r
CB&Q 17, Chicago (3:30p Sat)
    to Denver (8:20a Sun)
D&RGW 17, Denver (8:40a)
    to Salt Lake City (10:20p)
WP 17, Salt Lake City (10:40p)
    to Oakland (4:15p Mon)
Ferry to San Francisco (4:50p Mon)

NEW YORK to SAN FRANCISCO
CZ11: 6dbr-10r
    Alternate days to Chicago between:
PRR 1, New York (6:45p Fri)
    to Chicago (10:25a Sat)
NYC 19, New York (6:30p Fri)
    to Chicago (12:30p Sat)
As above, Chicago (3:30p)
    to San Francisco (4:50p Mon)

(Chicago to Oakland: dining car, Vista dome buffet club car. New York to Chicago: diner, lounge. The New York Central's Lake Shore Limited and the Pennsy's Pennsylvania Limited deliver the New York Pullmans to Chicago.)

California Zephyr, No. 18
Vista Dome streamliner
San Francisco to Chicago
Western Pacific
15 miles west of Wells

SAN FRANCISCO to CHICAGO
CZ10: 3dbr-dr-Vista Dome-obs.-lounge
CZ12: 16s   CZ14: 6dbr-10r
CZ15: 6dbr-10r
Ferry from San Francisco (9a Sun)
WP 18, Oakland (9:26a)
    to Salt Lake City (5:20a Mon)
D&RGW 18, Salt Lake City (5:40a) to Denver (7p)
CB&Q 18, Denver (7:15p) to Chicago (1:30p Tue)

SAN FRANCISCO to NEW YORK
CZ11: 6dbr-10r
As above, San Francisco (9a Sun)
    to Chicago (1:30p Tue)
Alternate days from Chicago:
NYC 22, Chicago (6:30p)
    to New York (12:30p Wed)
PRR 70, Chicago (5:30p)
    to New York (11:45a Wed)

(Oakland to Chicago: dining car, Vista Dome buffet club car; Chicago to New York: dining car, lounge on both the NYC's Lake Shore Limited and PRR's Admiral. The other east- and westbound Zephyrs are now in Nebraska.)

The *Zephyr* won justifiable acclaim for being the first western train to exploit the benefits of the dome car. Thirty Vista Dome cars went into the *Zephyr's* six consists, generally three in each train for coach passengers, and one car and the dome portion of another reserved exclusively for those in Pullmans.

San Francisco Overland, No. 27
Chicago to San Francisco
Southern Pacific
15 miles west of Wells

CHICAGO to SAN FRANCISCO
4dbr-4c-2dr   6s-6r-4dbr
10r-6dbr
C&NW 27, Chicago (3:30p Sat)
    to Omaha (12:30a Sun)
UP 27, Omaha (12:55a) to Ogden (8:33p)
SP 27, Ogden (9p) to Oakland (1:45p Mon)
Ferry to San Francisco (2:20p Mon)

NEW YORK to SAN FRANCISCO
10r-6dbr
    Alternate days to Chicago:
NYC 17, New York (6:15p Fri)
    to Chicago (11:15a Sat)
PRR 1, New York (6:45p Fri)
    to Chicago (10:25a Sat)
As above, Chicago (3:30p)
    to San Francisco (2:20p Mon)

ST. LOUIS to SAN FRANCISCO
6s-6r-4dbr
Wabash 9, St. Louis (4p Sat) to Kansas City (9p)
UP 9, Kansas City (9:30p)
    to Green River, Wyo. (3:35p)
UP 27, Green River (4:40p Sun) to Ogden (8:33p)
As above, Ogden (9p)
    to San Francisco (2:20p Mon)

SALT LAKE CITY to SAN FRANCISCO
6s-6r-4dbr
UP 29, Salt Lake City (8p) to Ogden (8:50p)
As above, Ogden (9p) to San Francisco (2:20p)

(New York to Chicago, Chicago to Oakland, and St. Louis to Green River: dining car, lounge. East of Chicago, the New York Central's Wolverine and the PRR's Pennsylvania Limited handle the accommodations. The other westbound Overland is in Iowa.)

The origins of the *Overland Limited*, as this venerable train was known until fairly late in its career, go back to 1887.

San Francisco Overland, No. 28
San Francisco to Chicago
Southern Pacific
80 miles west of Wells

SAN FRANCISCO to CHICAGO
4dbr-4c-2dr   6s-6r-4dbr
10r-6dbr
Ferry from San Francisco (11a Sun)
SP 28, Oakland (11:30a Sun) to Ogden (6:25a Mon)
UP 28, Ogden (6:55a) to Omaha (3:20a Tue)
C&NW 28, Omaha (3:45a) to Chicago (1p Tue)

SAN FRANCISCO to NEW YORK
10r-6dbr
As above, San Francisco (11a Sun)
    to Chicago (1p Tue)
Alternate days to New York:
PRR 70, Chicago (5:30p)
    to New York (11:45a Wed)
NYC 22, Chicago (6:30p) to
    New York (12:30p Wed)

SAN FRANCISCO to ST. LOUIS
6s-6r-4dbr
As above, San Francisco (11a Sun)
    to Ogden (6:25a Mon)
UP 10, Ogden (6:50a) to Kansas City (6:45a Tue)
Wabash 10, Kansas City (7a)
    to St. Louis (11:59a Tue)

SAN FRANCISCO to SALT LAKE CITY
6s-6r-4dbr
As above, San Francisco (11a Sun)
    to Ogden (6:25a Mon)
UP 30, Ogden (6:40a) to Salt Lake City (7:35a)

(Oakland to Chicago, Ogden to St. Louis, and Chicago to New York: dining car, lounge. The New York connections are the Pennsylvania's Admiral and the New York Central's Lake Shore Limited. Another eastbound Overland is in Nebraska.)

Shortly after the turn of the century, the *Overland Limited* was offering in-station telephone service from its observation car. A uniformed attendant placed the calls.

City of San Francisco, No. 101
Extra-fare streamliner
Chicago to San Francisco
Southern Pacific
28 miles west of Winnemucca

CHICAGO to SAN FRANCISCO   Consist varies:
4c-3dr   12s
10r-6dbr   5dbr-12 dup/sb
4dbr-4c-2dr   6s-6r-4dbr
C&NW 101, Chicago (7p Sat)
    to Omaha (2:45a Sun)
UP 101, Omaha (2:55a) to Ogden (6:05p)
SP 101, Ogden (6:22p) to Oakland (8:40a Mon)
Ferry to San Francisco (9:15a Mon)

(Dining car and first-class lounge with radio, Chicago to Oakland. Another westbound City of San Francisco is now in Iowa, 24 hours behind this one.)

On January 13, 1952, the *City of San Francisco* became snowbound in the Sierra Nevada, and the heroic three-day rescue efforts were reported on newscasts nationwide. Six of the more than 200 passengers suffered heart attacks, and others were overcome by carbon monoxide fumes emitted by propane engines underneath the cars. A snowplow fighting its way to the scene was struck by an avalanche, and the engineer buried alive. Passengers were finally evacuated to a lodge on January 16 to await a rescue train.

Crossing the SP's Great Salt Lake fill in 1968, the eastbound *City of San Francisco* meets a waiting freight near Promontory Point.

Nevada
(continued)

City of San Francisco, No. 102
Extra-fare streamliner
San Francisco to Chicago
Southern Pacific
8 miles east of Reno

SAN FRANCISCO to CHICAGO   Consist varies:

4c-3dr	12s
10r-6dbr	5dbr-12 dup/sb
4dbr-4c-2dr	6s-6r-4dbr

Ferry from San Francisco (5p Sun)
SP 102, Oakland (5:27p) to Ogden (9:40a Mon)
UP 102, Ogden (9:55a) to Omaha (2:50a Tue)
C&NW 102, Omaha (3a) to Chicago (10:45a Tue)

(Oakland to Chicago: diner and first-class lounge. Another eastbound City of San Francisco *is now in Nebraska.)*

City of Los Angeles, No. 104
Extra-fare streamliner
Los Angeles to Chicago
Union Pacific
4 miles east of Las Vegas

LOS ANGELES to CHICAGO   Consist varies:

11dbr	12s
10r-6dbr	18r
13r	4dbr-4c-2dr
4c-3dr	6s-6r-4br
2dr-c-dbr-obs.-lounge	4dbr-obs.-lounge
Obs.-buffet-lounge	

UP 104, Los Angeles (5p Sun)
      to Omaha (2:40a Tue)
C&NW 104, Omaha (2:50a)
      to Chicago (10:45a Tue)

(Los Angeles to Chicago: diner, first-class lounge. An earlier edition of the City of Los Angeles *is now in Nebraska, 24 hours ahead of this train.)*

City of Los Angeles, No. 103
Extra-fare streamliner
Chicago to Los Angeles
Union Pacific
South of Caliente

11dbr	12s
10r-6dbr	18r
13r	4dbr-4c-2dr
4c-3dr	6s-6r-4br
2dr-c-dbr-obs.-lounge	4dbr-obs.-lounge
Obs.-buffet-lounge	

C&NW 103, Chicago (7:15p Sat)
      to Omaha (3a Sun)
UP 103, Omaha (3:10a) to Los Angeles (9a Mon)

(Dining car, first-class lounge with bath, Chicago to Los Angeles. The City of Los Angeles *that will be near Caliente on the following midnight is now in Iowa.)*

The *City of Los Angeles* made its debut in May 1936, and late the next year a new seventeen-car train-set was added. Post-war equipment deliveries in 1947 gave the *City* the distinction of being the first 39¾-hour Chicago-Los Angeles train to go into daily operation.

New Hampshire
See map on page **14**

Red Wing/New Englander, No. 325
Boston To Montreal
Boston & Maine
29 miles southeast of White River Jct.

BOSTON to MONTREAL
    CP6: 4dbr-c-buffet-lounge   CP7: 12s-2dbr
B&M 325, Boston (9p) to White River Jct. (12:55a)
B&M, White River Jct. (1:45a) to Wells River (3a)
CPR 209, Wells River (3:10a) to Montreal (8:30a)

BOSTON to MONTREAL
    CN1: 12s-2dbr
B&M 325, Boston (9p) to White River Jct. (12:55a)
CV 325-21, White River Jct. (3:22a)
      to St. Albans (6:02a)
CNR 325-21, St. Albans (6:08a) to Montreal (8:15a)

BOSTON to ST. ALBANS, VT. Winter only
    V1: 10s-2dbr-c
B&M 325, Boston (9p) to White River Jct. (12:55a)
CV 303, White River Jct. (4:55a) to St. Albans (9a)

(The Canadian Pacific's Red Wing *and Canadian National* New Englander *run together to White River Jct. The latter carries a buffet-lounge.)*

The *Red Wing*, introduced in 1926, carried Quebec sleepers also. The *New Englander* included Pullmans for Burlington in its consists.

New Jersey

See map on page **45**

Pacific Express, No. 7
New York to Chicago
Erie
Standing in Jersey City

NEW YORK to HORNELL, N.Y.
  10s-dr-2dbr
  Ferry from New York
  Erie 7, Jersey City (12:55a) to Hornell (10:18a)

(Binghamton to Marion, diner-lounge)

The *Pacific Express* carried New York sleepers to Binghamton, Buffalo, and Chicago and picked up an Albany-Hornell car from the D&H. Other service on the Erie from Jersey City included Meadville, Bradford, and Oil City, Pa.; and Marion and Cincinnati, Ohio. The Erie turned around some of its Pullmans in Hornell rather than sending them on a long day trip to Chicago.

The Star, No. 11
New York to Buffalo
Lehigh Valley
Departing Newark

NEW YORK to ITHACA
  55: s-dbr-dr
  LV 11, New York (11:45p) to Ithaca (7:56a)

NEW YORK to BUFFALO
  17: s-dbr-dr
  LV 11, New York (11:45p) to Buffalo (10:45a)

(New York to Buffalo, cafe-lounge)

The *Star* handled a dedicated sleeper from New York to Wilkes-Barre.

Metropolitan Special, No. 11
New York to St. Louis
Baltimore & Ohio
Standing in Jersey City

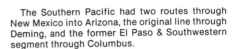

NEW YORK to BALTIMORE
  8s-5dbr
  Ferry or bus from New York
  B&O 11, Jersey City (12:50a) to Baltimore (5:55a)

NEW YORK to WASHINGTON
  12s-dr   8s-5dbr
  B&O 11 Jersey City (12:50a) to Washington (7a)

NEW YORK to ST. LOUIS
  12s-dr
  B&O 11, Jersey City (12:50a Mon)
        to St. Louis (7:40a Tue)

(New York to Washington, lunch counter and lounge; Washington to Cincinnati, dining car. Baltimore sleeper parked in Camden Station.)

The B&O scheduled a Newark-Washington sleeper on this line.

Owl, No. 15
New York to Buffalo
Via Paterson
Lackawanna
Standing in Hoboken

NEW YORK to BINGHAMTON
  151: 10r-6dbr
  Ferry from New York
  DL&W 15, Hoboken (1:05a) to Binghamton (6:20a)

NEW YORK to ELMIRA
  150: 10r-6dbr
  DL&W 15, Hoboken (1:05a) to Elmira (8:55a)

NEW YORK to SYRACUSE
  152: 10s-dr-2dbr
  DL&W 15, Hoboken (1:05a) to Binghamton (6:20a)
  DL&W 1915, Binghamton (8a)
        to Syracuse (10:15a)

(Diner-lounges: Binghamton to Buffalo, Binghamton to Syracuse)

Iron City Express, No. 37
New York to Pittsburgh
Pennsylvania Railroad
27 miles south of Newark

SPRINGFIELD to PITTSBURGH
  8s-buffet-lounge
  NH 59, Springfield (7p) to New Haven (8:26p)
  NH 189, New Haven (9:15p) to New York (11p)
  PRR 37, New York (11:20p) to Pittsburgh (7:50a)
  (Mon-Fri; Sunday schedule different)

BOSTON to PITTSBURGH
  12r-2sbr-3dbr   12s-dr
  NH 189, Boston (6p) to New York (11p)
  As above, New York (11:20p) to Pittsburgh (7:50a)

NEW YORK to PITTSBURGH
  6dbr-lounge   12s-dr
  12s-dr   12s-dr
  21r   12dup/r-4dbr (Sat)
  PRR 37, New York (11:20p) to Pittsburgh (7:50a)

NEW YORK to BIRMINGHAM
  10s-2c-dr
  PRR 37, New York (11:20p Sun)
        to Pittsburgh (7:50a Mon)
  PRR 205, Pittsburgh (10:15a) to Cincinnati (5:40p)
  L&N 1, Cincinnati (7:05p)
        to Birmingham (7:20a Tue)

(Dining cars: No. 189, the William Penn, *Boston to New York; No. 205, Pittsburgh to Columbus; No. 1, the* Azalean, *Cincinnati to Louisville)*

The *Iron City Express* also carried Boston sleepers for Cincinnati and St. Louis.

Gotham Limited, No. 55
New York to Chicago
Pennsylvania Railroad
Departing Newark

NEW YORK to CHICAGO
  5dbr-lounge   21r
  12s-dr
  PRR 55, New York (11:40p) to Chicago (3:40p)

(Diner: New York to Chicago; not open leaving New York)

New Mexico

See map on page **259**

The Southern Pacific had two routes through New Mexico into Arizona, the original line through Deming, and the former El Paso & Southwestern segment through Columbus.

Sunset Limited, No. 1
Streamliner, via Columbus
New Orleans to Los Angeles
Southern Pacific
20 miles west of El Paso

NEW ORLEANS to LOS ANGELES
  15: 10r-6dbr   16: 10r-6dbr
  17: 10r-6dbr   19: 10r-6dbr
  SP 1, New Orleans (12:30a Sun)
        to Los Angeles (4:30p Mon)

DALLAS to LOS ANGELES
  14: 10r-6dbr
  T&P 1, Dallas (7:50a Sun) to El Paso (10:15p)
  SP 1, El Paso (11:40p) to Los Angeles (4:30p Mon)

(New Orleans to Los Angeles: dining car, first-class lounge with shower; Dallas to El Paso, on T&P 1, the Texas Eagle, *dining-lounge car. Another edition of the westbound* Sunset Limited *is now standing in New Orleans, awaiting departure.)*

Texas Zephyr, No. 2
Streamliner
Dallas to Denver
Burlington (FtW&D/C&S)
40 miles northwest of Clayton

DALLAS/FORT WORTH to DENVER
  27: Std: s-dbr   28: Std: s-dr-dbr
  29: Std: s-dr
  FtW&D 2, Dallas (1p), Fort Worth (2p)
        to Denver (7a)

(Diner-lounge, Dallas to Denver)

California Limited, No. 4
Los Angeles to Chicago
Via Raton Pass, Topeka
Atchison, Topeka & Santa Fe
Standing in Las Vegas

LOS ANGELES to CHICAGO
  12s-dr   8s-dr-2c
  Santa Fe 4, Los Angeles (6:15p Sat)
        to Chicago (8:30a Tue)

SAN FRANCISCO to CHICAGO
  10s-dr-2c
  Bus from San Francisco (9a Sat)
  Santa Fe 4, Oakland (9:25a) to Barstow (10:10p)
  Santa Fe 4, Barstow (10:45p)
        to Chicago (8:30a Tue)

PHOENIX to DENVER
  8s-dr-2c
  Santa Fe 42, Phoenix (5p Sat) to Ash Fork (11:15p)
  Santa Fe 124, Ash Fork (2:10a Sun)
        to Albuquerque (10:25a)
  Santa Fe 4, Albuquerque (8a) to La Junta (6a Mon)
  Santa Fe 1, La Junta (8a) to Pueblo (9:45a)
  Santa Fe 102, Pueblo (9:55a) to Denver (1p Mon)

(Dining or cafe cars, all segments except west of La Junta on No. 4; lounge facilities on all trains except Oakland to Barstow; cafe-observation car, La Junta to Denver. The California Limited, *west of La Junta, makes meal stops at Fred Harvey dining rooms in Seligman and Winslow, Ariz., and Albuquerque. Although nominally from Phoenix, the Denver Pullman is more an accommodation for Albuquerque residents traveling to the Mile High City. Other eastbound editions of the* California Limited *are now in California and Missouri.)*

New Mexico   *(continued)*

No. 26
Carlsbad to Clovis
Atchison, Topeka & Santa Fe
27 miles south of Clovis

WAYNOKA, OKLA. to ALBUQUERQUE
  10s-dr-2c
Santa Fe 105, Waynoka (5:40p Sat)
    to Clovis (2a Sun)
Santa Fe 25, Clovis (3a) to Carlsbad (8:15a)
Santa Fe 26, Carlsbad (7:30p)
    to Clovis (12:45a Mon)
Santa Fe 105, Clovis (1:35a)
    to Albuquerque (7:45a Mon)

ALBUQUERQUE to WAYNOKA
  10s-dr-2c
Santa Fe 106, Albuquerque (9p Sat)
    to Clovis (2:35a Sun)
As above, Clovis (3a)
    to Carlsbad, Clovis (12:45a Mon)
Santa Fe 106, Clovis (4:15a)
    to Waynoka (12:30p Mon)

(Just as it does with its Grand Canyon operation, the Santa Fe shuttles east- and westbound sleepers, brought to Clovis on the Scout, *to this station near Carlsbad Caverns, and both cars lay over for the day while passengers sightsee. On the* Scout's *schedule, the Pullmans are conveniently turned around at Waynoka.)*

Waynoka is remembered as being one of the air/rail interchange points of the short-lived Transcontinental Air Transport's joint service with the Pennsylvania Railroad and the Santa Fe. After flying all day in 1929-vintage trimotors, westbound passengers landed at Waynoka to wait for the Santa Fe's *Missionary* and its overnight Pullman comforts, before embarking on their next flight the following day to the West Coast.

Pullmans from Albuquerque for Roswell traveled this line south from Clovis, as did a Kansas City-Carlsbad car.

Imperial, No. 39
Chicago to Los Angeles
Via Columbus, Phoenix, El Centro
Southern Pacific
Entering Carrizozo

CHICAGO to LOS ANGELES
  398: 6s-6dbr   391: 10s-dr-2c
RI 39, Chicago (8:15p Sat)
    to Tucumcari (8:26p CST Sun)
SP 39, Tucumcari (8:10p MST)
    to Los Angeles (11:30p Mon)

CHICAGO to PHOENIX   *Winter only*
  394: 12r-sbr-4dbr
RI 39, Chicago (8:15p Sat)
    to Tucumcari (8:26p CST Sun)
SP 39, Tucumcari (8:10p MST)
    to Phoenix (12:25p Mon)

MEMPHIS to LOS ANGELES
  1115: 10s-dr-2c
RI 111, Memphis (8p Sat)
    to Tucumcari (8p CST Sun)
SP 39, Tucumcari (8:10p MST)
    to Los Angeles (11:30p Mon)

LITTLE ROCK to LOS ANGELES
  1117: 8s-dr-2c
RI 111, Little Rock (11:35p Sat)
    to Tucumcari (8:10p MST)
SP 39, Tucumcari (8p CST Sun)
    to Los Angeles (11:30p Mon)

(Chicago to Los Angeles, dining car and first-class lounge. To Tucumcari on No. 111, the Cherokee: *lounge from Memphis, diner from Little Rock.)*

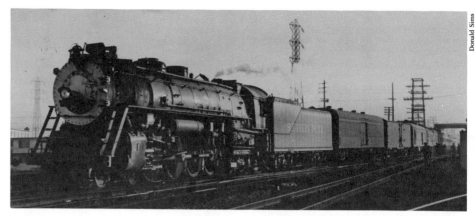

Southern Pacific's No. 43 from Tucumcari carried a Calexico sleeper to Los Angeles.

No. 13
Albuquerque to El Paso
Atchison, Topeka & Santa Fe
62 miles south of Belen

ALBUQUERQUE to EL PASO
  10s-dr-2c
Santa Fe 13, Albuquerque (9p) to El Paso (5:30a)

Service on this line once included sleepers to Deming and Silver City.

Imperial, No. 40
Los Angeles to Chicago
Via El Centro, Phoenix, Columbus
Southern Pacific
58 miles north of El Paso

LOS ANGELES to CHICAGO
  406: 6s-6dbr   405: 10s-dr-2c
SP 40, Los Angeles (11p Sat)
    to Tucumcari (5:50a MST Mon)
RI 40, Tucumcari (7:20a CST)
    to Chicago (8:40a Tue)

PHOENIX to CHICAGO   *Winter only*
  401: 12r-sbr-4dbr
SP 40, Phoenix (12:25p Sun)
    to Tucumcari (5:50a MST Mon)
RI 40, Tucumcari (7:20a CST)
    to Chicago (8:40a Tue)

LOS ANGELES to LITTLE ROCK
  403: 8s-dr-2c
SP 40, Los Angeles (11p Sat)
    to Tucumcari (5:50a MST Mon)
RI 112, Tucumcari (7:20a CST)
    to Little Rock (3a Tue)

LOS ANGELES to MEMPHIS
  404: 10s-dr-2c
SP 40, Los Angeles (11p Sat)
    to Tucumcari (5:50a MST Mon)
RI 112, Tucumcari (7:20a CST)
    to Memphis (6:50a Tue)

(Los Angeles to Chicago, diner and first-class lounge. From Tucumcari on No. 112, the Cherokee: *diner to Little Rock, lounge to Memphis. Two editions of the* Imperial, *both east- and westbound, are now in Iowa, and another, bound for Chicago, is still in California.)*

No. 14
El Paso to Albuquerque
Atchison, Topeka & Santa Fe
20 Miles north of El Paso

EL PASO to ALBUQUERQUE
  10s-dr-2c
Santa Fe 14, El Paso (11:30p)
    to Albuquerque (7:45a)

The Scout, No. 106
Albuquerque to Newton, Kans.
Atchison, Topeka & Santa Fe
10 miles west of Vaughn

LOS ANGELES to CHICAGO
  10s-dr-2c
Santa Fe 4, Los Angeles (6:15p Sat)
    to Albuquerque (6:45p Sun)
Santa Fe 13, Albuquerque (9p) to Belen (9:40p)
Santa Fe 106, Belen (9:55p)
    to Newton (5:25p Mon)
Santa Fe 6, Newton (6:10p)
    to Kansas City (10:15p)
Santa Fe 4, Kansas City (11p)
    to Chicago (8:30a Tue)

ALBUQUERQUE to WAYNOKA, OKLA.
  10s-dr-2c
Santa Fe 106, Albuquerque (9p Sun)
    to Clovis (2:35a Mon)
Santa Fe 25, Clovis (3a) to Carlsbad (8:15a)
Santa Fe 26, Carlsbad (7:30p)
    to Clovis (12:45a Tue)
Santa Fe 106, Clovis (4:15a)
    to Waynoka (12:30p Tue)

(Dining car on No. 106 east of Amarillo, after 7:15 a.m. Tuesday, and see California Limited, No. 4, *also on these pages. The* Scout *name applies to Nos. 13-106-6-4, even though it is a separate train only between Albuquerque and Newton. The* Scout *serves as the Santa Fe's southern route counterpart to the* California Limited *in much the same manner as the two Grand Canyon sections complement each other.)*

The *Scout* was inaugurated in early 1916, discontinued in 1931, but restored after a five-year hiatus. Working the southern route, it once operated all the way between Chicago and Los Angeles, carrying sleepers for Oklahoma and Texas cities and from New Orleans as well. The *Scout* also handled the Clovis-Waynoka Pullman that Transcontinental Air Transport passengers rode on the first night of their eastbound air/rail journey across the country. For a time the train also carried special cars reserved for women traveling with small children, prompting its description as a "rolling nursery" from Lucius Beebe, who probably never rode it.

Albuquerque had dedicated sleeping car service to Clovis and Amarillo on the southern route of the Santa Fe.

New York

See map on page **40**

Walter L. Greene's classic railroad painting *Eastward, Westward* (1929) showed the east- and westbound *Twentieth Century Limiteds* briefly pausing in the middle of the night at Buffalo's new Central Terminal. As the midpoint and nerve center of the New York Central system, Buffalo was naturally the stage for much nocturnal activity as the trains operating between East and Midwest called here all through the night. Either arriving, departing, or standing in Buffalo at this scheduling were the *Southwestern Limited* (No. 12), *Ohio State Limited* (15), both *New England States* (27 and 28), *Cleveland Limited* (58), and the *Buffalo-Pittsburgh Express* (279).

But with the Central's four-track main line, meetings of kindred flyers were taking place all night long across the Empire State. By the following schedules, no fewer than 28 name trains of the Great Steel Fleet were on the rails this midnight between the Pennsylvania line on the Erie lakeshore and the Hudson River valley at Albany.

From west to east, and apart from those just named, the others included:

West of Buffalo: *Advance Commodore Vanderbilt (66), Knickerbocker (41), Paul Revere (78), Ohio State Limited (16),* and *Advance Commodore Vanderbilt (65).*

Between Buffalo and Rochester: *Genesee (82), Detroiter (48), New Yorker (8),* and *Commodore Vanderbilt (67).*

Between Rochester and Syracuse: *Fifth Avenue Special (6), New England Wolverine (33), New York Special (44), Twentieth Century Limited (25), Interstate Express (46),* and *Wolverine (17).*

Between Syracuse and Albany: *Cayuga (34), Lake Shore Limited (19), Detroiter (47), Southwestern Limited (11), Cleveland Limited (57), North Star (21),* and *Easterner (52).*

It was a concentration like none other. Only the Pennsy hoisting its passenger cargoes over the Appalachians came close.

(Many eastern trains that primarily served business travelers operated six days a week, omitting Saturdays, when overnight traffic was generally light. Other trains operated all seven days but cut back the sleeping car consists considerably on Saturdays. A number of New York Central and Pennsylvania Railroad trains adhered to these patterns. Following are representative, full weekday schedules.)

Owl, No. 2
Pullman only
New York to Boston
New Haven Railroad
Standing in New York City

NEW YORK to BOSTON
Ten cars, Nos. 20-29, Std: *s-dr-c-r-sbr-dbr*
NH 2, New York (12:30a) to Boston (6:15a)

NEW YORK to PROVIDENCE
2000: Std. *s-dr-dbr-r*   2001: Std. *s-dr-dbr-r*
NH 2, New York (12:30a) to Providence (4:58a)

(New Haven trains operate from Grand Central Terminal, except for the through Washington-Boston PRR/NH trains, which use Pennsylvania Station)

For a brief time early in the century the New Haven experimented by furnishing sleeping spaces in the *Owl* with brass beds. Over the years, Pullmans departed New York on the New Haven for the resorts of Maine, Vermont, and New Hampshire and Canadian cities from Montreal to Halifax.

Pocono Express, No. 2
Buffalo to New York
Lackawanna
Standing in Buffalo

DETROIT to NEW YORK
440: 10s-3dbr
NYC 44, Detroit (5:10p) to Buffalo (9:50p)
DL&W 2, Buffalo (4:20a) to Hoboken (2:40p)
Ferry to New York (2:55p)

BUFFALO to NEW YORK
10s-3dbr
DL&W 2, Buffalo (4:20a) to Hoboken (2:40p)

(Diner, Detroit to Buffalo and Scranton to Hoboken; lounge, Detroit to Buffalo; diner-lounge, Elmira to Binghamton. NYC 44 is the New York Special.)

The DL&W scheduled a Buffalo-Scranton sleeper.

No. 4
Montreal to Utica
New York Central
26 miles north of Thendara

MONTREAL to BUFFALO
10s-buffet-lounge
NYC 4, Montreal (6:30p) to Utica (2:10a)
NYC 35, Utica (3:41a) to Buffalo (7:25a)

LAKE PLACID to NEW YORK
6dbr-buffet-lounge   8s-dr-3dbr
NYC 104, Lake Placid (9:35p)
        to Lake Clear Junction (10:15p)
NYC 4, Lake Clear Junction (10:32p)
        to Utica (2:10a)
NYC 44, Utica (2:30a) to New York (7:15a)

MALONE to NEW YORK   Summertime only
8s-dr-2c
NYC 4, Malone (8p) to Utica (1:20a)
NYC 44, Utica (2:30a) to New York (7:15a)

THENDARA to NEW YORK   Summer, Sun Only
8s-dr-2c
NYC 4, Thendara (11:52p) to Utica (1:20a)
NYC 44, Utica (2:30a) to New York (7:15a)

(No. 4 operates over the New York Central's Adirondack Division. No. 35 is the Iroquois, *and No. 44 the* New York Special. *The Central's trains to New York use Grand Central Terminal. More Lake Placid cars are added to summertime consists.)*

By these schedules, Lake Placid still had summertime sleepers to Chicago; previous service included Buffalo, Boston, and Philadelphia (PRR).

No. 4
Buffalo to New York
Lehigh Valley
7 miles south of Ithaca

BUFFALO to NEW YORK
12: Std. *s-dr-dbr*
LV 4, Buffalo (8:50p) to New York (7:55a)

BUFFALO to PHILADELPHIA
30: Std. *s-dbr-dr*
LV 4, Buffalo (8:50p) to Bethlehem (5:22a)
RDG 356, Bethlehem (5:40a)
        to Philadelphia (7:32a)

(Cafe-lounge, Buffalo to New York. In an arrangement dating back to World War I, Lehigh Valley trains use New York's Pennsylvania Station. Weekend schedules for Philadelphia sleeper are different.)

No. 4 carried sleepers from Wilkes-Barre to New York, and from Rochester and Wilkes-Barre to Philadelphia. The Lehigh Valley also offered a Buffalo-Washington sleeper with RDG/B&O. Rochester and Geneva sleepers for New York traveled this line.

Twilight, No. 5
New York to Buffalo
Lackawanna
20 miles northwest of Corning

NEW YORK to DETROIT
50: 10s-3dbr
Ferry from New York (4:28p)
DL&W 5, Hoboken (4:50p) to Buffalo (2:05a)
NYC 47, Buffalo (ntg) to Detroit (8a)

(Diners, Hoboken to Scranton, Buffalo to Detroit; lounges Hoboken to Binghamton, Buffalo to Detroit No. 47 is the streamlined, all-Pullman Detroiter.)

No. 5 carried Chicago sleepers that were moved west of Buffalo by either Michigan Central or Wabash. The DL&W also scheduled Pullmans from New York to Cleveland (NKP) and St. Louis (either NKP or Wabash).

Lake Cities, No 5
New York to Chicago
Erie Railroad
Hancock

NEW YORK to CHICAGO
10r-6dbr
Ferry or bus from New York
Erie 5, Jersey City (8:15p) to Chicago (4:45p)

NEW YORK to CLEVELAND
6r-6s-4dbr
Erie 5, Jersey City (8:15p) to Youngstown (10:01a)
Erie 605, Youngstown (10:20a)
        to Cleveland (11:42a)

NEW YORK to SALAMANCA
4s-6r-4dbr
Erie 5, Jersey City (8:15p) to Salamanca (6:14a)

(Diner-lounge, Jersey City to Chicago)

The *Lake Cities* once carried Pullmans from New York to Buffalo and Bradford and Meadville, Pa. It was the last long-distance passenger train of the Erie Lackawanna, ending service on Jan. 5, 1970.

Fifth Avenue Special, No. 6
Chicago to New York
New York Central
Departing Rochester

CHICAGO to NEW YORK
8s-buffet-lounge   10r-5dbr
14s
NYC 6, Chicago (12:01p) to New York (7a)

CLEVELAND to NEW YORK
10r-6dbr
NYC 6, Cleveland (7:16p) to New York (7a)

NIAGARA FALLS to NEW YORK
6dbr-buffet-lounge   17r
NYC 246, Niagara Falls (8:17p) to Buffalo (9:30p)
NYC 6, Buffalo (10:55p) to New York (7a)

ROCHESTER to NEW YORK
13dbr   22r
8s-5dbr
NYC 6, Rochester (11:59p) to New York (7a)

(Diner, Chicago to Buffalo)

No. 7
Albany to Rouses Point
Delaware & Hudson
Standing in Albany

NEW YORK to PLATTSBURG
218: 8s-5dbr
NYC 21, New York (8p) to Albany (11p)
D&H 7, Albany (12:10a) to Plattsburgh (5:38a)

(No. 21 is the North Star, *which carries a sleeper-lounge)*

New York   *(continued)*

Lake Cities, No. 6
Chicago to New York
Erie Railroad
21 miles east of Olean

CHICAGO to NEW YORK
10r-6dbr
Erie 6, Chicago (10:25a) to Jersey City (8:25a)
Ferry or bus to New York

CLEVELAND to NEW YORK
6r-6s-4dbr
Erie 606, Cleveland (5:25p) to Youngstown (6:51p)
Erie 6, Youngstown (7:13p) to Jersey City (8:25a)

SALAMANCA to NEW YORK
4s-6r-4dbr
Erie 6, Salamanca (11:08p) to Jersey City (8:25a)

(Diner-lounge, Chicago to Jersey City)

The Erie carried Pullmans to New York from Cincinnati, Marion (Ohio), Oil City (Pa.), Buffalo, and Binghamton.

Maple Leaf, No. 8
Toronto/Buffalo to New York
Lehigh Valley, via Ithaca
10 miles east of Buffalo

TORONTO to NEW YORK
36: Std. s-dbr
CNR 89, Toronto (8:35p) to Hamilton (9:30p)
CNR 90, Hamilton (9:40p)
   to Suspension Bridge (10:55p)
LV 8, Suspension Bridge (11:05p)
   to New York (9a)

TORONTO to PHILADELPHIA
16: Std. s-c-dr
As above, Toronto (8:35p)
   to Suspension Bridge (10:55p)
LV 8, Suspension Bridge (11:05p)
   to Bethlehem (6:52a)
RDG 312, Bethlehem (7:40a)
   to Philadelphia (9:18a)

BUFFALO to NEW YORK
25: Std. s-c-dr
LV 8, Buffalo (11:35p) to New York (9a)

(Toronto to Suspension Bridge, parlor-buffet; Lehighton, Pa., to New York, cafe-lounge. Weekend schedule of Philadelphia sleeper different. The Toronto and Buffalo sections of the Maple Leaf *are joined at Depew, 10 miles east of Buffalo.)*

The Canadian National provided Buffalo with sleeping cars to Chicago, Detroit and several resort areas in Ontario.

No. 8
Rouses Point to Troy
Delaware & Hudson
Departing Westport

PLATTSBURGH to NEW YORK
130: 8s-5dbr
D&H 8, Plattsburgh (10:35p) to Troy (3:40a)
NYC 72, Troy (4:20a) to New York (7:35a)

FORT TICONDEROGA to NEW YORK
135: 12s-dr   Summers, Mon only
D&H 8, Fort Ticonderoga (12:09a) to Troy (3a)
NYC 72, Troy (3:30a) to New York (6:40a)

(NYC 72 is the Mount Royal. *Notice how the northbound Plattsburgh sleeper operates via Albany, and this southbound one via Troy.)*

No. 8 handled Pullmans from Montreal and Plattsburgh to Albany, and Fort Edward to New York.

Maple leaf, No. 8
Lehigh Valley
Ithaca setout
Standing in Ithaca

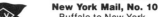

ITHACA to NEW YORK
10s-dr-dbr
LV 8, Ithaca (2:07a) to New York (9a)

New Yorker, No. 8
Chicago to New York
Lackawanna
62 miles east of Buffalo

CHICAGO to NEW YORK
80: 10r-6dbr
NKP 8, Chicago (10:25a) to Buffalo (10:35p)
DL&W 8, Buffalo (11p) to Hoboken (7:30a)
Ferry to New York (7:55a)

BUFFALO to NEW YORK
83: 10r-6dbr
DL&W 8, Buffalo (11p) to Hoboken (7:30a)

(Chicago to Buffalo, diner-lounge)

The Nickel Plate brought sleepers to Buffalo from Chicago and Hot Springs, Va. (C&O).

New Yorker, No. 8
Lackawanna
Elmira setout
Standing in Elmira

ELMIRA to NEW YORK
84: 10r-6dbr
DL&W 8, Elmira (1:38a) to Hoboken (7:30a)
Ferry to New York (7:55a)

New Yorker, No. 8
Lackawanna
Syracuse setout
Standing in Binghamton

SYRACUSE to NEW YORK
85: 10s-dr-2dbr
DL&W 1910, Syracuse (9:20p)
   to Binghamton (11:30p)
DL&W 8, Binghamton (2:48a) to Hoboken (7:30a)
Ferry to New York (7:55a)

(No. 1910, which also moves a Philadelphia sleeper, carries a buffet-lounge)

Wolverine, No. 8
New York Central
Buffalo setout
Standing in Buffalo

BUFFALO to NEW YORK
12s-dr
NYC 8, Buffalo (1:25a) to New York (9:20a)

(The Wolverine, *with its consist of Chicago and Detroit Pullmans, is some 60 miles west of Buffalo, in Ontario. It carries a diner and sleeper-lounge.)*

Montreal Limited, No. 10
Montreal to New York
Napierville Junction Railway
Entering Rouses Point

MONTREAL to NEW YORK

26: 6c-3dr	27: 6dbr-buffet-lounge
28: 22r	29: 8s-dr-3dbr
30: 22r	31: 6s-6dbr
32: 14sbr	

NJR 10, Montreal (10:50p)
   to Rouses Point (12:03a)
D&H 10, Rouses Point (12:08a) to Troy (4:47a)
NYC 62, Troy (5a) to New York (8:05a)

(The Napierville Junction Railway is a short-line subsidiary of Delaware & Hudson, connecting the D&H at Rouses Point with the two major Canadian railroads. Diner, Montreal to Whitehall.)

The *Montreal Limited* was inaugurated in 1924.

New York Mail, No. 10
Buffalo to New York
Lackawanna
Departing Binghamton

CHICAGO to NEW YORK
60: 10r-6dbr
NKP 6, Chicago (11:30p Sat)
   to Buffalo (12:40p Sun)
DL&W 10, Buffalo (5:45p) to Hoboken (5:30a Mon)
Ferry to New York (5:55a Mon)

BINGHAMTON to NEW YORK
100: 10r-6dbr
DL&W 10, Binghamton (11:59p)
   to Hoboken (5:30a)

(Chicago to Buffalo, diner-lounge; Buffalo to Elmira, buffet-lounge. No. 6 is the Nickel Plate Limited.)*

No. 10 carried Pullmans from Ithaca and Scranton to New York. The Lackawanna also scheduled sleepers to New York from Oswego, Richfield Springs, and Utica.

Southwestern Limited, No. 11
New York/Boston to St. Louis
New York Central
Outskirts of Utica

NEW YORK to ST. LOUIS

5dbr-obs.-buffet-lounge	14s
10r-6dbr	10r-6dbr

NYC 11, New York (7:30p) to St. Louis (4:10p)

BOSTON to ST. LOUIS

10r-6dbr	14s

NYC 11, Boston (4:30p) to Albany (9:50p)
NYC 11, Albany (10:18p) to St. Louis (4:10p)

(Dining cars: New York and Boston to Albany, Cleveland to St. Louis. Saturday consist is different.)

The *Southwestern Limited*, whose illustrious history goes back to the 1890's, was the Central's long-time flagship on its St. Louis line. For many years an all-Pullman train, the *Limited* carried Peoria sleepers and eventually acquired a heavy consist of the post-war cars to Oklahoma, Texas, and Mexico City, for a time running these in a separate section.

Southwestern Limited, No. 12
St. Louis to New York
New York Central
Departing Buffalo

ST. LOUIS to NEW YORK

5dbr-obs.-buffet-lounge	10r-6dbr
10r-6dbr	14s

NYC 12, St. Louis (9:30a) to New York (7:45a)

TORONTO to NEW YORK

10r-6dbr	10r-6dbr
10r-6dbr	13dbr
12s-dr	

CPR 821, Toronto (8:30p) to Hamilton (9:30p)
TH&B 382, Hamilton (9:35p) to Welland (10:37p)
NYC 382, Welland (10:40p) to Buffalo (11:45p)
NYC 12, Buffalo (11:59p) to New York (7:45a)

(St. Louis-Cleveland, dining car. No. 382 is the Ontarian. *Saturday consist is different.)*

Jim Shaughnessy

The eastbound *Twentieth Century Limited* drifts down Albany Hill behind two new E-7's in April 1948. Note pusher also descending.

Ohio State Limited, No. 15
Streamliner
New York to Cincinnati
New York Central
Departing Buffalo

NEW YORK to COLUMBUS
10r-6dbr
NYC 15, New York (4p) to Columbus (5:45a)

NEW YORK to CINCINNATI
10r-6dbr   10r-6dbr
5dbr-obs.-buffet-lounge 22r
NYC 15, New York (4p) to Cincinnati (8:35a)

BOSTON to CINCINNATI
10r-6dbr
NYC 49, Boston (11:30a) to Albany (5p)
NYC 41, Albany (5:15p) to Buffalo (10:40p)
NYC 15, Buffalo (11:59p) to Cincinnati (8:35a)

BUFFALO to CINCINNATI
22r
NYC 15, Buffalo (11:59p) to Cincinnati (8:35a)

NEW YORK to CINCINNATI
10r-6dbr
NYC 41, New York (2:30p) to Buffalo (10:40p)
NYC 15, Buffalo (11:59p) to Cincinnati (8:35a)

(Diners on Nos. 15, 49, and 41, the Knickerbocker*)*

Ohio State Limited, No. 16
Streamliner
Cincinnati to New York
New York Central
26 miles west of Buffalo

CINCINNATI to NEW YORK
5dbr-obs.-buffet-lounge 10r-6dbr
10r-6dbr   10r-6dbr
NYC 16, Cincinnati (4p) to New York (8:15a)

COLUMBUS to NEW YORK
10r-6dbr
NYC 16, Columbus (6:35p) to New York (8:15a)

(Dining car. Saturday schedule different)

Once an all-sleeper train, the eastbound *Ohio State Limited* ran as the second No. 26, the *Twentieth Century Limited.*

Wolverine, No. 17
New York to Chicago
Via Detroit
New York Central
20 miles west of Syracuse

NEW YORK to CHICAGO
6dbr-buffet-lounge 10r-6dbr
10r-6dbr   22r
NYC 17, New York (6:15p) to Chicago (11:15a)

NEW YORK to GRAND RAPIDS
10r-6dbr
NYC 17, New York (6:15p)
      to Jackson, Mich. (8:30a)
NYC 325, Jackson (8:40a)
      to Grand Rapids (11:05a)

MASSENA to PITTSBURGH Mon, Wed, Fri only
8s-dr-3dbr
NYC 8, Massena (6:20p) to Syracuse (11:25p)
NYC 17, Syracuse (11:40p) to Buffalo (2:12a)
NYC 279, Buffalo (2:30a) to Youngstown (6:35a)
P&LE 38, Youngstown (6:45a)
      to Pittsburgh (8:30a)

NEW YORK to SAN FRANCISCO Alt. days
10r-6dbr
NYC 17, New York (6:15p Sun)
      to Chicago (11:15a Mon)
C&NW 27, Chicago (3:30p Mon)
      to Omaha (12:30a Tue)
UP 27, Omaha (12:55a) to Ogden (8:33p)
SP 27, Ogden (9p) to Oakland pier (1:45p Wed)
Ferry to San Francisco (2:20p Wed)

(Dining cars, New York-Chicago, Youngstown-Pittsburgh, and Chicago-Oakland. Nos. 27, the San Francisco Overland, *also carry lounge facilities.)*

The formerly all-Pullman *Wolverine* also carried a New York-Bay City, Mich., sleeper.

New England States, No. 27
New York Central
Buffalo setout
Standing in Buffalo

BUFFALO to CHICAGO
10r-6dbr
NYC 27, Buffalo (12:20a) to Chicago (8:20a)

Lake Shore Limited, No. 19
New York to Chicago
Via Cleveland
New York Central
13 miles east of Syracuse

NEW YORK to CHICAGO
6dbr-buffet-lounge 12s-dr
8s-dr-2c   10r-6dbr
NYC 19, New York (6:30p) to Chicago (12:30p)

NEW YORK to TOLEDO
10r-6dbr
NYC 19, New York (6:30p) to Toledo (8:35a)

BOSTON to CHICAGO
6s-6dbr
NYC 33, Boston (3p) to Albany (7:55p)
NYC 19, Albany (9:35p) to Chicago (12:30p)

NEW YORK to SAN FRANCISCO Alt. days
10r-6dbr
NYC 19, New York (6:30p Sun)
      to Chicago (12:30p Mon)
CB&Q 17, Chicago (3:30p) to Denver (8:20a Tue)
D&RGW 17, Denver (8:40a)
      to Salt Lake City (10:20p)
WP 17, Salt Lake City (10:40p)
      to Oakland (4:15p Wed)
Ferry to San Francisco (4:50p)

(Dining car. Both No. 33, the New England Wolverine, *and Nos. 17, the* California Zephyr, *carry lounge and dining facilities.)*

The New York Central's original flagship train to Chicago before the *Twentieth Century Limited,* the *Lake Shore Limited* as of 1952 still had a significant consist that included one of the West Coast Pullmans.

New England States, No. 27
Streamliner
Boston to Chicago
New York Central
Approaching Buffalo

BOSTON to CHICAGO
5dbr-obs.buffet-lounge 10r-6dbr
10r-6dbr   10r-6dbr
10r-6dbr   22r
NYC 27, Boston (2:30p) to Chicago (8:20a)

(Dining car)

The *New England States* was inaugurated in 1938 to take the place of the Boston section of the *Twentieth Century Limited.*

New York   *(continued)*

North Star, No. 21
New York to Cleveland
New York Central
14 miles west of Schenectady

NEW YORK to CLEVELAND
  6dbr-buffet-lounge   10r-6dbr
NYC 21, New York (8p) to Cleveland (9:50a)

NEW YORK to TORONTO
  13dbr   12s-dr
  10r-6dbr   10r-6dbr
NYC 21, New York (8p) to Buffalo (4:35a)
NYC 371, Buffalo (5:15a) to Welland (6:20a)
TH&B 371, Welland (6:25a) to Hamilton (7:25a)
CPR 712, Hamilton (7:35a) to Toronto (8:35a)

BOSTON to TORONTO
  10r-6dbr
NYC 11, Boston (4:30p) to Albany (9:50p)
NYC 21, Albany (11:20p) to Buffalo (4:35a)
As above, Buffalo (5:15a) to Toronto (8:35a)

NEW YORK to LAKE PLACID
  6dbr-buffet-lounge   8s-dr-3dbr
NYC 21, New York (8p) to Utica (12:58a)
NYC 5, Utica (2:45a) to Lake Clear Jct. (6:46a)
NYC 105, Lake Clear Jct. (7:10a)
        to Lake Placid (8a)

NEW YORK to THENDARA   *Summer, Fri only*
  8s-dr-2c
NYC 21, New York (8p) to Utica (12:58a)
NYC 5, Utica (2:15a) to Thendara (4:04a)

NEW YORK to MALONE   *Summertime only*
  8s-dr-2c
NYC 21, New York (8p) to Utica (12:58a)
NYC 5, Utica (2:15a) to Malone (7:40a)

(Diner and lounge facilities, Boston to Albany, and Buffalo to Toronto. No. 11 is the Boston section of the Southwestern Limited. No. 5 carries additional Lake Placid sleepers in the summertime.)

Knickerbocker, No. 24
New York Central
Pittsburgh setout
Standing in Buffalo

PITTSBURGH to ALBANY
  10r-5dbr
P&LE 33, Pittsburgh (5p) to Youngstown (6:25p)
NYC 284, Youngstown (6:28p) to Buffalo (10:35p)
NYC 24, Buffalo (3:04a) to Albany (8:08a)

(The Knickerbocker is now departing Cleveland with a consist of sleepers for Toronto, Boston, and New York, and a diner. P&LE 33, the Empire Express, carries a diner).

The *Knickerbocker* previously was all-Pullman. The Central brought Albany its own sleepers from Chicago, Cleveland, Detroit, and Rochester.

New England Wolverine, No. 33
New York Central
Buffalo setout
Standing in Buffalo

BUFFALO to DETROIT
  22r
NYC 33, Buffalo (1:50a) to Detroit (6:30a)

Twentieth Century Limited, No. 25
All-Pullman streamliner
New York to Chicago
New York Central
East of Rochester

NEW YORK to CHICAGO
  5dbr-obs.-buffet-lounge   4c-4dbr-2dr
  4c-4dbr-2dr   4c-4dbr-2dr
  12dbr   12dbr
  12dbr   10r-6dbr
  10r-6dbr
NYC 25, New York (6p) to Chicago (9a)

NEW YORK to LOS ANGELES
  4c-4dbr-2dr   10r-6dbr
NYC 25, New York (6p Sun) to Chicago (9a Mon)
AT&SF 19, Chicago (1:30p)
      to Los Angeles (8:30a Wed)

(Dining cars and lounges, New York-Chicago-Los Angeles. No. 19 is the Santa Fe's all-Pullman, streamlined Chief. Both it and the Twentieth Century Limited are extra-fare trains.)

New England States, No. 28
Streamliner
Chicago to Boston
New York Central
Departing Buffalo

CHICAGO to BOSTON
  5dbr-obs.-buffet-lounge   22r
  10r-6dbr   10r-6dbr
  10r-6dbr   10r-6dbr
NYC 28, Chicago (2:20p) to Boston (9:40a)

DETROIT to BOSTON
  10r-6dbr
NYC 48, Detroit (7p) to Buffalo (ntg)
NYC 28, Buffalo (ntg) to Boston (9:40a)

PITTSBURGH to BOSTON
  10r-6dbr
P&LE 33, Pittsburgh (5p) to Youngstown (6:25p)
NYC 284, Youngstown (6:28p) to Buffalo (10:35p)
NYC 28, Buffalo (ntg) to Boston (9:40a)

(No. 48 is the all-Pullman, streamlined Detroiter. Dining cars all segments)

New England Wolverine, No. 33
Boston to Detroit
New York Central
15 miles east of Rochester

BOSTON to DETROIT
  6dbr-buffet-lounge   10r-6dbr
NYC 33, Boston (3p) to Detroit (6:30a)

BOSTON to CHICAGO
  6s-6dbr
NYC 33, Boston (3p) to Detroit (6:30a)
NYC 17, Detroit (7:10a) to Chicago (11:15a)

BOSTON to CHICAGO
  6s-6dbr
NYC 33, Boston (3p) to Buffalo (1:40a)
NYC 19, Buffalo (2:54a) to Chicago (12:30p)

BOSTON to PITTSBURGH
  10r-6dbr
NYC 33, Boston (3p) to Buffalo (1:40a)
NYC 279, Buffalo (2:30a) to Youngstown (6:35a)
P&LE 38, Youngstown (6:45a)
      to Pittsburgh (8:30a)

ALBANY to PITTSBURGH
  10r-5dbr
NYC 33, Albany (8:10p) to Buffalo (1:40a)
As above, Buffalo (2:30a) to Pittsburgh (8:30a)

(Dining service: Boston to Albany, Detroit and Buffalo to Chicago, and Youngstown to Pittsburgh. The first Chicago Pullman finishes its trip on No. 17, the Wolverine; the other Chicago car becomes part of the Lake Shore Limited west of Buffalo. Both trains carry lounges.)

Cayuga, No. 34
Syracuse to New York
New York Central
Departing Syracuse

SYRACUSE to NEW YORK
  10r-6dbr   22r
NYC 34, Syracuse (11:55p) to New York (6:15a)

ROCHESTER to NEW YORK
  10s-3dbr
NYC 18, Rochester (7:10p) to Syracuse (11:10p)
NYC 34, Syracuse (11:55p) to New York (6:15a)

(No. 18 travels the alternate, Auburn Road line to Syracuse. The Cayuga will pick up some St. Lawrence Division sleepers from Massena, Ogdensburg, and Watertown at Utica in a little more than an hour.)

Iroquois, New York section, No. 35
New York to Albany/Troy
New York Central
Approaching Poughkeepsie

NEW YORK to WATERTOWN
  10r-5dbr
NYC 35, New York (10:30p) to Utica (3:30a)
NYC 59, Utica (4:15a) to Watertown (6:50a)

NEW YORK to OGDENSBURG
  8s-4dbr
As above, New York (10:30p) to Watertown (6:50a)
NYC 59, Watertown (7:05a) to Ogdensburg (9:10a)

NEW YORK to MASSENA, N.Y.
  10r-5dbr
As Above, New York (10:30p) to Watertown (6:50a)
NYC 59, Watertown (7:05a) to Philadelphia (7:32a)
NYC 7, Philadelphia (7:49a) to Massena (10:10a)

NEW YORK to MONTREAL
  8s-5dbr
NYC 35, New York (10:30p) to Troy (1:50a)
B&M 5600, Troy (2:05a)
      to North Bennington, Vt. (2:58a)
RUT 51, North Bennington (3a)
      to Rouses Point (7:25a)
CNR 51, Rouses Point (7:28a) to Montreal (8:55a)

NEW YORK to RUTLAND   *Summers, Fri only*
  12s-dr
As above, New York (10p)
      to North Bennington (2:20a)
RUT 51, North Bennington (2:27a)
      to Rutland (3:40a)

(Weekend schedules are different.)

Iroquois, Boston section, No. 35
Boston to Chicago
New York Central
39 miles east of Albany

BOSTON to BUFFALO
  22r
NYC 35, Boston (7:35p) to Buffalo (7:25a)

BOSTON to CHICAGO
  6dbr-buffet-lounge   10r-5dbr
  8s-4dbr
NYC 35, Boston (7:35p) to Chicago (5:20p)

BOSTON to DETROIT
  10r-6dbr
NYC 35, Boston (7:35p) to Buffalo (7:25a)
NYC 335, Buffalo (8:25a) to Detroit (1:10p)

LAKE PLACID to CHICAGO *Summers, Sun, Wed*
  8s-dr-2c
NYC 104, Lake Placid (8:40p)
      to Lake Clear Jct. (9:20p)
NYC 4, Lake Clear Jct. (9:30p) to Utica (1:20a)
NYC 35, Utica (3:09a) to Chicago (5p)

(Dining cars: Buffalo to Chicago, Buffalo to Detroit. The two sections of the Iroquois are joined in Albany, and the Upstate New York cars ride the train as far as Utica.)

Knickerbocker, No. 41
New York to St. Louis
New York Central
66 miles west of Buffalo

NEW YORK to ST. LOUIS
10r-6dbr   22r
6dbr-buffet-lounge
NYC 41, New York (2:30p) to St. Louis (11:30a)

BOSTON to ST. LOUIS
8s-5dbr
NYC 49, Boston (11:30a) to Albany (5p)
NYC 41, Albany (5:15p) to St. Louis (11:30a)

(The Knickerbocker cut out its Cincinnati Pullmans in Buffalo. Down the line it will pick up sleepers for St. Louis from Buffalo, Cleveland, and Detroit. Dining cars on all segments.)

No. 42
Albany to Boston
New York Central
Standing in Albany

ALBANY to BOSTON
8s-4dbr
NYC 42, Albany (1:05a) to Boston (7:10a)

(No. 42 awaits a Boston Pullman from Buffalo; en route it will pick up a Worcester sleeper from the New Haven at Springfield)

New York Special, No. 44
Chicago to New York
Via Detroit
New York Central
17 miles east of Rochester

CHICAGO to NEW YORK
10r-6dbr
NYC 44, Chicago (9:45a) to New York (7:15a)

GRAND RAPIDS to NEW YORK
10r-6dbr
NYC 328, Grand Rapids (12:50p) to Jackson (3p)
NYC 44, Jackson (3:15p) to New York (7:15a)

DETROIT to NEW YORK
10r-5dbr
NYC 44, Detroit (5:10p) to New York (7:15a)

(Chicago to Detroit, diner-lounge; Detroit to Buffalo, diner; Detroit to Albany, lounge-sleeper. No. 44 will pick up a Lake Placid-New York sleeper in Utica. Another Detroit-New York Pullman already left the train at Buffalo for its final leg to Hoboken on the Lackawanna.)

Detroiter, No. 47
All-Pullman streamliner
New York to Detroit
New York Central
21 miles east of Syracuse

NEW YORK to DETROIT
6dbr-buffet-lounge   6dbr-buffet-lounge
22r   22r
4c-4dbr-2dr   4c-4dbr-2dr
12br   12br
12br   10r-6dbr
NYC 47, New York (7p) to Detroit (8a)

(Dining car)

The business travel generated by the close workings of Detroit's automobile industry and New York's financial and advertising communities is signaled in the solid-Pullman consist of the *Detroiter*, still going strong four decades after its inauguration in 1911.

Interstate Express, No. 46
Chicago to Boston
Via Cleveland
New York Central
30 miles east of Rochester

CHICAGO to BOSTON
6dbr-buffet-lounge   6s-6dbr
12s-dr
NYC 46, Chicago (9:40a) to Boston (9:10a)

CHICAGO to BOSTON   *via Detroit*
6s-6dbr
NYC 44, Chicago (9:45a) to Buffalo (9:50p)
NYC 46, Buffalo (10:20p) to Boston (9:10a)

DETROIT to BOSTON
6dbr-buffet-lounge   10r-6dbr
NYC 44, Detroit (5:10p) to Buffalo (9:50p)
NYC 46, Buffalo (10:20p) to Boston (9:10a)

TORONTO to BOSTON
10r-6dbr
CPR 801, Toronto (6:10p) to Hamilton (7:05p)
TH&B 380, Hamilton (7:10p) to Welland (8:10p)
NYC 380, Welland (8:15p) to Buffalo (9:15p)
NYC 46, Buffalo (10:20p) to Boston (9:10a)

BUFFALO to BOSTON
22r   12s-dr
NYC 46, Buffalo (10:20p) to Boston (9:10a)

BUFFALO to MONTREAL
10s-buffet-lounge
NYC 46, Buffalo (10:20p) to Utica (1:46a)
NYC 5, Utica (2:45a) to Montreal (11:30a)

CHICAGO to LAKE PLACID   *Summers, Tue, Fri*
8s-dr-2c
NYC 46, Chicago (9:50a) to Utica (1:41a)
NYC 5, Utica (2:15a) to Lake Clear Jct. (6:16a)
NYC 105, Lake Clear Jct. (6:40a)
          to Lake Placid (7:30a)

(Dining service on No. 44, Chicago to Buffalo; on No. 46, Chicago to Buffalo and Albany to Boston; and Toronto to Buffalo. No. 44 is the New York Special.)

Detroiter, No. 48
All-Pullman streamliner
Detroit to New York
New York Central
35 miles east of Buffalo

DETROIT to NEW YORK
6dbr-buffet-lounge   6dbr-buffet-lounge
22r   22r
4c-4dbr-2dr   4c-4dbr-2dr
12br   12br
12br
NYC 48, Detroit (7p) to New York (7:25a)

(Dining car)

Easterner, No. 52
Cleveland to New York
New York Central
Arriving in Albany

BUFFALO to BOSTON
12s-dr
NYC 52, Buffalo (4:45p) to Albany (12:05a)
NYC 42, Albany (1:05a) to Boston (7:10a)

BUFFALO to NEW YORK
22r
NYC 52, Buffalo (4:45p) to New York (4:30a)

(Diner-Lounge, Buffalo to Syracuse. For people who had to be in New York "the first thing in the morning," the Easterner was the earliest by far.)

No. 56
New York to Springfield
New Haven Railroad
Standing in New York

NEW YORK to SPRINGFIELD
561: Std. s-dr
NH 56, New York (12:10a) to Springfield (4:20a)

NEW YORK to WHITE RIVER JUNCTION, VT.
562: Std. s-dr-c
NH 56, New York (12:10a) to Springfield (4:20a)
B&M 703, Springfield, (5a)
          to White River Junction (8:35a)

NEW YORK to WORCESTER
563: Std. s-r-sbr-dbr
NH 56, New York (12:10a) to Springfield (4:20a)
NYC 42, Springfield (4:45a) to Worcester (5:53a)

The *Knickerbocker* races up the Hudson with unique center-door diner-lounge *Toreador*.

Howard W. Ameling collection

New York   *(continued)*

Easterner, No. 52
New York Central
Albany setout
Standing in Albany

ALBANY to NEW YORK
12s-2dbr
NYC 52, Albany (12:45a) to New York (4:30a)

Cleveland Limited, No. 57
New York to Cleveland
New York Central
21 miles east of Utica

NEW YORK to CLEVELAND
6dbr-buffet-lounge   6dbr-buffet-lounge
4c-4br-2dr   13dbr
10r-6dbr   22r
22r
NYC 57, New York (7:50p) to Cleveland (7:30a)

BOSTON to CLEVELAND
6dbr-buffet-lounge   22r
NYC 57, Boston (4:30p) to Albany (9:50p)
NYC 57, Albany (10:46p) to Cleveland (7:30a)

No. 57 was another of the Central's once-large fleet of all-Pullman limiteds. The eastbound edition was still sleeper-only in 1952.

Cleveland Limited, No. 58
All-Pullman
Cleveland to New York
New York Central
Buffalo

CLEVELAND to NEW YORK
6dbr-buffet-lounge   6dbr-buffet-lounge
4c-4br-2dr   13dbr
10r-6dbr   10r-6dbr
22r   22r
NYC 58, Cleveland (9p) to New York (7:55a)

(The Cleveland Limited *makes no scheduled stops between East Cleveland and Harmon, where the change to electric power is made for the trip into Grand Central Terminal.)*

Chicagoan, No. 59
New York to Chicago
New York Central
11 miles north of Harmon

NEW YORK to BUFFALO
6dbr-buffet-lounge   10r-6dbr
10r-5dbr   22r
12s-dr
NYC 59, New York (11p) to Buffalo (6:50a)

NEW YORK to CLEVELAND
10r-6dbr
NYC 59, New York (11p) to Cleveland (10:12a)

NEW YORK to CHICAGO
6dbr-buffet-lounge   8s-buffet-lounge
10r-5dbr   22r
12s-dr
NYC 59, New York (11p) to Chicago (3:30p)

(Buffalo to Chicago, dining car)

Cleveland Limited, No. 57
New York Central
Toronto Pullman
Standing in Buffalo

TORONTO to CLEVELAND
8s-5dbr
CPR 821, Toronto (8:30p) to Hamilton (9:30p)
TH&B, Hamilton (9:35p) to Welland (10:37p)
NYC 382, Welland (10:40p) to Buffalo (11:45p)
NYC 57, Buffalo (ntg) to Cleveland (7:30a)

Montreal Limited, No. 61
New York to Montreal
New York Central
20 miles north of New York

NEW YORK to MONTREAL
6dbr-buffet-lounge   6c-3dr
22r   22r
14sbr   6s-6dbr
8s-dr-3dbr
NYC 61, New York (11:30p) to Troy (2:25a)
D&H 9, Troy (2:35a) to Rouses Point (7:14a)
NJR 9, Rouses Point (7:19a) to Montreal (8:40a)

(Whitehall to Montreal, dining car)

As an all-Pullman train, the *Montreal Limited* carried sleepers for Ottawa and Quebec, as well.

Pittsburgher, No. 61
All-Pullman
New York to Pittsburgh
Pennsylvania Railroad
Departing New York

NEW YORK to PITTSBURGH
3dbr-dr-lounge   6dbr-lounge
21r   21r
21r   12dup/r-4dbr
12dup/r-4dbr   12dup/r-4dbr
10r-6dbr   4c-2dr-4dbr
PRR 61, New York (11:59p) to Pittsburgh (9a)

NEW YORK to WILLIAMSPORT, PA.
10s-3dbr
PRR 61, New York (11:59p) to Harrisburg (3:30a)
PRR 577, Harrisburg (4:05a)
          to Williamsport (7:45a)

(Diner serving breakfast into Pittsburgh. The train is now easing out of the 21-track Pennsylvania Station. This majestic structure occupying two square blocks opened in 1910, giving Manhattan Island direct rail access to the west via tunnels under the Hudson River.)

Advance Commodore Vanderbilt, No. 65
All-Pullman
New York to Chicago
New York Central
25 miles west of Buffalo

NEW YORK to CHICAGO
6dbr-buffet-lounge   10r-6dbr
10r-6dbr   10r-6dbr
22r   14s
NYC 65, New York (3:30p) to Chicago (7:30a)

NEW YORK to LOS ANGELES
10r-6dbr
NYC 65, New York (3:30p Sun)
          to Chicago (7:30a Mon)
C&NW 1, Chicago (12:01p) to Omaha (9p)
UP 1, Omaha (9:25p) to Los Angeles (10:40a Wed)

(Dining cars and lounges all segments. Although the Advance Commodore *was all-Pullman, it operated combined with the all-coach, streamlined* Pacemaker, *which had been introduced for the New York World's Fair of 1939. No. 1 is the* Los Angeles Limited.*)*

The *Advance Commodore Vanderbilt* began service on the eve of World War II, in 1941.

Advance Commodore Vanderbilt, No. 66
All-Pullman
Chicago to New York
New York Central
66 miles west of Buffalo

CHICAGO to NEW YORK
6dbr-buffet-lounge   10r-6dbr
10r-6dbr   22r
NYC 66, Chicago (3:15p) to New York (8:45a)

(Diner. The eastbound Advance Commodore *also runs combined with the* Pacemaker.*)*

Commodore Vanderbilt, No. 67
All-Pullman streamliner
New York to Chicago
New York Central
West of Rochester

NEW YORK to CHICAGO
4dbr-obs.-buffet-lounge 6dbr-buffet-lounge
4c-4dbr-2dr   4c-4dbr-2dr
13dbr   13dbr
10r-6dbr   10r-6dbr
22r   22r
22r   12dbr
NYC 67, New York (5:30p) to Chicago (8:30a)

(Diner. After Harmon, the Commodore Vanderbilt *makes no scheduled stops before Toledo.)*

Paul Revere, No. 78
Cleveland to Boston
New York Central
61 miles west of Buffalo

CLEVELAND to BOSTON
6dbr-buffet-lounge   22r
NYC 78, Cleveland (9:50p) to Boston (12:05p)

ST. LOUIS to BOSTON
14s   10r-6dbr
NYC 12, St. Louis (9:30a) to Cleveland (8:32p)
NYC 78, Cleveland (9:50p) to Boston (12:05p)

CINCINNATI to BOSTON
10r-6dbr
NYC 16, Cincinnati (4p) to Cleveland (9:07p)
NYC 78, Cleveland (9:50p) to Boston (12:05p)

(Diners all segments except Cleveland to Albany. Lounges all segments. No. 12 is the Southwestern Limited, *No. 16 the* Ohio State Limited.*)*

Paul Revere, No 78
New York Central
Pittsburgh Pullman
Standing in Buffalo

PITTSBURGH to MASSENA   Tue, Thu, Sun only
8s-dr-3dbr
P&LE 33, Pittsburgh (5p) to Youngstown (6:25p)
NYC 284, Youngstown (6:28p) to Buffalo (10:35p)
NYC 78, Buffalo (1:30a) to Syracuse (4a)
NYC 7, Syracuse (4:20a) to Massena (10:10a)

(No. 33, the Empire Express, *carries a diner)*

Genesee, No. 82
All-Pullman
Buffalo to New York
New York Central
30 miles east of Buffalo

BUFFALO to NEW YORK
6dbr-buffet-lounge   10r-6dbr
22r
NYC 82, Buffalo (11:30p) to New York (7:55a)

The *Genesee* formerly handled other Upstate sleepers to New York, such as cars from Rochester, Lockport, Oswego, and Syracuse.

No. 80
Ogdensburg to Utica
New York Central
35 miles north of Utica

MASSENA to NEW YORK
10r-5dbr
NYC 8, Massena (6:20p) to Philadelphia (8:10p)
NYC 80, Philadelphia (9:18p) to Utica (1a)
NYC 34, Utica (1:11a) to New York (6:15a)

OGDENSBURG to NEW YORK
8s-4dbr
NYC 80, Ogdensburg (7:55p)
to Philadelphia (9:17p)
As above, Philadelphia (9:18p)
to New York (6:15a)

WATERTOWN to NEW YORK
10r-5dbr
NYC 80, Watertown (10:10p) to Utica (1a)
NYC 34, Utica (1:11a) to New York (6:15a)

(All these sleepers from the St. Lawrence Division will go into No. 34, the Cayuga, at Utica. The Central schedules additional Pullmans from Ogdensburg and Watertown in the summertime.)

Serving both summer and winter vacation areas, the New York Central carried numerous Pullmans between New York and Upstate communities: Cape Vincent, Clayton, Lockport, Loon Lake, Newton Falls, Oswego, Paul Smith's Hotel, Raquette Lake, Saranac Lake, Thendara, Tupper Lake Junction, and Sacandaga Park; and with D&H from Lake George, North Creek, Rouses Point, Saratoga Springs, and Westport.

Tuscarora, No. 99
New York to Buffalo
New York Central
10 miles north of Grand Central Terminal

NEW YORK to SYRACUSE
10r-6dbr   *22r*
NYC 99, New York (11:45p) to Syracuse (5:03a)

NEW YORK to ROCHESTER
13dbr   *22r*
8s-5dbr
NYC 99, New York (11:45p) to Rochester (6:49a)

NEW YORK to ROCHESTER
10s-3dbr
NYC 99, New York (11:45p) to Syracuse (5:03a)
NYC 3, Syracuse (5:25a) to Rochester (10:05a)

NEW YORK to NIAGARA FALLS
6dbr-buffet-lounge   *17r*
NYC 99, New York (11:45p) to Buffalo (8:10a)
NYC 207, Buffalo (8:30a) to Niagara Falls (9:31a)

NEW YORK to DETROIT
10r-5dbr
NYC 99, New York (11:45p) to Buffalo (8:10a)
NYC 335, Buffalo (8:25a) to Detroit (1:10p)

NEW YORK to TORONTO
10r-6dbr
NYC 99, New York (11:45p) to Buffalo (8:10a)
NYC 377, Buffalo (9a) to Welland (10:10a)
TH&B 377, Welland (10:15a) to Hamilton (11:15a)
CPR 772, Hamilton (11:20a) to Toronto (12:20p)

(Diner, Buffalo to Detroit. The one Rochester sleeper travels via the Auburn Road.)

No. 101
New York to Washington
Pennsylvania Railroad
Standing in New York

NEW YORK to RICHMOND
10r-6dbr   *10r-6dbr*
PRR 101, New York (12:30a)
to Washington (5:15a)
RF&P 23, Washington (5:35a) to Richmond (8a)

NEW YORK to NORFOLK
10r-6dbr
As above, New York (12:30a) to Richmond (8a)
ACL 29, Richmond (8:30a) to Petersburg (9:05a)
N&W 22, Petersburg (9:15a) to Norfolk (10:55a)

(Washington to Richmond, parlor-diner)

Edison, No. 103
All-Pullman
New York to Washington
Pennsylvania Railroad
Standing in New York

NEW YORK to BALTIMORE
Std: s-dr-c-r
PRR 103, New York (12:55a) to Baltimore (5:50a)

NEW YORK to WASHINGTON
Std: s-dr-c-sbr-dbr-r-dup/r
PRR 103, New York (12:55a)
to Washington (7:15a)

NEW YORK to MEMPHIS
14r-4dbr
PRR 103, New York (12:55a) to Washington (7:15a)
SOU 45, Washington (8a)
to Lynchburg, Va. (11:55a)
N&W 45, Lynchburg (12:05p) to Bristol, Va. (5:20p)
SOU 45, Bristol (5:30p) to Memphis (7:45a)

(No. 45 is the streamlined Tennessean. Dining car, Washington to Knoxville; diner-lounge, Chattanooga to Memphis; lounge, Washington to Memphis. Sunday schedules differ.)

Day and night, trains departed Pennsylvania Station with sleepers bound for more than 100 destinations.

Night Cape Codder, No. 110
New York to Cape Cod
Summertime, Fridays only
New Haven Railroad
Standing in New York

NEW YORK to HYANNIS
3 Pullmans: s-dr-c-sbr
NH 110, New York (12:40a)
to Buzzards Bay, Mass. (7a)
NH 831, Buzzards Bay (7:10a) to Hyannis (7:57a)

NEW YORK to WOODS HOLE
4 Pullmans: s-dr-c-sbr-buffet-lounge
NH 110, New York (12:40a) to Buzzards Bay (7a)
NH 833, Buzzards Bay (ntg) to Woods Hole (7:56a)

WASHINGTON to HYANNIS
682: Std. s-dr-c
PRR 168, Washington (4:10p) to New York (8p)
NH 168, New York (8:25p) to New Haven (9:59p)
NH 110, New Haven (2:35a) to Buzzards Bay (7a)
NH 831 Buzzards Bay (7:10a) to Hyannis (7:57a)

(No. 168, the Montrealer, carries a diner to New York. Because the Night Cape Codder uses Grand Central Terminal, the Washington sleeper operates via Hell Gate Bridge and joins the train in New Haven.)

No. 163
New York to Albany
New York Central
Departing New York

NEW YORK to ALBANY
12s-2dbr
NYC 163, New York (11:58p) to Albany (3:45a)

NEW YORK to SYRACUSE
22r
NYC 163, New York (11:58p) to Albany (3:45a)
NYC 43, Albany (5:20a) to Syracuse (9:25a)

The New York Central offered Pullmans from Syracuse to Chicago, Boston, and Springfield, Mass.

The Federal, No. 172
New Haven Railroad
New York setout
Standing in New York

NEW YORK to BOSTON
1721: Std. s-dbr
NH 172, New York (3:10a) to Boston (8:10a)

(The Federal is now 18 miles north of Baltimore)

Buffalo-Pittsburgh Express, No. 279
Buffalo to Youngstown
New York Central
Standing in Buffalo

TORONTO to PITTSBURGH
6s-6dbr
CPR 821, Toronto (8:30p) to Hamilton (9:30p)
TH&B 382, Hamilton (9:35p) to Welland (10:37p)
NYC 382, Welland (10:40p) to Buffalo (11:45p)
NYC 279, Buffalo (2:30a) to Youngstown (6:35a)
P&LE 38, Youngstown (6:45a)
to Pittsburgh (8:30a)

BUFFALO to PITTSBURGH
8s-buffet-lounge   *22r*
As above, Buffalo (2:30a) to Pittsburgh (8:30a)

(Youngstown to Pittsburgh, dining car)

Interstate Express, No. 1306
Syracuse to Philadelphia
Lackawanna
Standing in Binghamton

SYRACUSE to PHILADELPHIA
130: 12s-dr
DL&W 1910, Syracuse (9:20p)
to Binghamton (11:30p)
DL&W 1306, Binghamton (12:20a)
to Scranton (ntg)
CNJ 306, Scranton (1:35a) to Bethlehem (ntg)
RDG 312, Bethlehem (7:40a)
to Philadelphia (9:18a)

(Syracuse to Binghamton, buffet-lounge. Weekend schedules different.)

The *Interstate Express* carried Syracuse-Washington and Binghamton-Philadelphia sleepers.

North Carolina

See map on page **118**

Sunland, No. 7
Washington to Florida
Seaboard Air Line
12 miles south of Hamlet

WASHINGTON to JACKSONVILLE
R360: 10s-dr-2c
RF&P 107, Washington (3:15p)
    to Richmond (6:05p)
SAL 7, Richmond (6:15p) to Jacksonville (7:55a)

PORTSMOUTH to TAMPA
B17: 20s-dr-2c
SAL 17, Portsmouth (4:30p) to Norlina (7:45p)
SAL 7, Norlina (8:11p) to Wildwood, Fla. (11a)
SAL 107, Wildwood (11:30a) to Tampa (1:30p)

(Dining car, Washington to Hamlet; parlor-dining car, Jacksonville to Wildwood)

Raleigh, the principal North Carolina city on the Seaboard's main line, had Pullman service to New York (RF&P/PRR).

Peach Queen, No. 30
Atlanta to New York
Southern Railway
10 miles south of Danville, Va.

ATLANTA to NEW YORK
S61: 10s-dr-2c
SOU 30, Atlanta (1:55p) to Washington (6:40a)
PRR 122, Washington (7:20a)
    to New York (11:30a)

ASHEVILLE to WASHINGTON
9: 10s-3dbr
SOU 16, Asheville (3:45p) to Greensboro (9:25p)
SOU 30, Greensboro (11:15p)
    to Washington (6:40a)

CHARLOTTE to RICHMOND
69: 12s-dr
SOU 30, Charlotte (8:40p) to Danville (12:13a)
SOU 12, Danville (4a) to Richmond (8a)

(Diner, Atlanta to Charlotte)

Over its busy main line, the Southern Railway brought Pullmans to Charlotte from Washington, Portsmouth (Va.), Raleigh, Savannah, Birmingham, and Columbia (Ga.).

Silver Comet, No. 33
New York to Birmingham
Seaboard Air Line
10 miles north of Hamlet

NEW YORK to BIRMINGHAM
S119: 10r-6dbr   S120: 10s-lounge
PRR 125, New York (12:30p)
    to Washington (4:25p)
RF&P 33, Washington (5p) to Richmond (7:38p)
SAL 33, Richmond (7:38p) to Birmingham (10:45a)

WASHINGTON to BIRMINGHAM
R366: 10s-dr-2c
As above, Washington (5p)
    to Birmingham (10:45a)

WASHINGTON to ATLANTA
R365: 10r-6dbr
As above, Washington (5p) to Richmond (7:38p)
SAL 33, Richmond (7:38p) to Atlanta (7:15a)

(Dining car and tavern observation, New York to Birmingham)

Silver Comet, No. 33
Seaboard Air Line
Portsmouth setout
Standing in Hamlet

PORTSMOUTH, Va. to ATLANTA
B3: 10s-dr-2c
SAL 17, Portsmouth (4:30p) to Norlina (7:45p)
SAL 7, Norlina (8:11p) to Hamlet (11:20p)
SAL 33, Hamlet (12:30a) to Atlanta (7:15a)

Washington-Atlanta-New Orleans Express, No. 35
Washington to New Orleans
Southern Railway
7 miles south of Charlotte

WASHINGTON to NEW ORLEANS
S74: 12s-dr
SOU 35, Washington (1:30p) to Atlanta (6:30a)
WPR 35, Atlanta (6:50a) to Montgomery (10:30a)
L&N 1, Montgomery (10:50a)
    to New Orleans (7:30p)

WASHINGTON to ATLANTA
57: 10s-c-2dbr
As above, Washington (1:30p) to Atlanta (6:30a)

CHARLOTTE to ATLANTA
60: 10s-dr-2c
SOU 35, Charlotte (11:50p) to Atlanta (6:30a)

(Dining cars: Monroe, Va., to Greensboro; Montgomery to New Orleans)

Well-served by crack Southern trains, Greensboro had dedicated sleepers to Atlanta, Jacksonville, Portsmouth, and New York (PRR). Pullmans from Norfolk and Richmond served Salisbury.

New Yorker, No. 40
Atlanta to New York
Southern Railway
25 miles south of Charlotte

ATLANTA to NEW YORK
NY2: 8s-5dbr   NY1: 12r-2sbr-3dbr
NY4: 2dr-3c-lounge
SOU 40, Atlanta (7p) to Washington (8:55a)
PRR 148, Washington (9:35a) to New York (1:25p)

ATLANTA to WASHINGTON
NY3: 8s-dr-3dbr
As above, Atlanta (7p) to Washington (8:55a)

(Diners: Atlanta to Washington, Washington to New York)

Orange Blossom Special, No. 45
All-Pullman
New York to Miami, winters
Seaboard Air Line
30 miles north of Aberdeen

NEW YORK to MIAMI
OB20: 6s-6dbr   OB21: 12r-sbr-4dbr
OB22: 2c-dr-buffet-lounge OB23: 6c-3dr
OB24: 6c-3dr   OB25: 6c-3dr
OB26: 6c-3dr   OB28: 6c-3dr
PRR 181, New York (1:45p) to Washington (5:40p)
RF&P 45, Washington (6:05p) to Richmond (ntg)
SAL 45, Richmond (8:38p) to Miami (3:10p)

BOSTON to MIAMI
MS7: 8s-dr-3dbr
NH 175, Boston (8:30a) to Washington (5:15p)
As above, Washington (6:05p) to Miami (3:10p)

WASHINGTON to MIAMI
R361: 6s-4r-4dbr   R362: 3c-dr-buffet-lounge
As above, Washington (6:05p) to Miami (3:10p)

(Dining cars: New York to Miami, and Washington to Wildwood)

The luxurious *Orange Blossom Special* was introduced in 1925, and before World War II operated in separate east and west coast sections. This 1952 carding was for its second-to-last season.

Silver Meteor, No. 57
New York to Florida
Seaboard Air Line
Leaving Raleigh

NEW YORK to MIAMI
S101: 10r-6dbr   S102: 10r-6dbr
S103: 10r-6dbr   S104: 6dbr-buffet-lounge
PRR 113, New York (2:25p) to Washington (6:25p)
RF&P 57, Washington (6:45p) to Richmond (ntg)
SAL 57, Richmond (9:18p) to Miami (3:55p)

NEW YORK to SARASOTA/VENICE
S105: 10r-6dbr
As above, New York (2:25p) to Richmond (ntg)
SAL 57, Richmond (9:18p)
    to Wildwood, Fla. (11:05a)
SAL 157, Wildwood (11:20a) to Tampa (1:10p)
SAL 26-257, Tampa (1:45p) to Sarasota (3:45p)
    and Venice (4:40p)

NEW YORK to ST. PETERSBURG
S106: 10r-6dbr   S107: 10r-6dbr
As above, New York (2:25p) to Wildwood (11:05a)
SAL 157, Wildwood (11:20a)
    to St. Petersburg (2:55p)

(Dining car and tavern observation, New York to Miami; diner, New York to St. Petersburg.)

The *Silver Meteor* pioneered streamlined train service to Florida on Feb. 2, 1939, with one-in-three-day operation. Pullmans were added in 1941.

ACL's *Palmetto*, offering Augusta-New York Pullman service, pauses at Richmond's Broad Street Station in November 1963.

William E. Griffin Jr. collection

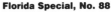

Silver Meteor, No. 58
Florida to New York
Seaboard Air Line
33 miles south of Raleigh

MIAMI to NEW YORK
B50: *10r-6dbr*   B51: *10r-6dbr*
B52: *10r-6dbr*   B53: *6dbr-buffet-lounge*
SAL 58, Miami (9a) to Richmond (3:30a)
RF&P 58, Richmond (ntg) to Washington (6:10a)
PRR 114, Washington (6:30a)
　　　　　to New York (10:30a)

VENICE/SARASOTA to NEW YORK
B48: *10r-6dbr*
SAL 258-25, Venice (8:20a) and
　　　　Sarasota (9:05a) to Tampa (11:05a)
SAL 158, Tampa (11:30a) to Wildwood (1:15p)
SAL 58, Wildwood (1:40p) to Richmond (3:30a)
As above, Richmond (ntg) to New York (10:30a)

ST. PETERSBURG to NEW YORK
B58: *10r-6dbr*   B59: *10r-6dbr*
SAL 158, St. Petersburg (9:45a)
　　　　to Wildwood (1:15p)
As above, Wildwood (1:40p) to New York (10:30a)

*(Diner and tavern observation, Miami to New York;
dining car, St. Petersburg to New York)*

Palmetto, No. 77
New York to Georgia
Atlantic Coast Line
8 miles north of Rocky Mount

NEW YORK to SAVANNAH
A158: *10s-2c-dr*
PRR 129, New York (2:30p) to Washington (6:45p)
RF&P 77, Washington (7:25p)
　　　　to Richmond (9:45p)
ACL 77, Richmond (10p) to Savannah (11:30a)

NEW YORK to AUGUSTA
A160: *10s-dr-2c*   A159: *10r-6dbr*
As above, New York (2:30p) to Richmond (9:45p)
ACL 77, Richmond (10p)
　　　　to Florence, S.C. (5:35a)
ACL 51, Florence (5:55a) to Augusta (10:30a)

WASHINGTON to SAVANNAH
R500: *8s-2c-dr*
As above, Washington (7:25p)
　　　　to Savannah (11:30a)

WASHINGTON to WILMINGTON
R501: *10s-2c-dr*
As above, Washington (7:25p) to Richmond (9:45p)
ACL 77, Richmond (10p)
　　　　to Rocky Mount (12:10a)
ACL 41, Rocky Mount (2:30a)
　　　　to Wilmington (7:30a)

(Dining cars: New York to Washington, Washington to Savannah)

Palmetto, No. 78
Georgia to New York
Atlantic Coast Line
Arriving in Wilson

SAVANNAH to NEW YORK
A158: *10s-2c-dr*
ACL 78, Savannah (3p) to Richmond (4:10a)
RF&P 78, Richmond (4:20a) to Washington (7a)
PRR 122, Washington (7:20a)
　　　　to New York (11:30a)

AUGUSTA to NEW YORK
A159: *10r-6dbr*   A160: *10s-dr-2c*
ACL 50, Augusta (4:10p) to Florence (8:25p)
ACL 78, Florence (8:45p) to Richmond (4:10a)
As above, Richmond (4:20a) to New York (11:30a)

SAVANNAH to WASHINGTON
R500: *8s-2c-dr*
As above, Savannah (3p) to Washington (7a)

(Dining cars: Savannah to Washington, Washington to New York)

Vacationer, No. 73
All-Pullman train
New York to Florida, winter
Atlantic Coast Line
15 miles north of Rocky Mount

NEW YORK to MIAMI
VA1: *6c-3dr*   VA2: *6c-3dr*
VA3: *8s-dr-3dbr*   VA4: *12r-sbr-4dbr*
VA5: *10s-c-2dbr*   VA6: *6c-3dr*
VA7: *sbr-dr-buffet-lounge*
PRR 117, New York (3:05p) to Washington (7:05p)
RF&P 73, Washington (7:35p)
　　　　to Richmond (9:55p)
ACL 73, Richmond (10:10p) to Jacksonville (9a)
FEC 73, Jacksonville (9:50a) to Miami (5:15p)

NEW YORK to JACKSONVILLE
VA8: *10r-6dbr, every third day*
As above, New York (3:05p) to Jacksonville (9a)

WASHINGTON to MIAMI
R504: *8s-5dbr*
As above, Washington (7:35p) to Miami (5:15p)

BOSTON to MIAMI
MA11: *6c-3dr*   MA12: *6s-4r-4dbr*
NH 177, Boston (11a) to Washington (7:15p)
As above, Washington (7:35p) to Miami (5:15p)

BOSTON to ST. PETERSBURG Alt. days
PA14: *10s-2c-dr*
As above, Boston (11a) to Jacksonville (9a)
ACL 191, Jacksonville (9:50a)
　　　　to St. Petersburg (5:05p)

BOSTON to TAMPA/SARASOTA Alt. days
RA14: *10s-2c-dr*
As above, Boston (11a) to Jacksonville (9a)
ACL 91, Jacksonville (9:50a) to Sarasota (5:25p)

(Diner, New York to Miami)

Palmetto, No. 78
Atlantic Coast Line
Wilmington setout
Standing in Rocky Mount

WILMINGTON to WASHINGTON
R501: *10s-2c-2dr*
ACL 42, Wilmington (7:20p)
　　　　to Rocky Mount (11:55p)
ACL 78, Rocky Mount (1:08a) to Richmond (4:10a)
RF&P 78, Richmond (4:20a) to Washington (7a)

Wilmington depended on the Atlantic Coast Line for most of its Pullman service, which included cars to Columbia, Norfolk, and, with Southern, Greensboro and Raleigh.

Florida Special, No. 87
All-Pullman
New York to Miami, winters
Atlantic Coast Line
10 miles north of Fayetteville

NEW YORK to MIAMI
FS21: *10r-6dbr*   FS22: *10r-6dbr*
FS23: *10r-6dbr*   FS24: *10r-6dbr*
FS25: *10r-6dbr*   FS26: *6dbr-lounge*
FS27: *6dbr-lounge*   FS28: *10r-6dbr*
FS29: *21r*   FS30 *14r-2dr*
FS31: *14r-2dr*   FS32: *10r-6dbr*
FS33: *21r*
PRR 193, New York (1:50p) to Washington (5:50p)
RF&P 87, Washington (6:15p)
　　　　to Richmond (8:30p)
ACL 87, Richmond (8:45p) to Jacksonville (7:15a)
FEC 87, Jacksonville (7:40a) to Miami (2:40p)

NEW YORK to JACKSONVILLE
J15: *10r-6dbr*
As above, New York (1:50p)
　　　　to Jacksonville (7:15a)

(Two dining cars, New York to Miami.)

The *Florida Special* was the patriarch of New York-Florida luxury service, having made its debut in 1888.

Florida Special, No. 88
All-Pullman
Miami to New York, winters
Atlantic Coast Line
35 miles north of Fayetteville

MIAMI to NEW YORK
FS21: *10r-6dbr*   FS22: *10r-6dbr*
FS23: *10r-6dbr*   FS24: *10r-6dbr*
FS25: *10r-6dbr*   FS26: *6dbr-lounge*
FS27: *6dbr-lounge*   FS28: *10r-6dbr*
FS29: *21r*   FS30: *14r-2dr*
FS31: *14r-2dr*   FS32: *10r-6dbr*
FS33: *21r*
FEC 88, Miami (9:25a) to Jacksonville (4:25p)
ACL 88, Jacksonville (4:40p) to Richmond (3:10a)
RF&P 88, Richmond (3:20a) to Washington (5:40a)
PRR 192, Washington (6a) to New York (10:15a)

JACKSONVILLE to NEW YORK
J15: *10r-6dbr*
As above, Jacksonville (4:40p)
　　　　to New York (10:15a)

(Two dining cars, Miami to New York.)

The All-Pullman *Florida Special* commonly operated in several sections, setting a record of seven in 1936.

West Coast Champion, No. 91
New York to Florida Gulf Coast
Atlantic Coast Line
8 miles south of Weldon

NEW YORK to TAMPA
A167: *10r-6dbr*   A168: *21r*
PRR 195, New York (3:40p) to Washington (7:40p)
RF&P 91, Washington (8p) to Richmond (10:20p)
ACL 91, Richmond (10:35p)
　　　　to Jacksonville (9:20a)
ACL 91, Jacksonville (9:50a) to Tampa (3:30p)

NEW YORK to SARASOTA
A169: *10r-6dbr*   A180: *8s-2c-dr*
As above, New York (3:40p)
　　　　to Jacksonville (9:20a)
ACL 91, Jacksonville (9:50a) to Sarasota (5:25p)

NEW YORK to ST. PETERSBURG
A170: *10r-6dbr*   A171: *10r-6dbr*
A172: *14r-2dr*   A173: *8s-2c-dr*
As above, New York (3:40p)
　　　　to Jacksonville (9:20a)
ACL 191, Jacksonville (9:50a)
　　　　to St. Petersburg (5:05p)

WASHINGTON to ST. PETERSBURG
R512: *6dbr-lounge*
As above, Washington (8p)
　　　　to St. Petersburg (5:05p)

WASHINGTON to TAMPA/SARASOTA
R513: *10r-6dbr*
As above, Washington (8p) to Tampa (3:30p)
　　　　and Sarasota (5:25p)

(Dining cars: New York to Tampa, Jacksonville to St. Petersburg; lounge New York to Tampa.)

No. 111
Goldsboro to Greensboro
Southern Railway
7 miles west of Durham

RALEIGH to ASHEVILLE
78: *10s-lounge*
SOU 111, Raleigh (11:05p) to Greensboro (1:10a)
SOU 15, Greensboro (2:40a) to Asheville (9:15a)

No. 112
Greensboro to Goldsboro
Southern Railway
Standing in Greensboro

ASHEVILLE to RALEIGH
78: *10s-lounge*
SOU 16, Asheville (3:45p) to Greensboro (9:25p)
SOU 112, Greensboro (1:30a) to Raleigh (4:05a)

North Dakota

See map on page 282

Ohio

See map on page 70

Empire Builder, No. 2
Streamliner, via New Rockford
Seattle and Portland to Chicago
Great Northern
80 miles west of Fargo

SEATTLE to CHICAGO
Cars 24-27, Std: s-r-dup/r-dbr-c-dr
20: Obs.-lounge-sleeper
GN 2, Seattle (3p Sat) to St. Paul (7a Mon)
CB&Q 44, St. Paul (7:15a) to Chicago (2p Mon)

PORTLAND to CHICAGO
Cars 21-22, Std: s-r-dup/r-dbr-c-dr
SP&S 2, Portland (3p Sat) to Spokane (10:35p)
GN 2, Spokane (11:15p) to St. Paul (7a Mon)
CB&Q 44, St. Paul (7:15a) to Chicago (2p Mon)

(Dining car, coffee shop/lounge, Seattle to Chicago; diner and buffet lounge sleeper, Portland to Spokane. This crack eastbound Builder is in its second night; tomorrow's edition is still in Washington, having departed from Spokane within the last hour.)

Soo-Dominion, No. 4
Moose Jaw, Sask., to St. Paul
Except summers; see No. 14-4
Soo Line
20 miles southeast of Enderlin

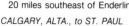

CALGARY, ALTA., to ST. PAUL
8s-dr-c
CPR 8, Calgary (8:10p Sat)
    to Moose Jaw (6:40a Sun)
CPR 14, Moose Jaw (8:20a)
    to N. Portal, Sask. (1:30p MST)
SOO 4, Portal (3:20p CST)
    to Minneapolis (6:30a Mon)
    and St. Paul (7:15a)

MOOSE JAW to ST. PAUL
16s (tourist)
As above, Moose Jaw (8:20a) to St. Paul (7:15a)

(Diner, Portal to Enderlin; buffet-parlor, Moose Jaw to North Portal. Through sleepers are exchanged at Moose Jaw with the Canadian Pacific's Montreal-Vancouver Dominions Nos. 7 and 8.)

The Soo scheduled sleepers to the Twin Cities from Enderlin, Oakes, Minot, and Adams.

Mountaineer, No. 14-4
Vancouver to St. Paul
Summertime only
Soo Line
36 miles southeast of Enderlin

VANCOUVER to ST. PAUL
6s-4dbr-4r   2c-dr-lounge
6c-3dr   8s-dr-2c
14s tourist
CPR 14-4, Vancouver (6:30p Fri)
    to N. Portal (11:15a Sun)
SOO 14-4, Portal (1:15p CST)
    to Minneapolis (6:30a Mon)
    and St. Paul (7:15a Mon)

BANFF to ST. PAUL
6s-6dbr
CPR 14-4, Banff (4:30p Sat)
    to N. Portal (11:15a Sun)
SOO 14-4, Portal (1:15p CST)
    to Minneapolis (6:30a Mon)
    and St. Paul (7:15a Mon)

(Diner, Vancouver to Enderlin. Open observation car, Vancouver to Calgary. The Mountaineer is the summertime version of the Soo-Dominion.)

Erie Limited, No. 1
New York to Chicago
Erie Railroad
Standing in Niles

NEW YORK to CHICAGO
10r-6dbr
Ferry or bus from New York
Erie 1, Jersey City (9:30a) to Chicago (7:50a)

SALAMANCA, N.Y. to CHICAGO
6r-6s-4dbr
Erie 1, Salamanca (7:40p) to Chicago (7:50a)

YOUNGSTOWN to CHICAGO
12r-2sbr-3dbr
Erie 1, Youngstown (11:48p) to Chicago (7:50a)

(Jersey City to Chicago, diner-lounge. The Erie commonly terminated sleepers in Salamanca or Meadville, Pa., to get a same-day turn-around of the car for increased productivity.)

The *Erie Limited* was inaugurated in June 1929, with an extra fare of $4.80 (compared to $10 on the *Twentieth Century Limited*), even though it was almost nine hours slower than its bigger rivals.

Erie Limited, No. 1
Erie Railroad
Akron setout
Standing in Akron

AKRON to CHICAGO
10r-6dbr
Erie 1, Akron (1:15a) to Chicago (7:50a)

Erie Limited, No. 2
Chicago to New York
Erie Railroad
Kenton

CHICAGO to NEW YORK
10r-6dbr
Erie 2, Chicago (6:05p) to Jersey City (6:55p)
Ferry or bus to New York

CHICAGO to SALAMANCA
6r-6s-4dbr
Erie 2, Chicago (6:05p) to Salamanca (8:56a)

(Diner-lounge, Chicago to Jersey City)

Pennsylvania Limited, No. 2
Chicago to New York
Pennsylvania Railroad
14 miles west of Mansfield

CHICAGO to NEW YORK
12dup/r-5dbr   21r
12s-dr   12s-dr
3dbr-dr-lounge
PRR 2, Chicago (6:30p) to New York (11:59a)

CHICAGO to WASHINGTON
8s-2c-dr
PRR 2, Chicago (6:30p) to Harrisburg (8:39a)
PRR 502, Harrisburg (8:50a) to Washington (12:10p)

LOS ANGELES to NEW YORK
10r-6dbr
UP 2, Los Angeles (12:01p Fri) to Omaha (3:50a Sun)
C&NW 2, Omaha (4:30a) to Chicago (2p Sun)
PRR 2, Chicago (6:30p) to New York (11:59a Mon)

(Dining car and lounge, Los Angeles to Chicago and Chicago to New York; cafe-coach, Harrisburg to Washington. UP/C&NW 2 is the Los Angeles Limited.)

Though since surpassed in fame and prestige by the *Broadway Limited*, the *Pennsylvania Limited* was the railroad's premier train of the 19th Century.

Pocahontas, No. 4
Cincinnati to Norfolk
Norfolk & Western
20 miles east of Cincinnati

CHICAGO to NORFOLK
2088: 10s-lounge   2089: 10r-6dbr
PRR 208, Chicago (1:30p) to Cincinnati (9:05p)
N&W 4, Cincinnati (11:25p) to Norfolk (5:30p)

CINCINNATI to WINSTON-SALEM
121: 10r-6dbr
N&W 4, Cincinnati (11:25p) to Roanoke (11:30a)
N&W 21, Roanoke (12:15p)
        to Winston-Salem (4:25p)

(Chicago to Cincinnati, dining car; Williamson, W.Va., to Norfolk, lounge and diner)

Diplomat, No. 4
St. Louis to New York
Baltimore & Ohio
19 miles east of Chillicothe

ST. LOUIS to NEW YORK
8s-buffet-lounge   8s-4dbr
B&O 4, St. Louis (1:20p) to Jersey City (3:25p)
Bus or ferry to New York

LOUISVILLE to NEW YORK
12s-dr
B&O 52, Louisville (5:10p)
        to North Vernon, Ind. (6:20p)
B&O 4, North Vernon (6:40p) to Jersey City (3:25p)

(St. Louis to Jersey City, dining car)

No. 5
Buffalo to Chicago
New York Central
Standing in Cleveland

BUFFALO to CHICAGO
10r-5dbr   12s-dr
NYC 5, Buffalo (8p) to Chicago (7:10a)

(No. 5 has just arrived in Cleveland with sleepers from Buffalo for St. Louis and Miami, now being switched to No. 41, the Knickerbocker, and No. 417, Midnight Special, respectively)

Nickel Plate Limited, No. 5
New York to Chicago
Nickel Plate
Departing Cleveland

NEW YORK to CHICAGO
30: 10r-6dbr
Ferry from New York
DL&W 5, Hoboken (10:30a) to Buffalo (6:45p)
NKP 5, Buffalo (7:30p) to Chicago (7:40a)

CLEVELAND to CHICAGO
51: 10r-6dbr   52: 10r-6dbr
53: Std: dbr   54: 10r-6dbr
NKP 5, Cleveland (11:55p) to Chicago (7:40a)

CLEVELAND to FORT WAYNE
501: 10r-6dbr
NKP 5, Cleveland (11:55p) to Fort Wayne (3:35a)

(This New York-Chicago train—actually, two trains with Buffalo as a common terminus—is known by its eastern parent as the Phoebe Snow and in the Midwest as the Nickel Plate Limited. The Phoebe carries an observation-diner-lounge; a diner-lounge runs west of Buffalo.)

The *Nickel Plate Limited*, inaugurated in 1928, originally paired up with the *Lackawanna Limited* for through service between Chicago and New York.

The Nickel Plate, more formally the New York, Chicago and St. Louis Railroad, was at home in Cleveland, with headquarters in Terminal Tower. To the east, the NKP provided Cleveland with dedicated Pullman service (with DL&W) to New York and Scranton and (with C&O) to White Sulphur Springs, W.Va., Hot Springs, and Clifton Forge, Va.

Richard J. Cook

An elephant-eared 4-8-4 hauls the NYC's *Iroquois* through Beria, Ohio, in March 1946.

Capitol Limited, No. 6
All-Pullman to Washington
Chicago to New York
Baltimore & Ohio
22 miles east of Akron

CHICAGO to WASHINGTON
10r-6dbr	10r-6dbr
10r-6dbr	10r-6dbr
12s-dr	14s
5dbr-obs.-lounge	8s-buffet-lounge
5r-sbr-3dr Strata Dome	

B&O 6, Chicago (4:30p) to Washington (8:55a)

SAN DIEGO to WASHINGTON
10r-6dbr
Santa Fe 71, San Diego (7:45a Fri)
    to Los Angeles (10:30a)
Santa Fe 20, Los Angeles (12:30p)
    to Chicago (10:30a Sun)
B&O 6, Chicago (4:30p) to Washington (8:55a Mon)

CHICAGO to NEW YORK
14r-4dbr	8s-5dbr

B&O 6, Chicago (4:30p) to Jersey City (1:29p)
Bus or ferry to New York

(Dining cars, all segments; lounges, all segments except San Diego to Los Angeles. Santa Fe 20 is the all-Pullman Chief. The Capitol Limited *is preceded by the all-coach* Columbian, *which departs Chicago at 4 p.m. The two will merge in Washington.)*

The *Capitol Limited*, born in May 1923, was an air-conditioning pioneer in 1932 and became the first diesel-powered train from the East to Chicago five years later. For a time the train carried Chicago-Philadelphia cars.

Chicago Mail, No. 9
Cleveland to Chicago
New York Central
Standing in Toledo

TOLEDO to CHICAGO
22r
NYC 9, Toledo (1:35a) to Chicago (5:55a)

The New York Central provided Pullman service from Toledo, not just to Chicago, but in all directions: Pittsburgh, Buffalo, Boston, and New York; with Southern to Jacksonville and Miami (SAL); Indianapolis, and Mackinaw City.

Washington-Pittsburgh-Chicago Express, No. 9
Washington to Chicago
Baltimore & Ohio
8 miles east of Akron

WASHINGTON to CHICAGO
8s-buffet-lounge	10s-2c-dr

B&O 9, Washington (1:45p) to Chicago (6:45a)

(Washington to Pittsburgh, diner)

Washington-Pittsburgh-Chicago Express, No. 9
Baltimore & Ohio
Akron setout
Standing in Akron

AKRON to CHICAGO
8s-4dbr
B&O 9, Akron (12:15a) to Chicago (6:45a)

(No. 9 will also pick up a Chicago sleeper from Wheeling at 2 a.m. in Willard.)

The B&O scheduled Pullmans from Akron to Detroit, Pittsburgh, and Washington.

Mohawk, No. 10
Chicago to New York
New York Central
Entering Toledo

CHICAGO to NEW YORK
10r-6dbr
NYC 10, Chicago (5:40p) to New York (6p)

CHICAGO to BUFFALO
12s
NYC 10, Chicago (5:40p) to Buffalo (7:20a)

(Chicago to Toledo, dining car; Buffalo to Albany, diner-lounge)

Mohawk, No. 10
New York Central
Cincinnati Pullman
Standing in Cleveland

CINCINNATI to BUFFALO
22r
NYC 424, Cincinnati (5p) to Cleveland (10:50p)
NYC 10, Cleveland (2:55a) to Buffalo (7:20a)

(No. 424 is the Cincinnati Mercury, *carrying diner and coach-buffet-lounge)*

Metropolitan Special, No. 11
New York to St. Louis
Baltimore & Ohio
Departing Cincinnati

NEW YORK to ST. LOUIS
12s-dr
Bus or ferry from New York
B&O 11, Jersey City (12:50a Mon)
    to St. Louis (7:40a Tue)

CINCINNATI to ST. LOUIS
8s-4dbr
B&O 11, Cincinnati (11:59p) to St. Louis (7:40a)

(New York to Washington, lunch counter; Washington to Cincinnati, diner-lounge)

The B&O predecessor companies gave the modern railroad a strong presence in Cincinnati, making the city a key hub of operations. Over this main line connecting the upper Ohio and Mississippi River valleys, the railroad scheduled Pullmans from Cincinnati east to Columbus, Wheeling, Pittsburgh, Washington, Baltimore, Philadelphia, New York, and Norfolk (N&W); in cooperation with the IC at Louisville, interline cars departed for Gulfport and New Orleans, and the SP carried tourist sleepers on to Los Angeles and San Francisco.

Lake Shore Limited, No. 22
Chicago to New York
New York Central
Approaching Sandusky

CHICAGO to NEW YORK
10r-6dbr	10r-5dbr
22r	12s-dr
6dbr-buffet-lounge	

NYC 22, Chicago (6:30p) to New York (12:30p)

CHICAGO to BOSTON
10r-5dbr	8s-4dbr

NYC 22, Chicago (6:30p) to Albany (9:25a)
NYC 22, Albany (9:45a) to Boston (2:55p)

LOS ANGELES to NEW YORK
10r-6dbr
UP 2, Los Angeles (12:01p Fri) to Omaha (3:50a Sun)
C&NW 2, Omaha (4:30a) to Chicago (2p Sun)
NYC 22, Chicago (6:30p) to New York (12:30p Mon)

SAN FRANCISCO to NEW YORK alternate days
10r-6dbr
Ferry from San Francisco (11a Fri)
SP 28, Oakland pier (11:30a) to Ogden (6:25a Sat)
UP 28, Ogden (6:55a) to Omaha (3:20a Sun)
C&NW 28, Omaha (3:45a) to Chicago (1p)
NYC 22, Chicago (6:30p) to New York (12:30p Mon)

SAN FRANCISCO to NEW YORK alternate days
10r-6dbr
Ferry from San Francisco (9a Fri)
WP 18, Oakland pier (9:26a)
    to Salt Lake City (5:20a Sat)
D&RGW 18, Salt Lake City (5:40a) to Denver (7p)
CB&Q 18, Denver (7:15p) to Chicago (1:30p Sun)
NYC 22, Chicago (6:30p) to New York (12:30p Mon)

(Dining and lounge facilities, all segments. No. 2 is the Los Angeles Limited, *No. 28 is the* San Francisco Overland, *No. 18 the* California Zephyr.)

The once all-sleeper *Lake Shore Limited* was the premier train of the New York Central until the *Twentieth Century Limited* acquired the mantle.

No. 27
Pittsburgh to St. Louis
Pennsylvania Railroad
25 miles west of Columbus

PITTSBURGH to ST. LOUIS
12s-dr
PRR 27, Pittsburgh (6:20p) to St. Louis (8:10a)

PITTSBURGH to LOUISVILLE
8s-buffet-lounge
PRR 27, Pittsburgh (6:20p) to Indianapolis (2:25a)
PRR 306, Indianapolis (3:40a) to Louisville (6:10a)

Ohio   (continued)

Knickerbocker, No. 24
St. Louis to New York
New York Central
Departing Cleveland

ST. LOUIS to NEW YORK
  10r-6dbr   10r-6dbr
  12s-dr   6dbr-buffet-lounge
NYC 24, St. Louis (12:45p) to New York (11a)

ST. LOUIS to BOSTON
  8s-5dbr
NYC 24, St. Louis (12:45p) to Albany (8:08a)
NYC 22, Albany (9:45a) to Boston (2:55p)

CINCINNATI to NEW YORK
  22r
NYC 424, Cincinnati (5p) to Cleveland (10:50p)
NYC 24, Cleveland (11:59p) to New York (11a)

CLEVELAND to NEW YORK
  10r-6dbr
NYC 24, Cleveland (11:59p) to New York (11a)

CLEVELAND to TORONTO
  8s-5dbr
NYC 24, Cleveland (11:59p) to Buffalo (2:54a)
NYC 371, Buffalo (5:15a) to Welland (6:20a)
TH&B 371, Welland (6:25a) to Hamilton (7:25a)
CPR 712, Hamilton (7:35a) to Toronto (8:35a)

(The Knickerbocker *is one of the New York Central's marvelously ecumenical trains providing a vital segment for several Pullman lines. A St. Louis-Richmond, Va., sleeper left the consist in Indianapolis for the* Carolina Special *and* Sportsman, *and a Pittsburgh-Albany Pullman will join the train in Buffalo. No. 22, the* Lake Shore Limited, *carries a diner to Boston. No. 424, the* Cincinnati Mercury, *and No. 371, the* Ontarian, *both offer dining and lounge facilities. The* Knickerbocker *carries a diner.)*

By way of the New York Central's Great Steel Fleet, Cleveland had dedicated Pullman service east to Buffalo, Rochester, and Albany, and south, via Cincinnati and the Southern Railway, to New Orleans, Jacksonville, St. Augustine, and St. Petersburg.

Twentieth Century Limited, No. 26
All-Pullman streamliner
Chicago to New York
New York Central
Approaching Ashtabula

CHICAGO to NEW YORK
  4c-4dbr-2dr   4c-4dbr-2dr
  4c-4dbr-2dr   12dbr
  12dbr   12dbr
  10r-6dbr   10r-6dbr
  5dbr-obs.-buffet-lounge
NYC 26, Chicago (4:30p) to New York (9:30a)

LOS ANGELES to NEW YORK
  4c-4dbr-2dr   10r-6dbr
Santa Fe 20, Los Angeles (12:30p Fri)
      to Chicago (10:30a Sun)
NYC 26, Chicago (4:30p) to New York (9:30a Mon)

(Dining cars and lounges, all segments. Both the Twentieth Century Limited *and* Santa Fe 20, *the* Chief, *are extra-fare, sleeper-only trains.)*

Its floral bill alone ran to thousands of dollars a year, and its well-trained staff included barbers, maids, valets, and secretaries, in addition to the legions employed in room and table service. It commonly moved in two or more separate sections—sometimes as many as seven—and it pioneered the modern all-room sleeping car train. The New York Central advertised its flagship *Twentieth Century Limited* as a fleet of trains, but it could just as well have called the service a way of life for the standards it maintained in the care of its passengers.

Broadway Limited, No. 28
All-Pullman streamliner
Chicago to New York
Pennsylvania Railroad
6 miles east of Alliance

CHICAGO to NEW YORK
  2dbr-lounge   21r
  4c-2dr-4dbr   4c-2dr-4dbr
  10r-6dbr   10r-6dbr
  10r-6dbr   10r-6dbr
  2mr-dbr-obs.-lounge   12dup/r-4dbr
PRR 28, Chicago (4:30p) to New York (9:30a)

LOS ANGELES to NEW YORK
  4c-2dr-4dbr
Santa Fe 20, Los Angeles (12:30p Fri)
      to Chicago (10:30a Sun)
PRR 28, Chicago (4:30p) to New York (9:30a Mon)

(Both the Broadway *and No. 20, the* Chief, *carry dining cars and lounges. At this scheduling, the* Chief *is the Santa Fe's primary train for the purpose of transporting the top-echelon transcontinental sleepers. Schedule differences made the obvious choice of the* Super Chief *unworkable, although in the twilight of coast-to-coast service, later in the decade, the train will carry some of the Los Angeles Pullmans. Note how both the* Broadway *and* Twentieth Century Limited *are partners with the* Chief, *for each eastern train has its own loyal following. The larger load, however, belongs to the* Century, *as usual.)*

The *Broadway Limited,* the flagship of the Pennsylvania Railroad's vast passenger operation, evolved from the *Pennsylvania Special,* which entered service in 1902. Later renamed the *Broad Way Limited,* in honor of the road's wide swath of trackage leading to New York, the train was the one-on-one challenger to the supremacy of the New York Central's *Twentieth Century Limited.* Though neither train ever was wanting in style or luxury, the *Broadway* was never able to match the volume or profitability of the multi-section *Century* fleet, although it did have the consolation of outliving its noble rival.

"Spirit of St. Louis," No. 30
All-Pullman
St. Louis to New York
Pennsylvania Railroad
28 miles east of Dennison

ST. LOUIS to NEW YORK
  14s   6dbr-lounge
  21r   4c-2dr-4dbr
  12dup/r-4dbr
PRR 30, St. Louis (12:30p) to New York (9:25a)

ST. LOUIS to WASHINGTON
  14s   10r-6dbr
  2dr-c-dbr-obs.-lounge
PRR 30, St. Louis (12:30p) to Harrisburg (6:09a)
PRR 530, Harrisburg (6:49a) to Washington (9:50a)

INDIANAPOLIS to NEW YORK
  21r   10r-6dbr
PRR 30, Indianapolis (4:40p) to New York (9:25a)

INDIANAPOLIS to WASHINGTON
  10r-5dbr
PRR 30, Indianapolis (4:40p) to Harrisburg (6:09a)
PRR 530, Harrisburg (6:49a) to Washington (9:50a)

(Dining cars: St. Louis to New York and Harrisburg to Washington. The all-coach, streamliner Jeffersonian *leaves St. Louis 30 minutes after the departure of the all-Pullman* "Spirit of St. Louis.")*

Although the *"Spirit of St. Louis"* dated from 1909, it was renamed in 1927 for the plane in which Colonel Charles Lindbergh made his historic flight across the Atlantic Ocean to Paris. If the Pennsylvania and New York Central put their foremost efforts into the New York-Chicago standard-bearers, the principal St. Louis trains were next in importance. As of this scheduling the *"Spirit of St. Louis,"* was still an all-Pullman consist in a field of a dozen or so competitors.

St. Louisan, No. 33
New York to St. Louis
Pennsylvania Railroad
32 miles west of Dennison

NEW YORK to ST. LOUIS
  6dbr-lounge   21r
  6s-6dbr
PRR 33, New York (12:55p) to St.Louis (8:20a)

NEW YORK to LOUISVILLE
  6s-6dbr
PRR 33, New York (12:55p) to Indianapolis (3:58a)
PRR 326, Indianapolis (5:50a) to Louisville (8:15a)

WASHINGTON to ST. LOUIS
  12s-dr
PRR 533, Washington (1:10p) to Harrisburg (4:05p)
PRR 33, Harrisburg (4:22p) to St. Louis (8:20a)

(New York to St. Louis, dining car; Indianapolis to Louisville, parlor-cafe)

No. 34
Columbus to Portsmouth
Norfolk & Western
6 miles north of Chillicothe

COLUMBUS to ROANOKE
  103: 10r-6dbr
N&W 34, Columbus (10:55p) to Portsmouth (1:55a)
N&W 4, Portsmouth (2:40a) to Roanoke (11:30a)

(No. 4, the Pocahontas, *carries lounge and dining cars from Williamson to Norfolk)*

Knickerbocker, No. 41
New York Central
Buffalo, Cleveland Pullmans
Standing in Cleveland

BUFFALO to ST. LOUIS
  8s-dbr
NYC 5, Buffalo (8p) to Cleveland (11:55p)
NYC 41, Cleveland (2:18a) to St.Louis (11:30a)

CLEVELAND to ST. LOUIS
  6s-6dbr
NYC 41, Cleveland (2:18a) to St. Louis (11:30a)

(The Knickerbocker, *carrying diner and sleeper lounge, is 66 miles west of Buffalo)*

Sportsman, No. 46
Detroit to Newport News
Chesapeake & Ohio
37 miles south of Columbus

DETROIT to NEWPORT NEWS (PHOEBUS)
  461: 10r-6dbr
C&O 46, Detroit (5:50p) to Ashland, Ky. (1:43a)
C&O 4-46, Ashland (2:16a) to Charlottesville (11:05a)
C&O 46, Charlottesville (11:30a) to Phoebus (5:30p)

DETROIT to HINTON, W.VA.
  464: 10r-6dbr
C&O 46, Detroit (5:50p) to Ashland (1:43a)
C&O 4-46, Ashland (2:16a) to Huntington (2:41a)
C&O 104, Huntington (7a) to Hinton (2:30p)

CLEVELAND to CLIFTON FORGE, VA.
  4451: 10r-6dbr
NYC 445, Cleveland (6p) to Columbus (9:30p)
C&O 46, Columbus (11:15p) to Clifton Forge (8:35a)

(Diners: Detroit to Columbus, Ashland to Richmond, and on NYC 445, the Capital City Special)

Cleveland had a Pullman car through to Charlottesville, Va., via the C&O.

Sportsman, No. 47
Chesapeake & Ohio
Columbus setout
Standing in Columbus

COLUMBUS to DETROIT
  473: 10r-6dbr
C&O 47, Columbus (2:55a) to Detroit (8:20a)

(The Sportsman *is 8 miles northwest of Ashland, Ky.)*

No. 58
Louisville to Detroit
Baltimore & Ohio
Departing Cincinnati

LOUISVILLE to DETROIT
   8s-4dbr
B&O 58, Louisville (6:50p) to Detroit (7:20a)

CINCINNATI to DETROIT
   8s-4dbr
B&O 58, Cincinnati (11:45p) to Detroit (7:20a)

ST. PETERSBURG to DETROIT
   10s-2c-dr
ACL 132, St. Petersburg (6:20p Sat)
   to Trilby, Fla. (8:35p)
ACL 32, Trilby (9:10p) to Albany, Ga. (3:35a Sun)
CofG 32-33, Albany (3:45a) to Atlanta (8:40a)
L&N 32, Atlanta (9:25a) to Cincinnati (9:25p)
B&O 58, Cincinnati (11:45p) to Detroit (7:20a Mon)

(Louisville to Cincinnati, dining-lounge car; Louisville to Detroit, lunch counter. The Florida Pullman travels to Cincinnati on the Southland, which carries dining cars, except between Thomasville and Macon, Ga., and a sleeper-lounge north of Atlanta.)

The B&O line from Cincinnati to Toledo handled both wintertime Pullman traffic to Florida and the summertime flow to northern Michigan resorts. Cincinnati had dedicated sleepers to Toledo and, with Pere Marquette, to Bay View and Harbor Springs, Mich.

No. 58
Baltimore & Ohio
Dayton setout
Standing in Dayton

DAYTON to DETROIT
   8s-4dbr
B&O 58, Dayton (1:25a) to Detroit (7:20a)

Forest City, No. 89
Cleveland to Chicago
New York Central
Standing in Cleveland

CLEVELAND to CHICAGO
6dbr-buffet-lounge	10r-6dbr
10r-6dbr	10r-6dbr
22r	22r

NYC 89, Cleveland (12:30a) to Chicago (7:10a)

(Though no longer all-Pullman, the Forest City still carries an impressive consist of six sleepers)

Admiral, No. 70
Chicago to New York
Pennsylvania Railroad
5 miles east of Mansfield

CHICAGO to NEW YORK
| 6dbr-lounge | 21r |
| 12s-dr | 6s-6dbr |

PRR 70, Chicago (5:30p) to New York (11:45a)

SAN FRANCISCO to NEW YORK alternate days
   10r-6dbr
Ferry from San Francisco (11a Fri)
SP 28, Oakland (11:30a) to Ogden (6:25a Sat)
UP 28, Ogden (6:55a) to Omaha (3:20a Sun)
C&NW 28, Omaha (3:45a) to Chicago (1p)
PRR 70, Chicago (5:30p) to New York (11:45a Mon)

SAN FRANCISCO to NEW YORK alternate days
   10r-6dbr
Ferry from San Francisco (9a Fri)
WP 18, Oakland (9:26a) to Salt Lake City (5:20a Sat)
D&RGW 18, Salt Lake City (5:40a) to Denver (7p)
CB&Q 18, Denver (7:15p) to Chicago (1:30p Sun)
PRR 70, Chicago (5:30p) to New York (11:45a Mon)

(Chicago to New York, dining car. No. 28, the San Francisco Overland, and No. 18, the California Zephyr, both carry dining and lounge facilities.)

The *Admiral* joined its comrade in arms, the *General*, before World War II, as defense preparations helped reverse the railroads' traffic slump that had finally touched bottom in the depths of the Depression.

Ohioan, No. 109
Columbus to Chicago
Pennsylvania Railroad
Departing Columbus

COLUMBUS to CHICAGO
| 10s-lounge | 12dup/r-5dbr |
| 18r | |

PRR 109, Columbus (11:50p) to Chicago (7:30a)

The Pennsylvania scheduled Pullmans from Columbus to Indianapolis and, in summers, Mackinaw City, Mich.

Southland, No. 201
Florida to Chicago
Pennsylvania Railroad
4 miles north of Norwood (Cincinnati)

CINCINNATI to CHICAGO
   10r-5dbr
PRR 201, Cincinnati (11:35p) to Chicago (6:40a)

JACKSONVILLE to CHICAGO
   R301: 8s-buffet-lounge
ACL 94, Jacksonville (9p Sat) to Albany (2a Sun)
CofG 94-5, Albany (2:10a) to Atlanta (8a)
L&N 32, Atlanta (9:25a) to Cincinnati (9:25p)
PRR 201, Cincinnati (11:35p) to Chicago (6:40a Mon)

ST. PETERSBURG to CHICAGO
   R303: 8s-2c-dr   R304: 13dbr or 4c-4dr
ACL 132, St. Petersburg (6:20p Sat) to Trilby (8:35p)
ACL 32, Trilby (9:10p) to Albany (3:35a Sun)
CofG 32-33, Albany (3:45a) to Atlanta (8:40a)
L&N 32, Atlanta (9:25a) to Cincinnati (9:25p)
PRR 201, Cincinnati (11:35p) to Chicago (6:40a Mon)

SARASOTA/TAMPA to CHICAGO
   R302: 8s-dr-3dbr
ACL 76, Sarasota (4:45p Sat) to Tampa (7:10p)
ACL, Tampa (7:35p) to Trilby (8:40p)
As above, Trilby (9:10p) to Chicago (6:40a Mon)

(Dining cars: St. Petersburg to Thomasville, Ga., Macon to Cincinnati. In Richmond, Ind., the Southland will pick up Chicago-bound sleepers from Springfield and Dayton.)

No. 202
Cincinnati to Pittsburgh
Pennsylvania Railroad
24 miles northeast of Cincinnati

NEW ORLEANS to PITTSBURGH
   10s-dr-2dbr
L&N 98, New Orleans (9p Sat)
   to Cincinnati (9:15p Sun)
PRR 202, Cincinnati (11:20p) to Pittsburgh (7a Mon)

CINCINNATI to PITTSBURGH
   18r
PRR 202, Cincinnati (11:20p) to Pittsburgh (7a)

CINCINNATI to NEW YORK
   12s-dr
PRR 202, Cincinnati (11:20p) to Pittsburgh (7a)
PRR 32, Pittsburgh (8a) to New York (4:30p)

(L&N 98, the Pan-American, carries to Cincinnati a diner from Montgomery and lounge sleeper from New Orleans. No. 32, the St. Louisan, has dining and lounge facilities.)

Cincinnati had dedicated sleepers via the Pennsy to Cleveland, New Castle, Pa., Washington, Atlantic City, and Boston (NH).

No. 202
Pennsylvania Railroad
Columbus setout
Standing in Columbus

COLUMBUS to PITTSBURGH
   12s-dr
PRR 202, Columbus (2:20a) to Pittsburgh (7a)

Pittsburgh-Buffalo Express, No. 272
Pittsburgh to Buffalo
New York Central
15 miles north of Youngstown

PITTSBURGH to BUFFALO
| 22r | 8s-lounge |
P&LE 7, Pittsburgh (9:50p) to Youngstown (11:30p)
NYC 272, Youngstown (11:40p) to Buffalo (4:30a)

PITTSBURGH to TORONTO
   6s-6dbr
As above, Pittsburgh (9:50p) to Buffalo (4:30a)
NYC 371, Buffalo (5:15a) to Welland (6:20a)
TH&B 371, Welland (6:25a) to Hamilton (7:25a)
CPR 712, Hamilton (7:35a) to Toronto (8:35a)

(No. 371, the Ontarian, carries a diner-lounge)

B&O's *Washington Express*, Tiffin, Ohio, October 1963. Pittsburgh sleepers will be added.

Howard W. Ameling collection

Ohio
(continued)

No. 203
Pittsburgh to Cincinnati
Pennsylvania Railroad
Standing in Steubenville

PITTSBURGH to COLUMBUS
12s-dr
PRR 203, Pittsburgh (10:40p) to Columbus (3:15a)

PITTSBURGH to CINCINNATI
18r
PRR 203, Pittsburgh (10:40p) to Cincinnati (6:45a)

PITTSBURGH to NEW ORLEANS
10s-2dbr-dr
PRR 203, Pittsburgh (10:40p Sun)
        to Cincinnati (6:45a Mon)
L&N 99, Cincinnati (9a) to New Orleans (7:10a Tue)

(L&N 99, the Pan-American, carries from Cincinnati a diner to Birmingham and sleeper-lounge to New Orleans)

Columbus had limited Pullman service of its own on this line: from Washington and New Castle, and to Cincinnati.

Chicago Night Express, No. 245
Wheeling to Chicago
Baltimore & Ohio
10 miles south of Mansfield

WHEELING to CHICAGO
10s-2c-dr
B&O 245, Wheeling (7p) to Willard (1:15a)
B&O 9, Willard (2a) to Chicago (6:45a)

(Wheeling to Newark, diner; No. 9 carries a sleeper-lounge)

Lake Cities Special, No. 287
Cleveland to Toledo
New York Central
Standing in Cleveland

CLEVELAND to DETROIT
22r
NYC 287, Cleveland (3:55a) to Toledo (5:50a)
NYC 304, Toledo (6:15a) to Detroit (7:45a)

Michigan Special, No. 302
Cincinnati to Detroit
New York Central
Arriving in Middletown

CINCINNATI to DETROIT
12s-dr   10r-6dbr
NYC 302, Cincinnati (11:15p) to Detroit (7a)

JACKSONVILLE to DETROIT
8s-dr-2c
SOU 4, Jacksonville (10p Sat)
        to Cincinnati (9:55p Sun)
NYC 302, Cincinnati (11:15p) to Detroit (7a Mon)

MIAMI to DETROIT
4c-4dbr-2dr   10r-6dbr
10r-5dbr   13dbr
5dbr-obs.-buffet-lounge
FEC 10, Miami (6:30p Sat)
        to Jacksonville (1:30a Sun)
SOU 6, Jacksonville (1:50a) to Cincinnati (10:15p)
NYC 302, Cincinnati (11:15p) to Detroit (7a Mon)

(Southern 4, the Royal Palm, carries a sleeper-lounge from Jacksonville and diner from Atlanta. The New Royal Palm, No. 6, offers dining and lounge facilities from Miami.)

Pullmans departed Cincinnati on this line for Columbus, Toledo, and Mackinaw City.

Detroit Night Express, No. 304
Indianapolis to Detroit
New York Central
7 miles west of Sidney

INDIANAPOLIS to DETROIT
22r
NYC 304, Indianapolis (9p) to Detroit (7:45a)

Indianapolis Express, No. 307
Detroit to Indianapolis
New York Central
Approaching Toledo

DETROIT to INDIANAPOLIS
22r
NYC 307, Detroit (10:25p) to Indianapolis (6:05a)

DETROIT to ST. LOUIS
10r-5dbr
NYC 307, Detroit (10:25p) to Indianapolis (6:05a)
NYC 41, Indianapolis (6:45a) to St. Louis (11:30a)

(No. 41, the Knickerbocker, carries lounge and diner)

Midnight Special, No. 417
Cleveland to Cincinnati
New York Central
Standing in Cleveland

CLEVELAND to COLUMBUS
12s-dr
NYC 417, Cleveland (12:10a) to Columbus (3:20a)

CLEVELAND to DAYTON
12s-dr
NYC 417, Cleveland (12:10a) to Dayton (5:35a)

CLEVELAND to CINCINNATI
12s-dr   12s-dr
22r
NYC 417, Cleveland (12:10a) to Cincinnati (7:10a)

CLEVELAND to MIAMI
10r-6dbr
NYC 417, Cleveland (12:10a Mon)
        to Cincinnati (7:10a)
SOU 5, Cincinnati (8:30a) to Jacksonville (4:05a Tue)
FEC 9, Jacksonville (4:25a) to Miami (11:25a Tue)

BUFFALO to MIAMI
14r-4dbr
NYC 5, Buffalo (8p Sun) to Cleveland (11:55p)
As above, Cleveland (12:10a Mon)
        to Miami (11:25a Tue)

(The Florida Pullmans travel on the streamlined New Royal Palm, with diner and lounge facilities through to Miami)

Columbus had its own Pullman service to St. Petersburg and Miami (Southern/SAL).

Gateway, No. 427
Cleveland to St. Louis
New York Central
14 miles west of Bellefontaine

CLEVELAND to ST. LOUIS
8s-buffet-lounge   10r-5dbr
NYC 427, Cleveland (8:35p) to St. Louis (8a)

(The Gateway will pick up a Cincinnati-St. Louis Pullman in Indianapolis)

Midnight Special, No. 442
New York Central
Dayton setout
Standing in Dayton

DAYTON to CLEVELAND
12s-dr
NYC 442, Dayton (1a) to Cleveland (6:20a)

Midnight Special, No. 442
New York Central
Columbus setout
Standing in Columbus

COLUMBUS to CLEVELAND
12s-dr
NYC 442, Columbus (3:10a) to Cleveland (6:20a)

C&O *Sportsman* at Fostoria, Ohio, 1948.

No. 437
Cincinnati to Chicago
New York Central
Departing Cincinnati

CINCINNATI to CHICAGO
6dbr-buffet-lounge   22r
22r
NYC 437, Cincinnati (11:55p) to Chicago (7:30a)

CINCINNATI to ST. LOUIS
10r-5dbr
NYC 437, Cincinnati (11:55p) to Indianapolis (1:35a)
NYC 427, Indianapolis (1:47a) to St. Louis (8a)

MIAMI to CHICAGO
10r-6dbr   10r-6dbr
FEC 10, Miami (6:30p Sat)
        to Jacksonville (1:30a Sun)
SOU 6, Jacksonville (1:50a) to Cincinnati (10:15p)
NYC 437, Cincinnati (11:55p) to Chicago (7:30a Mon)

JACKSONVILLE to CHICAGO
10s-dr-2c
SOU 4, Jacksonville (10p Sat)
        to Cincinnati (9:55p Sun)
NYC 437, Cincinnati (11:55p) to Chicago (7:30a Mon)

(Southern 6 is the New Royal Palm, with diner and lounge from Miami, and Southern 4 is the Royal Palm, carrying a diner from Atlanta and sleeper-lounge from Jacksonville)

A Cincinnati-Peoria Pullman traveled this route as far as Indianapolis.

Midnight Special, No. 442
Cincinnati to Cleveland
New York Central
19 miles north of Cincinnati

CINCINNATI to CLEVELAND
12s-dr   22r
NYC 442, Cincinnati (11:30p) to Cleveland (6:20a)

MIAMI to CLEVELAND
10r-6dbr
FEC 10, Miami (6:30p Sat)
        to Jacksonville (1:30a Sun)
SOU 6, Jacksonville (1:50a) to Cincinnati (10:15p)
NYC 442, Cincinnati (11:30p)
        to Cleveland (6:20a Mon)

MIAMI to BUFFALO
14r-4dbr
As above, Miami (6:30p Sat)
        to Cleveland (6:20a Mon)
NYC 90, Cleveland (6:35a) to Buffalo (10:20a Mon)

(NYC 442 is the Cleveland-Buffalo frond of the northbound New Royal Palm, which carries diner and lounge from Miami. NYC 90, the Chicagoan, has a diner and sleeper-lounge.)

Oklahoma

See map on page 246

Texas Special, No. 2
Streamliner
San Antonio to St. Louis
Missouri-Kansas-Texas
14 miles north of Wagoner

SAN ANTONIO to ST. LOUIS *via Dallas*
20: 12dr
MKT 2, San Antonio (12:01p) to Muskogee (11:15p) and Vinita (ntg)
SLSF 2, Vinita (ntg) to Monett (2:05a) and St. Louis (8:10a)

DALLAS TO ST. LOUIS
24: 14r-4br *26: 14r-4br*
MKT 2, Dallas (6:10p) to Muskogee (11:15p)
As above, Muskogee (11:30p) to St. Louis (8:10a)

FORT WORTH to ST. LOUIS
28: 14r-4br
MKT 12-2, Fort Worth (5:10p) to Denison (7:45p)
MKT 2, Denison (8:35p) to Muskogee (11:15p)
As above, Muskogee (11:30p) to St. Louis (8:10a)

SAN ANTONIO to WASHINGTON
22: 14r-4br
As above, San Antonio (12:01p Sun) to St. Louis (8:10a)
B&O 2, St. Louis (10:05a Mon) to Washington (7:25a Tue)

SAN ANTONIO to NEW YORK
25: 14r-4br, or 10r-5br
As above, San Antonio (12:01p Sun) to St. Louis (8:10a)
PRR 4, St. Louis (10a Mon) to New York (7:25 a Tue)

(Dining cars: San Antonio to Muskogee, via Dallas; Fort Worth to Denison; Springfield to St. Louis. Observation-lounge, San Antonio to St. Louis 2 days of 3; midtrain full lounge 3rd day. B&O 2 is the National Limited, *PRR 4 the* Penn Texas; *both carry diners and lounges. The* Texas Special, *in an arrangement dating to 1917, operates on the Katy south of Vinita and on Frisco tracks north of there.)*

Kansas Cityan, No. 11
Streamliner
Chicago to Oklahoma City
Atchison, Topeka & Santa Fe
3 miles south of Guthrie

CHICAGO to DALLAS
8s-2c-2dbr
Santa Fe 11, Chicago (9:30a) to Oklahoma City (12:35a)
Santa Fe 111, Oklahoma City (12:50a) to Fort Worth (6:50a) and Dallas (8a)

(Chicago to Oklahoma City: dining car, parlor-observation, and chair-lounge)

Despite its name, the *Kansas Cityan* was created as an Oklahoma City train in April 1938. The eastbound edition was the *Chicagoan*. Over the line to Dallas, the Santa Fe carried Oklahoma City-Ardmore sleepers.

Choctaw, No. 51
Memphis to Amarillo
Rock Island
7 miles west of El Reno

OKLAHOMA CITY to AMARILLO
517: 12s-dr
RI 51, Oklahoma City (10:45p) to Amarillo (7:30a)

Choctaw, No. 52
Amarillo to Memphis
Rock Island
3 miles east of Texas line

AMARILLO to OKLAHOMA CITY
527: 12s-dr
RI 52, Amarillo (8:30p) to Oklahoma City (6a)

As the steam-era *Oil Specials*, Nos. 51 and 52 handled Oklahoma City-Fort Worth-Dallas sleepers exchanged with the *Mid-Continent Special* at El Reno, as well as a St. Louis-Amarillo Pullman in cooperation with the Frisco at Oklahoma City. Over this Choctaw line the Rock Island scheduled Oklahoma City sleepers to Memphis, Little Rock, Muskogee, McAlester, and Amarillo. Via El Reno, Pullmans traveled to Wichita and Kansas City.

No. 111
Oklahoma City to Dallas
Atchison, Topeka & Santa Fe
Standing in Oklahoma City

OKLAHOMA CITY to DALLAS
10s-lounge
Santa Fe 111, Oklahoma City (12:50a) to Fort Worth (6:50a) and Dallas (8a)

(No. 111 will cut in the Chicago-Dallas sleeper after the arrival of No. 11, the Kansas Cityan, *at 12:35 a.m.*

Oklahoman, No. 112
Tulsa to Kansas City
Frisco
37 miles northeast of Tulsa

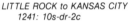

TULSA to KANSAS CITY
161: 10s-dr-2c
SLSF 112, Tulsa (11p) to Kansas City (7a)

Rainbow Special, No. 124
Little Rock to Kansas City
Missouri Pacific
3 miles north of Claremore

LITTLE ROCK to KANSAS CITY
1241: 10s-dr-2c
MP 124, Little Rock (4p) to Osawatomie, Kans. (5:15a)
MP 16, Osawatomie (5:45a) to Kansas City (7:40a)

(Little Rock to Coffeyville, grill-coach)

In addition to Kansas City service, the *Rainbow Special* and other trains on this crescent-shaped line carried sleepers from Little Rock to Fort Smith, Ark., Coffeyville, Kans., Monett, Mo., and Omaha.

Twin Star Rocket, No. 508
Houston to Minneapolis
Rock Island
3 miles south of Waurika

HOUSTON to MINNEAPOLIS
583: 8dup/r-6r-4dbr
RI 508, Houston (5p) to St. Paul (6p) and Minneapolis (6:30p)

HOUSTON to KANSAS CITY
581: 6s-6r-4dbr
RI 508, Houston (5p) to Kansas City (8:40a)

(Houston to Minneapolis, dining car and parlor-lounge-observation)

Black Gold, No. 517
Tulsa to Fort Worth
Frisco
18 miles south of Okmulgee

TULSA to DALLAS
50: 12r-2sb-3dbr
SLSF 517, Tulsa (10:30p) to Dallas (7a)

TULSA to FORT WORTH
52: 8s-dr-2c
SLSF 517, Tulsa (10:30p) to Fort Worth (8:20a)

(Tulsa to Fort Worth, buffet-lounge)

Oregon

See map on page 302

Cascade, No. 11
Streamliner, via Cascade line
Portland to San Francisco
Southern Pacific
15 miles south of Klamath Falls

PORTLAND to SAN FRANCISCO
Four all-room sleepers
SP 11, Portland (4:45p) to Oakland Pier (8:35a)
Ferry to San Francisco (9:15a)

SEATTLE to SAN FRANCISCO
112: 22r *114: 6dbr-10r*
113: 6dbr-10r *115: 4c-4dbr-2dr*
Pool 408, Seattle (12:30p) to Portland (4:30p)
As above, Portland (4:45p) to San Francisco (9:15a)

(Seattle to Portland, diner; Portland to Oakland, dining car and full lounge in an articulated, triple unit. Service between Seattle and Portland is pooled by the UP, GN, & NP.)

Portland Rose, No. 17
Denver to Portland
Union Pacific
Standing in Hinkle

DENVER to PORTLAND
6s-6r-4dbr
UP 17, Denver (5:40p Sat) to Portland (6a Mon)

KANSAS CITY to PORTLAND
10s-dr-2c
UP 37, Kansas City (11:59p Fri) to Denver (3:30p Sat)
As above, Denver (5:40p) to Portland (6a Mon)

CHICAGO to PORTLAND
10s-dr-c
C&NW 23, Chicago (8p Fri) to Omaha (7:30a Sat)
UP 23, Omaha (8:20a) to Green River, Wyo. (1:35a Sun)
UP 17, Green River (3:25a) to Portland (6a Mon)

(Denver to Portland, first-class lounge; Denver to Hinkle, dining car.)

Portland Rose, No. 18
Portland to Denver
Union Pacific
9 miles east of The Dalles

PORTLAND to DENVER
6s-6r-4dbr
UP 18, Portland (9:45p Sun) to Denver (8:10a Tue)

PORTLAND to KANSAS CITY
10s-dr-c
UP 18, Portland (9:45p Sun) to Denver (8:10a Tue)
UP 38, Denver (9a) to Kansas City (10:30p Tue)

PORTLAND to CHICAGO
10s-dr-2c
UP 18, Portland (9:45p Sun) to Green River (11:05p Mon)
UP 24, Green River (12:15a Tue) to Omaha (7:50p)
C&NW 24, Omaha (8:30p) to Chicago (8:15a Wed)

(Portland to Denver, first-class lounge; Hinkle to Denver, diner. The other east- and westbound Portland Roses *are now in Wyoming.)*

Klamath, No. 19
Portland to San Francisco
Southern Pacific
Entering Salem

PORTLAND to SAN FRANCISCO
Std: s-dr-c
SP 19, Portland (10:10p) to Oakland Pier (8:45p)
Ferry to San Francisco (9:20p)

(Portland to Oakland, diner/coffee shop)

Oregon   (continued)

The Spokane, No. 20
Portland to Spokane
Union Pacific
Standing in The Dalles

PORTLAND to SPOKANE
8s-buffet-lounge   10r-6dbr
UP 20, Portland (10p) to Spokane (7:20a)

PORTLAND to WALLA WALLA
10s-3dbr
UP 20, Portland (10p) to Hinkle (1:06a)
UP 72, Hinkle (2:55a) to Walla Walla (5:25a)

PORTLAND to LEWISTON, IDA.
8s-coach
UP 20, Portland (10p) to Ayer, Wash. (4:40a)
UP 74, Ayer (5:30a) to Lewiston (8:25a)

PORTLAND to YAKIMA
6s-coach
UP 20, Portland (10p) to Hinkle (1:06a)
UP 72, Hinkle (2:55a) to Wallula, Wash. (3:40a)
UP 64, Wallula (3:55a) to Yakima (7:50a)

The Columbia and Snake River waterways gave Portland early access to eastern Washington, and the commercial ties were sustained into the railroad era. *The Spokane* did a good Pullman business with commercial travelers through this territory.

The Spokane, No. 20
Salt Lake City sleeper
Union Pacific
Standing in Hinkle

SALT LAKE CITY to SPOKANE
10s-3dbr
UP 33, Salt Lake City (11:30p Sat)
    to Pocatello (4:15a)
UP 17, Pocatello (9:40a Sun) to Hinkle (11:20p)
UP 20, Hinkle (2:40a Mon) to Spokane (7:20a Mon)

(No. 17, lounge and diner)

City of Portland, No. 105
Streamliner
Chicago to Portland
Union Pacific   Logo 50
50 miles northwest of Huntington

CHICAGO to PORTLAND
2dr-4c-4dbr   6s-6r-4dbr
6s-6r-4dbr
C&NW 105, Chicago (5:30p Sat)
    to Omaha (1:30a Sun)
UP 105, Omaha (1:40a) to Portland (7:30a Mon)

CHICAGO to SEATTLE
12r-4dbr
As above Chicago (5:30p Sat)
    to Portland (7:30a Mon)
Pool 457, Portland (8a) to Seattle (11:59a Mon)

(Chicago to Portland, dining car and first-class lounge with bath. Portland to Seattle, Astra Dome observation-lounge and Astra Dome dining car, equipment from General Motors' former Train of Tomorrow, *purchased by UP in 1950.)*

When inaugurated in June 1935, the *City of Portland* broke the forty-hour barrier, shaving eighteen hours off the next best Chicago-Portland carding.

Rogue River, No. 330
Ashland to Portland
Southern Pacific
105 miles north of Grants Pass

ASHLAND to PORTLAND
Std: s-c-dr
SP 330, Ashland (6p) to Portland (7:25a)

(Ashland to Portland, snack-lounge)

The last eastbound *City of Portland* is serviced at LaGrande, Ore., April 30, 1971.

City of Portland, No. 106
Streamliner
Portland to Chicago
Union Pacific
18 miles east of La Grande

PORTLAND to CHICAGO
2dr-4c-4dbr   6s-6r-4dbr
6s-6r-4dbr
UP 106, Portland (5:30p Sun) to Omaha (3a Tue)
C&NW 106, Omaha (3:10a) to Chicago (11:30a Tue)

SEATTLE to CHICAGO
12r-4dbr
Pool 408, Seattle (12:30p Sun) to Portland (4:30p)
As above, Portland (5:30p) to Chicago (11:30a Tue)

(Seattle to Portland, diner; Portland to Chicago, dining car, first-class lounge with bath. The other eastbound edition of the City of Portland *is now in Nebraska, and the westbound one in Iowa.)*

Rogue River, No. 329
Portland to Ashland
Southern Pacific
4 miles south of Eugene

PORTLAND to ASHLAND
Std: s-c-dr
SP 329, Portland (7p) to Ashland (9:10a)

(Portland to Ashland, snack-lounge)

The *Rogue River* is the surviving night train on the Siskiyou line, which handled all the SP's California traffic before the opening of the Cascade line in the late 1920's. Sleepers for Eugene and Marshfield rode this line from Portland.

No. 401
Portland to Seattle
Pool
Leaving Portland

PORTLAND to SEATTLE
10s-dr-c   10s-dr-c
12s-dr
Pool 401, Portland (11:45p) to Seattle (6:45a)

PORTLAND to TACOMA
10s-dr-c
Pool 401, Portland (11:45p) to Tacoma (5a)

Pennsylvania
See map on page 50

At midnight, the cream of the Pennsylvania Railroad's western fleet was found on the sinuous main line between Pittsburgh and Harrisburg. Twenty name trains were among those that crested the Appalachians within a few hours of one another, either blasting their way up Horseshoe Curve westward to the summit with one or more K4 helpers, or, with sparking brake shoes, easing off the mountains down into the great marshaling yards of Altoona before continuing to the east. These were the night trains that worked this remarkable 245-mile stretch of trackage and their relative locations at midnight, from west to east (even numbers eastbound, odd westbound):

Between Pittsburgh and Greensburg: *Red Arrow* (68), *Cincinnati Limited* (40), *Clevelander* (38), *Philadelphia Night Express* (36).

Between Greensburg and Johnstown: *Statesman* (50), *General* (49), *Pittsburgher* (60), *Penn Texas* (4), *Liberty Limited* (59), *Red Arrow* (69).

Between Johnstown and Altoona: *Iron City Express* (16), *Cincinnati Limited* (41), *Manhattan Limited* (22), *Broadway Limited* (29), *American* (66), *"Spirit of St. Louis"* (31).

Between Altoona and Harrisburg: *Pennsylvania Limited* (1), *Penn Texas* (3), *Admiral* (71), *American* (67).

Westerner, No. 7 | Lackawanna Railroad |
New York to Chicago
Lackawanna
23 miles north of Scranton

NEW YORK to BUFFALO
71: 10r-6dbr
Ferry from New York
DL&W 7, Hoboken (8:15p) to Buffalo (4:45a)

NEW YORK to CHICAGO
70: 10r-6dbr
DL&W 7, Hoboken (8:15p) to Buffalo (4:45a)
NKP 7, Buffalo (5:20a) to Chicago (3:40p)

(New York to Scranton, buffet-lounge; Buffalo to Chicago, diner-lounge. Approaching Nicholson, the Westerner *is about to cross the picturesque Tunkhannock Viaduct, the largest reinforced concrete structure of its kind.)*

No. 7 carried a Scranton-Buffalo sleeper. The Lackawanna also provided Scranton with dedicated Pullman service to New York and, with Nickel Plate, to Cleveland and Chicago.

Pennsylvania Limited, No. 1
New York to Chicago
Pennsylvania Railroad
24 miles east of Altoona

NEW YORK to CHICAGO
6dbr-lounge	21r
6s-6dbr	12s-dr
12s-dr	

PRR 1, New York (6:45p) to Chicago (10:25a)

WASHINGTON to CHICAGO
8s-2c-dr
PRR 531, Washington (6:15p) to Harrisburg (9:20p)
PRR 1, Harrisburg (10:08p) to Chicago (10:25a)

NEW YORK to SAN FRANCISCO   alternate days
10r-6dbr
PRR 1, New York (6:45p Sun)
     to Chicago (10:25a Mon)
CB&Q 17, Chicago (3:30p) to Denver (8:20a Tue)
D&RGW 17, Denver (8:40a)
     to Salt Lake City (10:20p)
WP 17, Salt Lake City (10:40p)
     to Oakland (4:15p Wed)
Ferry to San Francisco (4:50p Wed)

NEW YORK to SAN FRANCISCO alternate days
10r-6dbr
PRR 1, New York (6:45p Sun)
     to Chicago (10:25a Mon)
C&NW 27, Chicago (3:30p) to Omaha (12:30a Tue)
UP 27, Omaha (12:55a) to Ogden (8:33p)
SP 27, Ogden (9p) to Oakland pier (1:45p Wed)
Ferry to San Francisco (2:20p Wed)

(Dining cars: New York to Chicago, and Washington to Harrisburg. Nos. 27, the San Francisco Overland, and Nos. 17, the California Zephyr, have dining car and lounge facilities, and the Zephyr has Vista Domes.)

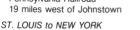

Penn Texas, No. 4
St. Louis to New York
Pennsylvania Railroad
19 miles west of Johnstown

ST. LOUIS to NEW YORK
10r-5dbr	12s-dr

PRR 4, St. Louis (10a) to New York (7:25a)

EL PASO to NEW YORK
14r-4dbr
T&P 2, El Paso (12:50a Sat) to Texarkana (10:30p)
MP 2, Texarkana (10:40p) to St. Louis (8:10a Sun)
PRR 4, St. Louis (10a) to New York (7:25a Mon)

HOUSTON to NEW YORK
10r-6dbr
MP 26, Houston (4p Sat) to Palestine (6:55p)
MP 22, Palestine (7:20p) to Longview (9p)
T&P 222, Longview (9:05p) to Texarkana (10:40p)
MP 22, Texarkana (10:50p) to St. Louis (8:20a Sun)
PRR 4, St. Louis (10a) to New York (7:25a Mon)

SAN ANTONIO to NEW YORK
10r-6dbr
MP 22, San Antonio (2:10p Sat) to Longview (9p)
As above, Longview (9:05p)
     to New York (7:25a Mon)

HOUSTON to WASHINGTON
10r-6dbr
As above, Houston (4p Sat) to St. Louis (8:20a Sun)
PRR 4, St. Louis (10a) to Harrisburg (3:52a Mon)
PRR 504, Harrisburg (4:15a)
     to Washington (7:35a Mon)

SAN ANTONIO to NEW YORK
14r-4dbr, or	10r-6dbr

MKT/SLSF 2, San Antonio (12:01p Sat)
     to St. Louis (8:10a Sun)
PRR 4, St. Louis (10a) to New York (7:25a Mon)

(MP 2 and 22 are the streamlined Texas Eagles; MKT/SLSF 2 is the streamlined Texas Special. Lounge facilities all segments; diners all but Muskogee to Springfield on the Texas Special and on PRR 504.)

Penn Texas, No. 3
New York to St. Louis
Pennsylvania Railroad
56 miles west of Harrisburg

NEW YORK to COLUMBUS
12dup/r-4dbr
PRR 3, New York (7:35p) to Columbus (8a)

NEW YORK to ST. LOUIS
10r-5dbr	12s-dr

PRR 3, New York (7:35p) to St. Louis (3p)

NEW YORK to EL PASO
14r-4dbr
PRR 3, New York (7:35p Sun) to St. Louis (3p Mon)
MP 1, St. Louis (5:30p) to Texarkana (3:05a Tue)
T&P 1, Texarkana (3:25a) to El Paso (10:15p Tue)

NEW YORK to HOUSTON
10r-6dbr
PRR 3, New York (7:35p Sun) to St. Louis (3p Mon)
MP 21, St. Louis (5:30p) to Texarkana (3a Tue)
T&P 221, Texarkana (3:10a)
     to Longview, Tex. (4:45a)
MP 21, Longview (4:50a) to Palestine (6:30a)
MP 25, Palestine (6:50a) to Houston (10a Tue)

NEW YORK to SAN ANTONIO
10r-6dbr
As above, New York (7:35p Sun)
     to Palestine (6:30a Tue)
MP 21, Palestine (6:45a) to San Antonio (11:40a Tue)

NEW YORK to SAN ANTONIO
14r-4dbr, or	10r-6dbr

PRR 3, New York (7:35p Sun) to St. Louis (3p Mon)
SLSF/MKT 1, St. Louis (5:30p)
     to San Antonio (1:55p Tue)

WASHINGTON to HOUSTON
10r-6dbr
PRR 575, Washington (7:45p Sun)
     to Harrisburg, Pa. (10:44p)
PRR 3, Harrisburg (10:53p) to St. Louis (3p Mon)
As above, St. Louis (5:30p) to Houston (10a Tue)

(MP 1 and 21 are the streamlined Texas Eagles; SLSF/MKT 1 is the streamlined Texas Special. Lounges carried on all segments; dining cars except between Washington and Harrisburg, and on the Texas Special between Springfield, Mo., and Muskogee, Okla.,

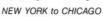

Capitol Limited, No. 5
New York to Chicago
Baltimore & Ohio
Arriving in Pittsburgh

NEW YORK to CHICAGO
14r-4dbr	8s-5dbr

Bus or ferry from New York
B&O 5, Jersey City (12:45p) to Chicago (8a)

WASHINGTON to CHICAGO
5dbr-observation-lounge	5r-sbr-3dr Strata Dome
8s-lounge	14s
12s-dr	

B&O 5, Washington (5:30p) to Chicago (8a)

WASHINGTON to SAN DIEGO
10r-6dbr
B&O 5, Washington (5:30p Sun) to Chicago (8a Mon)
Santa Fe 19, Chicago (1:30p)
     to Los Angeles (8:30a Wed)
Santa Fe 74, Los Angeles (11:30a)
     to San Diego (2:15p Wed)

(Diners on all segments; lounge facilities on all but No. 74. The Capitol Limited and overnight coach train Columbian operate together to Washington. There the Capitol Limited becomes all-Pullman, departing for Chicago 10 minutes before the Columbian. See B&O 25, also in Pennsylvania. No. 19 is the Santa Fe Chief.)

The *Capitol Limited* once carried Philadelphia-Chicago Pullmans, but by the time of this scheduling, all the B&O's dedicated service to Philadelphia had ended, including sleepers to Cincinnati, Cleveland, Detroit, Louisville, Parkersburg, W.Va., Pittsburgh, St. Louis, Washington, and New York.

Maple Leaf, No. 7
New York to Buffalo, Toronto
Lehigh Valley
22 miles south of Wilkes Barre

NEW YORK to BUFFALO
23: Std: s-c-dr
LV 7, New York (8:10p) to Buffalo (5:37a)

NEW YORK to TORONTO
20: Std: s-dbr
LV 7, New York (8:10p)
     to Suspension Bridge, N.Y. (6:20a)
CNR 93, Suspension Bridge (6:30a)
     to Hamilton (7:55a)
CNR 94, Hamilton (8a) to Toronto (9a)

PHILADELPHIA to TORONTO
329: Std: s-c-dr
RDG 329, Philadelphia (8:20p) to Bethlehem (10p)
LV 7, Bethlehem (10:21p)
     to Suspension Bridge (6:20a)
As above, Suspension Bridge (6:30a)
     to Toronto (9a)

(New York to Lehighton, cafe-lounge; Suspension Bridge to Toronto, buffet-parlor)

No. 7, previously known as the *New Yorker*, carried New York-Chicago, and Philadelphia-Detroit sleepers in cooperation with the Canadian National system. Lehigh Valley passenger service ended with the final run of the *Maple Leaf* in February 1961.

Express, No. 10
Chicago to Washington
Baltimore & Ohio
3 miles west of Connellsville

CHICAGO to WASHINGTON
8s-buffet-lounge	10s-2c-dr

B&O 10, Chicago (11a) to Washington (6:25a)

PITTSBURGH to NEW YORK
8s-dr-3dbr
B&O 10, Pittsburgh (10:35p) to Washington (6:25a)
B&O 2, Washington (8a)
     to Jersey City (12:50p)
Bus or ferry to New York

(Chicago to Pittsburgh, diner. No. 2, the National Limited, carries dining and lounge facilities.)

The B&O scheduled Pullmans from Pittsburgh to Baltimore and Richmond (RF&P).

Iron City Express, No. 16
Pittsburgh to New York
Pennsylvania Railroad
3 miles east of Johnstown

PITTSBURGH to NEW YORK
6dbr-lounge	12s-dr
21r	

PRR 16, Pittsburgh (10:10p) to New York (7:15a)

BIRMINGHAM to NEW YORK
10s-2c-dr
L&N 4, Birmingham (9:10p Sat)
     to Cincinnati (12:05p Sun)
PRR 204, Cincinnati (1p) to Pittsburgh (8:30p)
PRR 16, Pittsburgh (10:10p)
     to New York (7:15a Mon)

MEMPHIS to NEW YORK
10r-6dbr
L&N 104, Memphis (8:15p Sat)
     to Bowling Green, Ky. (3:40a Sun)
L&N 4, Bowling Green (4:15a) to Cincinnati (12:05p)
As above, Cincinnati (1p) to New York (7:15a Mon)

PITTSBURGH to PHILADELPHIA
8s-2c-dr
PRR 16, Pittsburgh (10:10p) to Harrisburg (3:47a)
PRR 36, Harrisburg (5:35a) to Philadelphia (7:35a)

(Diners, Louisville to Cincinnati and Columbus to Pittsburgh; lounge facilities, Memphis to Cincinnati. No. 36, the Philadelphia Night Express, carries cafe-coach. L&N 4 is the Azalean.)

The Pennsy's No. 72, *Juniata*, makes time near Losh's Run, Pa., on July 20, 1940.

Pennsylvania   (continued)

Cleveland Night Express, No. 18
Cleveland to Baltimore
Baltimore & Ohio
22 miles north of Pittsburgh

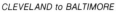

CLEVELAND to BALTIMORE
10r-5dbr   8s-dr-3dbr
8s-2c-dr
B&O 18, Cleveland (9p) to Baltimore (8:30a)

(Cleveland to Baltimore Camden Station, snack bar-lounge)

Cleveland Night Express, No. 18
Baltimore & Ohio
Pittsburgh setouts
Standing in Pittsburgh

PITTSBURGH to WASHINGTON
10s-3dbr
17r, or   12r-sbr-4dbr
B&O 18, Pittsburgh (12:40a) to Washington (7:35a)

(In Cumberland, Md., No. 18 will pick up Washington sleepers from Wheeling and Parkersburg, W. Va.)

Ambassador, No. 19
Baltimore to Detroit
Baltimore & Ohio
25 miles northwest of Connellsville

BALTIMORE to DETROIT
8s-2c-dr   8s-2c-dr
10r-6dbr   14sbr
5dbr-observation-lounge
B&O 19, Baltimore (4:45p) to Detroit (7:50a)

(Baltimore Camden Station to Detroit, dining car)

Pittsburgh had dedicated sleeping car service on the B&O to Akron and Detroit.

Ambassador, No. 20
Detroit to Baltimore
Baltimore & Ohio
10 miles south of New Castle

DETROIT to BALTIMORE
8s-2c-dr   8s-2c-dr
10r-6dbr   14sbr
5dbr-observation-lounge
B&O 20, Detroit (5:45p) to Baltimore (9:25a)

AKRON to BALTIMORE
12s-dr
B&O 20, Akron (10:20p) to Baltimore (9:25a)

(Detroit to Baltimore Camden Station, dining car)

Manhattan Limited, No. 22
Chicago to New York
Pennsylvania Railroad
9 miles east of Johnstown

CHICAGO to NEW YORK
3dbr-dr-lounge   10r-5dbr
12s-dr   12s-dr
PRR 22, Chicago (12:01p) to New York (6:45a)

CHICAGO to WASHINGTON
8s-2c-dr
PRR 22, Chicago (12:01p) to Harrisburg (3:03a)
PRR 504, Harrisburg (4:15a) to Washington (7:35a)

(Chicago to Pittsburgh, diner)

The *Manhattan Limited* was once all-Pullman.

Manhattan Limited, No. 23
New York to Chicago
Pennsylvania Railroad
12 miles west of Pittsburgh

NEW YORK to CHICAGO
3dbr-dr-lounge   10r-5dbr
12s-dr
PRR 23, New York (2:50p) to Chicago (7:20a)

(Diner, New York to Chicago. Saturday consist is different.)

Columbian, No. 25
Streamliner
New York to Chicago
Baltimore & Ohio
9 miles east of Pittsburgh

(Washington to Chicago, diner and both Strata Dome and observation lounges. Commonly considered an all-coach train, the Columbian will nonetheless pick up two Chicago Pullmans at Pittsburgh.)

The *Columbian* was introduced in the fall of 1929 as a New York-Washington day train. Two years later, it became the nation's first train to be totally air-conditioned.

Columbian, No. 25
Baltimore & Ohio
Pittsburgh setouts
Standing in Pittsburgh

PITTSBURGH to CHICAGO
8s-2c-dr   10r-5dbr
B&O 25, Pittsburgh (12:30a) to Chicago (8:20a)

Over the years, the B&O scheduled a considerable number of Pullmans to Pittsburgh, especially from West Virginia: Charleston, Clarksburg, Elkins (WM), Fairmont, Hinton (C&O), Kenova, and Parkersburg. Trains from the southwest brought sleepers from St. Louis, Louisville, Cincinnati, and Columbus. And the Buffalo, Rochester & Pittsburgh, eventually part of the B&O, offered service to the two New York cities.

Broadway Limited, No. 29
All-Pullman streamliner
New York to Chicago
Pennsylvania Railroad
10 miles east of Johnstown

NEW YORK to CHICAGO
2dbr-lounge   21r
4c-2dr-4dbr   4c-2dr-4dbr
10r-6dbr   10r-6dbr
10r-6dbr   10r-6dbr
12dup/r-4dbr   2mr-dbr-obs.-lounge
PRR 29, New York (6p) to Chicago (9a)

NEW YORK to LOS ANGELES
4c-2dr-4dbr
PRR 29, New York (6p Sun) to Chicago (9a Mon)
Santa Fe 19, Chicago (1:30p)
    to Los Angeles (8:30a Wed)

(Diners and lounges, all segments. No. 19 is the all-Pullman, streamlined *Chief*.)

"Spirit of St. Louis," No. 31
All-Pullman
New York to St. Louis
Pennsylvania Railroad
At Horseshoe Curve

NEW YORK to INDIANAPOLIS
10r-6dbr   21r
PRR 31, New York (6:10p) to Indianapolis (8:45a)

NEW YORK to ST. LOUIS
12dup/r-4dbr   6dbr-lounge
14s   21r
4c-2dr-4dbr
PRR 31, New York (6:10p) to St. Louis (1:10p)

WASHINGTON to INDIANAPOLIS
10r-5dbr
PRR 531, Washington (6:15p) to Harrisburg (9:20p)
PRR 31, Harrisburg (9:22p) to Indianapolis (8:45a)

WASHINGTON to ST. LOUIS
14s   10r-6dbr
2dr-c-dbr-observation-lounge
PRR 531, Washington (6:15p) to Harrisburg (9:20p)
PRR 31, Harrisburg (9:22p) to St. Louis (1:10p)

(Dining cars: New York to St. Louis, Washington to Harrisburg. The all-coach *Jeffersonian* leaves New York's Pennsylvania Station for St. Louis five minutes after the departure of the "Spirit of St. Louis.")

Pittsburgh Night Express, No. 35

Philadelphia to Pittsburgh
Pennsylvania Railroad
20 miles west of Philadelphia

PHILADELPHIA to PITTSBURGH
6sbr-buffet-lounge	12dup/r-5dbr
12s-dr	21r

PRR 35, Philadelphia (11:25p) to Pittsburgh (7:35a)

PHILADELPHIA to PITTSBURGH
8s-2c-dr
PRR 35, Philadelphia (11:25p) to Harrisburg (1:38a)
PRR 37, Harrisburg (2:42p) to Pittsburgh (7:50a)

(No. 35 departs from Broad Street Station. Saturday schedules differ. No. 37, the Iron City Express, carries a Birmingham sleeper, to which Philadelphia passengers can make intrain transfers.)

Pennsylvania trains to the west carried Pullmans from Philadelphia to Altoona, Cincinnati, and Chicago.

Philadelphia Night Express, No. 36

Pittsburgh to Philadelphia
Pennsylvania Railroad
5 miles west of Greensburg

PITTSBURGH to PHILADELPHIA
6sbr-buffet-lounge	12dup/r-5dbr
12s-dr	21r

PRR 36, Pittsburgh (11:25p) to Philadelphia (7:35a)

PITTSBURGH to HARRISBURG
12s-dr
PRR 36, Pittsburgh (11:25p) to Harrisburg (4:27a)

CLEVELAND to PHILADELPHIA
10r-5dbr	6s-6dbr

PRR 38, Cleveland (8:30p) to Harrisburg (4:35a)
PRR 36, Harrisburg (5:35a) to Philadelphia (7:35a)

(Cleveland to Harrisburg, diner-lounge; Harrisburg to Philadelphia, cafe-coach. No. 38 is the Clevelander. Philadelphia arrivals at Broad Street Station. Schedules on Saturday differ.)

No. 36 handled Pittsburgh's Pullman to Atlantic City.

Clevelander, No. 38

Cleveland to New York
Via Salem
Pennsylvania Railroad
15 miles east of Pittsburgh

CLEVELAND to NEW YORK
6dbr-lounge	12s-dr
21r	

PRR 38, Cleveland (8:30p) to New York (8:10a)

CLEVELAND to WASHINGTON
10r-5dbr	8s-2c-dr

PRR 38, Cleveland (8:30p) to Harrisburg (4:35a)
PRR 50, Harrisburg (4:50a) to Washington (7:50a)

AKRON to NEW YORK
3dbr-dr-lounge	10r-5dbr
12s-dr	12dup/r-5dbr

PRR 10, Akron (8:10p) to Pittsburgh (11:15p)
PRR 38, Pittsburgh (11:40p) to New York (8:10a)

(Diner-lounges: Akron to Pittsburgh, Cleveland to New York; buffet-lounge, Harrisburg to Washington)

The eastbound *Clevelander* was formerly known as the *Buckeye Limited*. In both directions the train was all-Pullman.

Clevelander, No. 39

New York to Cleveland
Via Salem
Pennsylvania Railroad
Arriving in Harrisburg

NEW YORK to CLEVELAND
6dbr-lounge	6s-6dbr
12s-dr	21r
10r-5dbr	

PRR 39, New York (8:45p) to Cleveland (8:30a)

(Diner-lounge)

Clevelander, No. 39

Pennsylvania Railroad
Washington Pullmans
Standing in Harrisburg

WASHINGTON to CLEVELAND
10r-5dbr	8s-2c-dr

PRR 539, Washington (8:40p) to Harrisburg (11:35p)
PRR 39, Harrisburg (12:15a) to Cleveland (8:30a)

Cincinnati Limited, No. 40

Cincinnati to New York
Pennsylvania Railroad
10 miles east of Pittsburgh

COLUMBUS to NEW YORK
12dup/r-4dbr
PRR 40, Columbus (7:58p) to New York (8:20a)

CINCINNATI to NEW YORK
6dbr-lounge	12s-dr
21r	21r
12dup/r-4dbr	2dr-c-dbr-obs.-lounge

PRR 40, Cincinnati (5:15p) to New York (8:20a)

LOUISVILLE to NEW YORK
10r-6dbr
L&N 8, Louisville (12:40p) to Cincinnati (4:55p)
PRR 40, Cincinnati (5:15p) to New York (8:20a)

NASHVILLE to NEW YORK
10r-6dbr
L&N 8, Nashville (6:40a) to Cincinnati (4:55p)
PRR 40, Cincinnati (5:15p) to New York (8:20a)

(Diners: Louisville to Cincinnati, Cincinnati to New York)

Cincinnati Limited, No. 41

New York to Cincinnati
Pennsylvania Railroad
6 miles east of Johnstown

NEW YORK to CINCINNATI
12s-dr	6dbr-lounge
21r	21r
12dup/r-4dbr	2dr-c-dbr-obs.-lounge

PRR 41, New York (5:25p) to Cincinnati (8:30a)

NEW YORK to LOUISVILLE
10r-6dbr
PRR 41, New York (5:25p) to Cincinnati (8:30a)
L&N 99, Cincinnati (9a) to Louisville (11:10a)

NEW YORK to NASHVILLE
10r-6dbr
PRR 41, New York (5:25p) to Cincinnati (8:30a)
L&N 99, Cincinnati (9a) to Nashville (3:20p)

NEW YORK to MEMPHIS
10r-6dbr
PRR 41, New York (5:25p) to Cincinnati (8:30a)
L&N 99, Cincinnati (9a) to Bowling Green (1:35p)
L&N 199, Bowling Green (1:55p) to Memphis (8:40p)

(Diner, New York to Cincinnati. No. 99, the Pan-American, carries a diner and sleeper-lounge. L&N 199 had a diner from Guthrie, Ky., to Memphis.)

New Englander, No. 46
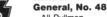
Pittsburgh to New York
Pennsylvania Railroad
Arriving North Philadelphia

PITTSBURGH to BOSTON
12r-2sbr-3dbr	12s-dr

PRR 46, Pittsburgh (4p) to New York (1:50a)
NH 186, New York (2:25a) to Boston (7:45a)

PITTSBURGH to SPRINGFIELD, MASS.
8s-buffet-lounge
PRR 46, Pittsburgh (4p) to New York (1:50a)
NH 186, New York (2:25a) to New Haven (3:56a)
NH 412, New Haven (6:55a) to Springfield (8:45a)

(Pittsburgh to Harrisburg, dining car. New Haven 186 is the Quaker.)

General, No. 48

All-Pullman
Chicago to New York
Pennsylvania Railroad
7 miles west of Pittsburgh

CHICAGO to NEW YORK
3dbr-dr-lounge	10r-6dbr
21r	12dup/r-4dbr
14s	

PRR 48, Chicago (3p) to New York (8:25a)

(At this scheduling, the General is combined with the all-coach Trail Blazer. Diner, Chicago to New York.)

General, No. 49

All-Pullman
New York to Chicago
Pennsylvania Railroad
9 miles east of Greensburg

NEW YORK to CHICAGO
3dbr-dr-lounge	10r-6dbr
12dup/r-4dbr	14s
21r	

PRR 49, New York (5p) to Chicago (8:20a)

NEW YORK to LOS ANGELES
10r-6dbr
PRR 49, New York (5p Sun) to Chicago (8:20a Mon)
C&NW 1, Chicago (12:01p) to Omaha (9p)
UP 1, Omaha (9:25p) to Los Angeles (10:40a Wed)

(Dining cars and lounges, all segments. The General operates with the all-coach Trail Blazer. C&NW/UP 1 is the Los Angeles Limited.)

The *General* was ranked second only to the *Broadway Limited* upon its introduction in April 1937. The pre-war traffic buildup resulted in creation of the *Advance General* four years later.

Statesman, No. 50

Pittsburgh to Washington
Pennsylvania Railroad
9 miles east of Greensburg

PITTSBURGH to WASHINGTON
8s-buffet-lounge	10r-5dbr
8s-2c-dr	8s-5dbr

PRR 50, Pittsburgh (11:05p) to Washington (7:50a)

Liberty Limited, No. 58

Chicago to Washington
Pennsylvania Railroad
Approaching Beaver Falls

CHICAGO to WASHINGTON
6dbr-lounge	12dup/r-4dbr
21r	21r
4c-2dr-4dbr	14s
14s	2dr-c-dbr-obs.-lounge

PRR 58, Chicago (3:40p) to Washington (8:35a)

(Dining car. Important Washington trains such as the Pennsy's Liberty Limited and B&O's Capitol Limited still carry a large proportion of sectional berths because of rules requiring government employees to travel overnight in the least costly sleeper accommodations.)

Liberty Limited, No. 59
Washington to Chicago
Pennsylvania Railroad
18 miles west of Johnstown

WASHINGTON to CHICAGO
6dbr-lounge	21r
21r	4c-2dr-4dbr
12dup/r-4dbr	14s
14s	2dr-c-dbrs-obs.-lounge

PRR 59, Washington (5:30p) to Chicago (8:30a)

(Diner)

The Pennsy operated a Washington section of the *Broadway Limited* until introducing the capital's own, all-Pullman *Liberty Limited* in 1925 as its principal Chicago train.

Pennsylvania   (continued)

Pittsburgher, No. 60
All-Pullman
Pittsburgh to New York
Pennsylvania Railroad
12 miles east of Greensburg

PITTSBURGH to NEW YORK

3dbr-dr-lounge	6dbr-lounge
21r	21r
21r	12dup/r-4dbr
12dup/r-4dbr	12dup/r-4dbr
4c-2dr-2dbr	10r-6dbr

PRR 60, Pittsburgh (11p) to New York (8a)

(Diner, open a.m.)

Golden Triangle, No. 63
Pittsburgh to Chicago
Pennsylvania Railroad
16 miles west of Beaver Falls

PITTSBURGH to CHICAGO

3dbr-dr-lounge	10r-6dbr
12dup/r-5dbr	12s-dr
4c-2dr-4dbr	12 dup/r-4dbr
21r	

PRR 63, Pittsburgh (11p) to Chicago (7:10a)

WASHINGTON to CHICAGO
   8s-2c-dr
PRR 533, Washington (1:10p) to Harrisburg (4:05p)
PRR 33, Harrisburg (4:22p) to Pittsburgh (9:10p)
PRR 63, Pittsburgh (11p) to Chicago (7:10a)

(No. 33, the St. Louisan, *carries dining and lounge cars)*

American, No. 66
St. Louis to New York
Pennsylvania Railroad
23 miles west of Altoona

ST. LOUIS to NEW YORK

3dbr-dr-lounge	21r
12s-dr	

PRR 66, St. Louis (9a) to New York (6:35a)

ST. LOUIS to WASHINGTON
   8s-5dbr
PRR 66, St. Louis (9a) to Harrisburg (2:55a)
PRR 504, Harrisburg (4:15a) to Washington (7:35a)

INDIANAPOLIS to WASHINGTON
   12s-dr
PRR 66, Indianapolis (1:02p) to Harrisburg (2:55a)
PRR 504, Harrisburg (4:15a) to Washington (7:35a)

WHEELING to NEW YORK
   10r-5dbr
PRR 702, Wheeling (7:50p)
       to Weirton Junction, W.Va. (ntg)
PRR 66, Weirton Junction (ca. 8:45p)
       to New York (6:35a)

(St. Louis to Pittsburgh, dining car)

Red Arrow, No. 68
Detroit to New York
Pennsylvania Railroad
Departing Pittsburgh

DETROIT to NEW YORK

6dbr-lounge	12s-dr
21r	21r
12dup/r-4dbr	

PRR 68, Detroit (5:30p) to New York (8:55a)

DETROIT to WASHINGTON

8s-lounge	10r-5dbr

PRR 68, Detroit (5:30p) to Harrisburg (5a)
PRR 574, Harrisburg (5:15a) to Washington (8:25a)

(Dining cars: Detroit to New York, Harrisburg to Washington)
The *Red Arrow* was formerly an all-sleeper train.

American, No. 67
New York to St. Louis
Pennsylvania Railroad
9 miles west of Harrisburg

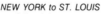

NEW YORK to ST. LOUIS

3dbr-dr-lounge	12s-dr
21r	

PRR 67, New York (8:25p) to St. Louis (4:10p)

WASHINGTON to ST. LOUIS
   8s-5dbr
PRR 539, Washington (8:40p) to Harrisburg (11:35p)
PRR 67, Harrisburg (11:50p) to St. Louis (4:10p)

NEW YORK to WHEELING
   10r-5dbr
PRR 67, New York (8:25p) to Weirton Junction (ntg)
PRR 701, Weirton Junction (ca. 6:15a)
       to Wheeling (7:20a)

(Pittsburgh to St. Louis, dining car)

The *American* was temporarily co-opted as the all-Pullman *Airway Limited* in July 1929, when the Pennsylvania and early TWA set up their transcontinental air/rail service. The *American/Airway Limited* departed New York at dinner time and deposited Pullman passenger at Port Columbus, Ohio, the next morning for the first leg of their journey west in tri-motor airliners.

Commodore Vanderbilt, No. 68
All-Pullman streamliner
Chicago to New York
New York Central
Approaching Erie

CHICAGO to NEW YORK

4dbr-obs.-lounge	6dbr-buffet-lounge
4c-4dbr-2dr	4c-4dbr-2dr
12dbr	13dbr
13dbr	10r-6dbr
22r	22r
22r	

NYC 68, Chicago (4p) to New York (9a)

(Dining car)

The New York Central inaugurated the *Commodore Vanderbilt* in 1929 as a luxury train to take some of the load off the *Twentieth Century Limited*, which was then averaging three sections a day in each direction. The *Commodore* was scheduled to depart after the regular and advance sections of the *Century*. By this 1952 scheduling, the *Commodore* was operating in two all-Pullman sections in advance of the *Century*. Like Rolls-Royce and Bentley automobiles, each train had its own loyal clientele.

Red Arrow, No. 69
New York to Detroit
Pennsylvania Railroad
Departing Johnstown

NEW YORK to DETROIT

6dbr-lounge	12s-dr
12dup/r-4dbr	21r
21r	

PRR 69, New York (5:10p) to Detroit (8a)

WASHINGTON to DETROIT

8s-lounge	10r-5dbr

PRR 569, Washington (5:20p) to Harrisburg (8:16p)
PRR 69, Harrisburg (8:27p) to Detroit (8a)

(Diners: New York to Detroit, Washington to Harrisburg)

Detroit Express, No. 105
Pittsburgh to Detroit
Pennsylvania Railroad
3 miles west of Beaver Falls

PITTSBURGH to DETROIT

6s-6dbr	10r-5dbr

PRR 105, Pittsburgh (11:10p) to Detroit (7:40a)

Admiral, No. 71
New York to Chicago
Pennsylvania Railroad
27 miles west of Harrisburg

NEW YORK to CHICAGO

3dbr-dr-lounge	21r
12s-dr	12dup/r-5dbr

PRR 71, New York (8p) to Chicago (1:05p)

NEW YORK to AKRON

3dbr-dr-lounge	10r-5dbr
12dup/r-5dbr	12s-dr

PRR 71, New York (8p) to Pittsburgh (4:30a)
PRR 9, Pittsburgh (5a) to Akron (8:15a)

WASHINGTON to CHICAGO
   12s-dr
PRR 575, Washington (7:45p) to Harrisburg (10:45p)
PRR 71, Harrisburg (11:28p) to Chicago (1:05p)

(Dining cars: New York to Chicago, Pittsburgh to Akron; buffet-lounge, Washington to Harrisburg. No. 9 is the Akronite.)

No. 87
Pittsburgh to Youngstown
Pittsburgh & Lake Erie
Standing in Pittsburgh

PITTSBURGH to DETROIT
   10r-6dbr
P&LE 87, Pittsburgh (12:40a) to Youngstown (2a)
Erie 687, Youngstown (2:10a) to Cleveland (3:35a)
NYC 287, Cleveland (3:55a) to Toledo (5:50a)
NYC 304, Toledo (6:15a) to Detroit (7:45a)

The P&LE, controlled by the New York Central, teamed with the Erie to provide Pittsburgh with Pullman connections to Chicago, Cleveland, Toledo (NYC), and St. Louis (NYC). An early pairing with Western Maryland resulted in service to Baltimore.

Quaker, No. 186
Philadelphia to Boston
Pennsylvania Railroad
Standing in Broad Street Station

PHILADELPHIA to BOSTON

14r-4dbr	14sbr
8s-dr-3dbr	

PRR 186, Philadelphia (12:10a) to New York (2a)
NH 186, New York (2:25a) to Boston (7:45a)

PHILADELPHIA to PROVIDENCE   *Monday only*
   12s-dr
PRR 186, Philadelphia (12:10a) to New York (2a)
NH 186, New York (2:25a) to Providence (6:32a)

(In New York, the Quaker *will pick up Pittsburgh sleepers for Boston and Springfield, delivered by the* New Englander.)

Philadelphia had dedicated Pullman service to New York, Springfield (NH), and Lake Placid (NYC). The summertime *Bar Harbor Express* was carrying sleepers for Portland, Rockland, and Ellsworth, Me., (NH/B&M/MEC) as of this scheduling, and the choice once included Pullmans to the original Bar Harbor gateway of Mount Desert Ferry, Me., and Bretton Woods, N.H. (NH/B&M/MEC).

Interstate Express, No. 301
Philadelphia to Syracuse
Reading
28 miles north of Philadelphia

PHILADELPHIA to SYRACUSE
   12s-dr
RDG 301, Philadelphia (11:15p) to Bethlehem (ntg)
CNJ 301, Bethlehem (ntg) to Scranton (5:33a)
DL&W 1301, Scranton (ntg) Binghamton (6:42a)
DL&W 1915, Binghamton (8a) to Syracuse (10:15a)

(Binghamton to Syracuse, buffet-lounge)

The *Interstate Limited* once carried sleepers from Philadelphia to Scranton and Binghamton.

Howard W. Ameling collection

Nickel Plate No. 7, the *Westerner*, under Alco power at Cleveland, July 10, 1948.

Havana Special/Piedmont Limited, No. 141
New York to Florida/New Orleans
Pennsylvania Railroad
Departing Philadelphia

NEW YORK to MIAMI
10s-c-2dbr
PRR 141, New York (10:15p Sun)
    to Washington (2:10a Mon)
RF&P 75, Washington (3:05a) to Richmond (5:50a)
ACL 75, Richmond (6:30a) to Jacksonville (8:30p)
FEC 35, Jacksonville (10:30p) to Miami (7a Tue)

NEW YORK to TAMPA
10s-c-dr
As above, New York (10:15p Sun)
    to Jacksonville (8:30p Mon)
ACL 75, Jacksonville (11p) to Tampa (7a Tue)

NEW YORK to FORT MYERS
10s-2c-dr
As above, New York (10:15p Sun)
    to Jacksonville (8:30p Mon)
ACL 275, Jacksonville (10:30p)
    to Fort Myers (8:30a Tue)

NEW YORK to WILMINGTON, N.C.
10s-2c-dr
As above, New York (10:15p) to Richmond (5:50a)
ACL 75, Richmond (6:30a)
    to Rocky Mount, N.C. (8:55a)
ACL 49, Rocky Mount (9:35a) to Wilmington (1:15p)

NEW YORK to NEW ORLEANS
8s-5dbr   10s-2c-dr
PRR 141, New York (10:15p Sun)
    to Washington (2:10a Mon)
SOU 33, Washington (2:40a) to Atlanta (6:30p)
WPR 33, Atlanta (7:15p) to Montgomery (10:40p)
L&N 33, Montgomery (11:10p)
    to New Orleans (8:05a Tue)

NEW YORK to MOBILE
10s-2c-dr
As above, New York (10:15p Sun)
    to Montgomery (10:40p Mon)
L&N 33, Montgomery (11:10p) to Mobile (4:10a Tue)

NEW YORK to CHARLOTTE
10s-2c-dr
PRR 141, New York (10:15p) to Washington (2:10a)
SOU 33, Washington (2:40a) to Charlotte (12:10p)

(The Havana Special carries a diner from Washington to Jacksonville; the Piedmont Limited has diners on all segments south of Monroe, Va.)

The *Havana Special* carried a sleeper from Key West to Philadelphia before the Oversea service was lost to the 1935 hurricane.

No. 331
Philadelphia to Bethlehem
Reading
14 miles north of Philadelphia

PHILADELPHIA to BUFFALO
331: 10s-2br-dr
RDG 331, Philadelphia (11:30p) to Bethlehem (1:29a)
LV 11, Bethlehem (2:08a) to Buffalo (10:45a)

(Bethlehem to Buffalo, cafe-lounge)

No. 331 formerly offered service from Philadelphia to Rochester as well as a through sleeper to Chicago (LV/Michigan Central). Other joint Reading/LV service included Pullmans from Philadelphia to Wilkes Barre and Niagara Falls and Auburn, N.Y.

No. 431
Philadelphia to Washington
Pennsylvania Railroad
Standing in Broad Street Station

PHILADELPHIA to WASHINGTON
Std: s-dr
PRR 431, Philadelphia (1:10a) to Washington (4:45a)

PHILADELPHIA to RICHMOND
10r-6dbr
PRR 431, Philadelphia (1:10a) to Washington (4:45a)
RF&P 23, Washington (5:35a) to Richmond (8a)

(Washington to Richmond, parlor-diner)

The Pennsy scheduled Philadelphia-Baltimore sleepers.

Dominion Express, No. 574
Buffalo to Washington
Pennsylvania Railroad
2 miles north of Emporium

BUFFALO to WASHINGTON
8s-lounge   18r
8s-5dbr
PRR 574, Buffalo (8:30p) to Washington (8:25a)

BUFFALO to PHILADELPHIA
10r-5dbr   12s-dr
PRR 574, Buffalo (8:30p) to Harrisburg (4:55a)
PRR 36, Harrisburg (5:35a) to Philadelphia (7:35a)

(In Harrisburg, the Dominion Express cuts in a diner as well as two Detroit-Washington sleepers delivered by the Red Arrow. No. 36, the Philadelphia Night Express from Pittsburgh, carries a cafe-coach. Sunday schedules differ.)

The *Dominion Express* once carried a consist of Buffalo sleepers for Florida and Atlantic City, as well as Toronto-Philadelphia (CPR/TH&B) cars.

Dominion Express, No. 575
Washington to Buffalo
Pennsylvania Railroad
29 miles north of Harrisburg

WASHINGTON to BUFFALO
8s-buffet-lounge   18r
8s-5dbr
PRR 575, Washington (7:45p) to Buffalo (8a)

PHILADELPHIA to BUFFALO
12s-dr   10r-5dbr
PRR 581, Philadelphia (8:10p) to Harrisburg (10:40p)
PRR 575, Harrisburg (11:20p) to Buffalo (8a)

WASHINGTON to CANANDAIGUA, N.Y.
12s-dr
PRR 575, Washington (7:45p) to Williamsport (1:25a)
PRR 595, Williamsport (3:05a)
    to Canandaigua (7:50a)

(The Dominion Express is also the Washington connection for the Admiral and Penn Texas' Chicago and Houston Pullmans, respectively. Philadelphia to Harrisburg, cafe-coach)

Southern Express, No. 580
Erie to Harrisburg
Pennsylvania Railroad
26 miles south of Emporium

ERIE to PHILADELPHIA
10s-3dbr
PRR 580, Erie (5:45p) to Harrisburg (4:40a)
PRR 36, Harrisburg (5:35a) to Philadelphia (7:35a)

ERIE to WASHINGTON
10s-3dbr
PRR 580, Erie (5:45p) to Harrisburg (4:40a)
PRR 574, Harrisburg (5:15a) to Washington (8:25a)

ERIE to NEW YORK
8s-5dbr
PRR 580, Erie (5:45p) to Harrisburg (4:40a)
PRR 68, Harrisburg (5:30a) to New York (8:55a)

(No. 574, the Dominion Express, carries a diner from Harrisburg to Washington. No. 36, the Philadelphia Night Express, has a cafe-coach; No. 68, the Red Arrow, a diner.)

The *Southern Express* handled Oil City-New York, and Red Bank-Philadelphia sleepers.

Southern Express, No. 580
Pennsylvania Railroad
Williamsport setout
Standing in Williamsport

WILLIAMSPORT to NEW YORK
10s-3dbr
PRR 580, Williamsport (2:21a) to Harrisburg (4:40a)
PRR 68, Harrisburg (5:30a) to New York (8:55a)

The Pennsylvania once scheduled Williamsport-Washington sleepers, and the Reading/CNJ provided competing service to New York.

Pennsylvania   *(continued)*

(continued)

Northern Express, No. 581
Philadelphia to Erie
Pennsylvania Railroad
45 miles north of Harrisburg

PHILADELPHIA to ERIE
10s-3dbr
PRR 581, Philadelphia (8:10p) to Erie (10a)

NEW YORK to ERIE
8s-5dbr
PRR 1, New York (6:45p) to Harrisburg (10p)
PRR 581, Harrisburg (11p) to Erie (10a)

WASHINGTON to ERIE
10s-3dbr
PRR 575, Washington (7:45p) to Harrisburg (10:45p)
PRR 581, Harrisburg (11p) to Erie (10a)

NEW YORK to EMPORIUM   Mon and Thu
12s-dr
PRR 1, New York (6:45p) to Harrisburg (10p)
PRR 581, Harrisburg (11p) to Emporium (3:59a)

(Philadelphia to Erie, cafe-coach. No. 1, the Pennsylvania Limited, carries a diner and sleeper-lounge. Washington to Harrisburg, buffet-lounge.)

No. 596
Canandaigua to Williamsport
Pennsylvania Railroad
54 miles north of Williamsport

CANANDAIGUA to WASHINGTON
12s-dr
PRR 596, Canandaigua (8:55p)
    to Williamsport (1:47a)
PRR 580, Williamsport (2:21a) to Harrisburg (4:40a)
PRR 574, Harrisburg (5:15a) to Washington (8:25a)

(Diner, Harrisburg to Washington)

No. 645
Harrisburg to Hagerstown
Pennsylvania Railroad
14 miles south of Harrisburg

NEW YORK to ROANOKE
10r-6dbr   10s-c-dr
PRR 3, New York (7:35p) to Harrrisburg (10:53p)
PRR 645, Harrisburg (11:35p) to Hagerstown (2a)
N&W 1, Hagerstown, Md. (2:15a) to Roanoke (9:55a)

(Diner and lounge: on No. 3, the Penn Texas, and on N&W 1, from Shenandoah, Va., to Roanoke.)

Rhode Island
See map on page **14**

Owl, No. 3
New Haven Railroad
Providence setout
Standing in Providence

PROVIDENCE to NEW YORK
3000: Std-s-dr-dbr-r   3001: Std-s-dr-dbr-r
NH 3, Providence (1:48a) to New York (6:20a)

(The Owl is standing in Boston)

The New Haven also originated Pullman service from Providence to Philadelphia.

The Federal, No. 173
Boston to Washington
New Haven Railroad
Arriving in Providence

BOSTON to WASHINGTON
6sbr-buffet-lounge   2c-dr-2dbr-2sbr
14s   12s-dr
13dbr   14r-4dbr (Sun Tue Thu)
18r   18r
NH 173, Boston (11p) to New York (3:50a)
PRR 173, New York (4:02a) to Washington (8:20a)

(The Federal will pick up Providence and Springfield sleepers for Washington in New Haven)

Before the opening of New York's Hell Gate Bridge, the *Federal* bypassed New York City in favor of the Poughkeepsie Bridge route. In January 1953 it made the news as the runaway train that crashed into Washington Union Station and dropped through the floor of the huge concourse.

South Carolina
See map on page **124**

Miamian, No. 7
New York to Miami
Atlantic Coast Line
South of Charleston

NEW YORK to MIAMI
A173: 7c-lounge   A174: 10s-2c-dr
A175: 6c-3dbr   A176: 10s-2c-dr
A177: 12r-sbr-4dbr
PRR 133, New York (10:25a) to Washington (2:25p)
RF&P 7, Washington (2:45p) to Richmond (5p)
ACL 7, Richmond (5:15p) to Jacksonville (4a)
FEC 7, Jacksonville (4:30a) to Miami (11:45a)

WASHINGTON to MIAMI
R507: 6s-6dbr
As above, Washington (2:45p) to Miami (11:45a)

(Diner and lounge, New York to Miami)

The *Miamian*, another of the great trains inaugurated in the late 1920's, operated then all-Pullman in two sections. The first included a twice-weekly Quebec-Miami sleeper.

The ACL brought dedicated Pullmans to Charleston from Savannah, Atlanta (Georgia Railroad), Washington (RF&P) and New York (RF&P/PRR).

No. 17
Columbia to Greenville
Southern Railway
17 miles north of Greenwood

COLUMBIA to ATLANTA
31: 12s-dr
SOU 17, Columbia (8:30p) to Greenville (1:45a)
SOU 35, Greenville (2:50a) to Atlanta (6:30a)

The *Southerner*, shown here at Birmingham in 1970, was SR's crack, mostly coach streamliner between New York and New Orleans.

Silver Star, No. 21
Streamliner
New York to Florida
Seaboard Air Line
35 miles south of Columbia

NEW YORK to MIAMI
S109: *10r-6db*   S110: *8s-5dbr*
S111: *6c-buffet-lounge*
PRR 127, New York (9:50a) to Washington (1:50p)
RF&P 21, Washington (2:10p) to Richmond (4:43p)
SAL 21, Richmond (4:43p) to Miami (11a)

NEW YORK to PORT BOCA GRANDE
S112: *6c-3dr*
As above, New York (9:50a) to Richmond (4:43p)
SAL 21, Richmond (4:43p) to Wildwood, Fla. (5:55a)
SAL 121, Wildwood (6:10a) to Tampa (8a)
SAL 321, Tampa (8:25a)
to Port Boca Grande (11:55a)

NEW YORK to ST. PETERSBURG
S113: *8s-5dbr*   S114: *10r-6dbr*
As above, New York (9:50a) to Wildwood (5:55a)
SAL 121, Wildwood (6:10a) to St. Petersburg (9:45a)

(Dining cars, New York to St. Petersburg and Miami; tavern coach and observation coach, New York to Miami)

Columbia, the main stop for SAL trains in South Carolina, had its own Pullman service to Washington (RF&P) and New York (RF&P/PRR).

Skyland Special, No. 23
North Carolina to Jacksonville
Southern Railway
30 miles south of Columbia

ASHEVILLE to JACKSONVILLE
S124: *8s-dr-3dbr*
SOU 10, Asheville (4:45p) to Columbia (10:35p)
SOU 23, Columbia (11:10p) to Jacksonville (7:30a)

CHARLOTTE to JACKSONVILLE
S126: *12s-dr*
SOU 23, Charlotte (7:50p) to Columbia (11p)
SOU 23, Columbia (11:10p) to Jacksonville (7:30a)

(Dinette coach, Asheville to Jacksonville)

The *Skyland Special* formerly originated in Cincinnati. Both Spartanburg and Columbia had dedicated sleepers from Cincinnati on this line through Asheville, and another car, from Chicago (NYC), came to Columbia.

Silver Comet, No. 34
Streamliner
Birmingham to New York
Seaboard Air Line
5 miles northeast of Chester

BIRMINGHAM to NEW YORK
B72: *10r-6dbr*   B74: *10s-lounge*
SAL 34, Birmingham (2p) to Richmond (6:45a)
RF&P 34, Richmond (6:45a) to Washington (9:20a)
PRR 196, Washington (9:45a) to New York (1:35p)

BIRMINGHAM to WASHINGTON
B71: *10s-dr-2c*
As above, Birmingham (2p) to Washington (9:20a)

ATLANTA to WASHINGTON
B70: *10r-6dbr*
SAL 34, Atlanta (7:10p) to Richmond (6:45a)
RF&P 34, Richmond (6:45a) to Washington (9:20a)

ATLANTA to PORTSMOUTH
B4: *10s-dr-2c*
SAL 34, Atlanta (7:10p) to Hamlet, N.C. (1:55a)
SAL 8, Hamlet (5:45a) to Norlina, N.C. (9a)
SAL 18, Norlina (9a) to Portsmouth (12:25p)

(Dining car and tavern-observation, Birmingham to New York)

Orange Blossom Special, No. 46
All-Pullman
Miami to New York
Seaboard Air Line
55 miles northeast of Columbia

MIAMI to BOSTON
B29: *8s-dr-3dbr*
SAL 46, Miami (11a) to Richmond (5:55a)
RF&P 46, Richmond (ntg) to Washington (8:30a)
PRR 180, Washington (8:50a) to New York (12:50p)
NH 174, New York (2p) to Boston (6:45p)

MIAMI to NEW YORK
B20: *6s-6dbr*   B21: *12r-4dbr*
B22: *2c-dr-buffet-lounge*   B23: *6c-3dr*
B24: *6c-3dr*   B25: *6c-3dr*
B26: *6c-3dr*   B28: *6c-3dr*
As above, Miami (11a) to New York (12:50p)

MIAMI to WASHINGTON
B27: *3c-dr-buffet-lounge*   B30: *6s-4r-4dbr*
As above, Miami (11a) to Washington (8:30a)

(Dining cars: Miami to New York, Wildwood to Washington)

Like other luxury, vacation trains, the *Orange Blossom* was suspended during World War II. Before its business was taken over by the *Silver Meteor* and *Silver Star*, the *Orange Blossom* carried the gray and maroon Pullman cars that in summers traveled on the *Bar Harbor Express.*

Southerner, No. 48
Streamliner
New Orleans to New York
Southern Railway
12 miles southwest of Spartanburg

BIRMINGHAM to NEW YORK
S3: *10r-6dbr*
SOU 48, Birmingham (3:20p) to Washington (9:10a)
PRR 148, Washington (9:35a) to New York (1:25p)

(Diner, observation coach lounge, and coach lounge, New Orleans to New York)

Although inaugurated as an all-coach streamliner, the *Southerner* was later assigned a Birmingham-New York Pullman.

No. 55
Wilmington, N.C., to Augusta, Ga.
Atlantic Coast Line
18 miles northeast of Robbins

WILMINGTON to ATLANTA
A53: *10s-2c-dr*
ACL 55, Wilmington (4:30p) to Augusta (1:20a)
Georgia 3, Augusta (2:20a) to Atlanta (7a)

FLORENCE to ATLANTA
A6: *10s-c-dr*
ACL 55, Florence (8:25p) to Augusta (1:20a)
Georgia 3, Augusta (2:20a) to Atlanta (7a)

(Cafe-lounge, Wilmington to Florence)

Vacationer, No. 74
All-Pullman
Florida to New York
Atlantic Coast Line
North of Hardeeville

MIAMI to NEW YORK
VA1: *6c-3dr*   VA2: *6c-3dr*
VA3: *8s-dr-3dr*   VA4: *12r-sb-4dbr*
VA5: *10s-c-2dbr*   VA6: *6c-3dr*
VA7: *sb-dr-buffet-lounge*
FEC 74, Miami (1p) to Jacksonville (8:20p)
ACL 74, Jacksonville (8:55p) to Richmond (7:45a)
RF&P 74, Richmond (7:55a) to Washington (10:15a)
PRR 116, Washington (10:40a) to New York (2:30p)

MIAMI to WASHINGTON
F504: *8s-5dbr*
As above, Miami (1p) to Washington (10:15a)

JACKSONVILLE to NEW YORK every third day
VA8: *10r-6dbr*
As above, Jacksonville (8:55p) to New York (2:30p)

MIAMI to BOSTON
MB11: *6c-3dr*   MB12: *6s-4r-4dbr*
As above, Miami (1p) to Washington (10:15a)
PRR 176, Washington (12n) to New York (3:45p)
NH 176, New York (4p) to Boston (8:40p)

ST. PETERSBURG to BOSTON alternate days
PA14: *10s-2c-dr*
ACL 192, St. Petersburg (11a) to Jacksonville (6:20p)
As above, Jacksonville (8:55p)
to Washington (10:15a)
As above, Washington (12n) to Boston (8:40p)

SARASOTA to BOSTON   alternate days
RA14: *10s-2c-dr*
ACL 92, Sarasota (10:30a) to Jacksonville (6:25p)
As above, Jacksonville (8:55p) to Boston (8:40p)

(Diners, Miami to New York, St. Petersburg to Jacksonville; diner and lounge, Tampa to Jacksonville)

West Coast Champion, No. 92
Streamliner
Florida Gulf Coast to New York
Atlantic Coast Line
37 miles south of Florence

SARASOTA to NEW YORK
A169: *10r-6dbr*   A180: *8s-2c-dr*
ACL 92, Sarasota (10:30a) to Jacksonville (6:25p)
ACL 92, Jacksonville (6:55p) to Richmond (5:40a)
RF&P 92, Richmond (5:50a) to Washington (8:15a)
PRR 194, Washington (8:40a) to New York (12:30p)

TAMPA to NEW YORK
A167: *10r-6dbr*   A168: *21r*
ACL 92, Tampa (12:40p) to Jacksonville (6:25p)
As above, Jacksonville (6:55p) to New York (12:30p)

ST. PETERSBURG to NEW YORK
A170: *10r-6dbr*   A171: *10r-6dbr*
A172: *14r-2dr*   A173: *8s-2c-dr*
ACL 192, St. Petersburg (11a) to Jacksonville (6:20p)
As above, Jacksonville (6:55p) to New York (12:30p)

ST. PETERSBURG to WASHINGTON
R512: *6dbr-lounge*
As above, St. Petersburg (11a) to Washington (8:15a)

SARASOTA to WASHINGTON
R513: *10r-6dbr*
As above, Sarasota (10:30a) to Washington (8:15a)

(Diner and lounge, Tampa to New York; diner, St. Petersburg to Jacksonville)

The ACL *Champion*, introduced in 1939 to meet the competition of the SAL's *Silver Meteor*, became so popular that two sections of the train, one for each coast of Florida, were put into service.

South Dakota

See map on page 298

Olympian Hiawatha, No. 16
Streamliner
Tacoma/Seattle to Chicago
Milwaukee Road
77 miles west of Aberdeen

TACOMA/SEATTLE to CHICAGO
  161: 10r-6dbr   162: 8dbr-lounge
  A16: 14s MILW Touralux   B16: 14s Touralux
MILW 16, Tacoma (1:30p Sat) and Seattle (2:45p)
  to Chicago (1:45p Mon)

(Diner, Tip Top grill car. Another edition of the two-night Olympian Hiawatha is now in Idaho.)

The *Columbian* carried a sleeper to Aberdeen from Minneapolis.

Minnesota and Black Hills Express, No. 514
Rapid City to Mankato
Chicago & North Western
Standing in Pierre

RAPID CITY to CHICAGO
  8s-dr-2c
C&NW 514, Rapid City (7p Sun)
  to Huron (4:38a Mon)
C&NW 518, Huron (5:10a) to Mankato, Minn. (9:45a)
C&NW 514, Mankato (4:20p) to Elroy, Wis. (ntg Tue)
C&NW 514, Elroy (1:30a) to Chicago (7:05a Tue)

(Huron to Mankato, lounge car; Mankato to Elroy, cafe-coach. This two-night scheduling effectively makes Mankato, with the Pullman's long layover there, an intermediate destination and originating point. In Elroy, No. 514—east of Mankato known as the Rochester Minnesota Special—is combined with No. 514 from Minneapolis, The Victory, to complete the journey to Chicago. No. 518 is the Dakota 400.)

Dakota 400, No. 519
Streamliner
Chicago to Huron
Chicago & North Western
11 miles east of Huron

CHICAGO to RAPID CITY
  8s-dr-2c
C&NW 515, Chicago (9p Sat)
  to Mankato (11:42a Sun)
C&NW 519, Mankato (7:19p)
  to Huron, S.D. (12:14a Mon)
C&NW 515, Huron (12:35a)
  to Rapid City (9:15a Mon)

(Elroy to Mankato, cafe-coach; Mankato to Huron, lounge car. This unusual Pullman routing puts the standard Rapid City sleeper aboard the North Western's Nos. 518 and 519, the Dakota 400, the only streamliner in the famous fleet of day trains to carry a sleeping car. Nos. 515 are the Rochester-Minnesota Special and Minnesota and Black Hills Express, east and west of Mankato, respectively.)

The *Badger State Express*, a Chicago-Pierre train, served Huron with Pullmans to Minneapolis/St. Paul and Rochester.

Tennessee

See map on page 104

The Louisville & Nashville operated two main lines south of Nashville, both of which had passenger service. The principal passenger trains, however, operated by way of Lewisburg, as did all of the trains carrying Pullmans south of Nashville in the schedules below.

Ponce de Leon, No. 2
Jacksonville to Cincinnati
Southern Railway
2 miles south of Harriman Jct.

TAMPA to CLEVELAND
  B123: 10s-dr-2c
SAL 2-10, Tampa (11:10p Sat)
  to Jacksonville (6a Sun)
SOU 2, Jacksonville (9a) to Cincinnati (7:55a Mon)
NYC 426, Cincinnati (9:15a) to Cleveland (3:45p Mon)

JACKSONVILLE to CINCINNATI
  170: 10s-diner-lounge
SOU 2, Jacksonville (9a) to Cincinnati (7:55a)

(NYC 426, the Cleveland Special, carries a diner-lounge)

Chattanooga had its own Pullmans to Jacksonville and Miami on the *Royal Palm*, which followed this same route, as well as service on other trains to Cincinnati, Louisville, Atlanta, Mobile (M&O), New Orleans, and Shreveport (IC). Knoxville had dedicated sleepers to Cincinnati, Louisville, and St. Louis.

No. 2-102
Nashville to Memphis
Nashville, Chattanooga & St. Louis
48 miles west of Nashville

NASHVILLE to MEMPHIS
  7: 12s-dr
NC&StL 2, Nashville (10:30p) to Bruceton (1:30a)
NC&StL 102, Bruceton (1:45a) to Memphis (6:50a)

The NC&StL once scheduled sleepers from Nashville to both St. Louis and Chicago, via the IC at Martin.

No. 3
Nashville to Atlanta
Nashville, Chattanooga & St. Louis
64 miles northwest of Chattanooga

NASHVILLE to ATLANTA
  1: 14s
NC&StL 3, Nashville (9p) to Atlanta (8:10a)

NASHVILLE to BRISTOL
  K38: 10s-c-2dbr
NC&StL 3, Nashville (9p) to Chattanooga (3:30a)
SOU 46, Chattanooga (4:35a) to Bristol (10:40a)

As they did with the Bristol line, the NC&StL and Southern cooperated on Pullman service from Nashville to Chattanooga, Knoxville, Asheville, and, with N&W, Washington, and New York. Another sleeper for New York also traveled on just the Southern via Asheville.

Azalean, No. 4
New Orleans to Cincinnati
Louisville & Nashville
34 miles south of Lewisburg

BIRMINGHAM to NEW YORK
  L3: 10s-2c-dr
L&N 4, Birmingham (9:10p Sun)
  to Cincinnati (12:05p Mon)
PRR 204, Cincinnati (1p) to Pittsburgh (8:30p)
PRR 16, Pittsburgh (10:10p) to New York (7:15a Tue)

(Dining cars: New Orleans to Birmingham, Louisville to Cincinnati, Columbus to Pittsburgh. PRR 16 is the Iron City Express, carrying a sleeper-lounge; Saturday night schedule from Pittsburgh slightly different.)

Panama Limited, No. 5
Illinois Central
Memphis setout
Standing in Memphis

MEMPHIS to NEW ORLEANS
  551: 6s-4r-4dbr
IC 5, Memphis (2:28a) to New Orleans (9:30a)

(The all-Pullman Panama Limited is now 122 miles to the north, just departing from Fulton, Ky. This Memphis setout will be the eleventh sleeper in the streamliner's consist, which includes lounge and twin diners.)

Over its latticework of lines through Mississippi, the IC carried dedicated sleepers from Memphis to Greenville, Greenwood, Gulfport, Jackson, Natchez, and Yazoo City. Service in the other direction included cars to Louisville and, with B&O, Washington and New York.

Humming Bird, No. 6
Streamliner
New Orleans to Cincinnati
Louisville & Nashville
7 miles north of Nashville

NEW ORLEANS to CINCINNATI
  6: 6s-6dbr   7: 12s-dr
L&N 6, New Orleans (10:55a) to Cincinnati (7:45a)

NASHVILLE to CINCINNATI
  2: 12s-dr
L&N 6, Nashville (11:50p) to Cincinnati (7:45a)

(New Orleans to Cincinnati, lounge and dining cars)

Panama Limited, No. 6
All-Pullman streamliner
New Orleans to Chicago
Illinois Central
21 miles north of Memphis

NEW ORLEANS to CHICAGO
  610: 10r-6dbr   612: 2dr-4c-4dbr
  614: 10r-6dbr   616: 10r-5dbr
  618: 6s-6r-4dbr   620: 6s-6r-4dbr
  602: 2dbr-dr-2c-observation
IC 6, New Orleans (5p) to Chicago (9:30a)

NEW ORLEANS to ST. LOUIS
  606: 6s-6r-4dbr   608: 10r-6dbr
IC 6, New Orleans (5p) to Carbondale, Ill. (4:08a)
IC 16, Carbondale (4:50a) to St. Louis (7:47a)

JACKSON, MISS., to CHICAGO
  604: 6s-6r-4dbr
IC 6, Jackson (8p) to Chicago (9:30a)

(New Orleans to Chicago: lounge, twin diners. Carbondale to St. Louis, buffet-lounge.)

Gulf Coast Rebel, No. 15
St. Louis to Mobile
Gulf Mobile & Ohio
21 miles north of Jackson

CHICAGO to MOBILE
  1: 4s-8r-c-3dbr
GM&O 1, Chicago (11:45a) to St. Louis (4:55p)
GM&O 15, St. Louis (6p) to Mobile (10:55a)

ST. LOUIS to MOBILE
  3: 8s-dr-3dbr
GM&O 15, St. Louis (6p) to Mobile (10:55a)

(St. Louis to Mobile, diner-lounge. No. 1 is the Alton Limited, carrying parlor-lounge and dining car.)

The Mobile & Ohio's *Gulf Coast Special* became the *Gulf Coast Rebel* in October 1940, upon the railroad's merger with the Gulf Mobile & Northern. The IC continued to deliver the Chicago sleeper to St. Louis until after World War II, when the Alton became part of the GM&O system, finally giving the railroad its own access to Chicago.

Two E3's lead MoPac's Memphis-bound No. 220 out of Little Rock in June 1960.

Seminole, No. 10
Florida to Chicago
Illinois Central
20 miles south of Jackson

BIRMINGHAM to CHICAGO
101: 8s-2c-dr
IC 10, Birmingham (6:30p) to Chicago (10:45a)

ST. PETERSBURG to CHICAGO   alternate days
907: 8s-2c-dr
ACL 38, St. Petersburg (6:30p Sat)
to Jacksonville (6a Sun)
ACL 18, Jacksonville (7:45a) to Albany, Ga. (12:25p)
CofG 9, Albany (12:30p) to Birmingham (6:10p)
IC 10, Birmingham (6:30p) to Chicago (10:45a Mon)

SARASOTA/TAMPA to CHICAGO   alternate days
903: 8s-2c-dr
ACL 76 Sarasota (4:45p Sat) and Tampa (10:30p)
to Jacksonville, Fla. (6:30a Sun)
As above, Jacksonville (7:45a)
to Chicago (10:45a Mon)

SARASOTA/TAMPA to ST. LOUIS   alternate days
905: 8s-2c-dr
As above, Sarasota (4:45p Sat) and Tampa (10:30p)
to Birmingham (6:10p Sun)
IC 10, Birmingham (6:30p)
to Carbondale (4:35a Mon)
IC 16, Carbondale (4:50a) to St. Louis (7:47a Mon)

ST. PETERSBURG to ST. LOUIS   alternate days
909: 8s-2c-dr
As above, St. Petersburg (6:30p Sat)
to Birmingham (6:10p Sun)
As above, Birmingham (6:30p)
to St. Louis (7:47a Mon)

(The Florida sleepers alternate, so that one departs every day from both Sarasota/Tampa and St. Petersburg, and one of which is bound for Chicago, and the other for St. Louis. Dining car, Jacksonville to Chicago. Buffet-lounge, Carbondale to St. Louis. The Seminole at this point is operating on former Mobile & Ohio tracks, which arrangement the IC had negotiated when putting together its route to Birmingham in 1908. In fact, the GM&O's northbound Gulf Coast Rebel is just 20 miles behind the Seminole on this midnight. This joint usage presages the IC and GM&O merger of 1972, except by then the passenger trains will be just a memory.)

Chickasaw, No. 16
Memphis to St. Louis
Illinois Central
48 miles north of Memphis

MEMPHIS to ST. LOUIS
162: 8s-dr-3dbr
IC 16, Memphis (11p) to St. Louis (7:47a)

(The Chickasaw arrives in Carbondale at 4:15a; there it takes on the St. Louis sleepers from both the northbound Panama Limited and Seminole, as well as a buffet-lounge)

Over the years, Memphis travelers had four choices overnight to St. Louis: this one and the Frisco's, both still operating in 1952, in addition to the Missouri Pacific, and a 19th Century L&N/IC lashup via Milan.

Gulf Coast Rebel, No. 16
Mobile to St. Louis
Gulf Mobile & Ohio
40 miles south of Jackson

MOBILE to ST. LOUIS
4: 8s-dr-3dbr
GM&O 16, Mobile (3:15p) to St. Louis (8:15a)

MOBILE to CHICAGO
2: 4s-8r-c-3dbr
GM&O 16, Mobile (3:15p) to St. Louis (8:15a)
GM&O 2, St. Louis (8:58a) to Chicago (2:08p)

(Mobile to St. Louis, diner-lounge. GM&O 2 is the Abraham Lincoln, which carries an observation-lounge, parlor-lounge, and dining car, in addition to the Oklahoma City-Chicago Pullman that the Frisco's Meteor delivers to St. Louis. The well-appointed Abraham Lincoln was a favorite train of lawmakers and lobbyists traveling to and from the Illinois capital, Springfield.)

Both the southern lines that formed the GM&O converged at Jackson. One of them, the Gulf Mobile & Northern, scheduled sleepers from the western Tennessee city to Jackson, Miss., and Mobile.

Birmingham Special, No. 18
Birmingham to Washington
Southern Railway
2 miles east of Morristown

BIRMINGHAM to NEW YORK
S8: 10r-6dbr
SOU 18, Birmingham (3:30p) to Bristol (2:10a)
N&W 18, Bristol (2:20a) to Lynchburg, Va. (7:15a)
SOU 18, Lynchburg (7:15a) to Washington (11:25a)
PRR 176, Washington (12n) to New York (3:45p)

CHATTANOOGA to NEW YORK
NW15: 10s-dr-2dbr
SOU 18, Chattanooga (8p) to Bristol (2:10a)
As above, Bristol (2:20a) to New York (3:45p)

BIRMINGHAM to WASHINGTON
S46: 8s-diner-lounge
As above, Birmingham (3:30p)
to Washington (11:25a)

(Roanoke to Washington, diner. PRR 176, the Senator, carries dining car and parlor-lounge.)

The *Birmingham Special* once traveled Southern all the way to Washington, via Atlanta. The present routing, however, forever enriched American popular music, for it was a trip on this train, in the opposite direction, that inspired Harry Warren to write the 1941 hit *Chattanooga Choo Choo*.

Both Chattanooga and Knoxville had their own sleepers to Salisbury, N.C., and Washington, and Chattanooga had dedicated service to Asheville.

Flamingo, No. 18
Jacksonville to Cincinnati
Louisville & Nashville
Standing in Knoxville

ATLANTA to CINCINNATI
18: 8s-dr-3dbr
L&N 18, Atlanta (7:50p) to Cincinnati (8:05a)

ATLANTA to LOUISVILLE
22: 8s-dr-3dbr, or   10s-dr-2dbr
L&N 18, Atlanta (7:50p) to Corbin, Ky. (3:04a)
L&N 24, Corbin (3:40a) to Louisville (7:20a)

KNOXVILLE to CINCINNATI
8: 10s-2c-dr
L&N 18, Knoxville (12:10a) to Cincinnati (8:05a)

(Atlanta to Etowah, diner)

Carolina Special, No. 28
Carolinas to Chicago
Southern Railway
35 miles north of Harriman Jct.

CHARLESTON, S.C. to CHICAGO
S4: 8s-5dbr
SOU 27, Charleston (7:30a) to Asheville (5p)
SOU 28, Asheville (5:30p) to Cincinnati (7a)
NYC 415, Cincinnati (9:45a) to Chicago (3:30p)

GREENSBORO, N.C. to CHICAGO
S49: 10s-dr-2c
SOU 21, Greensboro (10a) to Asheville (5:05p)
As above, Asheville (5:30p) to Chicago (3:30p)

ASHEVILLE to CINCINNATI
47: 10s-diner-lounge
SOU 28, Asheville (5:30p) to Cincinnati (7a)

COLUMBIA to LOUISVILLE
21: 12s-dr
SOU 27, Columbia (11:20a) to Asheville (5p)
SOU 28, Asheville (5:30p) to Danville, Ky. (3a)
SOU 24, Danville (4:30a) to Louisville (7:25a)

(Dining cars: Columbia to Knoxville, Cincinnati to Chicago; dinette-coach, Goldsboro to Asheville. In timetables, the Carolina Special is listed as 27-28. Going to Asheville, the train is considered westbound, hence the odd number. The Asheville-Cincinnati leg is considered northbound, thus the even number. Whichever the direction, the train is therefore listed as 27-28.)

Tennessean, No. 45
Streamliner
Washington to Memphis
Southern Railway
Arriving in Chattanooga

NEW YORK to MEMPHIS
NA72: 10r-4dbr
PRR 103, New York (12:55a) to Washington (7:15a)
SOU 45, Washington (8a) to Lynchburg (11:55a)
N&W 45, Lynchburg (12:05p) to Bristol (5:20p)
SOU 45, Bristol (5:30p) to Memphis (7:45a)

WASHINGTON to MEMPHIS
S66: 14r-4dbr
As above, Washington (8a) to Memphis (7:45a)

BRISTOL to NASHVILLE
S38: 10s-c-2dbr
SOU 45, Bristol (5:30p) to Chattanooga (12:01a)
NC&StL 4, Chattanooga (1:45a) to Nashville (6:30a)

KNOXVILLE to MEMPHIS
S20: 10s-c-2dbr
SOU 45, Knoxville (9:15p) to Memphis (7:45a)

(Dining car, Washington to Knoxville; diner-lounge, Chattanooga to Memphis; lounge, Washington to Memphis. Sunday schedules differ.)

The *Tennessean* made its bow in streamlined livery in May 1941. This is one of six routings that Pullman cars traveled between New York and Memphis. Among the others: the PRR/L&N via Cincinnati, B&O/IC via Louisville, both PRR/RF&P/SAL/SLSF and PRR/SOU/SLSF via Birmingham, and a PRR/N&W/SOU line via the Cumberland Valley.

A nine-car *Georgian* zips along C&EI iron near Martinton, Ill., in June 1948.

Tennessee   (continued)

Tennessean, No. 45

Southern Railway
Chattanooga setout
Standing in Chattanooga

CHATTANOOGA to MEMPHIS
   62: 8s-c-diner-lounge
   SOU 45, Chattanooga (12:35a) to Memphis (7:45a)

Humming Bird, No. 54
Streamliner
New Orleans to Chicago/St. Louis
Louisville & Nashville
Standing in Nashville

NEW ORLEANS to CHICAGO
   L8: 8s-2c-dr
   L&N 6, New Orleans (10:55a) to Nashville (11:35p)
   L&N 54, Nashville (12:05a) to Evansville (3:25a)
   C&EI 18, Evansville (3:45a) to Chicago (9:15a)

MONTGOMERY to ST. LOUIS
   32: 10s-c-dr
   L&N 6, Montgomery (5:45p) to Nashville (11:35p)
   L&N 54, Nashville (12:05a) to Evansville (3:25a)
   L&N 56, Evansville (3:50a) to St. Louis (7:55a)

MONTGOMERY to CHICAGO
   L15: 6s-4r-4dbr
   As above, Montgomery (5:45p) to Evansville (3:25a)
   C&EI 18, Evansville (3:45a) to Chicago (9:15a)

NASHVILLE to ST. LOUIS
   48: 12s-dr
   As above, Nashville (12:05a) to St. Louis (7:55a)

ATLANTA to ST. LOUIS
   800: 8s-2c-dr
   NC&StL 80, Atlanta (6:30p) to Nashville (11:40p)
   As above, Nashville (12:05a) to St. Louis (7:55a)

(Dining and lounge cars: New Orleans to Nashville, Evansville to Chicago; diner, Evansville to St. Louis. The Atlanta sleeper is fresh off the Georgian, which carries both lounge and diner.)

No. 101-1
Memphis to Nashville
Nashville, Chattanooga & St. Louis
43 miles east of Memphis

MEMPHIS to NASHVILLE
   7: 12s-dr
   NC&StL 101, Memphis (10:30p) to Bruceton (3:30a)
   NC&StL 1, Bruceton (3:50a) to Nashville (7a)

   Around the turn of the century, another sleeper operated to Nashville on the L&N/NC&StL via McKenzie.

Georgian, No. 80

Streamliner
Atlanta to Chicago
Louisville & Nashville
Departing Nashville

ATLANTA to CHICAGO
   801: 12r-2sbr-dbr-dr   802: 8s-dr-3dbr
   803: 8s-2c-dr
   NC&StL 80, Atlanta (6:30p) to Nashville (11:40p)
   L&N 80, Nashville (11:59p) to Evansville (3:15a)
   C&EI 8, Evansville (3:25a) to Chicago (8:55a)

NASHVILLE to CHICAGO
   L14: 10s-c-dr
   As above, Nashville (11:59p) to Chicago (8:55a)

(Atlanta to Chicago, lounge and dining cars)

   The *Georgian* and *Humming Bird* were two post-war streamliners whose paths crossed in Nashville. The *Georgian* started out as a St. Louis-Atlanta train, faltered, then revived after Chicago was made its northern terminus.

Georgian, No. 81
Streamliner
Chicago to Atlanta
Louisville & Nashville
39 miles north of Nashville

CHICAGO to ATLANTA
   811: 12r-2sbr-dbr-dr   812: 8s-dr-3dbr
   813: 8s-2c-dr
   C&EI 7, Chicago (4p) to Evansville (9:20p)
   L&N 81, Evansville (9:40p) to Nashville (12:55a)
   NC&StL 81, Nashville (1:20a) to Atlanta (8:35a)

ST. LOUIS to ATLANTA
   810: 8s-2c-dr
   L&N 93, St. Louis (4:35p) to Evansville (9:15p)
   As above, Evansville (9:40p) to Atlanta (8:35a)

(Chicago to Atlanta, lounge and dining car. No. 93 is the St. Louis section of the Humming Bird, which carries a diner.)

No. 104
Memphis to Bowling Green, Ky.
Louisville & Nashville
17 miles northeast of McKenzie

MEMPHIS to CINCINNATI
   74: 10s-lounge
   L&N 104, Memphis (8:15p) to Bowling Green (3:40a)
   L&N 4, Bowling Green (4:15a) to Cincinnati (12:05p)

MEMPHIS to NEW YORK
   L4: 10r-6dbr
   As above, Memphis (8:15p Sun)
       to Cincinnati (12:05p)
   PRR 204, Cincinnati (1p) to Pittsburgh (8:30p)
   PRR 16, Pittsburgh (10:10p) to New York (7:15a Tue)

(Dining cars: Louisville to Cincinnati; Columbus to Pittsburgh. No. 4 is the Azalean. PRR 16 is the Iron City Express. Saturday night schedules from Pittsburgh slightly different.)

   Memphis had Pullman service to Louisville and, via the Pennsy's *Sea Gull*, to Atlantic City.

Texas

See map on page 252

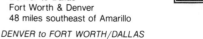

Texas Zephyr, No. 1

Streamliner
Denver to Dallas
Fort Worth & Denver
48 miles southeast of Amarillo

DENVER to FORT WORTH/DALLAS
   17: Std: s-dbr   18: Std: s-dr-dbr
   FtW&D 1, Denver (12:01p)
       to Fort Worth (6a) and Dallas (7:15a)

AMARILLO to FORT WORTH/DALLAS
   100: Std: s-dr
   FtW&D 1, Amarillo (11:05p) to Fort Worth (6a)
       and Dallas (7:15a)

(Diner-lounge. The Fort Worth & Denver is part of the Burlington system.)

Southern Belle, No. 2
New Orleans to Kansas City
Streamliner, via Coushatta, La.
Kansas City Southern
18 miles south of Texarkana

NEW ORLEANS to KANSAS CITY
   2: 14r-4dbr
   KCS 2, New Orleans (4p) to Kansas City (10:15a)

PORT ARTHUR to KANSAS CITY
   4: 14r-4dbr
   KCS 102, Port Arthur (4:30p) to Shreveport (10:10p)
   KCS 2, Shreveport (11p) to Kansas City (10:15a)

(Dining cars over all segments; observation-lounge from New Orleans to Kansas City)

   The KCS scheduled a Port Arther-St. Louis Pullman in cooperation with the Missouri Pacific.

Texas Eagle, No. 2
Streamliner
El Paso to St. Louis
Texas & Pacific
Standing in El Paso

EL PASO to NEW YORK
   PR9: 14r-4dbr
   T&P 2, El Paso (12:50a Mon) to Texarkana (10:30p)
   MP 2, Texarkana (10:40p) to St. Louis (8:10a Tue)
   PRR 4, St. Louis (10a) to New York (7:25a Wed)

(Diner-lounge, El Paso to Fort Worth; dining and lounge cars, Fort Worth-St. Louis-New York. PRR 4 is the Penn Texas.)

   The *Texas Eagles*, the post-war successors to the 1915-vintage *Sunshine Specials*, operated, like their predecessors, as two separate trains in Texas. This one is the West Texas section, serving El Paso, Fort Worth, and Dallas. The South Texas section accommodated Houston, Austin, and San Antonio. The El Paso line was part of the Sunshine Route to California, over which Pullmans from St. Louis and Memphis were once scheduled.

Sunset Limited, No. 2

Streamliner
Los Angeles to New Orleans
Southern Pacific
64 miles west of Del Rio

LOS ANGELES to NEW ORLEANS
   26: 10r-6dbr   27: 10r-6dbr
   29: 10r-6dbr
   SP 2, Los Angeles (8p Sat) to New Orleans (4p Mon)

LOS ANGELES to SAN ANTONIO
   25: 10r-6dbr
   SP 2, Los Angeles (8p Sat)
       to San Antonio (4:30a Mon)

(Dining car and first-class lounge)

Sunset Limited, No. 2
Southern Pacific
San Antonio setout
Standing in San Antonio

SAN ANTONIO to NEW ORLEANS
25: 10r-6dbr
SP 2, San Antonio (4:45a) to New Orleans (4p)

The *Sunset Limited* once carried a San Antonio-San Francisco Pullman.

No. 4
Fort Worth to St. Louis
Texas & Pacific
27 miles west of Longview

FORT WORTH to LITTLE ROCK
41: 10s-dr-2c
T&P 4, Fort Worth (8p) to Texarkana (3:40a)
MP 4, Texarkana (3:55a) to Little Rock (7:30a)

Katy Flyer, No. 5
St. Louis and Kansas City
to San Antonio, via Dallas
Missouri-Kansas-Texas
22 miles south of Dallas

KANSAS CITY to SAN ANTONIO
52: 12s-dr
MKT 25, Kansas City (8:40a)
  to Parsons, Kans. (11:40a)
MKT 25, Parsons (12:10p) to Denison (7:10p)
MKT 5, Denison (7:40p) to Waco (1:50a)
MKT 5, Waco (2:10a) to San Antonio (7:50a)

DALLAS to SAN ANTONIO
53: 12s-dr
MKT 5, Dallas (11:30p) to Waco (1:50a)
MKT 5, Waco (2:10a) to San Antonio (7:50a)

(Diner, Parsons to San Antonio)

The *Katy Flyer* and other earlier MKT trains handled numerous dedicated Pullman cars operating between the larger cities of Texas and Missouri and beyond. Sleepers originated in Dallas, Fort Worth, Waco, Austin, San Antonio, Corpus Christi, Galveston, Houston, and Wichita Falls for overnight trips among these cities as well as to St. Louis, Kansas City, Shreveport (KCS), Tulsa, Minneapolis (CGW), Chicago (Alton) and Denver (UP).

Argonaut, No. 5
New Orleans to Los Angeles
Via Maricopa, Ariz.
Southern Pacific
49 miles east of San Antonio

NEW ORLEANS to LOS ANGELES
53: 8s-5dbr   54: Std: s-dr-c
SP 5, New Orleans (11a Sun)
  to Los Angeles (4p Tue)

NEW ORLEANS to SAN ANTONIO
55: Std: s-dr
SP 5, New Orleans (11a) to San Antonio (1:30a)

HOUSTON to SAN ANTONIO
56: Std: s-dr-c
SP 5, Houston (7:50p) to San Antonio (1:30a)

(Dining car and first-class lounge)

Argonaut, No. 5
Southern Pacific
San Antonio setout
Standing in San Antonio

SAN ANTONIO to LOS ANGELES
55: Std: s-dr
SP 5, San Antonio (2:05a Mon)
  to Los Angeles (4p Tue)

The *Argonaut* was an important source of San Antonio's Pullman connections. A sleeper came from Chicago on the Illinois Central to be cut into the *Argonaut* at New Orleans. Other service included Pullmans to Atlanta (L&N/WPR) and Mexico City (NdeM via Eagle Pass).

No. 5
Kansas City to Galveston
Santa Fe
48 miles south of Fort Worth

CHICAGO to GALVESTON
10s-2dbr-c
Santa Fe 3, Chicago (8:45p Sat)
  to Kansas City (7:15a Sun)
Santa Fe 5, Kansas City (8:50a)
  to Galveston (9:50a Mon)

FORT WORTH to GALVESTON
6s-dr-4dbr
Santa Fe 5, Fort Worth (10:45p) to Galveston (9:50a)

(Dining cars, Chicago-Kansas City-Fort Worth; lounge cars, Chicago to Newton, Kans. No. 3 is the California Limited.)

Argonaut, No. 6
Los Angeles to New Orleans
Via Bowie and Maricopa, Ariz.
Southern Pacific
Approaching Sierra Blanca

LOS ANGELES to NEW ORLEANS
64: 8s-5dbr   65: Std: s-dr-c
66: Std: s-dr
SP 6, Los Angeles (8:30p Sat)
  to New Orleans (7a Tue)

(Diner, first-class lounge)

With early evening departures of the *Argonaut* from El Paso, through Pullman service sufficed. The West Texas metropolis, however, formerly had dedicated sleepers to Douglas, Globe, Tucson, and Phoenix, Ariz., Mexico City (SPdeM at Nogales), and Los Angeles (tourist car).

Argonaut, No. 6
Los Angeles to New Orleans
Via Bowie and Maricopa, Ariz.
Southern Pacific
7 miles east of Beaumont

LOS ANGELES to NEW ORLEANS
64: 8s-5dbr   65: Std: s-dr-c
66: Std: s-dr
SP 6, Los Angeles (8:30p Fri)
  to New Orleans (7a Mon)

HOUSTON to NEW ORLEANS
67: 8s-5dbr
SP 6, Houston (10p) to New Orleans (7a)

(Diner, first-class lounge. As testament to the vastness of Texas, two editions of the eastbound Argonaut—though 24 hours apart in schedule—are still within the state's boundaries. The one near Beaumont will soon be in Louisiana; the one in West Texas approaches Sierra Blanca, where SP and T&P tracks came together in 1881 to form America's second transcontinental rail line.)

SP trains provided Houston with dedicated sleeper service to Los Angeles and Chicago (IC).

No. 6
Galveston to Kansas City
Santa Fe
8 miles south of Brenham

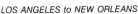

GALVESTON to CHICAGO
10s-2dbr-c
Santa Fe 6, Galveston (7:30p Sun)
  to Kansas City (10:15p)
Santa Fe 4, Kansas City (11p Mon)
  to Chicago (8:30a Tue)

GALVESTON to FORT WORTH
6s-dr-4dbr
Santa Fe 6, Galveston (7:30p) to Fort Worth (6:45a)

(Dining cars: Fort Worth-Kansas City-Chicago; lounge cars: Newton, Kans.,-Kansas City-Chicago. No. 4 is the California Limited.)

Westerner, No. 6
El Paso to Dallas
Texas & Pacific
7 miles west of Big Spring

EL PASO to DALLAS
88: Std: r-br   86: Std: s-dr
T&P 6, El Paso (3:30p) to Dallas (8:45a)

LOS ANGELES to DALLAS
24: 10r-6dbr
SP 2, Los Angeles (8p Sat) to El Paso (3p Sun)
T&P 6, El Paso (3:30p) to Dallas (8:45a Mon)

(Dining-lounge, El Paso to Big Spring; grill coach, El Paso to Dallas. SP 2, the extra-fare Sunset Limited, carries a first-class lounge and diner.)

On this state-spanning line of the T&P, Pecos, Big Spring, Sweetwater, and Abilene all had dedicated Pullman service to Fort Worth/Dallas, as did Amarillo and Lubbock with the cooperation of the Santa Fe, and San Angelo with a Santa Fe predecessor line. Schedules of 1909 show New Orleans sleepers terminating in Abilene, and St. Louis cars in Big Spring.

Katy Flyer, No. 6
San Antonio to Kansas City
and St. Louis, via Dallas
Missouri-Kansas-Texas
42 miles north of San Antonio

SAN ANTONIO to KANSAS CITY
62: 12s-dr
MKT 6, San Antonio (11p) to Denison (10:50a)
MKT 6-26, Denison (11:20a) to Parsons (6:05p)
MKT 26, Parsons (6:25p) to Kansas City (9:25p)

SAN ANTONIO to DALLAS
63: 12s-dr
MKT 6, San Antonio (11p) to Dallas (6:50a)

SAN ANTONIO to FORT WORTH
64: 12s-dr
MKT 6, San Antonio (11p) to Waco (4a)
MKT 26, Waco (4:30a) to Fort Worth (7:10a)

(San Antonio to Parsons, dining car. The Katy serves Dallas and Fort Worth with separate trains between Waco and Denison. No. 26 is the Fort Worth train, carrying the sleeper, and a coach to Denison. At Parsons, the trains separate again, with No. 6 carrying a chair car to St. Louis.)

Westerner, No. 7
Texarkana to El Paso
Texas & Pacific
45 miles west of Fort Worth

TEXARKANA to EL PASO
73: 12s-dr
T&P 7, Texarkana (3p) to El Paso (1:35p)

DALLAS to EL PASO
72: Std: r-br
T&P 7, Dallas (9:40p) to El Paso (1:35p)

(Diner-lounge: Marshall to Fort Worth, Big Spring to El Paso)

No. 8
Dallas to Denver
Fort Worth & Denver
62 miles north of Fort Worth

DALLAS to AMARILLO
800: Std: s-dr
FtW&D 8, Dallas (9:15p) and
  Fort Worth (10:35p) to Amarillo (8:10a)

(Fort Worth to Amarillo, dining car)

No. 8 previously carried sleepers from both Dallas and Fort Worth to Wichita Falls. Other trains handled the Lubbock and Plainview Pullmans and a car that ran from Fort Worth to Colorado Springs. The Fort Worth & Denver also scheduled an Amarillo-Wichita Falls sleeper.

Texas   *(continued)*

Bluebonnet, No. 8-28
San Antonio to Kansas City
Missouri-Kansas-Texas
Standing in Denison

SAN ANTONIO to KANSAS CITY   *via Dallas*
    81: 8s-5br
    MKT 8, San Antonio (12:25p) to Waco (5:40p)
    MKT 8, Waco (6:05p) to Denison (11:40p)
    MKT 8-28, Denison (12:01a) to Kansas City (8:35a)

WACO to KANSAS CITY   *via Fort Worth*
    82: 10s-dr-2c
    MKT 28, Waco (6p) to Denison (11:25p)
    MKT 8-28, Denison (12:01a) to Kansas City (8:35a)

DALLAS to KANSAS CITY
    83: 8s-6r-2br
    MKT 8, Dallas (9:20p) to Denison (11:40p)
    MKT 8-28, Denison (12:01a) to Kansas City (8:35a)

(San Antonio to Kansas City, via Dallas: diner and lounge)

The *Bluebonnet*, named for the Texas state flower, was inaugurated in 1927 as a joint Frisco-MKT train, originally connecting the Texas cities with Tulsa and St. Louis. By this scheduling, it's solely a Katy operation to and from Kansas City.

Houstonian, No. 10
Houston to New Orleans
Missouri Pacific
13 miles east of Beaumont

HOUSTON to NEW ORLEANS
    95: 8s-5dbr
    MP 10, Houston (9:35p) to New Orleans (7:05a)

Pioneer, No. 15
Houston to Brownsville
Missouri Pacific
104 miles southwest of Houston

HOUSTON to CORPUS CHRISTI
    155: 10s-dr-2c
    MP 15, Houston (9:20p) to Odem (3:10a)
    MP accom., Odem (3:45a) Corpus Christi (4:35a)

HOUSTON to BROWNSVILLE
    150: 10s-dr-2c
    MP 15, Houston (9:20p) to Brownsville (8a)

Pioneer, No. 16
Brownsville to Houston
Missouri Pacific
25 miles south of Odem

MISSION to SAN ANTONIO
    1160: 12s-dr
    MP 116, Mission (6:30p) to Harlingen (8:50p)
    MP 16, Harlingen (9:15p) to Odem (12:40a)
    MP 216, Odem (1:15a) to San Antonio (6:45a)

BROWNSVILLE to HOUSTON
    160: 10s-dr-2c
    MP 16, Brownsville (8:15p) to Houston (7:30a)

In the mid-1920's, the Missouri Pacific scheduled sleepers on this route from Brownsville to New Orleans and Chicago (IC).

No. 16 accommodation
Corpus Christi to Odem
Missouri Pacific
Approaching Odem

CORPUS CHRISTI to HOUSTON
    2161: 10s-dr-2c
    MP accom., Corpus Christi (11:45p)
        to Odem (12:15a)
    MP 16, Odem (1:05a) to Houston (7:30a)

Owl, No. 17
Houston to Dallas
Southern Pacific
25 miles north of Houston

HOUSTON to DALLAS
    171: 8s-5br
    SP 17, Houston (11:30p) to Dallas (6:55a)

In its days of heavier consists, the *Owl* also carried sleepers from Galveston to Dallas, St. Louis, and Denver (FtW&D), from Houston to Austin, Waco, Fort Worth, and St. Louis (SSW), and a New Orleans-Denver Pullman (FtW&D).

Owl, No. 18
Dallas to Houston
Southern Pacific
20 miles south of Dallas

DALLAS to HOUSTON
    181: 8s-5br
    SP 18, Dallas (11:30p) to Houston (6:30a)

In addition to Galveston and Houston cars, other sleepers traveled this line to San Antonio and Los Angeles.

Katy Flyer, No. 25-5
Kansas City and St. Louis
to San Antonio, via Fort Worth
Missouri-Kansas-Texas
32 miles south of Fort Worth

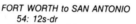

FORT WORTH to SAN ANTONIO
    54: 12s-dr
    MKT 25, Fort Worth (10:55p) to Waco (1:35a)
    MKT 5, Waco (2:10a) to San Antonio (7:50a)

No. 76
Sweetwater to Brownwood
Santa Fe
27 miles west of Brownwood

LOS ANGELES to DALLAS
    10r-5dbr
    Santa Fe 24, Los Angeles (1:40p Sat)
        to Clovis (2:30p)
    Santa Fe 98, Clovis (4p CST Sun) to Texico (ntg)
    Santa Fe 95, Texico (4:15p) to Sweetwater (9:50p)
    Santa Fe 76, Sweetwater (9:50p)
        to Brownwood (12:45a Mon)
    Santa Fe 78, Brownwood (1:10a)
        to Fort Worth (6:25a)
    Santa Fe 111, Fort Worth (7:15a) to Dallas (8a Mon)

CLOVIS to DALLAS
    10s-dr-2c
    As above, Clovis (4p) to Dallas (8a)

CLOVIS to HOUSTON
    8s-dr-2c
    As above, Clovis (4p) to Sweetwater (9:50p)
    Santa Fe 76, Sweetwater (9:50p) to Temple (3:55a)
    Santa Fe 65, Temple (4:07a) to Houston (8:15a)

SAN FRANCISCO to NEW ORLEANS
    247: 6s-6r-4dbr
    Bus from San Francisco (7:45a Sat)
    Santa Fe 60, Oakland (8:07a) to Bakersfield (2:05p)
    Santa Fe 24, Bakersfield (2:15p) to Barstow (6p)
    Santa Fe 24, Barstow (6:25p) to Clovis (2:30p Sun)
    As above, Clovis (4p) to Houston (8:15a Mon)
    MP 4, Houston (9:05a) to New Orleans (7:15p Mon)

SAN ANGELO to HOUSTON
    10s-lounge
    Santa Fe 78, San Angelo (8:50p)
        to Brownwood (12:25a)
    Santa Fe 76, Brownwood (1a) to Temple (3:55a)
    Santa Fe 65, Temple (4:07a) to Houston (8:15a)

(Dining and lounge facilities on all segments except into Dallas and out of San Angelo. Santa Fe 24 is the Grand Canyon; No. 60 is a Golden Gate; MP 4 is the Orleanean. Probably no branch line train in the country carries a larger, more cosmopolitan consist of Pullmans than No. 76 as it trundles in the night across Coleman County.)

No. 112
Dallas to Oklahoma City
Santa Fe
58 miles north of Fort Worth

DALLAS to CHICAGO
    8s-2c-2dbr
    Santa Fe 112, Dallas (8:15p)
        to Oklahoma City (4:40a)
    Santa Fe 12, Oklahoma City (5a) to Chicago (8p)

DALLAS to OKLAHOMA CITY
    10s-lounge
    Santa Fe 112, Dallas (8:15p)
        to Oklahoma City (4:40a)

(Oklahoma City to Chicago: diner, parlor-observation, chair-lounge. No. 12 is the Chicagoan.)

Southerner, No. 38
Texas to St. Louis
Missouri Pacific
8 miles east of Taylor

SAN ANTONIO to LITTLE ROCK
    10s-dr-2c
    MP 38, San Antonio (8:15p) to Longview (8:10a)
    T&P 238, Longview (8:30a) to Texarkana (11:30a)
    MP 8, Texarkana (12:05p) to Little Rock (3:05p)

(Longview to St. Louis, diner-lounge. In Little Rock, No. 8 will exchange the present sleeper for the Hot Springs-Chicago Pullman.)

No. 75
Temple to Sweetwater
Santa Fe
54 miles west of Temple

NEW ORLEANS to SAN FRANCISCO
    237: 6s-6r-4dbr
    MP 3, New Orleans (8:35a Sun) to Houston (6:10p)
    Santa Fe 66, Houston (6:45p) to Temple (10:40p)
    Santa Fe 75, Temple (10:50p)
        to Sweetwater (4:45a Mon.)
    Santa Fe 94, Sweetwater (5:05a) to Texico (ntg)
    Santa Fe 97, Texico (11:05a) to Clovis, N.M. (11:30a)
    Santa Fe 23, Clovis (11a MST)
        to Barstow (5:55a Tue)
    Santa Fe 23, Barstow (6:20a) to Bakersfield (10:50a)
    Santa Fe 61, Bakersfield (11:20a) to Oakland (5p)
    Bus to San Francisco (5:25p Tue)

HOUSTON to CLOVIS
    8s-dr-2c
    As above, Houston (6:45p) to Clovis (11:30a)

HOUSTON to SAN ANGELO
    10s-lounge
    Santa Fe 66, Houston (6:45p) to Temple (10:40p)
    Santa Fe 75, Temple (10:50p) to Brownwood (1:40a)
    Santa Fe 77, Brownwood (2:30a)
        to San Angelo (6:10a)

(Houston to Clovis, dining car and lounge. MoPac No. 3, the Orleanean, carries a diner-lounge. Santa Fe No. 23, the Grand Canyon, has a lounge and diner. No. 61, a Golden Gate, carries a chair lounge and lunch counter dining car.)

No. 78
San Angelo to Brownwood
Santa Fe
12 miles west of Brownwood

SAN ANGELO to HOUSTON
    10s-lounge
    Santa Fe 78, San Angelo (8:50p)
        to Brownwood (12:25a)
    Santa Fe 76, Brownwood (1a) to Temple (3:55a)
    Santa Fe 65, Temple (4:07a) to Houston (8:15a)

SAN ANGELO to DALLAS
    10s-2dr
    Santa Fe 78, San Angelo (8:50p)
        to Fort Worth (6:25a)
    Santa Fe 111, Fort Worth (7:15a)
        to Dallas (8a)

No. 77
Fort Worth to Brownwood
Santa Fe
47 miles northeast of Brownwood

DALLAS to LOS ANGELES
 10r-5dbr
 Santa Fe 112, Dallas (8:15p Sun) to Fort Worth (9p)
 Santa Fe 77, Fort Worth (9:20p)
  to Brownwood (1:40a Mon)
 Santa Fe 75, Brownwood (2:05a)
  to Sweetwater (4:45a)
 Santa Fe 94, Sweetwater (5:05a) to Texico (ntg)
 Santa Fe 97, Texico (11:05a) to Clovis (11:30a)
 Santa Fe 23, Clovis (11a MST)
  to Los Angeles (10:40a Tue)

DALLAS to CLOVIS
 10s-dr
 As above, Dallas (8:15p) to Clovis (11:30a)

DALLAS to SAN ANGELO
 10s-2dr
 As above, Dallas (8:15p) to Brownwood (1:40a)
 Santa Fe 77, Brownwood (2:30a)
  to San Angelo (6:10a)

(Dining car and lounge: Brownwood to Clovis and on No. 23, the Grand Canyon, Clovis to Los Angeles.

Scout, No. 105
Newton, Kans., to Albuquerque
Santa Fe
28 miles west of Amarillo

CHICAGO to LOS ANGELES
 10s-dr-2c
 Santa Fe 3, Chicago (8:45p Sat)
  to Kansas City (7:15a)
 Santa Fe 5, Kansas City (8:50a Sun)
  to Newton (12:10p)
 Santa Fe 105, Newton (12:55p)
  to Belen, N.M. (6:35a Mon)
 Santa Fe 14, Belen (6:55a) to Albuquerque (7:45a)
 Santa Fe 3, Albuquerque (8:30a)
  to Los Angeles (7a Tue)

WAYNOKA, OKLA. to ALBUQUERQUE
 10s-dr-2c
 Santa Fe 105, Waynoka (5:40p Sun)
  to Clovis (2a Mon)
 Santa Fe 25, Clovis (3a) to Carlsbad, N.M. (8:15a)
 Santa Fe 26, Carlsbad (7:30p) to Clovis (12:45a Tue)
 Santa Fe 105, Clovis (1:35a MST) to Belen 6:35a)
 Santa Fe 14, Belen (6:55a)
  to Albuquerque (7:45a Tue)

(Diner: Chicago-Kansas City-Amarillo; lounge: Chicago-Kansas City-Newton. No. 3, the California Limited, makes meal stops at Fred Harvey dining rooms in Gallup, N.M., and Williams, Ariz. The Scout acts as a southern route counterpart to No. 3, its most notable service handling the Pullmans full of tourists bound for Carlsbad Caverns. Just as at the Grand Canyon, the Carlsbad sleeper lays over during the day while the passengers sightsee.)

No. 28
Houston to Shreveport
Southern Pacific
45 miles north of Houston

HOUSTON to SHREVEPORT
 280: Std: s-dr
 SP 28, Houston (10:45p) to Shreveport (6:15a)

Southerner, No. 37
St. Louis to Texas
Missouri Pacific
64 miles southwest of Palestine

ST. LOUIS to SAN ANTONIO
 75: 10s-dr-2c
 MP 37, St. Louis (12:45a Sun) to Texarkana (2:30p)
 T&P 237, Texarkana (4:15p) to Longview (6:50p)
 MP 37, Longview (7:10p) to San Antonio (7:10a Mon)

(St. Louis to Longview, diner-lounge)

No. 215
San Antonio to Odem
Missouri Pacific
37 miles south of San Antonio

SAN ANTONIO to MISSION
 2150: 12s-dr
 MP 215, San Antonio (10:40p) to Odem (2:45a)
 MP 15, Odem (3:30a) to Harlingen (6:45a)
 MP 115, Harlingen (7:05a) to Mission (8:55a)

Border Limited, No. 303
Houston to McAllen
Southern Pacific
77 miles southwest of Houston

HOUSTON to CORPUS CHRISTI
 3030: Std: s-dr
 SP 303, Houston (9:45p) to Skidmore (3:23a)
 SP 309, Skidmore (3:50a) to Corpus Christi (5:15a)

HOUSTON to McALLEN
 3031: Std: s-dr
 SP 303, Houston (9:45p) to McAllen (7:50a)

Border Limited, No. 304
McAllen to Houston
Southern Pacific
17 miles southwest of Skidmore

McALLEN to HOUSTON
 3120: Std: s-dr
 SP 304, McAllen (8:30p) to Houston (6:50a)

No. 310
Corpus Christi to Skidmore
Southern Pacific
13 miles southeast of Skidmore

CORPUS CHRISTI to HOUSTON
 3040: Std: s-dr
 SP 310, Corpus Christi (11p) to Skidmore (12:20a)
 SP 304, Skidmore (12:55a) to Houston (6:50a)

Black Gold, No. 518
Texas to Tulsa
Frisco
37 miles north of Dallas

FORT WORTH to TULSA
 152: 8s-dr-2c
 SLSF 518, Fort Worth (9:35p) to Tulsa (8:05a)

DALLAS to TULSA
 150: 12r-2sb-3dbr
 SLSF 518, Dallas (11p) to Tulsa (8:05a)

(Fort Worth to Tulsa, buffet-lounge)

Utah

See map on page 200

City of St. Louis, No. 9
Streamliner
St. Louis to Los Angeles
Union Pacific
100 miles south of Salt Lake City

ST. LOUIS to LOS ANGELES
 94: 12r-4dbr
 Wabash 9, St. Louis (4p Sat) to Kansas City (9p)
 UP 9, Kansas City (9:30p) to Los Angeles (3p Mon)

SPOKANE to LOS ANGELES *Friday only, winter*
 6s-6dbr
 UP 19, Spokane (9p Sat) to Hinkle, Ore. (1:15a Sun)
 UP 18, Hinkle (2:10a) to Pocatello (5:05p)
 UP 34, Pocatello (5:35p) to Salt Lake City (10p)
 UP 9, Salt Lake City (10:10p)
  to Los Angeles (3p Mon)

SALT LAKE CITY to CEDAR CITY *Summertime only*
 8s-dr-2c
 UP 9, Salt Lake City (10:10p) to Milford (1:45a)
 UP 309, Milford (4:15a) to Cedar City (6:30a)

(Diner and first-class lounge on No. 9 and No. 18, the Portland Rose; Cafe-lounge on No. 34, the Northwest Special. The "Saturday" departure of the Spokane sleeper is used to make it consistent with this schedule format and is obviously not a true schedule.)

The UP carried dedicated sleepers from Salt Lake City to Caliente and Las Vegas, Nev., and Los Angeles.

City of St. Louis, No. 10
Streamliner
Los Angeles to St. Louis
Union Pacific
8 miles west of Lund

LOS ANGELES to ST. LOUIS
 104: 12r-4dbr
 UP 10, Los Angeles (10:30a Sun)
  to Kansas City (6:45a Tue)
 Wabash 10, Kansas City (7a)
  to St. Louis (11:59a Tue)

LOS ANGELES to CEDAR CITY *Summer season*
 10s-dr-2c
 UP 10, Los Angeles (10:30a) to Milford (12:45a)
 UP 309, Milford (4:15a) to Cedar City (6:30a)

(Los Angeles to St. Louis, diner and first-class lounge. Cedar City is the gateway to Zion and Bryce Canyon National Parks and surrounding recreational areas. Other east- and westbound editions of the City of St. Louis are now in Kansas.)

UP's westbound *Gold Coast* notches 75 mph with 14 cars crossing Nebraska in July 1953.

Utah   (continued)

Los Angeles Limited, No. 2

Los Angeles to Chicago
Union Pacific
19 miles southwest of Lund

LOS ANGELES to CHICAGO
2dr-4c-4dbr   12r-4dbr
14s
UP 2, Los Angeles (12:01p Sun)
    to Omaha (3:50a Tue)
C&NW 2, Omaha (4:30a) to Chicago (2p Tue)

LOS ANGELES to NEW YORK
10r-6dbr
As above, Los Angeles (12:01p Sun)
    to Chicago (2p Tue)
NYC 22, Chicago (6:30p) to New York (12:30p Wed)

LOS ANGELES to NEW YORK
10r-6dbr
As above, Los Angeles (12:01p Sun)
    to Chicago (2p Tue)
PRR 2, Chicago (6:30p) to New York (11:59a Wed)

LOS ANGELES to ST. PAUL/MINNEAPOLIS
6s-6r-4dbr
UP 2, Los Angeles (12:01p Sun)
    to Omaha (3:50a Tue)
C&NW 204, Omaha (9:25a) to St. Paul (6:15p Tue)
    and Minneapolis (6:50p)

LOS ANGELES to OMAHA
14s
UP 2, Los Angeles (12:01p Sun)
    to Omaha (3:50a Tue)

(Dining cars and lounges, Los Angeles to Chicago and Chicago to New York; cafe-lounge Omaha to Minneapolis on C&NW 204, the North American. NYC 22 is the Lake Shore Limited, and PRR 2 is the Pennsylvania Limited. Another eastbound Los Angeles Limited is now in Nebraska.)

Northwest Special, No. 33

Salt Lake City to Ashton, Ida.
Union Pacific
15 miles south of Ogden

SALT LAKE CITY to SPOKANE
10s-3dbr
UP 33, Salt Lake City (11:30p Sun)
    to Pocatello (4:15a)
UP 17, Pocatello (9:40a Mon) to Hinkle (11:20p)
UP 20, Hinkle (2:40a Tue) to Spokane (7:20a Tue)

(No. 17, the Portland Rose, carries diner and first-class lounge; No. 20 is the Spokane)

The UP's trains to the Northwest carried sleepers from Salt Lake City to Ashton, Buhl, and Idaho Falls, Ida., and Huntington, Ore.

No. 310

Cedar City to Milford
Union Pacific
4 miles south of Milford

CEDAR CITY to SALT LAKE CITY   Summertime only
8s-dr-2c
UP 310, Cedar City (10:05p) to Milford (12:05a)
UP 10, Milford (12:55a) to Salt Lake City (4:40a)

CEDAR CITY to LOS ANGELES Summertime only
10s-dr-2c
UP 310, Cedar City (10:05p) to Milford (12:05a)
UP 9, Milford (1:55a) to Los Angles (3p)

(These sleepers are carried to and from Milford on the City of St. Louis, Nos. 9 and 10)

Vermont

See map on page 14

Washingtonian, No. 20

Montreal to Washington
Central Vermont
Departing Montpelier

MONTREAL to NEW YORK
129: 8s-5dbr   127: 14sbr
CNR 20, Montreal (8:35p) to St. Albans (10:36p)
CV 20, St. Albans (10:41p) to White River Jct. (1:25a)
B&M 732, White River Jct. (1:40a)
    to Springfield (4:40a)
NH 169, Springfield (4:55a) to New York (7:55a)

MONTREAL to WASHINGTON
131: 8s-5dbr   133: 5c-buffet-lounge
As above, Montreal (8:35p) to New York (7:55a)
PRR 169, New York (8:30a) to Washington (12:15p)

MONTREAL to BOSTON
10s-2dbr-c
As above, Montreal (8:35p)
    to White River Jct. (1:25a)
B&M 302, White River Jct. (4:15a) to Boston (8:05a)

(New York to Washington, diner; White River Jct. to Boston, buffet-lounge. No 302 is the New Englander. The Washingtonian is the southbound version of the Montrealer. The CV logo accompanying this schedule is of the modern design created for the railroads in the Canadian National family.)

Sleepers traveled the CV/CNR route from Boston to Montpelier and, before World War I, all the way to Chicago.

Mount Royal, No. 52

Montreal to New York
Rutland Railway
16 miles north of Rutland

MONTREAL to NEW YORK
8s-5dbr
CNR 52, Montreal (8:05p)
    to Rouses Point, N.Y. (9:27p)
RUT 52, Rouses Point (9:30p)
    to North Bennington (2:28a)
B&M 5603, North Bennington (2:30a) to Troy (3:20a)
NYC 8-72, Troy (4:20a) to New York (7:35a)

RUTLAND to NEW YORK   Summers, Monday only
12s-dr
RUT 52, Rutland (1:05a) to North Bennington (2:27a)
As above, North Bennington (2:30a) to New York (6:40a)

The *Mount Royal* carried Alburgh-New York Pullmans and also had a Boston section with sleepers from both Montreal and Ogdensburg, N.Y.

Night White Mountains, No. 76

New Hampshire to New York
Summertime, Sundays only
Boston & Maine
21 miles south of Wells River

BRETTON WOODS-FABYAN to NEW YORK
760: 10s-dr-2c   761: 6c-3dr
762: 6c-buffet-lounge   763: 14sbr
B&M 76, Bretton Woods (9:30p) to Springfield (ntg)
NH 167, Springfield (ntg) to New York (7:45a)

BRETTON WOODS-FABYAN to WASHINGTON
764: 6s-6dbr
B&M 76, Bretton Woods (9:30p)
    to White River Jct. (12:35a)
B&M 732, White River Jct. (1:40a)
    to Springfield (4:40a)
NH 169, Springfield (4:55a) to New York (7:55a)
PRR 169, New York (8:30a) to Washington (12:15p)

WOODSVILLE to NEW YORK
765: 10s-dr-2c
B&M 76, Woodsville (11:20p) to Springfield (ntg)
NH 167, Springfield (ntg) to New York (7:45a)

(Because the Night White Mountains arrives in New York's Grand Central Terminal, the Washington sleeper must be turned over to a Hell Gate Bridge route train beforehand. In this case, Nos. 732/169, the Washingtonian, obliges at White River Jct. Diner, New York to Washington.)

Red Wing, No. 210

Montreal to Boston
Canadian Pacific
46 miles north of St. Johnsbury

MONTREAL to BOSTON
12s-2dbr   4dbr-c-obs.-buf-lounge
CPR 210, Montreal (9p) to Wells River (2:10a)
B&M, Wells River (2:20a) to White River Jct. (3:20a)
B&M 302, White River Jct. (4:15a) to Boston (8:05a)

Vermonter, No. 304

St. Albans to White River Jct.
Central Vermont
16 miles north of White River Jct.

ST. ALBANS to NEW YORK
125: 10s-3dbr
CV 304, St. Albans (8:30p)
    to White River Jct. (12:30a)
B&M 732, White River Jct. (1:40a)
    to Springfield (4:40a)
NH 169, Springfield (4:55a) to New York (7:55a)

ST. ALBANS to BOSTON
B1: 10s-2dbr-c
CV 304, St. Albans (8:30p)
    to White River Jct. (12:30a)
B&M 302, White River Jct. (4:15a) to Boston (8:05a)

(Buffet-lounges: White River Jct. to Boston and New York. No. 302 is the New Englander. Sunday schedules different.)

B&M/CPR *Red Wing,* with observation-buffet-compartment *Mt. Huber,* arriving in Montreal.

Otto C. Perry/Denver Public Library

Virginia

See map on page **110**

George Washington, No. 1

New York/Phoebus
to Cincinnati/Louisville
Chesapeake & Ohio
10 miles west of Covington

WASHINGTON to CINCINNATI
　19: 10r-6dbr
　C&O 1, Washington (6:01p) to Cincinnati (8:45a)

WASHINGTON to LOUISVILLE
　115: 10r-6dbr
　C&O 1, Washington (6:01p) to Ashland, Ky. (5:15a)
　C&O 21, Ashland (5:45a) to Louisville (11a)

NEW YORK to LOUISVILLE
　C51: 10r-5dbr
　PRR 125, New York (12:30p) to Washington (4:25p)
　As above, Washington (6:01p) to Louisville (11a)

PHOEBUS to CINCINNATI
　10: 5dbr-observation-buffet-lounge
　C&O 41, Phoebus (2:35p) to Charlottesville (8:10p)
　C&O 1, Charlottesville (8:42p) to Cincinnati (8:45a)

PHOEBUS to CHICAGO
　C16: 10r-6dbr
　As above, Phoebus (2:35p) to Cincinnati (8:45a)
　NYC 415, Cincinnati (9:45a) to Chicago (3:30p)

RICHMOND to LOUISVILLE
　11: 10r-6dbr
　C&O 41, Richmond (5:25p) to Charlottesville (8:10p)
　C&O 1, Charlottesville (8:42p) to Ashland (5:15a)
　C&O 21, Ashland (5:45a) to Louisville (11a)

RICHMOND to ST. LOUIS
　C18: 10r-6dbr
　C&O 41, Richmond (5:25p) to Charlottesville (8:10p)
　C&O 1, Charlottesville (8:42p) to Cincinnati (8:45a)
　NYC 415, Cincinnati (9:45a) to Indianapolis (11:10a)
　NYC 11, Indianapolis (11:25a) to St. Louis (4:10p)

(Diners: Washington and Richmond to Charlottesville, Ashland to Cincinnati and Louisville. Phoebus is one of several stations in the Newport News-Hampton area used by the C&O.)

The *George Washington* garnered kudos in 1932 for being the first air-conditioned Pullman train. The *GW* and other C&O trains carried sleepers from Newport News to Huntington, Ashland, Louisville, St. Louis (NYC), Chicago (NYC, via Toledo), and New York (RF&P/PRR). Richmond had dedicated Pullmans to Lynchburg, Clifton Forge, White Sulphur Springs, Huntington, Ashland, Cincinnati, Detroit, Grand Rapids, and New York (RF&P/PRR).

At Charlottesville, the C&O combined the sections from Washington and Newport News. This picturesque home of the University of Virginia was served by Pullmans to Cincinnati, Louisville, Cleveland (NYC), Detroit, and Chicago (NYC)

No. 11

Richmond to Danville
Southern Railway
75 miles northeast of Danville

RICHMOND to CHARLOTTE
　69: 12s-dr
　SOU 11, Richmond (10:15p) to Danville (2:15a)
　SOU 29, Danville (4:18a) to Charlotte (7:45a)

(Dining car, Danville to Charlotte)

The Southern Railway scheduled Pullmans from Richmond to Danville, Asheville, Salisbury, Winston-Salem, Atlanta, Birmingham, and Memphis.

Fast Flying Virginian (F.F.V.), No. 3

New York/Phoebus
to Cincinnati/Louisville
Chesapeake & Ohio
7 miles south of Manassas

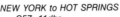

NEW YORK to HOT SPRINGS
　C57: 11dbr
　PRR 151, New York (6:25p) to Washington (10:20p)
　C&O 3, Washington (10:55p) to Covington (5:20a)
　C&O 303, Covington (6:15a) to Hot Springs (8a)

NEW YORK to WHITE SULPHUR SPRINGS
　C56: 11dbr
　PRR 151, New York (6:25p) to Washington (10:20p)
　C&O 3, Washington (10:55p) to Ashland (11:49a)
　　　to White Sulphur Springs (6a)

WASHINGTON to HUNTINGTON
　32: 10r-6dbr
　C&O 3, Washington (10:55p) to Huntington (11:05a)

NEW YORK to HUNTINGTON, W.VA.
　C52: 10r-5dbr
　As above, New York (6:25p) to Huntington (11:05a)

NEW YORK to ASHLAND, KY.
　C54: 10r-6dbr
　PRR 151, New York (6:25p) to Washington (10:20p)
　C&O 3, Washington (10:55p) to Ashland (11:49a)

NEW YORK to LEXINGTON, KY.
　C53: 10r-6dbr
　As above, New York (6:25p) to Ashland (11:49a)
　C&O 23, Ashland (12:10p) to Lexington (2:45p)

PHOEBUS to CINCINNATI
　35: 10r-6dbr
　C&O 43, Phoebus (7:30p) to Charlottesville (1:25a)
　C&O 3, Charlottesville (2:13a) to Cincinnati (3:10p)

(Dining service: Clifton Forge to Huntington, Ashland to Cincinnati, and Ashland to Lexington. Pullman service to Hot Springs and White Sulphur Springs three times weekly in the winter, daily during warm weather.)

The distinguished *F.F.V.*, one of the nation's premier name trains, was inaugurated in 1889 with an eye-catching consist of wooden coaches in orange and dark red livery.

Fast Flying Virginian (F.F.V.), No. 6

Cincinnati to Washington
Chesapeake & Ohio
20 miles west of Charlottesville

HUNTINGTON to WASHINGTON
　60: 10r-6dbr
　C&O 6, Huntington (3:45p) to Washington (4:10a)

ASHLAND to NEW YORK
　C65: 10r-6dbr
　C&O 6, Ashland (2:52p) to Washington (4:10a)
　PRR 112, Washington (5:05a) to New York (9:20a)

HUNTINGTON to NEW YORK
　C64: 10r-5dbr
　As above, Huntington (3:45p) to New York (9:20a)

WHITE SULPHUR SPRINGS to NEW YORK
　C67: 11dbr
　C&O 6, White Sulphur Springs (8:55p)
　　　to Washington (4:10a)
　PRR 112, Washington (5:05a) to New York (9:20a)

HOT SPRINGS to NEW YORK
　C63: 11dbr
　C&O 306, Hot Springs (8:15p)
　　　to Clifton Forge (9:40p)
　C&O 6, Clifton Forge (10:05p) to Washington (4:10a)
　PRR 112, Washington (5:05a) to New York (9:20a)

HINTON, W.VA. to RICHMOND
　61: 10r-6dbr
　C&O 6, Hinton (7:40p) to Charlottesville (12:40a)
　C&O 42, Charlottesville (6:45a) to Richmond (8:45a)

(Dining cars: Huntington to Clifton Forge, Washington to New York. Pullman service from White Sulphur Springs and Hot Springs thrice weekly in winter, daily during warm weather.)

No. 2

Roanoke to Hagerstown
Norfolk & Western
3 miles north of Front Royal

ROANOKE to NEW YORK
　RN1: 10s-c-dr　　　　W6: 10r-6dbr
　N&W 2, Roanoke (6:40p) to Hagerstown (1:45a)
　PRR 638, Hagerstown (1:53a) to Harrisburg (3:55a)
　PRR 60, Harrisburg (4:10a) to New York (8a)

(Diner-lounge from Roanoke to Shenandoah. Dining car and sleeper-lounge from Harrisburg to New York.)

Palmland, No. 10-2

Miami to New York
Seaboard Air Line
7 miles south of Petersburg

HAMLET, N.C. to NEW YORK　(via Southern Pines)
　B63: 8s-dr-3dbr　　　　B64: 12r-sbr-4dbr
　SAL 10-2, Hamlet (6:35p) to Richmond (1:10a)
　RF&P 110, Richmond (1:10a) to Washington (4a)
　PRR 112, Washington (5:05a) to New York (9:20a)

HAMLET to WASHINGTON
　B&1: 10s-dr-2c
　As above, Hamlet (6:35p) to Washington (4a)

MIAMI to NEW YORK
　B122: 8s-5dbr
　SAL 10-2, Miami (9p Sat) to Richmond (1:10a Mon)
　As above, Richmond (1:10a)
　　　to New York (9:20a Mon)

(Dining car, Jacksonville to Raleigh; diner-lounge, Washington to New York)

Cavalier, No. 15

Norfolk to Cincinnati
Norfolk & Western
In Petersburg

NORFOLK to ROANOKE
　99: 10s-lounge
　N&W 15, Norfolk (10p) to Roanoke (5:25a)

RICHMOND to BRISTOL
　100: 12s-dr
　ACL 33, Richmond (11:15p) to Petersburg (11:55p)
　N&W 15, Petersburg (12:25a) to Roanoke (5:25a)
　N&W 41, Roanoke (6a) to Bristol (10a)

(Diner, Roanoke to Bristol)

Peach Queen, No. 29

New York to Atlanta
Southern Railway
18 miles south of Manassas

NEW YORK to ATLANTA
　SR1: 2dr-3c-lounge　　　SR2: 12r-2sbr-3dbr
　SR3: 8s-5dbr　　　　　　SR4: 10s-dr-2c
　PRR 151, New York (6:25p) to Washington (10:20p)
　SOU 29, Washington (10:45p) to Atlanta (2:30p)

NEW YORK to CHARLOTTE
　SR5: 10s-dr-2c
　PRR 151, New York (6:25p) to Washington (10:20p)
　SOU 29, Washington (10:45p) to Charlotte (7:45a)

NEW YORK to SALISBURY
　SR6: 8s-5dbr
　PRR 151, New York (6:25p) to Washington (10:20p)
　SOU 29, Washington (10:45p) to Greensboro (5:16a)
　SOU 34, Greensboro (6:25a) to Salisbury (7:43a)

NEW YORK to WINSTON-SALEM
　SR31: 10r-6dbr
　As above, New York (6:25p) to Greensboro (5:16a)
　SOU 1, Greensboro (7:05a)
　　　to Winston-Salem (7:55a)

NEW YORK to DURHAM-RALEIGH
　SR32: 10r-6dbr
　As above, New York (6:25p) to Greensboro (5:16a)
　SOU 16, Greensboro (5:50a) to Durham (7:40a)
　　　and Raleigh (8:40a)

WASHINGTON to ATLANTA
　61: 8s-dr-3dbr
　SOU 29, Washington (10:45p) to Atlanta (2:30p)

Virginia   (continued)

Cavalier, No. 16
Cincinnati to Norfolk
Norfolk & Western
8 miles west of Roanoke

WILLIAMSON, W.VA. to NEW YORK
 WN1: *10r-6dbr*
 N&W 16, Williamson (5:50p) to Roanoke (12:20a)
 N&W 42, Roanoke (12:45a) to Lynchburg (2a)
 SOU 42, Lynchburg (2:10a) to Washington (6:45a)
 PRR 122, Washington (7:20a) to New York (11:30a)

(Diner-lounge, Williamson to Roanoke; dining car, Washington to New York)

Cavalier, No. 16
Norfolk & Western
Roanoke setout
Standing in Roanoke

ROANOKE to NORFOLK
 99: *10s-lounge*
 N&W 16, Roanoke (12:50a) to Norfolk (8a)

 Roanoke had dedicated sleeping cars from Cleveland (NYC) and Charleston, W. Va. (C&O).

Birmingham Special, No. 17
New York to Birmingham
Norfolk & Western
35 miles northeast of Bristol

NEW YORK to BIRMINGHAM
 S8: *10r-6dbr*
 PRR 123, New York (11:30a) to Washington (3:25p)
 SOU 17, Washington (3:50p) to Lynchburg (7:50p)
 N&W 17, Lynchburg (7:55p) to Bristol (12:55a)
 SOU 17, Bristol (1:05a) to Birmingham (9:30a)

NEW YORK to KNOXVILLE
 N77: *10s-dr-2dbr*
 PRR 123, New York (11:30a) to Washington (3:25p)
 As above, Washington (3:50p) to Bristol (12:55a)
 SOU 17, Bristol (1:05a) to Knoxville (4:15a)

WASHINGTON to BIRMINGHAM
 S46: *8s-diner-lounge*
 As above, Washington (3:50p)
  to Birmingham (9:30a)

(Buffet-lounge, dining car from New York to Washington; diner, Washington to Roanoke)

**Aiken-Augusta Special and
Asheville Special, No. 31**
New York to Aiken,
Augusta, and Asheville
Southern Railway
17 miles north of Danville

NEW YORK to AUGUSTA
 SR17: *12r-2sbr-3dbr*
 PRR 149, New York (2:55p) to Washington (6:50p)
 SOU 31, Washington (7:20p) to Augusta (11:15a)

WASHINGTON to AUGUSTA
 43: *10s-dr-2c*
 SOU 31, Washington (7:20p) to Augusta (11:15a)

NEW YORK to ASHEVILLE
 SR18: *10s-dr-2c*   SR19: *10r-6dbr*
 PRR 149, New York (2:55p) to Washington (6:50p)
 SOU 31, Washington (7:20p) to Greensboro (1:55a)
 SOU 15, Greensboro (2:40a) to Asheville (9:15a)

WASHINGTON to ASHEVILLE
 9: *10s-3dbr*
 As above, Washington (7:20p) to Asheville (9:15a)

(Diners: New York to Washington, Washington to Greensboro and Augusta, and Greensboro to Asheville.)

 Augusta was a favorite golfing resort for President Dwight D. Eisenhower, whose trips contributed to its popularity.

The northbound *Florida Special* pauses at Richmond Broad Street, Feb. 16, 1963.

Piedmont Limited, No. 34
New Orleans to New York
Southern Railway
23 miles south of Manassas

NEW ORLEANS to NEW YORK
 L42: *10s-dr-2c*   L43: *8s-5dbr*
 L&N 34, New Orleans (5p Sat)
  to Montgomery (2a Sun)
 WPR 34, Montgomery (2:35a) to Atlanta (8:20a)
 SOU 34, Atlanta (9a) to Washington (1:25a Mon)
 PRR 108, Washington (2:10a)
  to New York (6:25a Mon)

MOBILE to NEW YORK
 L44: *10s-dr-2c*
 L&N 34, Mobile (8:45p Sat) to Montgomery (2a Sun)
 As above, Montgomery (2:35a)
  to New York (6:25a Mon)

NEW ORLEANS to WASHINGTON
 L40: *10s-lounge*
 As above, New Orleans (5p Sat)
  to Washington (1:25a Mon)

BIRMINGHAM to NEW YORK
 S45: *10s-dr-2c*
 SOU 12, Birmingham (10:30p Sat)
  to Atlanta (6a Sun)
 As above, Atlanta (9a) to New York (6:25a Mon)

SALISBURY, N.C. to NEW YORK
 S41: *8s-5dbr*
 SOU 34, Salisbury (5:05p) to Washington (1:25a)
 PRR 108, Washington (2:10a) to New York (6:25a)

(Dining car, New Orleans to Monroe, Va.)

Pelican, No. 42
Norfolk & Western
Roanoke setout
Standing in Roanoke

ROANOKE to WASHINGTON
 W3: *10r-6dbr*
 N&W 42, Roanoke (12:45a) to Lynchburg (2a)
 SOU 42, Lynchburg (2:10a) to Washington (6:45a)

F.F.V. accommodation, No. 43
Phoebus to Charlottesville
Chesapeake & Ohio
25 miles east of Gordonsville

PHOEBUS to CINCINNATI
 35: *10r-6dbr*
 C&O 43, Phoebus (7:30p) to Charlottesville (1:25a)
 C&O 3, Charlottesville (2:13a) to Cincinnati (3:10p)

**Aiken-Augusta Special and
Asheville Special, No. 32**
Augusta, Aiken, and
Asheville to New York
Southern Railway
13 miles south of Lynchburg

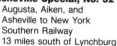
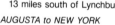

AUGUSTA to NEW YORK
 S14: *12r-2sbr-3dbr*
 SOU 32, Augusta (1:45p) to Washington (4:20a)
 PRR 118, Washington (4:55a) to New York (9:15a)

AUGUSTA to WASHINGTON
 43: *10s-dr-2c*
 SOU 32, Augusta (1:45p) to Washington (4:20a)

CHARLOTTE to NEW YORK
 S33: *10s-dr-2c*
 SOU 32, Charlotte (7:30p) to Washington (4:20a)
 PRR 118, Washington (4:55a) to New York (9:15a)

RALEIGH to NEW YORK
 S64: *10r-6dbr*
 SOU 13, Raleigh (6p) to Greensboro (9p)
 SOU 32, Greensboro (9:50p) to Washington (4:20a)
 PRR 118, Washington (4:55a) to New York (9:15a)

ASHEVILLE to NEW YORK
 S83: *10r-6dbr*   S88: *10s-dr-2c*
 SOU 16, Asheville (3:45p) to Greensboro (9:25p)
 As above, Greensboro (9:50p) to New York (9:15a)

WINSTON-SALEM to NEW YORK
 S34: *10r-6dbr*
 SOU 16, Winston-Salem (8:45p)
  to Greensboro (9:25p)
 As above, Greensboro (9:50p) to New York (9:15a)

(Diners: Augusta to Greensboro and Washington, Asheville to Greensboro, and Washington to New York)

Southerner, No. 47
Streamliner
New York to New Orleans
Southern Railway
12 miles north of Lynchburg

NEW YORK to BIRMINGHAM
 S3: *10r-6dbr*
 PRR 157, New York (4:35p) to Washington (8:35p)
 SOU 47, Washington (8:50p) to Birmingham (1:25p)

(Diner, observation-coach-lounge, and coach-lounge. The Birmingham sleeper was the one exception to the all-coach consist.)

The Crescent, No. 38
All-Pullman streamliner
New Orleans to New York
Southern Railway
Standing in Lynchburg

NEW ORLEANS to NEW YORK
L24: 2dr-mr-buffet-lounge   L25: 10r-6dbr
L26: 10r-6dbr   L27: 10r-6dbr
L28: 5dbr-observation-lounge
L&N 38, New Orleans (11p Sat)
            to Montgomery (7:20a)
WPR 38, Montgomery (7:40a Sun) to Atlanta (1:15p)
SOU 38, Atlanta (1:45p) to Washington (4:15a Mon)
PRR 118, Washington 4:55a)
            to New York (9:15a Mon)

NEW ORLEANS to WASHINGTON
L19: 10r-6dbr
As above, New Orleans (11p Sat)
            to Washington (4:15a Mon)

ATLANTA to NEW YORK
S20: 10r-6dbr   S21: 10r-6dbr
S22: 10r-6dbr   S23: 10r-6dbr
As above, Atlanta (1:45p) to New York (9:15a)

ATLANTA to WASHINGTON
18: 10r-6dbr
As above, Atlanta (1:45p) to Washington (4:15a)

(Diner, New Orleans to New York; coaches, New Orleans to Atlanta.)

The coaches were a later compromise: *The Crescent Limited*, as it was called upon its introduction in 1925, was an all-Pullman, extra-fare train. Its luxurious, heavyweight sleepers were easily identified by their signature color scheme of two tones of green.

Pelican, No. 42
New Orleans to New York
Norfolk & Western
Entering Roanoke

NEW ORLEANS to NEW YORK
N9: 10r-6dbr
SOU 42, New Orleans (9p Sat) to Bristol (7:40p Sun)
N&W 42, Bristol (7:55p) to Lynchburg (2a Mon)
SOU 42, Lynchburg (2:10a) to Washington (6:45a)
PRR 122, Washington (7:20a)
            to New York (11:30a Mon)

KNOXVILLE to NEW YORK
S67: 10r-6dbr
SOU 42, Knoxville (3:55p) to Bristol (7:40p)
As above, Bristol (7:55p) to New York (11:30a)

SHREVEPORT to NEW YORK
2264: 10s-2c-dr
IC 226, Shreveport (4:30p Sat) to Meridian (2a Sun)
SOU 42, Meridian (2:55a) to Bristol (7:40p)
As above, Bristol (7:55p) to New York (11:30a Mon)

BRISTOL to NEW YORK
BN1: 10r-6dbr
As above, Bristol (7:55p) to New York (11:30a)

NEW ORLEANS to WASHINGTON
S44: 10s-lounge
As above, New Orleans (9p Sat)
            to Washington (6:45a Mon)

BRISTOL to RICHMOND
101: 12s-dr
N&W 42, Bristol (7:55p) to Roanoke (12:05a)
N&W 16, Roanoke (12:50a) to Petersburg (5:25a)
ACL 22, Petersburg (6:35a) to Richmond (7:20a)

(Diners: Birmingham to Roanoke, Washington to New York)

In its varied consist, the *Pelican* had the distinction of carrying the only Pullman operating to New York from west of the Mississippi River (excluding St. Louis and the post-war Southwestern and West Coast expansion), the Shreveport sleeper.

The N&W/ACL interline Bristol-Richmond sleeper was the sole survivor of service that had included Richmond cars from Columbus (Ohio), Roanoke, and Lynchburg.

Havana Special, No. 76
Florida to New York
Richmond, Fredericksburg
& Potomac
13 miles north of Fredericksburg

MIAMI to NEW YORK
F161: 10s-c-2dbr
FEC 76, Miami (10p Sat) to Jacksonville (7a Sun)
ACL 76, Jacksonville (8:10a) to Richmond (10:05p)
RF&P 76, Richmond (10:30p)
            to Washington (1:10a Mon)
PRR 110, Washington (2:15a)
            to New York (7:05a Mon)

TAMPA to NEW YORK
A162: 10s-c-dr
ACL 76, Tampa (10:30p Sat)
            to Jacksonville (6:30a Sun)
As above, Jacksonville (8:10a)
            to New York (7:05a Mon)

FORT MYERS to NEW YORK
A163: 10s-dr-2c
ACL 276, Fort Myers (6:45p Sat)
            to Jacksonville (5:45a)
As above, Jacksonville (8:10a)
            to New York (7:05a Mon)

ORLANDO to WASHINGTON
R508: 8s-2c-dr
ACL 76, Orlando (1:35a) to Jacksonville (6:30a)
As above, Jacksonville (8:10a) to Washington (1:10a)

WILMINGTON, N.C. to NEW YORK
A165: 10s-2c-dr
ACL 48, Wilmington (3:40p) to Rocky Mount (7p)
ACL 76, Rocky Mount (7:30p) to Richmond (10:05p)
As above, Richmond (10:30p) to New York (7:05a)

RICHMOND to NEW YORK
Two cars: 10r-6dbr
RF&P 76, Richmond (10:30p) to Washington (1:10a)
PRR 108, Washington (2:10a) to New York (6:25a)

RICHMOND to PHILADELPHIA
10r-6dbr
RF&P 76, Richmond (10:30p) to Washington (1:10a)
PRR 108, Washington (2:10a) to Philadelphia (5:30a)

NORFOLK to NEW YORK
10r-6dbr
N&W 21, Norfolk (7:35p) to Petersburg (9:15p)
ACL 20, Petersburg (9:20p) to Richmond (10p)
As above, Richmond (10:30p) to New York (7:05a)

(Dining car, Jacksonville to Washington; lounge, Jacksonville to Richmond.)

The *Havana Special* was the principal New York-Key West train until the Oversea Extension of the Florida East Coast Railway was destroyed by the great Labor Day hurricane of 1935.

Richmond also had sleepers to Washington and Pittsburgh (B&O).

The Crescent, No. 37
All-Pullman streamliner
New York to New Orleans
Southern Railway
Entering Danville

NEW YORK to NEW ORLEANS
SR24: 2dr-mr-buffet-lounge SR25: 10r-6dbr
SR26: 10r-6dbr   SR27: 10r-6dbr
SR28: 5dbr-observation-lounge
PRR 149, New York (2:55p) to Washington (6:50p)
SOU 37, Washington (7:15p) to Atlanta (9:05a)
WPR 37, Atlanta (9:40a) to Montgomery (12:35p)
L&N 37, Montgomery (12:45p)
            to New Orleans (7:55p)

WASHINGTON to NEW ORLEANS
S35: 10r-6dbr
As above, Washington (7:15p)
            to New Orleans (7:55p)

NEW YORK to ATLANTA
SR20: 10r-6dbr   SR21: 10r-6dbr
SR22: 10r-6dbr   SR23: 10r-6dbr
As above, New York (2:55p) to Atlanta (9:05a)

WASHINGTON to ATLANTA
85: 10r-6dbr
As above, Washington (7:15p) to Atlanta (9:05a)

(Diner, New York to New Orleans. Coaches carried south of Atlanta.)

The *Crescent's* patrician ancestry dated to the 1891 *Vestibuled Limited*. This modern version, streamlined in 1941, carried eleven stainless steel Pullmans, under power of green, white and gray diesel units.

Cavalier, No. 468
Cape Charles to New York
Pennsylvania Railroad
25 miles north of Cape Charles

CAPE CHARLES to NEW YORK
12s-c-dr   8s-4dbr
18r
PRR 468, Cape Charles (11:25p)
            to New York (7:30a)

CAPE CHARLES to PHILADELPHIA
10s-3dbr
PRR 468, Cape Charles (11:25p)
            to Philadelphia (6:15a)

(Ferries departed Norfolk at 7:45 p.m. and Old Point Comfort at 8:40 p.m. for 10:35 p.m. arrival at Cape Charles.)

September 1938, and Southern's northbound *Crescent Limited* steams into Greenville, S.C.

W.H. Thrall/Frank Ardrey collection

Washington

See map on page 292

Empire Builder, No. 1
Streamliner
Chicago to Seattle
Great Northern
Leaving Spokane

CHICAGO to SEATTLE
   10: Observation-buffet-sleeper-lounge
   Cars 14-17, with 7 types of accommodations
   CB&Q 49, Chicago (1p Sat) to St. Paul (7:45p)
   GN 1, St. Paul (8:15p) to Seattle (8a Mon)

(Dining car, coffee shop-lounge. The other westbound
Empire Builder is now in Minnesota.)

The *Empire Builder*, honoring the Great Northern's
James Hill, was the GN's premier train after its inauguration in 1929. The 1952 version was the second
postwar streamlining of the train.

Empire Builder, No. 1
Portland connection
Spokane, Portland & Seattle
Standing in Spokane

CHICAGO to PORTLAND
   Cars 11-12: Standard accommodations
   CB&Q 49, Chicago (1p Sat) to St. Paul (7:45p)
   GN 1, St. Paul (8:15p) to Spokane (11:30p Sun)
   SP&S 1, Spokane (12:06a Mon)
     to Portland (7:30a Mon)

SPOKANE to PORTLAND
   10: 6r-3dbr-buffet-lounge
   SP&S 1, Spokane (12:06a) to Portland (7:30a)

(Diner and coffee shop-lounge, Chicago to Spokane)

North Coast Limited, No. 1
Chicago to Seattle
Northern Pacific
Leaving Pasco

CHICAGO to SEATTLE
   252: 3br-c-6r-8dup/r   BN: 14s tourist
   254: 4br-c-observation-lounge
   CB&Q 51, Chicago (11p Fri) to St. Paul (8a Sat)
   NP 1, St. Paul (9a) to Seattle (7:30a Mon)

SPOKANE to SEATTLE
   250: 3br-c-6r-8dup/r
   NP 1, Spokane (8:40p) to Seattle (7:30a)

WALLA WALLA to SEATTLE
   32: 10s-dr-c
   NP 348, Walla Walla (8:55p) to Pasco (11:20p)
   NP 1, Pasco (11:59p) to Seattle (7:30a)

(Dining car. In summers, sleeper 253: 4dbr-sb-12r add-
ed, Chicago to Seattle. By the above schedule, the
train has just cut out its Portland cars at Pasco, where
the SP&S No. 3, now approaching town with the Port-
land cars of the GN's Western Star, will add the North
Coast Limited's cars to its consist. Other westbound
editions of the North Coast are now in Illinois and
Montana.)

The Northern Pacific scheduled dedicated sleepers
for Seattle from St. Paul and Gardiner, Mont., its Yellowstone Park station, and, with the Burlington, from
Denver, Kansas City, Lincoln, and St. Louis. Cars on
the NP for Spokane included Tacoma, Yakima, Walla
Walla, Butte, and Garrison, Mont.

North Coast Limited, No. 1
Northern Pacific
Yakima setout
Standing in Yakima

YAKIMA to SEATTLE
   Y: 10s (NP tourist sleeper)
   NP 1, Yakima (2:30a) to Seattle (7:30a)

North Coast Limited, No. 2
Seattle to Chicago
Northern Pacific
40 miles east of Auburn

SEATTLE to CHICAGO
   262: 3br-c-6r-8dup/r   NB: 14s tourist
   264: 4brc-observation-lounge
   NP 2, Seattle (10p Sun) to St. Paul (10p Tue)
   CB&Q 50, St. Paul (10:45p) to Chicago (7:45a Wed)

SEATTLE to SPOKANE
   260: 3br-c-6r-8dup/r
   NP 2, Seattle (10p) to Spokane (8:25a)

SEATTLE to WALLA WALLA
   31: 10s-dr-c
   NP 2, Seattle (10p) to Pasco (5:15a)
   NP 347, Pasco (5:30a) to Walla Walla (7:50a)

SEATTLE to YAKIMA
   S: 10s NP tourist sleeper
   NP 2, Seattle (10p) to Yakima (3a)

(Dining car. In summer, sleeper 263: 4dbr-sb-12r, Seat-
tle to Chicago, added. A three-night train, the North
Coast Limited also has eastbound editions now in
Montana and Wisconsin.)

Although it was just a summer train when inaugurated in 1900, the *North Coast Limited* eventually went on
fulltime schedules and came to be the oldest name
train between Chicago and the Pacific Northwest. In
1952, the year of this schedule, the *North Coast* was
re-equipped with handsome train sets in two shades of
green designed by Raymond Loewy. Vista dome cars,
including sleepers, were added in 1954.

Empire Builder, No. 2
Streamliner
Seattle and Portland to Chicago
Great Northern
30 miles northeast of Spokane

SEATTLE to CHICAGO
   20: Observation-sleeper-lounge
   Cars 24-27, with 7 types of accommodations
   GN 2, Seattle (3p Sun) to St. Paul (7a Tue)
   CB&Q 44, St. Paul (7:15a) to Chicago (2p Tue)

PORTLAND to CHICAGO
   Cars 21-22: standard accommodations
   SP&S 2, Portland (3p Sun) to Spokane (10:35p)
   GN 2, Spokane (11:15p) to St. Paul (7a Tue)
   CB&Q 44, St. Paul (7:15a) to Chicago (2p Tue)

(Dining car, coffee shop-lounge Seattle to Chicago;
diner and sleeper-buffet-lounge Portland to Spokane.
The crack Empire Builder is one of only two Seattle
trains making the trek to Chicago in two nights. The
eastbound edition that left the previous day is now in
North Dakota.)

Western Star, No. 3
Streamliner
Chicago to Seattle
Great Northern
92 miles west of Spokane

CHICAGO to SEATTLE
   34: 4s-8dup/r-4dbr   35: 16dup/r-4dbr
   39: dr-2dbr-observation-buffet-lounge
   CB&Q 53, Chicago (11:15p Fri)
     to St. Paul (8:15a Sat)
   GN 3, St. Paul (8:45a) to Seattle (7:30a Mon)

SPOKANE to SEATTLE
   33: 16dup/r-4dbr
   GN 3, Spokane (9:45p) to Seattle (7:30a)

(Diner, coffee shop. Stops at Glacier National Park,
Mont., in summertime, when additional sleeper, 36:
16dup/r-4dbr, added Chicago to Seattle. Other edi-
tions of the train are now in Montana and Illinois.)

The *Western Star* came into being in 1951, when the
Empire Builder got its second postwar refurbishing.
From the older, 1947 equipment, the new train was
formed.

Western Star (Portland section) No. 3
Spokane to Portland
Spokane, Portland & Seattle
48 miles north of Pasco

CHICAGO to PORTLAND
   32: 4s-4dbr-8dup/r
   CB&Q 53, Chicago (11:15p Fri)
     to St. Paul (8:15a Sat)
   GN 3, St. Paul (8:45a) to Spokane (9:30p Sun)
   SP&S 3, Spokane (9:45p) to Portland (7a Mon)

SPOKANE to PORTLAND
   30: 8s-observation-buffet-lounge
   SP&S 3, Spokane (9:45p) to Portland (7a)

North Coast Limited (Portland section) No. 3
Pasco to Portland
Spokane, Portland & Seattle
Standing in Pasco

CHICAGO to PORTLAND
   251: 3br-c-6r-8dup/r   BR: 14s tourist
   CB&Q 51, Chicago (11p Fri) to St. Paul (8a Sat)
   NP 1, St. Paul (9a) to Pasco (11:40p Sun)
   SP&S 3, Pasco (1:35a Mon) to Portland (7a Mon)

North Coast Limited/Western Star, No. 4
(Portland sections)
Portland to Pasco/Spokane
Spokane, Portland & Seattle
3 miles west of Wishram

PORTLAND to CHICAGO
   42: 4s-4dbr-8dup/r
   SP&S 4, Portland (9:15p Sun)
     to Spokane (6:35a Mon)
   GN 4, Spokane (7:20a) to St. Paul (10:30p Tue)
   CB&Q 54, St. Paul (11p) to Chicago (8a Wed)

PORTLAND to SPOKANE
   40: 8s-observation-buffet-lounge
   SP&S 4, Portland (9:15p) to Spokane (6:35a)

PORTLAND to CHICAGO
   261: 3br-c-6r-8dup/r   PN: 14s tourist
   SP&S 4, Portland (9:15p Sun)
     to Pasco, Wash. (2:30a Mon)
   NP 2, Pasco (5:25a) to St. Paul (10p Tue)
   CB&Q 50, St. Paul (10:45p) to Chicago (7:45a Wed)

(Carrying the sleepers of its parent NP and GN along
the north bank of the Columbia River, SP&S No. 4 is
being shadowed across the dark water by the UP's
eastbound Portland Rose and Spokane. Their west-
bound counterparts will illuminate the Oregon side sev-
eral hours later, then, with the coming of dawn, the City
of Portland will make its appearance.)

Western Star, No. 4
Streamliner
Seattle to Chicago
Great Northern
30 miles east of Everett

SEATTLE to CHICAGO
   44: 4s-8dup/r-4dbr   45: 16dup/r-4dbr
   49: dr-2dbr-observation-buffet-lounge
   GN 4, Seattle (10:15p Sun) to St. Paul (10:30p Tue)
   CB&Q 54, St. Paul (11p) to Chicago (8a Wed)

SEATTLE to SPOKANE
   43: 16dup/r-4dbr
   GN 4, Seattle (10:15p) to Spokane (6:50a)

(Diner, coffee shop car; additional Seattle-Chicago
sleeper, 46: 16dup/r-4dbr, carried in summer. East-
bound editions of the train are now in both Montana
and Wisconsin.)

The *Western Star* became the Great Northern's No.
2 train in 1951, at which time its venerable predecessor, the *Oriental Limited*, was retired.

Columbian, No. 17

Chicago to Seattle/Tacoma
Milwaukee Road
25 miles east of Othello

CHICAGO to SEATTLE/TACOMA
171: 6s-6dbr
K17: 14s MILW tourist
MILW 17, Chicago (10p Fri)
to Seattle (7:30a Mon) and Tacoma (9:05a)

SPOKANE to SEATTLE/TACOMA
178: 10s-buffet-lounge
MILW 17, Spokane (9:15p)
to Seattle (7:30a) and Tacoma (9:05a)

(Diner-lounge. This edition of the Columbian is now in its third night; others follow in Montana and Wisconsin.)

The Milwaukee Road's *Columbian* was introduced in 1911, upon completion of the railroad's extension to the Pacific Coast. Always the maid-of-all-work, the *Columbian* vanished from the Depression-era schedules but enjoyed a decade's reprise after World War II.

Columbian, No. 18

Tacoma/Seattle to Chicago
Milwaukee Road
37 miles east of Black River

TACOMA/SEATTLE to CHICAGO
181: 6s-6dbr
K18: 14s MILW tourist
MILW 18, Tacoma (8:45p Sun) and
Seattle (10:30p Sun) to Chicago (8:45a Wed)

TACOMA/SEATTLE to SPOKANE
189: 10s-buffet-lounge
MILW 18, Tacoma (8:45p) and
Seattle (10:30p) to Spokane (8a)

(Diner-lounge. This eastbound Columbian is preceded by editions of the train now in Montana and Minnesota. The schedules for the train at its western termini reflects the Milwaukee Road's practice of using Tacoma as both origin and end of the line and backing its trains between that city and Seattle.)

In the late 1920's, the Milwaukee Road scheduled seasonal sleepers from Seattle to its Yellowstone Park station, Gallatin Gateway. The *Olympian* carried a Spokane-Chicago Pullman.

Spokane, No. 19

Spokane to Portland
Union Pacific
22 miles north of Wallula

SPOKANE to PORTLAND
10r-6dbr
8s-buffet-lounge
UP 19, Spokane (9p) to Portland (6:30a)

SPOKANE to SALT LAKE CITY
10s-3dbr
UP 19, Spokane (9p) to Hinkle, Ore. (1:15a)
UP 18, Hinkle (2:10a) to Pocatello (5:05p)
UP 34, Pocatello (5:35p) to Salt Lake City (10p)

SPOKANE to LOS ANGELES   *Fridays, winter only*
6s-6dbr
As above, Spokane (9p Sun)
to Salt Lake City (10p Mon)
UP 9, Salt Lake City (10:10p) to Los Angeles (3p Tue)

LEWISTON, IDA. to PORTLAND
8s-coach
Camas Prairie 73, Lewiston (7:30p) to Ayer (10:15p)
UP 19, Ayer (11:20p) to Portland (6:30a)

(The "Sunday" departure of the Los Angeles sleeper is used to conform to the format of these schedules but is obviously not actual schedule information)

The UP formerly carried Pullmans from Spokane to Boise, Walla Walla, and San Francisco.

Fast Mail, No. 27

St. Paul/Minneapolis to Seattle
Great Northern
25 miles east of Cascade Tunnel

WENATCHEE to SEATTLE
270: 12s-dr
GN 27, Wenatchee (11:15p) to Seattle (5a)

Fast Mail, No. 28

Seattle to Minneapolis/St. Paul
Great Northern
Leaving Everett

SEATTLE to WENATCHEE
280: 12s-dr
GN 28, Seattle (10:45p) to Wenatchee, (4:20a)

No. 71

Walla Walla to Pendleton
Union Pacific
3 miles south of Wallula

WALLA WALLA to PORTLAND
10s-3dbr
UP 71, Walla Walla (10:15p) to Hinkle (12:40a)
UP 19, Hinkle (1:45a) to Portland (6:30a)

YAKIMA to PORTLAND
6s-coach
UP 63, Yakima (8p) to Wallula (11:40p)
UP 71, Wallula (11:55p) to Hinkle (12:40a)
UP 19, Hinkle (1:45a) to Portland (6:30a)

No. 402

Seattle to Portland
Pool service
9 miles north of Auburn

SEATTLE to PORTLAND
4022: 10s-dr-c
4023: 10s-dr-c
4024: 12s-dr
Pool 402, Seattle (11:45p) to Portland (6:45a)

SEATTLE to ST. LOUIS
4020: 6s-6r-4dbr
Pool 402, Seattle (11:45p Sun)
to Portland (6:45a Mon)
UP 12, Portland (8:10a)
to Green River, Wyo. (9:50a Tue)
UP 10, Green River (10:45a)
to Kansas City (6:45a Wed)
Wabash 10, Kansas City (7a)
to St. Louis (11:59a Wed)

No. 402
 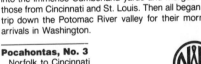
Pool service
Tacoma setout
Standing in Tacoma

TACOMA to PORTLAND
4021: 10s-dr-c
Pool 402, Tacoma (1:15a) to Portland (6:45a)

Jim Fredrickson

NP's venerable *North Coast Limited* prepares to depart Seattle King Street in 1969.

West Virginia
See map on page 94

The Baltimore & Ohio, in terms of Pullman numbers, dominated the nighttime scenario in West Virginia, and the railroad's historic and scenic main line, which generally paralleled the Potomac River between Harpers Ferry and Cumberland, accommodated a magnificent flow of traffic. The westbound parade of royal blue started early in the evening, marshaled by the all-Pullman *Capitol Limited*. Ten minutes later came the *Columbian*, then *Ambassador*, *National Limited*, *Diplomat*, *Cleveland Night Express*, and, finally, as midnight approached, the *West Virginian* and *Shenandoah*. With the coming of dawn, the process was reversed as the main-line trains from Pittsburgh and Chicago eased into the immense Cumberland yards and merged with those from Cincinnati and St. Louis. Then all began the trip down the Potomac River valley for their morning arrivals in Washington.

Pocahontas, No. 3

Norfolk to Cincinnati
Norfolk & Western
39 miles west of Bluefield

NORFOLK to CHICAGO
111: 10s-lounge
M1: 10r-6dbr
N&W 3, Norfolk (1:45p) to Cincinnati (7:45a)
PRR 215, Cincinnati (8:30a) to Chicago (1:45p)

WINSTON-SALEM to CINCINNATI
104: 10r-6dbr
N&W 22, Winston-Salem (2p) to Roanoke (6:20p)
N&W 3, Roanoke (7:50p) to Cincinnati (7:45a)

ROANOKE to COLUMBUS
102: 10r-6dbr
N&W 3, Roanoke (7:50p)
to Portsmouth, Ohio (4:25a)
N&W 33, Portsmouth (5a) to Columbus (7:40a)

(Diner-lounge, Norfolk to Williamson; diner, Cincinnati to Chicago)

The *Pocahontas* made its first run in November 1926.

Both Bluefield and Williamson had Pullman service to Washington and New York, via N&W/Southern.

West Virginia   *(continued)*

National Limited, No. 1
Washington to St. Louis
Baltimore & Ohio
12 miles east of Grafton

WASHINGTON to ST. LOUIS
  3c-dr-obs.-buf-lounge   14s
  B&O 1, Washington (6:30p) to St. Louis (1p)

NEW YORK to ST. LOUIS
  10s-2c-dr
  Ferry or bus from New York City
  B&O 1, Jersey City (1:55p) to St. Louis (1p)

NEW YORK to LOUISVILLE
  8s-5dbr
  Ferry or bus from New York City
  B&O 1, Jersey City (1:55p) to Cincinnati (7:15a)
  B&O 57, Cincinnati (7:35a) to Louisville (10:05a)

WASHINGTON to SAN ANTONIO
  14r-4dbr
  As above, Washington (6:30p Sun)
      to St. Louis (1p Mon)
  SLSF/MKT 1, St. Louis (5:30p)
      to San Antonio (1:55p Tue)

WASHINGTON to FORT WORTH
  14r-4dbr
  As above, Washington (6:30p Sun)
      to St. Louis (1p Mon)
  MP 1, St. Louis (5:30p) to Texarkana (3:05a Tue)
  T&P 1, Texarkana (3:25a) to Dallas (7:30a)
      and Fort Worth (8:30a Tue)

(New York to St. Louis, dining car. The San Antonio sleeper travels on the Texas Special, *which carries lounge facilities and dining cars from St. Louis to Springfield, Mo., and Muskogee to San Antonio. The* Texas Eagle *takes the Fort Worth Pullman; a sleeper-lounge and dining car are included in the consist.)*

National Limited, No. 2
St. Louis to Washington
Baltimore & Ohio
19 miles west of Clarksburg

ST. LOUIS to WASHINGTON
  3c-dr-obs.-buf.-lounge   14s
  B&O 2, St. Louis (10:05a) to Washington (7:25a)

ST. LOUIS to NEW YORK
  14r-4dbr   10s-2c-dr
  B&O 2, St. Louis (10:05a) to Jersey City (12:10p)
  Ferry or bus to New York

LOUISVILLE to NEW YORK
  8s-5dbr
  B&O 50, Louisville (1:55p) to North Vernon (3:07p)
  B&O 2, North Vernon (3:28p) to Jersey City (12:10p)
  Ferry or bus to New York

SAN ANTONIO to WASHINGTON
  14r-4dbr
  MKT/SLSF 2, San Antonio (12:01p Sat)
      to St. Louis (8:10a Sun)
  B&O 2, St. Louis (10:05a)
      to Washington (7:25a Mon)

FORT WORTH to WASHINGTON
  14r-4dbr
  T&P 2, Fort Worth (5:15p Sat) and
     Dallas (6:15p) to Texarkana (10:30p)
  MP 2, Texarkana (10:40p) to St. Louis (8:10a Sun)
  B&O 2, St. Louis (10:05a)
      to Washington (7:25a Mon)

(T&P/MP 2, the Texas Eagle, *carries a dining car and sleeper-lounge. MKT/SLSF 2, the* Texas Special, *has dining cars from San Antonio to Muskogee, and Springfield to St. Louis, and lounge facilities the full distance. St. Louis to New York, diner and lounge.)*

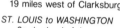

C&O's *Greenbrier* (11dbr) on the park track at White Sulphur Springs, W.Va.

George Washington, No. 2
Louisville/Cincinnati to New York/Phoebus
Chesapeake & Ohio
42 miles west of Hinton

ASHLAND to NEW YORK
  C21: 10r-6dbr
  C&O 2, Ashland (9:07p) to Washington (8:45a)
  PRR 148, Washington (9:35a) to New York (1:25p)

CINCINNATI to WASHINGTON
  27: 10r-6dbr
  C&O 2, Cincinnati (6:05p) to Washington (8:45a)

CINCINNATI to PHOEBUS
  20: 5dbr-obs.-buf.-lounge   29: 10r-6dbr
  C&O 2, Cincinnati (6:05p) to Charlottesville (5:30a)
  C&O 2-42, Charlottesville (6:45a)
      to Phoebus (12:10p)

LEXINGTON to RICHMOND
  28: 10r-6dbr
  C&O 22, Lexington (4:05p) to Ashland (7:25p)
  C&O 2, Ashland (9:07p)
      to Charlottesville (5:30a)
  C&O 2-42, Charlottesville (6:45a)
      to Richmond (8:45a)

LOUISVILLE to NEW YORK
  C24: 10r-5dbr
  C&O 22, Louisville (1:25p) to Ashland (7:25p)
  As above, Ashland (9:07p) to New York (1:25p)

LOUISVILLE to WASHINGTON
  23: 10r-6dbr
  C&O 22, Louisville (1:25p) to Ashland (7:25p)
  C&O 2, Ashland (9:07p) to Washington (8:45a)

(Dining cars all segments, except Ashland to Charlottesville)

In the Depression year of its birth, 1932, the *George Washington* took the honor of being the first air-conditioned Pullman train and thereupon assumed the title of the C&O's flagship from the venerable *Fast Flying Virginian.*

The C&O provided most of the Pullman service in and out of Huntington and Charleston. At this scheduling, Charleston's dedicated service was gone, but it once included sleepers from Bluefield, Roanoke, Washington, Columbus, Cincinnati, Toledo, and Detroit. Pullmans came to Huntington from the latter three and Cleveland, Phoebus, and Richmond.

Cleveland Night Express, No. 17
Baltimore to Cleveland
Baltimore & Ohio
30 miles east of Cumberland, Md.

BALTIMORE to CLEVELAND
  8s-dr-2c   10r-5dbr
  8s-dr-3dbr
  B&O 17, Baltimore (8:30p) to Cleveland (8:10a)

BALTIMORE to AKRON
  12s-dr
  B&O 17, Baltimore (8:30p) to Akron (7:02a)

(Baltimore to Cleveland, snack bar-lounge)

West Virginian, No. 23
Washington to Parkersburg
Baltimore & Ohio
4 miles east of Martinsburg

WASHINGTON to PARKERSBURG
  12s-dr
  B&O 23, Washington (10:25p) to Parkersburg (9:10a)

WASHINGTON to WHEELING
  10s-2c-dr
  B&O 23, Washington (10:25p) to Grafton (5:30a)
  B&O 343, Grafton (5:55a) to Wheeling (9:50a)

(Lunch counter service, all segments)

Both Parkersburg and Fairmont had dedicated Pullman service from New York, Baltimore, Washington, Pittsburgh, and Chicago.

West Virginian, No. 24
Parkersburg to Cumberland
Baltimore & Ohio
Entering Grafton

PARKERSBURG to WASHINGTON
  12s-dr
  B&O 24, Parkersburg (9p) to Cumberland (3:30a)
  B&O 18, Cumberland (4:32a) to Washington (7:35a)

(No. 24 is about to pick up the Wheeling setout. Both sleepers will ride No. 18, the Cleveland Night Express, *into Washington.)*

West Virginian, No. 24
Baltimore & Ohio
Wheeling setout
Standing in Grafton

WHEELING to WASHINGTON
  10s-2c-dr
  B&O 344, Wheeling (8:25p) to Grafton (11:50p)
  B&O 24, Grafton (12:10a) to Cumberland (3:30a)
  B&O 18, Cumberland (4:32a) to Washington (7:35a)

Wheeling also had sleepers to Baltimore and New York.

No. 77
Pittsburgh to Kenova
Baltimore & Ohio
6 miles south of Wheeling

PITTSBURGH to HUNTINGTON
  12s-dr
  B&O 77, Pittsburgh (9:30p) to Huntington (6:35a)

This Ohio River line once saw Pullman service from Pittsburgh to Kenova and Charleston (C&O); Huntington to both Parkersburg and Wheeling; and Charleston to Clarksburg (C&O).

No. 78
Kenova to Pittsburgh
Baltimore & Ohio
20 miles north of Huntington

HUNTINGTON to PITTSBURGH
  12s-dr
  B&O 78, Huntington (11:30p) to Pittsburgh (8:20a)

Wisconsin

See map on page 270

Pioneer Limited, No. 1
Milwaukee Road
Milwaukee setouts
Standing in Milwaukee

MILWAUKEE to ST. PAUL/MINNEAPOLIS
17: 8dup/r-6r-4db
MILW 1, Milwaukee (12:55a) to St Paul (8a)
and Minneapolis (8:40a)

MILWAUKEE to MINOCQUA
19: 10s-c-dr
MILW 1, Milwaukee (12:55a) to New Lisbon (3:34a)
MILW 217, New Lisbon (4:15a) to Minocqua (10:10a)

(Tip Top Tap diner. The Pioneer Limited *is 39 miles north of Chicago.)*

The *Pioneer Limited*, the Milwaukee Road's oldest name train, evolved out of service that began in 1872, although the *Pioneer* title didn't come into use until 1898. Before World War I, the train ran in two sections, one solid with the railroad's own sleeping cars.

Copper Country Limited, No. 2
Champion, Mich., to Chicago
Milwaukee Road
35 miles north of Green Bay

CALUMET, MICH., to CHICAGO
20: 10s-c-dr
DSS&A 10, Calumet (5:15p) to Champion (7:55p)
MILW 2, Champion (8:05p) to Milwaukee (4:10a)
MILW 56, Milwaukee (4:35a) to Chicago (6:20a)

SAULT STE. MARIE, MICH., to CHICAGO
22: 10s-2c
SOO 7, Sault Ste. Marie (3:50p) to Pembine (10:10p)
MILW 2, Pembine (10:50p) to Milwaukee (4:10a)
MILW 56, Milwaukee (4:35a) to Chicago (6:20a)

(Soo 7 carries a sleeper-diner)

The *Copper Country Limited*, another venerable Milwaukee Road train dating to the 19th Century, once offered sleeping accommodations from Channing, Mich., to Milwaukee and from Green Bay to Chicago. Other service between Upper Michigan and Milwaukee included cars from Champion, Iron River, and Ontonagon. Until early in the Depression, the Sault Ste. Marie Pullman traveled to and from Chicago on the North Western.

No. 7
Sault Ste. Marie to Minneapolis
Soo Line
25 miles east of Rhinelander

SAULT STE. MARIE to ST. PAUL/MINNEAPOLIS
8s-restaurant
SOO 7, Sault Ste. Marie, (3:50p) to St. Paul (7:35a)
and Minneapolis (8:15a)

The former *Minneapolis-St. Paul-Montreal Express*, No. 7 once offered through sleepers to the Twin Cities both from Montreal and, before World War I, from Boston. By 1952, through passenger cars no longer crossed the international bridge, but travelers could still avail themselves of Canadian Pacific sleepers operating between Toronto and Sault Ste. Marie, Ont.

No. 8
Minneapolis to Sault Ste. Marie
Soo Line
3 miles west of Ladysmith

MINNEAPOLIS/ST. PAUL to SAULT STE MARIE
8s-restaurant
SOO 8, Minneapolis (7:25p) and
St. Paul (8:15p) to Sault Ste. Marie (11:45p)

No. 8
Duluth to Marquette, Mich.
Duluth, South Shore & Atlantic
Departing Ashland

DULUTH to MARQUETTE   *Tue, Thu, Sun only*
Standard accommodations
DSS&A 8, Duluth (8:15p) to Marquette (7:30a)

The "South Shore" over the years scheduled sleepers from Duluth to the Michigan communites of Calumet, Marquette, Mackinaw City, and Sault Ste. Marie. In the first decade of the Twentieth Century, however, service was briefly extended, via Canadian Pacific, from Duluth to Montreal on the railroads' *Boston Express.*

Copper Country Limited, No. 9
Chicago to Champion
Milwaukee Road
Approaching Green Bay

CHICAGO to CALUMET
90: 10s-c-dr
MILW 9, Chicago (7:45p) to Champion (6a)
DSS&A 9, Champion (6:10a) to Calumet (8:55a)

CHICAGO to SAULT STE. MARIE
92: 10s-2c
MILW 9, Chicago (7:45p) to Pembine (3:05a)
SOO 8, Pembine (5:22a) to Sault Ste. Marie (11:45a)

(Soo 8 carries a sleeper-diner)

Sioux, No. 11
Chicago to Canton, S.D.
Milwaukee Road
21 miles south of Madison

CHICAGO to CANTON
110: 10s-c-dr
MILW 11, Chicago (9:30p) to Canton (2:40p)

CHICAGO to AUSTIN, MINN.
112: 8s-2c-dr
MILW 11, Chicago (9:30p) to Calmar, Iowa (5:20a)
MILW 103, Calmar (5:50a) to Austin (7:35a)

The consist of the *Sioux* has dwindled from the days when it was a sleeper-laden Rapid City train, and as of this scheduling, Sioux Falls passengers were even finishing their journey via pre-paid taxi from Canton. The lineup, however, once included dedicated sleeping cars to Jackson, Minn., St. Paul/Minneapolis, Mason City, Sioux Falls, and Rapid City, in addition to those still carried.

Laker, No. 17
Chicago to Duluth
Soo Line
Standing in Neenah

CHICAGO to DULUTH
10s-3db   2dr-3db-8r
SOO 17 Chicago (6:30p) to Duluth (8:15a)

CHICAGO to ST. PAUL/MINNEAPOLIS
8s-diner-lounge)
SOO 17, Chicago (6:30p) to Owen (4a)
SOO 5, Owen (4:35a) to St. Paul (8:20a)
and Minneapolis (9a)

CHICAGO to ASHLAND
8s-2c-dr
SOO 17, Chicago (6:30p) to Spencer (ntg)
SOO 117, Spencer (3:50a) to Ashland (8:40a)

(Chicago to Duluth, diner-lounge)

Named the *Laker* after World War II, No. 17 once offered Milwaukee-Duluth sleepers before service ceased in the late 1930s and Soo passengers were conveyed by interurban between Milwaukee and Waukesha. A Chicago-Stevens Point sleeper was also scheduled, returning from Neenah southbound. When traffic to the Twin Cities waned to the point where a separate train was uneconomical, No. 17 took on those cars, giving it the three-way deployment of this scheduling. By then the Milwaukee-Twin Cities sleeper was also a memory.

Columbian, No. 17
Chicago to Seattle/Tacoma
Milwaukee Road
18 miles west of Milwaukee

CHICAGO to SEATTLE/TACOMA
171: 6s-6db
MILW 17, Chicago (10p Sun) to Seattle (7:30a Wed)
and Tacoma (9:05a Wed)

CHICAGO to ST. PAUL/MINNEAPOLIS
176: 10s-c-dr
MILW 17, Chicago (10p) to St. Paul (7:15a)
and Minneapolis (7:55a)

(The Columbian, *which carries a diner-lounge, is on the road three nights between Chicago and Puget Sound. Other editions are now in Montana and Washington.)*

Laker, No. 18
Soo Line
Duluth to Chicago
Switching in Spencer

DULUTH to CHICAGO
10s-3db   2dr-3db-8r
SOO 18, Duluth (7p) to Chicago (8:25a)

MINNEAPOLIS/ST. PAUL to CHICAGO
8s-diner-lounge
SOO 6, Minneapolis (6:15p) and
St. Paul (7p) to Owen (10:50p)
SOO 18, Owen (11:25p) to Chicago (8:25a)

ASHLAND to CHICAGO
8s-2c-dr
SOO 118, Ashland (7p) to Spencer (11:40p)
SOO 18, Spencer (12:10a) to Chicago (8:25a)

(Duluth to Chicago, diner-lounge)

North Coast Limited, No. 50
Seattle/Portland to Chicago
Burlington
64 miles north of La Crosse

SEATTLE to CHICAGO
262: 3br-c-6r-8dup/r   264: 4br-c-obs.-lounge
NP 2, Seattle (10p Fri) to St. Paul (10p Sun)
CB&Q 50, St. Paul (10:45p Sun)
to Chicago (7:45a Mon)

PORTLAND to CHICAGO
261: 3br-c-6r-8dup/r
SP&S 4, Portland (9:15p Fri)
to Pasco, Wash. (2:30a Sat)
NP 2, Pasco (5:25a) to St. Paul (10p Sun)
CB&Q 50, St. Paul (10:45p Sun) to Chicago (7:45a Mon)

BILLINGS to CHICAGO
263: 3br-c-6r-8dup/r
NP 2, Billings, Mont. (1a Sun) to St. Paul (10p)
CB&Q 50, St. Paul (10:45p) to Chicago (7:45a Mon)

CODY to CHICAGO   *Summertime only*
B38: 10s-c-dr
CB&Q 28, Cody (7:45p Sat) to Billings (11p)
As above, Billings (1a Sun) to Chicago (7:45a Mon)

(Seattle to Chicago: diner with luncheon counter, coach-buffet-lounge. The patriarch of name-train travel to and from the Pacific Northwest, the North Coast Limited *makes the trip on a comfortable three-night schedule. The train's other eastbound editions are now in Washington and Montana.)*

No. 65
Duluth to St. Paul/Minneapolis
Northern Pacific
Entering Superior

DULUTH to ST. PAUL/MINNEAPOLIS
650: 10s-c-dr
NP 65, Duluth (11:30p) to St. Paul (5:55a)
and Minneapolis (6:40a)

Wisconsin   (continued)

Western Star/Black Hawk, No. 54

Streamliner
Seattle/Portland to Chicago
Burlington
51 miles south of St. Paul

SEATTLE to CHICAGO
    44: 4s-8dup/r-4db   45: 16dup/r-4dbr
    49: dr-2dbr-observation-buffet-lounge
    GN 4, Seattle (10:15p Fri) to St. Paul (10:30p Sun)
    CB&Q 54, St. Paul (11p Sun) to Chicago (8a Mon)

PORTLAND to CHICAGO
    42: 4s-8dup/r-8dbr
    SP&S 4, Portland (9:15p Fri) to Spokane (6:35a Sat)
    GN 4, Spokane (7:20a) to St. Paul (10:30p Sun)
    CB&Q 54, St. Paul (11p) to Chicago (8a Mon)

GREAT FALLS, MONT. to CHICAGO
    48: 16dup/r-4dbr
    GN 4, Great Falls (9:35p Sat) to St. Paul (10:30p Sun)
    CB&Q 54, St. Paul (11p) to Chicago (8a Mon)

MINNEAPOLIS/ST. PAUL to CHICAGO
    Std: s-dbr
    CB&Q 54, Minneapolis (10:15p) and
        St. Paul (11p) to Chicago (8a)

MINNEAPOLIS/ST. PAUL to ROCK ISLAND, ILL.
    10s-c-dr
    CB&Q 54, Minneapolis (10:15p) and
        St. Paul (11p) to Savanna, Ill. (4:35a)
    CB&Q 48, Savanna (5:12a) to Rock Island, Ill. (8:05a)

(Seattle to Chicago: dining car, coffee shop; Portland to Spokane, observation-sleeper-lounge. Other eastbound editions of the three-night Western Star *are now in Washington and Montana. Note that the Portland cars for both the* North Coast Limited *and* Western Star *begin their journey aboard the same SP&S train.)*

The *Black Hawk* was the Burlington's principal overnight train from the Twin Cities, one of the road's three new "80th Anniversary Fleet" trains introduced in 1930.

Both the Burlington and Rock Island railroads once competed for the Twin Cities-Rock Island, Ill., trade, just as they did for the Minneapolis-Peoria overnight business. The latter included a third contestant, the Minneapolis & St. Louis.

Fast Mail, No. 56

Minneapolis to Chicago
Milwaukee Road
Standing in La Crosse

MINNEAPOLIS/ST. PAUL to CHICAGO
    560: 10s-c-dr
    MILW 56, Minneapolis (8:20p) and
        St. Paul (9p) to Chicago (6:20a)

(Soon there will be more to the Fast Mail than the long string of head-end cars and single sleeper from the Twin Cities now standing in this major rail junction on the Upper Mississippi. Down the line at New Lisbon, No. 56 will meet No. 256 and take on the Pullmans from Minocqua for both Milwaukee and Chicago. At Milwaukee the consist will swell even further as the Copper Country Limited's sleepers from Sault Ste. Marie and Calumet are attached.)

Unlike No. 56, the westbound *Fast Mail* carried no passengers at this scheduling, but it was nonetheless a well remembered train. On June 12, 1924, the *Fast Mail* was stopped at Rondout, Ill., by gun-wielding bandits aboard the train, who stole a shipment from the Chicago Federal Reserve Bank of three million dollars in cash and securities. It was the largest robbery loss in the history of rail travel. The case was quickly solved, however, after the trail led to a crooked postal inspector, mastermind of the scheme.

New Year's Day 1959, C&NW's *Chicago Limited* (left) and DW&P's No. 619 prepare to depart from Duluth. In pre-RDC days, the "Peg" train carried a Winnipeg sleeper.

William D. Middleton

Wisconsin Lakes Special, No. 123

Chicago to Watersmeet, Mich.
Summer Fridays only
Chicago & North Western
4 miles south of Oshkosh

CHICAGO to WATERSMEET
    10s-dr-2c   6c-3dr
    12s-dr
    C&NW 123, Chicago (7:30p) to Monico (5:25a)
    C&NW 11, Monico (6a) to Watersmeet (8:35a)

(Cafe-lounge. No. 123 operates via Fond du Lac and New London. No. 11 also carries Chicago-Watersmeet sleepers scheduled year-round on the Ashland Limited*.)*

Wisconsin Lakes Special, No. 124

Watersmeet to Chicago
Summer Sundays only
Chicago & North Western
2 miles north of New London

WATERSMEET to CHICAGO
    10s-dr-2c   6c-3dr
    12s-dr
    C&NW 12, Watersmeet (7:25p) to Monico (9:10p)
    C&NW 124, Monico (9:15p) to Chicago (6:40a)

(Cafe-lounge. No. 124 operates via New London and Fond du Lac. No. 12 also carries Watersmeet-Chicago Pullmans scheduled year-round in the Ashland Limited*.)*

The Milwaukee Road's counterpart to the *Wisconsin Lakes Special* was called the *North Woods Fisherman* and served the resorts around Minocqua, just to the west. A daily summer train in the 1920s, it later ran on weekends but ceased operation in 1948. The *Tomahawk*, however, still was carrying Minocqua sleepers year-round as of this scheduling and was no doubt more than equal to the demand.

Iron and Copper Country Express, No. 161
Chicago to Ishpeming, Mich.
Via Fond du Lac
Chicago & North Western
18 miles north of Neenah

CHICAGO to ISHPEMING
    12s-dr
    C&NW 161, Chicago (6:30p) to Ishpeming (7:20a)

(Chicago to Green Bay, cafe-lounge)

The *Iron and Copper Country Express* was an old and honored heavyweight among North Western trains. It once carried dedicated sleepers from Chicago to the Upper Peninsula communities of Calumet, Marquette, Escanaba, and Sault Ste. Marie. A Milwaukee-Iron Mountain Pullman operated on this route.

Ashland Limited, No. 161-211
Milwaukee to Ashland, via Manitowoc
Chicago & North Western
Standing in Green Bay

CHICAGO to ASHLAND
    10s-dr-2c
    C&NW 161, Chicago (6:30p) to Milwaukee (8:30p)
    C&NW 211, Milwaukee (9:10p) to Ashland (10:15a)

CHICAGO to WATERSMEET
    10s-dr-2c
    C&NW 161, Chicago (6:30p) to Milwaukee (8:30p)
    C&NW 211, Milwaukee (9:10p) to Monico (5:15a)
    C&NW 11, Monico (6a) to Watersmeet (8:35a)

(Chicago to Green Bay, cafe-lounge)

Another senior north woods train, the *Ashland Limited* once counted in its consist Pullmans from Chicago for Rhinelander and Marenisco, Mich. Seasonal cars to Phelps moved over this route, and Milwaukee had dedicated service to Ironwood and Watersmeet, Mich., as well as Ashland.

Iron and Copper Country Express, No. 162
Ishpeming to Chicago, via Fond du Lac
Chicago & North Western
Standing in Green Bay

ISHPEMING to CHICAGO
    12s-dr
    C&NW 162, Ishpeming (6:30p) to Chicago (6:45a)

The southbound *Iron and Copper Country Express* formerly cut in one of the Green Bay sleepers for Chicago, as well as one that moved earlier from Menominee to Oshkosh, giving Fox River Valley residents more convenient accommodations to Chicago.

Ashland Limited, No. 212
Ashland to Chicago, via Manitowoc
Chicago & North Western
29 miles east of Eland

ASHLAND to CHICAGO
    10s-dr-2c
    C&NW 212, Ashland (5:15p) to Chicago (7a)

WATERSMEET to CHICAGO
    10s-dr-2c
    C&NW 12, Watersmeet (7:25p) to Monico (9:10p)
    C&NW 212, Monico (9:40p) to Chicago (7a)

ASHLAND to MILWAUKEE   Summertime, Sunday only
    12s-dr
    C&NW 212, Ashland (5:15p) to Milwaukee (4:35a)

(Green Bay to Chicago, cafe-lounge. In summertime, No. 212 carries additional Ashland-Chicago sleepers.)

When Wausau had its own sleeper on the North Western to Chicago, the *Ashland Limited* handled it.

Ashland Limited, No. 212
Chicago & North Western
Green Bay setout
Standing in Green Bay

GREEN BAY to CHICAGO
  10s-dr-c
  C&NW 212, Green Bay (1:20a) to Chicago (7a)

(Milwaukee to Chicago, cafe-lounge)

No. 224
Escanaba, Mich., to Chicago
Via Fond du Lac
Chicago & North Western
20 miles south of Fond du Lac

GREEN BAY to CHICAGO
  8s-dr-4db
  C&NW 224, Green Bay (8:40p) to Milwaukee (1:05a)
  C&NW 212, Milwaukee (4:45a) to Chicago (7a)

(Milwaukee to Chicago, cafe-lounge)

Tomahawk, No. 256-56
Minocqua to Chicago
Milwaukee Road
12 miles north of New Lisbon

MINOCQUA to CHICAGO
  2560: 10s-c-dr
  MILW 256 Minocqua (6:30p) to New Lisbon (12:20a)
  MILW 56 New Lisbon (1:28a) to Chicago (6:20a)

MINOCQUA to MILWAUKEE
  2561: 10s-c-dr
  MILW 256, Minocqua (6:30p) to New Lisbon (12:20a)
  MILW 56, New Lisbon (1:28a) to Milwaukee (4:22a)

Predecessors of the *Tomahawk* carried Wausau sleepers for both Chicago and Minneapolis.

North Western Limited, No. 405
Chicago & North Western
Milwaukee setout
Standing in Milwaukee

MILWAUKEE to MINNEAPOLIS
  3db-c-16dup/r
  C&NW 405 Milwaukee (1a) to St. Paul (7:40a)
      and Minneapolis (8:15a)

(Chicago to Minneapolis, dining car. The North Western Limited *is now 7 miles south of Kenosha.)*

The *North Western Limited*, the railroad's flagship, was so named several years before the turn of the century. Originally serving Madison before the C&NW's new line from Milwaukee was built, the *Limited* broke the drab Pullman-green tradition long before the streamliner era with a sprightly yellow and green livery that anticipated that of the *400* fleet. Among the *Limited's* earlier duties were cutting in sleepers for the Twin Cities from Eau Claire and Fond du Lac, the latter of which traveled a circuitous route via Green Bay and Wausau.

Duluth-Superior Limited, No. 511
Chicago to Duluth
Via Beloit and Madison
Chicago & North Western
4 miles north of Elroy

CHICAGO to DULUTH
  c-3db-16dup/r   6s-6dbr
  C&NW 511, Chicago (6:30p) to Duluth (8a)

CHICAGO to ST. PAUL/MINNEAPOLIS
  16s-dr
  C&NW 511, Chicago (6:30p) to Altoona (2:05a)
  C&NW 405, Altoona (2:35a) to St. Paul (7:40a)
      and Minneapolis (8:15a)

CHICAGO to DRUMMOND   Summertime, Fri. only
  10s-dr-2c
  C&NW 511, Chicago (6:30p) to Spooner (5:30a)
  C&NW 311, Spooner (6a) to Drummond (7:20a)

(Chicago to Madison, diner; Chicago to Elroy, buffet-parlor; Altoona to Duluth, cafe-lounge.)

Chicago Limited, No. 510-406
Duluth to Chicago, via Milwaukee
Chicago & North Western
4 miles north of Chippewa Falls

DULUTH to CHICAGO
  c-3db-16dup/r
  C&NW 510, Duluth (8:15p) to Altoona (ntg)
  C&NW 406, Altoona (1:51a) to Chicago (8a)

DRUMMOND to CHICAGO   Summertime, Sunday only
  10s-dr-2c
  C&NW 310, Drummond (8:50p) to Spooner (10:15p)
  C&NW 510, Spooner (10:30p) to Altoona (ntg)
  C&NW 406, Altoona (1:51) to Chicago (8:05a)

(No. 510 carries a buffet-lounge; No. 406, the North Western Limited, *has a full dining car)*

Milwaukee had dedicated Pullman service from Duluth/Superior and Rochester.

Minnesota and Black Hills Express, No. 514
Rapid City, S.D., to Chicago
Via Madison and Clinton Jct.
Chicago & North Western
24 miles north of Elroy

RAPID CITY to CHICAGO
  10s-dr-2c
  C&NW 514, Rapid City (7p Sat)
      to Huron, S.D. (4:38a Sun)
  C&NW 518, Huron (5:10a) to Mankato, Minn. (9:45a)
  C&NW 514, Mankato (4:20p) to Elroy (ntg)
  C&NW 514, Elroy (1:30a Mon)
      to Chicago (7:05a Mon)

ROCHESTER to CHICAGO
  12r-2sbr-3dbr
  C&NW 514, Rochester (7:45p) to Elroy (ntg)
  C&NW 514, Elroy (1:30a) to Chicago (7:05a)

(The Minnesota and Black Hills Express *and the* Victory *both carry the number 514 and are combined into one train at Elroy. No. 518, the* Dakota 400, *carries a lounge; a cafe-coach is available in No. 514 between Mankato and Elroy.)*

Victory, No. 514
Minneapolis to Chicago
Via Madison, Clinton Jct.
Chicago & North Western
8 miles north of Wyeville

DULUTH to CHICAGO
  10s-dr-c
  C&NW 512, Duluth (4:15p) to Altoona (ntg)
  C&NW 514, Altoona (10:05p) to Elroy (12:45a)
  C&NW 514, Elroy (1:30a) to Chicago (7:05a)

MINNEAPOLIS/ST. PAUL to MADISON
  12s-dr
  C&NW 514, Minneapolis (6:15p) and
      St. Paul (7p) to Madison (3:15a)

Rochester-Minnesota Special and Minnesota and Black Hills Express, No. 515
Chicago to Rapid City
Via Clinton Jct., Madison
Chicago & North Western
Standing in Madison

CHICAGO to RAPID CITY
  8s-dr-2c
  C&NW 515, Chicago (9p Sun)
      to Mankato (11:42a Mon)
  C&NW 519, Mankato (7:30p) to Huron (12:14a Tue)
  C&NW 515, Huron (12:35a) to Rapid City (9:15a Tue)

CHICAGO to ROCHESTER
  12r-2sbr-3dbr
  C&NW 515, Chicago (9p) to Rochester (8:05a)

(Elroy to Mankato, cafe coach. No. 519, the Dakota 400, *carries a lounge car. Holding over the Rapid City sleeper for the much swifter* Dakota 400 *to move on to Huron had the effect of improving the Pullman schedules for the intermediate market of Mankato, and marginally serving customers in nearby Rochester and the Twin Cities. It certainly wasn't a very attractive schedule for travelers going the distance to Rapid City.)*

Wyoming

See map on page **197**

On this night in Wyoming, a trio of westbound Overland trains meet their eastbound counterparts heading for Sherman Hill: *Portland Rose, Gold Coast,* and *Pony Express.* The schedules for all three of them are illustrative of the Union Pacific's massive switching of Pullmans among the Chicago and Denver/Kansas City trains on one hand, and the Los Angeles, San Francisco, and Portland ones on the other. While the *City* streamliners were dedicated to their namesake destinations, these secondary trains worked closely with one another, exchanging sleepers in either Green River or Ogden.

Portland Rose, No. 17
Denver to Portland
Union Pacific
Approaching Rawlins

DENVER to PORTLAND
  6s-6r-4dbr
  UP 17, Denver (5:40p Sun) to Portland (6a Tue)

KANSAS CITY to PORTLAND
  10s-dr-2c
  UP 37, Kansas City (11:59p Sat)
      to Denver (3:30p Sun)
  UP 17, Denver (5:40p) to Portland (6a Tue)

(Denver to Portland, first-class lounge; Denver to Hinkle, Ore., dining car; No. 37, the Pony Express, *carries a diner; No. 23,* Gold Coast, *has both diner and first-class lounge)*

Cheyenne had dedicated Pullmans to Portland and Omaha.

Gold Coast, No. 23
Chicago to San Francisco
Union Pacific
50 miles east of Rock Springs

CHICAGO to SAN FRANCISCO
  12s-dr cars as needed
  C&NW 23, Chicago (8p Sat) to Omaha (7:30a Sun)
  UP 23, Omaha (8:20a) to Ogden (6:15a Mon)
  SP 23, Ogden (8:15a) to Oakland pier (7a Tue)
  Ferry to San Francisco (7:35a Tue)

CHICAGO to PORTLAND
  10s-dr-c
  C&NW 23, Chicago (8p Sat) to Omaha (7:30a Sun)
  UP 23, Omaha (8:20a)
      to Green River, Wyo. (1:35a Mon)
  UP 17, Green River (3:25a) to Portland (6a Tue)

CHICAGO to LOS ANGELES
  10s-3dbr   14s
  As above, Chicago (8p Sat) to Ogden (6:15a Mon)
  UP 37 Ogden (7a) to Los Angeles (7a Tue)

(From Chicago and Denver, to Oakland, Portland, and Los Angeles: dining and lounge service. Breakfast not served into Portland on No. 17, the Portland Rose, *or into Los Angeles on No. 37, the* Pony Express.)*

Introduced in 1926 as a Chicago-San Francisco train, the *Gold Coast* also served other West Coast termini. Suspended during the Depression, it was revived in 1947 as a maid of all work with numerous conditional stops scheduled in Wyoming and other states.

The *North Shore Limited,* which travels MC via Ontario, takes water at Tivoli, N.Y.

Wyoming   (continued)

Portland Rose, No. 18

Portland to Denver
Union Pacific
11 miles east of Rock Springs

PORTLAND to DENVER
6s-6r-4dbr
UP 18, Portland (9:45p Sat) to Denver (8:10a Mon)

PORTLAND to KANSAS CITY
10s-dr-c
UP 18, Portland (9:45p Sat) to Denver (8:10a Mon)
UP 38, Denver (9a) to Kansas City (10:30p Mon)

(First-class lounge from Portland, diner east of Pendleton, Ore., on No. 18; No. 38, the Pony Express, carries a diner. The other east- and westbound Portland Roses are now in Oregon.)

No. 30
Billings to Denver
Burlington
18 miles north of Wendover

BILLINGS to DENVER
B305: Std: s-dr
CB&Q 30, Billings (12:45p) to Wendover (12:30a)
C&S 30, Wendover (12:35a) to Denver (7:40a)

(Diner, Lovell to Casper)

No. 30 once carried Pullmans from Casper to both Cheyenne and Denver. Cheyenne had service to Billings. Summertime travel to Yellowstone Park brought numerous sleepers to Cody, as of this scheduling from Chicago (NP/Burlington) and in previous years from Billings, Denver, Edgemont, S.D., Lincoln, Omaha, and Glacier Park, Mont. (GN). The Edgemont (Black Hills) and Glacier Park cars facilitated travel for tourists visiting several scenic locations in sequence.

Pony Express, No. 38

Los Angeles to Kansas City
Union Pacific
23 miles east of Rock Springs

LOS ANGELES to KANSAS CITY
12s-dr
UP 38, Los Angeles (7:30p Sat)
to Kansas City (10:30p Mon)

SAN FRANCISCO to DENVER
12s-dr
Ferry from San Francisco (6p Sat)
SP 24, Oakland (6:30p) to Ogden (6:30p Sun)
UP 38, Ogden (7:10p) to Denver (8a Mon)

SALT LAKE CITY to DENVER
6s-6r-4dbr
UP 38, Salt Lake City (5:45p) to Denver (8a)

(Los Angeles to Denver, first-class lounge; Las Vegas to Kansas City, diner; SP 24, the Gold Coast, carries both facilities. Another edition of the Pony Express, 24 hours behind this one, is now in California.)

Gold Coast, No. 24
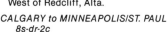
San Francisco to Chicago
Union Pacific
Switching in Green River

SAN FRANCISCO to CHICAGO
12s-dr cars as needed
Ferry from San Francisco (6p Sat)
SP 24, Oakland pier (6:30p) to Odgen (6:30p Sun)
UP 24, Ogden (7:15p) to Omaha (7:50p Mon)
C&NW 24, Omaha (8:30p) to Chicago (8:15a Tue)

PORTLAND to CHICAGO
10s-dr-c
UP 18, Portland (9:45p Sat)
to Green River (11:05p Sun)
UP 24, Green River (12:15a Mon) to Omaha (7:50p)
C&NW 24, Omaha (8:30p) to Chicago (8:15a Tue)

LOS ANGELES to CHICAGO
10s-3dbr   14s
UP 38, Los Angeles (7:30p Sat)
to Ogden (6:40p Sun)
As above, Ogden (7:15p) to Chicago (8:15a Tue)

(First-class lounge service to Chicago from Los Angeles, Portland, and San Francisco; dining service east of Las Vegas on No. 38, the Pony Express, and east of Pendleton on No. 18, the Portland Rose. East- and westbound editions of the Gold Coast are now in both Iowa and California.)

Pony Express, No. 37

Kansas City to Los Angeles
Union Pacific
13 miles west of Rawlins

KANSAS CITY to LOS ANGELES
12s-dr
UP 37, Kansas City (11:59p Sat)
to Los Angeles (7a Tue)

DENVER to SAN FRANCISCO
12s-dr
UP 37, Denver (5:35p Sun) to Ogden (6:35a Mon)
SP 23, Ogden (8:15a) to Oakland (7a Tue)
Ferry to San Francisco (7:35a)

DENVER to SALT LAKE CITY
6s-6r-4dbr
UP 37, Denver (5:35p) to Salt Lake City (8a)

(Kansas City to Las Vegas, dining car; Denver to Los Angeles, first-class lounge. SP No. 23, the Gold Coast, carries first-class lounge and diner. Other westbound editions of the Pony Express are now in Missouri and California.)

Canada

Mountaineer, No. 3-13

St. Paul to Vancouver, B.C.
Summers only
Canadian Pacific
(See below for location)

ST. PAUL to VANCOUVER
6s-4dbr-4r   2c-dr-lounge
6c-3dr   8s-dr-2c
14s tourist
Soo 3-13, St. Paul (9:35p Sat) and
Minneapolis (10:25p)
to Portal, N.D. (2:30p Sun)
CPR 3-13, North Portal (2:20p MT)
to Vancouver (7:30a Tue)

ST. PAUL to BANFF
6s-6dbr
As above, St. Paul (9:35p Sat) to Portal (2:30p Sun)
CPR 3-13, North Portal (2:20p MT)
to Banff (10:10a Mon)

(Dining car, Enderlin, N.D., to Vancouver; open observation car, Calgary to Vancouver. Two editions of the westbound Mountaineer are now in Canada, one west of Swift Current, Sask., the other west of Kamloops, B.C. A three-night train, the Mountaineer has another edition in Minnesota.)

Soo-Dominion, No. 7

St. Paul to Calgary
Canadian Pacific
West of Swift Current

ST. PAUL/MINNEAPOLIS to CALGARY
8s-dr-2c
Soo 3, St. Paul (9:35p Sat)
and Minneapolis (10:25p)
to Portal (2p Sun)
CPR 13, North Portal (2:20p MST)
to Moose Jaw, Sask. (7:15p)
CPR 7, Moose Jaw (8:50p) to Calgary (8:30a Mon)

(Diners: Enderlin to Portal; Moose Jaw to Calgary. The Soo-Dominion is the off-season version of the Mountaineer.)

Soo-Dominion, No. 8

Calgary to St. Paul
Canadian Pacific
West of Redcliff, Alta.

CALGARY to MINNEAPOLIS/ST. PAUL
8s-dr-2c
CPR 8, Calgary (8:10p Sun)
to Moose Jaw (6:40a Mon)
CPR 14, Moose Jaw (8:20a)
to North Portal (1:30p MST)
SOO 4, Portal (3:20p CST)
to Minneapolis (6:30a Tue)
and St. Paul (7:15a Tue)

(Dining cars: Calgary to Moose Jaw, Portal to Enderlin)

Wolverine, No. 8
Chicago to New York
New York Central
Canfield Jct., Ont.

CHICAGO to NEW YORK
10r-6dbr   10r-6dbr
22r   6dbr-buffet-lounge
NYC 8, Chicago (2:30p) to New York (9:20a)

DETROIT to NEW YORK
10r-6dbr
NYC 8, Detroit (8:50p) to New York (9:20a)

(Dining car. Saturday consist different.)

No. 9
Toronto to Detroit
Canadian National
Departing Toronto

TORONTO to DETROIT
25: 12s-2dbr
CNR 9, Toronto (11:59p) to Detroit (8:05a)

No. 10
Detroit to Toronto
Canadian National
Windsor, Ont.

DETROIT to TORONTO
26: 12s-2dbr
CNR 10, Detroit (10:30p) to Toronto (8a)

Mountaineer, No. 14-4
Vancouver to St. Paul
Canadian Pacific
Summertime only

VANCOUVER to ST. PAUL
6s-4dbr-4r   2c-dr-lounge
6c-3dr   8s-dr-2c
14s tourist
CPR 14-4, Vancouver (6:30p Sat)
     to North Portal (11:15a Mon)
SOO 14-4, Portal (1:15p CT)
     to Minneapolis (6:30a Tue)
     and St. Paul (7:15a Tue)

BANFF to ST. PAUL
6s-6dbr
CPR 14-4, Banff (4:30p Sun)
     to North Portal (11:15a Mon)
SOO 14-4, Portal (1:15p CT)
     to Minneapolis (6:30a Tue)
     and St. Paul (7:15a Tue)

(Diner, Vancouver to Enderlin. Open observation car, Vancouver to Calgary. The Mountaineer, *a three-night train, has two eastbound editions now in Canada: one east of North Bend, B.C., and one east of Medicine Hat, Alta. A third edition is in North Dakota.)*

International Limited, No. 15
Toronto to Chicago
Canadian National
East of Princeton, Ont.

TORONTO to CHICAGO
1: 24dup/r   17: 8s-dr-2c
3: 2c-dr-3sbr-buffet-lounge
CNR/GTW 15, Toronto (10p) to Chicago (8a)

(Dining car: Battle Creek, Mich., to Chicago)

Inter-City Limited, No. 17
Montreal to Chicago
Canadian National
Departing Coteau, Que.

MONTREAL to CHICAGO
93: 8s-dr-2c
CNR 17, Montreal (11p) to Chicago (8:20a)

MONTREAL to DETROIT
95: 12s-2dbr
CNR 17, Montreal (11p) to London (11:42a)
CNR 117, London (12:08p) to Detroit (3:05p)

(Toronto to Port Huron, cafe-parlor car; Lansing to Chicago, dining car; Durand, Mich., to Chicago, buffet-parlor car)

Overseas, No. 22
Chicago to Montreal
Canadian Pacific
Entering Oshawa, Ont.

DETROIT to MONTREAL
10r-5dbr
CPR 22, Detroit (4:15p) to Montreal (7:25a)

(Diners: Chicago to Detroit, Windsor to Toronto)

Maple Leaf, No. 16
Detroit to Montreal
Canadian National
West of Pickering, Ont.

CHICAGO to MONTREAL
94: 8s-dr-2c
GTW/CNR 20, Chicago (9:40a) to Toronto (10:10p)
CNR 16, Toronto (11:30p) to Montreal (7:30a)

DETROIT to MONTREAL
96: 12s-2dbr
CNR 16, Detroit (3:45p) to Toronto (10:25p)
CNR 16, Toronto (11:30p) to Montreal (7:30a)

(No. 20: Chicago to Lansing, dining car; Port Huron, Mich., to Toronto, cafe car. No. 16: Windsor to Toronto, diner.)

The *Maple Leaf* formerly carried Chicago-New York sleepers it turned over to the Lehigh Valley.

Canadian, No. 19
Montreal to Chicago
Canadian Pacific
Departing Woodstock, Ont.

TORONTO to CHICAGO
12s-dr   10r-6dbr
CPR 19, Toronto (10p) to Detroit (3a)
NYC 39, Detroit (3:15a) to Chicago (8:10a)

(Detroit to Chicago, dining car. NYC 39 is the North Shore Limited.)

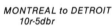

Chicago Express, No. 21
Montreal to Chicago
Canadian Pacific
10 miles east of Green Valley, Ont.

MONTREAL to DETROIT
10r-5dbr
CPR 21, Montreal (11p) to Detroit (2:40p)

(Montreal to Toronto, buffet-lounge; Toronto to Windsor, Detroit to Chicago, dining cars)

North Shore Limited, No. 39
New York to Chicago
New York Central
East of La Salette, Ont.

NEW YORK to CHICAGO
10r-6dbr
NYC 39, New York (12 noon) to Chicago (8:10a)

BUFFALO to CHICAGO
10r-5dbr   8s-buffet-lounge
NYC 39, Buffalo (10:10p) to Chicago (8:10a)

(Detroit to Chicago, diner-lounge. No. 39 will cut in a Toronto-Chicago Pullman, delivered to Detroit by the CPR's Canadian *at 3 a.m.)*

No. 364
Detroit to Buffalo
New York Central
East of Tillsonburg, Ont.

DETROIT to NEW YORK
12s-dr
NYC 364, Detroit (9p) to Buffalo (2:40a)
NYC 22, Buffalo (4:10a) to New York (12:30p)

(No. 22 is the Lake Shore Limited, *carrying dining and lounge facilities)*

Michigan, No. 635
Toronto to Chicago
Canadian Pacific
Departing West Toronto

TORONTO to DETROIT
8s-4dbr
CPR 635, Toronto (11:30p) to Detroit (7:45a)

(Detroit to Chicago, dining car)

Mexico

Aztec Eagle, No. 1
Mexico City to Nuevo Laredo
National Railways of Mexico
South of Saltillo
NdeM

MEXICO CITY to SAN ANTONIO
10s-2c-dr   8s-5dbr
NdeM 1, Mexico City (8:20a) to Nuevo Laredo (8a)
MP 22, Laredo (9:05a) to San Antonio (1:30p)

(Diner-lounge)

Aztec Eagle, No. 2
Nuevo Laredo to Mexico City
National Railways of Mexico
South of Monterrey
NdeM

SAN ANTONIO to MEXICO CITY
10s-2c-dr   8s-5dbr
MP 21, San Antonio (12:05p Sun) to Laredo (4:20p)
NdeM 2, Nuevo Laredo (6p)
     to Mexico City (8p Mon)

(Diner-lounge)

No. 7
Mexico City to El Paso
National Railways of Mexico
Location: See below
NdeM

MEXICO CITY to EL PASO
10s-dr-2br
NdeM 7, Mexico City (7:40p Sun)
     to El Paso (3:30p Tue)

(Two editions of No. 7 are now in Mexico: one south of San Juan del Rio, the other south of Escalon. The train carries a diner-lounge)

No. 8
El Paso to Mexico City
National Railways of Mexico
Location: See below
NdeM

EL PASO to MEXICO CITY
10s-dr-2br
NdeM 8, El Paso (11:40a Sun)
     to Mexico City (9:20a Tue)

(One edition of No. 8 approaches Santa Rosalia, and the other has just departed Leon. Diner-lounge)

No. 9
Guadalajara to Nogales
Southern Pacific of Mexico
Location: see below
SPdeM

GUADALAJARA to NOGALES
12s-c-dr   6s-cafe
SPdeM 9, Guadalajara (10:40a Fri)
     to Nogales (4:40p Sun)

(Two northbound editions of No. 9 are under way in Mexico at midnight, one departing Ruiz, another approaching Navojoa)

No. 10
Nogales to Guadalajara
Southern Pacific of Mexico
Location: see below
SPdeM

NOGALES to GUADALAJARA
12s-c-dr   6s-cafe
SPdeM 10, Nogales (10a Sun)
     to Guadalajara (5p Tue)

(One of the two southbound editions of No. 10 is approaching Ciudad Obregon, the other Acaponeta)

Bibliography

For the reader who wants a general education in passenger trains, the works of these authors in the bibliography below are highly recommended: Arthur D. Dubin and William W. Kratville for the trains, John H. White Jr. for the cars, George H. Drury for the railroad history, Lucius Beebe for descriptive color.

Books

Alexander, Edwin P. *Down at the Depot.* New York: Clarkson N. Potter, 1970.

Archer, Robert F. *Lehigh Valley Railroad.* Berkeley, Calif.: Howell-North Books, 1977.

Athearn, Robert G. *Rebel of the Rockies: A History of the Denver and Rio Grande Western Railroad.* New Haven: Yale University Press, 1962.

Barger, Ralph L. *A Century of Pullman Cars.* Vol. I. Sykesville, Md.: Greenberg Publishing Company, 1988.

Beebe, Lucius. *Mr. Pullman's Elegant Palace Car.* New York: Doubleday, 1961.

———. *The Overland Limited.* Berkeley: Howell-North, 1963.

———. *Twentieth Century.* Berkeley: Howell-North Books, 1962.

Beebe, Lucius and Charles Clegg. *Hear the Train Blow: A Pictorial Epic of America in the Railroad Age.* New York: Grosset & Dunlap, 1952.

———. *Narrow Gauge in the Rockies.* Berkeley: Howell-North Books, 1958.

———. *Rio Grande: Mainline of the Rockies.* Berkeley: Howell-North, 1962.

———. *The Trains We Rode.* Vol. I. Berkeley: Howell-North, 1965.

———. *The Trains We Rode.* Vol. II. Berkeley: Howell-North, 1966.

———. *Virginia & Truckee: A Story of Virginia City and Comstock Times.* Oakland: Grahame H. Hardy, 1949.

Bramson, Seth. *Speedway to Sunshine: The Story of the Florida East Coast Railway.* Erin, Ont.: Boston Mills Press, 1984.

Bryant, Keith L. *History of the Atchison, Topeka and Santa Fe Railway.* New York: Macmillan, 1974.

Burgess, George H. and Miles C. Kennedy. *Centennial History of the Pennsylvania Railroad Company.* Philadelphia: Pennsylvania Railroad Company, 1949.

Casey, Robert J. and W.A.S. Douglas. *Pioneer Railroad: The Story of the Chicago and North Western System.* New York: Whittlesey House (McGraw-Hill), 1948.

Condit, Carl W. *The Port of New York.* Vols. I & II. Chicago: University of Chicago Press, 1980.

———. *The Railroad and the City.* Columbus: Ohio State University Press, 1977.

Cook, Roger and Karl Zimmermann. *The Western Maryland Railway.* San Diego: Howell-North Books, 1981.

Corliss, Carlton J. *Main Line of Mid-America.* New York: Creative Age Press, 1950.

Crump, Spencer. *Redwoods, Iron Horses, and the Pacific.* Los Angeles: Trans-Anglo Books, 1971.

Culp, Edwin D. *Stations West.* Caldwell, Idaho: Caxton Printers, 1972.

Davis, Burke. *The Southern Railway: Road of the Innovators.* Chapel Hill: University of North Carolina Press, 1985.

Derleth, August. *The Milwaukee Road.* New York: Creative Age Press, 1948.

Dixon, Thomas W. Jr. *Chessie: The Railroad Kitten.* Sterling, Va.: TLC Publishing Co., 1988.

Donovan, Frank P. Jr. *Mileposts On the Prairie.* New York: Simmons-Boardman Publishing Corp., 1950.

Dorin, Patrick C. *Canadian Pacific Railway.* Seattle: Superior Publishing Co., 1974.

———. *Everywhere West: The Burlington Route.* Seattle: Superior Publishing Co., 1976.

———. *The Grand Trunk Western Railroad.* Seattle: Superior Publishing Co., 1977.

———. *The Milwaukee Road East.* Seattle: Superior Publishing Co., 1978.

Droege, John A. *Passenger Terminals and Trains.* Milwaukee: Kalmbach, 1969. (Reprint of original, New York: McGraw-Hill, 1916)

Drury, George H. *The Historical Guide to North American Railroads.* Milwaukee: Kalmbach, 1985.

———. *The Train-Watcher's Guide to North American Railroads.* Milwaukee: Kalmbach, second printing, 1985.

Dubin, Arthur D. *More Classic Trains.* Milwaukee: Kalmbach, 1974.

———. *Some Classic Trains.* Milwaukee: Kalmbach, 1964.

Duke, Donald and Stan Kistler. *Santa Fe: Steel Rails Through California.* San Marino, Calif.: Golden West, 1963.

Dunbar, Willis Frederick. *All Aboard! A History of Railroads in Michigan.* Grand Rapids: William B. Eerdmans Publishing Company, 1969.

Dunscomb, Guy L. and Fred A. Stindt. *Western Pacific Steam Locomotives, Passenger Trains and Cars.* Modesto, Calif.

Fair, James R. Jr. *The North Arkansas Line.* Berkeley: Howell-North Books, 1969.

Federal Writers' Project of the Works Progress Administration. *Maine: A Guide Down East.* Boston: Houghton Mifflin, 1937.

———. *New Jersey: A Guide to Its Present and Past.* New York: Hastings House, 1946.

———. *West Virginia: A Guide to the Mountain State.* New York: Oxford University Press, 1941.

———. *The WPA Guide to Illinois.* New York: Pantheon, 1983.

Grant, Roger H. *The Corn Belt Route.* DeKalb, Ill.: Northern Illinois University Press, 1984.

Griffin, William E. Jr. *One Hundred Fifty Years of History Along the Richmond, Fredericksburg and Potomac Railroad.* Richmond, Va.: Whittet & Shepperson, 1984.

Harlow, Alvin F. *The Road of the Century: The Story of the New York Central.* New York: Creative Age Press, 1947.

Hayes, William Edward. *Iron Road to Empire: The History of 100 Years of the Progress and Achievements of the Rock Island Lines.* H. Wolff Book Manufacturing Co., Inc., 1953.

Henwood, James N.J. *A Short Haul to the Bay: A History of the Narragansett Pier Railroad.* Brattleboro, Vt.: The Stephen Greene Press, 1969.

Herr, Kincaid A. *Louisville & Nashville Railroad, 1850-1963.* Revised edition. Louisville: Louisville & Nashville Railroad, 1964.

Hilton, George W. *Monon Route.* Berkeley: Howell-North Books, 1978, second printing.

Hofsommer, Don L. *The Southern Pacific, 1901-1985.* College Station, Texas: Texas A&M University Press, 1986.

Hofsommer, Donovan L., editor. *Railroads in Oklahoma.* Oklahoma City: Oklahoma Historical Society, 1977.

Holbrook, Stewart H. *The Story of American Railroads.* New York: Crown, 1947.

Hubbard, Freeman. *Railroad Avenue: Great Stories and Legends of American Railroading.* Revised edition. San Marino, Calif.: Golden West Books, 1964.

Hull, Clifton E. *Shortline Railroads of Arkansas.* Norman, Okla.: University of Oklahoma Press, 1969.

Hungerford, Edward. *The Story of the Baltimore & Ohio Railroad.* New York: G.P. Putnam's Sons, 1928.

Husband, Joseph. *The Story of the Pullman Car.* Chicago: A.C. McClurg & Co., 1917.

Johnson, James D. *The Lincoln Land Traction.* Wheaton, Ill.: The Traction Orange Co., 1965.

Kratville, William W. *Steam, Steel & Limiteds.* Omaha: Barnhart Press, 1967.

Krause, John with Donald Duke. *American Narrow Gauge.* San Marino Calif.: Golden West Books, 1978.

Lemly, James Hutton. *The Gulf, Mobile and Ohio.* Homewood, Ill.: Richard D. Irwin, Inc., 1953.

Lind, Alan R. *Limiteds Along the Lakefront: The Illinois Central in Chicago.* Park Forest, Ill.: Transport History Press, 1986.

Marshall, James. *Santa Fe: The Railroad That Built an Empire.* New York: Random House, 1945.

Middleton, William D. *Grand Central.* San Marino, Calif.: Golden West Books, 1977.

Myrick, David F. *Railroads of Arizona.* Vol. I. San Diego: Howell-North, 1981, second printing.

————. *Railroads of Arizona.* Vol. II. San Diego: Howell-North Books, 1980.

————. *Railroads of Nevada and Eastern California.* Vol. II. Burbank, Calif.: Howell-North Books, 1963.

Nolan, Edward W. *Northern Pacific Views.* Helena: Montana Historical Society Press, 1983.

Overton, Richard C. *Burlington Route: A History of the Burlington Lines.* New York: Knopf, 1968.

Parks, Pat. *The Railroad That Died at Sea: The Florida East Coast's Key West Extension.* Brattleboro, Vt.: The Stephen Greene Press, 1968.

Prince, Richard E. *Atlantic Coast Line Railroad Steam Locomotives, Ships, and History.* Green River, Wyo.: Richard E. Prince, 1966.

————. *Nashville, Chattanooga & St. Louis Railway.* Green River, Wyo.: Richard E. Prince, 1967.

————. *Seaboard Air Line Railway.* Green River, Wyo.: Richard E. Prince, 1969.

————. *Steam Locomotives and History: Georgia Railroad and West Point Route.* Green River, Wyo.: Richard E. Prince, Second printing, 1972.

Rehor, John A. *The Nickel Plate Story.* Milwaukee: Kalmbach, 1965.

Runte, Alfred. *Trains of Discovery: Western Railroads and the National Parks.* Flagstaff, Ariz.: Northland Press, 1984.

Scribbins, Jim. *The 400 Story.* Park Forest, Ill.: PTJ Publishing, 1982.

————. *The Hiawatha Story.* Milwaukee: Kalmbach, 1978.

Serling, Robert. *Howard Hughes' Airline: An Informal History of TWA.* New York: St. Martin's/Marek, 1983.

Shaughnessy, Jim. *Delaware & Hudson.* Berkeley: Howell-North, 1967.

Stilgoe, John R. *Metropolitan Corridor: Railroads and the American Scene.* New Haven: Yale University Press, 1983.

Stover, John F. *History of the Illinois Central Railroad.* New York: Macmillan, 1975.

Suprey, Leslie V. *Steam Trains of the Soo.* Mora, Minn.: B&W Printers & Publishers, Third revised edition, 1962.

Taylor, George Rogers and Irene Neu. *The American Railroad Network, 1861-1890.* Cambridge, Mass.: Harvard University Press, 1956.

Turner, Charles W. *Chessie's Road.* Richmond, Va.: Garrett & Massie, 1956.

Wagner, Jack R. *Short Line Junction.* Fresno: Academy Library Guild, 1956.

Wallin, Richard R., Paul H. Stringham, and John Szwajkart. *Chicago & Illinois Midland.* San Marino, Calif.: Golden West Books, 1979.

Waters, L.L. *Steel Trails to Santa Fe.* Lawrence, Kans.: University of Kansas Press, 1950.

Weller, John L. *The New Haven Railroad: Its Rise and Fall.* New York: Hastings House, 1969.

White, John H. Jr. *The American Railroad Passenger Car.* Parts 1 & 2. Baltimore: Johns Hopkins University Press, softcover editions, 1985.

Wilson, Neill C. and Frank J. Taylor. *Southern Pacific: The Roaring Story of a Fighting Railroad.* New York: McGraw-Hill Book Company, 1952.

Wood, Charles and Dorothy. *Spokane Portland and Seattle Railway.* Seattle: Superior Publishing Co., 1974.

Wurm, Ted and Harre W. Demoro. *The Silver Short Line.* Glendale, Calif.: Trans-Anglo Books, 1983.

Ziel, Ron and George H. Foster. *Steel Rails to the Sunrise.* New York: Duell, Sloan and Pearce, 1965.

The Complete Roster of Heavyweight Pullman Cars. New York: Wayner Publications, 1985.

Pamphlets

Short History of the Baltimore & Ohio Railroad. Baltimore: Baltimore & Ohio Railroad, 1936.

The Bangor and Aroostook 1891-1966. Bangor and Aroostook Railroad. Concord, N.H.: Rumford Press, 1966.

Periodicals

Willard V. Anderson. "The Nickel-Plated Railroad." *Trains*, 8, No. 7 (May 1948), pp. 20-29.

Phil Borleske. "Soo Line Sojourns." *Passenger Train Journal*, (January 1985), pp. 21-33.

———. "Last Days of Great Western Passenger Trains." *Passenger Train Journal*, (October 1983), pp. 20-30.

Thomas W. Dixon Jr. "Along Came the F.F.V." *Chesapeake and Ohio Historical Newsletter*, XI, No. 12 (December 1979), pp. 4-22.

———. "The George Washington: The Most Wonderful Train in the World." *Chesapeake and Ohio Historical Newsletter*, IX, No. 5 (May 1977), pp. 14-31.

Richard G. Durnin. "West Shore Railroad." *National Railway Historical Society Bulletin*, (Third Quarter, 1962), pp. 22-25.

Charles E. Fisher. "Through Car Service from New England." *Bulletin of the Railway and Locomotive Historical Society*, No. 87 (October 1952), pp. 47-58.

Robert A. Le Massena. "All Off at Denver." *Trains*, 30, No. 8 (June 1970), pp. 40-49.

Frank E. Shaffer. "Pullman Prolificacy." *Trains*, 27, No. 12 (October 1967), pp. 24-28.

Jim Shaughnessy. "Case of the Ambidextrous Short Line." *Trains*, 19, No. 5 (March 1959), pp. 40-48.

Richard H. Steinmetz. "Mainline Hot Spot." *Trains*, 5, No. 4 (February 1945), pp. 16-22.

Lawrence M. Thomas. "Prairie State Sentinels." *Passenger Train Journal*, Vol. 18, No. 1 (January-February 1987), pp. 17-27.

The Official Guide of the Railways: June 1869 (reprint), December 1890, December 1895, August 1907, March 1909, August 1909, September 1910, September 1913, July 1917, August 1919, September 1925, July 1928, January 1930 (reprint), June 1938, September 1946, July 1951, March 1952, January 1965, April 1971, May 1971.

The Pullman News Chicago, The Pullman Company. 1922.

Folios

Proposed Lake Front Passenger Terminal, Twelfth Street Boulevard and Indian Avenue, Chicago, Illinois, Illinois Central Railroad Company, 1925. A folio of architectural renderings in the collection of the Chicago Historical Society.

Dissertation

Patterson, William H. "Through the Heart of the South: A History of the Seaboard Air Line Railroad Company, 1832-1950." Diss. University of South Carolina 1951.

Index

Page numbers in **bold** indicate photographs